BIO103: Human Biology

BIO103: Human Biology

LUMEN LEARNING & OPENSTAX

Edits to the original text were made in this version by Teresa Burke, Elizabeth Justin, and Gordon Lake.

Published by SUNY OER Services

SUNY Office of Library and Information Services
10 N Pearl St
Albany, NY 12207

ISBN: 978-1-64176-065-2

Contents

Chapter 5 — Reproductive System

Chapter 6 — Development and Inheritance

Chapter 7 — The Skeletal System

Chapter 8 — Muscular System

Chapter 9 — The Urinary System

Chapter 10 — Blood

Chapter 11 — The Cardiovascular System

Chapter 12 — The Respiratory System

Chapter 13 — The Nervous System

Chapter 14 — Sensory Systems

Chapter 15 — The Lymphatic and Immune System

About This Course

Welcome to BIO103: Human Biology

BIO103: Human Biology is an introductory course which concerns the structure and function of the human body and the maintenance of homeostasis. The course is designed for non-science majors and does not fulfill the elective requirement of the LAX student.

Upon successful completion of the course, students will:

1. List and define, in order, the main steps of the scientific method, relate these steps to an experiment and critically analyze data obtained from laboratory experiments.
2. Relate basic structure, function and properties of cells.
3. Focus and identify histology slides under high power (40X) objective of a compound light microscope.
4. Locate and identify major organs of various organ systems in the human body, their parts and associated functions.
5. Recognize the concept of homeostasis in the human body.

Schedule of topics and labs:

Week	Lecture	Lab & Histology drawings
1	**Introduction and Organization** Basic characteristics of living things, Levels of biological organization, Scientific method.	**Microscope** Cheek cell smear : Squamous epithelial cells
2	**Basic Chemistry of Life** Atomic Structure, bonding, acids, bases, salts, pH, carbohydrates, lipids, proteins, nucleic acids, dehydration synthesis, hydrolysis, Enzymes.	**Chemistry of Life** Starch granules in potato cells
3	**The Cell** Compare prokaryotic and eukaryotic cell, cell size and microscopy, cell structure and functions, organelles, plasma membrane and transport of materials, basics of protein synthesis, cellular respiration and ATP synthesis.	**Diffusion and Osmosis** Red blood cell in isotonic, hypotonic, and hypertonic solutions
4	**Body Organization and Homeostasis** Cells to organs systems, body cavities, homeostasis, positive and negative feedback, thermoregulation Digestive System Organs and functions, digestive enzymes, nutrient absorption. Glucose Homeostasis with insulin and glucagon.	**Body Organization** C.S of small intestine showing columnar epithelium
5	**Reproductive System** Male and female reproductive organs and functions, menstrual cycle, oxytocin, prolactin, FSH, testosterone, estrogen, progesterone, birth control, sexually transmitted diseases	**Lab Exam I**
6	**Development and Heredity** Fertilization, prenatal development, zygote, embryo, fetus, fetal circulation, and birth	**Reproductive System** C.S of testis and ovary
7	**Skeletal System** Bone structure and function, bone cells, bone development and remodeling , axial and appendicular skeleton, joints, hormonal and nutritional effects on bones, calcium homeostasis with thyroid hormone and parathyroid hormone,	**Skeletal System** Compact bone tissue and hyaline cartilage
8	**Muscular System** Function and characteristics of muscles. mechanism of muscular contraction, twitch, summation, tetanus, fatigue, energy and oxygen supply, aerobic and anaerobic respiration, oxygen debt and ATP	**Muscular System** Muscle fiber drawings; smooth muscle, skeletal muscle, and cardiac muscle
9	**Urinary System** Organs and functions, nephron, urine production, basic water and electrolyte balance, ADH, Aldosterone	**Urinary System** Kidney - glomerular capsule
10	**Blood** Function and composition of blood. Blood types, erythropoietin, blood clotting, blood cell disorders.	**Lab Exam II**
11	**Cardiovascular System and Lymphatic System** Heart, blood vessels, blood flow, cardiac cycle, EKG, blood pressure, cardiovascular diseases, lymphatic system (basic)	**Heart and Blood** Human blood smear slide
12	**Respiratory System** Structures of respiratory system, mechanism of breathing, gas exchange and transport of gases, hemoglobin, respiratory centers in brain, respiratory disorders.	**Respiratory System** Normal Lung Pseudo stratified ciliated columnar epithelium
13	**Nervous System** Neuron, basics of nerve impulse conduction and synaptic transmission, CNS: brain, spinal cord, Hormonal function of hypothalamus and Pituitary gland, PNS: nerves, reflex arc, autonomic nervous system.	**Nervous System** Multipolar neuron
14	**Senses organs** Sensory receptors of skin, vision, hearing, balance, smell and taste.	**Senses Organs** No drawings.
15	**FINAL EXAM WEEK**	**Lab exam III**

CHAPTER 1 — SCIENTIFIC METHOD AND CHARACTERISTICS OF LIFE

The Process of Science

LEARNING OBJECTIVE

By the end of this section, you will be able to:

- Discuss the fundamental relationship between anatomy and physiology
- Identify the shared characteristics of the natural sciences
- Understand the process of scientific inquiry
- Identify and describe the properties of life
- Define homeostasis and three parts of homeostatic feedback regulation system.
- Describe positive and negative feedback mechanisms with examples.
- Describe the levels of organization among living things

Human **anatomy** is the scientific study of the body's structures. Some of these structures are very small and can only be observed and analyzed with the assistance of a microscope. Other larger structures can readily be seen, manipulated, measured, and weighed. The word "anatomy" comes from a Greek root that means "to cut apart." Human anatomy was first studied by observing the exterior of the body and observing the wounds of soldiers and other injuries. Later, physicians were allowed to dissect bodies of the dead to augment their knowledge. When a body is dissected, its structures are cut apart in order to observe their physical attributes and their relationships to one another. Dissection is still used in medical schools, anatomy courses, and in pathology labs. In order to observe structures in living people, however, a number of imaging techniques have been developed. These techniques allow clinicians to visualize structures inside the living body such as a cancerous tumor or a fractured bone.

Like most scientific disciplines, anatomy has areas of specialization. **Gross anatomy** is the study of the larger structures of the body, those visible without the aid of magnification. Macro- means "large," thus, gross anatomy is also referred to as **macroscopic anatomy**. In contrast, micro- means "small," and microscopic anatomy is the study of structures that can be observed only with the use of a microscope or other magnification devices. Microscopic anatomy includes cytology, the study of cells and histology, the study of tissues. As the technology of microscopes has advanced, anatomists have been able to observe smaller and smaller structures of the body, from slices of large structures like the heart, to the three-dimensional structures of large molecules in the body.

Anatomists take two general approaches to the study of the body's structures: regional and systemic. **Regional anatomy** is the study of the interrelationships of all of the structures in a specific body region, such as the abdomen. Studying regional anatomy helps us appreciate the interrelationships of body structures, such as how muscles, nerves, blood vessels, and other structures work together to serve a particular body region. In contrast, **systemic anatomy** is the study of the structures that make up a discrete body system—that is, a group of structures that work together to perform a unique body function. For example, a systemic anatomical study of the muscular system would consider all of the skeletal muscles of the body.

Whereas anatomy is about structure, physiology is about function. Human **physiology** is the scientific study of the chemistry and physics of the structures of the body and the ways in which they work together to support the functions of life. Much of the study of physiology centers on the body's tendency toward homeostasis. **Homeostasis** is the state of steady internal conditions maintained by living things. The study of physiology certainly includes observation, both with the naked eye and with microscopes, as well as manipulations and measurements. However, current advances in physiology usually depend on carefully designed laboratory experiments that reveal the functions of the many structures and chemical compounds that make up the human body.

Like anatomists, physiologists typically specialize in a particular branch of physiology. For example, neurophysiology is the study of the brain, spinal cord, and nerves and how these work together to perform functions as complex and diverse as vision, movement, and thinking. Physiologists may work from the organ level (exploring, for example, what different parts of the brain do) to the molecular level (such as exploring how an electrochemical signal travels along nerves).

Form is closely related to function in all living things. For example, the thin flap of your eyelid can snap down to clear away dust particles and almost instantaneously slide back up to allow you to see again. At the microscopic level, the arrangement and function of the nerves and muscles that serve the eyelid allow for its quick action and retreat. At a smaller level of analysis, the function of these nerves and muscles likewise relies on the interactions of specific molecules and ions. Even the three-dimensional structure of certain molecules is essential to their function.

Your study of anatomy and physiology will make more sense if you continually relate the form of the structures you are studying to their function. In fact, it can be somewhat frustrating to attempt to study anatomy without an understanding of the physiology that a body structure supports. Imagine, for example, trying to appreciate the unique arrangement of the bones of the human hand if you had no conception of the function of the hand. Fortunately, your understanding of how the human hand manipulates tools—from pens to cell phones—helps you appreciate the unique alignment of the thumb in opposition to the four fingers, making your hand a structure that allows you to pinch and grasp objects and type text messages.

The Nature of Science

Science is a very specific way of learning, or knowing, about the world. The history of the past 500 years demonstrates that science is a very powerful way of knowing about the world; it is largely responsible for the technological revolutions that have taken place during this time. There are however, areas of knowledge and human experience that the methods of science cannot be applied to. These include such things as answering purely moral questions, aesthetic questions, or what can be generally categorized as spiritual questions. Science has cannot investigate these areas because they are outside the realm of material phenomena, the phenomena of matter and energy, and cannot be observed and measured.

The scientific method is a method of research with defined steps that include experiments and careful observation. The steps of the scientific method will be examined in detail later, but one of the most important aspects of this method is the testing of hypotheses. A hypothesis is a suggested explanation for an event, which can be tested. Hypotheses, or tentative explanations, are generally produced within the context of a scientific theory. A scientific theory is a generally accepted, thoroughly tested and confirmed explanation for a set of observations or phenomena. Scientific theory is the foundation of scientific knowledge. In addition, in many scientific disciplines (less so in biology) there are scientific laws, often expressed in mathematical formulas, which describe how elements of nature will behave under certain specific conditions. There is not an evolution of hypotheses through theories to laws as if they represented some increase in certainty about the world. Hypotheses are the day-to-day material that scientists work with and they are developed within the context of theories. Laws are concise descriptions of parts of the world that are amenable to formulaic or mathematical description.

Natural Sciences

What would you expect to see in a museum of natural sciences? Frogs? Plants? Dinosaur skeletons? Exhibits about how the brain functions? A planetarium? Gems and minerals? Or maybe all of the above? Science includes such diverse fields as astronomy, biology, computer sciences, geology, logic, physics, chemistry, and mathematics. However, those fields of science related to the physical world and its phenomena and processes are considered natural sciences. Thus, a museum of natural sciences might contain any of the items listed above.

There is no complete agreement when it comes to defining what the natural sciences include. For some experts, the natural sciences are astronomy, biology, chemistry, earth science, and physics. Other scholars choose to divide natural sciences into life sciences, which study living things and include biology, and physical sciences, which study nonliving matter and include astronomy, physics, and chemistry. Some disciplines such as biophysics and biochemistry build on two sciences and are interdisciplinary.

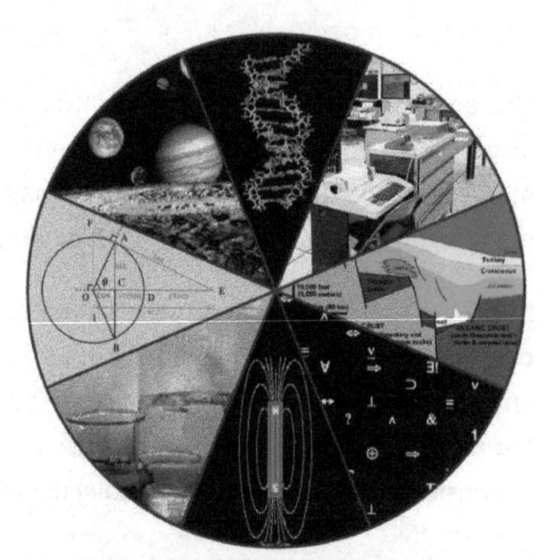

Some fields of science include astronomy, biology, computer science, geology, logic, physics, chemistry, and mathematics. (credit: "Image Editor"/Flickr)

Scientific Inquiry

One thing is common to all forms of science: an ultimate goal "to know." Curiosity and inquiry are the driving forces for the development of science. Scientists seek to understand the world and the way it operates. Two methods of logical thinking are used: inductive reasoning and deductive reasoning.

Inductive reasoning is a form of logical thinking that uses related observations to arrive at a general conclusion. This type of reasoning is common in descriptive science. A life scientist such as a biologist makes observations and records them. These data can be qualitative (descriptive) or quantitative (consisting of numbers), and the raw data can be supplemented with drawings, pictures, photos, or videos. From many observations, the scientist can infer conclusions (inductions) based on evidence. Inductive reasoning involves formulating generalizations inferred from careful observation and the analysis of a large amount of data. Brain studies often work this way. Many brains are observed while people are doing a task. The part of the brain that lights up, indicating activity, is then demonstrated to be the part controlling the response to that task.

Deductive reasoning or deduction is the type of logic used in hypothesis-based science. In deductive reasoning, the pattern of thinking moves in the opposite direction as compared to inductive reasoning. Deductive reasoning is a form of logical thinking that uses a general principle or law to forecast specific results. From those general principles, a scientist can extrapolate and predict the specific results that would be valid as long as the general principles are valid. For example, a prediction would be that if the climate is becoming warmer in a region, the distribution of plants and animals should change. Comparisons have been made between distributions in the past and the present, and the many changes that have been found are consistent with a warming climate. Finding the change in distribution is evidence that the climate change conclusion is a valid one.

Both types of logical thinking are related to the two main pathways of scientific study: descriptive science and hypothesis-based science. Descriptive (or discovery) science aims to observe, explore, and discover, while hypothesis-based science begins with a specific question or problem and a potential answer or solution that can be tested. The boundary between these two forms of study is often blurred, because most scientific endeavors combine both approaches. Observations lead to questions, questions lead to forming a hypothesis as a possible answer to those questions, and then the hypothesis is tested. Thus, descriptive science and hypothesis-based science are in continuous dialogue.

Hypothesis Testing

Scientists study the living world by posing questions about it and seeking science-based responses. This approach is common to other sciences as well and is often referred to as the scientific method. The scientific method was used even in ancient times, but it was first documented by England's Sir Francis Bacon (1561–1626) , who set up inductive methods for scientific inquiry. The scientific method is not exclusively used by biologists but can be applied to almost anything as a logical problem-solving method.

The scientific process typically starts with an observation (often a problem to be solved) that leads to a question. Let's think about a simple problem that starts with an observation and apply the scientific method to solve the problem. One Monday morning, a student arrives at class and quickly discovers that the classroom is too warm. That is an observation that also describes a problem: the classroom is too warm. The student then asks a question: "Why is the classroom so warm?"

Recall that a hypothesis is a suggested explanation that can be tested. To solve a problem, several hypotheses may be proposed. For example, one hypothesis might be, "The classroom is warm because no one turned on the air conditioning." But there could be other responses to the question, and therefore other hypotheses may be proposed. A second hypothesis might be, "The classroom is warm because there is a power failure, and so the air conditioning doesn't work."

Once a hypothesis has been selected, a prediction may be made. A prediction is similar to a hypothesis but it typically has the format "If . . . then . . . ." For example, the prediction for the first hypothesis might be, "*If* the student turns on the air conditioning, *then* the classroom will no longer be too warm."

A hypothesis must be testable to ensure that it is valid. For example, a hypothesis that depends on what a bear thinks is not testable, because it can never be known what a bear thinks. It should also be falsifiable, meaning that it can be disproven by experimental results. An example of an unfalsifiable hypothesis is "Botticelli's *Birth of Venus* is beautiful." There is no experiment that might show this statement to be false. To test a hypothesis, a researcher will conduct one or more experiments designed to eliminate one or more of the hypotheses. This is important. A hypothesis can be disproven, or eliminated, but it can never be proven. Science does not deal in proofs like mathematics. If an experiment fails to disprove a hypothesis, then we find support for that explanation, but this is not to say that down the road a better explanation will not be found, or a more carefully designed experiment will be found to falsify the hypothesis.

Each experiment will have one or more variables and one or more controls. A variable is any part of the experiment that can vary or change during the experiment. A control is a part of the experiment that does not change. Look for the variables and controls in the example that follows. As a simple example, an experiment might be conducted to test the hypothesis that phosphate limits the growth of algae in freshwater ponds. A series of artificial ponds are filled with water and half of them are treated by adding phosphate each week, while the other half are treated by adding a salt that is known not to be used by algae. The variable here is the phosphate (or lack of phosphate), the experimental or treatment cases are the ponds with added phosphate and the control ponds are those with something inert added, such as the salt. Just adding something is also a control against the possibility that adding extra matter to the pond has an effect. If the treated ponds show lesser growth of algae, then we have found support for our hypothesis. If they do not, then we reject our hypothesis. Be aware that rejecting one hypothesis does not determine whether or not the other hypotheses can be accepted; it simply eliminates one hypothesis that is not valid. Using the scientific method, the hypotheses that are inconsistent with experimental data are rejected.

SCIENTIFIC METHOD

A flow chart shows the steps in the scientific method. In step 1, an observation is made. In step 2, a question is asked about the observation. In step 3, an answer to the question, called a hypothesis, is proposed. In step 4, a prediction is made based on the hypothesis. In step 5, an experiment is done to test the prediction. In step 6, the results are analyzed to determine whether or not the hypothesis is supported. If the hypothesis is not supported, another hypothesis is made. In either case, the results are reported.

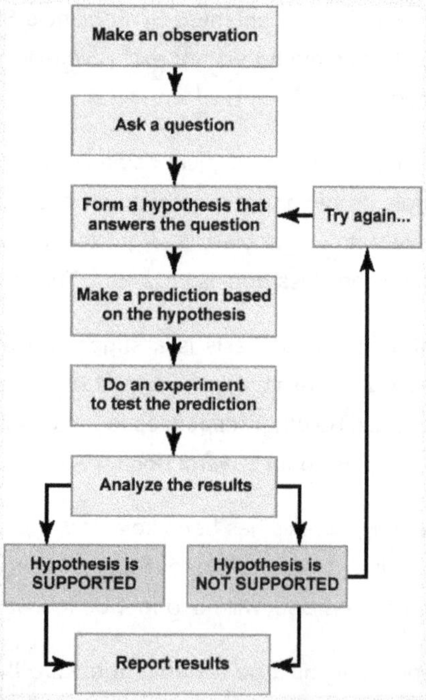

Make an observation
Ask a question
Form a hypothesis that answers the question
Try again...
Make a prediction based on the hypothesis
Do an experiment to test the prediction
Analyze the results
Hypothesis is SUPPORTED
Hypothesis is NOT SUPPORTED
Report results

In the example below, the scientific method is used to solve an everyday problem. Which part in the example below is the hypothesis? Which is the prediction? Based on the results of the experiment, is the hypothesis supported? If it is not supported, propose some alternative hypotheses.

1. My toaster doesn't toast my bread.
2. Why doesn't my toaster work?
3. There is something wrong with the electrical outlet.
4. If something is wrong with the outlet, my coffeemaker also won't work when plugged into it.
5. I plug my coffeemaker into the outlet.
6. My coffeemaker works.

In practice, the scientific method is not as rigid and structured as it might at first appear. Sometimes an experiment leads to conclusions that favor a change in approach; often, an experiment brings entirely new scientific questions to the puzzle. Many times, science does not operate in a linear fashion; instead, scientists continually draw inferences and make generalizations, finding patterns as their research proceeds. Scientific reasoning is more complex than the scientific method alone suggests.

Basic and Applied Science

The scientific community has been debating for the last few decades about the value of different types of science. Is it valuable to pursue science for the sake of simply gaining knowledge, or does scientific knowledge only have worth if we can apply it to solving a specific problem or bettering our lives? This question focuses on the differences between two types of science: basic science and applied science.

Basic science or "pure" science seeks to expand knowledge regardless of the short-term application of that knowledge. It is not focused on developing a product or a service of immediate public or commercial value. The immediate goal of basic science is knowledge for knowledge's sake, though this does not mean that in the end it may not result in an application.

In contrast, applied science or "technology," aims to use science to solve real-world problems, making it possible, for example, to improve a crop yield, find a cure for a particular disease, or save animals threatened by a natural disaster. In applied science, the problem is usually defined for the researcher.

Some individuals may perceive applied science as "useful" and basic science as "useless." A question these people might pose to a scientist advocating knowledge acquisition would be, "What for?" A careful look at the history of science, however, reveals that basic knowledge has resulted in many remarkable applications of great value. Many scientists think that a basic understanding of science is necessary before an application is developed; therefore, applied science relies on the results generated through basic science. Other scientists think that it is time to move on from basic science and instead to find solutions to actual problems. Both approaches are valid. It is true that there are problems that demand immediate attention; however, few solutions would be found without the help of the knowledge generated through basic science.

One example of how basic and applied science can work together to solve practical problems occurred after the discovery of DNA structure led to an understanding of the molecular mechanisms governing DNA replication. Strands of DNA, unique in every human, are found in our cells, where they provide the instructions necessary for life. During DNA replication, new copies of DNA are made, shortly before a cell divides to form new cells. Understanding the mechanisms of DNA replication enabled scientists to develop laboratory techniques that are now used to identify genetic diseases, pinpoint individuals who were at a crime scene, and determine paternity. Without basic science, it is unlikely that applied science would exist.

Another example of the link between basic and applied research is the Human Genome Project, a study in which each human chromosome was analyzed and mapped to determine the precise sequence of DNA subunits and the exact location of each gene. (The gene is the basic unit of heredity; an individual's complete collection of genes is his or her genome.) Other organisms have also been studied as part of this project to gain a better understanding of human chromosomes. The Human Genome Project relied on basic research carried out with non-human organisms and, later, with the human genome. An important end goal eventually became using the data for applied research seeking cures for genetically related diseases.

Reporting Scientific Work

Whether scientific research is basic science or applied science, scientists must share their findings for other researchers to expand and build upon their discoveries. Communication and collaboration within and between sub disciplines of science are key to the advancement of knowledge in science. For this reason, an important aspect of a scientist's work is disseminating results and communicating with peers. Scientists can share results by presenting them at a scientific meeting or conference, but this approach can reach only the limited few who are present. Instead, most scientists present their results in peer-reviewed articles that are published in scientific journals. Peer-reviewed articles are scientific papers that are reviewed, usually anonymously by a scientist's colleagues, or peers. These colleagues are qualified individuals, often experts in the same research area, who judge whether or not the scientist's work is suitable for publication. The process of peer review helps to ensure that the research described in a scientific paper or grant proposal is original, significant, logical, and thorough. Grant proposals, which are requests for research funding, are also subject to peer review. Scientists publish their work so other scientists can reproduce their experiments under similar or different conditions to expand on the findings. The experimental results must be consistent with the findings of other scientists.

There are many journals and the popular press that do not use a peer-review system. A large number of online open-access journals, journals with articles available without cost, are now available many of which use rigorous peer-review systems, but some of which do not. Results of any studies published in these forums without peer review are not reliable and should not form the basis for other scientific work. In one exception, journals may allow a researcher to cite a personal communication from another researcher about unpublished results with the cited author's permission.

Characteristics of Life

Biology is the science that studies life, but what exactly is life? This may sound like a silly question with an obvious response, but it is not always easy to define life. For example, a branch of biology called virology studies viruses, which exhibit some of the characteristics of living entities but lack others. It turns out that although viruses can attack living organisms, cause diseases, and even reproduce, they do not meet the criteria that biologists use to define life. Consequently, virologists are not biologists, strictly speaking. Similarly, some biologists study the early molecular evolution that gave rise to life; since the events that preceded life are not biological events, these scientists are also excluded from biology in the strict sense of the term.

From its earliest beginnings, biology has wrestled with three questions: What are the shared properties that make something "alive"? And once we know something is alive, how do we find meaningful levels of organization in its structure? And, finally, when faced with the remarkable diversity of life, how do we organize the different kinds of organisms so that we can better understand them? As new organisms are discovered every day, biologists continue to seek answers to these and other questions.

Properties of Life

All living organisms share several key characteristics or functions: order, sensitivity or response to the environment, reproduction, adaptation, growth and development, regulation, homeostasis, energy processing, and evolution. When viewed together, these nine characteristics serve to define life.

Order

Organisms are highly organized, coordinated structures that consist of one or more cells. Even very simple, single-celled organisms are remarkably complex: inside each cell, atoms make up molecules; these in turn make up cell organelles and other cellular inclusions. In multicellular organisms, similar cells form tissues. Tissues, in turn, collaborate to create organs (body structures with a distinct function). Organs work together to form organ systems.

Sensitivity or Response to Stimuli

Organisms respond to diverse stimuli. For example, plants can bend toward a source of light, climb on fences and walls, or respond to touch. Even tiny bacteria can move toward or away from chemicals (a process called *chemotaxis*) or light (*phototaxis*). Movement toward a stimulus is considered a positive response, while movement away from a stimulus is considered a negative response.

Reproduction

Single-celled organisms reproduce by first duplicating their DNA, and then dividing it equally as the cell prepares to divide to form two new cells. Multicellular organisms often produce specialized reproductive germline cells that will form new individuals. When reproduction occurs, genes containing DNA are passed along to an organism's offspring. These genes ensure that the offspring will belong to the same species and will have similar characteristics, such as size and shape.

Growth and Development

Organisms grow and develop following specific instructions coded for by their genes. These genes provide instructions that will direct cellular growth and development, ensuring that a species' young will grow up to exhibit many of the same characteristics as its parents.

Regulation

Even the smallest organisms are complex and require multiple regulatory mechanisms to coordinate internal functions, respond to stimuli, and cope with environmental stresses. Two examples of internal functions regulated in an organism are nutrient transport and blood flow. Organs (groups of tissues working together) perform specific functions, such as carrying oxygen throughout the body, removing wastes, delivering nutrients to every cell, and cooling the body.

Homeostasis

What is Homeostasis?

Living Things Must Maintain Homeostasis. **Homeostasis** means **"steady state".** Homeostasis is the tendency of an organism or cell to **maintain a constant internal environment**. Living things constantly adjust to internal and external changes. Homeostasis means to maintain **dynamic equilibrium** in the body. It is dynamic because it is constantly adjusting to the changes that the body's systems encounter. It is equilibrium because body functions are kept within specific ranges or normal limits.

Maintaining homeostasis requires that the body continuously monitor its internal conditions. From body temperature to blood pressure to levels of certain nutrients, each physiological condition has a particular set point. A **set point** is the physiological value around which the normal range fluctuates. A **normal range** is the restricted set of values that is optimally healthful and stable. For example, the set point for normal human body temperature is approximately 37°C (98.6°F) Physiological parameters, such as body temperature and blood pressure, tend to fluctuate within a normal range a few degrees above and below that point. Even an animal that is apparently inactive is maintaining this homeostatic equilibrium. An inability to maintain homeostasis may lead to death or a disease, a condition known as homeostatic imbalance.

Homeostasis, in a general sense, refers to stability, balance, or equilibrium. Physiologically, it is the body's attempt to maintain a constant and balanced internal environment, which requires persistent monitoring and adjustments as conditions change. Adjustment of physiological systems within the body is called homeostatic regulation, which involves three parts or mechanisms:

- **Sensory receptors**
- **Control center – (Brain)**
- **Effector organs (muscles / glands)**
- The **sensory receptor** receives information that something in the environment is changing. A change in the internal or external environment is called a **stimulus** and is detected by a receptor. sensory receptor monitors a physiological value. This value is reported to the control center.
- The **control center or integration center (Brain)** receives and processes information from the sensory receptor.
- The **effector organ** responds to the commands of the control center by either opposing or enhancing the stimulus. The effector is a **muscle** (that contracts or relaxes) or a **gland** that secretes.

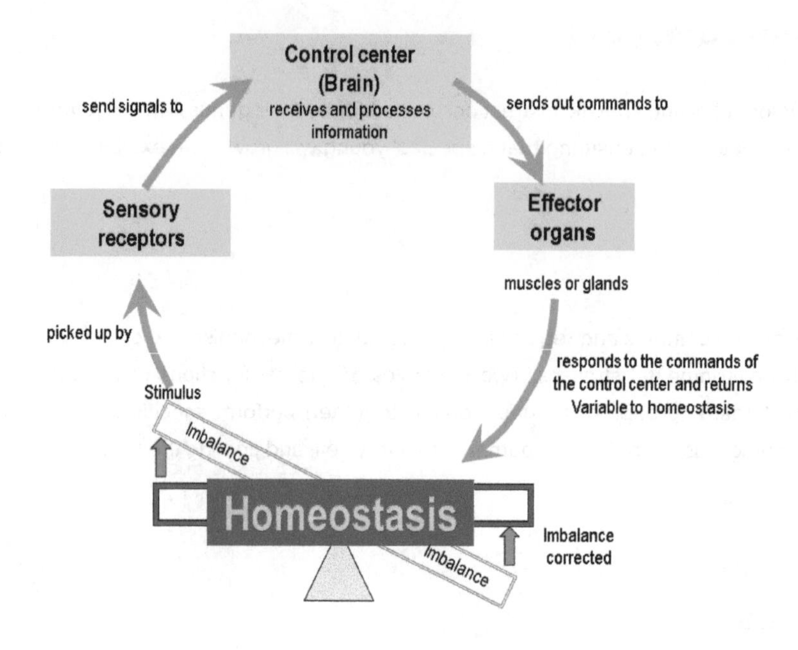

Components of a feedback loop

Maintenance of Homeostasis by Feedback loops

When a change occurs in an animal's environment, an adjustment must be made. The receptor senses the change in the environment, then sends a signal to the control center (in most cases, the brain) which in turn generates a response that is signaled to an effector. Homeostasis is controlled by the nervous and endocrine system of mammals. **Homeostatsis is maintained by negative feedback loops.** Positive feedback loops actually push the organism further out of homeostasis, but may be necessary for life to occur.

Types of homeostatic feedback systems / mechanisms:

- Negative Feedback Mechanism
- Positive Feedback Mechanism

Examples for Homeostasis

Negative Feedback Mechanism maintains	**Positive Feedback Mechanism**
Body Temperature	Childbirth
Blood pressure	Blood Clotting
Blood Sugar	Ovulation
Blood pH	Lactation

Negative Feedback Mechanisms

Any homeostatic process that changes the direction of the stimulus is a **negative feedback loop**. It may either increase or decrease the stimulus, but the stimulus is not allowed to continue as it did before the receptor sensed it. In other words, if a level is too high, the body does something to bring it down, and conversely, if a level is too low, the body does something to make it go up. Hence the term negative feedback.

Thermoregulation: :

Thermoregulation is a process that allows your body to maintain its core internal temperature. The set point for normal human body temperature is approximately 37°C (98.6°F). Body temperature affects body activities. Body proteins, including enzymes, begin to denature and lose their function with high heat (around 50°C for mammals). Enzyme activity will decrease by half for every ten degree centigrade drop in temperature, to the point of freezing, with a few exceptions.

During body temperature regulation, **temperature receptors in the skin (sensory receptor)** communicate information to the **brain (the control center)** which signals the **blood vessels and sweat glands in the skin (effectors).** As the internal and external environment of the body are constantly changing, adjustments must be made continuously to stay at or near a specific value: the **set point**. (approximately 37°C / 98.6°F)

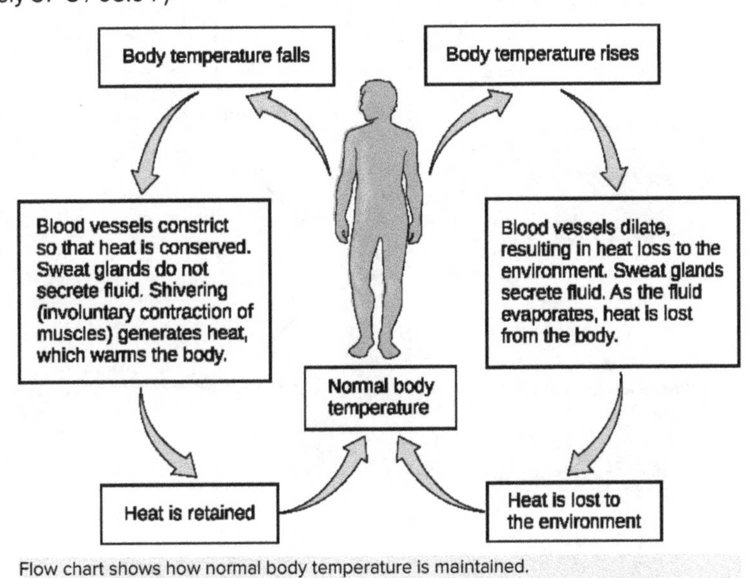

Flow chart shows how normal body temperature is maintained.

When body temperature exceeds its normal range:

When the brain's temperature regulation center receives data from the sensors indicating that the body's temperature exceeds its normal range, the brain (the control center) which signals the effectors: blood vessels and sweat glands in the skin. This stimulation has three major effects:

- Blood vessels in the skin begin to dilate allowing more blood from the body core to flow to the surface of the skin allowing the heat to radiate into the environment.
- As blood flow to the skin increases, sweat glands are activated to increase their output. As the sweat evaporates from the skin surface into the surrounding air, it takes heat with it.
- The depth of respiration increases, and a person may breathe through an open mouth instead of through the nasal passageways. This further increases heat loss from the lungs.

When body temperature reduces below its normal range:

In contrast, activation of the brain's heat-gain center by exposure to cold

- reduces blood flow to the skin, and blood returning from the limbs is diverted into a network of deep veins. This arrangement traps heat closer to the body core and restricts heat loss.
- If heat loss is severe, the brain triggers an increase in random signals to skeletal muscles, causing them to contract and producing shivering. The muscle contractions of shivering release heat while using up ATP.
- The brain triggers the thyroid gland in the endocrine system to release thyroid hormone, which increases metabolic activity and heat production in cells throughout the body. The brain also signals the adrenal glands to release epinephrine (adrenaline), a hormone that causes the breakdown of glycogen into glucose, which can be used as an energy source. The breakdown of glycogen into glucose also results in increased metabolism and heat production.

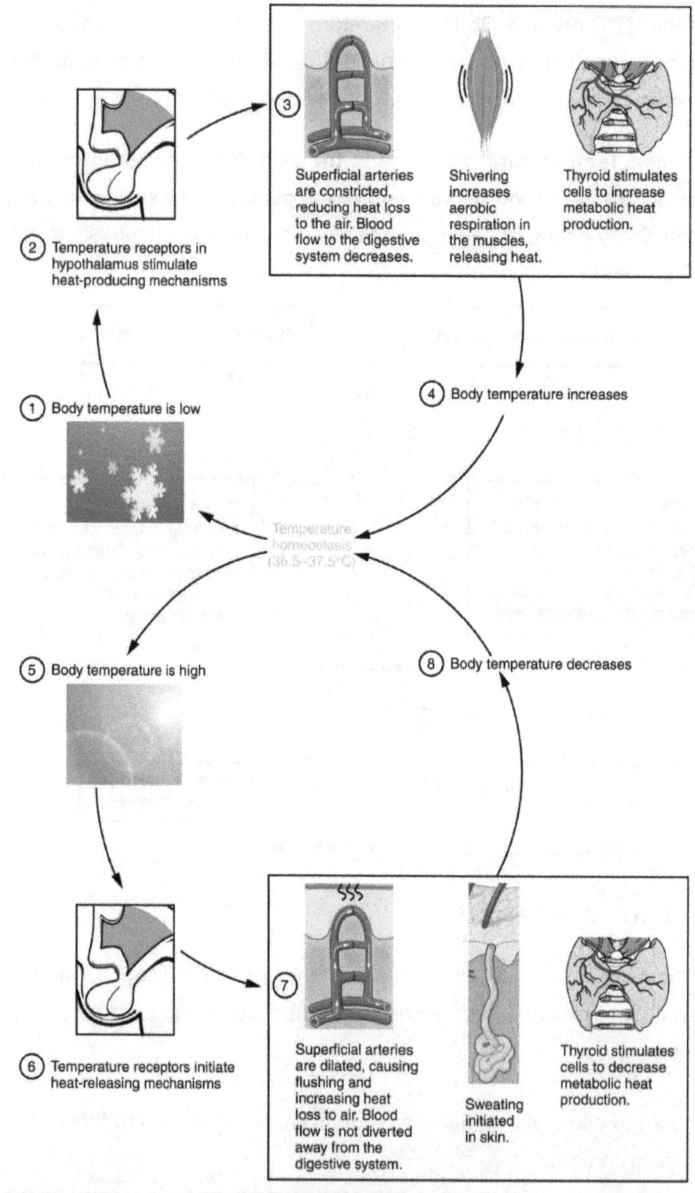

The hypothalamus in Brain controls thermoregulation.

Positive Feedback Mechanism

Positive feedback intensifies a change in the body's physiological condition rather than reversing it, maintains the direction of the stimulus, possibly accelerating it. A deviation from the normal range results in more change, and the system moves farther away from the normal range. Positive feedback in the body is normal only when there is a definite end point. Childbirth and the body's response to blood loss are two examples of positive feedback loops that are normal but are activated only when needed.

Example 1: Labor and Delivery During Childbirth:

The extreme muscular work of labor and delivery are the result of a positive feedback system . The first contractions of labor (the stimulus) push the baby toward the cervix (the lowest part of the uterus). The cervix contains **stretch-sensitive nerve cells (the sensory receptors)** that monitor the degree of stretching . These nerve cells send messages to the brain, which in turn causes the **pituitary gland at the base of the brain (control center)** to release the hormone **oxytocin** into the bloodstream. Oxytocin causes stronger contractions of the smooth muscles in of the **uterus (the effectors),** pushing the baby further down the birth canal. This causes even greater stretching of the cervix. The cycle of stretching, oxytocin release, and increasingly more forceful contractions stops only when the baby comes out of the uterus. At this point, the stretching of the cervix halts, stopping the release of oxytocin.

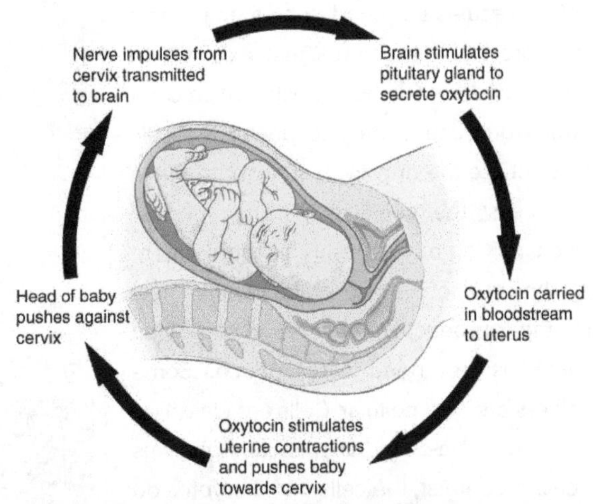

Positive Feedback Loop. Normal childbirth is driven by a positive feedback loop. A positive feedback loop results in a change in the body's status, rather than a return to homeostasis.

Example 2: Steps in Blood Clotting

A second example of positive feedback centers on reversing extreme damage to the body. Following a penetrating wound, the most immediate threat is excessive blood loss. Less blood circulating means reduced blood pressure and reduced perfusion (penetration of blood) to the brain and other vital organs. If perfusion is severely reduced, vital organs will shut down and the person will die. The body responds to this potential catastrophe by releasing substances in the injured blood vessel wall that begin the process of blood clotting. **As each step of clotting occurs, it stimulates the release of more clotting substances.** This accelerates the processes of clotting and sealing off the damaged area. Clotting is contained in a local area based on the tightly controlled availability of clotting proteins. This is an adaptive, life-saving cascade of events.

Energy Processing

All organisms use a source of energy for their metabolic activities. Some organisms capture energy from the sun and convert it into chemical energy in food; others use chemical energy in molecules they take in as food.

Levels of Organization of Living Things

Living things are highly organized and structured, following a hierarchy that can be examined on a scale from small to large. The atom is the smallest and most fundamental unit of matter. It consists of a nucleus surrounded by electrons. Atoms form molecules. A molecule is a chemical structure consisting of at least two atoms held together by one or more chemical bonds. Many molecules that are biologically important are macromolecules, large molecules that are typically formed by polymerization (a polymer is a large molecule that is made by combining smaller units called monomers, which are simpler than macromolecules). An example of a macromolecule is deoxyribonucleic acid (DNA), which contains the instructions for the structure and functioning of all living organisms.

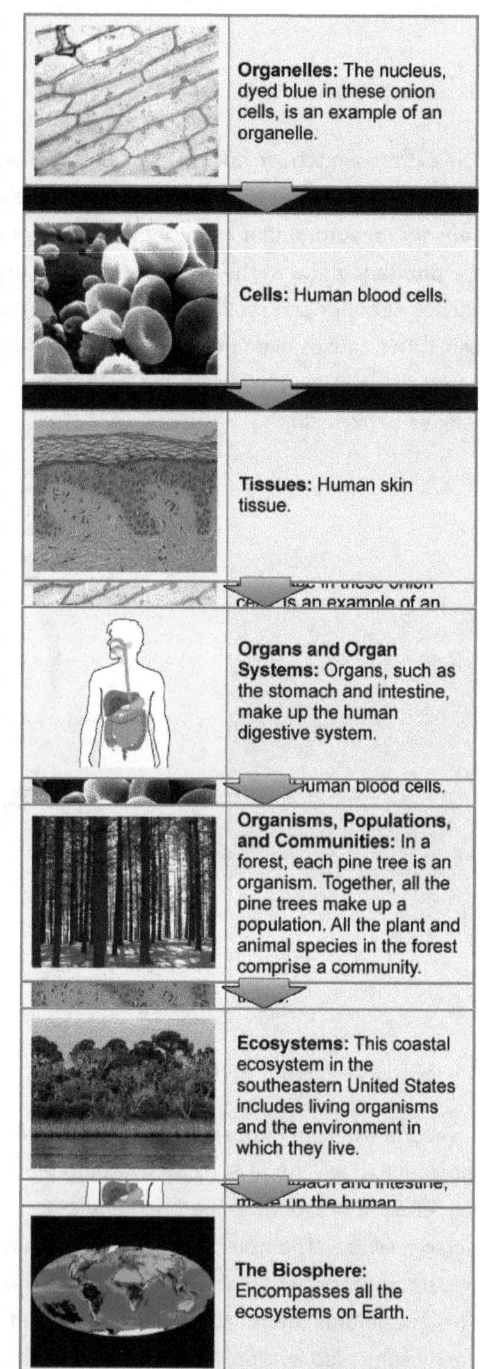

The biological levels of organization of living things are shown. From a single organelle to the entire biosphere, living organisms are parts of a highly structured hierarchy. (credit "organelles": modification of work by Umberto Salvagnin; credit "cells": modification of work by Bruce Wetzel, Harry Schaefer/ National Cancer Institute; credit "tissues": modification of work by Kilbad; Fama Clamosa; Mikael Häggström; credit "organs": modification of work by Mariana Ruiz Villareal; credit "organisms": modification of work by "Crystal"/Flickr; credit "ecosystems": modification of work by US Fish and Wildlife Service Headquarters; credit "biosphere": modification of work by NASA)

Some cells contain aggregates of macromolecules surrounded by membranes; these are called organelles. Organelles are small structures that exist within cells. Examples of organelles include mitochondria and chloroplasts, which carry out indispensable functions: mitochondria produce energy to power the cell, while chloroplasts enable green plants to utilize the energy in sunlight to make sugars. All living things are made of cells; the cell itself is the smallest fundamental unit of structure and function in living organisms. (This requirement is why viruses are not considered living: they are not made of cells. To make new viruses, they have to invade and hijack the reproductive mechanism of a living cell; only then can they obtain the materials they need to reproduce.) Some organisms consist of a single cell and others are multicellular. Cells are classified as prokaryotic or eukaryotic. Prokaryotes are single-celled or colonial organisms that do not have membrane-bound nuclei; in contrast, the cells of eukaryotes do have membrane-bound organelles and a membrane-bound nucleus.

In larger organisms, cells combine to make tissues, which are groups of similar cells carrying out similar or related functions. Organs are collections of tissues grouped together performing a common function. Organs are present not only in animals but also in plants. An organ system is a higher level of organization that consists of functionally related organs. Mammals have many organ systems. For instance, the circulatory system transports blood through the body and to and from the lungs; it includes organs such as the heart and blood vessels. Organisms are individual living entities. For example, each tree in a forest is an organism. Single-celled prokaryotes and single-celled eukaryotes are also considered organisms and are typically referred to as microorganisms.

All the individuals of a species living within a specific area are collectively called a population. For example, a forest may include many pine trees. All of these pine trees represent the population of pine trees in this forest. Different populations may live in the same specific area. For example, the forest with the pine trees includes populations of flowering plants and also insects and microbial populations. A community is the sum of populations inhabiting a particular area. For instance, all of the trees, flowers, insects, and other populations in a forest form the forest's community. The forest itself is an ecosystem. An ecosystem consists of all the living things in a particular area together with the abiotic, non-living parts of that environment such as nitrogen in the soil or rain water. At the highest level of organization, the biosphere is the collection of all ecosystems, and it represents the zones of life on earth. It includes land, water, and even the atmosphere to a certain extent.

Which of the following statements is false?

1. Tissues exist within organs which exist within organ systems.
2. Communities exist within populations which exist within ecosystems.
3. Organelles exist within cells which exist within tissues.
4. Communities exist within ecosystems which exist in the biosphere.

The Diversity of Life

The fact that biology, as a science, has such a broad scope has to do with the tremendous diversity of life on earth. The source of this diversity is evolution, the process of gradual change during which new species arise from older species. Evolutionary biologists study the evolution of living things in everything from the microscopic world to ecosystems.

The evolution of various life forms on Earth can be summarized in a phylogenetic tree. A phylogenetic tree is a diagram showing the evolutionary relationships among biological species based on similarities and differences in genetic or physical traits or both. A phylogenetic tree is composed of nodes and branches. The internal nodes represent ancestors and are points in evolution when, based on scientific evidence, an ancestor is thought to have diverged to form two new species. The length of each branch is proportional to the time elapsed since the split.

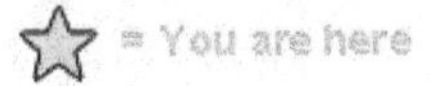

This phylogenetic tree was constructed by microbiologist Carl Woese using data obtained from sequencing ribosomal RNA genes. The tree shows the separation of living organisms into three domains: Bacteria, Archaea, and Eukarya. Bacteria and Archaea are prokaryotes, single-celled organisms lacking intracellular organelles. (credit: Eric Gaba; NASA Astrobiology Institute)

Evolution Connection

Carl Woese and the Phylogenetic TreeIn the past, biologists grouped living organisms into five kingdoms: animals, plants, fungi, protists, and bacteria. The organizational scheme was based mainly on physical features, as opposed to physiology, biochemistry, or molecular biology, all of which are used by modern systematics. The pioneering work of American microbiologist Carl Woese in the early 1970s has shown, however, that life on Earth has evolved along three lineages, now called domains—Bacteria, Archaea, and Eukarya. The first two are prokaryotic cells with microbes that lack membrane-enclosed nuclei and organelles. The third domain contains the eukaryotes and includes unicellular microorganisms together with the four original kingdoms (excluding bacteria). Woese defined Archaea as a new domain, and this resulted in a new taxonomic tree. Many organisms belonging to the Archaea domain live under extreme conditions and are called extremophiles. To construct his tree, Woese used genetic relationships rather than similarities based on morphology (shape).

Woese's tree was constructed from comparative sequencing of the genes that are universally distributed, present in every organism, and conserved (meaning that these genes have remained essentially unchanged throughout evolution). Woese's approach was revolutionary because comparisons of physical features are insufficient to differentiate between the prokaryotes that appear fairly similar in spite of their tremendous biochemical diversity and genetic variability. The comparison of homologous DNA and RNA sequences provided Woese with a sensitive device that revealed the extensive variability of prokaryotes, and which justified the separation of the prokaryotes into two domains: bacteria and archaea.

Concept Check

1. **How** is anatomy different from physiology?
2. **Define** the scientific method.
3. **List** the steps of the scientific method and **briefly describe** of what happens at each step.
4. **Compare and contrast** inductive reasoning and deductive reasoning.
5. **What** is the difference between descriptive science and hypothesis-based science?
6. **What** is a hypothesis and **how** is it different from a prediction?
7. **What** is a variable?
8. **Why** is a control an important part of an experiment?
9. **How** is basic science different from applied science?
10. **Why** is it so important to share scientific work?
11. **List and describe** the properties of life.
12. **What** is homeostasis? **Give** one example of each negative feedback and positive feedback.
13. **Define** each level of organization in living things.
14. **What** is a phylogenetic tree used for?

CHAPTER 2 — CHEMISTRY OF LIFE

Chemistry

LEARNING OBJECTIVES

- Describe the fundamental composition of matter
- Distinguish between ionic bonds, covalent bonds, and hydrogen bonds
- Compare and contrast the four important classes of organic (carbon-based) compounds—proteins, carbohydrates, lipids and nucleic acids—according to their composition and functional importance to human life
- Discuss the relationships between matter, mass, elements, compounds, atoms, and subatomic particles
- Compare and contrast ionic, covalent (polar and nonpolar), and hydrogen bonds
- Identify the properties of water that make it essential to life

The smallest, most fundamental material components of the human body are basic chemical elements. In fact, chemicals called nucleotide bases are the foundation of the genetic code with the instructions on how to build and maintain the human body from conception through old age. There are about three billion of these base pairs in human DNA.

Human chemistry includes organic molecules (carbon-based) and biochemicals (those produced by the body). Human chemistry also includes elements. In fact, life cannot exist without many of the elements that are part of the earth. All of the elements that contribute to chemical reactions, to the transformation of energy, and to electrical activity and muscle contraction—elements that include phosphorus, carbon, sodium, and calcium, to name a few—originated in stars.

These elements, in turn, can form both the inorganic and organic chemical compounds important to life, including, for example, water, glucose, and proteins. This chapter begins by examining elements and how the structures of atoms, the basic units of matter, determine the characteristics of elements by the number of protons, neutrons, and electrons in the atoms. The chapter then builds the framework of life from there.

Look around you. Everything is made of chemicals of one sort or another. Life is chemistry organized into astonishing complexity and intricacy. To make sense of this organization we can look at life's chemistry as a hierarchy—levels of organization. From simple elemental ions, to simple organic molecules, complexity rises with increasingly larger macromolecules.

A person is between 1–2 meters (m) tall, but there are many length scales and biological levels of detail which are important for understanding anatomy and physiology. For perspective on size difference, consider an atom. Atoms are basic units of matter. Atoms contain a positive center (nucleus) surrounded by a cloud of electrons that allow interatomic interactions. We can't really grasp how small atoms are, but think big instead of small. A length of 10^{10} m is more than the distance from the Earth to the moon.

Table 1. Units of Measurement

Elements	Unit
Atoms and ions	Å = 10^{-10} m
Molecules	nm = 10^{-9} m
Cells	μm 10^{-6} m
Tissues	mm= 10^{-3} m
Organs	cm = 10^{-2} m

The smallest length scale that we will cover is the size of individual atoms, but the movement of subatomic particles called electrons, can change atomic charge. Ions are atoms that carry either a positive or negative charge from altered numbers of electrons, and many atoms and molecules exist in the body as ions.

Ionic chemistry is important in human medicine and health. Ions play an essential role in physiological processes, particularly as they move across cell membranes. Appropriate intracellular and extracellular concentrations of sodium, potassium and calcium ions are required for nerve impulses and heart beats, enable cell-to-cell communication and initiate cellular processes. For example, release of insulin by beta cells of the pancreas is mediated by ions. Transport of ions across membranes may occur by passive diffusion, through ion channels, or through pumps. Pumps often move ions against a concentration gradient. Ionic chemistry is important in human medicine. Anesthetic drugs such as Novocain block sodium channels. Neurotoxins from some snakes and puffer fish work by blocking ion movements that mormally occur in nerve transmission. Malfunctions in ionic channel or pump molecules can result in serious physiological ailments, including cystic fibrosis (mutation in a gene that codes for cell membrane chloride channel) and epilepsy.

Even subatomic particles, which are too small to see with the best microscopes in the world, play an extremely important role in maintaining proper physiology.

The substance of the universe—from a grain of sand to a star—is called **matter**. Scientists define matter as anything that occupies space and has mass. An object's mass and its weight are related concepts, but not quite the same. An object's mass is the amount of matter contained in the object, and the object's mass is the same whether that object is on Earth or in the zero-gravity environment of outer space. An object's weight, on the other hand, is its mass as affected by the pull of gravity. Where gravity strongly pulls on an object's mass its weight is greater than it is where gravity is less strong. An object of a certain mass weighs less on the moon, for example, than it does on Earth because the gravity of the moon is less than that of Earth. In other words, weight is variable, and is influenced by gravity. A piece of cheese that weighs a pound on Earth weighs only a few ounces on the moon.

Elements and Compounds

All matter in the natural world is composed of one or more of the 92 fundamental substances called elements. An **element** is a pure substance that is distinguished from all other matter by the fact that it cannot be created or broken down by ordinary chemical means. While your body can assemble many of the chemical compounds needed for life from their constituent elements, it cannot make elements. They must come from the environment. A familiar example of an element that you must take in is calcium (Ca^{++}). Calcium is essential to the human body; it is absorbed and used for a number of processes, including strengthening bones. When you consume dairy products your digestive system breaks down the food into components small enough to cross into the bloodstream. Among these is calcium, which, because it is an element, cannot be broken down further. The elemental calcium in cheese, therefore, is the same as the calcium that forms your bones. Some other elements you might be familiar with are oxygen, sodium, and iron. The elements in the human body are shown in the table below, beginning with the most abundant: oxygen (O), carbon (C), hydrogen (H), and nitrogen (N). Each element's name can be replaced by a one- or two-letter symbol; you will become familiar with some of these during this course. All the elements in your body are derived from the foods you eat and the air you breathe.

***Elements of the Human Body.* The main elements that compose the human body are shown from most abundant to least abundant.**

Element	Symbol	Percentage in Body	At a Look
Oxygen	O	65.0	
Carbon	C	18.5	
Hydrogen	H	9.5	
Nitrogen	N	3.2	
Calcium	Ca	1.5	
Phosphorus	P	1.0	
Potassium	K	0.4	
Sulfur	S	0.3	
Sodium	Na	0.2	
Chlorine	Cl	0.2	
Magnesium	Mg	0.1	
Trace elements include boron (B), chromium (Cr), cobalt (Co), copper (Cu), fluorine (F), iodine (I), iron (Fe), manganese (Mn), molybdenum (Mo), selenium (Se), silicon (Si), tin (Sn), vanadium (V), and zinc (Zn)		less than 1.0	

In nature, elements rarely occur alone. Instead, they combine to form compounds. A *compound* is a substance composed of two or more elements joined by chemical bonds. For example, the compound glucose is an important body fuel. It is always composed of the same three elements: carbon, hydrogen, and oxygen. Moreover, the elements that make up any given compound always occur in the same relative amounts. In glucose, there are always six carbon and six oxygen units for every twelve hydrogen units. But what, exactly, are these “units” of elements?

Atoms and Subatomic Particles

An **atom** is the smallest quantity of an element that retains the unique properties of that element. In other words, an atom of hydrogen is a unit of hydrogen—the smallest amount of hydrogen that can exist. As you might guess, atoms are almost unfathomably small. The period at the end of this sentence is millions of atoms wide.

Atomic Structure and Energy

Atoms are made up of even smaller subatomic particles, three types of which are important: the **proton**, **neutron**, and ***electron***. The number of positively-charged protons and non-charged (“neutral”) neutrons, gives mass to the atom, and the number of each in the nucleus of the atom determine the element. The number of negatively-charged electrons that “spin” around the nucleus at close to the speed of light equals the number of protons. An electron has about 1/2000th the mass of a proton or neutron.

Figure 1 shows two models that can help you imagine the structure of an atom—in this case, helium (He). In the planetary model, helium’s two electrons are shown circling the nucleus in a fixed orbit depicted as a ring. Although this model is helpful in visualizing atomic structure, in reality, electrons do not travel in fixed orbits, but whiz around the nucleus erratically in a so-called electron cloud.

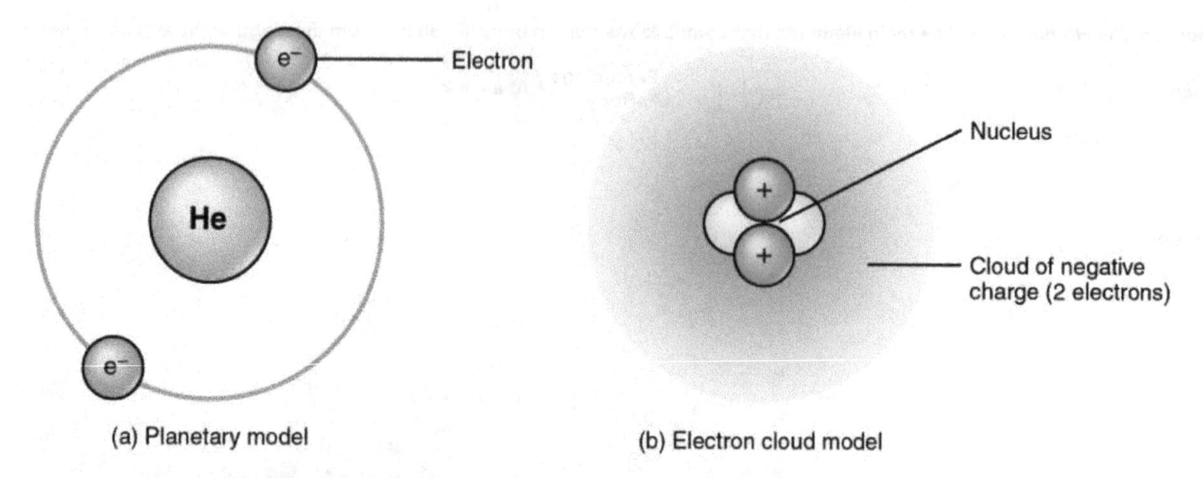

Figure 1. Two Models of Atomic Structure. (a) In the planetary model, the electrons of helium are shown in fixed orbits, depicted as rings, at a precise distance from the nucleus, somewhat like planets orbiting the sun. (b) In the electron cloud model, the electrons of carbon are shown in the variety of locations they would have at different distances from the nucleus over time.

An atom's protons and electrons carry electrical charges. Protons, with their positive charge, are designated p^+. Electrons, which have a negative charge, are designated e^-. An atom's neutrons have no charge: they are electrically neutral. Just as a magnet sticks to a steel refrigerator because their opposite charges attract, the positively charged protons attract the negatively charged electrons. This mutual attraction gives the atom some structural stability. The attraction by the positively charged nucleus helps keep electrons from straying far. The number of protons and electrons within a neutral atom are equal, thus, the atom's overall charge is balanced.

Atomic Number and Mass Number

An atom of carbon is unique to carbon, but a proton of carbon is not. One proton is the same as another, whether it is found in an atom of carbon, sodium (Na), or iron (Fe). The same is true for neutrons and electrons. So, what gives an element its distinctive properties—what makes carbon so different from sodium or iron? The answer is the unique quantity of protons each contains. Carbon by definition is an element whose atoms contain six protons. No other element has exactly six protons in its atoms. Moreover, *all* atoms of carbon, whether found in your liver or in a lump of coal, contain six protons. Thus, the ***atomic number***, which is the number of protons in the nucleus of the atom, identifies the element. Because an atom usually has the same number of electrons as protons, the atomic number identifies the usual number of electrons as well.

In their most common form, many elements also contain the same number of neutrons as protons. The most common form of carbon, for example, has six neutrons as well as six protons, for a total of 12 subatomic particles in its nucleus. An element's **mass number** is the sum of the number of protons and neutrons in its nucleus. So the most common form of carbon's mass number is 12. (Electrons have so little mass that they do not appreciably contribute to the mass of an atom.) Carbon is a relatively light element. Uranium (U), in contrast, has a mass number of 238 and is referred to as a heavy metal. Its atomic number is 92 (it has 92 protons) but it contains 146 neutrons; it has the most mass of all the naturally occurring elements.

The **periodic table of the elements**, shown in Figure 2, is a chart identifying the 92 elements found in nature, as well as several larger, unstable elements discovered experimentally. The elements are arranged in order of their atomic number, with hydrogen and helium at the top of the table, and the more massive elements below. The periodic table is a useful device because for each element, it identifies the chemical symbol, the atomic number, and the mass number, while organizing elements according to their propensity to react with other elements. The number of protons and electrons in an element are equal. The number of protons and neutrons may be equal for some elements, but are not equal for all.

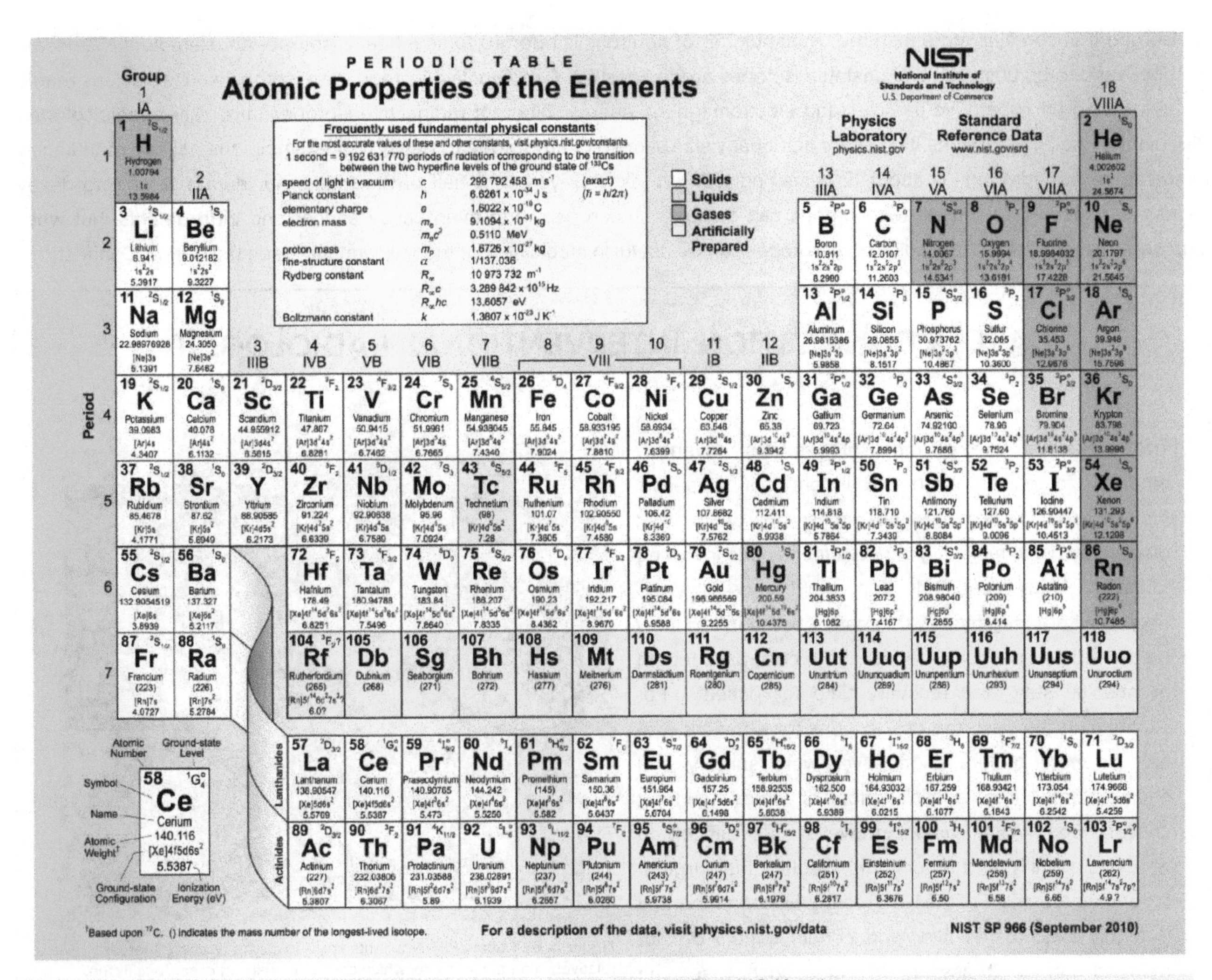

Figure 2. The Periodic Table of the Elements. (credit: R.A. Dragoset, A. Musgrove, C.W. Clark, W.C. Martin)

Isotopes

Although each element has a unique number of protons, it can exist as different isotopes. An *isotope* is one of the different forms of an element, distinguished from one another by different numbers of neutrons. The standard isotope of carbon is ^{12}C, commonly called carbon twelve. ^{12}C has six protons and six neutrons, for a mass number of twelve. All of the isotopes of carbon have the same number of protons; therefore, ^{13}C has seven neutrons, and ^{14}C has eight neutrons. The different isotopes of an element can also be indicated with the mass number hyphenated (for example, C-12 instead of ^{12}C). Hydrogen has three common isotopes, shown in Figure 3.

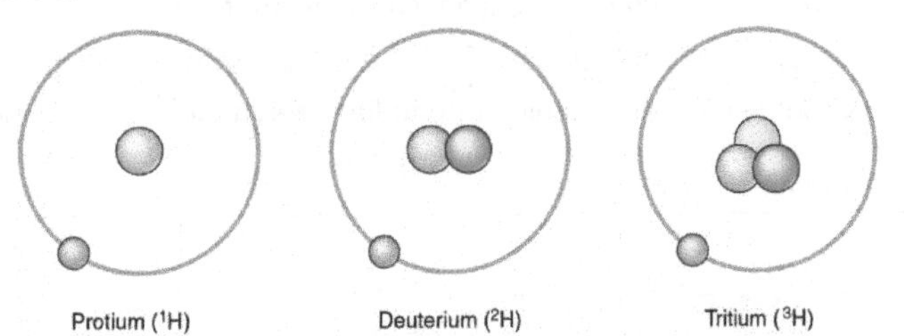

Figure 3. Isotopes of Hydrogen. Protium, designated ^1H, has one proton and no neutrons. It is by far the most abundant isotope of hydrogen in nature. Deuterium, designated ^2H, has one proton and one neutron. Tritium, designated ^3H, has two neutrons.

An isotope that contains more than the usual number of neutrons is referred to as a heavy isotope. An example is ^{14}C. Heavy isotopes tend to be unstable, and unstable isotopes are radioactive. A *radioactive isotope* is an isotope whose nucleus readily decays, giving off subatomic particles and electromagnetic energy. Different radioactive isotopes (also called radioisotopes) differ in their half-life, the time it takes for half of any size sample of an isotope to decay. For example, the half-life of tritium—a radioisotope of hydrogen—is about 12 years, indicating it takes 12 years for half of the tritium nuclei in a sample to decay. Excessive exposure to radioactive isotopes can damage human cells and even cause cancer and birth defects, but when exposure is controlled, some radioactive isotopes can be useful in medicine. For more information, see the Career Connections.

CAREER CONNECTION: INTERVENTIONAL RADIOLOGIST

The controlled use of radioisotopes has advanced medical diagnosis and treatment of disease. Interventional radiologists are physicians who treat disease by using minimally invasive techniques involving radiation. Many conditions that could once only be treated with a lengthy and traumatic operation can now be treated non-surgically, reducing the cost, pain, length of hospital stay, and recovery time for patients. For example, in the past, the only options for a patient with one or more tumors in the liver were surgery and chemotherapy (the administration of drugs to treat cancer). Some liver tumors, however, are difficult to access surgically, and others could require the surgeon to remove too much of the liver. Moreover, chemotherapy is highly toxic to the liver, and certain tumors do not respond well to it anyway. In some such cases, an interventional radiologist can treat the tumors by disrupting their blood supply, which they need if they are to continue to grow. In this procedure, called radioembolization, the radiologist accesses the liver with a fine needle, threaded through one of the patient's blood vessels. The radiologist then inserts tiny radioactive "seeds" into the blood vessels that supply the tumors. In the days and weeks following the procedure, the radiation emitted from the seeds destroys the vessels and directly kills the tumor cells in the vicinity of the treatment.

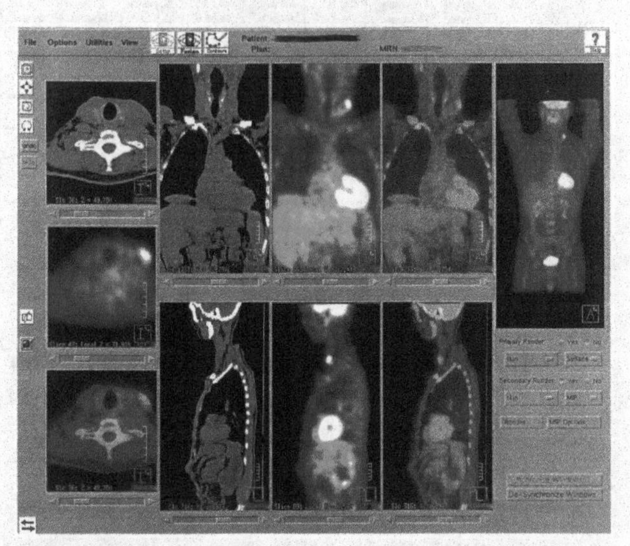

Figure 4. PET Scan. PET highlights areas in the body where there is relatively high glucose use, which is characteristic of cancerous tissue. This PET scan shows sites of the spread of a large primary tumor to other sites.

Radioisotopes emit subatomic particles that can be detected and tracked by imaging technologies. One of the most advanced uses of radioisotopes in medicine is the positron emission tomography (PET) scanner, which detects the activity in the body of a very small injection of radioactive glucose, the simple sugar that cells use for energy. The PET camera reveals to the medical team which of the patient's tissues are taking up the most glucose. Thus, the most metabolically active tissues show up as bright "hot spots" on the images (Figure 4). PET can reveal some cancerous masses because cancer cells consume glucose at a high rate to fuel their rapid reproduction.

The Behavior of Electrons

In the human body, atoms do not exist as independent entities. Rather, they are constantly reacting with other atoms to form and to break down more complex substances. To fully understand anatomy and physiology you must grasp how atoms participate in such reactions. The key is understanding the behavior of electrons.

Although electrons do not follow rigid orbits a set distance away from the atom's nucleus, they do tend to stay within certain regions of space called electron shells. An *electron shell* is a layer of electrons that encircle the nucleus at a distinct energy level.

The atoms of the elements found in the human body have from one to five electron shells, and all electron shells hold eight electrons except the first shell, which can only hold two. This configuration of electron shells is the same for all atoms. The precise number of shells depends on the number of electrons in the atom. Hydrogen and helium have just one and two electrons, respectively. If you take a look at the periodic table of the elements, you will notice that hydrogen and helium are placed alone on either sides of the top row; they are the only elements that have just one electron shell (Figure 5). A second shell is necessary to hold the electrons in all elements larger than hydrogen and helium.

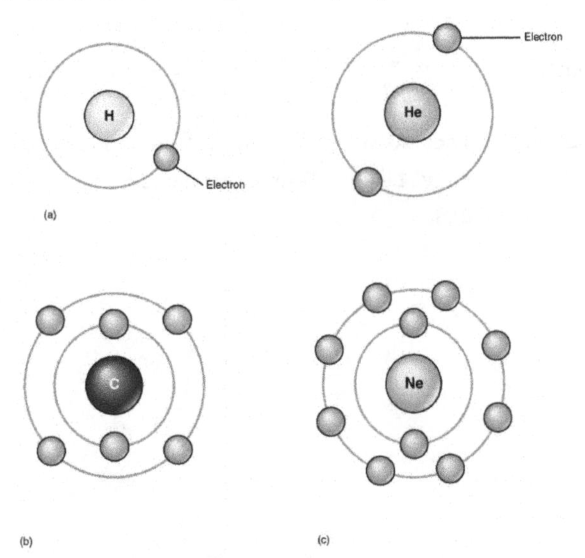

Figure 5. Electron Shells. Electrons orbit the atomic nucleus at distinct levels of energy called electron shells. (a) With one electron, hydrogen only half-fills its electron shell. Helium also has a single shell, but its two electrons completely fill it. (b) The electrons of carbon completely fill its first electron shell, but only half-fills its second. (c) Neon, an element that does not occur in the body, has 10 electrons, filling both of its electron shells.

Lithium (Li), whose atomic number is 3, has three electrons. Two of these fill the first electron shell, and the third spills over into a second shell. The second electron shell can accommodate as many as eight electrons. Carbon, with its six electrons, entirely fills its first shell, and half-fills its second. With ten electrons, neon (Ne) entirely fills its two electron shells. Again, a look at the periodic table reveals that all of the elements in the second row, from lithium to neon, have just two electron shells. Atoms with more than ten electrons require more than two shells. These elements occupy the third and subsequent rows of the periodic table.

The factor that most strongly governs the tendency of an atom to participate in chemical reactions is the number of electrons in its valence shell. A *valence shell* is an atom's outermost electron shell. If the valence shell is full, the atom is stable; meaning its electrons are unlikely to be pulled away from the nucleus by the electrical charge of other atoms. If the valence shell is not full, the atom is reactive; meaning it will tend to react with other atoms in ways that make the valence shell full. Consider hydrogen, with its one electron only half-filling its valence shell. This single electron is likely to be drawn into relationships with the atoms of other elements, so that hydrogen's single valence shell can be stabilized.

All atoms (except hydrogen and helium with their single electron shells) are most stable when there are exactly eight electrons in their valence shell. This principle is referred to as the octet rule, and it states that an atom will give up, gain, or share electrons with another atom so that it ends up with eight electrons in its own valence shell. For example, oxygen, with six electrons in its valence shell, is likely to react with other atoms in a way that results in the addition of two electrons to oxygen's valence shell, bringing the number to eight. When two hydrogen atoms each share their single electron with oxygen, covalent bonds are formed, resulting in a molecule of water, H_2O.

In nature, atoms of one element tend to join with atoms of other elements in characteristic ways. For example, carbon commonly fills its valence shell by linking up with four atoms of hydrogen. In so doing, the two elements form the simplest of organic molecules, methane, which also is one of the most abundant and stable carbon-containing compounds on Earth. As stated above, another example is water; oxygen needs two electrons to fill its valence shell. It commonly interacts with two atoms of hydrogen, forming H_2O. Incidentally, the name "hydrogen" reflects its contribution to water (hydro- = "water"; -gen = "maker"). Thus, hydrogen is the "water maker."

Chemical Bonds

Atoms separated by a great distance cannot link; rather, they must come close enough for the electrons in their valence shells to interact. But do atoms ever actually touch one another? Most physicists would say no, because the negatively charged electrons in their valence shells repel one another. No force within the human body—or anywhere in the natural world—is strong enough to overcome this electrical repulsion. So when you read about atoms linking together or colliding, bear in mind that the atoms are not merging in a physical sense.

Instead, atoms link by forming a chemical bond. A *bond* is a weak or strong electrical attraction that holds atoms in the same vicinity. The new grouping is typically more stable—less likely to react again—than its component atoms were when they were separate. A more or less stable grouping of two or more atoms held together by chemical bonds is called a **molecule**. The bonded atoms may be of the same element, as in the case of H_2, which is called molecular hydrogen or hydrogen gas. When a molecule is made up of two or more atoms of different elements, it is called a chemical **compound**. Thus, a unit of water, or H_2O, is a compound, as is a single molecule of the gas methane, or CH_4.

Three types of chemical bonds are important in human physiology, because they hold together substances that are used by the body for critical aspects of homeostasis, signaling, and energy production, to name just a few important processes. These are ionic bonds, covalent bonds, and hydrogen bonds.

Ions and Ionic Bonds

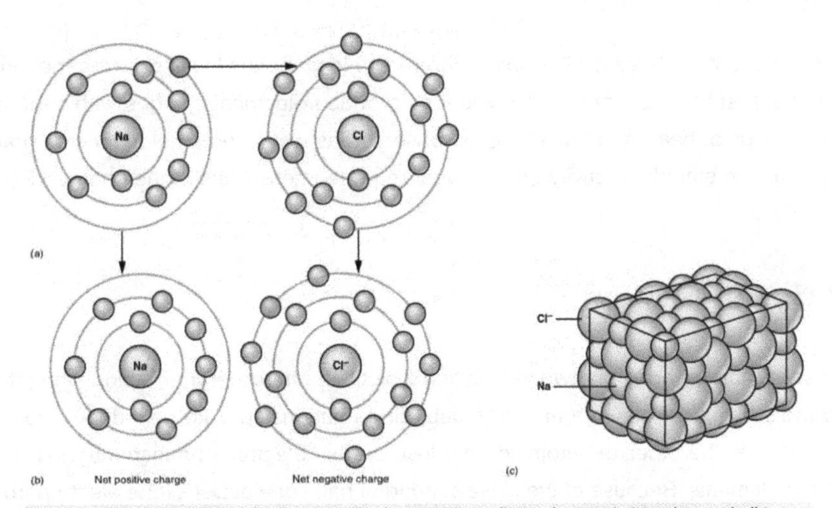

Figure 1. Ionic Bonding. (a) Sodium readily donates the solitary electron in its valence shell to chlorine, which needs only one electron to have a full valence shell. (b) The opposite electrical charges of the resulting sodium cation and chloride anion result in the formation of a bond of attraction called an ionic bond. (c) The attraction of many sodium and chloride ions results in the formation of large groupings called crystals.

Recall that an atom typically has the same number of positively charged protons and negatively charged electrons. As long as this situation remains, the atom is electrically neutral. But when an atom participates in a chemical reaction that results in the donation or acceptance of one or more electrons, the atom will then become positively or negatively charged. This happens frequently for most atoms in order to have a full valence shell, as described previously. This can happen either by gaining electrons to fill a shell that is more than half-full, or by giving away electrons to empty a shell than is less than half-full, thereby leaving the next smaller electron shell as the new, full, valence shell. An atom that has an electrical charge—whether positive or negative—is an **ion**.

Potassium (K), for instance, is an important element in all body cells. Its atomic number is 19. It has just one electron in its valence shell. This characteristic makes potassium highly likely to participate in chemical reactions in which it donates one electron. (It is easier for potassium to donate one electron than to gain seven electrons.) The loss will cause the positive charge of potassium's protons to be more influential than the negative charge of potassium's electrons. In other words, the resulting potassium ion will be slightly positive. A potassium ion is written K^+, indicating that it has lost a single electron. A positively charged ion is known as a **cation**.

Now consider fluorine (F), a component of bones and teeth. Its atomic number is nine, and it has seven electrons in its valence shell. Thus, it is highly likely to bond with other atoms in such a way that fluorine accepts one electron (it is easier for fluorine to gain one electron than to donate seven electrons). When it does, its electrons will outnumber its protons by one, and it will have an overall negative charge. The ionized form of fluorine is called fluoride, and is written as F^-. A negatively charged ion is known as an **anion**.

Atoms that have more than one electron to donate or accept will end up with stronger positive or negative charges. A cation that has donated two electrons has a net charge of +2. Using magnesium (Mg) as an example, this can be written Mg^{++} or Mg^{2+}. An anion that has accepted two electrons has a net charge of –2. The ionic form of selenium (Se), for example, is typically written Se^{2-}.

The opposite charges of cations and anions exert a moderately strong mutual attraction that keeps the atoms in close proximity forming an ionic bond. An **ionic bond** is an ongoing, close association between ions of opposite charge. The table salt you sprinkle on your food owes its existence to ionic bonding. As shown in Figure 1, sodium commonly donates an electron to chlorine, becoming the cation Na^+. When chlorine accepts the electron, it becomes the chloride anion, Cl^-. With their opposing charges, these two ions strongly attract each other.

Water is an essential component of life because it is able to break the ionic bonds in salts to free the ions. In fact, in biological fluids, most individual atoms exist as ions. These dissolved ions produce electrical charges within the body. The behavior of these ions produces the tracings of heart and brain function observed as waves on an electrocardiogram (EKG or ECG) or an electroencephalogram (EEG). The electrical activity that derives from the interactions of the charged ions is why they are also called electrolytes.

Covalent Bonds

Unlike ionic bonds formed by the attraction between a cation's positive charge and an anion's negative charge, molecules formed by a **covalent bond** share electrons in a mutually stabilizing relationship. Like next-door neighbors whose kids hang out first at one home and then at the other, the atoms do not lose or gain electrons permanently. Instead, the electrons move back and forth between the elements. Because of the close sharing of pairs of electrons (one electron from each of two atoms), covalent bonds are stronger than ionic bonds.

Nonpolar Covalent Bonds

Figure 2 shows several common types of covalent bonds. Notice that the two covalently bonded atoms typically share just one or two electron pairs, though larger sharings are possible. The important concept to take from this is that in covalent bonds, electrons in the outermost valence shell are shared to fill the valence shells of both atoms, ultimately stabilizing both of the atoms involved. In a single covalent bond, a single electron is shared between two atoms, while in a double covalent bond, two pairs of electrons are shared between two atoms. There even are triple covalent bonds, where three atoms are shared.

(a) A single covalent bond: hydrogen gas (H—H). Two atoms of hydrogen each share their solitary electron in a single covalent bond.

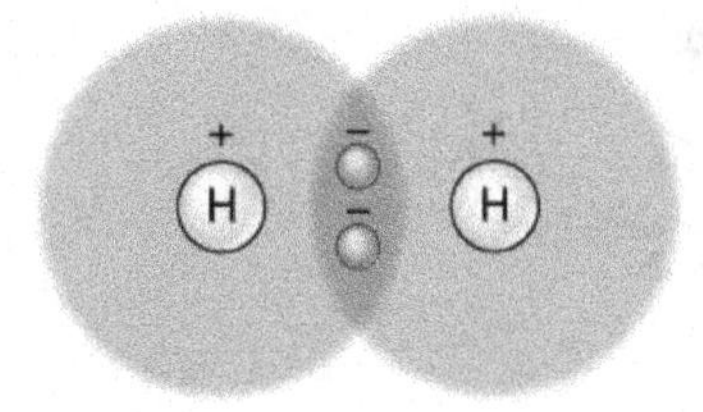

(b) A double covalent bond: oxygen gas (O=O). An atom of oxygen has six electrons in its valence shell; thus, two more would make it stable. Two atoms of oxygen achieve stability by sharing two pairs of electrons in a double covalent bond.

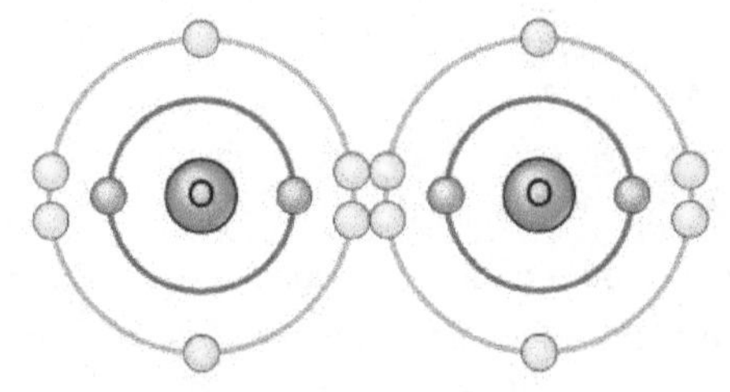

Molecule of oxygen gas (O_2)

(c) Two double covalent bonds: carbon dioxide (O=C=O). An atom of carbon has four electrons in its valence shell; thus, four more would make it stable. An atom of carbon and two atoms of oxygen achieve stability by sharing two electron pairs each, in two double covalent bonds.

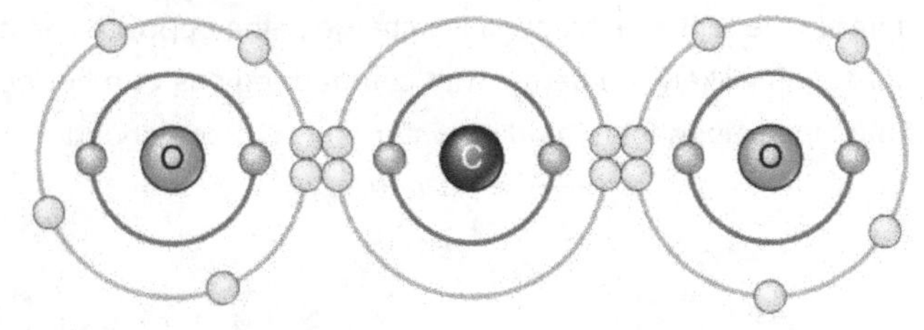

Figure 2. Covalent Bonding

You can see that the covalent bonds shown in Figure 2 are balanced. The sharing of the negative electrons is relatively equal, as is the electrical pull of the positive protons in the nucleus of the atoms involved. This is why covalently bonded molecules that are electrically balanced in this way are described as nonpolar; that is, no region of the molecule is either more positive or more negative than any other.

Polar Covalent Bonds

Groups of legislators with completely opposite views on a particular issue are often described as "polarized" by news writers. In chemistry, a *polar molecule* is a molecule that contains regions that have opposite electrical charges. Polar molecules occur when atoms share electrons unequally, in polar covalent bonds.

The most familiar example of a polar molecule is water (Figure 3). The molecule has three parts: one atom of oxygen, the nucleus of which contains eight protons, and two hydrogen atoms, whose nuclei each contain only one proton. Because every proton exerts an identical positive charge, a nucleus that contains eight protons exerts a charge eight times greater than a nucleus that contains one proton. This means that the negatively charged electrons present in the water molecule are more strongly attracted to the oxygen nucleus than to the hydrogen nuclei. Each hydrogen atom's single negative electron therefore migrates toward the oxygen atom, making the oxygen end of their bond slightly more negative than the hydrogen end of their bond.

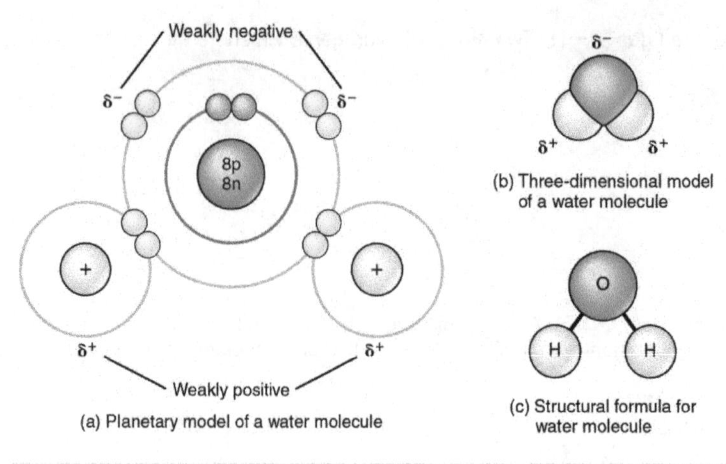

Figure 3. Polar Covalent Bonds in a Water Molecule

What is true for the bonds is true for the water molecule as a whole; that is, the oxygen region has a slightly negative charge and the regions of the hydrogen atoms have a slightly positive charge. These charges are often referred to as "partial charges" because the strength of the charge is less than one full electron, as would occur in an ionic bond. As shown in Figure 3, regions of weak polarity are indicated with the Greek letter delta (∂) and a plus (+) or minus (–) sign.

Even though a single water molecule is unimaginably tiny, it has mass, and the opposing electrical charges on the molecule pull that mass in such a way that it creates a shape somewhat like a triangular tent (see Figure 3b). This dipole, with the positive charges at one end formed by the hydrogen atoms at the "bottom" of the tent and the negative charge at the opposite end (the oxygen atom at the "top" of the tent) makes the charged regions highly likely to interact with charged regions of other polar molecules. For human physiology, the resulting bond is one of the most important formed by water—the hydrogen bond.

Hydrogen Bonds

A **hydrogen bond** is formed when a weakly positive hydrogen atom already bonded to one electronegative atom (for example, the oxygen in the water molecule) is attracted to another electronegative atom from another molecule. In other words, hydrogen bonds always include hydrogen that is already part of a polar molecule.

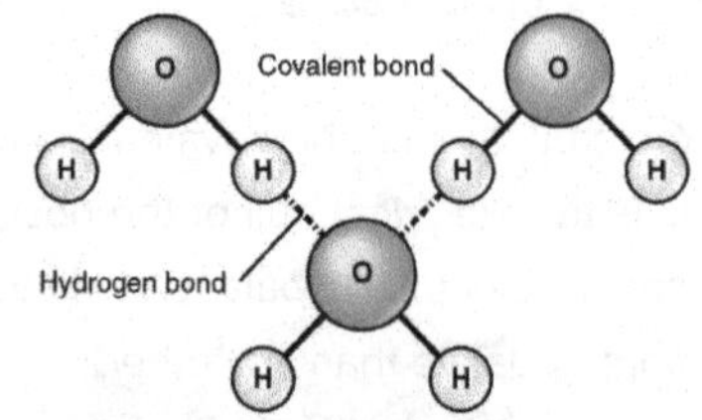

Figure 4. Hydrogen Bonds between Water Molecules. Notice that the bonds occur between the weakly positive charge on the hydrogen atoms and the weakly negative charge on the oxygen atoms. Hydrogen bonds are relatively weak, and therefore are indicated with a dotted (rather than a solid) line.

The most common example of hydrogen bonding in the natural world occurs between molecules of water. It happens before your eyes whenever two raindrops merge into a larger bead, or a creek spills into a river. Hydrogen bonding occurs because the weakly negative oxygen atom in one water molecule is attracted to the weakly positive hydrogen atoms of two other water molecules (Figure 4).

Water molecules also strongly attract other types of charged molecules as well as ions. This explains why "table salt," for example, actually is a molecule called a "salt" in chemistry, which consists of equal numbers of positively-charged sodium (Na^+) and negatively-charged chloride (Cl^-), dissolves so readily in water, in this case forming dipole-ion bonds between the water and the electrically-charged ions (electrolytes). Water molecules also repel molecules with nonpolar covalent bonds, like fats, lipids, and oils. You can demonstrate this with a simple kitchen experiment: pour a teaspoon of vegetable oil, a compound formed by nonpolar covalent bonds, into a glass of water. Instead of instantly dissolving in the water, the oil forms a distinct bead because the polar water molecules repel the nonpolar oil.

Water

The concepts you have learned so far in this chapter govern all forms of matter, and would work as a foundation for geology as well as biology. This section of the chapter narrows the focus to the chemistry of human life; that is, the compounds important for the body's structure and function. In general, these compounds are either inorganic or organic.

- An **inorganic compound** is a substance that does not contain both carbon and hydrogen. A great many inorganic compounds do contain hydrogen atoms, such as water (H_2O) and the hydrochloric acid (HCl) produced by your stomach. In contrast, only a handful of inorganic compounds contain carbon atoms. Carbon dioxide (CO_2) is one of the few examples.
- An **organic compound**, then, is a substance that contains both carbon and hydrogen. Organic compounds are synthesized via covalent bonds within living organisms, including the human body. Recall that carbon and hydrogen are the second and third most abundant elements in your body. You will soon discover how these two elements combine in the foods you eat, in the compounds that make up your body structure, and in the chemicals that fuel your functioning.

The following section examines the three groups of inorganic compounds essential to life: water, salts, acids, and bases. Organic compounds are covered later in the chapter.

Water

As much as 70 percent of an adult's body weight is water. This water is contained both within the cells and between the cells that make up tissues and organs. Its several roles make water indispensable to human functioning.

Water as a Lubricant and Cushion

Water is a major component of many of the body's lubricating fluids. Just as oil lubricates the hinge on a door, water in synovial fluid lubricates the actions of body joints, and water in pleural fluid helps the lungs expand and recoil with breathing. Watery fluids help keep food flowing through the digestive tract, and ensure that the movement of adjacent abdominal organs is friction free.

Water also protects cells and organs from physical trauma, cushioning the brain within the skull, for example, and protecting the delicate nerve tissue of the eyes. Water cushions a developing fetus in the mother's womb as well.

Water as a Heat Sink

A heat sink is a substance or object that absorbs and dissipates heat but does not experience a corresponding increase in temperature. In the body, water absorbs the heat generated by chemical reactions without greatly increasing in temperature. Moreover, when the environmental temperature soars, the water stored in the body helps keep the body cool. This cooling effect happens as warm blood from the body's core flows to the blood vessels just under the skin and is transferred to the environment. At the same time, sweat glands release warm water in sweat. As the water evaporates into the air, it carries away heat, and then the cooler blood from the periphery circulates back to the body core.

Water as a Component of Liquid Mixtures

A mixture is a combination of two or more substances, each of which maintains its own chemical identity. In other words, the constituent substances are not chemically bonded into a new, larger chemical compound. The concept is easy to imagine if you think of powdery substances such as flour and sugar; when you stir them together in a bowl, they obviously do not bond to form a new compound. The room air you breathe is a gaseous mixture, containing three discrete elements—nitrogen, oxygen, and argon—and one compound, carbon dioxide. There are three types of liquid mixtures, all of which contain water as a key component. These are solutions, colloids, and suspensions.

For cells in the body to survive, they must be kept moist in a water-based liquid called a solution. In chemistry, a liquid **solution** consists of a solvent that dissolves a substance called a solute. An important characteristic of solutions is that they are homogeneous; that is, the solute molecules are distributed evenly throughout the solution. If you were to stir a teaspoon of sugar into a glass of water, the sugar would dissolve into sugar molecules separated by water molecules. The ratio of sugar to water in the left side of the glass would be the same as the ratio of sugar to water in the right side of the glass. If you were to add more sugar, the ratio of sugar to water would change, but the distribution—provided you had stirred well—would still be even.

Water is considered the "universal solvent" and it is believed that life cannot exist without water because of this. Water is certainly the most abundant solvent in the body; essentially all of the body's chemical reactions occur among compounds dissolved in water. Because water molecules are polar, with regions of positive and negative electrical charge, water readily dissolves ionic compounds and polar covalent compounds. Such compounds are referred to as hydrophilic, or "water-loving." As mentioned above, sugar dissolves well in water. This is because sugar molecules contain regions of hydrogen-oxygen polar bonds, making it hydrophilic. Nonpolar molecules, which do not readily dissolve in water, are called hydrophobic, or "water-fearing."

The Role of Water in Chemical Reactions

Two types of chemical reactions involve the creation or the consumption of water: dehydration synthesis and hydrolysis.

- In dehydration synthesis, one reactant gives up an atom of hydrogen and another reactant gives up a hydroxyl group (OH) in the synthesis of a new product. In the formation of their covalent bond, a molecule of water is released as a byproduct (Figure 1). This is also sometimes referred to as a condensation reaction.
- In hydrolysis, a molecule of water disrupts a compound, breaking its bonds. The water is itself split into H and OH. One portion of the severed compound then bonds with the hydrogen atom, and the other portion bonds with the hydroxyl group.

These reactions are reversible, and play an important role in the chemistry of organic compounds (which will be discussed shortly).

(a) Dehydration synthesis

Monomers are joined by removal of OH from one monomer and removal of H from the other at the site of bond formation.

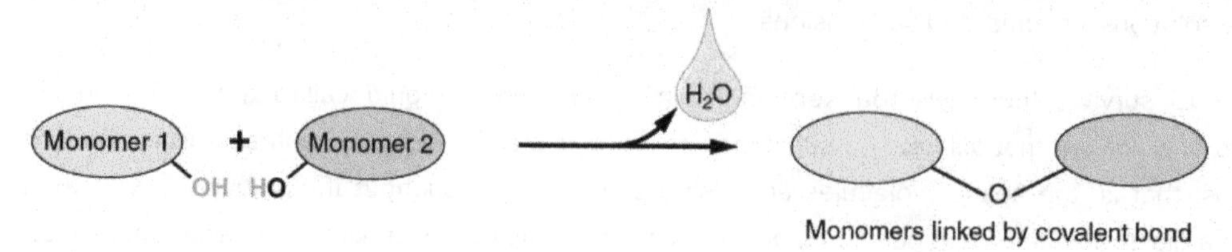

(b) Hydrolysis

Monomers are released by the addition of a water molecule, adding OH to one monomer and H to the other.

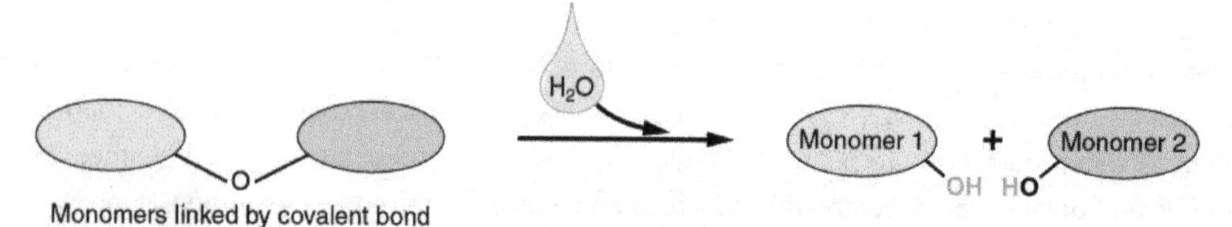

Figure 1. Dehydration Synthesis and Hydrolysis. Monomers, the basic units for building larger molecules, form polymers (two or more chemically-bonded monomers). (a) In dehydration synthesis, two monomers are covalently bonded in a reaction in which one gives up a hydroxyl group and the other a hydrogen atom. A molecule of water is released as a byproduct during dehydration reactions. (b) In hydrolysis, the covalent bond between two monomers is split by the addition of a hydrogen atom to one and a hydroxyl group to the other, which requires the contribution of one molecule of water.

Salts

Recall that salts are formed when ions form ionic bonds. In these reactions, one atom gives up one or more electrons, and thus becomes positively charged, whereas the other accepts one or more electrons and becomes negatively charged. You can now define a salt as a substance that, when dissolved in water, dissociates into ions other than H^+ or OH^-. This fact is important in distinguishing salts from acids and bases, discussed next.

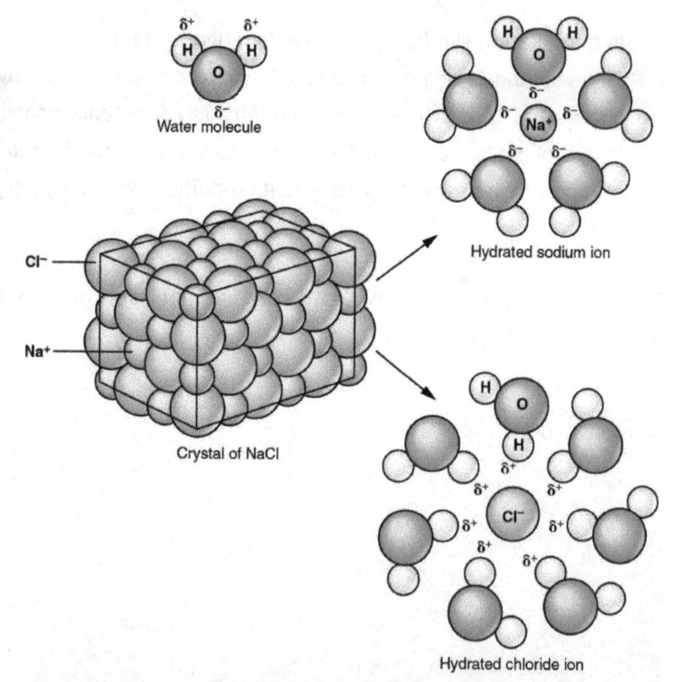

Figure 2. Dissociation of Sodium Chloride in Water. Notice that the crystals of sodium chloride dissociate not into molecules of NaCl, but into Na^+ cations and Cl^- anions, each completely surrounded by water molecules.

A typical salt, NaCl, dissociates completely in water (Figure 2). The positive and negative regions on the water molecule (the hydrogen and oxygen ends respectively) attract the negative chloride and positive sodium ions, pulling them away from each other. Again, whereas nonpolar and polar covalently bonded compounds break apart into molecules in solution, salts dissociate into ions. These ions are electrolytes; they are capable of conducting an electrical current in solution. This property is critical to the function of ions in transmitting nerve impulses and prompting muscle contraction.

Many other salts are important in the body. For example, bile salts produced by the liver help break apart dietary fats, and calcium phosphate salts form the mineral portion of teeth and bones.

Acids and Bases

Acids and bases, like salts, dissociate in water into electrolytes. Acids and bases can very much change the properties of the solutions in which they are dissolved.

Acids

An **acid** is a substance that releases hydrogen ions (H^+) in solution (Figure 3a). Because an atom of hydrogen has just one proton and one electron, a positively charged hydrogen ion is simply a proton. This solitary proton is highly likely to participate in chemical reactions. Strong acids are compounds that release all of their H^+ in solution; that is, they ionize completely. Hydrochloric acid (HCl), which is released from cells in the lining of the stomach, is a strong acid because it releases all of its H^+ in the stomach's watery environment. This strong acid aids in digestion and kills ingested microbes. Weak acids do not ionize completely; that is, some of their hydrogenions remain bonded within a compound in solution. An example of a weak acid is vinegar, or acetic acid; it is called acetate after it gives up a proton.

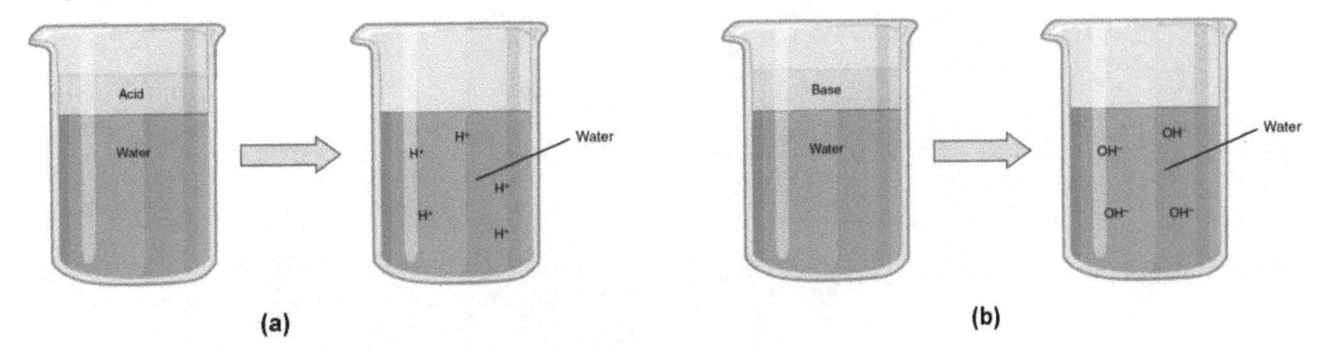

Figure 3. Acids and Bases. (a) In aqueous solution, an acid dissociates into hydrogen ions (H^+) and anions. Nearly every molecule of a strong acid dissociates, producing a high concentration of H^+. (b) In aqueous solution, a base dissociates into hydroxyl ions (OH^-) and cations. Nearly every molecule of a strong base dissociates, producing a high concentration of OH^-.

Bases

A **base** is a substance that releases hydroxyl ions (OH^-) in solution, or one that accepts H^+ already present in solution (see Figure 3b). The hydroxyl ions or other base combine with H^+ present to form a water molecule, thereby removing H^+ and reducing the solution's acidity. Strong bases release most or all of their hydroxyl ions; weak bases release only some hydroxyl ions or absorb only a few H^+. Food mixed with hydrochloric acid from the stomach would burn the small intestine, the next portion of the digestive tract after the stomach, if it were not for the release of bicarbonate (HCO_3^-), a weak base that attracts H^+. Bicarbonate accepts some of the H^+ protons, thereby reducing the acidity of the solution.

The Concept of pH

The relative acidity or alkalinity of a solution can be indicated by its pH. A solution's **pH** is the negative, base-10 logarithm of the hydrogen ion (H^+) concentration of the solution. As an example, a pH 4 solution has an H^+ concentration that is ten times greater than that of a pH 5 solution. That is, a solution with a pH of 4 is ten times more acidic than a solution with a pH of 5. The concept of pH will begin to make more sense when you study the pH scale, like that shown in Figure 4. The scale consists of a series of increments ranging from 0 to 14. A solution with a pH of 7 is considered neutral—neither acidic nor basic. Pure water has a pH of 7. The lower the number below 7, the more acidic the solution, or the greater the concentration of H^+. The concentration of hydrogen ions at each pH value is 10 times different than the next pH. For instance, a pH value of 4 corresponds to a proton concentration of 10^{-4} M, or 0.0001M, while a pH value of 5 corresponds to a proton concentration of 10^{-5} M, or 0.00001M. The higher the number above 7, the more basic (alkaline) the solution, or the lower the concentration of H^+. Human urine, for example, is ten times more acidic than pure water, and HCl is 10,000,000 times more acidic than water.

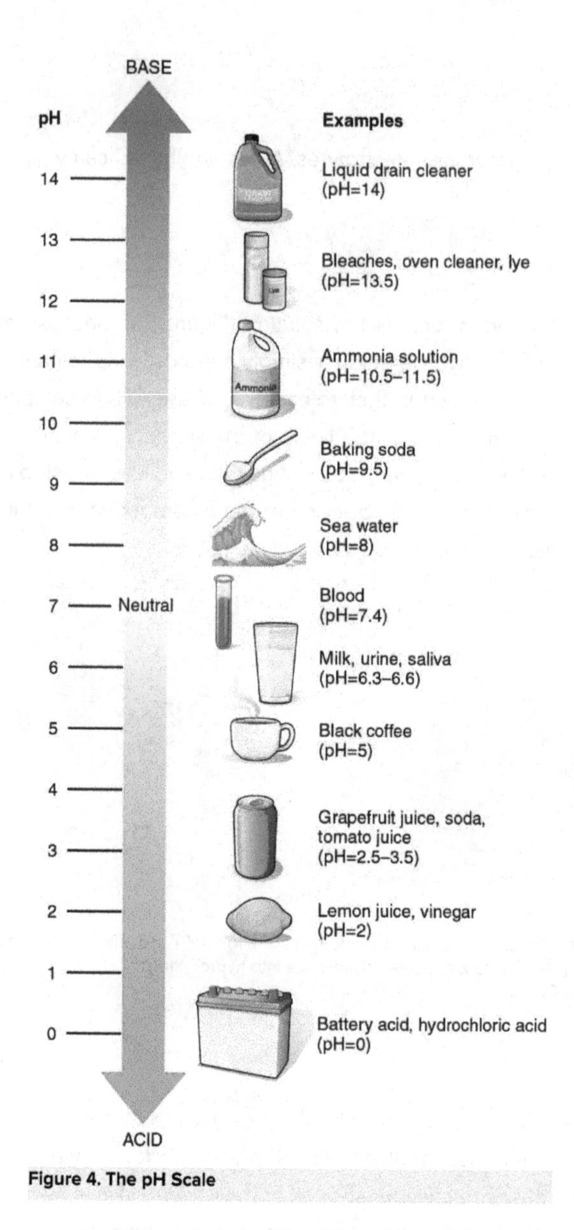

Figure 4. The pH Scale

Buffers

The pH of human blood normally ranges from 7.35 to 7.45, although it is typically identified as pH 7.4. At this slightly basic pH, blood can reduce the acidity resulting from the carbon dioxide (CO_2) constantly being released into the bloodstream by the trillions of cells in the body. Homeostatic mechanisms (along with exhaling CO_2 while breathing) normally keep the pH of blood within this narrow range. This is critical, because fluctuations—either too acidic or too alkaline—can lead to life-threatening disorders.

All cells of the body depend on homeostatic regulation of acid–base balance at a pH of approximately 7.4. The body therefore has several mechanisms for this regulation, involving breathing, the excretion of chemicals in urine, and the internal release of chemicals collectively called buffers into body fluids. A **buffer** is a solution of a weak acid and its conjugate base. A buffer can neutralize small amounts of acids or bases in body fluids. For example, if there is even a slight decrease below 7.35 in the pH of a bodily fluid, the buffer in the fluid—in this case, acting as a weak base—will bind the excess hydrogen ions. In contrast, if pH rises above 7.45, the buffer will act as a weak acid and contribute hydrogen ions.

HOMEOSTATIC IMBALANCES: ACIDS AND BASES

Excessive acidity of the blood and other body fluids is known as acidosis. Common causes of acidosis are situations and disorders that reduce the effectiveness of breathing, especially the person's ability to exhale fully, which causes a buildup of CO_2 (and H^+) in the bloodstream. Acidosis can also be caused by metabolic problems that reduce the level or function of buffers that act as bases, or that promote the production of acids. For instance, with severe diarrhea, too much bicarbonate can be lost from the body, allowing acids to build up in body fluids. In people with poorly managed diabetes (ineffective regulation of blood sugar), acids called ketones are produced as a form of body fuel. These can build up in the blood, causing a serious condition called diabetic ketoacidosis. Kidney failure, liver failure, heart failure, cancer, and other disorders also can prompt metabolic acidosis.

In contrast, alkalosis is a condition in which the blood and other body fluids are too alkaline (basic). As with acidosis, respiratory disorders are a major cause; however, in respiratory alkalosis, carbon dioxide levels fall too low. Lung disease, aspirin overdose, shock, and ordinary anxiety can cause respiratory alkalosis, which reduces the normal concentration of H^+.

Metabolic alkalosis often results from prolonged, severe vomiting, which causes a loss of hydrogen and chloride ions (as components of HCl). Medications also can prompt alkalosis. These include diuretics that cause the body to lose potassium ions, as well as antacids when taken in excessive amounts, for instance by someone with persistent heartburn or an ulcer.

Carbon and macromolecules

Organic compounds typically consist of groups of carbon atoms covalently bonded to hydrogen, usually oxygen, and often other elements as well. Created by living things, they are found throughout the world, in soils and seas, commercial products, and every cell of the human body. The four types most important to human structure and function are carbohydrates, lipids, proteins, and nucleotides. Before exploring these compounds, you need to first understand the chemistry of carbon.

The Chemistry of Carbon

What makes organic compounds ubiquitous is the chemistry of their carbon core. Recall that carbon atoms have four electrons in their valence shell, and that the octet rule dictates that atoms tend to react in such a way as to complete their valence shell with eight electrons. Carbon atoms do not complete their valence shells by donating or accepting four electrons. Instead, they readily share electrons via covalent bonds.

Commonly, carbon atoms share with other carbon atoms, often forming a long carbon chain referred to as a carbon skeleton. When they do share, however, they do not share all their electrons exclusively with each other. Rather, carbon atoms tend to share electrons with a variety of other elements, one of which is always hydrogen. Carbon and hydrogen groupings are called hydrocarbons. If you study the figures of organic compounds in the remainder of this chapter, you will see several with chains of hydrocarbons in one region of the compound.

Many combinations are possible to fill carbon's four "vacancies." Carbon may share electrons with oxygen or nitrogen or other atoms in a particular region of an organic compound. Moreover, the atoms to which carbon atoms bond may also be part of a functional group. A **functional group** is a group of atoms linked by strong covalent bonds and tending to function in chemical reactions as a single unit. You can think of functional groups as tightly knit "cliques" whose members are unlikely to be parted. Five functional groups are important in human physiology; these are the hydroxyl, carboxyl, amino, methyl and phosphate groups (Table 1).

Table 1. Functional Groups Important in Human Physiology

Functional group	Structural formula	Importance
Hydroxyl	—O—H	Hydroxyl groups are polar. They are components of all four types of organic compounds discussed in this chapter. They are involved in dehydration synthesis and hydrolysis reactions.
Carboxyl	O—C—OH	Carboxyl groups are found within fatty acids, amino acids, and many other acids.
Amino	$—N—H_2$	Amino groups are found within amino acids, the building blocks of proteins.
Methyl	$—C—H_3$	Methyl groups are found within amino acids.
Phosphate	$—P—O_4^{2-}$	Phosphate groups are found within phospholipids and nucleotides.

Carbon's affinity for covalent bonding means that many distinct and relatively stable organic molecules nevertheless readily form larger, more complex molecules. Any large molecule is referred to as **macromolecule** (macro- = "large"), and the organic compounds in this section all fit this description. However, some macromolecules are made up of several "copies" of single units called monomer (mono- = "one"; -mer = "part"). Like beads in a long necklace, these monomers link by covalent bonds to form long polymers (poly- = "many"). There are many examples of monomers and polymers among the organic compounds.

Monomers form polymers by engaging in dehydration synthesis. As was noted earlier, this reaction results in the release of a molecule of water. Each monomer contributes: One gives up a hydrogen atom and the other gives up a hydroxyl group. Polymers are split into monomers by hydrolysis (-lysis = "rupture"). The bonds between their monomers are broken, via the donation of a molecule of water, which contributes a hydrogen atom to one monomer and a hydroxyl group to the other.

Carbohydrates

The term carbohydrate means "hydrated carbon." Recall that the root hydro- indicates water. A **carbohydrate** is a molecule composed of carbon, hydrogen, and oxygen; in most carbohydrates, hydrogen and oxygen are found in the same two-to-one relative proportions they have in water. In fact, the chemical formula for a "generic" molecule of carbohydrate is $(CH_2O)_n$.

Carbohydrates are referred to as saccharides, a word meaning "sugars." Three forms are important in the body. Monosaccharides are the monomers of carbohydrates. Disaccharides (di- = "two") are made up of two monomers. **Polysaccharides** are the polymers, and can consist of hundreds to thousands of monomers.

Monosaccharides

A **monosaccharide** is a monomer of carbohydrates. Five monosaccharides are important in the body. Three of these are the hexose sugars, so called because they each contain six atoms of carbon. These are glucose, fructose, and galactose, shown in Figure 1a. The remaining monosaccharides are the two pentose sugars, each of which contains five atoms of carbon. They are ribose and deoxyribose, shown in Figure 1.

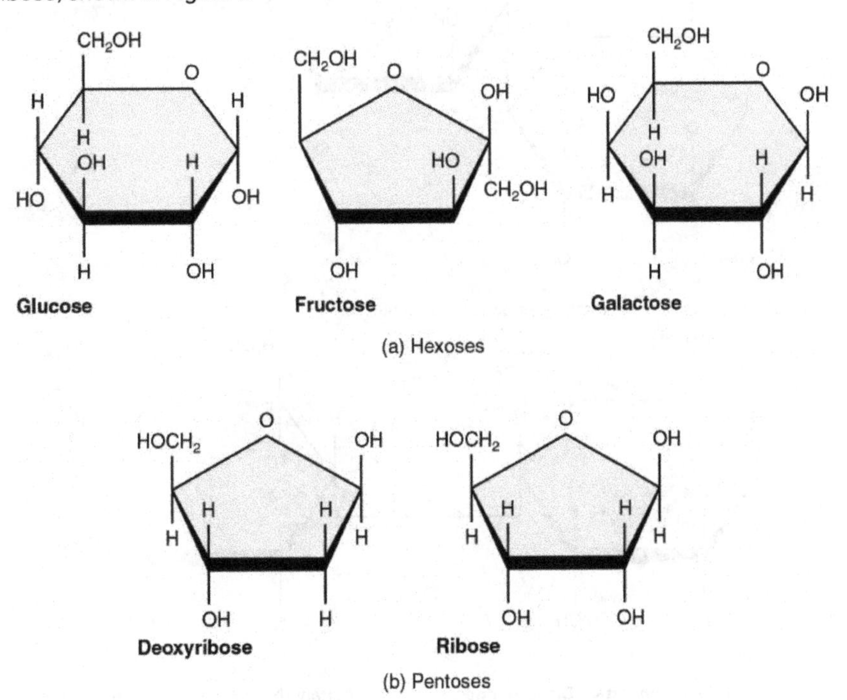

Figure 1. Five Important Monosaccharides.

Disaccharides

A **disaccharide** is a pair of monosaccharides. Disaccharides are formed via dehydration synthesis, and the bond linking them is referred to as a glycosidic bond (glyco- = "sugar"). Three disaccharides (shown in Figure 2) are important to humans. These are sucrose, commonly referred to as table sugar; lactose, or milk sugar; and maltose, or malt sugar. As you can tell from their common names, you consume these in your diet; however, your body cannot use them directly. Instead, in the digestive tract, they are split into their component monosaccharides via hydrolysis.

(a) The monosaccharides glucose and fructose bond to form sucrose

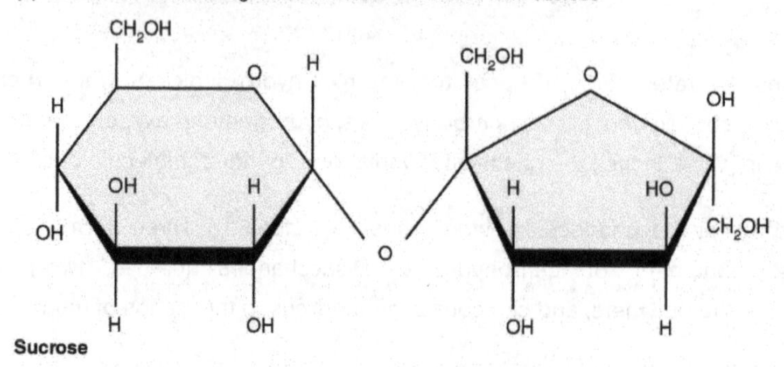

(b) The monosaccharides galactose and glucose bond to form lactose.

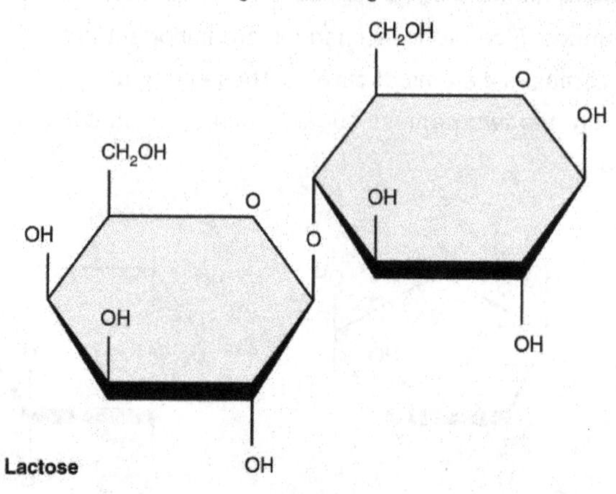

(c) Two glucose monosaccharides bond to form maltose.

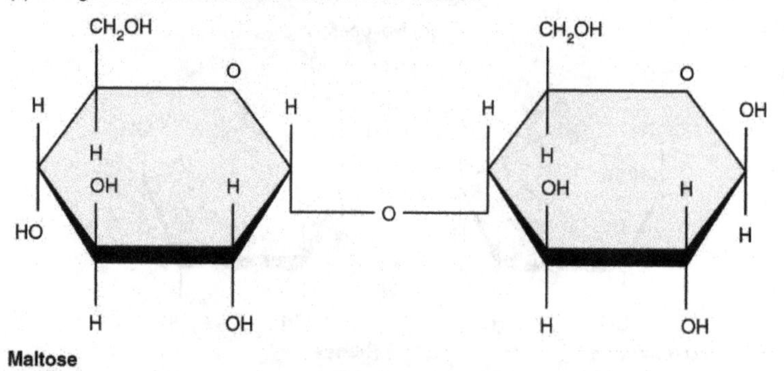

Figure 2. Three Important Disaccharides. All three important disaccharides form by dehydration synthesis.

Polysaccharides

Polysaccharides can contain a few to a thousand or more monosaccharides. Three are important to the body (Figure 3):

- Starches are polymers of glucose. They occur in long chains called amylose or branched chains called amylopectin, both of which are stored in plant-based foods and are relatively easy to digest.
- Glycogen is also a polymer of glucose, but it is stored in the tissues of animals, especially in the muscles and liver. It is not considered a dietary carbohydrate because very little glycogen remains in animal tissues after slaughter; however, the human body stores excess glucose as glycogen, again, in the muscles and liver.
- Cellulose, a polysaccharide that is the primary component of the cell wall of green plants, is the component of plant food referred to as "fiber". In humans, cellulose/fiber is not digestible; however, dietary fiber has many health benefits. It helps you feel full so you eat less, it promotes a healthy digestive tract, and a diet high in fiber is thought to reduce the risk of heart disease and possibly some forms of cancer.

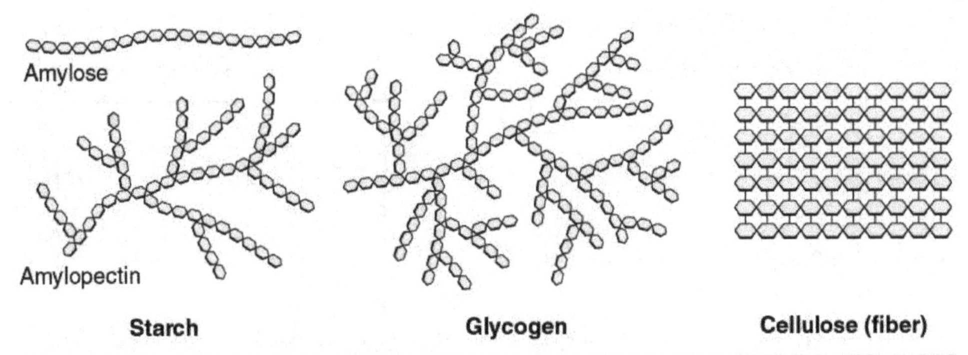

Figure 3. Three Important Polysaccharides. Three important polysaccharides are starches, glycogen, and fiber.

Functions of Carbohydrates

The body obtains carbohydrates from plant-based foods. Grains, fruits, and legumes and other vegetables provide most of the carbohydrate in the human diet, although lactose is found in dairy products.

Although most body cells can break down other organic compounds for fuel, all body cells can use glucose. Moreover, nerve cells (neurons) in the brain, spinal cord, and through the peripheral nervous system, as well as red blood cells, can use only glucose for fuel. In the breakdown of glucose for energy, molecules of adenosine triphosphate, better known as ATP, are produced. **Adenosine triphosphate (ATP)** is composed of a ribose sugar, an adenine base, and three phosphate groups. ATP releases free energy when its phosphate bonds are broken, and thus supplies ready energy to the cell. More ATP is produced in the presence of oxygen (O_2) than in pathways that do not use oxygen. The overall reaction for the conversion of the energy in glucose to energy stored in ATP can be written:

$$C_6H_{12}O_6 + 6\ O_2 \rightarrow 6\ CO_2 + 6\ H_2O + ATP$$

In addition to being a critical fuel source, carbohydrates are present in very small amounts in cells' structure. For instance, some carbohydrate molecules bind with proteins to produce glycoproteins, and others combine with lipids to produce glycolipids, both of which are found in the membrane that encloses the contents of body cells.

Lipids

A **lipid** is one of a highly diverse group of compounds made up mostly of hydrocarbons. The few oxygen atoms they contain are often at the periphery of the molecule. Their nonpolar hydrocarbons make all lipids hydrophobic. In water, lipids do not form a true solution, but they may form an emulsion, which is the term for a mixture of solutions that do not mix well.

Triglycerides

A **triglyceride** is one of the most common dietary lipid groups, and the type found most abundantly in body tissues. This compound, which is commonly referred to as a fat, is formed from the synthesis of two types of molecules (Figure 4):

- A glycerol backbone at the core of triglycerides, consists of three carbon atoms.
- Three fatty acids, long chains of hydrocarbons with a carboxyl group and a methyl group at opposite ends, extend from each of the carbons of the glycerol.

Three fatty acid chains are bound to glycerol by dehydration synthesis.

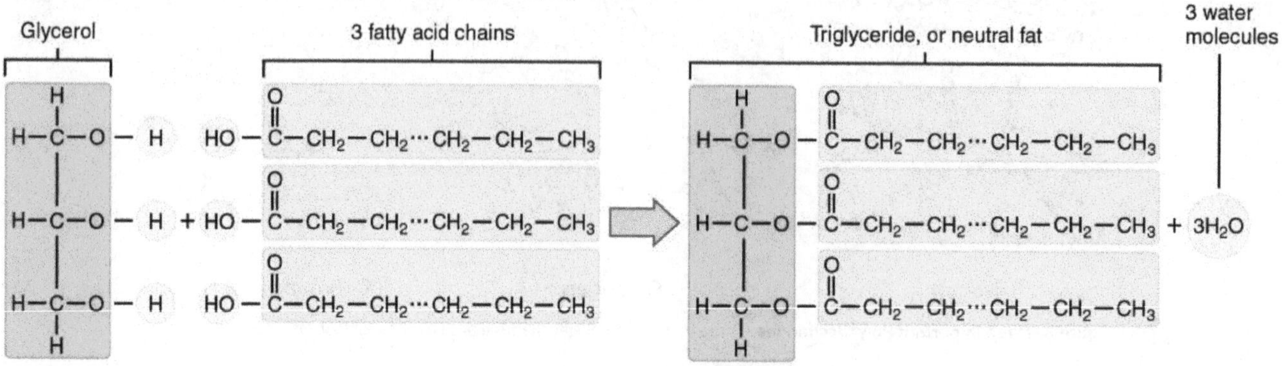

Figure 4. Triglycerides. Triglycerides are composed of glycerol attached to three fatty acids via dehydration synthesis. Notice that glycerol gives up a hydrogen atom, and the carboxyl groups on the fatty acids each give up a hydroxyl group.

Triglycerides form via dehydration synthesis. Glycerol gives up hydrogen atoms from its hydroxyl groups at each bond, and the carboxyl group on each fatty acid chain gives up a hydroxyl group. A total of three water molecules are thereby released.

Fatty acid chains that have no double carbon bonds anywhere along their length and therefore contain the maximum number of hydrogen atoms are called saturated fatty acids. These straight, rigid chains pack tightly together and are solid or semi-solid at room temperature (Figure 5a). Butter and lard are examples, as is the fat found on a steak or in your own body. In contrast, fatty acids with one double carbon bond are kinked at that bond (Figure 5b). These monounsaturated fatty acids are therefore unable to pack together tightly, and are liquid at room temperature. Polyunsaturated fatty acids contain two or more double carbon bonds, and are also liquid at room temperature. Plant oils such as olive oil typically contain both mono- and polyunsaturated fatty acids.

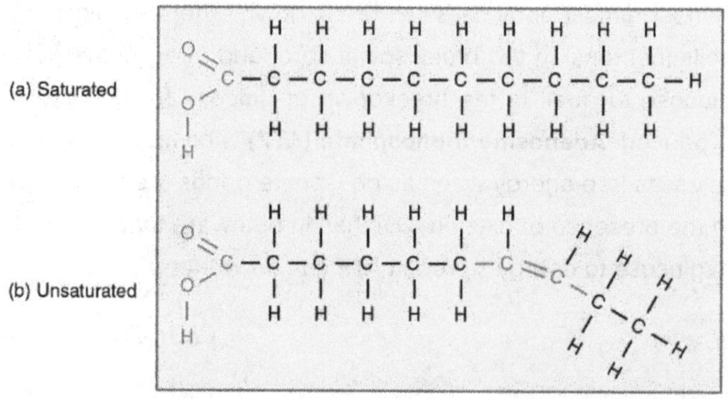

Figure 5. Fatty Acid Shapes. The level of saturation of a fatty acid affects its shape. (a) Saturated fatty acid chains are straight. (b) Unsaturated fatty acid chains are kinked.

Whereas a diet high in saturated fatty acids increases the risk of heart disease, a diet high in unsaturated fatty acids is thought to reduce the risk. This is especially true for the omega-3 unsaturated fatty acids found in cold-water fish such as salmon. These fatty acids have their first double carbon bond at the third hydrocarbon from the methyl group (referred to as the omega end of the molecule).

Finally, *trans* fatty acids found in some processed foods, including some stick and tub margarines, are thought to be even more harmful to the heart and blood vessels than saturated fatty acids. *Trans* fats are created from unsaturated fatty acids (such as corn oil) when chemically treated to produce partially hydrogenated fats.

As a group, triglycerides are a major fuel source for the body. When you are resting or asleep, a majority of the energy used to keep you alive is derived from triglycerides stored in your fat (adipose) tissues. Triglycerides also fuel long, slow physical activity such as gardening or hiking, and contribute a modest percentage of energy for vigorous physical activity. Dietary fat also assists the absorption and transport of the nonpolar fat-soluble vitamins A, D, E, and K. Additionally, stored body fat protects and cushions the body's bones and internal organs, and acts as insulation to retain body heat.

Fatty acids are also components of glycolipids, which are sugar-fat compounds found in the cell membrane. Lipoproteins are compounds in which the hydrophobic triglycerides are packaged in protein envelopes for transport in body fluids.

Phospholipids

As its name suggests, a **phospholipid** is a bond between the glycerol component of a lipid and a phosphorous molecule. In fact, phospholipids are similar in structure to triglycerides. However, instead of having three fatty acids, a phospholipid is generated from a diglyceride, a glycerol with just two fatty acid chains (Figure 6). The third binding site on the glycerol is taken up by the phosphate group, which in turn is attached to a polar "head" region of the molecule. Recall that triglycerides are nonpolar and hydrophobic. This still holds for the fatty acid portion of a phospholipid compound. However, the phosphate-containing group at the head of the compound is polar and thereby hydrophilic. In other words, one end of the molecule can interact with oil, and the other end with water. This makes phospholipids ideal emulsifiers, compounds that help disperse fats in aqueous liquids, and enables them to interact with both the watery interior of cells and the watery solution outside of cells as components of the cell membrane.

(a) Phospholipids

Two fatty acid chains and a phosphorus-containing group are attached to the glycerol backbone.

Example: Phosphatidylcholine

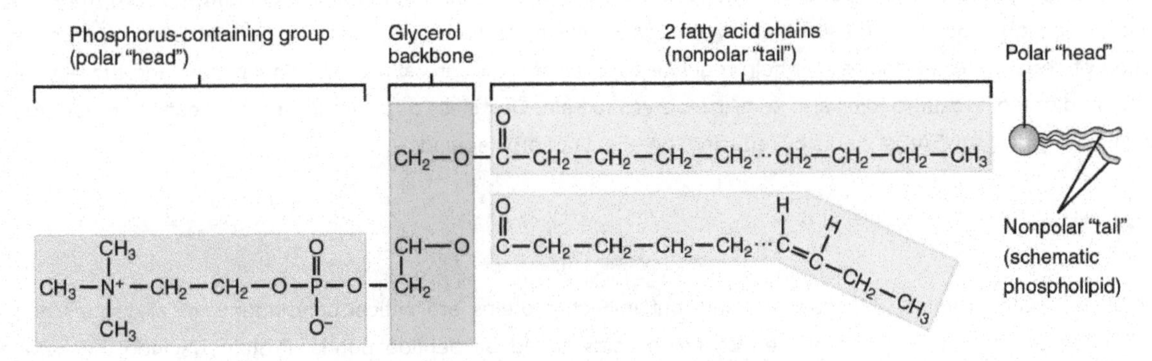

(b) Sterols

Four interlocking hydrocarbon rings from a steroid.

Example: Cholesterol (cholesterol is the basis for all steroids formed in the body)

(c) Prostaglandins

PGF2X

PGE2

Figure 6. Other Important Lipids. (a) Phospholipids are composed of two fatty acids, glycerol, and a phosphate group. (b) Sterols are ring-shaped lipids. Shown here is cholesterol. (c) Prostaglandins are derived from unsaturated fatty acids. Prostaglandin E2 (PGE2) includes hydroxyl and carboxyl groups.

Steroids

A **steroid** compound (referred to as a sterol) has as its foundation a set of four hydrocarbon rings bonded to a variety of other atoms and molecules (see Figure 6b). Although both plants and animals synthesize sterols, the type that makes the most important contribution to human structure and function is cholesterol, which is synthesized by the liver in humans and animals and is also present in most animal-based foods. Like other lipids, cholesterol's hydrocarbons make it hydrophobic; however, it has a polar hydroxyl head that is hydrophilic. Cholesterol is an important component of bile acids, compounds that help emulsify dietary fats. In fact, the word root *chole–* refers to bile. Cholesterol is also a building block of many hormones, signaling molecules that the body releases to regulate processes at distant sites. Finally, like phospholipids, cholesterol molecules are found in the cell membrane, where their hydrophobic and hydrophilic regions help regulate the flow of substances into and out of the cell.

Prostaglandins

Like a hormone, a **prostaglandin** is one of a group of signaling molecules, but prostaglandins are derived from unsaturated fatty acids (see Figure 6c). One reason that the omega-3 fatty acids found in fish are beneficial is that they stimulate the production of certain prostaglandins that help regulate aspects of blood pressure and inflammation, and thereby reduce the risk for heart disease. Prostaglandins also sensitize nerves to pain. One class of pain-relieving medications called nonsteroidal anti-inflammatory drugs (NSAIDs) works by reducing the effects of prostaglandins.

Proteins

You might associate proteins with muscle tissue, but in fact, proteins are critical components of all tissues and organs. A **protein** is an organic molecule composed of amino acids linked by peptide bonds. Proteins include the keratin in the epidermis of skin that protects underlying tissues, the collagen found in the dermis of skin, in bones, and in the meninges that cover the brain and spinal cord. Proteins are also components of many of the body's functional chemicals, including digestive enzymes in the digestive tract, antibodies, the neurotransmitters that neurons use to communicate with other cells, and the peptide-based hormones that regulate certain body functions (for instance, growth hormone). While carbohydrates and lipids are composed of hydrocarbons and oxygen, all proteins also contain nitrogen (N), and many contain sulfur (S), in addition to carbon, hydrogen, and oxygen.

Microstructure of Proteins

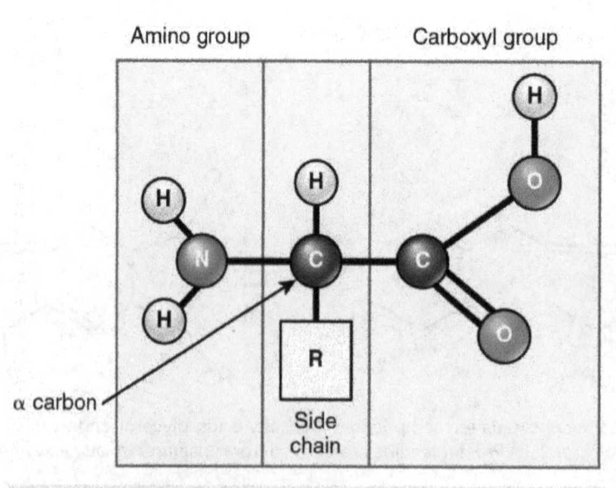

Figure 7. Structure of an Amino Acid

Proteins are polymers made up of nitrogen-containing monomers called amino acids. An **amino acid** is a molecule composed of an amino group and a carboxyl group, together with a variable side chain. Just 20 different amino acids contribute to nearly all of the thousands of different proteins important in human structure and function. Body proteins contain a unique combination of a few dozen to a few hundred of these 20 amino acid monomers. All 20 of these amino acids share a similar structure (Figure 7). All consist of a central carbon atom to which the following are bonded:

- a hydrogen atom
- an alkaline (basic) amino group NH_2 (see Table 1)
- an acidic carboxyl group COOH (see Table 1)
- a variable group

Notice that all amino acids contain both an acid (the carboxyl group) and a base (the amino group) (*amine* = "nitrogen-containing"). For this reason, they make excellent buffers, helping the body regulate acid–base balance. What distinguishes the 20 amino acids from one another is their variable group, which is referred to as a side chain or an R-group. This group can vary in size and can be polar or nonpolar, giving each amino acid its unique characteristics. For example, the side chains of two amino acids—cysteine and methionine—contain sulfur. Sulfur does not readily participate in hydrogen bonds, whereas all other amino acids do. This variation influences the way that proteins containing cysteine and methionine are assembled.

Amino acids join via dehydration synthesis to form protein polymers (Figure 8). The unique bond holding amino acids together is called a peptide bond. A **peptide bond** is a covalent bond between two amino acids that forms by dehydration synthesis. A peptide, in fact, is a very short chain of amino acids. Strands containing fewer than about 100 amino acids are generally referred to as polypeptides rather than proteins.

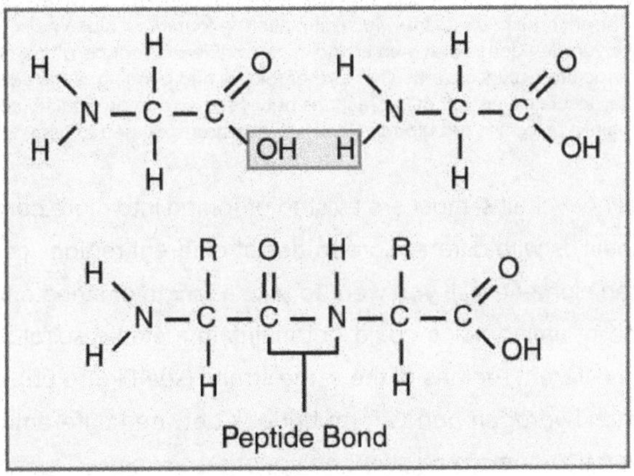

Figure 8. Peptide Bond. Different amino acids join together to form peptides, polypeptides, or proteins via dehydration synthesis. The bonds between the amino acids are peptide bonds.

The body is able to synthesize most of the amino acids from components of other molecules; however, nine cannot be synthesized and have to be consumed in the diet. These are known as the essential amino acids.

Free amino acids available for protein construction are said to reside in the amino acid pool within cells. Structures within cells use these amino acids when assembling proteins. If a particular essential amino acid is not available in sufficient quantities in the amino acid pool, however, synthesis of proteins containing it can slow or even cease.

Shape of Proteins

Just as a fork cannot be used to eat soup and a spoon cannot be used to spear meat, a protein's shape is essential to its function. A protein's shape is determined, most fundamentally, by the sequence of amino acids of which it is made (Figure 9a). The sequence is called the primary structure of the protein.

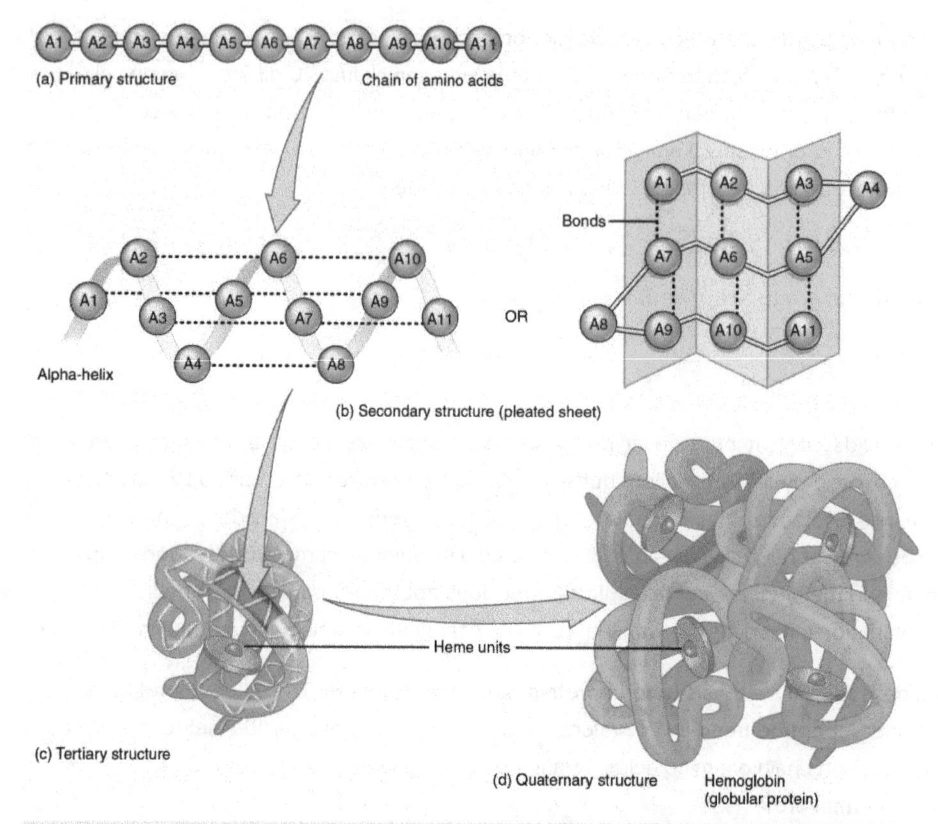

Figure 9. The Shape of Proteins. (a) The primary structure is the sequence of amino acids that make up the polypeptide chain. (b) The secondary structure, which can take the form of an alpha-helix or a beta-pleated sheet, is maintained by hydrogen bonds between amino acids in different regions of the original polypeptide strand. (c) The tertiary structure occurs as a result of further folding and bonding of the secondary structure. (d) The quaternary structure occurs as a result of interactions between two or more tertiary subunits. The example shown here is hemoglobin, a protein in red blood cells which transports oxygen to body tissues.

Although some polypeptides exist as linear chains, most are twisted or folded into more complex secondary structures that form when bonding occurs between amino acids with different properties at different regions of the polypeptide. The most common secondary structure is a spiral called an alpha-helix. If you were to take a length of string and simply twist it into a spiral, it would not hold the shape. Similarly, a strand of amino acids could not maintain a stable spiral shape without the help of hydrogen bonds, which create bridges between different regions of the same strand (see Figure 9b). Less commonly, a polypeptide chain can form a beta-pleated sheet, in which hydrogen bonds form bridges between different regions of a single polypeptide that has folded back upon itself, or between two or more adjacent polypeptide chains.

The secondary structure of proteins further folds into a compact three-dimensional shape, referred to as the protein's tertiary structure (see Figure 9c). In this configuration, amino acids that had been very distant in the primary chain can be brought quite close via hydrogen bonds or, in proteins containing cysteine, via disulfide bonds. A **disulfide bond** is a covalent bond between sulfur atoms in a polypeptide. Often, two or more separate polypeptides bond to form an even larger protein with a quaternary structure (see Figure 9d). The polypeptide subunits forming a quaternary structure can be identical or different. For instance, hemoglobin, the protein found in red blood cells is composed of four tertiary polypeptides, two of which are called alpha chains and two of which are called beta chains.

When they are exposed to extreme heat, acids, bases, and certain other substances, proteins will denature. *Denaturation* is a change in the structure of a molecule through physical or chemical means. Denatured proteins lose their functional shape and are no longer able to carry out their jobs. An everyday example of protein denaturation is the curdling of milk when acidic lemon juice is added.

The contribution of the shape of a protein to its function can hardly be exaggerated. For example, the long, slender shape of protein strands that make up muscle tissue is essential to their ability to contract (shorten) and relax (lengthen). As another example, bones contain long threads of a protein called collagen that acts as scaffolding upon which bone minerals are deposited. These elongated proteins, called fibrous proteins, are strong and durable and typically hydrophobic.

In contrast, globular proteins are globes or spheres that tend to be highly reactive and are hydrophilic. The hemoglobin proteins packed into red blood cells are an example (see Figure 9d); however, globular proteins are abundant throughout the body, playing critical roles in most body functions. Enzymes, introduced earlier as protein catalysts, are examples of this. The next section takes a closer look at the action of enzymes.

Proteins Function as Enzymes

If you were trying to type a paper, and every time you hit a key on your laptop there was a delay of six or seven minutes before you got a response, you would probably get a new laptop. In a similar way, without enzymes to catalyze chemical reactions, the human body would be nonfunctional. It functions only because enzymes function.

Enzymatic reactions—chemical reactions catalyzed by enzymes—begin when substrates bind to the enzyme. A **substrate** is a reactant in an enzymatic reaction. This occurs on regions of the enzyme known as active sites (Figure 10). Any given enzyme catalyzes just one type of chemical reaction. This characteristic, called specificity, is due to the fact that a substrate with a particular shape and electrical charge can bind only to an active site corresponding to that substrate.

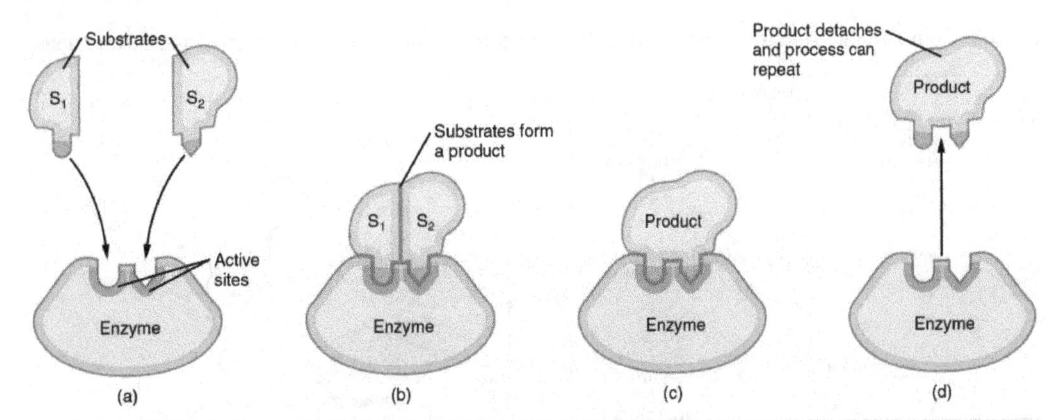

Figure 10. Steps in an Enzymatic Reaction. (a) Substrates approach active sites on enzyme. (b) Substrates bind to active sites, producing an enzyme–substrate complex. (c) Changes internal to the enzyme–substrate complex facilitate interaction of the substrates. (d) Products are released and the enzyme returns to its original form, ready to facilitate another enzymatic reaction.

Binding of a substrate produces an enzyme–substrate complex. It is likely that enzymes speed up chemical reactions in part because the enzyme–substrate complex undergoes a set of temporary and reversible changes that cause the substrates to be oriented toward each other in an optimal position to facilitate their interaction. This promotes increased reaction speed. The enzyme then releases the product(s), and resumes its original shape. The enzyme is then free to engage in the process again, and will do so as long as substrate remains.

Other Functions of Proteins

Advertisements for protein bars, powders, and shakes all say that protein is important in building, repairing, and maintaining muscle tissue, but the truth is that proteins contribute to all body tissues, from the skin to the brain cells. Also, certain proteins act as hormones, chemical messengers that help regulate body functions, For example, growth hormone is important for skeletal growth, among other roles.

As was noted earlier, the basic and acidic components enable proteins to function as buffers in maintaining acid–base balance, but they also help regulate fluid–electrolyte balance. Proteins attract fluid, and a healthy concentration of proteins in the blood, the cells, and the spaces between cells helps ensure a balance of fluids in these various "compartments." Moreover, proteins in the cell membrane help to transport electrolytes in and out of the cell, keeping these ions in a healthy balance. Like lipids, proteins can bind with carbohydrates. They can thereby produce glycoproteins or proteoglycans, both of which have many functions in the body.

The body can use proteins for energy when carbohydrate and fat intake is inadequate, and stores of glycogen and adipose tissue become depleted. However, since there is no storage site for protein except functional tissues, using protein for energy causes tissue breakdown, and results in body wasting.

Nucleotides

The fourth type of organic compound important to human structure and function are the nucleotides (Figure 11). A **nucleotide** is one of a class of organic compounds composed of three subunits:

- one or more phosphate groups
- a pentose sugar: either deoxyribose or ribose
- a nitrogen-containing base: adenine, cytosine, guanine, thymine, or uracil

Nucleotides can be assembled into nucleic acids (DNA or RNA) or the energy compound adenosine triphosphate.

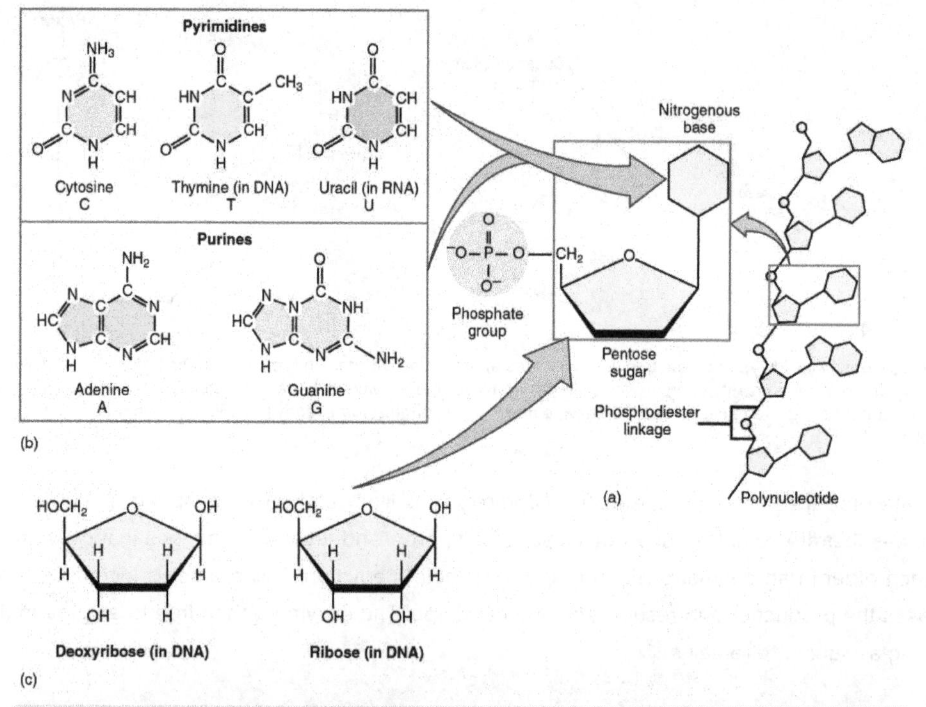

Figure 11. Nucleotides. (a) The building blocks of all nucleotides are one or more phosphate groups, a pentose sugar, and a nitrogen-containing base. (b) The nitrogen-containing bases of nucleotides. (c) The two pentose sugars of DNA and RNA.

Nucleic Acids

The nucleic acids differ in their type of pentose sugar. **Deoxyribonucleic acid (DNA)** is nucleotide that stores genetic information. DNA contains deoxyribose (so-called because it has one less atom of oxygen than ribose) plus one phosphate group and one nitrogen-containing base. The "choices" of base for DNA are adenine, cytosine, guanine, and thymine. **Ribonucleic acid (RNA)** is a ribose-containing nucleotide that helps manifest the genetic code as protein. RNA contains ribose, one phosphate group, and one nitrogen-containing base, but the "choices" of base for RNA are adenine, cytosine, guanine, and uracil.

The nitrogen-containing bases adenine and guanine are classified as purines. A **purine** is a nitrogen-containing molecule with a double ring structure, which accommodates several nitrogen atoms. The bases cytosine, thymine (found in DNA only) and uracil (found in RNA only) are pyramidines. A **pyramidine** is a nitrogen-containing base with a single ring structure

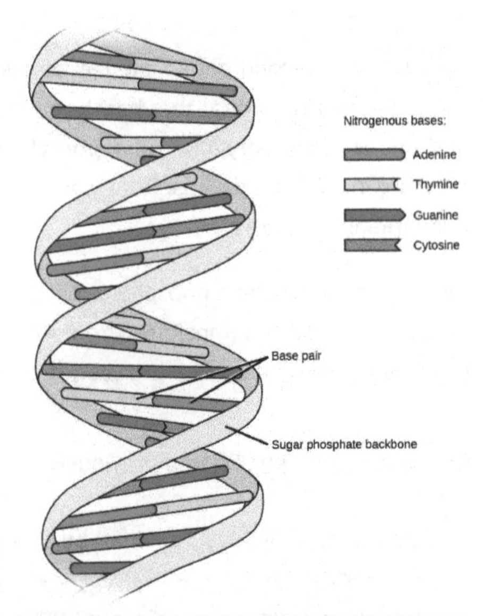

Figure 12. DNA. In the DNA double helix, two strands attach via hydrogen bonds between the bases of the component nucleotides.

Bonds formed by dehydration synthesis between the pentose sugar of one nucleic acid monomer and the phosphate group of another form a "backbone," from which the components' nitrogen-containing bases protrude. In DNA, two such backbones attach at their protruding bases via hydrogen bonds. These twist to form a shape known as a double helix (Figure 12). The sequence of nitrogen-containing bases within a strand of DNA form the genes that act as a molecular code instructing cells in the assembly of amino acids into proteins. Humans have almost 22,000 genes in their DNA, locked up in the 46 chromosomes inside the nucleus of each cell (except red blood cells which lose their nuclei during development). These genes carry the genetic code to build one's body, and are unique for each individual except identical twins.

In contrast, RNA consists of a single strand of sugar-phosphate backbone studded with bases. Messenger RNA (mRNA) is created during protein synthesis to carry the genetic instructions from the DNA to the cell's protein manufacturing plants in the cytoplasm, the ribosomes.

Adenosine Triphosphate

The nucleotide adenosine triphosphate (ATP), is composed of a ribose sugar, an adenine base, and three phosphate groups (Figure 13). ATP is classified as a high energy compound because the two covalent bonds linking its three phosphates store a significant amount of potential energy. In the body, the energy released from these high energy bonds helps fuel the body's activities, from muscle contraction to the transport of substances in and out of cells to anabolic chemical reactions.

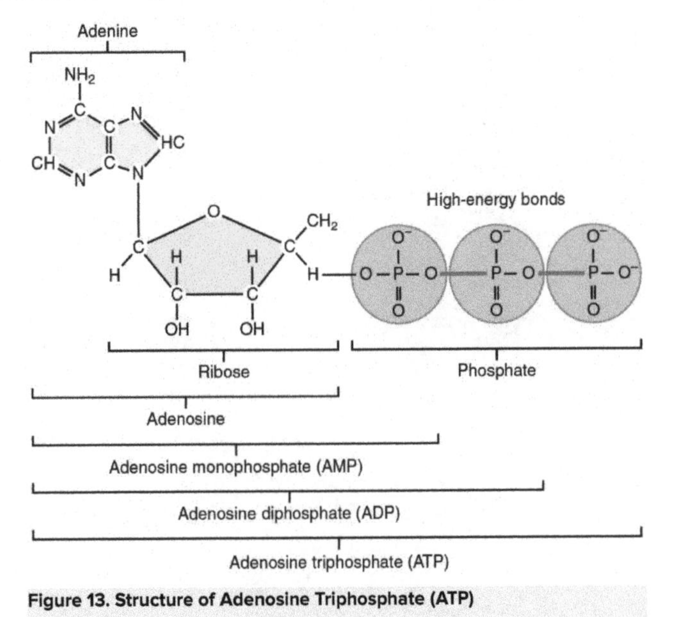

Figure 13. Structure of Adenosine Triphosphate (ATP)

When a phosphate group is cleaved from ATP, the products are adenosine diphosphate (ADP) and inorganic phosphate (P_i). This hydrolysis reaction can be written:

$$ATP + H_2O \rightarrow ADP + P_i + energy$$

Removal of a second phosphate leaves adenosine monophosphate (AMP) and two phosphate groups. Again, these reactions also liberate the energy that had been stored in the phosphate-phosphate bonds. They are reversible, too, as when ADP undergoes phosphorylation. **Phosphorylation** is the addition of a phosphate group to an organic compound, in this case, resulting in ATP. In such cases, the same level of energy that had been released during hydrolysis must be reinvested to power dehydration synthesis.

Cells can also transfer a phosphate group from ATP to another organic compound. For example, when glucose first enters a cell, a phosphate group is transferred from ATP, forming glucose phosphate ($C_6H_{12}O_6$—P) and ADP. Once glucose is phosphorylated in this way, it can be stored as glycogen or metabolized for immediate energy.

Genes to proteins: Central Dogma

Genes specify functional products (such as proteins)

A DNA molecule is divided up into functional units called **genes**. Each gene provides instructions for a functional product, that is, a molecule needed to perform a job in the cell. In many cases, the functional product of a gene is a protein. For example, Mendel's flower color gene provides instructions for a protein that helps make colored molecules (pigments) in flower petals.

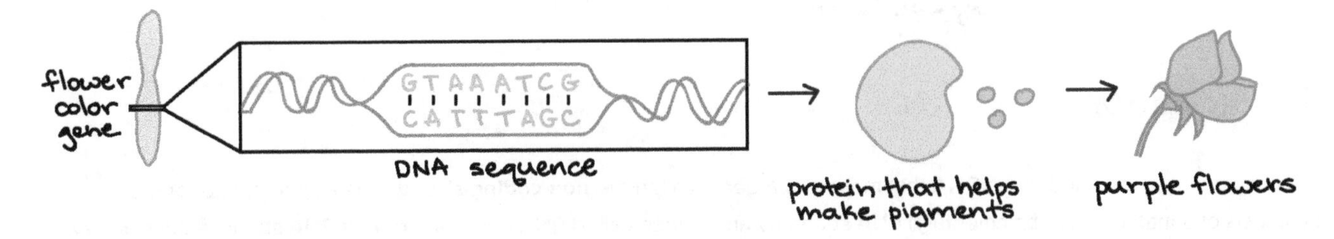

The flower color gene that Mendel studied consists of a stretch of DNA found on a chromosome. The DNA has a particular sequence; part of it, shown in this diagram, is GTAAATCG (upper strand), paired with the complementary sequence CATTTAGC (lower strand). The DNA of the gene specifies production of a protein that helps make pigments. When the protein is present and functional, pigments are produced, and the flowers of a plant have a purple color.

The functional products of most known genes are proteins, or, more accurately, polypeptides. **Polypeptide** is just another word for a chain of amino acids. Although many proteins consist of a single polypeptide, some are made up of multiple polypeptides. Genes that specify polypeptides are called **protein-coding** genes. Not all genes specify polypeptides. Instead, some provide instructions to build functional RNA molecules, such as the transfer RNAs and ribosomal RNAs that play roles in translation.

How does the DNA sequence of a gene specify a particular protein?

Many genes provide instructions for building polypeptides. How, exactly, does DNA direct the construction of a polypeptide? This process involves two major steps: transcription and translation.

- In **transcription**, the DNA sequence of a gene is copied to make an RNA molecule. This step is called *transcription* because it involves rewriting, or transcribing, the DNA sequence in a similar RNA "alphabet." In eukaryotes, the RNA molecule must undergo processing to become a mature **messenger RNA** (**mRNA**).

- In **translation**, the sequence of the mRNA is decoded to specify the amino acid sequence of a polypeptide. The name *translation* reflects that the nucleotide sequence of the mRNA sequence must be translated into the completely different "language" of amino acids.

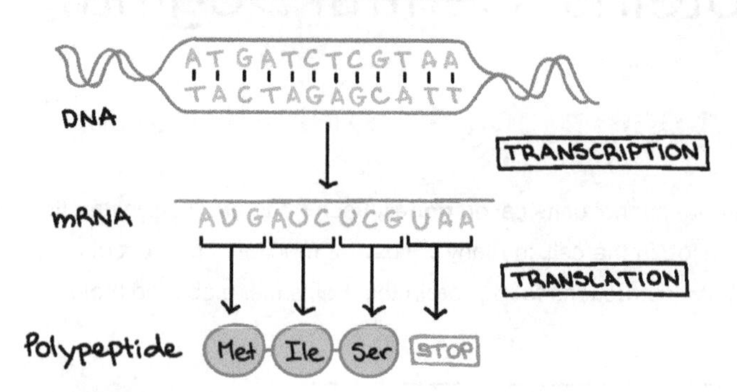

Transcription

In transcription, one strand of the DNA that makes up a gene, called the **non-coding strand**, acts as a template for the synthesis of a matching (complementary) RNA strand by an enzyme called RNA polymerase. This RNA strand is the **primary transcript**.

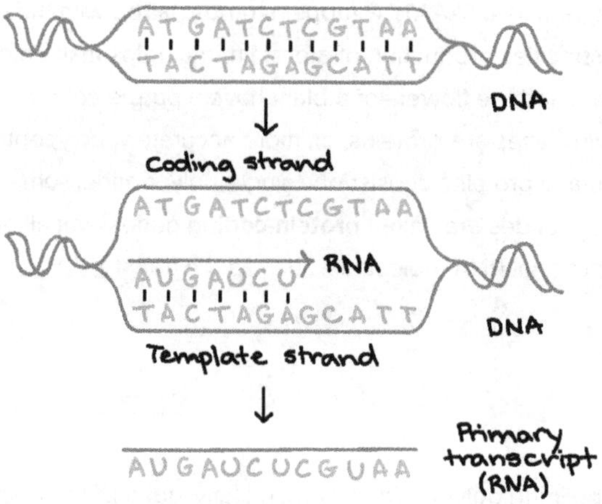

ATGATCTCGTAA
TACTAGAGCATT
DNA
Coding strand
ATGATCTCGTAA
AUGAUCU
RNA
TACTAGAGCATT
DNA
Template strand
AUGAUCUCGUAA
Primary transcript (RNA)

RNA polymerase

The main enzyme involved in transcription is **RNA polymerase**, which uses a single-stranded DNA template to synthesize a complementary strand of RNA. Specifically, RNA polymerase builds an RNA strand, adding each new nucleotide to the strand.

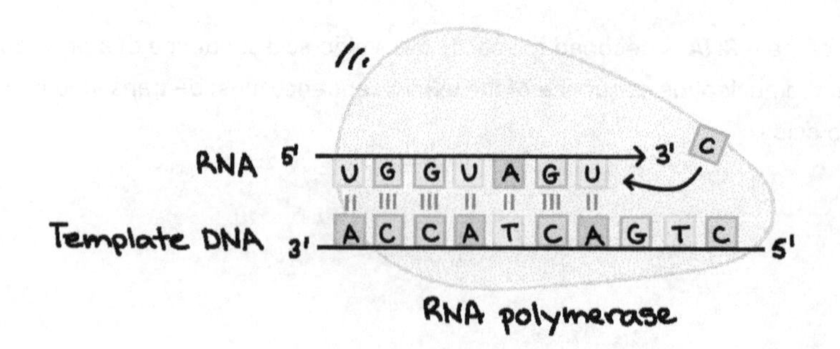

Stages of transcription

Transcription of a gene takes place in three stages: initiation, elongation, and termination. Here, we will briefly see how these steps happen.

1. **Initiation.** RNA polymerase binds to a sequence of DNA called the **promoter**, found near the beginning of a gene. Each gene has its own promoter. Once bound, RNA polymerase separates the DNA strands, providing the single-stranded template needed for transcription.

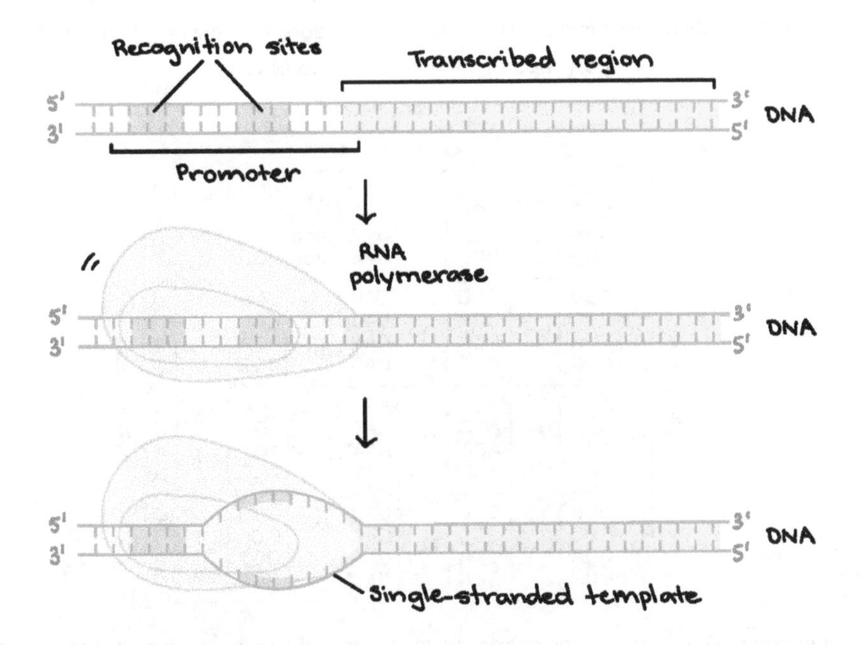

   The promoter region comes before (and slightly overlaps with) the transcribed region whose transcription it specifies. It contains recognition sites for RNA polymerase or its helper proteins to bind to. The DNA opens up in the promoter region so that RNA polymerase can begin transcription.

2. **Elongation.** One strand of DNA, the **template strand**, acts as a template for RNA polymerase. As it "reads" this template one base at a time, the polymerase builds an RNA molecule out of complementary nucleotides, making a chain. The RNA transcript carries the same information as the non-template (**coding**) strand of DNA, but it contains the base uracil (U) instead of thymine (T).
3. **Termination.** Sequences called **terminators** signal that the RNA transcript is complete. Once they are transcribed, they cause the transcript to be released from the RNA polymerase. An example of a termination mechanism involving formation of a hairpin in the RNA is shown below.

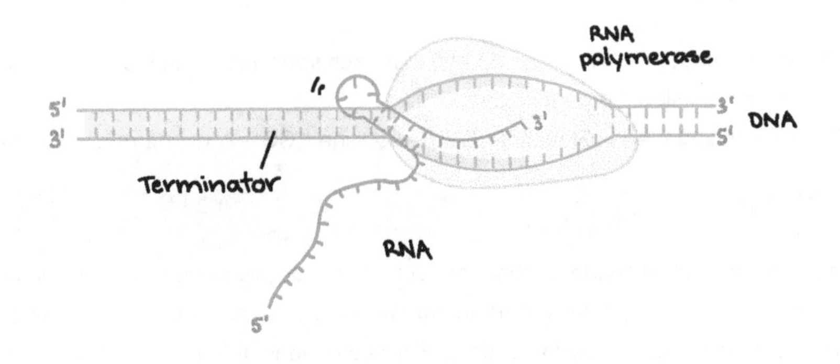

Translation

The genetic code

During translation, a cell "reads" the information in a messenger RNA (mRNA) and uses it to build a protein. Actually, to be a little more technical, an mRNA doesn't always encode—provide instructions for—a whole protein. Instead, what we can confidently say is that it always encodes a **polypeptide**, or chain of amino acids.

Second letter

First letter	U	C	A	G	Third letter
U	UUU Phe	UCU Ser	UAU Tyr	UGU Cys	U
	UUC Phe	UCC Ser	UAC Tyr	UGC Cys	C
	UUA Leu	UCA Ser	**UAA Stop**	**UGA Stop**	A
	UUG Leu	UCG Ser	**UAG Stop**	UGG Trp	G
C	CUU Leu	CCU Pro	CAU His	CGU Arg	U
	CUC Leu	CCC Pro	CAC His	CGC Arg	C
	CUA Leu	CCA Pro	CAA Gln	CGA Arg	A
	CUG Leu	CCG Pro	CAG Gln	CGG Arg	G
A	AUU Ile	ACU Thr	AAU Asn	AGU Ser	U
	AUC Ile	ACC Thr	AAC Asn	AGC Ser	C
	AUA Ile	ACA Thr	AAA Lys	AGA Arg	A
	AUG Met	ACG Thr	AAG Lys	AGG Arg	G
G	GUU Val	GCU Ala	GAU Asp	GGU Gly	U
	GUC Val	GCC Ala	GAC Asp	GGC Gly	C
	GUA Val	GCA Ala	GAA Glu	GGA Gly	A
	GUG Val	GCG Ala	GAG Glu	GGG Gly	G

Each three-letter sequence of mRNA nucleotides corresponds to a specific amino acid, or to a stop codon. UGA, UAA, and UAG are stop codons. AUG is the codon for methionine, and is also the start codon. In an mRNA, the instructions for building a polypeptide are RNA nucleotides (As, Us, Cs, and Gs) read in groups of three. These groups of three are called **codons**. One codon, AUG, specifies the amino acid methionine and also acts as a **start codon** to signal the start of protein construction. There are three more codons that do *not* specify amino acids. These **stop codons**, UAA, UAG, and UGA, tell the cell when a polypeptide is complete. All together, this collection of codon-amino acid relationships is called the **genetic code**, because it lets cells "decode" an mRNA into a chain of amino acids.

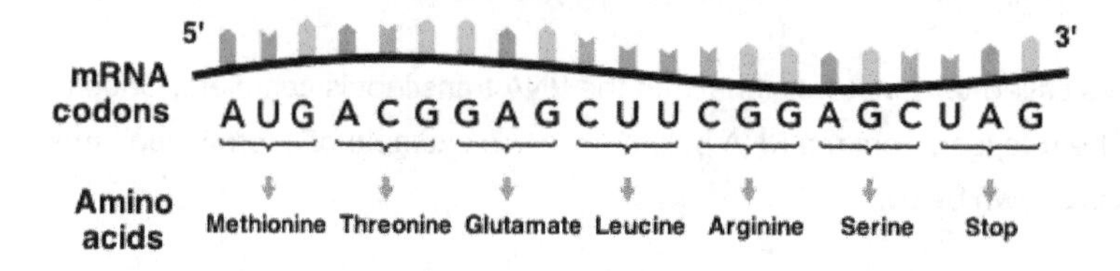

Each mRNA contains a series of codons (nucleotide triplets) that each specifies an amino acid. The correspondence between mRNA codons and amino acids is called the genetic code.
AUG – Methionine ACG – Threonine GAG – Glutamate CUU – Leucine CGG – Arginine AGC – Serine UAG – Stop

Transfer RNAs (tRNAs)

Transfer RNAs, or **tRNAs**, are molecular "bridges" that connect mRNA codons to the amino acids they encode. One end of each tRNA has a sequence of three nucleotides called an **anticodon**, which can bind to specific mRNA codons. The other end of the tRNA carries the amino acid specified by the codons. There are many different types of tRNAs. Each type reads one or a few codons and brings the right amino acid matching those codons.

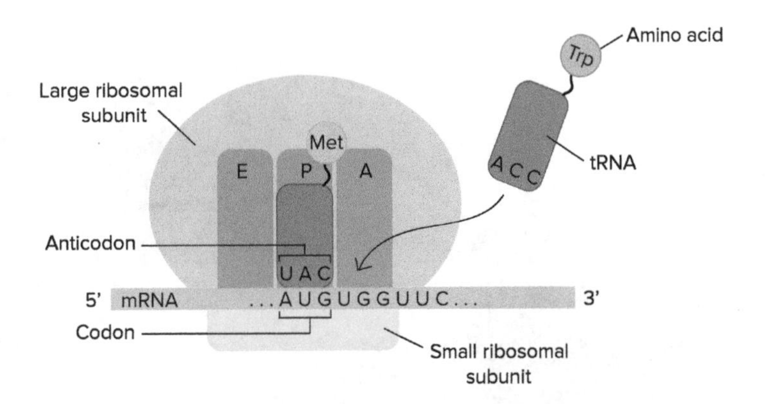

Ribosomes are composed of a small and large subunit and have three sites where tRNAs can bind to an mRNA (the A, P, and E sites). Each tRNA vcarries a specific amino acid and binds to an mRNA codon that is complementary to its anticodon.

Ribosomes

Ribosomes are the structures where polypeptides (proteins) are built. They are made up of protein and RNA (**ribosomal RNA**, or **rRNA**). Each ribosome has two subunits, a large one and a small one, which come together around an mRNA—kind of like the two halves of a hamburger bun coming together around the patty. The ribosome provides a set of handy slots where tRNAs can find their matching codons on the mRNA template and deliver their amino acids. These slots are called the A, P, and E sites. Not only that, but the ribosome also acts as an enzyme, catalyzing the chemical reaction that links amino acids together to make a chain.

Steps of translation

Your cells are making new proteins every second of the day. And each of those proteins must contain the right set of amino acids, linked together in just the right order. That may sound like a challenging task, but luckily, your cells (along with those of other animals, plants, and bacteria) are up to the job.
To see how cells make proteins, let's divide translation into three stages: initiation (starting off), elongation (adding on to the protein chain), and termination (finishing up).

Getting started: Initiation

In **initiation**, the ribosome assembles around the mRNA to be read and the first tRNA (carrying the amino acid methionine, which matches the start codon, AUG). This setup, called the initiation complex, is needed in order for translation to get started.

Extending the chain: Elongation

Elongation is the stage where the amino acid chain gets **long**er. In elongation, the mRNA is read one codon at a time, and the amino acid matching each codon is added to a growing protein chain.
Each time a new codon is exposed:

- A matching tRNA binds to the codon
- The existing amino acid chain (polypeptide) is linked onto the amino acid of the tRNA via a chemical reaction
- The mRNA is shifted one codon over in the ribosome, exposing a new codon for reading

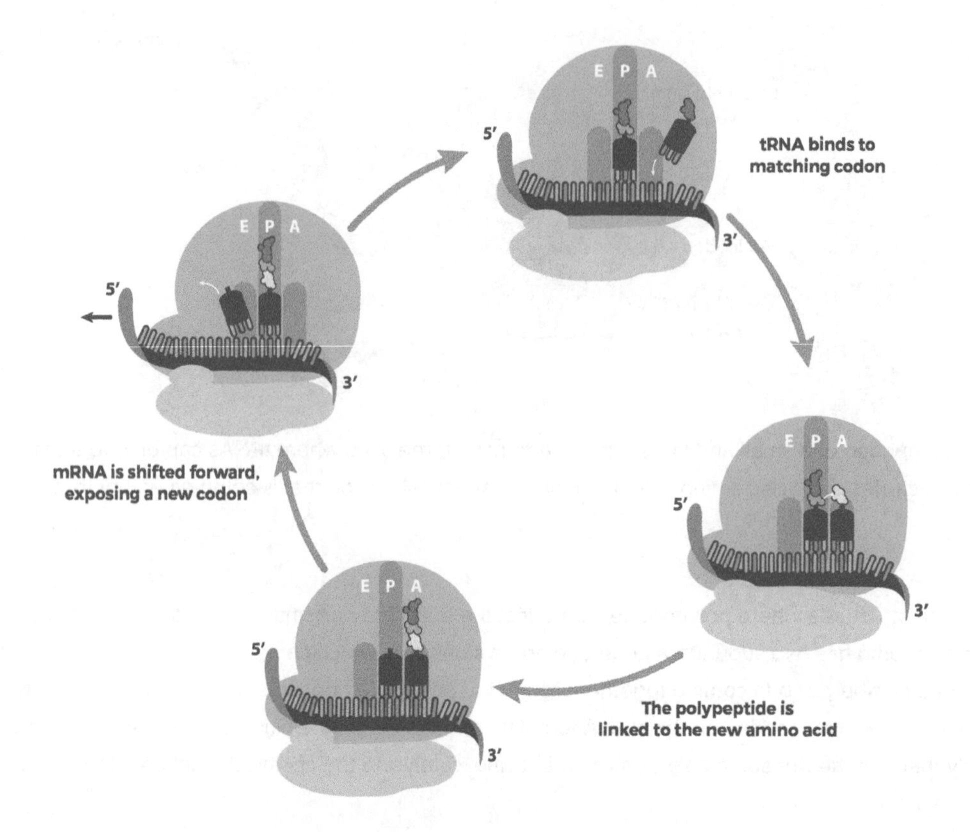

Elongation has three stages:

1) The anticodon of an incoming tRNA pairs with the mRNA codon exposed in the A site.

2) A peptide bond is formed between the new amino acid (in the A site) and the previously-added amino acid (in the P site), transferring the polypeptide from the P site to the A site.

3) The ribosome moves one codon down on the mRNA. The tRNA in the A site (carrying the polypeptide) shifts to the P site. The tRNA in the P site shifts to the E site and exits the ribosome.

During elongation, tRNAs move through the A, P, and E sites of the ribosome, as shown above. This process repeats many times as new codons are read and new amino acids are added to the chain.
For more details on the steps of elongation, see the stages of translation article.

Finishing up: Termination

Termination is the stage in which the finished polypeptide chain is released. It begins when a stop codon (UAG, UAA, or UGA) enters the ribosome, triggering a series of events that separate the chain from its tRNA and allow it to drift out of the ribosome. After termination, the polypeptide may still need to fold into the right 3D shape, undergo processing (such as the removal of amino acids), get shipped to the right place in the cell, or combine with other polypeptides before it can do its job as a functional protein.

Concept Check

1. **What** is the relationship between atoms, elements, and compounds?
2. **List and state** the charge and location of each subatomic particle.
3. **Distinguish** between atomic number and mass number.
4. **What** is an isotope and how can they be used in science?
5. **What** is a chemical bond and **why** do they form?
6. **Define** cation and anion.
7. **How** does an ionic bond form?
8. **What** are the different types of covalent bods?
9. **How** are ionic bonds different from covalent bonds and hydrogen bonds?
10. **How** are organic compounds different from inorganic compounds?
11. **List and explain** the importance of water.
12. **How** does salt dissolve in water?
13. **Compare and contrast** acids and bases.
14. **What** are buffers and **why** are they important to living things?
15. **How** many types can carbon bond? **Why** is this important?
16. **What** are some different functional groups and on **which** molecules are they typically found?
17. **List and explain** the importance of the four macromolecules.
18. **What** is the role of phopholipids?
19. **Why** are enzymes so important to living things?
20. **What** is ATP?

CHAPTER 3 — THE CELL

Studying Cells

LEARNING OBJECTIVES

- Describe the structure and function of the cell membrane, including its regulation of materials into and out of the cell
- Describe the functions of the various cytoplasmic organelles
- Explain the process by which a cell builds proteins using the DNA code
- List the stages of the cell cycle in order, including the steps of cell division in both somatic cells
- Describe the role of cells in organisms
- Compare and contrast light microscopy and electron microscopy
- Explain the structure of ATP and how this molecule is used in the process of chemical reactions
- Define the different steps of cellular respiration
- Describe the stages of the cell cycl

You developed from a single fertilized egg cell into the complex organism containing trillions of cells that you see when you look in a mirror. During this developmental process, early, undifferentiated cells differentiate and become specialized in their structure and function. These different cell types form specialized tissues that work in concert to perform all of the functions necessary for the living organism. Cellular and developmental biologists study how the continued division of a single cell leads to such complexity and differentiation.

Consider the difference between a structural cell in the skin and a nerve cell. A structural skin cell may be shaped like a flat plate (squamous) and live only for a short time before it is shed and replaced. Packed tightly into rows and sheets, the squamous skin cells provide a protective barrier for the cells and tissues that lie beneath. A nerve cell, on the other hand, may be shaped something like a star, sending out long processes up to a meter in length and may live for the entire lifetime of the organism. With their long winding appendages, nerve cells can communicate with one another and with other types of body cells and send rapid signals that inform the organism about its environment and allow it to interact with that environment.

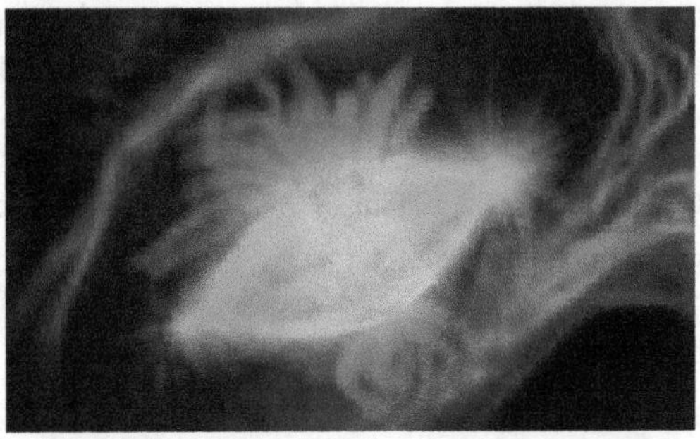

Figure 1. Flourescence-stained Cell Undergoing Mitosis. A lung cell from a newt, commonly studied for its similarity to human lung cells, is stained with fluorescent dyes. The green stain reveals mitotic spindles, red is the cell membrane and part of the cytoplasm, and the structures that appear light blue are chromosomes. This cell is in anaphase of mitosis. (credit: "Mortadelo2005"/Wikimedia Commons)

These differences illustrate one very important theme that is consistent at all organizational levels of biology: the form of a structure is optimally suited to perform particular functions assigned to that structure. Keep this theme in mind as you tour the inside of a cell and are introduced to the various types of cells in the body. A primary responsibility of each cell is to contribute to homeostasis.

Homeostasis is a term used in biology that refers to a dynamic state of balance within parameters that are compatible with life. For example, living cells require a water-based environment to survive in, and there are various physical (anatomical) and physiological mechanisms that keep all of the trillions of living cells in the human body moist. This is one aspect of homeostasis. When a particular parameter, such as blood pressure or blood oxygen content, moves far enough *out* of homeostasis (generally becoming too high or too low), illness or disease—and sometimes death—inevitably results.

The concept of a cell started with microscopic observations of dead cork tissue by scientist Robert Hooke in 1665. Without realizing their function or importance, Hook coined the term "cell" based on the resemblance of the small subdivisions in the cork to the rooms that monks inhabited, called cells. About ten years later, Antonie van Leeuwenhoek became the first person to observe living and moving cells under a microscope. In the century that followed, the theory that cells represented the basic unit of life would develop. These tiny fluid-filled sacs house components responsible for the thousands of biochemical reactions necessary for an organism to grow and survive. In this chapter, you will learn about the major components and functions of a prototypical, generalized cell and discover some of the different types of cells in the human body.

A cell is the smallest unit of a living thing. Whether comprised of one cell (like bacteria) or many cells (like a human), we call it an organism. Thus, cells are the basic building blocks of all organisms.

Several cells of one kind that interconnect with each other and perform a shared function form tissues. These tissues combine to form an organ (your stomach, heart, or brain), and several organs comprise an organ system (such as the digestive system, circulatory system, or nervous system). Several systems that function together form an organism (like a human being). Here, we will examine the structure and function of cells.

There are many types of cells, which scientists group into one of two broad categories: prokaryotic and eukaryotic. For example, we classify both animal and plant cells as eukaryotic cells; whereas, we classify bacterial cells as prokaryotic. Before discussing the criteria for determining whether a cell is prokaryotic or eukaryotic, we will first examine how biologists study cells.

Cell Theory

The microscopes we use today are far more complex than those that Dutch shopkeeper Antony van Leeuwenhoek, used in the 1600s. Skilled in crafting lenses, van Leeuwenhoek observed the movements of single-celled organisms, which he collectively termed "animalcules."

In the 1665 publication *Micrographia*, experimental scientist Robert Hooke coined the term "cell" for the box-like structures he observed when viewing cork tissue through a lens. In the 1670s, van Leeuwenhoek discovered bacteria and protozoa. Later advances in lenses, microscope construction, and staining techniques enabled other scientists to see some components inside cells.

By the late 1830s, botanist Matthias Schleiden and zoologist Theodor Schwann were studying tissues and proposed the unified cell theory, which states that:

1. One or more cells comprise all living things
2. The cell is the basic unit of life
3. New cells arise from existing cells.

Microscopy

Cells vary in size. With few exceptions, we cannot see individual cells with the naked eye, so scientists use microscopes (micro- = "small"; -scope = "to look at") to study them. A microscope is an instrument that magnifies an object. We photograph most cells with a microscope, so we can call these images micrographs.

The optics of a microscope's lenses change the image orientation that the user sees. A specimen that is right-side up and facing right on the microscope slide will appear upside-down and facing left when one views through a microscope, and vice versa. Similarly, if one moves the slide left while looking through the microscope, it will appear to move right, and if one moves it down, it will seem to move up. This occurs because microscopes use two sets of lenses to magnify the image. Because of the manner by which light travels through the lenses, this two lens system produces an inverted image (binocular, or dissecting microscopes, work in a similar manner, but include an additional magnification system that makes the final image appear to be upright).

Light Microscopes

To give you a sense of cell size, a typical human red blood cell is about eight millionths of a meter or eight micrometers (abbreviated as eight μm) in diameter. A pin head is about two thousandths of a meter (two mm) in diameter. That means about 250 red blood cells could fit on a pinhead.

Most student microscopes are light microscopes (Figure **1a**). Visible light passes and bends through the lens system to enable the user to see the specimen. Light microscopes are advantageous for viewing living organisms, but since individual cells are generally transparent, their components are not distinguishable unless they are colored with special stains. Staining, however, usually kills the cells.

Light microscopes that undergraduates commonly use in the laboratory magnify up to approximately 400 times. Two parameters that are important in microscopy are magnification and resolving power. Magnification is the process of enlarging an object in appearance. Resolving power is the microscope's ability to distinguish two adjacent structures as separate: the higher the resolution, the better the image's clarity and detail. When one uses oil immersion lenses to study small objects, magnification usually increases to 1,000 times. In order to gain a better understanding of cellular structure and function, scientists typically use electron microscopes.

(a) (b)

Figure 1. (a) Most light microscopes in a college biology lab can magnify cells up to approximately 400 times and have a resolution of about 200 nanometers. (b) Electron microscopes provide a much higher magnification, 100,000x, and a have a resolution of 50 picometers. (credit a: modification of work by "GcG"/Wikimedia Commons; credit b: modification of work by Evan Bench)

Electron Microscopes

In contrast to light microscopes, electron microscopes use a beam of electrons instead of a beam of light. Not only does this allow for higher magnification and, thus, more detail , it also provides higher resolving power. The method to prepare the specimen for viewing with an electron microscope kills the specimen. Electrons have short wavelengths (shorter than photons) that move best in a vacuum, so we cannot view living cells with an electron microscope.

In a scanning electron microscope, a beam of electrons moves back and forth across a cell's surface, creating details of cell surface characteristics. In a transmission electron microscope, the electron beam penetrates the cell and provides details of a cell's internal structures. As you might imagine, electron microscopes are significantly more bulky and expensive than light microscopes.

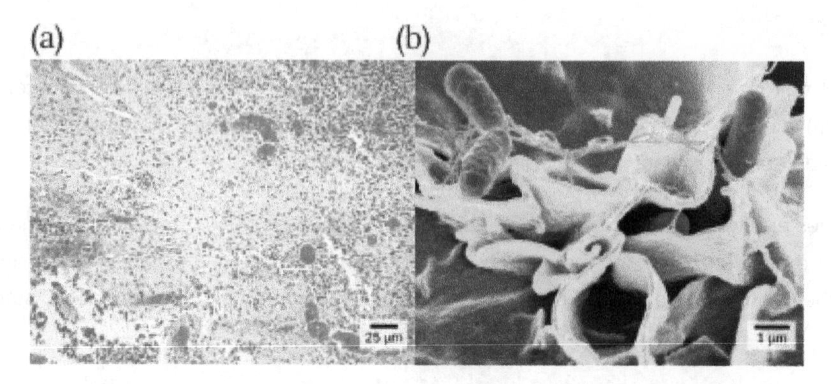

Figure 2. (a) These Salmonella bacteria appear as tiny purple dots when viewed with a light microscope. (b) This scanning electron microscope micrograph shows Salmonella bacteria (in red) invading human cells (yellow). Even though subfigure (b) shows a different Salmonella specimen than subfigure (a), you can still observe the comparative increase in magnification and detail. (credit a: modification of work by CDC/Armed Forces Institute of Pathology, Charles N. Farmer, Rocky Mountain Laboratories; credit b: modification of work by NIAID, NIH; scale-bar data from Matt Russell)

CAREER CONNECTION

Cytotechnologist

Have you ever heard of a medical test called a Pap smear (Figure 3)? In this test, a doctor takes a small sample of cells from the patient's uterine cervix and sends it to a medical lab where a cytotechnologist stains the cells and examines them for any changes that could indicate cervical cancer or a microbial infection.

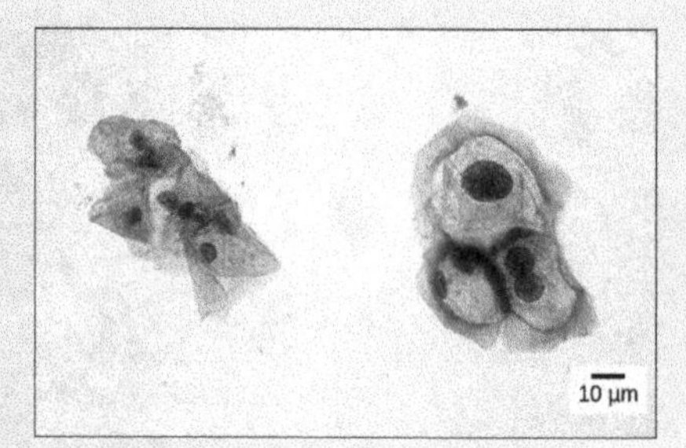

Figure 3. These uterine cervix cells, viewed through a light microscope, are from a Pap smear. Normal cells are on the left. The cells on the right are infected with human papillomavirus (HPV). Notice that the infected cells are larger. Also, two of these cells each have two nuclei instead of one, the normal number. (credit: modification of work by Ed Uthman, MD; scale-bar data from Matt Russell)

Cytotechnologists (cyto- = "cell") are professionals who study cells via microscopic examinations and other laboratory tests. They are trained to determine which cellular changes are within normal limits and which are abnormal. Their focus is not limited to cervical cells. They study cellular specimens that come from all organs. When they notice abnormalities, they consult a pathologist, a medical doctor who interprets and diagnoses changes that disease in body tissue and fluids cause.

Cytotechnologists play a vital role in saving people's lives. When doctors discover abnormalities early, a patient's treatment can begin sooner, which usually increases the chances of a successful outcome.

The Cytoplasm and Cellular Organelles

Now that you have learned that the cell membrane surrounds all cells, you can dive inside of a prototypical human cell to learn about its internal components and their functions. All living cells in multicellular organisms contain an internal cytoplasmic compartment, and a nucleus within the cytoplasm. **Cytosol**, the jelly-like substance within the cell, provides the fluid medium necessary for biochemical reactions. Eukaryotic cells, including all animal cells, also contain various cellular organelles. An **organelle** ("little organ") is one of several different types of membrane-enclosed bodies in the cell, each performing a unique function. Just as the various bodily organs work together in harmony to perform all of a human's functions, the many different cellular organelles work together to keep the cell healthy and performing all of its important functions. The organelles and cytosol, taken together, compose the cell's **cytoplasm**. The **nucleus** is a cell's central organelle, which contains the cell's DNA (Figure 1).

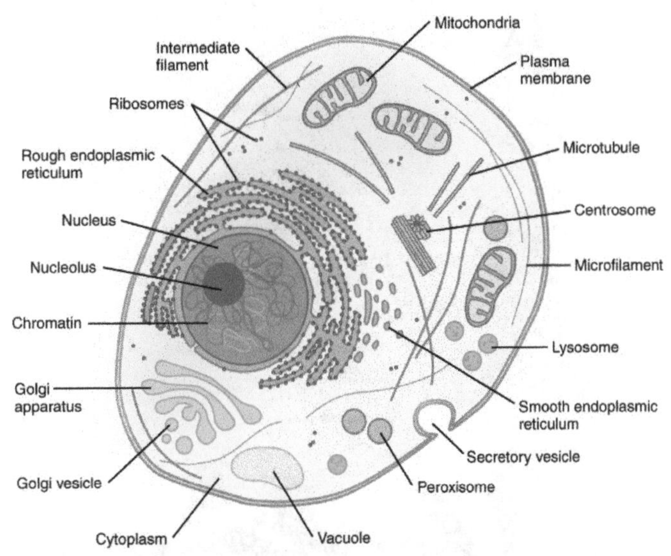

Figure 1. Prototypical Human Cell. While this image is not indicative of any one particular human cell, it is a prototypical example of a cell containing the primary organelles and internal structures.

The Nucleus

Typically, the nucleus is the most prominent organelle in a cell ((Figure 2)). The nucleus (plural = nuclei) houses the cell's DNA and directs the synthesis of ribosomes and proteins. Let's look at it in more detail.

The Nuclear Envelope

The nuclear envelope is a double-membrane structure that constitutes the nucleus' outermost portion ((Figure)). Both the nuclear envelope's inner and outer membranes are phospholipid bilayers. The nuclear envelope is punctuated with pores that control the passage of ions, molecules, and RNA between the nucleoplasm and cytoplasm. The nucleoplasm is the semi-solid fluid inside the nucleus, where we find the chromatin and the nucleolus.

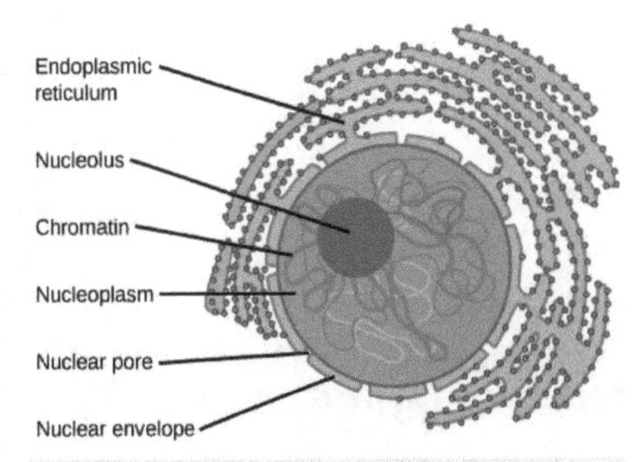

Figure 2. The nucleus stores chromatin (DNA plus proteins) in a gel-like substance called the nucleoplasm. The nucleolus is a condensed chromatin region where ribosome synthesis occurs. We call the nucleus' boundary the nuclear envelope. It consists of two phospholipid bilayers: an outer and an inner membrane. The nuclear membrane is continuous with the endoplasmic reticulum. Nuclear pores allow substances to enter and exit the nucleus.

Chromatin and Chromosomes

To understand chromatin, it is helpful to first explore chromosomes, structures within the nucleus that are made up of DNA, the hereditary material. You may remember that in prokaryotes, DNA is organized into a single circular chromosome. In eukaryotes, chromosomes are linear structures. Every eukaryotic species has a specific number of chromosomes in the nucleus of each cell. For example, in humans, the chromosome number is 46, while in fruit flies, it is eight.

Chromosomes are only visible and distinguishable from one another when the cell is getting ready to divide. When the cell is in the growth and maintenance phases of its life cycle, proteins attach to chromosomes, and they resemble an unwound, jumbled bunch of threads. We call these unwound protein-chromosome complexes chromatin ((Figure 3)). Chromatin describes the material that makes up the chromosomes both when condensed and decondensed.

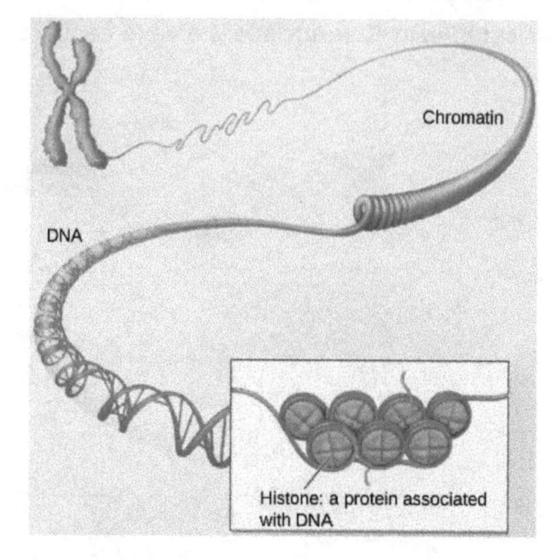

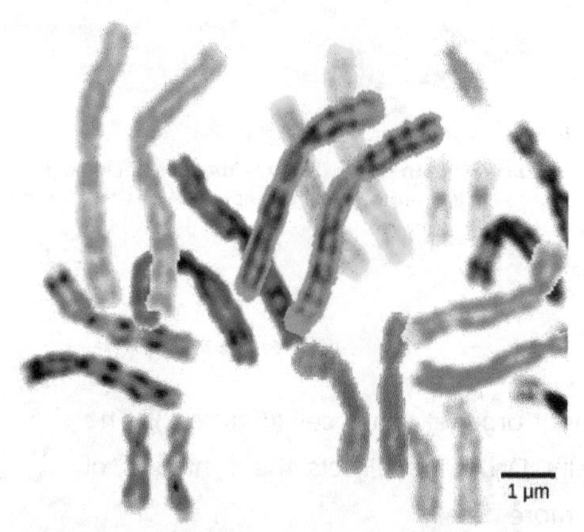

Figure 3. (a) This image shows various levels of chromatin's organization (DNA and protein). (b) This image shows paired chromosomes. (credit b: modification of work by NIH; scale-bar data from Matt Russell)

The Nucleolus

We already know that the nucleus directs the synthesis of ribosomes, but how does it do this? Some chromosomes have sections of DNA that encode ribosomal RNA. A darkly staining area within the nucleus called the nucleolus (plural = nucleoli) aggregates the ribosomal RNA with associated proteins to assemble the ribosomal subunits that are then transported out through the pores in the nuclear envelope to the cytoplasm.

Organelles of the Endomembrane System

A set of three major organelles together form a system within the cell called the endomembrane system. These organelles work together to perform various cellular jobs, including the task of producing, packaging, and exporting certain cellular products. The organelles of the endomembrane system include the endoplasmic reticulum, Golgi apparatus, and vesicles.

Endoplasmic Reticulum

The **endoplasmic reticulum (ER)** is a system of channels that is continuous with the nuclear membrane (or "envelope") covering the nucleus and composed of the same lipid bilayer material. The ER can be thought of as a series of winding thoroughfares similar to the waterway canals in Venice. The ER provides passages throughout much of the cell that function in transporting, synthesizing, and storing materials. The winding structure of the ER results in a large membranous surface area that supports its many functions (Figure 4).

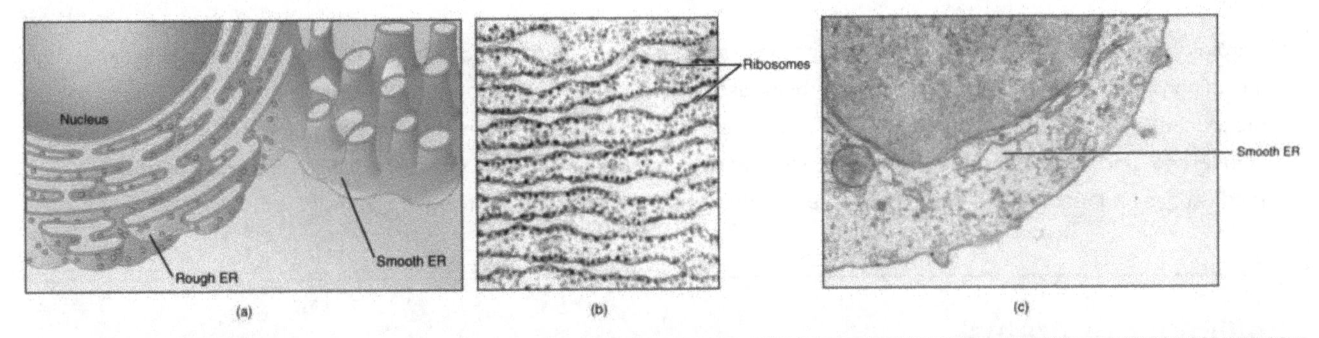

Figure 4. Endoplasmic Reticulum (ER). Click for a larger image. (a) The ER is a winding network of thin membranous sacs found in close association with the cell nucleus. The smooth and rough endoplasmic reticula are very different in appearance and function (source: mouse tissue). (b) Rough ER is studded with numerous ribosomes, which are sites of protein synthesis (source: mouse tissue). EM × 110,000. (c) Smooth ER synthesizes phospholipids, steroid hormones, regulates the concentration of cellular Ca^{++},metabolizes some carbohydrates, and breaks down certain toxins (source: mouse tissue). EM × 110,510. (Micrographs provided by the Regents of University of Michigan Medical School © 2012)

Endoplasmic reticulum can exist in two forms: rough ER and smooth ER. These two types of ER perform some very different functions and can be found in very different amounts depending on the type of cell. Rough ER (RER) is so-called because its membrane is dotted with embedded granules—organelles called ribosomes, giving the RER a bumpy appearance. A **ribosome** is an organelle that serves as the site of protein synthesis. It is composed of two ribosomal RNA subunits that wrap around mRNA to start the process of translation, followed by protein synthesis. Smooth ER (SER) lacks these ribosomes. One of the main functions of the smooth ER is in the synthesis of lipids. The smooth ER synthesizes phospholipids, the main component of biological membranes, as well as steroid hormones. For this reason, cells that produce large quantities of such hormones, such as those of the female ovaries and male testes, contain large amounts of smooth ER. In addition to lipid synthesis, the smooth ER also sequesters (i.e., stores) and regulates the concentration of cellular Ca^{++}, a function extremely important in cells of the nervous system where Ca^{++} is the trigger for neurotransmitter release. The smooth ER additionally metabolizes some carbohydrates and performs a detoxification role, breaking down certain toxins. In contrast with the smooth ER, the primary job of the rough ER is the synthesis and modification of proteins destined for the cell membrane or for export from the cell. For this protein synthesis, many ribosomes attach to the ER (giving it the studded appearance of rough ER). Typically, a protein is synthesized within the ribosome and released inside the channel of the rough ER, where sugars can be added to it (by a process called glycosylation) before it is transported within a vesicle to the next stage in the packaging and shipping process: the Golgi apparatus.

CAREER CONNECTION

Cardiologist

Heart disease is the leading cause of death in the United States. This is primarily due to our sedentary lifestyle and our high trans-fat diets.

Heart failure is just one of many disabling heart conditions. Heart failure does not mean that the heart has stopped working. Rather, it means that the heart can't pump with sufficient force to transport oxygenated blood to all the vital organs. Left untreated, heart failure can lead to kidney failure and other organ failure.

Cardiac muscle tissue comprises the heart's wall. Heart failure occurs when cardiac muscle cells' endoplasmic reticula do not function properly. As a result, an insufficient number of calcium ions are available to trigger a sufficient contractile force.

Cardiologists (cardi- = "heart"; -ologist = "one who studies") are doctors who specialize in treating heart diseases, including heart failure. Cardiologists can diagnose heart failure via a physical examination, results from an electrocardiogram (ECG, a test that measures the heart's electrical activity), a chest X-ray to see whether the heart is enlarged, and other tests. If the cardiologist diagnoses heart failure, he or she will typically prescribe appropriate medications and recommend a reduced table salt intake and a supervised exercise program.

The Golgi Apparatus

The **Golgi apparatus** is responsible for sorting, modifying, and shipping off the products that come from the rough ER, much like a post-office. The Golgi apparatus looks like stacked flattened discs, almost like stacks of oddly shaped pancakes. Like the ER, these discs are membranous. The Golgi apparatus has two distinct sides, each with a different role. One side of the apparatus receives products in vesicles. These products are sorted through the apparatus, and then they are released from the opposite side after being repackaged into new vesicles. If the product is to be exported from the cell, the vesicle migrates to the cell surface and fuses to the cell membrane, and the cargo is secreted (Figure 5).

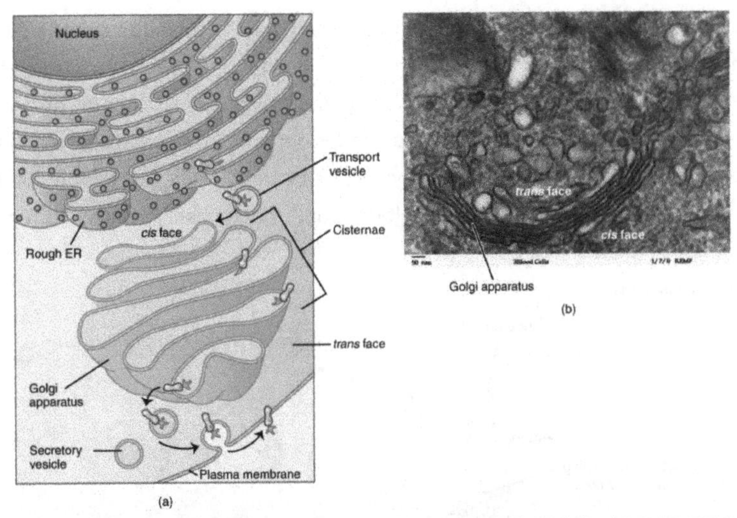

Figure 5. Golgi Apparatus. (a) The Golgi apparatus manipulates products from the rough ER, and also produces new organelles called lysosomes. Proteins and other products of the ER are sent to the Golgi apparatus, which organizes, modifies, packages, and tags them. Some of these products are transported to other areas of the cell and some are exported from the cell through exocytosis. Enzymatic proteins are packaged as new lysosomes (or packaged and sent for fusion with existing lysosomes). (b) An electron micrograph of the Golgi apparatus.

CAREER CONNECTION

Geneticist

Many diseases arise from genetic mutations that prevent synthesizing critical proteins. One such disease is Lowe disease (or oculocerebrorenal syndrome, because it affects the eyes, brain, and kidneys). In Lowe disease, there is a deficiency in an enzyme localized to the Golgi apparatus. Children with Lowe disease are born with cataracts, typically develop kidney disease after the first year of life, and may have impaired mental abilities.

A mutation on the X chromosome causes Lowe disease. The X chromosome is one of the two human sex chromosomes, as these chromosomes determine a person's sex. Females possess two X chromosomes while males possess one X and one Y chromosome. In females, the genes on only one of the two X chromosomes are expressed. Females who carry the Lowe disease gene on one of their X chromosomes are carriers and do not show symptoms of the disease. However, males only have one X chromosome and the genes on this chromosome are always expressed. Therefore, males will always have Lowe disease if their X chromosome carries the Lowe disease gene. Geneticists have identified the mutated gene's location, as well as many other mutation locations that cause genetic diseases. Through prenatal testing, a woman can find out if the fetus she is carrying may be afflicted with one of several genetic diseases.

Geneticists analyze prenatal genetic test results and may counsel pregnant women on available options. They may also conduct genetic research that leads to new drugs or foods, or perform DNA analyses for forensic investigations.

Lysosomes

Some of the protein products packaged by the Golgi include digestive enzymes that are meant to remain inside the cell for use in breaking down certain materials. The enzyme-containing vesicles released by the Golgi may form new lysosomes, or fuse with existing, lysosomes. A **lysosome** is an organelle that contains enzymes that break down and digest unneeded cellular components, such as a damaged organelle. (A lysosome is similar to a wrecking crew that takes down old and unsound buildings in a neighborhood.) **Autophagy** ("self-eating") is the process of a cell digesting its own structures. Lysosomes are also important for breaking down foreign material. For example, when certain immune defense cells (white blood cells) phagocytize bacteria, the bacterial cell is transported into a lysosome and digested by the enzymes inside. As one might imagine, such phagocytic defense cells contain large numbers of lysosomes. Under certain circumstances, lysosomes perform a more grand and dire function. In the case of damaged or unhealthy cells, lysosomes can be triggered to open up and release their digestive enzymes into the cytoplasm of the cell, killing the cell. This "self-destruct" mechanism is called **autolysis**, and makes the process of cell death controlled (a mechanism called "apoptosis").

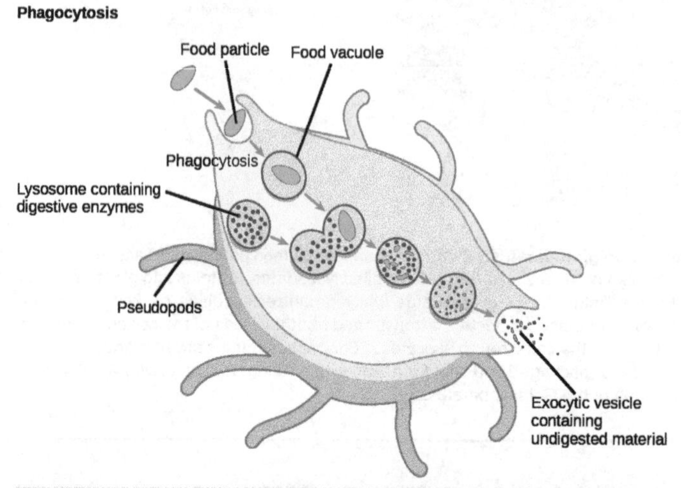

A macrophage has engulfed (phagocytized) a potentially pathogenic bacterium and then fuses with lysosomes within the cell to destroy the pathogen. Other organelles are present in the cell but for simplicity we do not show them.

Organelles for Energy Production and Detoxification

In addition to the jobs performed by the endomembrane system, the cell has many other important functions. Just as you must consume nutrients to provide yourself with energy, so must each of your cells take in nutrients, some of which convert to chemical energy that can be used to power biochemical reactions. Another important function of the cell is detoxification. Humans take in all sorts of toxins from the environment and also produce harmful chemicals as byproducts of cellular processes. Cells called hepatocytes in the liver detoxify many of these toxins.

Mitochondria

A **mitochondrion** (plural = mitochondria) is a membranous, bean-shaped organelle that is the "energy transformer" of the cell. Mitochondria consist of an outer lipid bilayer membrane as well as an additional inner lipid bilayer membrane. The inner membrane is highly folded into winding structures with a great deal of surface area, called cristae. It is along this inner membrane that a series of proteins, enzymes, and other molecules perform the biochemical reactions of cellular respiration. These reactions convert energy stored in nutrient molecules (such as glucose) into adenosine triphosphate (ATP), which provides usable cellular energy to the cell. Cells use ATP constantly, and so the mitochondria are constantly at work. Oxygen molecules are required during cellular respiration, which is why you must constantly breathe it in. One of the organ systems in the body that uses huge amounts of ATP is the muscular system because ATP is required to sustain muscle contraction. As a result, muscle cells are packed full of mitochondria. Nerve cells also need large quantities of ATP to run their sodium-potassium pumps. Therefore, an individual neuron will be loaded with over a thousand mitochondria. On the other hand, a bone cell, which is not nearly as metabolically-active, might only have a couple hundred mitochondria.

EVOLUTION CONNECTION

Endosymbiosis

We have mentioned that both mitochondria and chloroplasts contain DNA and ribosomes. Have you wondered why? Strong evidence points to endosymbiosis as the explanation.

Symbiosis is a relationship in which organisms from two separate species depend on each other for their survival. Endosymbiosis (endo- = "within") is a mutually beneficial relationship in which one organism lives inside the other. Endosymbiotic relationships abound in nature. We have already mentioned that microbes that produce vitamin K live inside the human gut. This relationship is beneficial for us because we are unable to synthesize vitamin K. It is also beneficial for the microbes because they are protected from other organisms and from drying out, and they receive abundant food from the environment of the large intestine.

Scientists have long noticed that bacteria, mitochondria, and chloroplasts are similar in size. We also know that bacteria have DNA and ribosomes, just like mitochondria and chloroplasts. Scientists believe that host cells and bacteria formed an endosymbiotic relationship when the host cells ingested both aerobic and autotrophic bacteria (cyanobacteria) but did not destroy them. Through many millions of years of evolution, these ingested bacteria became more specialized in their functions, with the aerobic bacteria becoming mitochondria and the autotrophic bacteria becoming chloroplasts.

Peroxisomes

Like lysosomes, a **peroxisome** is a membrane-bound cellular organelle that contains mostly enzymes (Figure 7). Peroxisomes perform a couple of different functions, including lipid metabolism and chemical detoxification. In contrast to the digestive enzymes found in lysosomes, the enzymes within peroxisomes serve to transfer hydrogen atoms from various molecules to oxygen, producing hydrogen peroxide (H_2O_2). In this way, peroxisomes neutralize poisons such as alcohol. In order to appreciate the importance of peroxisomes, it is necessary to understand the concept of reactive oxygen species.

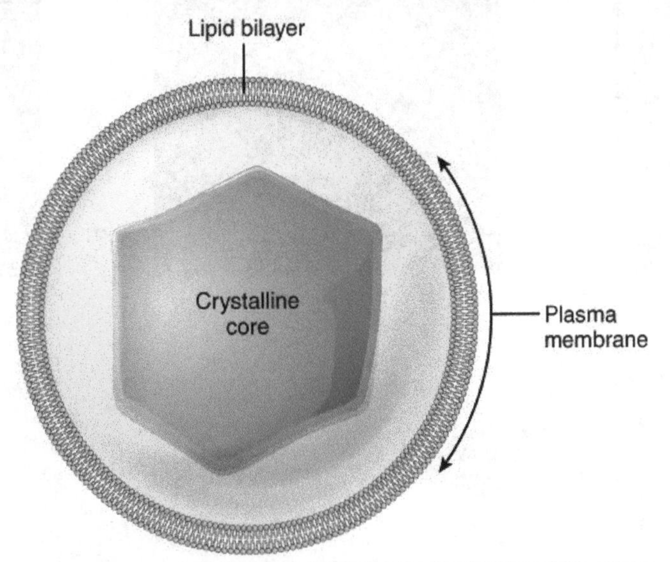

Figure 7. Peroxisome. Peroxisomes are membrane-bound organelles that contain an abundance of enzymes for detoxifying harmful substances and lipid metabolism.

AGING AND THE CELL: THE FREE RADICAL THEORY

The free radical theory on aging was originally proposed in the 1950s, and still remains under debate. Generally speaking, the free radical theory of aging suggests that accumulated cellular damage from oxidative stress contributes to the physiological and anatomical effects of aging. There are two significantly different versions of this theory: one states that the aging process itself is a result of oxidative damage, and the other states that oxidative damage causes age-related disease and disorders. The latter version of the theory is more widely accepted than the former. However, many lines of evidence suggest that oxidative damage does contribute to the aging process. Research has shown that reducing oxidative damage can result in a longer lifespan in certain organisms such as yeast, worms, and fruit flies. Conversely, increasing oxidative damage can shorten the lifespan of mice and worms. Interestingly, a manipulation called calorie-restriction (moderately restricting the caloric intake) has been shown to increase life span in some laboratory animals. It is believed that this increase is at least in part due to a reduction of oxidative stress. However, a long-term study of primates with calorie-restriction showed no increase in their lifespan. A great deal of additional research will be required to better understand the link between reactive oxygen species and aging.

The Cytoskeleton

Much like the bony skeleton structurally supports the human body, the cytoskeleton helps the cells to maintain their structural integrity. The **cytoskeleton** is a group of fibrous proteins that provide structural support for cells, but this is only one of the functions of the cytoskeleton. Cytoskeletal components are also critical for cell motility, cell reproduction, and transportation of substances within the cell. The cytoskeleton forms a complex thread-like network throughout the cell consisting of three different kinds of protein-based filaments: microfilaments, intermediate filaments, and microtubules (Figure 8).

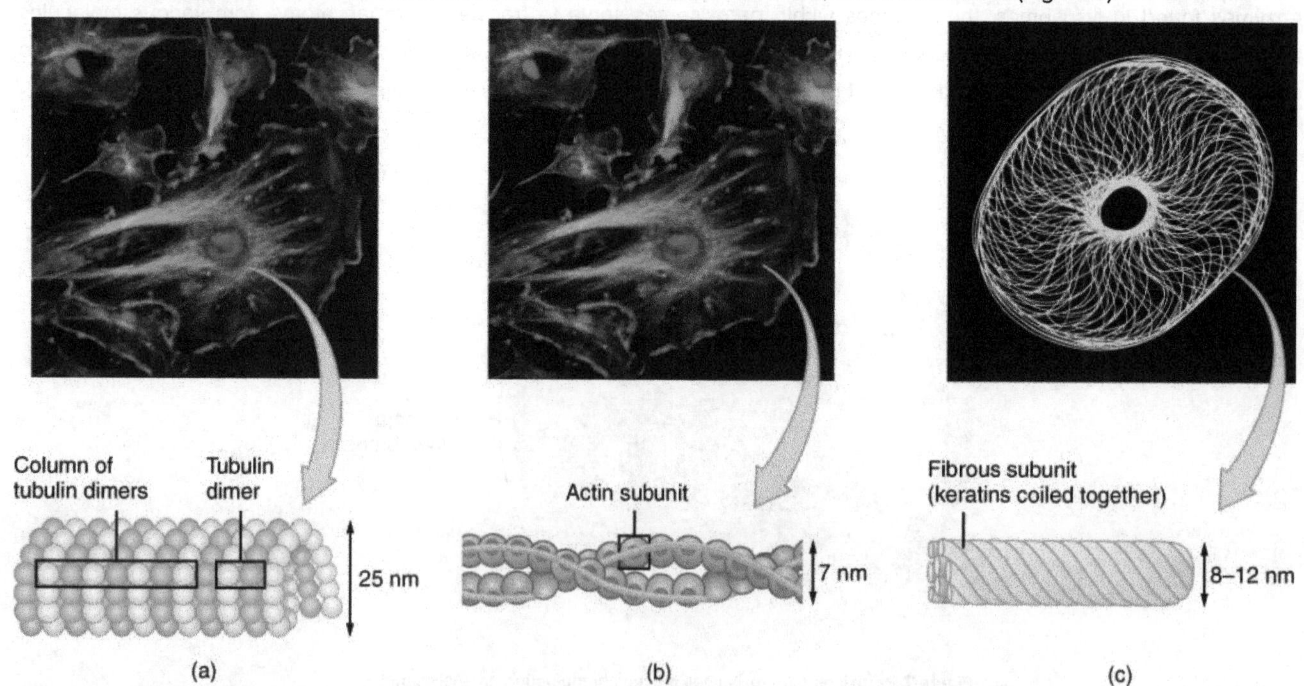

Figure 8. The Three Components of the Cytoskeleton. The cytoskeleton consists of (a) microtubules, (b) microfilaments, and (c) intermediate filaments. The cytoskeleton plays an important role in maintaining cell shape and structure, promoting cellular movement, and aiding cell division.

The thickest of the three components is the **microtubule**, a structural filament composed of subunits of a protein called tubulin. Microtubules maintain cell shape and structure, help resist compression of the cell, and play a role in positioning the organelles within the cell. Microtubules also make up two types of cellular appendages important for motion: cilia and flagella. **Cilia** are found on many cells of the body, including the epithelial cells that line the airways of the respiratory system. Cilia move rhythmically; they beat constantly, moving waste materials such as dust, mucus, and bacteria upward through the airways, away from the lungs and toward the mouth. Beating cilia on cells in the female fallopian tubes move egg cells from the ovary towards the uterus. A **flagellum** (plural = *flagella*) is an appendage larger than a cilium and specialized for cell locomotion. The only flagellated cell in humans is the sperm cell that must propel itself towards female egg cells.

A very important function of microtubules is to set the paths (somewhat like railroad tracks) along which the genetic material can be pulled (a process requiring ATP) during cell division, so that each new daughter cell receives the appropriate set of chromosomes. Two short, identical microtubule structures called centrioles are found near the nucleus of cells. A **centriole** can serve as the cellular origin point for microtubules extending outward as cilia or flagella or can assist with the separation of DNA during cell division. Microtubules grow out from the centrioles by adding more tubulin subunits, like adding additional links to a chain.

In contrast with microtubules, the **microfilament** is a thinner type of cytoskeletal filament. Actin, a protein that forms chains, is the primary component of these microfilaments. Actin fibers, twisted chains of actin filaments, constitute a large component of muscle tissue and, along with the protein myosin, are responsible for muscle contraction. Like microtubules, actin filaments are long chains of single subunits (called actin subunits). In muscle cells, these long actin strands, called thin filaments, are "pulled" by thick filaments of the myosin protein to contract the cell. Actin also has an important role during cell division. When a cell is about to split in half during cell division, actin filaments work with myosin to create a cleavage furrow that eventually splits the cell down the middle, forming two new cells from the original cell.

The final cytoskeletal filament is the intermediate filament. As its name would suggest, an **intermediate filament** is a filament intermediate in thickness between the microtubules and microfilaments. Intermediate filaments are made up of long fibrous subunits of a protein called keratin that are wound together like the threads that compose a rope. Intermediate filaments, in concert with the microtubules, are important for maintaining cell shape and structure. Unlike the microtubules, which resist compression, intermediate filaments resist tension—the forces that pull apart cells. There are many cases in which cells are prone to tension, such as when epithelial cells of the skin are compressed, tugging them in different directions. Intermediate filaments help anchor organelles together within a cell and also link cells to other cells by forming special cell-to-cell junctions.

Extracellular Matrix of Animal Cells

While cells in most multicellular organisms release materials into the extracellular space, animal cells will be discussed as an example. The primary components of these materials are proteins, and the most abundant protein is collagen. Collagen fibers are interwoven with proteoglycans, which are carbohydrate-containing protein molecules. Collectively, we call these materials the extracellular matrix ((Figure 9)). Not only does the extracellular matrix hold the cells together to form a tissue, but it also allows the cells within the tissue to communicate with each other. How can this happen?

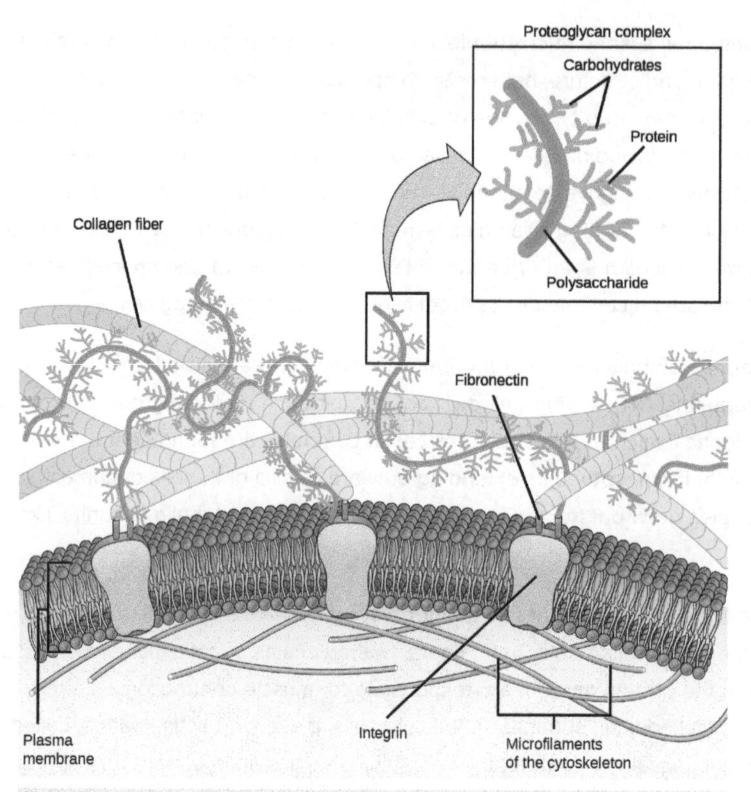

Figure 9. The extracellular matrix consists of a network of proteins and carbohydrates.

Cells have protein receptors on their plasma membranes' extracellular surfaces. When a molecule within the matrix binds to the receptor, it changes the receptor's molecular structure. The receptor, in turn, changes the microfilaments' conformation positioned just inside the plasma membrane. These conformational changes induce chemical signals inside the cell that reach the nucleus and turn "on" or "off" the transcription of specific DNA sections, which affects the associated protein production, thus changing the activities within the cell.

Blood clotting provides an example of the extracellular matrix's role in cell communication. When the cells lining a blood vessel are damaged, they display a protein receptor, which we call tissue factor. When tissue factor binds with another factor in the extracellular matrix, it causes platelets to adhere to the damaged blood vessel's wall, stimulates the adjacent smooth muscle cells in the blood vessel to contract (thus constricting the blood vessel), and initiates a series of steps that stimulate the platelets to produce clotting factors.

Intercellular Junctions

Cells can also communicate with each other via direct contact, or intercellular junctions. There are differences in the ways that plant and animal and fungal cells communicate. Plasmodesmata are junctions between plant cells; whereas, animal cell contacts include tight junctions, gap junctions, and desmosomes.

Tight Junctions

A tight junction is a watertight seal between two adjacent animal cells ((Figure 10)). Proteins (predominantly two proteins called claudins and occludins) tightly hold the cells against each other.

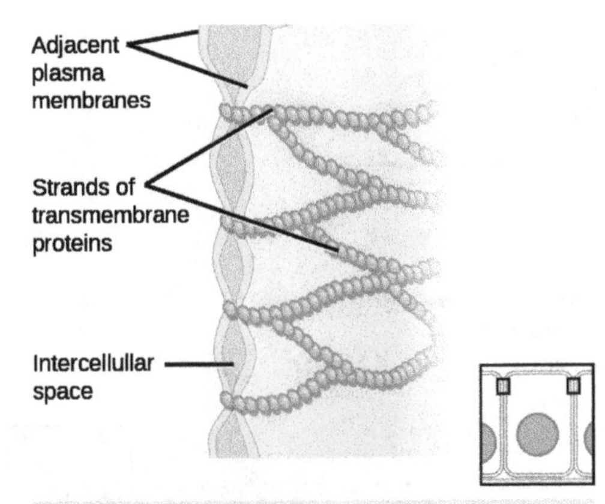

Figure 10. Tight junctions form watertight connections between adjacent animal cells. Proteins create tight junction adherence. (credit: modification of work by Mariana Ruiz Villareal)

This tight adherence prevents materials from leaking between the cells; tight junctions are typically found in epithelial tissues that line internal organs and cavities, and comprise most of the skin. For example, the tight junctions of the epithelial cells lining your urinary bladder prevent urine from leaking out into the extracellular space.

Desmosomes

Also only in animal cells are desmosomes, which act like spot welds between adjacent epithelial cells ((Figure 11)). Cadherins, short proteins in the plasma membrane connect to intermediate filaments to create desmosomes. The cadherins connect two adjacent cells and maintain the cells in a sheet-like formation in organs and tissues that stretch, like the skin, heart, and muscles.

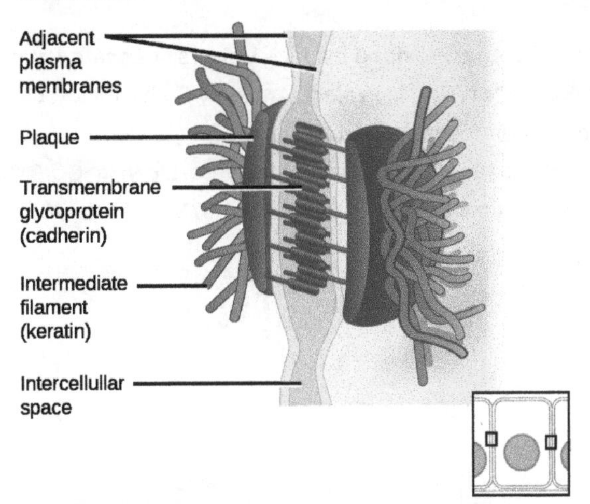

Figure 11. A desmosome forms a very strong spot weld between cells. Linking cadherins and intermediate filaments create it. (credit: modification of work by Mariana Ruiz Villareal)

Gap Junctions

Gap junctions in animal cells are channels between adjacent cells that allow for transporting ions, nutrients, and other substances that enable cells to communicate ((Figure 12)). Structurally, however, gap junctions and plasmodesmata differ.

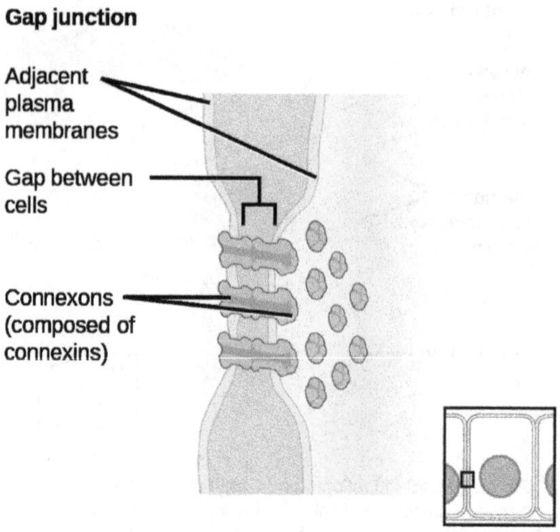

Figure 12. A gap junction is a protein-lined pore that allows water and small molecules to pass between adjacent animal cells. (credit: modification of work by Mariana Ruiz Villareal)

Gap junctions develop when a set of six proteins (connexins) in the plasma membrane arrange themselves in an elongated donut-like configuration – a connexon. When the connexon's pores ("doughnut holes") in adjacent animal cells align, a channel between the two cells forms. Gap junctions are particularly important in cardiac muscle. The electrical signal for the muscle to contract passes efficiently through gap junctions, allowing the heart muscle cells to contract in tandem.

Cellular energy

Scientists use the term bioenergetics to describe the concept of energy flow (Figure 1) through living systems, such as cells. Cellular processes such as the building and breaking down of complex molecules occur through stepwise chemical reactions. Some of these chemical reactions are spontaneous and release energy, whereas others require energy to proceed.

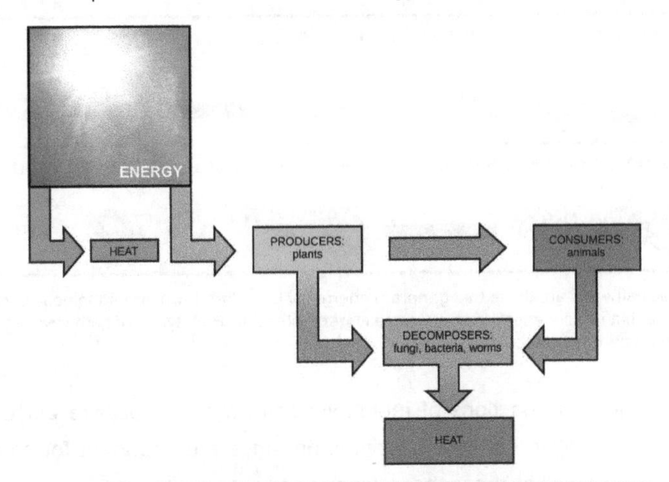

Figure 1. Ultimately, most life forms get their energy from the sun. Plants use photosynthesis to capture sunlight, and herbivores eat the plants to obtain energy. Carnivores eat the herbivores, and eventual decomposition of plant and animal material contributes to the nutrient pool.

Just as living things must continually consume food to replenish their energy supplies, cells must continually produce more energy to replenish that used by the many energy-requiring chemical reactions that constantly take place. Together, all of the chemical reactions that take place inside cells, including those that consume or generate energy, are referred to as the cell's metabolism.

Metabolic Pathways

Consider the metabolism of sugar. This is a classic example of one of the many cellular processes that use and produce energy. Living things consume sugars as a major energy source, because sugar molecules have a great deal of energy stored within their bonds. For the most part, photosynthesizing organisms like plants produce these sugars. During photosynthesis, plants use energy (originally from sunlight) to convert carbon dioxide gas (CO_2) into sugar molecules (like glucose: $C_6H_{12}O_6$). They consume carbon dioxide and produce oxygen as a waste product. This reaction is summarized as:

$$6CO_2+6H_2O\rightarrow C_6H_{12}O_6+6O_2$$

Because this process involves synthesizing an energy-storing molecule, it requires energy input to proceed. During the light reactions of photosynthesis, energy is provided by a molecule called adenosine triphosphate (ATP), which is the primary energy currency of all cells. Just as the dollar is used as currency to buy goods, cells use molecules of ATP as energy currency to perform immediate work. In contrast, energy-storage molecules such as glucose are consumed only to be broken down to use their energy. The reaction that harvests the energy of a sugar molecule in cells requiring oxygen to survive can be summarized by the reverse reaction to photosynthesis. In this reaction, oxygen is consumed and carbon dioxide is released as a waste product. The reaction is summarized as:

$$C_6H_{12}O_6\rightarrow 6H_2O+6CO_2$$

Both of these reactions involve many steps.

The processes of making and breaking down sugar molecules illustrate two examples of metabolic pathways. A metabolic pathway is a series of chemical reactions that takes a starting molecule and modifies it, step-by-step, through a series of metabolic intermediates, eventually yielding a final product. In the example of sugar metabolism, the first metabolic pathway synthesized sugar from smaller molecules, and the other pathway broke sugar down into smaller molecules. These two opposite processes—the first requiring energy and the second producing energy—are referred to as anabolic pathways (building polymers) and catabolic pathways (breaking down polymers into their monomers), respectively. Consequently, metabolism is composed of synthesis (anabolism) and degradation (catabolism) (Figure 2).

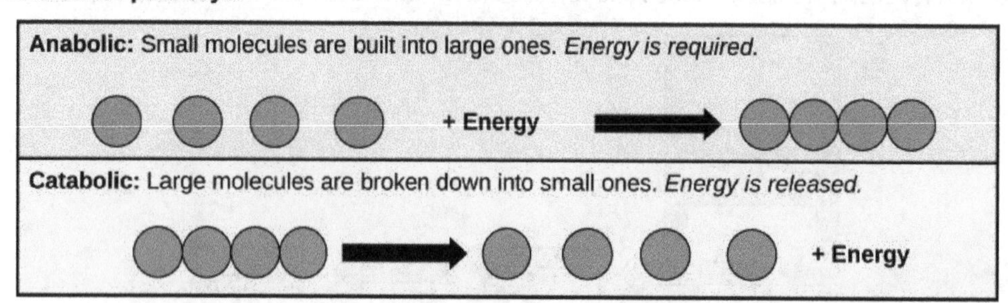

Figure 2. Catabolic pathways are those that generate energy by breaking down larger molecules. Anabolic pathways are those that require energy to synthesize larger molecules. Both types of pathways are required for maintaining the cell's energy balance.

It is important to know that the chemical reactions of metabolic pathways do not take place on their own. Each reaction step is facilitated, or catalyzed, by a protein called an enzyme. Enzymes are important for catalyzing all types of biological reactions—those that require energy as well as those that release energy.

ATP In Living Systems

All living things require energy to function. While different organisms acquire this energy in different ways, they store (and use it) in the same way. In this section, we'll learn about ATP—the energy of life. ATP is how cells store energy. These storage molecules are produced in the mitochondria, tiny organelles found in eukaryotic cells sometimes called the "powerhouse" of the cell.

MITOCHONDRIAL DISEASE PHYSICIAN

What happens when the critical reactions of cellular respiration do not proceed correctly? Mitochondrial diseases are genetic disorders of metabolism. Mitochondrial disorders can arise from mutations in nuclear or mitochondrial DNA, and they result in the production of less energy than is normal in body cells.

In type 2 diabetes, for instance, the oxidation efficiency of NADH is reduced, impacting oxidative phosphorylation but not the other steps of respiration. Symptoms of mitochondrial diseases can include muscle weakness, lack of coordination, stroke-like episodes, and loss of vision and hearing. Most affected people are diagnosed in childhood, although there are some adult-onset diseases.

Identifying and treating mitochondrial disorders is a specialized medical field. The educational preparation for this profession requires a college education, followed by medical school with a specialization in medical genetics. Medical geneticists can be board certified by the American Board of Medical Genetics and go on to become associated with professional organizations devoted to the study of mitochondrial diseases, such as the Mitochondrial Medicine Society and the Society for Inherited Metabolic Disease.

A living cell cannot store significant amounts of free energy. Excess free energy would result in an increase of heat in the cell, which would result in excessive thermal motion that could damage and then destroy the cell. Rather, a cell must be able to handle that energy in a way that enables the cell to store energy safely and release it for use only as needed. Living cells accomplish this by using the compound adenosine triphosphate (ATP). ATP is often called the "energy currency" of the cell, and, like currency, this versatile compound can be used to fill any energy need of the cell. How? It functions similarly to a rechargeable battery.

When ATP is broken down, usually by the removal of its terminal phosphate group, energy is released. The energy is used to do work by the cell, usually by the released phosphate binding to another molecule, activating it. For example, in the mechanical work of muscle contraction, ATP supplies the energy to move the contractile muscle proteins. Recall the active transport work of the sodium-potassium pump in cell membranes. ATP alters the structure of the integral protein that functions as the pump, changing its affinity for sodium and potassium. In this way, the cell performs work, pumping ions against their electrochemical gradients.

ATP Structure and Function

At the heart of ATP is a molecule of adenosine monophosphate (AMP), which is composed of an adenine molecule bonded to a ribose molecule and to a single phosphate group (Figure 3). Ribose is a five-carbon sugar found in RNA, and AMP is one of the nucleotides in RNA. The addition of a second phosphate group to this core molecule results in the formation of adenosine diphosphate (ADP); the addition of a third phosphate group forms adenosine triphosphate (ATP).

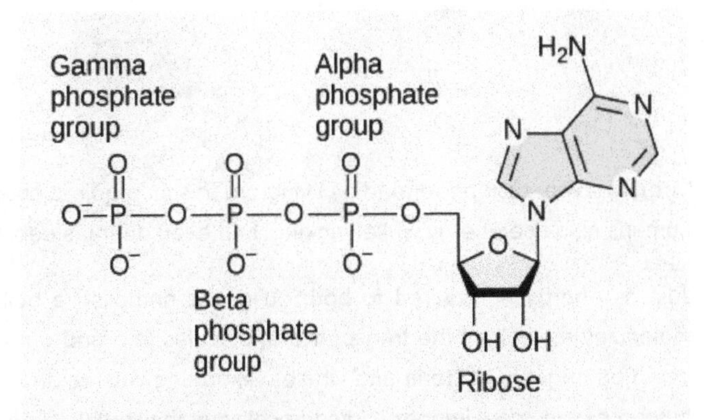

Figure 3. ATP (adenosine triphosphate) has three phosphate groups that can be removed by hydrolysis to form ADP (adenosine diphosphate) or AMP (adenosine monophosphate).The negative charges on the phosphate group naturally repel each other, requiring energy to bond them together and releasing energy when these bonds are broken.

The addition of a phosphate group to a molecule requires energy. Phosphate groups are negatively charged and thus repel one another when they are arranged in series, as they are in ADP and ATP. This repulsion makes the ADP and ATP molecules inherently unstable. The release of one or two phosphate groups from ATP, a process called **dephosphorylation**, releases energy.

Energy from ATP

Hydrolysis is the process of breaking complex macromolecules apart. During hydrolysis, water is split, or lysed, and the resulting hydrogen atom (H^+) and a hydroxyl group (OH^-) are added to the larger molecule. The hydrolysis of ATP produces ADP, together with an inorganic phosphate ion (P_i), and the release of free energy. To carry out life processes, ATP is continuously broken down into ADP, and like a rechargeable battery, ADP is continuously regenerated into ATP by the reattachment of a third phosphate group. Water, which was broken down into its hydrogen atom and hydroxyl group during ATP hydrolysis, is regenerated when a third phosphate is added to the ADP molecule, reforming ATP.

Obviously, energy must be infused into the system to regenerate ATP. Where does this energy come from? In nearly every living thing on earth, the energy comes from the metabolism of glucose. In this way, ATP is a direct link between the limited set of exergonic pathways of glucose catabolism and the multitude of endergonic pathways that power living cells.

Phosphorylation

Recall that, in some chemical reactions, enzymes may bind to several substrates that react with each other on the enzyme, forming an intermediate complex. An intermediate complex is a temporary structure, and it allows one of the substrates (such as ATP) and reactants to more readily react with each other; in reactions involving ATP, ATP is one of the substrates and ADP is a product. During an endergonic chemical reaction, ATP forms an intermediate complex with the substrate and enzyme in the reaction. This intermediate complex allows the ATP to transfer its third phosphate group, with its energy, to the substrate, a process called phosphorylation.

When the intermediate complex breaks apart, the energy is used to modify the substrate and convert it into a product of the reaction. The ADP molecule and a free phosphate ion are released into the medium and are available for recycling through cell metabolism.

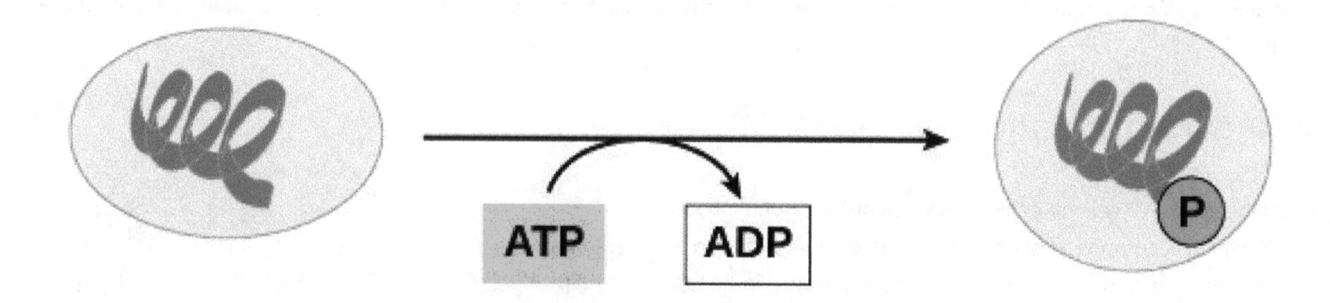

Virtually every task performed by living organisms requires energy. Energy is needed to perform heavy labor and exercise, but humans also use energy while thinking, and even during sleep. In fact, the living cells of every organism constantly use energy.

Just as energy is required to both build and demolish a building, energy is required for the synthesis and breakdown of molecules as well as the transport of molecules into and out of cells. In addition, processes such as ingesting and breaking down pathogenic bacteria and viruses, exporting wastes and toxins, and movement of the cell require energy. Nutrients and other molecules are imported into the cell, metabolized (broken down) and possibly synthesized into new molecules, modified if needed, transported around the cell, and possibly distributed to the entire organism. For example, the large proteins that make up muscles are built from smaller molecules imported from dietary amino acids. Complex carbohydrates are broken down into simple sugars that the cell uses for energy.

Cellular Respiration: Synthesis of ATP

Cellular respiration extracts the energy from the bonds in glucose and converts it into a form that all living things can use. Now let's take a more detailed look at how all eukaryotes—which includes humans!—make use of this stored energy. There are two types of cellular respiration: 1. aerobic and 2. anaerobic. Aerobic respiration requires oxygen and yields high amounts of ATP per one glucose molecule compared to anaerobic that does not require oxygen and yields low levels of ATP per glucose molecule. Both aerobic and anaerobic respiration begin with process called **glycolysis**.

Glycolysis

You have read that nearly all of the energy used by living things comes to them in the bonds of the sugar, glucose. **Glycolysis** is the first step in the breakdown of glucose to extract energy for cell metabolism. Many living organisms carry out glycolysis as part of their metabolism. Glycolysis takes place in the cytoplasm of most prokaryotic and all eukaryotic cells.

Glycolysis begins with the six-carbon, ring-shaped structure of a single glucose molecule and ends with two molecules of a three-carbon sugar called pyruvate. Glycolysis consists of two distinct phases. In the first part of the glycolysis pathway, energy is used to make adjustments so that the six-carbon sugar molecule can be split evenly into two three-carbon pyruvate molecules. In the second part of glycolysis, ATP and nicotinamide-adenine dinucleotide (NADH) are produced (Figure 5).

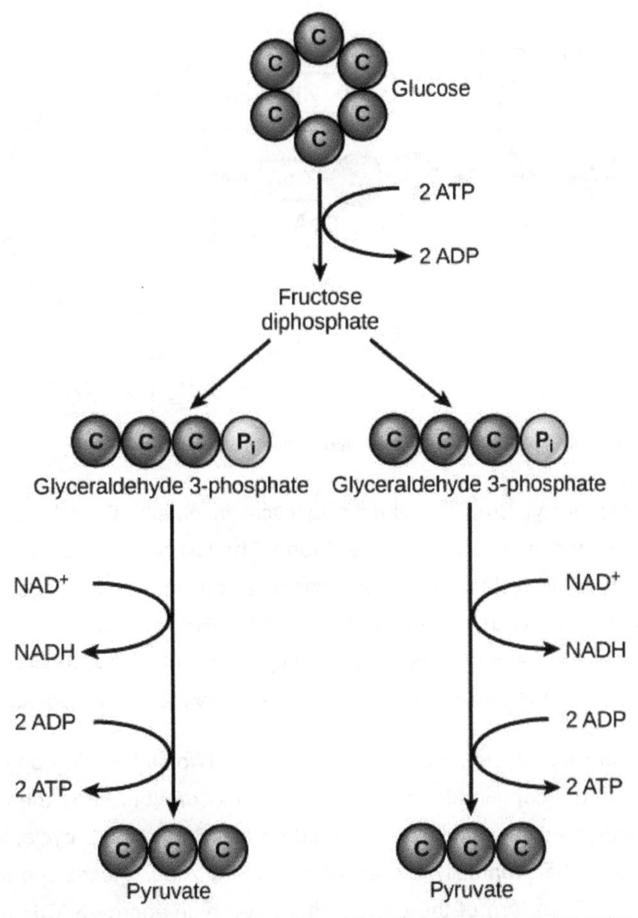

Figure 5. In glycolysis, a glucose molecule is converted into two pyruvate molecules.

If the cell cannot catabolize the pyruvate molecules further, it will harvest only two ATP molecules from one molecule of glucose. For example, mature mammalian red blood cells are only capable of glycolysis, which is their sole source of ATP. If glycolysis is interrupted, these cells would eventually die.

Aerobic Respiration: Citric Acid Cycle and Oxidative Phosphorylation

In eukaryotic cells, the pyruvate molecules produced at the end of glycolysis are transported into mitochondria, which are sites of cellular respiration. If oxygen is available, aerobic respiration will go forward. There are two main processes of aerobic respiration: The Citric Acid Cycle (or the Kreb's Cycle) and Oxidative Phosphorylation.

The Citric Acid Cycle

In mitochondria, pyruvate will be transformed into a two-carbon acetyl group (by removing a molecule of carbon dioxide) that will be picked up by a carrier compound called coenzyme A (CoA), which is made from vitamin B_5. The resulting compound is called acetyl CoA. (Figure 6). Acetyl CoA can be used in a variety of ways by the cell, but its major function is to deliver the acetyl group derived from pyruvate to the next pathway in glucose catabolism.

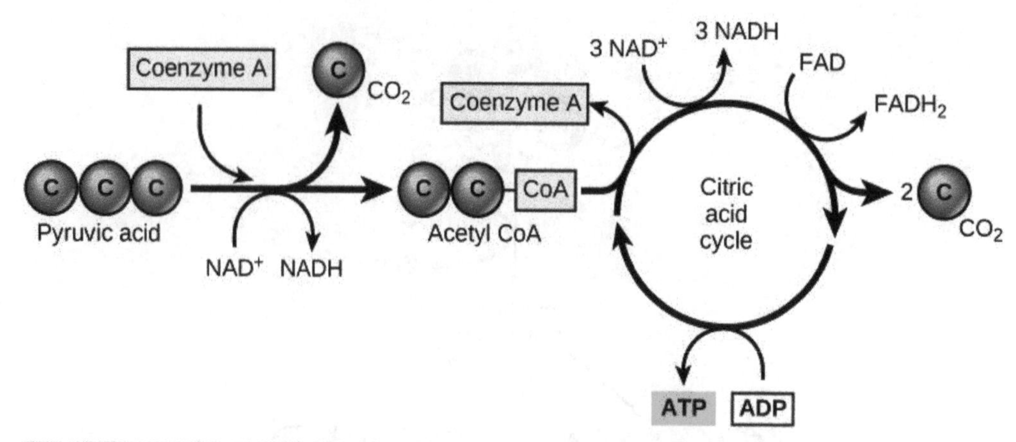

Figure 6. Pyruvate is converted into acetyl-CoA before entering the citric acid cycle.

Like the conversion of pyruvate to acetyl CoA, the citric acid cycle in eukaryotic cells takes place in the matrix of the mitochondria. Unlike glycolysis, the citric acid cycle is a closed loop: The last part of the pathway regenerates the compound used in the first step. The eight steps of the cycle are a series of chemical reactions that produces two carbon dioxide molecules, one ATP molecule (or an equivalent), and reduced forms (NADH and $FADH_2$) of NAD^+ and FAD^+, important coenzymes in the cell. Part of this is considered an aerobic pathway (oxygen-requiring) because the NADH and $FADH_2$ produced must transfer their electrons to the next pathway in the system, which will use oxygen. If oxygen is not present, this transfer does not occur.

Two carbon atoms come into the citric acid cycle from each acetyl group. Two carbon dioxide molecules are released on each turn of the cycle; however, these do not contain the same carbon atoms contributed by the acetyl group on that turn of the pathway. The two acetyl-carbon atoms will eventually be released on later turns of the cycle; in this way, all six carbon atoms from the original glucose molecule will be eventually released as carbon dioxide. It takes two turns of the cycle to process the equivalent of one glucose molecule. Each turn of the cycle forms three high-energy NADH molecules and one high-energy $FADH_2$ molecule. These high-energy carriers will connect with the last portion of aerobic respiration to produce ATP molecules. One ATP (or an equivalent) is also made in each cycle. Several of the intermediate compounds in the citric acid cycle can be used in synthesizing non-essential amino acids; therefore, the cycle is both anabolic and catabolic.

Oxidative Phosphorylation

You have just read about two pathways in glucose catabolism—glycolysis and the citric acid cycle—that generate ATP. Most of the ATP generated during the aerobic catabolism of glucose, however, is not generated directly from these pathways. Rather, it derives from a process that begins with passing electrons through a series of chemical reactions to a final electron acceptor, oxygen. These reactions take place in specialized protein complexes located in the inner membrane of the mitochondria of eukaryotic organisms and on the inner part of the cell membrane of prokaryotic organisms. The energy of the electrons is harvested and used to generate a electrochemical gradient across the inner mitochondrial membrane. The potential energy of this gradient is used to generate ATP. The entirety of this process is called oxidative phosphorylation.

The electron transport chain (Figure 7a) is the last component of aerobic respiration and is the only part of metabolism that uses atmospheric oxygen. Oxygen continuously diffuses into plants for this purpose. In animals, oxygen enters the body through the respiratory system. Electron transport is a series of chemical reactions that resembles a bucket brigade in that electrons are passed rapidly from one component to the next, to the endpoint of the chain where oxygen is the final electron acceptor and water is produced.

The **electron transport chain** are four complexes composed of proteins and accessory electron carriers. The electron transport chain is present in multiple copies in the inner mitochondrial membrane of eukaryotes. In each transfer of an electron through the electron transport chain, the electron loses energy, but with some transfers, the energy is stored as potential energy by using it to pump hydrogen ions across the inner mitochondrial membrane into the intermembrane space, creating an electrochemical gradient. This process is called **chemiosmosis** (7c).

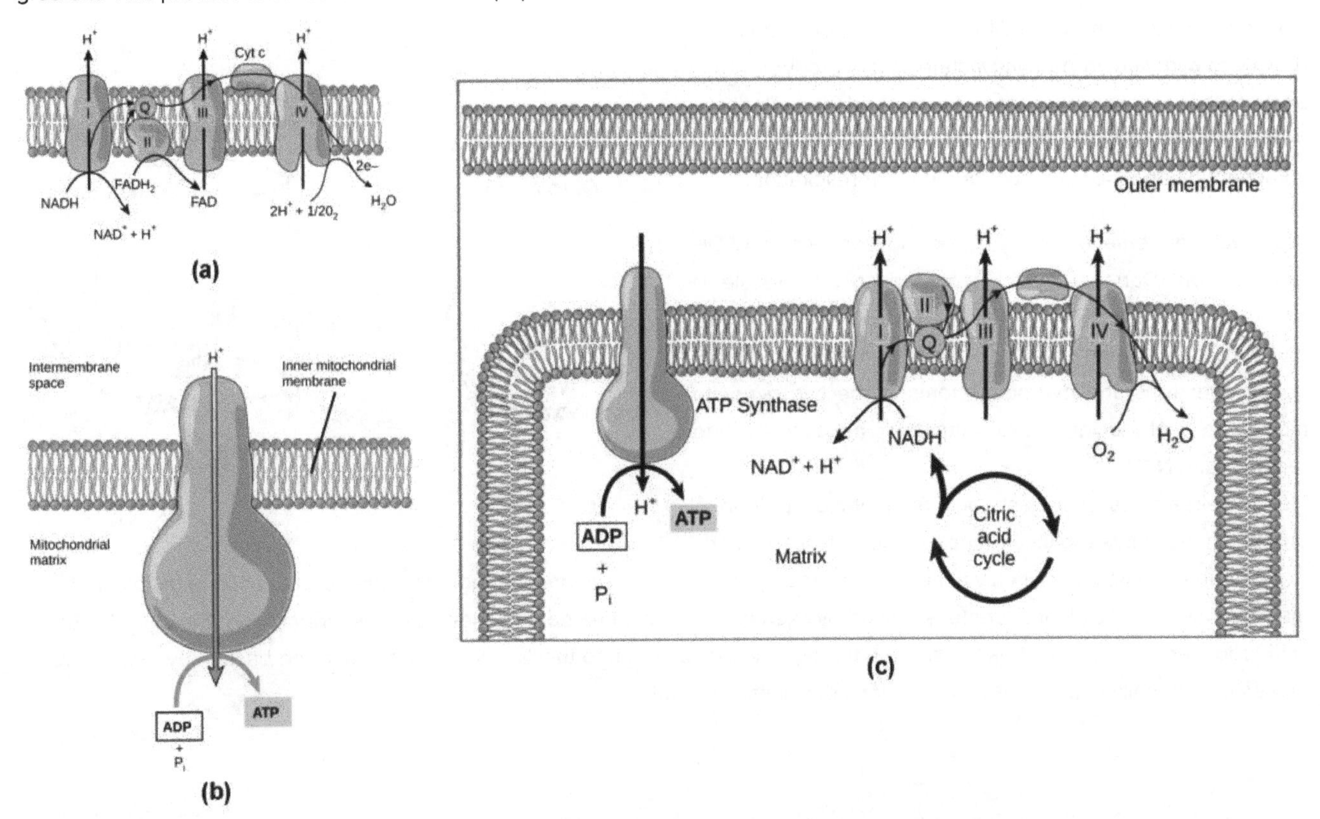

Figure 7. (a) The electron transport chain is a set of molecules that supports a series of oxidation-reduction reactions. (b) ATP synthase is a complex, molecular machine that uses an H+ gradient to regenerate ATP from ADP. (c) Chemiosmosis relies on the potential energy provided by the H+ gradient across the membrane.

Chemiosmosis

Chemiosmosis is the movement of ions across a semipermeable membrane, down their electrochemical gradient. Hydrogen ions, or protons, will diffuse from an area of high proton concentration to an area of lower proton concentration. This electrochemical concentration gradient of protons can be used to make ATP.

ATP synthase is the enzyme that makes ATP by chemiosmosis. It allows protons to pass through the membrane and uses the free energy difference to phosphorylate adenosine diphosphate (ADP), making ATP. The energy from the electron movement through electron transport chains to ATP synthase allows the proton to pass through ATP synthase to photophosphorylate ADP making ATP.

Anaerobic Respiration: Fermenation

Fermentation is another anaerobic (non-oxygen-requiring) pathway for breaking down glucose, one that's performed by many types of organisms and cells. In **fermentation**, the only energy extraction pathway is glycolysis, with one or two extra reactions tacked on at the end. Fermentation and cellular respiration begin the same way, with glycolysis. In fermentation, however, the pyruvate made in glycolysis does not continue through oxidation and the citric acid cycle, and the electron transport chain does not run.

Lactic acid fermentation

In **lactic acid fermentation**, NADH transfers its electrons directly to pyruvate, generating lactate as a byproduct. The bacteria that make yogurt carry out lactic acid fermentation, as do the red blood cells in your body, which don't have mitochondria and thus can't perform cellular respiration.

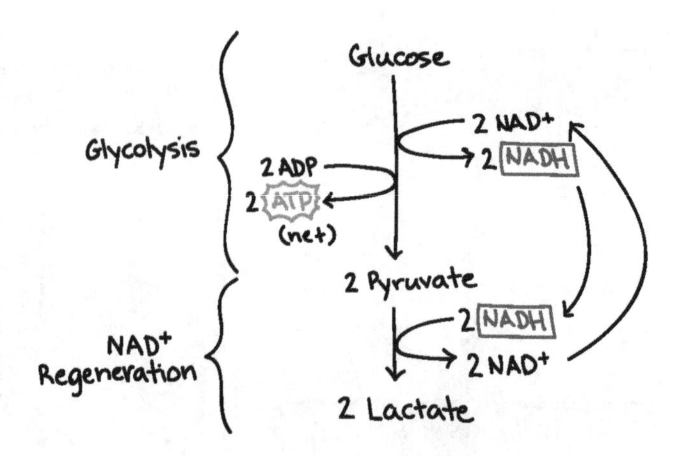

Lactic acid fermentation has two steps: glycolysis and NADH regeneration. During glycolysis, one glucose molecule is converted to two pyruvate molecules, producing two net ATP and two NADH. During NADH regeneration, the two NADH donate electrons and hydrogen atoms to the two pyruvate molecules, producing two lactate molecules and regenerating NAD+.

Muscle cells also carry out lactic acid fermentation, though only when they have too little oxygen for aerobic respiration to continue—for instance, when you've been exercising very hard. It was once thought that the accumulation of lactate in muscles was responsible for soreness caused by exercise, but recent research suggests this is probably not the case. Lactic acid produced in muscle cells is transported through the bloodstream to the liver, where it's converted back to pyruvate and processed normally in the remaining reactions of cellular respiration.

The Cell Membrane and Transport

Despite differences in structure and function, all living cells in multicellular organisms have a surrounding cell membrane. As the outer layer of your skin separates your body from its environment, the cell membrane (also known as the plasma membrane) separates the inner contents of a cell from its exterior environment. This cell membrane provides a protective barrier around the cell and regulates which materials can pass in or out.

Structure and Composition of the Cell Membrane

The **cell membrane** is an extremely pliable structure composed primarily of back-to-back phospholipids (a "bilayer"). Cholesterol is also present, which contributes to the fluidity of the membrane, and there are various proteins embedded within the membrane that have a variety of functions. A single phospholipid molecule has a phosphate group on one end, called the "head," and two side-by-side chains of fatty acids that make up the lipid tails (Figure 1). The phosphate group is negatively charged, making the head polar and hydrophilic—or "water loving."

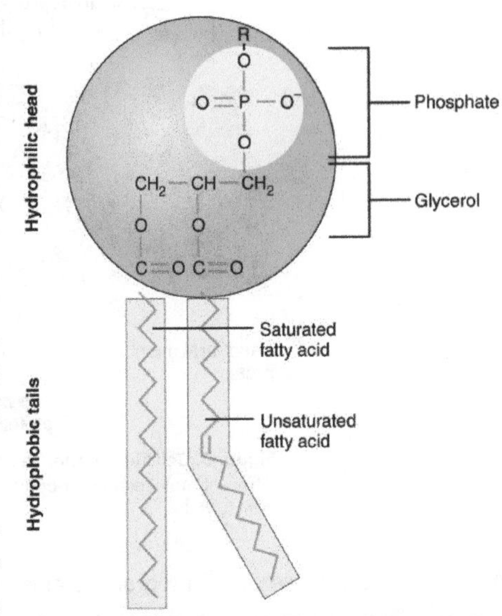

Figure 1. Phospholipid Structure. A phospholipid molecule consists of a polar phosphate "head," which is hydrophilic and a non-polar lipid "tail," which is hydrophobic. Unsaturated fatty acids result in kinks in the hydrophobic tails.

A **hydrophilic** molecule (or region of a molecule) is one that is attracted to water. The phosphate heads are thus attracted to the water molecules of both the extracellular and intracellular environments. The lipid tails, on the other hand, are uncharged, or nonpolar, and are hydrophobic—or "water fearing."

A **hydrophobic** molecule (or region of a molecule) repels and is repelled by water. Some lipid tails consist of saturated fatty acids and some contain unsaturated fatty acids. This combination adds to the fluidity of the tails that are constantly in motion. Phospholipids are thus amphipathic molecules.

An **amphipathic** molecule is one that contains both a hydrophilic and a hydrophobic region. In fact, soap works to remove oil and grease stains because it has amphipathic properties. The hydrophilic portion can dissolve in water while the hydrophobic portion can trap grease in micelles that then can be washed away.

The cell membrane consists of two adjacent layers of phospholipids. The lipid tails of one layer face the lipid tails of the other layer, meeting at the interface of the two layers. The phospholipid heads face outward, one layer exposed to the interior of the cell and one layer exposed to the exterior (Figure 2).

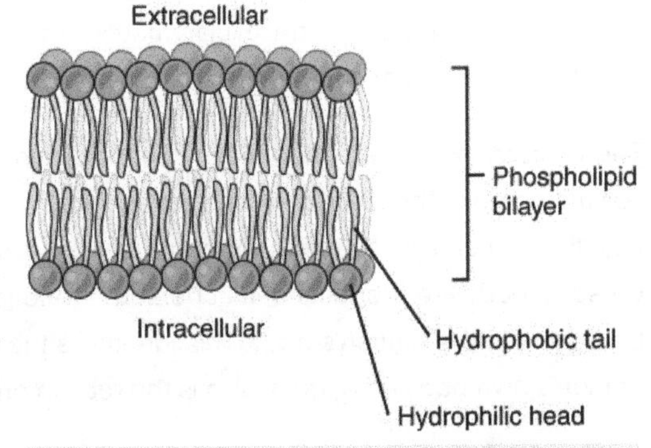

Figure 2. Phospolipid Bilayer. The phospholipid bilayer consists of two adjacent sheets of phospholipids, arranged tail to tail. The hydrophobic tails associate with one another, forming the interior of the membrane. The polar heads contact the fluid inside and outside of the cell.

Because the phosphate groups are polar and hydrophilic, they are attracted to water in the intracellular fluid. **Intracellular fluid (ICF)** is the fluid interior of the cell. The phosphate groups are also attracted to the extracellular fluid. **Extracellular fluid (ECF)** is the fluid environment outside the enclosure of the cell membrane. **Interstitial fluid (IF)** is the term given to extracellular fluid not contained within blood vessels. Because the lipid tails are hydrophobic, they meet in the inner region of the membrane, excluding watery intracellular and extracellular fluid from this space. The cell membrane has many proteins, as well as other lipids (such as cholesterol), that are associated with the phospholipid bilayer. An important feature of the membrane is that it remains fluid; the lipids and proteins in the cell membrane are not rigidly locked in place.

Membrane Proteins

The lipid bilayer forms the basis of the cell membrane, but it is peppered throughout with various proteins. Two different types of proteins that are commonly associated with the cell membrane are the integral proteins and peripheral protein (Figure 3). As its name suggests, an **integral protein** is a protein that is embedded in the membrane. A **channel protein** is an example of an integral protein that selectively allows particular materials, such as certain ions, to pass into or out of the cell.

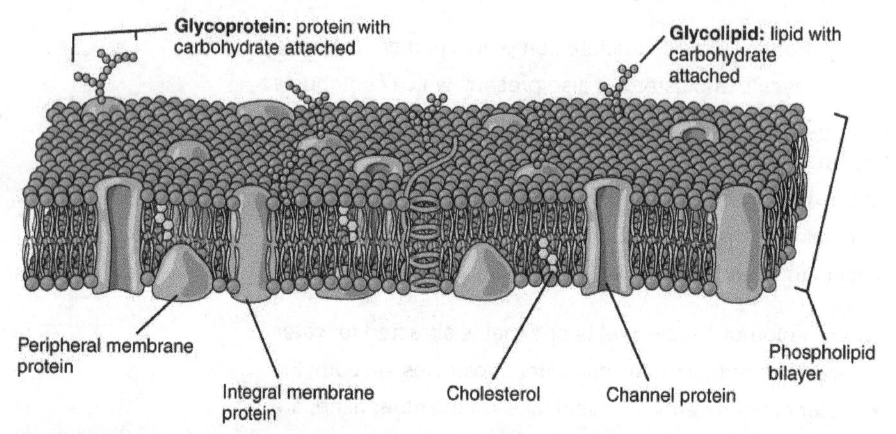

Figure 3. Cell Membrane. The cell membrane of the cell is a phospholipid bilayer containing many different molecular components, including proteins and cholesterol, some with carbohydrate groups attached.

Another important group of integral proteins are cell recognition proteins, which serve to mark a cell's identity so that it can be recognized by other cells. A **receptor** is a type of recognition protein that can selectively bind a specific molecule outside the cell, and this binding induces a chemical reaction within the cell. A **ligand** is the specific molecule that binds to and activates a receptor. Some integral proteins serve dual roles as both a receptor and an ion channel. One example of a receptor-ligand interaction is the receptors on nerve cells that bind neurotransmitters, such as dopamine. When a dopamine molecule binds to a dopamine receptor protein, a channel within the transmembrane protein opens to allow certain ions to flow into the cell.

Some integral membrane proteins are glycoproteins. A **glycoprotein** is a protein that has carbohydrate molecules attached, which extend into the extracellular matrix. The attached carbohydrate tags on glycoproteins aid in cell recognition. The carbohydrates that extend from membrane proteins and even from some membrane lipids collectively form the glycocalyx.

The **glycocalyx** is a fuzzy-appearing coating around the cell formed from glycoproteins and other carbohydrates attached to the cell membrane. The glycocalyx can have various roles. For example, it may have molecules that allow the cell to bind to another cell, it may contain receptors for hormones, or it might have enzymes to break down nutrients. The glycocalyces found in a person's body are products of that person's genetic makeup. They give each of the individual's trillions of cells the "identity" of belonging in the person's body. This identity is the primary way that a person's immune defense cells "know" not to attack the person's own body cells, but it also is the reason organs donated by another person might be rejected.

Peripheral proteins are typically found on the inner or outer surface of the lipid bilayer but can also be attached to the internal or external surface of an integral protein. These proteins typically perform a specific function for the cell. Some peripheral proteins on the surface of intestinal cells, for example, act as digestive enzymes to break down nutrients to sizes that can pass through the cells and into the bloodstream.

Transport across the Cell Membrane

One of the great wonders of the cell membrane is its ability to regulate the concentration of substances inside the cell. These substances include ions such as Ca^{2+}, Na^{+}, K^{+}, and Cl^{-}; nutrients including sugars, fatty acids, and amino acids; and waste products, particularly carbon dioxide (CO_2), which must leave the cell. The membrane's lipid bilayer structure provides the first level of control. The phospholipids are tightly packed together, and the membrane has a hydrophobic interior. This structure causes the membrane to be selectively permeable.

Selective Permeability

A membrane that has **selective permeability** allows only substances meeting certain criteria to pass through it unaided. In the case of the cell membrane, only relatively small, nonpolar materials can move through the lipid bilayer (remember, the lipid tails of the membrane are nonpolar). Some examples of these are other lipids, oxygen and carbon dioxide gases, and alcohol. However, water-soluble materials—like glucose, amino acids, and electrolytes—need some assistance to cross the membrane because they are repelled by the hydrophobic tails of the phospholipid bilayer. All substances that move through the membrane do so by one of two general methods, which are categorized based on whether or not energy is required.

Passive and Active Passage Through the Cell Membrane

Passive transport is the movement of substances across the membrane without the expenditure of cellular energy. In contrast, **active transport** is the movement of substances across the membrane using energy from adenosine triphosphate (ATP).

Passive Transport

Passive processes do not use ATP but do need some sort of driving force. It is usually from kinetic energy in the form of a concentration gradient. Molecules will tend to move from high to low concentrations by the random movement of molecules. There are 3 main types of passive processes.

1. Diffusion and facilitated diffusion
2. Osmosis and tonicity
3. Filtration

In order to understand *how* substances move passively across a cell membrane, it is necessary to understand concentration gradients and diffusion. A **concentration gradient** is the difference in concentration of a substance across a space. Molecules (or ions) will spread/diffuse from where they are more concentrated to where they are less concentrated until they are equally distributed in that space. (When molecules move in this way, they are said to move *down* their concentration gradient.) **Diffusion** is the movement of particles from an area of higher concentration to an area of lower concentration. A couple of common examples will help to illustrate this concept. Imagine being inside a closed bathroom. If a bottle of perfume were sprayed, the scent molecules would naturally diffuse from the spot where they left the bottle to all corners of the bathroom, and this diffusion would go on until no more concentration gradient remains. Another example is a spoonful of sugar placed in a cup of tea. Eventually the sugar will diffuse throughout the tea until no concentration gradient remains. In both cases, if the room is warmer or the tea hotter, diffusion occurs even faster as the molecules are bumping into each other and spreading out faster than at cooler temperatures. Having an internal body temperature around 98.6° F thus also aids in diffusion of particles within the body.

Whenever a substance exists in greater concentration on one side of a semipermeable membrane, such as the cell membranes, any substance that can move down its concentration gradient across the membrane will do so. Consider substances that can easily diffuse through the lipid bilayer of the cell membrane, such as the gases oxygen (O_2) and CO_2. O_2 generally diffuses into cells because it is more concentrated outside of them, and CO_2 typically diffuses out of cells because it is more concentrated inside of them. Neither of these examples requires any energy on the part of the cell, and therefore they use passive transport to move across the membrane. Before moving on, you need to review the gases that can diffuse across a cell membrane. Because cells rapidly use up oxygen during metabolism, there is typically a lower concentration of O_2 inside the cell than outside. As a result, oxygen will diffuse from the interstitial fluid directly through the lipid bilayer of the membrane and into the cytoplasm within the cell. On the other hand, because cells produce CO_2 as a byproduct of metabolism, CO_2 concentrations rise within the cytoplasm; therefore, CO_2 will move from the cell through the lipid bilayer and into the interstitial fluid, where its concentration is lower. This mechanism of molecules spreading from where they are more concentrated to where they are less concentration is a form of passive transport called simple diffusion (Figure 4).

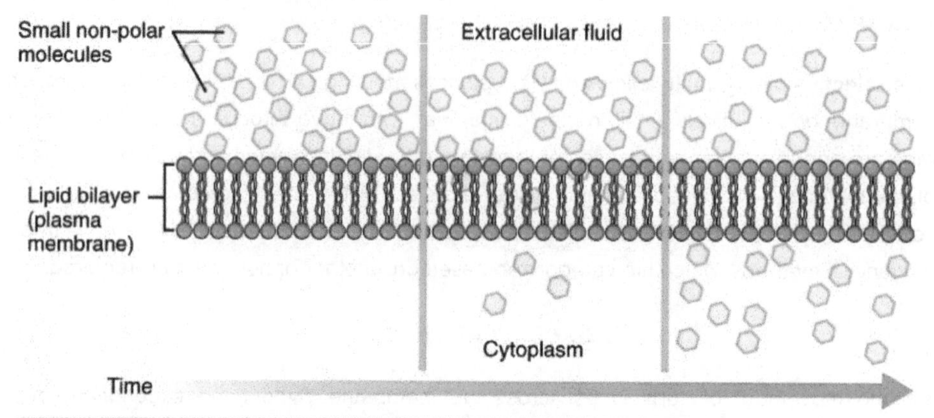

Figure 4. Simple Diffusion across the Cell (Plasma) Membrane. The structure of the lipid bilayer allows only small, non-polar substances such as oxygen and carbon dioxide to pass through the cell membrane, down their concentration gradient, by simple diffusion.

Solutes dissolved in water on either side of the cell membrane will tend to diffuse down their concentration gradients, but because most substances cannot pass freely through the lipid bilayer of the cell membrane, their movement is restricted to protein channels and specialized transport mechanisms in the membrane. **Facilitated diffusion** is the diffusion process used for those substances that cannot cross the lipid bilayer due to their size and/or polarity (Figure 5). A common example of facilitated diffusion is the movement of glucose into the cell, where it is used to make ATP. Although glucose can be more concentrated outside of a cell, it cannot cross the lipid bilayer via simple diffusion because it is both large and polar. To resolve this, a specialized carrier protein called the glucose transporter will transfer glucose molecules into the cell to facilitate its inward diffusion.

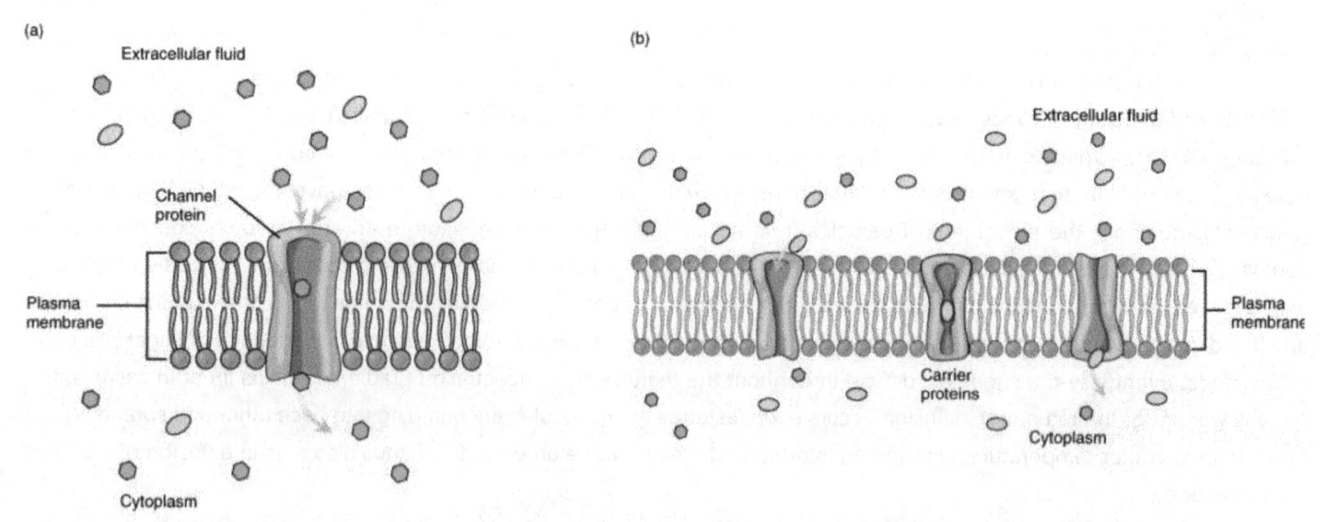

Figure 5. Facilitated Diffusion. (a) Facilitated diffusion of substances crossing the cell (plasma) membrane takes place with the help of proteins such as channel proteins and carrier proteins. Channel proteins are less selective than carrier proteins, and usually mildly discriminate between their cargo based on size and charge. (b) Carrier proteins are more selective, often only allowing one particular type of molecule to cross.

Even though sodium ions (Na^+) are highly concentrated outside of cells, these electrolytes are polarized and cannot pass through the nonpolar lipid bilayer of the membrane. Their diffusion is facilitated by membrane proteins that form sodium channels (or "pores"), so that Na^+ ions can move down their concentration gradient from outside the cells to inside the cells. There are many other solutes that must undergo facilitated diffusion to move into a cell, such as amino acids, or to move out of a cell, such as wastes. Because facilitated diffusion is a passive process, it does not require energy expenditure by the cell. Water also can move freely across the cell membrane of all cells, either through protein channels or by slipping between the lipid tails of the membrane itself.

Osmosis is the diffusion of water through a semipermeable membrane (Figure 6).

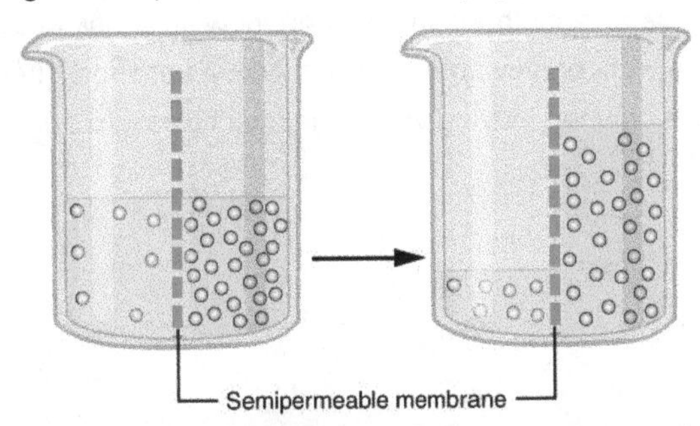

Figure 6. Osmosis. Osmosis is the diffusion of water through a semipermeable membrane down its concentration gradient. If a membrane is permeable to water, though not to a solute, water will equalize its own concentration by diffusing to the side of lower water concentration (and thus the side of higher solute concentration). In the beaker on the left, the solution on the right side of the membrane is hypertonic.

The movement of water molecules is not itself regulated by cells, so it is important that cells are exposed to an environment in which the concentration of solutes outside of the cells (in the extracellular fluid) is equal to the concentration of solutes inside the cells (in the cytoplasm). Two solutions that have the same concentration of solutes are said to be **isotonic** (equal tension). When cells and their extracellular environments are isotonic, the concentration of water molecules is the same outside and inside the cells, and the cells maintain their normal shape (and function). Osmosis occurs when there is an imbalance of solutes outside of a cell versus inside the cell. A solution that has a higher concentration of solutes than another solution is said to be **hypertonic**, and water molecules tend to diffuse into a hypertonic solution (Figure 7). Cells in a hypertonic solution will shrivel as water leaves the cell via osmosis. In contrast, a solution that has a lower concentration of solutes than another solution is said to be **hypotonic**, and water molecules tend to diffuse out of a hypotonic solution. Cells in a hypotonic solution will take on too much water and swell, with the risk of eventually bursting. A critical aspect of homeostasis in living things is to create an internal environment in which all of the body's cells are in an isotonic solution. Various organ systems, particularly the kidneys, work to maintain this homeostasis.

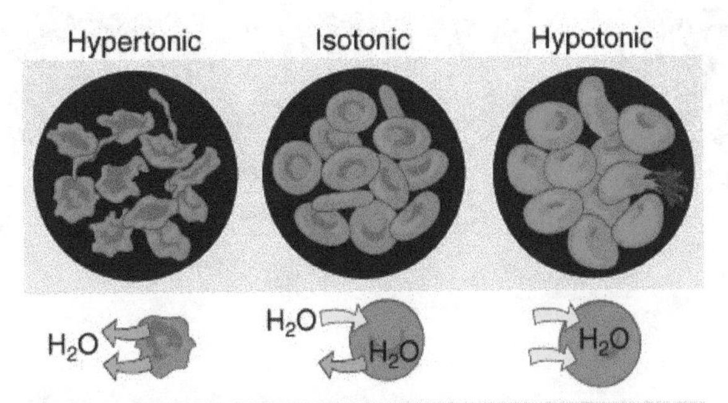

Figure 7. Concentration of Solutions. A hypertonic solution has a solute concentration higher than another solution. An isotonic solution has a solute concentration equal to another solution. A hypotonic solution has a solute concentration lower than another solution.

Another mechanism besides diffusion to passively transport materials between compartments is filtration. Unlike diffusion of a substance from where it is more concentrated to less concentrated, filtration uses a hydrostatic pressure gradient that pushes the fluid—and the solutes within it—from a higher pressure area to a lower pressure area. Filtration is an extremely important process in the body. For example, the circulatory system uses filtration to move plasma and substances across the endothelial lining of capillaries and into surrounding tissues, supplying cells with the nutrients. Filtration pressure in the kidneys provides the mechanism to remove wastes from the bloodstream.

Active Transport

You have just finished investigating the passive methods of transport, now let's look at active methods. In active methods the cell must expend energy (ATP) to do the work of moving molecules. Active transport often occurs when the molecule is being moved against its concentration gradient or when moving very large molecules into our out of the cell. There are 3 main types of active processes.

1. Primary Active Transport or Solute Pumps
2. Endocytosis
3. Exocytosis

During active transport, ATP is required to move a substance across a membrane, often with the help of protein carriers, and usually *against* its concentration gradient. One of the most common types of active transport involves proteins that serve as pumps. The word "pump" probably conjures up thoughts of using energy to pump up the tire of a bicycle or a basketball. Similarly, energy from ATP is required for these membrane proteins to transport substances—molecules or ions—across the membrane, usually against their concentration gradients (from an area of low concentration to an area of high concentration). The **sodium-potassium pump**, which is also called N^+/K^+ ATPase, transports sodium out of a cell while moving potassium into the cell. The Na^+/K^+ pump is an important ion pump found in the membranes of many types of cells. These pumps are particularly abundant in nerve cells, which are constantly pumping out sodium ions and pulling in potassium ions to maintain an electrical gradient across their cell membranes. An **electrical gradient** is a difference in electrical charge across a space. In the case of nerve cells, for example, the electrical gradient exists between the inside and outside of the cell, with the inside being negatively-charged (at around -70 mV) relative to the outside. The negative electrical gradient is maintained because each Na^+/K^+ pump moves three Na^+ ions out of the cell and two K^+ ions into the cell for each ATP molecule that is used (Figure 8). This process is so important for nerve cells that it accounts for the majority of their ATP usage.

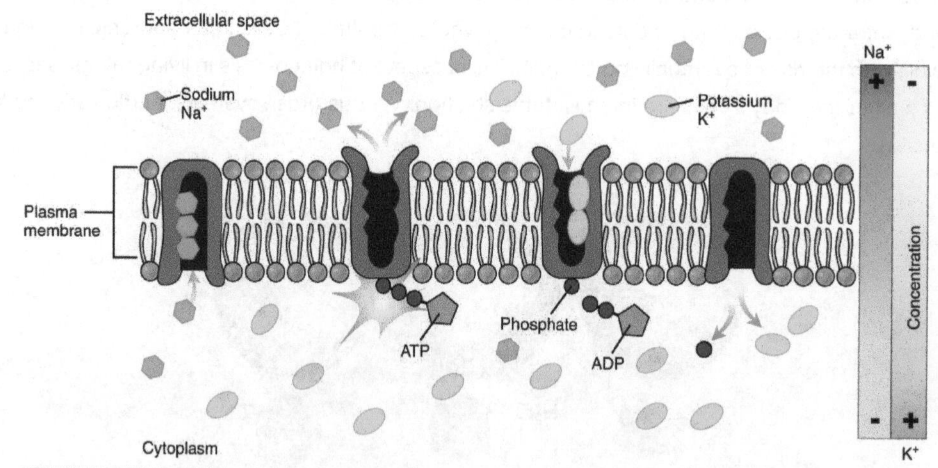

Figure 8. Sodium-Potassium Pump. The sodium-potassium pump is found in many cell (plasma) membranes. Powered by ATP, the pump moves sodium and potassium ions in opposite directions, each against its concentration gradient. In a single cycle of the pump, three sodium ions are extruded from and two potassium ions are imported into the cell.

Other forms of active transport do not involve membrane carriers. **Endocytosis** (bringing "into the cell") is the process of a cell ingesting material by enveloping it in a portion of its cell membrane, and then pinching off that portion of membrane (Figure 9).

Once pinched off, the portion of membrane and its contents becomes an independent, intracellular vesicle. A **vesicle** is a membranous sac—a spherical and hollow organelle bounded by a lipid bilayer membrane. Endocytosis often brings materials into the cell that must to be broken down or digested. **Phagocytosis** ("cell eating") is the endocytosis of large particles. Many immune cells engage in phagocytosis of invading pathogens. Like little Pac-men, their job is to patrol body tissues for unwanted matter, such as invading bacterial cells, phagocytize them, and digest them. In contrast to phagocytosis, **pinocytosis** ("cell drinking") brings fluid containing dissolved substances into a cell through membrane vesicles.

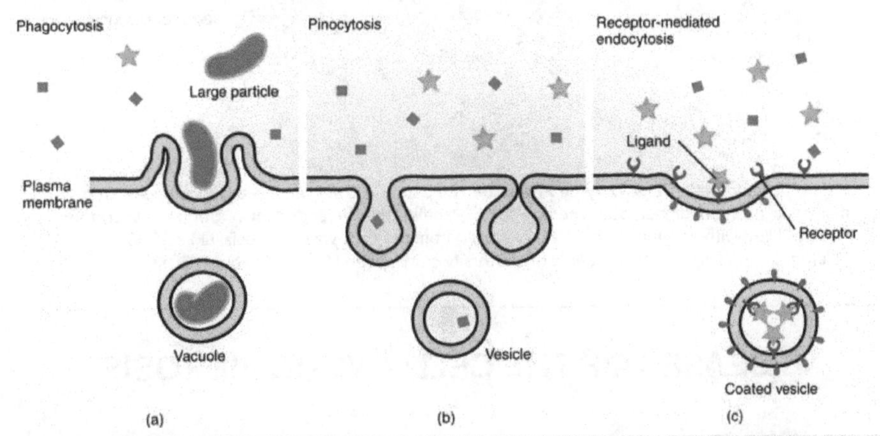

Figure 9. Three Forms of Endocytosis. Endocytosis is a form of active transport in which a cell envelopes extracellular materials using its cell membrane. (a) In phagocytosis, which is relatively nonselective, the cell takes in a large particle. (b) In pinocytosis, the cell takes in small particles in fluid. (c) In contrast, receptor-mediated endocytosis is quite selective. When external receptors bind a specific ligand, the cell responds by endocytosing the ligand.

Phagocytosis and pinocytosis take in large portions of extracellular material, and they are typically not highly selective in the substances they bring in. Cells regulate the endocytosis of specific substances via receptor-mediated endocytosis. **Receptor-mediated endocytosis** is endocytosis by a portion of the cell membrane that contains many receptors that are specific for a certain substance. Once the surface receptors have bound sufficient amounts of the specific substance (the receptor's ligand), the cell will endocytose the part of the cell membrane containing the receptor-ligand complexes. Iron, a required component of hemoglobin, is endocytosed by red blood cells in this way. Iron is bound to a protein called transferrin in the blood. Specific transferrin receptors on red blood cell surfaces bind the iron-transferrin molecules, and the cell endocytoses the receptor-ligand complexes. In contrast with endocytosis, **exocytosis** (taking "out of the cell") is the process of a cell exporting material using vesicular transport (Figure 10).

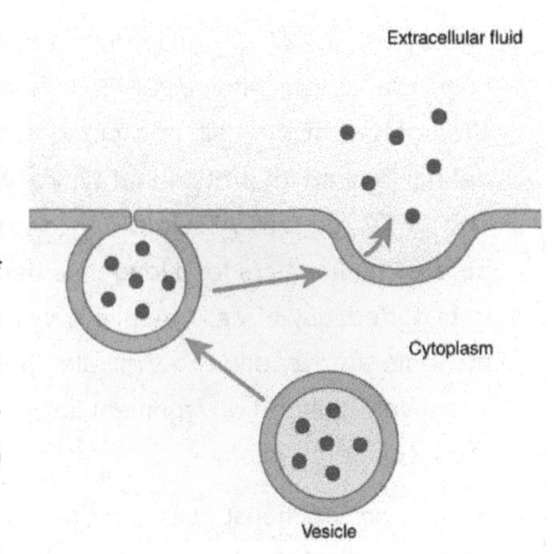

Figure 10. Exocytosis. Exocytosis is much like endocytosis in reverse. Material destined for export is packaged into a vesicle inside the cell. The membrane of the vesicle fuses with the cell membrane, and the contents are released into the extracellular space.

Many cells manufacture substances that must be secreted, like a factory manufacturing a product for export. These substances are typically packaged into membrane-bound vesicles within the cell. When the vesicle membrane fuses with the cell membrane, the vesicle releases it contents into the interstitial fluid. The vesicle membrane then becomes part of the cell membrane. Cells of the stomach and pancreas produce and secrete digestive enzymes through exocytosis (Figure 11). Endocrine cells produce and secrete hormones that are sent throughout the body, and certain immune cells produce and secrete large amounts of histamine, a chemical important for immune responses.

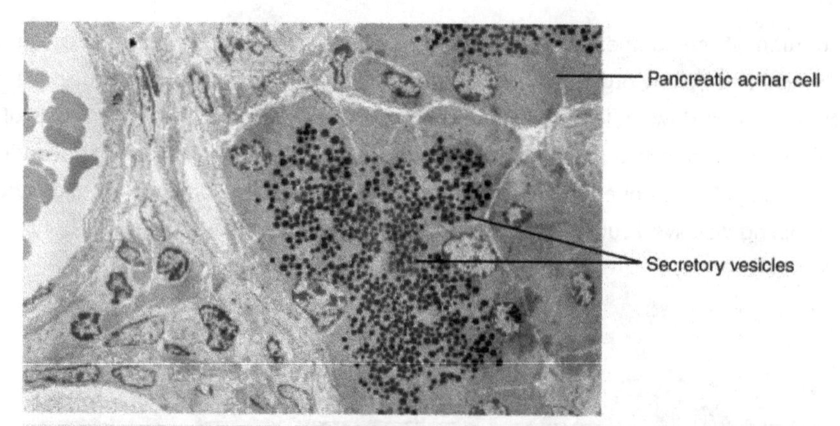

Figure 11. Pancreatic Cells' Enzyme Products. The pancreatic acinar cells produce and secrete many enzymes that digest food. The tiny black granules in this electron micrograph are secretory vesicles filled with enzymes that will be exported from the cells via exocytosis. LM × 2900. (Micrograph provided by the Regents of University of Michigan Medical School © 2012)

DISEASES OF THE CELL: CYSTIC FIBROSIS

Cystic fibrosis (CF) affects approximately 30,000 people in the United States, with about 1,000 new cases reported each year. The genetic disease is most well known for its damage to the lungs, causing breathing difficulties and chronic lung infections, but it also affects the liver, pancreas, and intestines. Only about 50 years ago, the prognosis for children born with CF was very grim—a life expectancy rarely over 10 years. Today, with advances in medical treatment, many CF patients live into their 30s.

The symptoms of CF result from a malfunctioning membrane ion channel called the cystic fibrosis transmembrane conductance regulator, or CFTR. In healthy people, the CFTR protein is an integral membrane protein that transports Cl^- ions out of the cell. In a person who has CF, the gene for the CFTR is mutated, thus, the cell manufactures a defective channel protein that typically is not incorporated into the membrane, but is instead degraded by the cell. The CFTR requires ATP in order to function, making its Cl^- transport a form of active transport. This characteristic puzzled researchers for a long time because the Cl^- ions are actually flowing *down* their concentration gradient when transported out of cells. Active transport generally pumps ions *against* their concentration gradient, but the CFTR presents an exception to this rule. In normal lung tissue, the movement of Cl^- out of the cell maintains a Cl^--rich, negatively charged environment immediately outside of the cell. This is particularly important in the epithelial lining of the respiratory system.

Respiratory epithelial cells secrete mucus, which serves to trap dust, bacteria, and other debris. A cilium (plural = *cilia*) is one of the hair-like appendages found on certain cells. Cilia on the epithelial cells move the mucus and its trapped particles up the airways away from the lungs and toward the outside. In order to be effectively moved upward, the mucus cannot be too viscous; rather it must have a thin, watery consistency. The transport of Cl^- and the maintenance of an electronegative environment outside of the cell attract positive ions such as Na^+ to the extracellular space. The accumulation of both Cl^- and Na^+ ions in the extracellular space creates solute-rich mucus, which has a low concentration of water molecules. As a result, through osmosis, water moves from cells and extracellular matrix into the mucus, "thinning" it out. This is how, in a normal respiratory system, the mucus is kept sufficiently watered-down to be propelled out of the respiratory system.

If the CFTR channel is absent, Cl^- ions are not transported out of the cell in adequate numbers, thus preventing them from drawing positive ions. The absence of ions in the secreted mucus results in the lack of a normal water concentration gradient. Thus, there is no osmotic pressure pulling water into the mucus. The resulting mucus is thick and sticky, and the ciliated epithelia cannot effectively remove it from the respiratory system. Passageways in the lungs become blocked with mucus, along with the debris it carries. Bacterial infections occur more easily because bacterial cells are not effectively carried away from the lungs.

Cell Growth and Division

While there are a few cells in the body that do not undergo cell division (such as gametes, red blood cells, most neurons, and some muscle cells), most somatic cells divide regularly. A **somatic cell** is a general term for a body cell, and all human cells, except for the cells that produce eggs and sperm (which are referred to as germ cells), are somatic cells. Somatic cells contain **two** copies of each of their chromosomes (one copy received from each parent).

A **homologous** pair of chromosomes is the two copies of a single chromosome found in each somatic cell. The human is a **diploid** organism, having 23 homologous pairs of chromosomes in each of the somatic cells. The condition of having pairs of chromosomes is known as diploidy. Cells in the body replace themselves over the lifetime of a person. For example, the cells lining the gastrointestinal tract must be frequently replaced when constantly "worn off" by the movement of food through the gut. But what triggers a cell to divide, and how does it prepare for and complete cell division? The **cell cycle** is the sequence of events in the life of the cell from the moment it is created at the end of a previous cycle of cell division until it then divides itself, generating two new cells.

The Cell Cycle

One "turn" or cycle of the cell cycle consists of two general phases: interphase, followed by mitosis and cytokinesis. **Interphase** is the period of the cell cycle during which the cell is not dividing. The majority of cells are in interphase most of the time. **Mitosis** is the division of genetic material, during which the cell nucleus breaks down and two new, fully functional, nuclei are formed. **Cytokinesis** divides the cytoplasm into two distinctive cells.

Interphase

A cell grows and carries out all normal metabolic functions and processes in a period called G_1 (Figure 1). **G_1 phase** (gap 1 phase) is the first gap, or growth phase in the cell cycle. For cells that will divide again, G_1 is followed by replication of the DNA, during the S phase. The **S phase** (synthesis phase) is period during which a cell replicates its DNA.

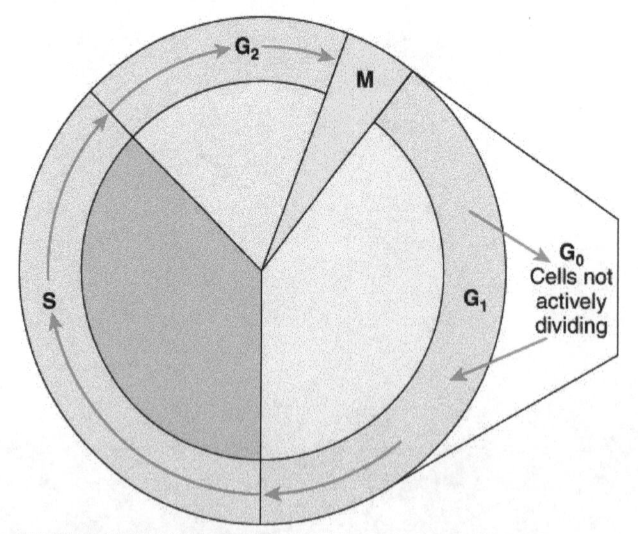

Figure 1. Cell Cycle. The two major phases of the cell cycle include mitosis (cell division), and interphase, when the cell grows and performs all of its normal functions. Interphase is further subdivided into G_1, S, and G_2 phases.

After the synthesis phase, the cell proceeds through the G_2 phase. The **G_2 phase** is a second gap phase, during which the cell continues to grow and makes the necessary preparations for mitosis. Between G_1, S, and G_2 phases, cells will vary the most in their duration of the G1 phase. It is here that a cell might spend a couple of hours, or many days. The S phase typically lasts between 8-10 hours and the G_2 phase approximately 5 hours. In contrast to these phases, the **G_0 phase** is a resting phase of the cell cycle. Cells that have temporarily stopped dividing and are resting (a common condition) and cells that have permanently ceased dividing (like nerve cells) are said to be in G_0.

The Structure of Chromosomes

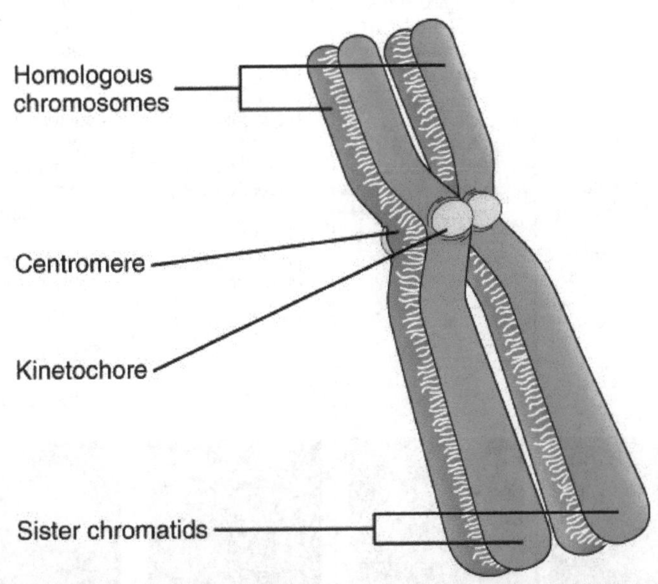

Figure 2. A Homologous Pair of Chromosomes with their Attached Sister Chromatids. The red and blue colors correspond to a homologous pair of chromosomes. Each member of the pair was separately inherited from one parent. Each chromosome in the homologous pair is also bound to an identical sister chromatid, which is produced by DNA replication, and results in the familiar "X" shape.

Billions of cells in the human body divide every day. During the synthesis phase (S, for DNA synthesis) of interphase, the amount of DNA within the cell precisely doubles. Therefore, after DNA replication but before cell division, each cell actually contains **two** copies of each chromosome. Each copy of the chromosome is referred to as a **sister chromatid** and is physically bound to the other copy. The **centromere** is the structure that attaches one sister chromatid to another. Because a human cell has 46 chromosomes, during this phase, there are 92 chromatids (46 × 2) in the cell. Make sure not to confuse the concept of a pair of chromatids (one chromosome and its exact copy attached during mitosis) and a homologous pair of chromosomes (two paired chromosomes which were inherited separately, one from each parent) (Figure 2).

Mitosis

The **mitotic phase** of the cell typically takes between 1 and 2 hours. During this phase, a cell undergoes two major processes. First, it completes mitosis, during which the contents of the nucleus are equitably pulled apart and distributed between its two halves. Cytokinesis then occurs, dividing the cytoplasm and cell body into two new cells. Mitosis is divided into four major stages that take place after interphase (Table 1) and in the following order: prophase, metaphase, anaphase, and telophase. The process is then followed by cytokinesis.

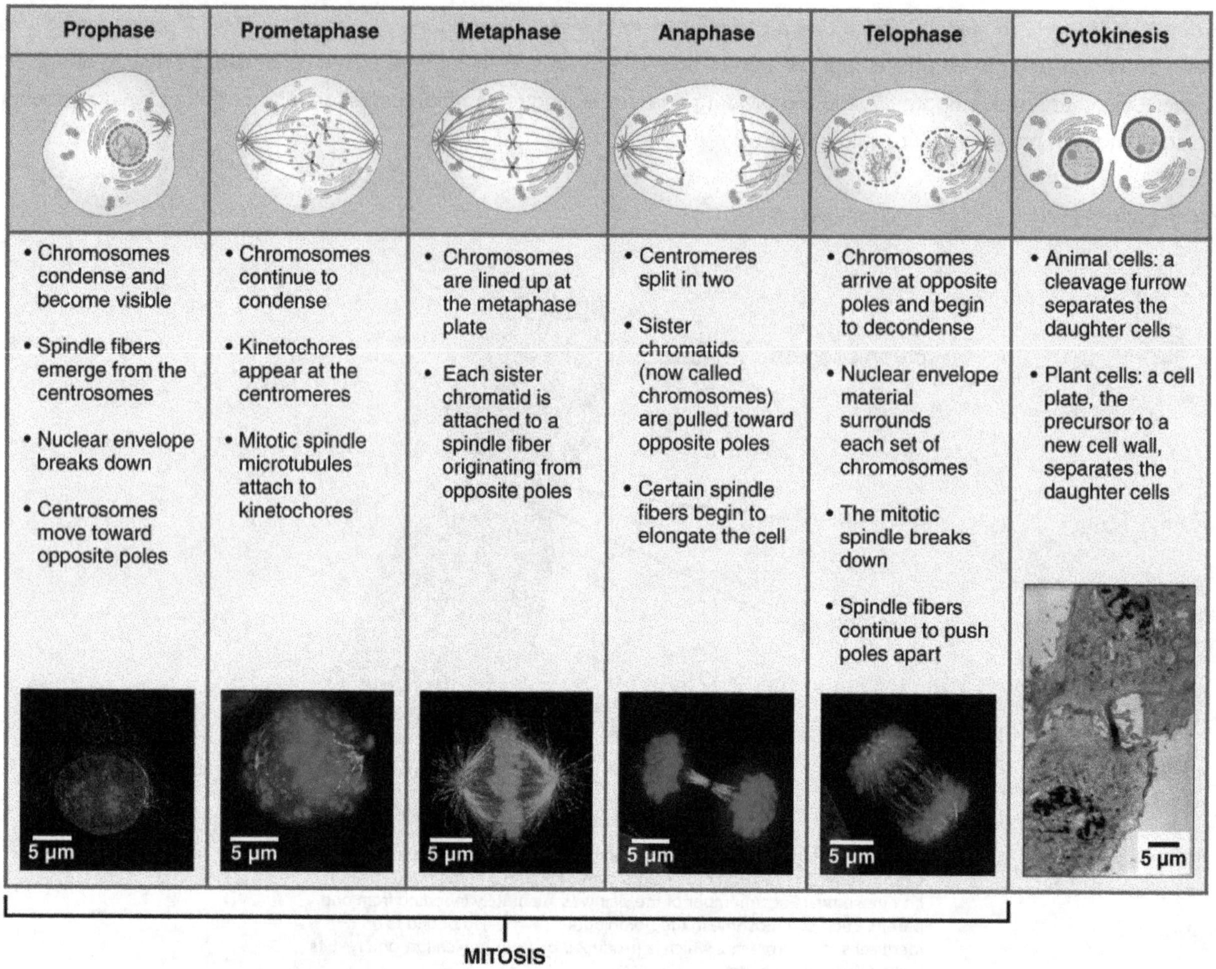

Prophase	Prometaphase	Metaphase	Anaphase	Telophase	Cytokinesis
• Chromosomes condense and become visible • Spindle fibers emerge from the centrosomes • Nuclear envelope breaks down • Centrosomes move toward opposite poles	• Chromosomes continue to condense • Kinetochores appear at the centromeres • Mitotic spindle microtubules attach to kinetochores	• Chromosomes are lined up at the metaphase plate • Each sister chromatid is attached to a spindle fiber originating from opposite poles	• Centromeres split in two • Sister chromatids (now called chromosomes) are pulled toward opposite poles • Certain spindle fibers begin to elongate the cell	• Chromosomes arrive at opposite poles and begin to decondense • Nuclear envelope material surrounds each set of chromosomes • The mitotic spindle breaks down • Spindle fibers continue to push poles apart	• Animal cells: a cleavage furrow separates the daughter cells • Plant cells: a cell plate, the precursor to a new cell wall, separates the daughter cells
5 μm	5 μm	5 μm	5 μm	5 μm	5 μm

MITOSIS

Prophase is the first phase of mitosis, during which the loosely packed chromatin coils and condenses into visible chromosomes. During prophase, each chromosome becomes visible with its identical partner attached, forming the familiar X-shape of sister chromatids. The nucleolus disappears early during this phase, and the nuclear envelope also disintegrates. A major occurrence during prophase concerns a very important structure that contains the origin site for microtubule growth. Recall the cellular structures called centrioles that serve as origin points from which microtubules extend. These tiny structures also play a very important role during mitosis. A **centrosome** is a pair of centrioles together. The cell contains two centrosomes side-by-side, which begin to move apart during prophase. As the centrosomes migrate to two different sides of the cell, microtubules begin to extend from each like long fingers from two hands extending toward each other. The **mitotic spindle** is the structure composed of the centrosomes and their emerging microtubules. Near the end of prophase there is an invasion of the nuclear area by microtubules from the mitotic spindle. The nuclear membrane has disintegrated, and the microtubules attach themselves to the centromeres that adjoin pairs of sister chromatids. The **kinetochore** is a protein structure on the centromere that is the point of attachment between the mitotic spindle and the sister chromatids. This stage is referred to as late prophase or "prometaphase" to indicate the transition between prophase and metaphase.

Metaphase is the second stage of mitosis. During this stage, the sister chromatids, with their attached microtubules, line up along a linear plane in the middle of the cell. A metaphase plate forms between the centrosomes that are now located at either end of the cell. The **metaphase plate** is the name for the plane through the center of the spindle on which the sister chromatids are positioned. The microtubules are now poised to pull apart the sister chromatids and bring one from each pair to each side of the cell.

Anaphase is the third stage of mitosis. Anaphase takes place over a few minutes, when the pairs of sister chromatids are separated from one another, forming individual chromosomes once again. These chromosomes are pulled to opposite ends of the cell by their kinetochores, as the microtubules shorten. Each end of the cell receives one partner from each pair of sister chromatids, ensuring that the two new daughter cells will contain identical genetic material.

Telophase is the final stage of mitosis. Telophase is characterized by the formation of two new daughter nuclei at either end of the dividing cell. These newly formed nuclei surround the genetic material, which uncoils such that the chromosomes return to loosely packed chromatin. Nucleoli also reappear within the new nuclei, and the mitotic spindle breaks apart, each new cell receiving its own complement of DNA, organelles, membranes, and centrioles. At this point, the cell is already beginning to split in half as cytokinesis begins.

Cytokinesis

The **cleavage furrow** is a contractile band made up of microfilaments that forms around the midline of the cell during cytokinesis. (Recall that microfilaments consist of actin.) This contractile band squeezes the two cells apart until they finally separate. Two new cells are now formed. One of these cells (the "stem cell") enters its own cell cycle; able to grow and divide again at some future time. The other cell transforms into the functional cell of the tissue, typically replacing an "old" cell there. Imagine a cell that completed mitosis but never underwent cytokinesis. In some cases, a cell may divide its genetic material and grow in size, but fail to undergo cytokinesis. This results in larger cells with more than one nucleus. Usually this is an unwanted aberration and can be a sign of cancerous cells.

Cell Cycle Control

A very elaborate and precise system of regulation controls direct the way cells proceed from one phase to the next in the cell cycle and begin mitosis. The control system involves molecules within the cell as well as external triggers. These internal and external control triggers provide "stop" and "advance" signals for the cell. Precise regulation of the cell cycle is critical for maintaining the health of an organism, and loss of cell cycle control can lead to cancer.

Mechanisms of Cell Cycle Control

As the cell proceeds through its cycle, each phase involves certain processes that must be completed before the cell should advance to the next phase. A **checkpoint** is a point in the cell cycle at which the cycle can be signaled to move forward or stopped. At each of these checkpoints, different varieties of molecules provide the stop or go signals, depending on certain conditions within the cell. A **cyclin** is one of the primary classes of cell cycle control molecules (Figure 3). A **cyclin-dependent kinase (CDK)** is one of a group of molecules that work together with cyclins to determine progression past cell checkpoints. By interacting with many additional molecules, these triggers push the cell cycle forward unless prevented from doing so by "stop" signals, if for some reason the cell is not ready. At the G_1 checkpoint, the cell must be ready for DNA synthesis to occur. At the G_2 checkpoint the cell must be fully prepared for mitosis. Even during mitosis, a crucial stop and go checkpoint in metaphase ensures that the cell is fully prepared to complete cell division. The metaphase checkpoint ensures that all sister chromatids are properly attached to their respective microtubules and lined up at the metaphase plate before the signal is given to separate them during anaphase.

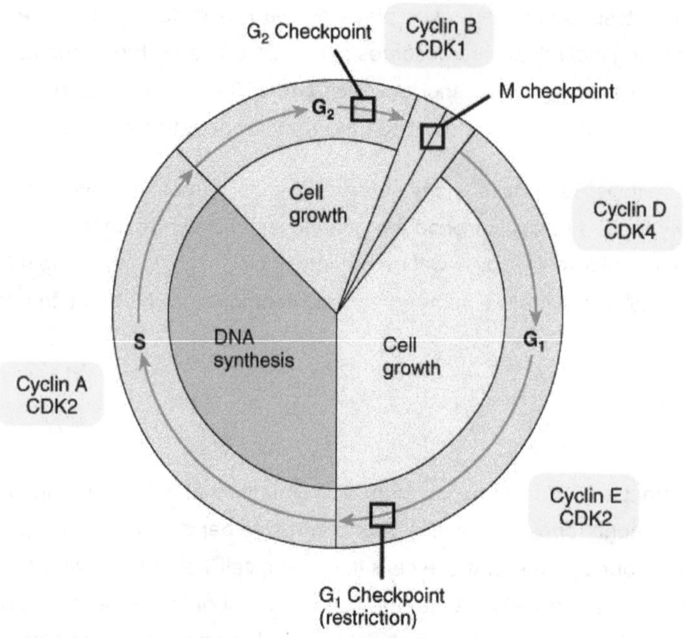

Figure 3. Control of the Cell Cycle. Cells proceed through the cell cycle under the control of a variety of molecules, such as cyclins and cyclin-dependent kinases. These control molecules determine whether or not the cell is prepared to move into the following stage.

The Cell Cycle Out of Control: Implications

Most people understand that cancer or tumors are caused by abnormal cells that multiply continuously. If the abnormal cells continue to divide unstopped, they can damage the tissues around them, spread to other parts of the body, and eventually result in death. In healthy cells, the tight regulation mechanisms of the cell cycle prevent this from happening, while failures of cell cycle control can cause unwanted and excessive cell division. Failures of control may be caused by inherited genetic abnormalities that compromise the function of certain "stop" and "go" signals. Environmental insult that damages DNA can also cause dysfunction in those signals. Often, a combination of both genetic predisposition and environmental factors lead to cancer. The process of a cell escaping its normal control system and becoming cancerous may actually happen throughout the body quite frequently. Fortunately, certain cells of the immune system are capable of recognizing cells that have become cancerous and destroying them. However, in certain cases the cancerous cells remain undetected and continue to proliferate. If the resulting tumor does not pose a threat to surrounding tissues, it is said to be benign and can usually be easily removed. If capable of damage, the tumor is considered malignant and the patient is diagnosed with cancer.

HOMEOSTATIC IMBALANCES: CANCER ARISES FROM HOMEOSTATIC IMBALANCES

Cancer is an extremely complex condition, capable of arising from a wide variety of genetic and environmental causes. Typically, mutations or aberrations in a cell's DNA that compromise normal cell cycle control systems lead to cancerous tumors. Cell cycle control is an example of a homeostatic mechanism that maintains proper cell function and health. While progressing through the phases of the cell cycle, a large variety of intracellular molecules provide stop and go signals to regulate movement forward to the next phase. These signals are maintained in an intricate balance so that the cell only proceeds to the next phase when it is ready.

The homeostatic control of the cell cycle can be thought of like a car's cruise control. Cruise control will continually apply just the right amount of acceleration to maintain a desired speed, unless the driver hits the brakes, in which case the car will slow down. Similarly, the cell includes molecular messengers, such as cyclins, that push the cell forward in its cycle. In addition to cyclins, a class of proteins that are encoded by genes called proto-oncogenes provide important signals that regulate the cell cycle and move it forward. Examples of proto-oncogene products include cell-surface receptors for growth factors, or cell-signaling molecules, two classes of molecules that can promote DNA replication and cell division.

In contrast, a second class of genes known as tumor suppressor genes sends stop signals during a cell cycle. For example, certain protein products of tumor suppressor genes signal potential problems with the DNA and thus stop the cell from dividing, while other proteins signal the cell to die if it is damaged beyond repair. Some tumor suppressor proteins also signal a sufficient surrounding cellular density, which indicates that the cell need not presently divide. The latter function is uniquely important in preventing tumor growth: normal cells exhibit a phenomenon called "contact inhibition;" thus, extensive cellular contact with neighboring cells causes a signal that stops further cell division.

These two contrasting classes of genes, proto-oncogenes and tumor suppressor genes, are like the accelerator and brake pedal of the cell's own "cruise control system," respectively. Under normal conditions, these stop and go signals are maintained in a homeostatic balance. Generally speaking, there are two ways that the cell's cruise control can lose control: a malfunctioning (overactive) accelerator, or a malfunctioning (underactive) brake. When compromised through a mutation, or otherwise altered, proto-oncogenes can be converted to oncogenes, which produce oncoproteins that push a cell forward in its cycle and stimulate cell division even when it is undesirable to do so.

For example, a cell that should be programmed to self-destruct (a process called apoptosis) due to extensive DNA damage might instead be triggered to proliferate by an oncoprotein. On the other hand, a dysfunctional tumor suppressor gene may fail to provide the cell with a necessary stop signal, also resulting in unwanted cell division and proliferation. A delicate homeostatic balance between the many proto-oncogenes and tumor suppressor genes delicately controls the cell cycle and ensures that only healthy cells replicate. Therefore, a disruption of this homeostatic balance can cause aberrant cell division and cancerous growths.

Concept check

1. **Define** a cell.
2. **Compare and contrast** the different types of microscopes.
3. **List** the three parts of the cell theory.
4. **Explain** the difference between hydrophilic and hydrophobic. **How** does this relate to a phospholipid?
5. **How** is a phospholipid bilayer formed?
6. **Where** are the following fluids located: intracellular fluid (ICF), extracellular fluid (ECF), and interstitial fluid (IF)?
7. **List** the different components of the cell membrane.
8. **Define** selectively permeable.
9. **Compare and contrast** passive transport with active transport.
10. **What** is osmosis and **what** is meant by tonicity?
11. **List and define** the different types of active transport.
12. **What** are the components of the cytoplasm?
13. **List and briefly** describe the functions of the organelles.
14. **What** are the components of the cytoskeleton?
15. **How** do the three types of cell junctions differ?
16. **Define** the cell cycle.
17. **List and describe** the stages of mitosis.
18. **Compare and contrast** meiosis and mitosis.

CHAPTER 4 — BODY ORGANIZATION AND DIGESTIVE SYSTEM

Introduction to the Tissue Level of Organization

LEARNING OBJECTIVES

- Define the terminology, both directional and anatomical, associated with anatomy of the human body
- Identify the main tissue types and discuss their roles in the human body
- Explain the process of tissue repair
- Describe the structure of the human body in terms of six levels of organization
- List the eleven organ systems of the human body and identify at least one organ and one major function of each
- Identify the major organs of the digestive system and state the functions.
- Describe how carbohydrates, proteins and lipids are digested and absorbed into the blood.
- Explain Glucose Homeostasis.

The body contains at least 200 distinct cell types. These cells contain essentially the same internal structures yet they vary enormously in shape and function. The different types of cells are not randomly distributed throughout the body; rather they occur in organized layers, a level of organization referred to as tissue. The micrograph that opens this chapter shows the high degree of organization among different types of cells in the tissue of the cervix. You can also see how that organization breaks down when cancer takes over the regular mitotic functioning of a cell.

The variety in shape reflects the many different roles that cells fulfill in your body. The human body starts as a single cell at fertilization. As this fertilized egg divides, it gives rise to trillions of cells, each built from the same blueprint, but organizing into tissues and becoming irreversibly committed to a developmental pathway.

Directional Terms

Certain directional anatomical terms appear throughout this and any other anatomy textbook (Figure 2). These terms are essential for describing the relative locations of different body structures. For instance, an anatomist might describe one band of tissue as "inferior to" another or a physician might describe a tumor as "superficial to" a deeper body structure. Commit these terms to memory to avoid confusion when you are studying or describing the locations of particular body parts.

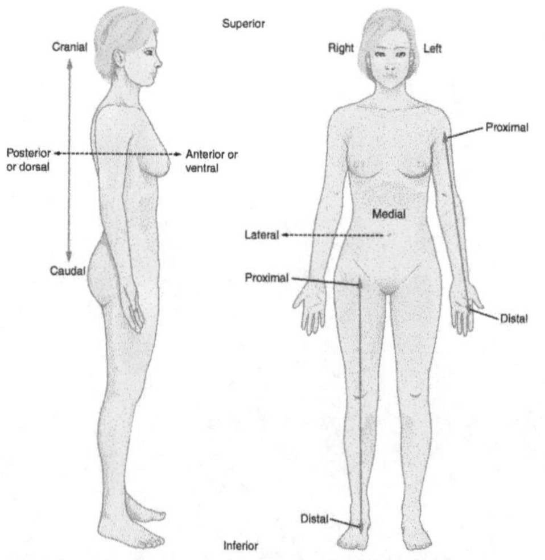

Figure 2. Directional Terms Applied to the Human Body. Paired directional terms are shown as applied to the human body.

- **Anterior** (or *ventral*) Describes the front or direction toward the front of the body. The toes are anterior to the foot.
- **Posterior** (or *dorsal*) Describes the back or direction toward the back of the body. The popliteus is posterior to the patella.
- **Superior** (or *cranial*) describes a position above or higher than another part of the body proper. The orbits are superior to the oris.
- **Inferior** (or *caudal*) describes a position below or lower than another part of the body proper; near or toward the tail (in humans, the coccyx, or lowest part of the spinal column). The pelvis is inferior to the abdomen.
- **Lateral** describes the side or direction toward the side of the body. The thumb (pollex) is lateral to the digits.

- **Medial** describes the middle or direction toward the middle of the body. The hallux is the medial toe.
- **Proximal** describes a position in a limb that is nearer to the point of attachment or the trunk of the body. The brachium is proximal to the antebrachium.
- **Distal** describes a position in a limb that is farther from the point of attachment or the trunk of the body. The crus is distal to the femur.
- **Superficial** describes a position closer to the surface of the body. The skin is superficial to the bones.
- **Deep** describes a position farther from the surface of the body. The brain is deep to the skull.

Body Planes

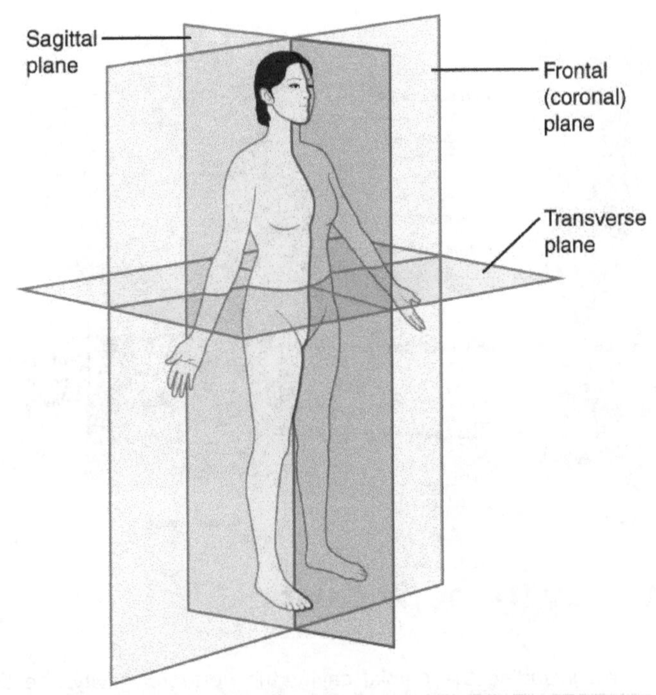

Figure 3. Planes of the Body. The three planes most commonly used in anatomical and medical imaging are the sagittal, frontal (or coronal), and transverse plane.

A **section** is a two-dimensional surface of a three-dimensional structure that has been cut. Modern medical imaging devices enable clinicians to obtain "virtual sections" of living bodies. We call these scans. Body sections and scans can be correctly interpreted, however, only if the viewer understands the plane along which the section was made. A **plane** is an imaginary two-dimensional surface that passes through the body. There are three planes commonly referred to in anatomy and medicine, as illustrated in Figure 3.

- The **sagittal plane** is the plane that divides the body or an organ vertically into right and left sides. If this vertical plane runs directly down the middle of the body, it is called the midsagittal or median plane. If it divides the body into unequal right and left sides, it is called a parasagittal plane or less commonly a longitudinal section.
- The **frontal plane** is the plane that divides the body or an organ into an anterior (front) portion and a posterior (rear) portion. The frontal plane is often referred to as a coronal plane. ("Corona" is Latin for "crown.")
- The **transverse plane** is the plane that divides the body or organ horizontally into upper and lower portions. Transverse planes produce images referred to as cross sections.

Body Cavities and Serous Membranes

By the broadest definition, a body cavity is any fluid-filled space in a multicellular organism. However, the term usually refers to the space where internal organs develop, located between the skin and the outer lining of the gut cavity. "The human body cavity," normally refers to the ventral body cavity because it is by far the largest one in volume. Blood vessels are not considered cavities but may be held within cavities. Most cavities provide room for the organs to adjust to changes in the organism's position. They usually contains protective membranes and sometimes bones that protect the organs.

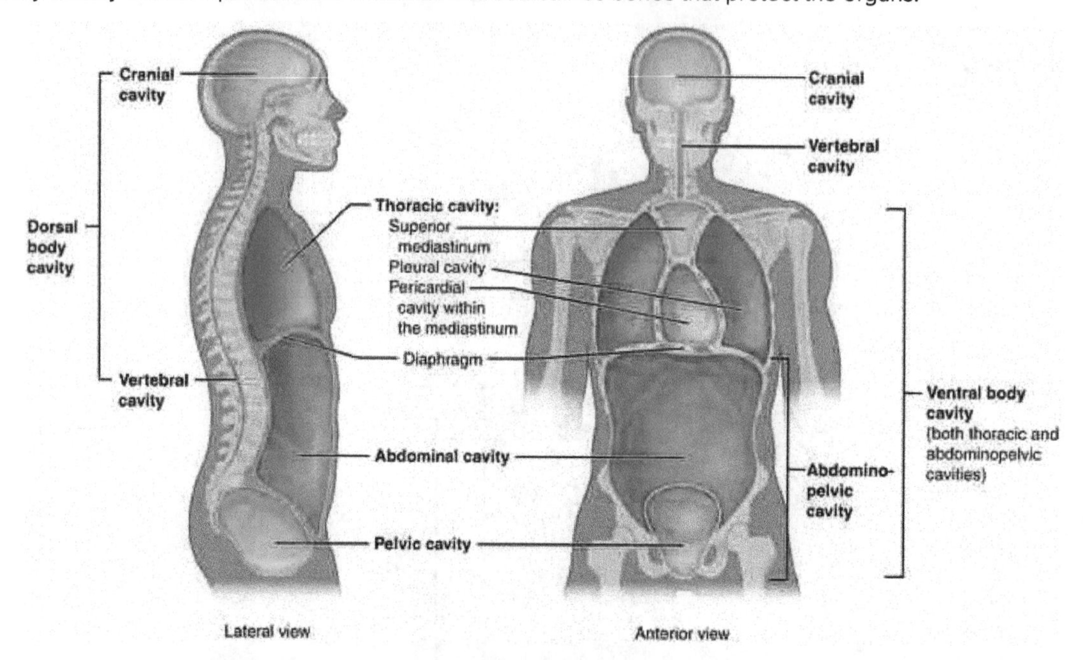

Anatomical terminology for body cavities

Humans have multiple body cavities, including the cranial cavity, the vertebral cavity, the thoracic cavity (containing the pericardial cavity and the pleural cavity), the abdominal cavity, and the pelvic cavity. In mammals, the diaphragm separates the thoracic cavity from the abdominal cavity.
Dorsal

The dorsal cavity is a continuous cavity located on the dorsal side of the body. It houses the organs of the upper central nervous system, including the brain and the spinal cord. The meninges is a multi-layered membrane within the dorsal cavity that envelops and protects the brain and spinal cord.

- Cranial: The cranial cavity is the anterior portion of the dorsal cavity consisting of the space inside the skull. This cavity contains the brain, the meninges of the brain, and cerebrospinal fluid.
- Vertebral: The vertebral cavity is the posterior portion of the dorsal cavity and contains the structures within the vertebral column. These include the spinal cord, the meninges of the spinal cord, and the fluid-filled spaces between them. This is the most narrow of all body cavities, sometimes described as threadlike.

Ventral

The ventral cavity, the interior space in the front of the body, contains many different organ systems. The organs within the ventral cavity are also called viscera. The ventral cavity has anterior and posterior portions divided by the diaphragm, a sheet of skeletal muscle found beneath the lungs.

- **Thoracic**: The thoracic cavity is the anterior ventral body cavity found within the rib cage in the torso. It houses the primary organs of the cardiovascular and respiratory systems, such as the heart and lungs, but also includes organs from other systems, such as the esophagus and the thymus gland. The thoracic cavity is lined by two types of mesothelium, a type of membrane tissue that lines the ventral cavity: the pleura lining of the lungs, and the pericadium lining of the heart.
- **Abdominopelvic:** The abdominoplevic cavity is the posterior ventral body cavity found beneath the thoracic cavity and diaphragm. It is generally divided into the abdominal and pelvic cavities. The abdominal cavity is not contained within bone and houses many organs of the digestive and renal systems, as well as some organs of the endocrine system, such as the adrenal glands. The pelvic cavity is contained within the pelvis and houses the bladder and reproductive system. The abdominopelvic cavity is lined by a type of mesothelium called the peritoneum.

Abdominal Regions and Quadrants

To promote clear communication, for instance about the location of a patient's abdominal pain or a suspicious mass, health care providers typically divide up the cavity into either nine regions or four quadrants (Figure 4).

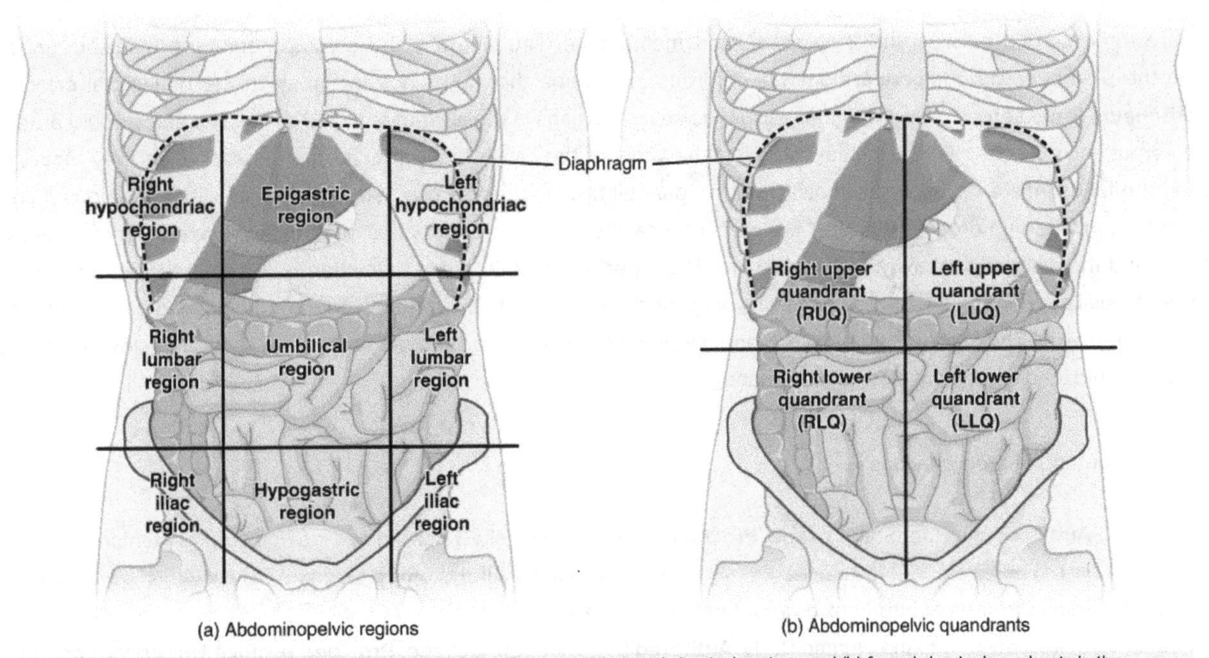

Figure 4. Regions and Quadrants of the Peritoneal Cavity. There are (a) nine abdominal regions and (b) four abdominal quadrants in the peritoneal cavity.

The more detailed regional approach subdivides the cavity with one horizontal line immediately inferior to the ribs and one immediately superior to the pelvis, and two vertical lines drawn as if dropped from the midpoint of each clavicle (collarbone). There are nine resulting regions. The simpler quadrants approach, which is more commonly used in medicine, subdivides the cavity with one horizontal and one vertical line that intersect at the patient's umbilicus (navel).

Membranes of the Anterior (Ventral) Body Cavity

A **serous membrane** (also referred to a serosa) is one of the thin membranes that cover the walls and organs in the thoracic and abdominopelvic cavities. The parietal layers of the membranes line the walls of the body cavity (pariet- refers to a cavity wall). The visceral layer of the membrane covers the organs (the viscera). Between the parietal and visceral layers is a very thin, fluid-filled serous space, or cavity (Figure 5).

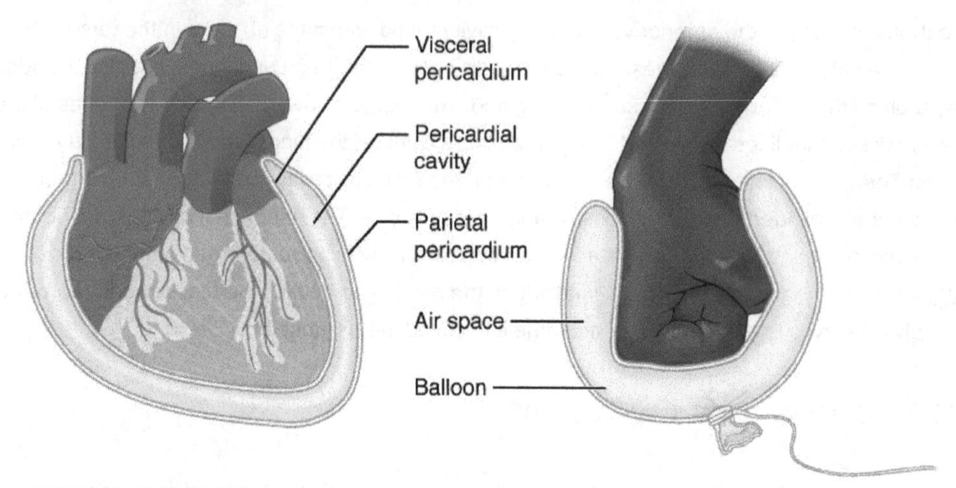

Figure 5. Serous Membrane. Serous membrane lines the pericardial cavity and reflects back to cover the heart—much the same way that an underinflated balloon would form two layers surrounding a fist.

There are three serous cavities and their associated membranes. The **pleura** is the serous membrane that surrounds the lungs in the pleural cavity; the **pericardium** is the serous membrane that surrounds the heart in the pericardial cavity; and the **peritoneum** is the serous membrane that surrounds several organs in the abdominopelvic cavity. The serous fluid produced by the serous membranes reduces friction between the walls of the cavities and the internal organs when they move, such as when the lungs inflate or the heart beats. Both the parietal and visceral serosa secrete the thin, slippery serous fluid that prevents friction when an organ slides past the walls of a cavity. In the pleural cavities, pleural fluid prevents friction between the lungs and the walls of the cavity. In the pericardial sac, pericardial fluid prevents friction between the heart and the walls of the pericardial sac. And in the peritoneal cavity, peritoneal fluid prevents friction between abdominal and pelvic organs and the wall of the cavity. The serous membranes therefore provide additional protection to the viscera they enclose by reducing friction that could lead to inflammation of the organs.

Tissue Types

The term **tissue** is used to describe a group of cells found together in the body. The cells within a tissue share a common embryonic origin. Microscopic observation reveals that the cells in a tissue share morphological features and are arranged in an orderly pattern that achieves the tissue's functions. From the evolutionary perspective, tissues appear in more complex organisms. For example, multicellular protists, ancient eukaryotes, do not have cells organized into tissues.

Although there are many types of cells in the human body, they are organized into four broad categories of tissues: epithelial, connective, muscle, and nervous. Each of these categories is characterized by specific functions that contribute to the overall health and maintenance of the body. A disruption of the structure is a sign of injury or disease. Such changes can be detected through **histology**, the microscopic study of tissue appearance, organization, and function.

The Four Types of Tissues

Epithelial tissue, also referred to as epithelium, refers to the sheets of cells that cover exterior surfaces of the body, lines internal cavities and passageways, and forms certain glands. **Connective tissue**, as its name implies, binds the cells and organs of the body together and functions in the protection, support, and integration of all parts of the body. **Muscle tissue** is excitable, responding to stimulation and contracting to provide movement, and occurs as three major types: skeletal (voluntary) muscle, smooth muscle, and cardiac muscle in the heart. **Nervous tissue** is also excitable, allowing the propagation of electrochemical signals in the form of nerve impulses that communicate between different regions of the body.

The next level of organization is the organ, where several types of tissues come together to form a working unit. Just as knowing the structure and function of cells helps you in your study of tissues, knowledge of tissues will help you understand how organs function. The epithelial and connective tissues are discussed in detail in this chapter. Muscle and nervous tissues will be discussed only briefly in this chapter.

Tissue Membranes

A **tissue membrane** is a thin layer or sheet of cells that covers the outside of the body (for example, skin), the organs (for example, pericardium), internal passageways that lead to the exterior of the body (for example, abdominal mesenteries), and the lining of the moveable joint cavities. There are two basic types of tissue membranes: connective tissue and epithelial membranes.

Epithelial Tissues

Epithelial tissues cover the outside of organs and structures in the body and line the lumens of organs in a single layer or multiple layers of cells. The types of epithelia are classified by the shapes of cells present and the number of layers of cells. Epithelia composed of a single layer of cells is called simple epithelia; epithelial tissue composed of multiple layers is called stratified epithelia. (Figure 4) summarizes the different types of epithelial tissues.

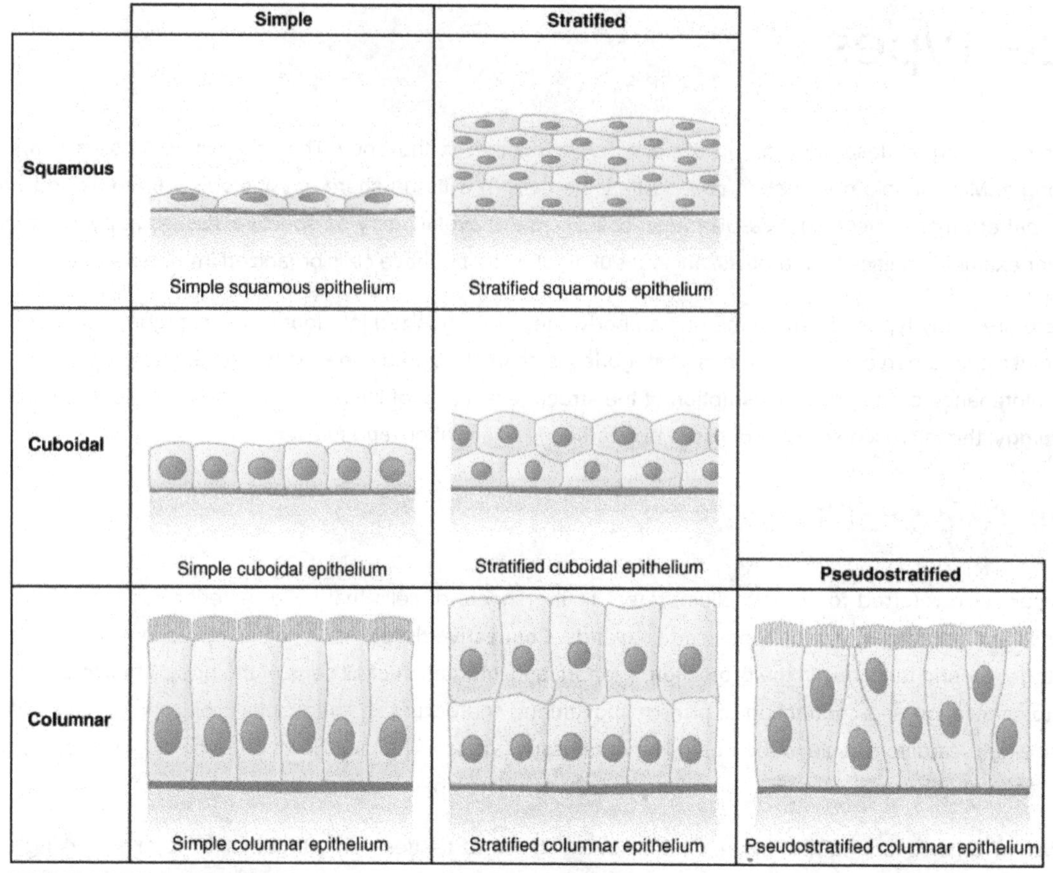

Figure 4. Cells of Epithelial Tissue. Simple epithelial tissue is organized as a single layer of cells and stratified epithelial tissue is formed by several layers of cells.

Different Types of Epithelial Tissues

Cell shape	Description	Location
squamous	flat, irregular round shape	simple: lung alveoli, capillaries; stratified: skin, mouth, vagina
cuboidal	cube shaped, central nucleus	glands, renal tubules
columnar	tall, narrow, nucleus toward base; tall, narrow, nucleus along cell	simple: digestive tract; pseudostratified: respiratory tract
transitional	round, simple but appear stratified	urinary bladder

Squamous Epithelia

Squamous epithelial cells are generally round, flat, and have a small, centrally located nucleus. The cell outline is slightly irregular, and cells fit together to form a covering or lining. When the cells are arranged in a single layer (simple epithelia), they facilitate diffusion in tissues, such as the areas of gas exchange in the lungs and the exchange of nutrients and waste at blood capillaries.

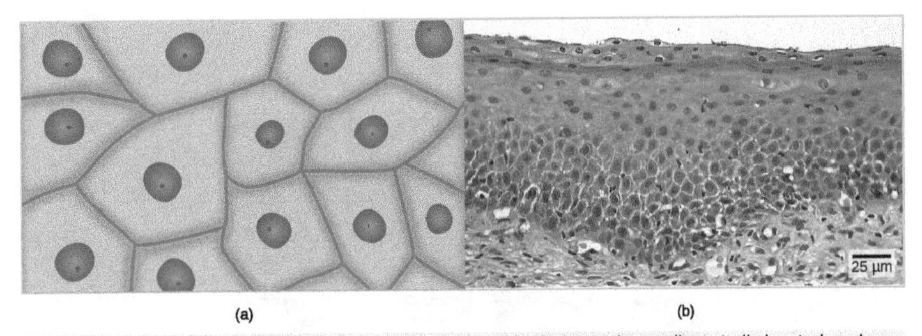

Figure 5. Squamous epithelia cells (a) have a slightly irregular shape, and a small, centrally located nucleus. These cells can be stratified into layers, as in (b) this human cervix specimen. (credit b: modification of work by Ed Uthman; scale-bar data from Matt Russell)

(Figure 5)**a** illustrates a layer of squamous cells with their membranes joined together to form an epithelium. Image (Figure 5)**b** illustrates squamous epithelial cells arranged in stratified layers, where protection is needed on the body from outside abrasion and damage. This is called a stratified squamous epithelium and occurs in the skin and in tissues lining the mouth and vagina.

Cuboidal Epithelia

Cuboidal epithelial cells, shown in (Figure 6), are cube-shaped with a single, central nucleus. They are most commonly found in a single layer representing a simple epithelia in glandular tissues throughout the body where they prepare and secrete glandular material. They are also found in the walls of tubules and in the ducts of the kidney and liver.

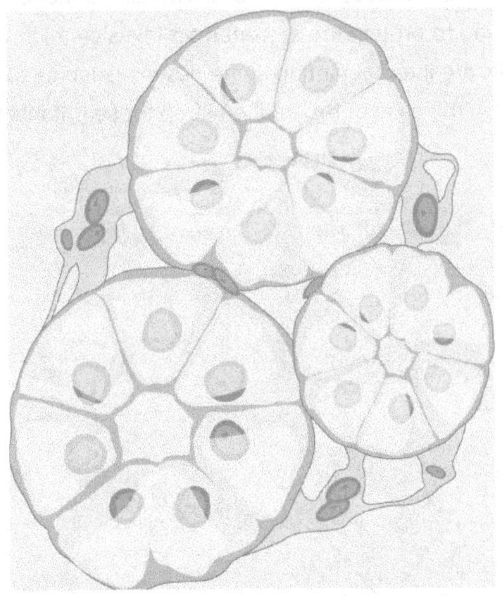

Figure 6. Simple cuboidal epithelial cells line tubules in the mammalian kidney, where they are involved in filtering the blood.

Columnar Epithelia

Columnar epithelial cells are taller than they are wide: they resemble a stack of columns in an epithelial layer, and are most commonly found in a single-layer arrangement. The nuclei of columnar epithelial cells in the digestive tract appear to be lined up at the base of the cells, as illustrated in (Figure 7). These cells absorb material from the lumen of the digestive tract and prepare it for entry into the body through the circulatory and lymphatic systems.

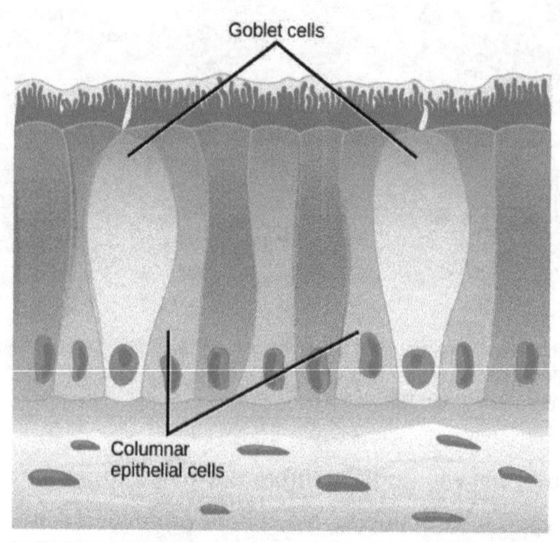

Figure 7. Simple columnar epithelial cells absorb material from the digestive tract. Goblet cells secret mucous into the digestive tract lumen.

Columnar epithelial cells lining the respiratory tract appear to be stratified. However, each cell is attached to the base membrane of the tissue and, therefore, they are simple tissues. The nuclei are arranged at different levels in the layer of cells, making it appear as though there is more than one layer, as seen in (Figure 8). This is called pseudostratified, columnar epithelia. This cellular covering has cilia at the apical, or free, surface of the cells. The cilia enhance the movement of mucous and trapped particles out of the respiratory tract, helping to protect the system from invasive microorganisms and harmful material that has been breathed into the body. Goblet cells are interspersed in some tissues (such as the lining of the trachea). The goblet cells contain mucous that traps irritants, which in the case of the trachea keep these irritants from getting into the lungs.

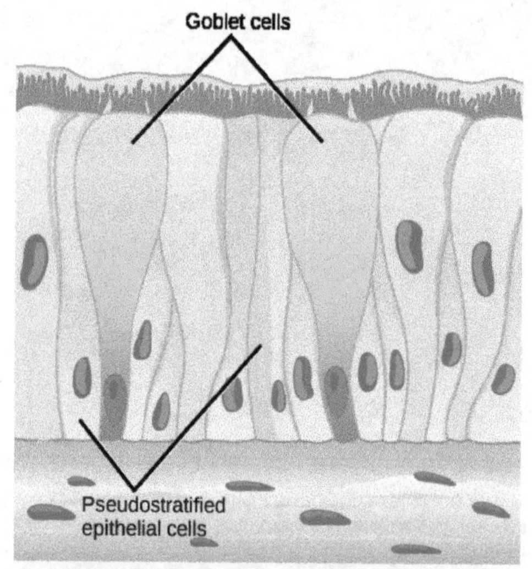

Figure 8. Pseudostratified columnar epithelia line the respiratory tract. They exist in one layer, but the arrangement of nuclei at different levels makes it appear that there is more than one layer. Goblet cells interspersed between the columnar epithelial cells secrete mucous into the respiratory tract.

Connective Tissues

Connective tissues are made up of a matrix consisting of living cells and a nonliving substance, called the ground substance. The ground substance is made of an organic substance (usually a protein) and an inorganic substance (usually a mineral or water). The principal cell of connective tissues is the fibroblast. This cell makes the fibers found in nearly all of the connective tissues.

Fibroblasts are motile, able to carry out mitosis, and can synthesize whichever connective tissue is needed. Macrophages, lymphocytes, and, occasionally, leukocytes can be found in some of the tissues. Some tissues have specialized cells that are not found in the others. The matrix in connective tissues gives the tissue its density. When a connective tissue has a high concentration of cells or fibers, it has proportionally a less dense matrix.

The organic portion or protein fibers found in connective tissues are either collagen, elastic, or reticular fibers. Collagen fibers provide strength to the tissue, preventing it from being torn or separated from the surrounding tissues. Elastic fibers are made of the protein elastin; this fiber can stretch to one and one half of its length and return to its original size and shape. Elastic fibers provide flexibility to the tissues. Reticular fibers are the third type of protein fiber found in connective tissues. This fiber consists of thin strands of collagen that form a network of fibers to support the tissue and other organs to which it is connected. The various types of connective tissues, the types of cells and fibers they are made of, and sample locations of the tissues is summarized in below.

Connective Tissues

Tissue	Cells	Fibers	Location
loose/areolar	fibroblasts, macrophages, some lymphocytes, some neutrophils	few: collagen, elastic, reticular	around blood vessels; anchors epithelia
dense, fibrous connective tissue	fibroblasts, macrophages	mostly collagen	irregular: skin; regular: tendons, ligaments
cartilage	chondrocytes, chondroblasts	hyaline: few: collagen fibrocartilage: large amount of collagen	shark skeleton, fetal bones, human ears, intervertebral discs
bone	osteoblasts, osteocytes, osteoclasts	some: collagen, elastic	vertebrate skeletons
adipose	adipocytes	few	adipose (fat)
blood	red blood cells, white blood cells	none	blood

Loose/Areolar Connective Tissue

Loose connective tissue, also called areolar connective tissue, has a sampling of all of the components of a connective tissue. As illustrated in (Figure 9), loose connective tissue has some fibroblasts; macrophages are present as well. Collagen fibers are relatively wide and stain a light pink, while elastic fibers are thin and stain dark blue to black. The space between the formed elements of the tissue is filled with the matrix. The material in the connective tissue gives it a loose consistency similar to a cotton ball that has been pulled apart. Loose connective tissue is found around every blood vessel and helps to keep the vessel in place. The tissue is also found around and between most body organs. In summary, areolar tissue is tough, yet flexible, and comprises membranes.

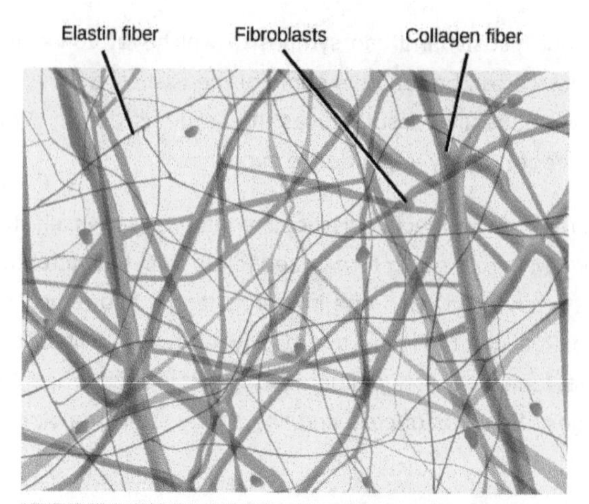

Figure 9. Loose connective tissue is composed of loosely woven collagen and elastic fibers. The fibers and other components of the connective tissue matrix are secreted by fibroblasts.

Fibrous Connective Tissue

Fibrous connective tissues contain large amounts of collagen fibers and few cells or matrix material. The fibers can be arranged irregularly or regularly with the strands lined up in parallel. Irregularly arranged fibrous connective tissues are found in areas of the body where stress occurs from all directions, such as the dermis of the skin. Regular fibrous connective tissue, shown in (Figure 10), is found in tendons (which connect muscles to bones) and ligaments (which connect bones to bones).

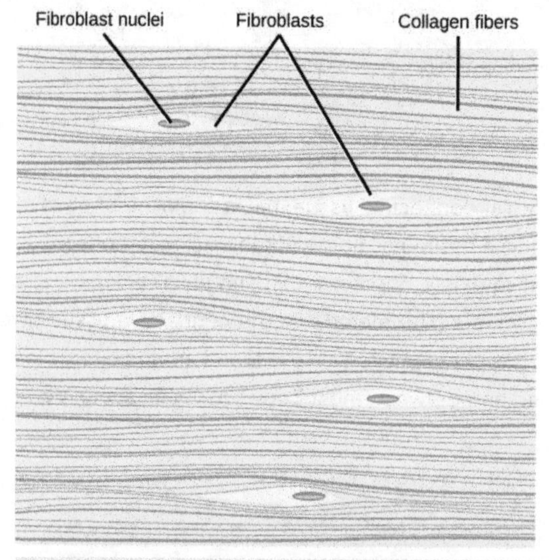

Figure 10. Fibrous connective tissue from the tendon has strands of collagen fibers lined up in parallel.

Cartilage

Cartilage is a connective tissue with a large amount of the matrix and variable amounts of fibers. The cells, called chondrocytes, make the matrix and fibers of the tissue. Chondrocytes are found in spaces within the tissue called lacunae.

The three main types of cartilage tissue are hyaline cartilage, fibrocartilage, and elastic cartilage. **Hyaline cartilage**, the most common type of cartilage in the body, consists of short and dispersed collagen fibers and contains large amounts of proteoglycans. Under the microscope, tissue samples appear clear. The surface of hyaline cartilage is smooth. Both strong and flexible, it is found in the rib cage and nose and covers bones where they meet to form moveable joints. It makes up a template of the embryonic skeleton before bone formation. A plate of hyaline cartilage at the ends of bone allows continued growth until adulthood. **Fibrocartilage** is tough because it has thick bundles of collagen fibers dispersed through its matrix. The knee and jaw joints and the the intervertebral discs are examples of fibrocartilage. **Elastic cartilage** contains elastic fibers as well as collagen and proteoglycans. This tissue gives rigid support as well as elasticity. Tug gently at your ear lobes, and notice that the lobes return to their initial shape. The external ear contains elastic cartilage.

Bone

Bone, or osseous tissue, is a connective tissue that has a large amount of two different types of matrix material. The organic matrix is similar to the matrix material found in other connective tissues, including some amount of collagen and elastic fibers. This gives strength and flexibility to the tissue. The inorganic matrix consists of mineral salts—mostly calcium salts—that give the tissue hardness. Without adequate organic material in the matrix, the tissue breaks; without adequate inorganic material in the matrix, the tissue bends.

There are three types of cells in bone: osteoblasts, osteocytes, and osteoclasts. Osteoblasts are active in making bone for growth and remodeling. Osteoblasts deposit bone material into the matrix and, after the matrix surrounds them, they continue to live, but in a reduced metabolic state as osteocytes. Osteocytes are found in lacunae of the bone. Osteoclasts are active in breaking down bone for bone remodeling, and they provide access to calcium stored in tissues. Osteoclasts are usually found on the surface of the tissue.

Bone can be divided into two types: compact and spongy. Compact bone is found in the shaft (or diaphysis) of a long bone and the surface of the flat bones, while spongy bone is found in the end (or epiphysis) of a long bone. Compact bone is organized into subunits called osteons, as illustrated in (Figure12). A blood vessel and a nerve are found in the center of the structure within the Haversian canal, with radiating circles of lacunae around it known as lamellae. The wavy lines seen between the lacunae are microchannels called canaliculi; they connect the lacunae to aid diffusion between the cells. Spongy bone is made of tiny plates called trabeculae; these plates serve as struts to give the spongy bone strength. Over time, these plates can break causing the bone to become less resilient. Bone tissue forms the internal skeleton of vertebrate animals, providing structure to the animal and points of attachment for tendons.

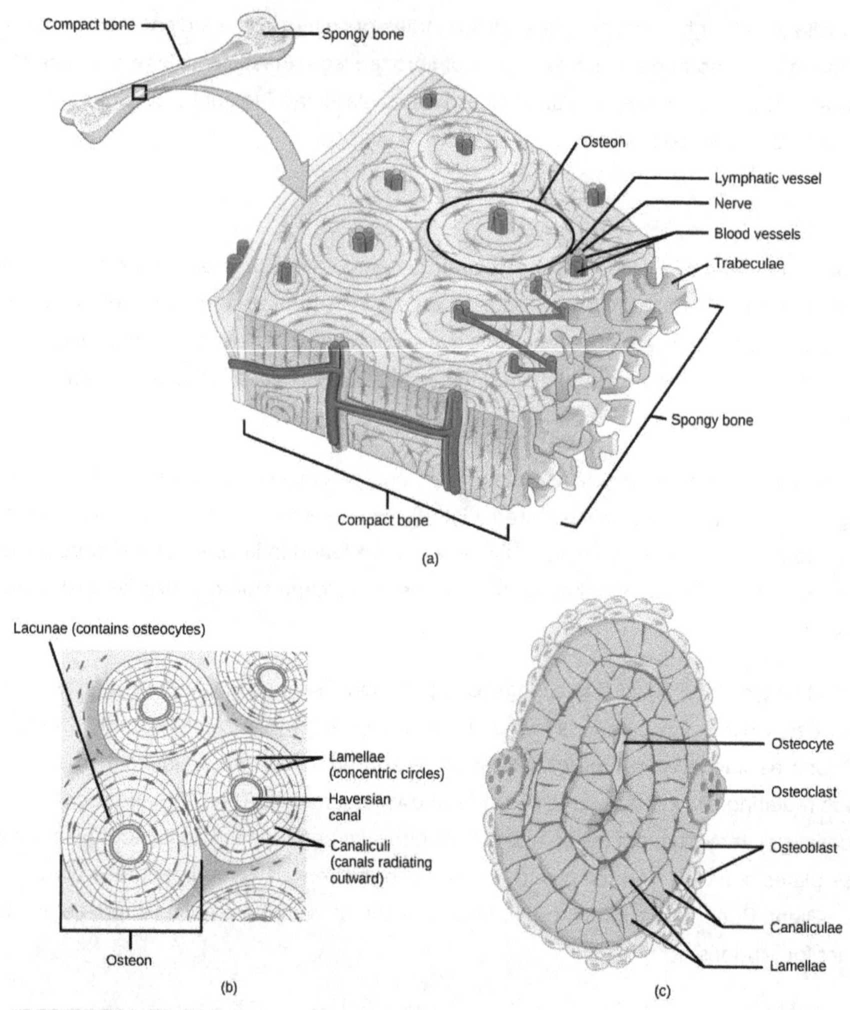

Figure 12. (a) Compact bone is a dense matrix on the outer surface of bone. Spongy bone, inside the compact bone, is porous with web-like trabeculae. (b) Compact bone is organized into rings called osteons. Blood vessels, nerves, and lymphatic vessels are found in the central Haversian canal. Rings of lamellae surround the Haversian canal. Between the lamellae are cavities called lacunae. Canaliculi are microchannels connecting the lacunae together. (c) Osteoblasts surround the exterior of the bone. Osteoclasts bore tunnels into the bone and osteocytes are found in the lacunae.

Adipose Tissue

Adipose tissue, or fat tissue, is considered a connective tissue even though it does not have fibroblasts or a real matrix and only has a few fibers. Adipose tissue is made up of cells called adipocytes that collect and store fat in the form of triglycerides, for energy metabolism. Adipose tissues additionally serve as insulation to help maintain body temperatures, allowing animals to be endothermic, and they function as cushioning against damage to body organs. Under a microscope, adipose tissue cells appear empty due to the extraction of fat during the processing of the material for viewing, as seen in. The thin lines in the image are the cell membranes, and the nuclei are the small, black dots at the edges of the cells.

Blood

Blood is considered a connective tissue because it has a matrix, as shown in (Figure 14). The living cell types are red blood cells (RBC), also called erythrocytes, and white blood cells (WBC), also called leukocytes. The fluid portion of whole blood, its matrix, is commonly called plasma.

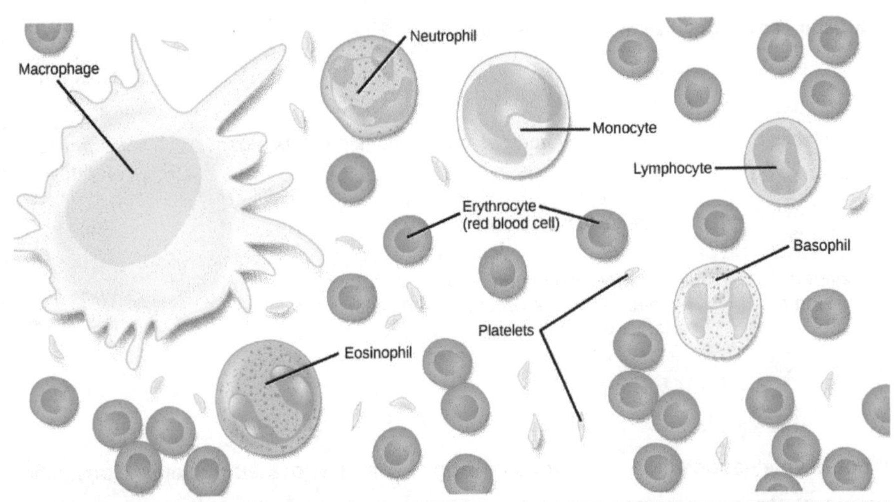

Figure 14. Blood is a connective tissue that has a fluid matrix, called plasma, and no fibers. Erythrocytes (red blood cells), the predominant cell type, are involved in the transport of oxygen and carbon dioxide. Also present are various leukocytes (white blood cells) involved in immune response.

The cell found in greatest abundance in blood is the erythrocyte. Erythrocytes are counted in millions in a blood sample: the average number of red blood cells in primates is 4.7 to 5.5 million cells per microliter. Erythrocytes are consistently the same size in a species, but vary in size between species. For example, the average diameter of a primate red blood cell is 7.5 µl, a dog is close at 7.0 µl, but a cat's RBC diameter is 5.9 µl. Sheep erythrocytes are even smaller at 4.6 µl. Mammalian erythrocytes lose their nuclei and mitochondria when they are released from the bone marrow where they are made. Fish, amphibian, and avian red blood cells maintain their nuclei and mitochondria throughout the cell's life. The principal job of an erythrocyte is to carry and deliver oxygen to the tissues.

Leukocytes are the predominant white blood cells found in the peripheral blood. Leukocytes are counted in the thousands in the blood with measurements expressed as ranges: primate counts range from 4,800 to 10,800 cells per µl, dogs from 5,600 to 19,200 cells per µl, cats from 8,000 to 25,000 cells per µl, cattle from 4,000 to 12,000 cells per µl, and pigs from 11,000 to 22,000 cells per µl.

Lymphocytes function primarily in the immune response to foreign antigens or material. Different types of lymphocytes make antibodies tailored to the foreign antigens and control the production of those antibodies. Neutrophils are phagocytic cells and they participate in one of the early lines of defense against microbial invaders, aiding in the removal of bacteria that has entered the body. Another leukocyte that is found in the peripheral blood is the monocyte. Monocytes give rise to phagocytic macrophages that clean up dead and damaged cells in the body, whether they are foreign or from the host animal. Two additional leukocytes in the blood are eosinophils and basophils—both help to facilitate the inflammatory response.

The slightly granular material among the cells is a cytoplasmic fragment of a cell in the bone marrow. This is called a platelet or thrombocyte. Platelets participate in the stages leading up to coagulation of the blood to stop bleeding through damaged blood vessels. Blood has a number of functions, but primarily it transports material through the body to bring nutrients to cells and remove waste material from them.

Muscle Tissues

There are three types of muscle in animal bodies: smooth, skeletal, and cardiac. They differ by the presence or absence of striations or bands, the number and location of nuclei, whether they are voluntarily or involuntarily controlled, and their location within the body.

Table 1. Comparison of Structure and Properties of Muscle Tissue Types

Tissue	Histology	Function	Location
Skeletal	Long cylindrical fiber, striated, many peripherally located nuclei	Voluntary movement, produces heat, protects organs	Attached to bones and around entrance points to body (e.g., mouth, anus)
Cardiac	Short, branched, striated, single central nucleus	Contracts to pump blood	Heart
Smooth	Short, spindle-shaped, no evident striation, single nucleus in each fiber	Involuntary movement, moves food, involuntary control of respiration, moves secretions, regulates flow of blood in arteries by contraction	Walls of major organs and passageways, visceral organs

Smooth Muscle

Smooth muscle does not have striations in its cells. It has a single, centrally located nucleus.. Constriction of smooth muscle occurs under involuntary, autonomic nervous control and in response to local conditions in the tissues. Smooth muscle tissue is also called non-striated as it lacks the banded appearance of skeletal and cardiac muscle. The walls of blood vessels, the tubes of the digestive system, and the tubes of the reproductive systems are composed of mostly smooth muscle.

Skeletal Muscle

Skeletal muscle has striations across its cells caused by the arrangement of the contractile proteins actin and myosin. These muscle cells are relatively long and have multiple nuclei along the edge of the cell. Skeletal muscle is under voluntary, somatic nervous system control and is found in the muscles that move bones.

Cardiac Muscle

Cardiac muscle is found only in the heart. Like skeletal muscle, it has cross striations in its cells, but cardiac muscle has a single, centrally located nucleus. Cardiac muscle is not under voluntary control but can be influenced by the autonomic nervous system to speed up or slow down. An added feature to cardiac muscle cells is a line than extends along the end of the cell as it abuts the next cardiac cell in the row. This line is called an intercalated disc: it assists in passing electrical impulse efficiently from one cell to the next and maintains the strong connection between neighboring cardiac cells.

Nervous Tissues

Nervous tissues are made of cells specialized to receive and transmit electrical impulses from specific areas of the body and to send them to specific locations in the body. Two main classes of cells make up nervous tissue: the **neuron** and **neuroglia**. The large structure with a central nucleus is the cell body of the neuron.

Projections from the cell body are either dendrites specialized in receiving input or a single axon specialized in transmitting impulses. Some neuroglial cells are also shown. Astrocytes regulate the chemical environment of the nerve cell, and oligodendrocytes insulate the axon so the electrical nerve impulse is transferred more efficiently. Other glial cells that are not shown support the nutritional and waste requirements of the neuron. Some of the glial cells are phagocytic and remove debris or damaged cells from the tissue. A nerve consists of neurons and glial cells.

Nervous tissue is characterized as being excitable and capable of sending and receiving electrochemical signals that provide the body with information. Neurons propagate information via electrochemical impulses, called action potentials, which are biochemically linked to the release of chemical signals. Neuroglia play an essential role in supporting neurons and modulating their information propagation.

Neurons display distinctive morphology, well suited to their role as conducting cells, with three main parts. The cell body includes most of the cytoplasm, the organelles, and the nucleus. Dendrites branch off the cell body and appear as thin extensions. A long "tail," the axon, extends from the neuron body and can be wrapped in an insulating layer known as **myelin**, which is formed by accessory cells. The synapse is the gap between nerve cells, or between a nerve cell and its target, for example, a muscle or a gland, across which the impulse is transmitted by chemical compounds known as neurotransmitters.

CAREER CONNECTIONS

Pathologist

A pathologist is a medical doctor or veterinarian who has specialized in the laboratory detection of disease in animals, including humans. These professionals complete medical school education and follow it with an extensive post-graduate residency at a medical center. A pathologist may oversee clinical laboratories for the evaluation of body tissue and blood samples for the detection of disease or infection. They examine tissue specimens through a microscope to identify cancers and other diseases. Some pathologists perform autopsies to determine the cause of death and the progression of disease.

Tissue Injury and Aging

Tissues of all types are vulnerable to injury and, inevitably, aging. In the former case, understanding how tissues respond to damage can guide strategies to aid repair. In the latter case, understanding the impact of aging can help in the search for ways to diminish its effects.

Tissue Injury and Repair

Inflammation is the standard, initial response of the body to injury. Whether biological, chemical, physical, or radiation burns, all injuries lead to the same sequence of physiological events. Inflammation limits the extent of injury, partially or fully eliminates the cause of injury, and initiates repair and regeneration of damaged tissue. **Necrosis**, or accidental cell death, causes inflammation. **Apoptosis** is programmed cell death, a normal step-by-step process that destroys cells no longer needed by the body. By mechanisms still under investigation, apoptosis does not initiate the inflammatory response. Acute inflammation resolves over time by the healing of tissue. If inflammation persists, it becomes chronic and leads to diseased conditions. Arthritis and tuberculosis are examples of chronic inflammation. The suffix "*-itis*" denotes inflammation of a specific organ or type, for example, peritonitis is the inflammation of the peritoneum, and meningitis refers to the inflammation of the meninges, the tough membranes that surround the central nervous system

The four cardinal signs of inflammation—redness, swelling, pain, and local heat—were first recorded in antiquity. Cornelius Celsus is credited with documenting these signs during the days of the Roman Empire, as early as the first century AD. A fifth sign, loss of function, may also accompany inflammation.

Upon tissue injury, damaged cells release inflammatory chemical signals that evoke local **vasodilation**, the widening of the blood vessels. Increased blood flow results in apparent redness and heat. In response to injury, mast cells present in tissue degranulate, releasing the potent vasodilator **histamine**. Increased blood flow and inflammatory mediators recruit white blood cells to the site of inflammation. The endothelium lining the local blood vessel becomes "leaky" under the influence of histamine and other inflammatory mediators allowing neutrophils, macrophages, and fluid to move from the blood into the interstitial tissue spaces. The excess liquid in tissue causes swelling, more properly called edema. The swollen tissues squeezing pain receptors cause the sensation of pain. Prostaglandins released from injured cells also activate pain neurons. Non-steroidal anti-inflammatory drugs (NSAIDs) reduce pain because they inhibit the synthesis of prostaglandins. High levels of NSAIDs reduce inflammation. Antihistamines decrease allergies by blocking histamine receptors and as a result the histamine response.

After containment of an injury, the tissue repair phase starts with removal of toxins and waste products. **Clotting** (coagulation) reduces blood loss from damaged blood vessels and forms a network of fibrin proteins that trap blood cells and bind the edges of the wound together. A scab forms when the clot dries, reducing the risk of infection. Sometimes a mixture of dead leukocytes and fluid called pus accumulates in the wound. As healing progresses, fibroblasts from the surrounding connective tissues replace the collagen and extracellular material lost by the injury. Angiogenesis, the growth of new blood vessels, results in vascularization of the new tissue known as granulation tissue. The clot retracts pulling the edges of the wound together, and it slowly dissolves as the tissue is repaired. When a large amount of granulation tissue forms and capillaries disappear, a pale scar is often visible in the healed area. A **primary union** describes the healing of a wound where the edges are close together. When there is a gaping wound, it takes longer to refill the area with cells and collagen. The process called **secondary union** occurs as the edges of the wound are pulled together by what is called **wound contraction**. When a wound is more than one quarter of an inch deep, sutures (stitches) are recommended to promote a primary union and avoid the formation of a disfiguring scar. Regeneration is the addition of new cells of the same type as the ones that were injured (Figure 1).

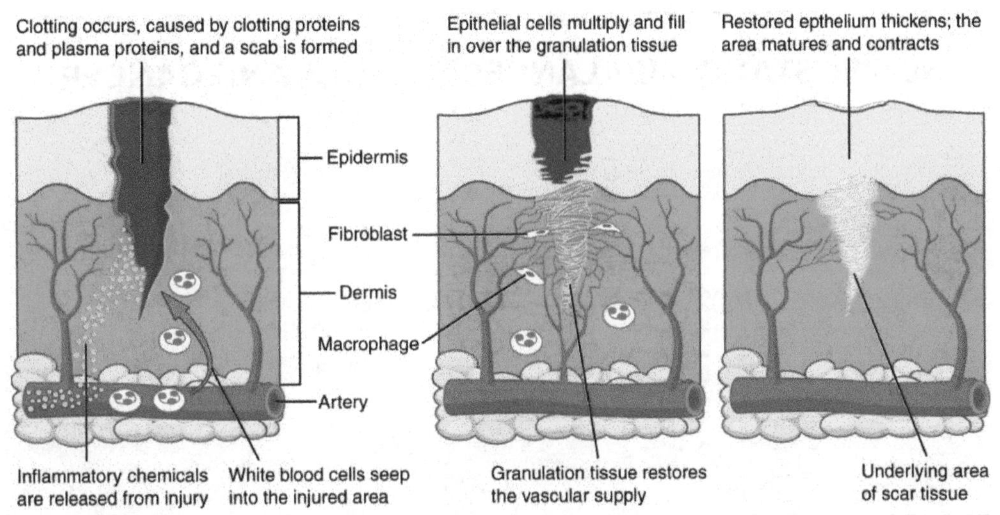

Figure 1. Tissue Healing. During wound repair, collagen fibers are laid down randomly by fibroblasts that move into repair the area.

Tissue and Aging

According to poet Ralph Waldo Emerson, "The surest poison is time." In fact, biology confirms that many functions of the body decline with age. All the cells, tissues, and organs are affected by senescence, with noticeable variability between individuals owing to different genetic makeup and lifestyles. The outward signs of aging are easily recognizable. The skin and other tissues become thinner and drier, reducing their elasticity, contributing to wrinkles and high blood pressure. Hair turns gray because follicles produce less melanin, the brown pigment of hair and the iris of the eye. The face looks flabby because elastic and collagen fibers decrease in connective tissue and muscle tone is lost. Glasses and hearing aids may become parts of life as the senses slowly deteriorate, all due to reduced elasticity. Overall height decreases as the bones lose calcium and other minerals. With age, fluid decreases in the fibrous cartilage disks intercalated between the vertebrae in the spine. Joints lose cartilage and stiffen. Many tissues, including those in muscles, lose mass through a process called **atrophy**. Lumps and rigidity become more widespread. As a consequence, the passageways, blood vessels, and airways become more rigid. The brain and spinal cord lose mass. Nerves do not transmit impulses with the same speed and frequency as in the past. Some loss of thought clarity and memory can accompany aging. More severe problems are not necessarily associated with the aging process and may be symptoms of underlying illness.

As exterior signs of aging increase, so do the interior signs, which are not as noticeable. The incidence of heart diseases, respiratory syndromes, and type 2 diabetes increases with age, though these are not necessarily age-dependent effects. Wound healing is slower in the elderly, accompanied by a higher frequency of infection as the capacity of the immune system to fend off pathogen declines.

Aging is also apparent at the cellular level because all cells experience changes with aging. Telomeres, regions of the chromosomes necessary for cell division, shorten each time cells divide. As they do, cells are less able to divide and regenerate. Because of alterations in cell membranes, transport of oxygen and nutrients into the cell and removal of carbon dioxide and waste products from the cell are not as efficient in the elderly. Cells may begin to function abnormally, which may lead to diseases associated with aging, including arthritis, memory issues, and some cancers.

The progressive impact of aging on the body varies considerably among individuals, but Studies indicate, however, that exercise and healthy lifestyle choices can slow down the deterioration of the body that comes with old age.

HOMEOSTATIC IMBALANCES: TISSUES AND CANCER

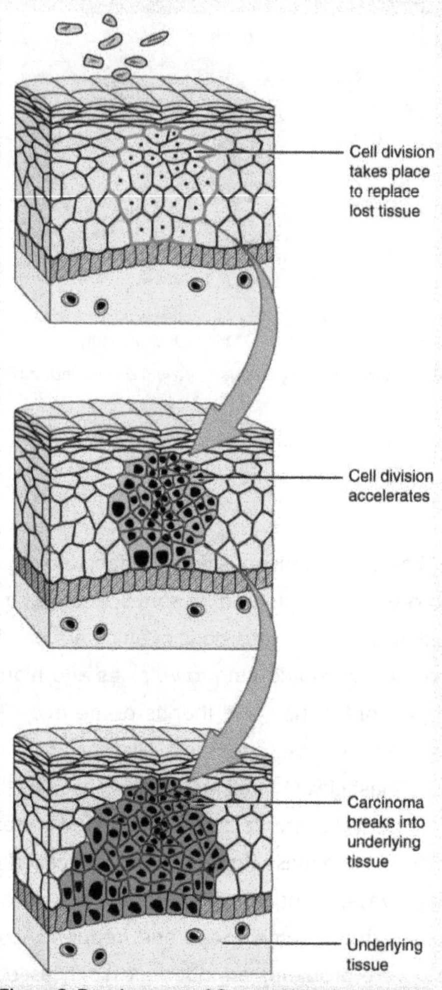

Figure 2. Development of Cancer. Note the change in cell size, nucleus size, and organization in the tissue.

Cancer is a generic term for many diseases in which cells escape regulatory signals. Uncontrolled growth, invasion into adjacent tissues, and colonization of other organs, if not treated early enough, are its hallmarks. Health suffers when tumors "rob" blood supply from the "normal" organs.

A mutation is defined as a permanent change in the DNA of a cell. Epigenetic modifications, changes that do not affect the code of the DNA but alter how the DNA is decoded, are also known to generate abnormal cells. Alterations in the genetic material may be caused by environmental agents, infectious agents, or errors in the replication of DNA that accumulate with age. Many mutations do not cause any noticeable change in the functions of a cell; however, if the modification affects key proteins that have an impact on the cell's ability to proliferate in an orderly fashion, the cell starts to divide abnormally.

As changes in cells accumulate, they lose their ability to form regular tissues. A tumor, a mass of cells displaying abnormal architecture, forms in the tissue. Many tumors are benign, meaning they do not metastasize nor cause disease. A tumor becomes malignant, or cancerous, when it breaches the confines of its tissue, promotes angiogenesis, attracts the growth of capillaries, and metastasizes to other organs (Figure 2).

The specific names of cancers reflect the tissue of origin. Cancers derived from epithelial cells are referred to as carcinomas. Cancer in myeloid tissue or blood cells form myelomas. Leukemias are cancers of white blood cells, whereas sarcomas derive from connective tissue. Cells in tumors differ both in structure and function. Some cells, called cancer stem cells, appear to be a subtype of cell responsible for uncontrolled growth. Recent research shows that contrary to what was previously assumed, tumors are not disorganized masses of cells, but have their own structures.

Cancer treatments vary depending on the disease's type and stage. Traditional approaches, including surgery, radiation, chemotherapy, and hormonal therapy, aim to remove or kill rapidly dividing cancer cells, but these strategies have their limitations. Depending on a tumor's location, for example, cancer surgeons may be unable to remove it. Radiation and chemotherapy are difficult, and it is often impossible to target only the cancer cells. The treatments inevitably destroy healthy tissue as well. To address this, researchers are working on pharmaceuticals that can target specific proteins implicated in cancer-associated molecular pathways.

Structural Organization of the Human Body

Before you begin to study the different structures and functions of the human body, it is helpful to consider its basic architecture; that is, how its smallest parts are assembled into larger structures. It is convenient to consider the structures of the body in terms of fundamental levels of organization that increase in complexity: subatomic particles, atoms, molecules, organelles, cells, tissues, organs, organ systems, and organism.

The Levels of Organization

To study the chemical level of organization, scientists consider the simplest building blocks of matter: subatomic particles, atoms and molecules. All matter in the universe is composed of one or more unique pure substances called elements, familiar examples of which are hydrogen, oxygen, carbon, nitrogen, calcium, and iron. The smallest unit of any of these pure substances (elements) is an atom. Atoms are made up of subatomic particles such as the proton, electron and neutron. Two or more atoms combine to form a molecule, such as the water molecules, proteins, and sugars found in living things. Molecules are the chemical building blocks of all body structures.

A **cell** is the smallest independently functioning unit of a living organism. Even bacteria, which are extremely small, independently-living organisms, have a cellular structure. Each bacterium is a single cell. All living structures of human anatomy contain cells, and almost all functions of human physiology are performed in cells or are initiated by cells.

A human cell typically consists of flexible membranes that enclose cytoplasm, a water-based cellular fluid together with a variety of tiny functioning units called **organelles**. In humans, as in all organisms, cells perform all functions of life. A **tissue** is a group of many similar cells (though sometimes composed of a few related types) that work together to perform a specific function. An **organ** is an anatomically distinct structure of the body composed of two or more tissue types. Each organ performs one or more specific physiological functions. **An organ system** is a group of organs that work together to perform major functions or meet physiological needs of the body.

This book covers eleven distinct organ systems in the human body (Figure 2 and Figure 3). Assigning organs to organ systems can be imprecise since organs that "belong" to one system can also have functions integral to another system. In fact, most organs contribute to more than one system.

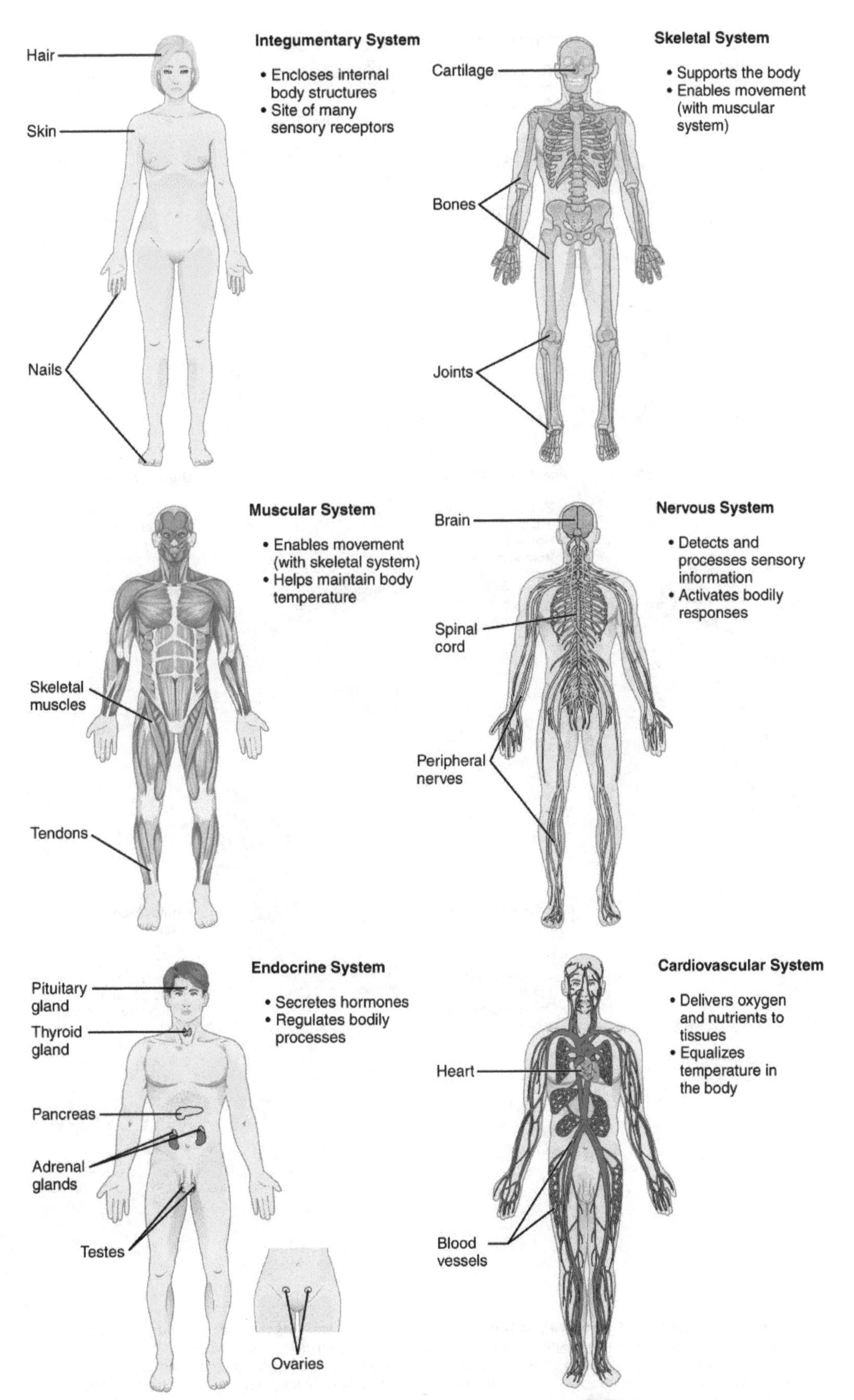

Figure 2. Organ Systems of the Human Body. Organs that work together are grouped into organ systems.

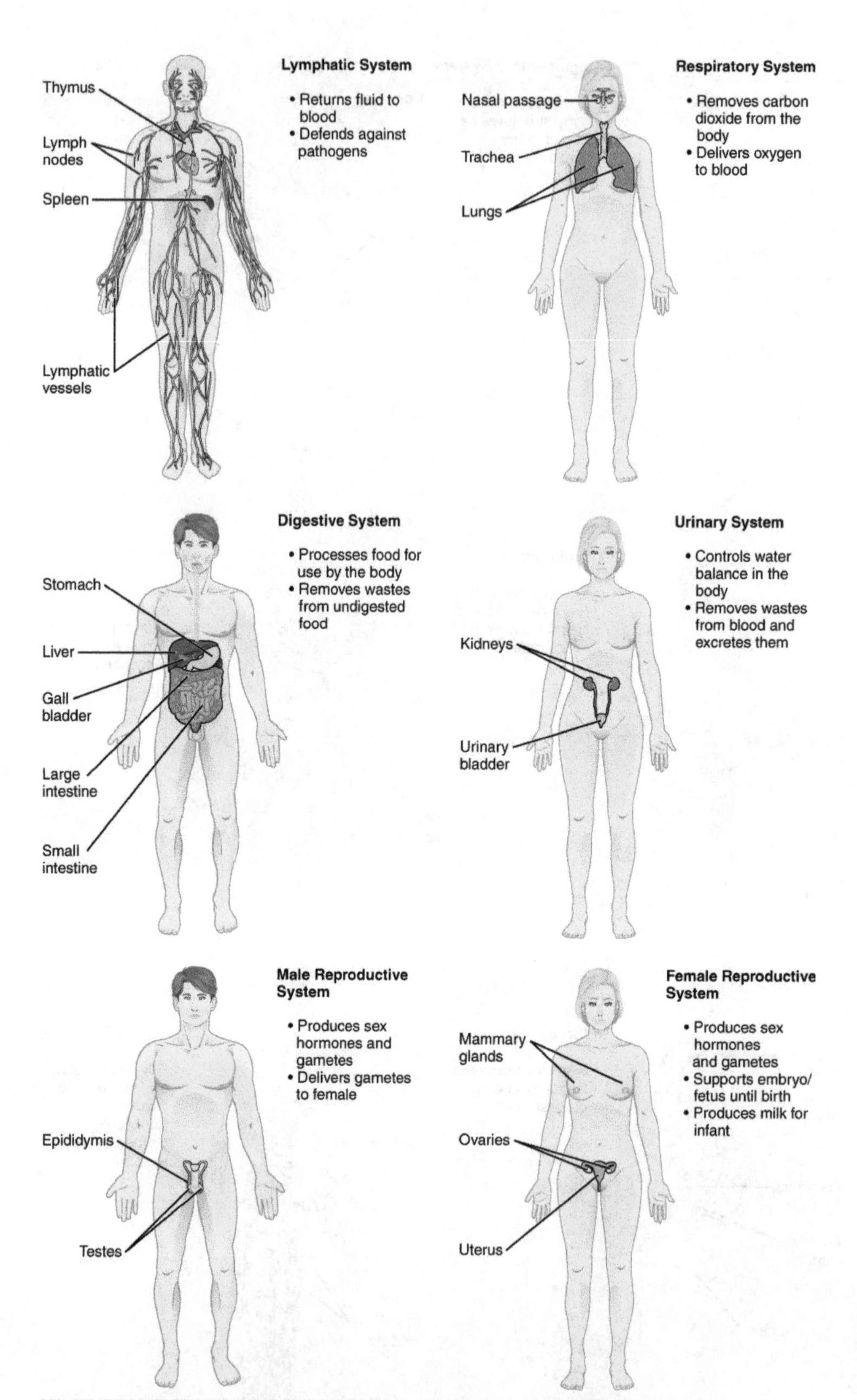

Figure 3. Organ Systems of the Human Body (continued). Organs that work together are grouped into organ systems. The organism level is the highest level of organization. An **organism** is a living being that has a cellular structure and that can independently perform all physiologic functions necessary for life. In multicellular organisms, including humans, all cells, tissues, organs, and organ systems of the body work together to maintain the life and health of the organism.

Digestive System : Introduction

The function of the digestive system is to break down the foods you eat, release their nutrients, and absorb those nutrients into the body. Although the small intestine is where the majority of digestion occurs, and where most of the released nutrients are absorbed into the blood or lymph, each of the digestive system organs makes a vital contribution to this process.

The processes of digestion include six activities:

- **1. Ingestion:** The first of these processes, ingestion, refers to the entry of food into the alimentary canal through the mouth. Chewing increases the surface area of the food and allows an appropriately sized bolus to be produced.
- **2. Propulsion:** Food leaves the mouth when the tongue and pharyngeal muscles propel it into the esophagus. This act of swallowing, the last voluntary act until defecation, is an example of propulsion, which refers to the movement of food through the digestive tract. It includes both the voluntary process of swallowing and the involuntary process of peristalsis. Peristalsis consists of sequential, alternating waves of contraction and relaxation of alimentary wall smooth muscles, which act to propel food along (Figure 1). These waves also play a role in mixing food with digestive juices. Peristalsis is so powerful that foods and liquids you swallow enter your stomach even if you are standing on your head.

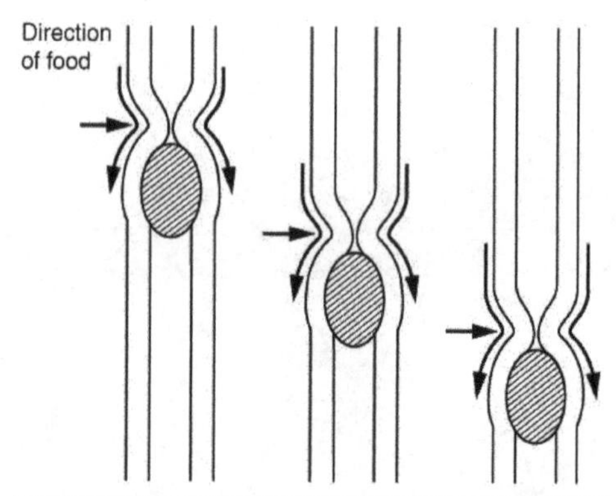

Figure 1. Peristalsis moves food through the digestive tract with alternating waves of muscle contraction and relaxation.

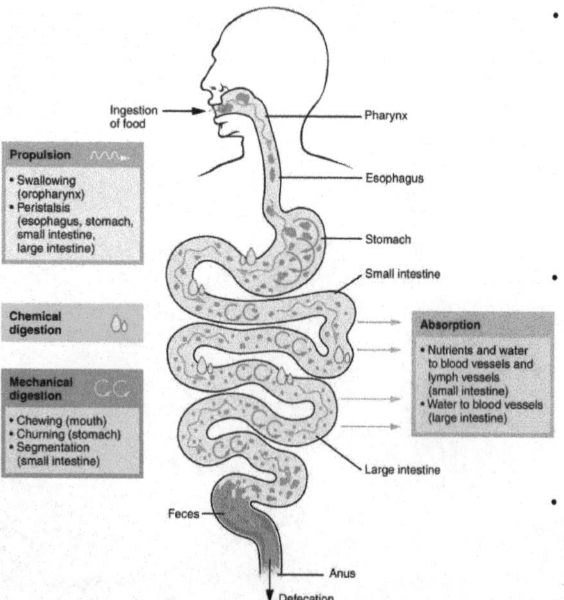

Figure 2. :The different processes involved in digestion.

- **3. Mechanical or physical digestion:** Digestion includes both mechanical and chemical processes. Mechanical digestion is a purely physical process that does not change the chemical nature of the food. Instead, it makes the food smaller to increase both surface area and mobility. It includes mastication **or chewing,** mechanical **churning of food in the stomach** and **s**egmentation in small intestine.
- **4. Chemical digestion :** Chemical digestion, starts in the mouth, digestive secretions break down complex food molecules into their chemical building blocks (for example, proteins into separate amino acids). These secretions vary in composition, but typically contain water, various enzymes, acids, and salts. Chemical digestion is completed in the small intestine.
- **5. Absorption:** Food that has been broken down is of no value to the body unless it enters the bloodstream and its nutrients are put to work. This occurs through the process of absorption, which takes place primarily within the small intestine.
- **6. Defecation / Elimination:** In defecation, the final step in digestion, undigested materials are removed from the body as feces.

Digestive System Anatomy:

The human digestive system consists of a digestive tube, **the alimentary canal,** and **accessory organs** that secrete digestive chemicals.

Alimentary Canal and Organs: Also called the **gastrointestinal (GI) tract** or gut, the **alimentary canal** (aliment- = "to nourish") is a one-way tube about 7.62 meters (25 feet) in length. The main function of the organs of the alimentary canal is to nourish the body. This tube begins at the mouth and terminates at the anus. Between those two points,the alimentary canal is divided into specialized digestive organs along its length :

mouth (oral cavity)→ pharynx→esophagus→ stomach→ small intestine→ large intestine (colon)→ rectum→ anus.

The inside space of the alimentary canal (tube) is called **lumen.** Food goes through the lumen. Both the mouth and anus are open to the external environment; thus, food and wastes within the alimentary canal are technically considered to be outside the body. Only through the process of absorption do the nutrients in food enter into and nourish the body's "inner space."

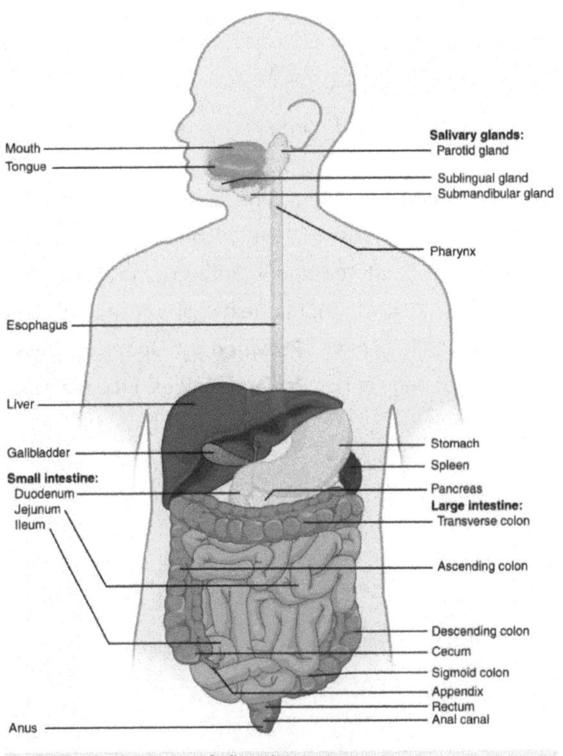

Figure 3. Mains organs of digestive system.

Wall of the Alimentary Canal

Throughout its length, the alimentary tract is composed of the same four tissue layers; the details of their structural arrangements vary to fit their specific functions. Starting from the lumen and moving outwards, these layers are the **mucosa, submucosa, muscularis, and serosa.**

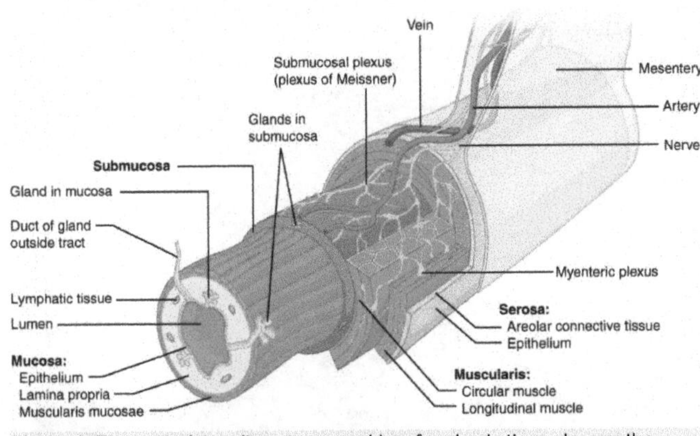

Figure 4. The wall of the alimentary canal has four basic tissue layers: the mucosa, submucosa, muscularis, and serosa.

The **mucosa** is referred to as a mucous membrane, because mucus production is a characteristic feature of gut epithelium. The membrane consists of epithelium, which is in direct contact with ingested food.

The **submucosa** lies immediately beneath the mucosa. A broad layer of dense connective tissue, it connects the overlying mucosa to the underlying muscularis. It includes blood and lymphatic vessels (which transport absorbed nutrients), and a scattering of submucosal glands that release digestive secretions. Additionally, it serves as a conduit for a dense branching network of nerves.

The **muscalaris** (also called the muscularis externa) is the third layer of the alimentary canal. The muscularis in the small intestine is made up of a **double layer of smooth muscle: an inner circular layer and an outer longitudinal layer.** The contractions of these layers promote mechanical digestion, and move the food along the canal. The basic two-layer structure found in the small intestine is modified in the organs proximal and distal to it. The **stomach** is equipped for its churning function by the addition of a **third layer, the oblique muscle.** While the colon has two layers like the small intestine, its longitudinal layer is segregated into three narrow parallel bands, the tenia coli, which make it look like a series of pouches rather than a simple tube.

The **serosa** is the portion of the alimentary canal superficial to the muscularis. Present only in the region of the alimentary canal within the abdominal cavity, it consists of a layer of visceral peritoneum overlying a layer of loose connective tissue. Instead of serosa, the mouth, pharynx, and esophagus have a dense sheath of collagen fibers called the adventitia. These tissues serve to hold the alimentary canal in place near the ventral surface of the vertebral column.

Anatomy of Organs of the Digestive System and Their Functions

Gastrointestinal tract, also called **digestive tract** or **alimentary canal**, pathway by which food enters the body and solid wastes are expelled. The gastrointestinal tract includes the mouth, pharynx, esophagus, stomach, small intestine, large intestine, and anus. *See* digestion.

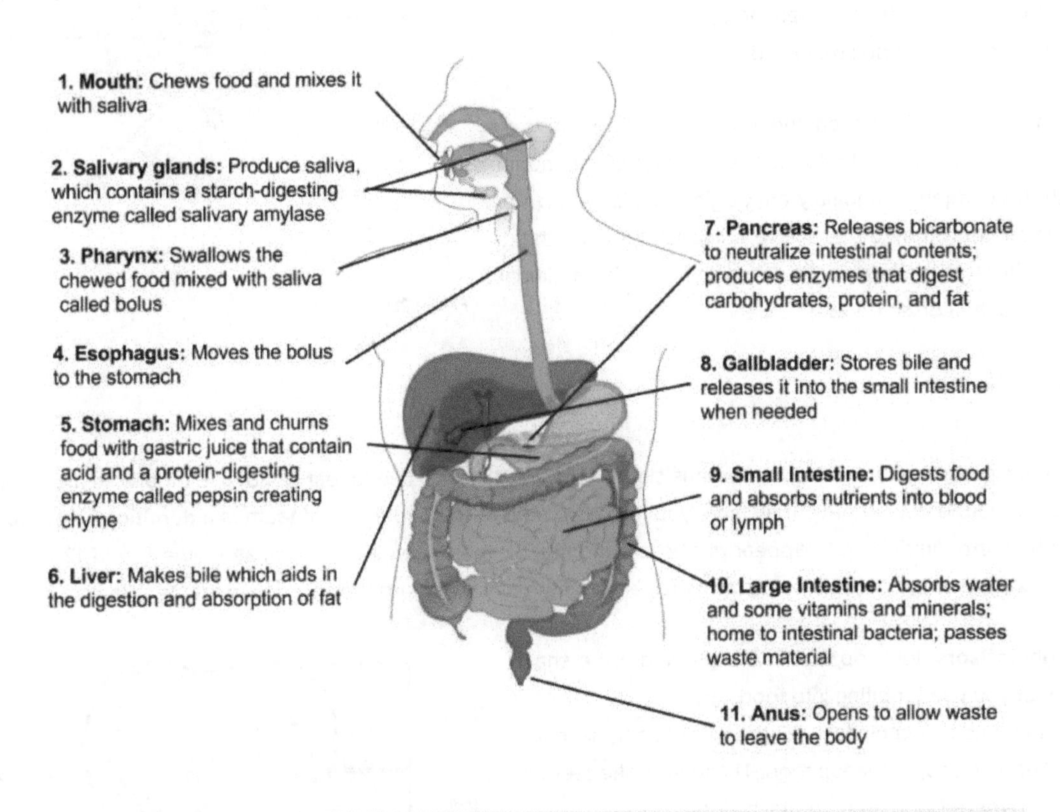

Figure 1: Digestive system organs and functions – overview.

The Oral Cavity or Mouth:

The oral cavity, or mouth, is the point of entry of food into the digestive system. The cheeks, tongue, and palate frame the mouth, which is also called the **oral cavity** (or buccal cavity). The arched shape of the roof of your mouth is called the **palate**. If you run your tongue along the roof of your mouth, the front part is the **hard palate**, and at the back, the roof becomes fleshier. This part of the palate, known as the **soft palate.** A fleshy bead of tissue called the **uvula** drops down from the center of the posterior edge of the soft palate. Although some have suggested that the uvula is a vestigial organ, it serves an important purpose. When you swallow, the soft palate and uvula move upward, helping to keep foods and liquid from entering the nasal cavity. Unfortunately, it can also contribute to the sound produced by snoring. Two muscular folds extend downward from the soft palate, on either side of the uvula. Between these two arches are the **palatine tonsils**, clusters of lymphoid tissue that protect the pharynx. The lingual tonsils are located at the base of the tongue. These glands make antibodies that help fight infection. **Tonsillitis** means that your tonsils are inflamed. Both bacteria and viruses can cause tonsillitis.

The Tongue

The tongue is a workhorse, facilitating ingestion, mechanical digestion, chemical digestion (lingual lipase), sensation (of taste, texture, and temperature of food), swallowing, and vocalization. The tongue perform three important digestive functions in the mouth:

- Position food for optimal chewing
- Gather food into a **bolus** (rounded mass)
- Position food so it can be swallowed

A fold of mucous membrane on the underside of the tongue, the **lingual frenulum**, tethers the tongue to the floor of the mouth. People with the congenital anomaly ankyloglossia, also known by the non-medical term "tongue tie," have a lingual frenulum that is too short or otherwise malformed. Severe ankyloglossia can impair speech and must be corrected with surgery.

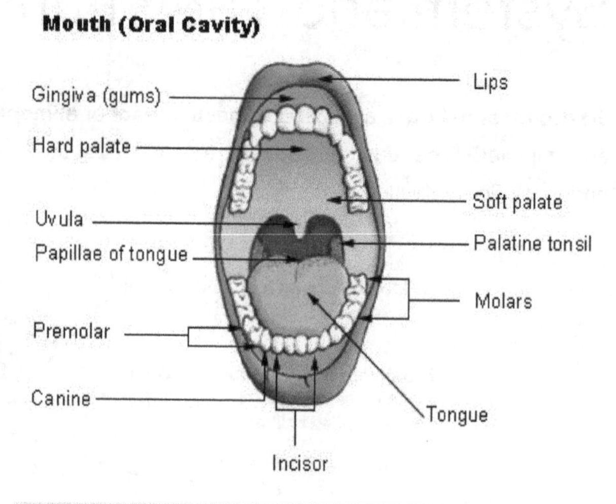

Figure 2: Oral cavity

The Teeth

The teeth, or **dentes** (singular = dens), are organs similar to bones that you use to tear, grind, and otherwise mechanically break down food. During the course of your lifetime, you have two sets of teeth (one set of teeth is a **dentition**). Your 20 **deciduous teeth**, or baby teeth, first begin to appear at about 6 months of age. Between approximately age 6 and 12, these teeth are replaced by 32 **permanent teeth**. Moving from the center of the mouth toward the side, these are as follows:

- The eight **incisors**, four top and four bottom, are the sharp front teeth you use for biting into food.
- The four **cuspids** (or canines) flank the incisors and have a pointed edge (cusp) to tear up food. These fang-like teeth are superb for piercing tough or fleshy foods.
- Posterior to the cuspids are the eight **premolars** (or bicuspids), which have an overall flatter shape with two rounded cusps useful for mashing foods.
- The most posterior and largest are the 12 **molars**, which have several pointed cusps used to crush food so it is ready for swallowing. The third members of each set of three molars, top and bottom, are commonly referred to as the wisdom teeth, because their eruption is commonly delayed until early adulthood. It is not uncommon for wisdom teeth to fail to erupt; that is, they remain impacted. In these cases, the teeth are typically removed by orthodontic surgery.

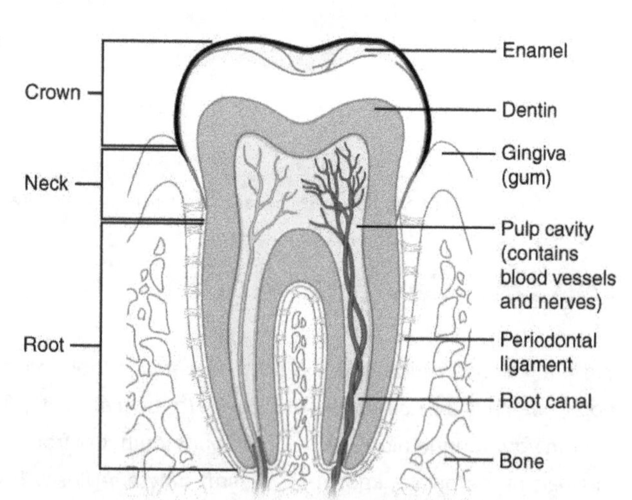

Figure 3: This longitudinal section through a molar in its alveolar socket shows the relationships between enamel, dentin, and pulp.

Anatomy of a Tooth: The teeth are secured in the alveolar processes (sockets) of the maxilla and the mandible. **Gingivae** (commonly called the gums) are soft tissues that line the alveolar processes and surround the necks of the teeth.

The two main parts of a tooth are the **crown**, which is the portion projecting above the gum line, and the **root**, which is embedded within the maxilla and mandible. Both parts contain an inner **pulp cavity**, containing loose connective tissue through which run nerves and blood vessels. The region of the pulp cavity that runs through the root of the tooth is called the root canal. Surrounding the pulp cavity is **dentin**, a bone-like tissue. In the root of each tooth, the dentin is covered by an even harder bone-like layer called **cementum**. In the crown of each tooth, the dentin is covered by an outer layer of **enamel**, the hardest substance in the body.

Although enamel protects the underlying dentin and pulp cavity, it is still nonetheless susceptible to mechanical and chemical erosion, or what is known as tooth decay. The most common form, **dental caries (cavities)** develops when colonies of bacteria feeding on sugars in the mouth release acids that cause soft tissue inflammation and degradation of the calcium crystals of the enamel.

PERIONDONTAL DISEASES / GUM DISEASES

The word periodontal means "around the tooth." Periodontal diseases, also called gum diseases, are serious bacterial infections that attack the gums and the surrounding tissues. If it's left untreated, the disease will continue and the underlying bone around the teeth will dissolve and will no longer be able to hold the teeth in place.

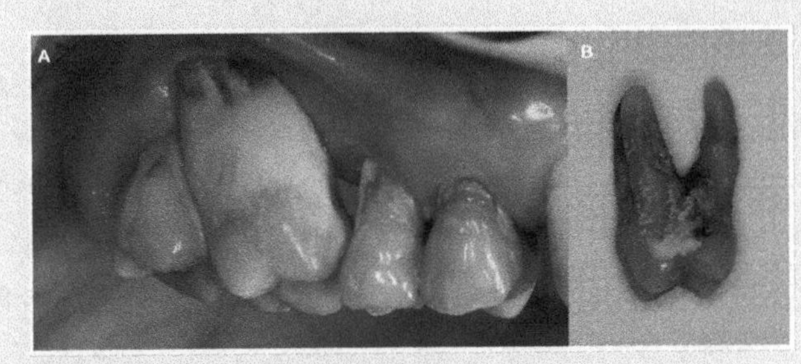

Figure 4:: Advanced periodontal disease

Gingivitis. This is the mildest form of periodontal disease. The gums are likely to become red, swollen, and tender. They may bleed easily during daily cleanings and flossing.

Moderate to advanced periodontitis. This most advanced stage of gum disease shows significant bone loss, deepening of periodontal pockets, and possibly receding gums surrounding the teeth. Teeth may loosen and need to be extracted.

The Salivary Glands

Many small **salivary glands** are housed within the mucous membranes of the mouth and tongue. These minor exocrine glands are constantly secreting **saliva,** either directly into the oral cavity or indirectly through ducts, even while you sleep.

The Major Salivary Glands

- The **submandibular glands**, which are in the floor of the mouth, secrete saliva into the mouth through the submandibular ducts.
- The **sublingual glands**, which lie below the tongue, use the lesser sublingual ducts to secrete saliva into the oral cavity.
- The **parotid glands** lie between the skin and the masseter muscle, near the ears

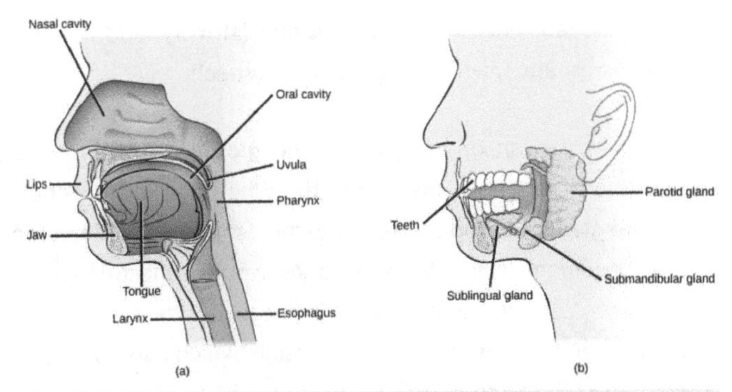

Figure 5:. Digestion of food begins in the (a) oral cavity. Food is masticated by teeth and moistened by saliva secreted from the (b) salivary glands. Enzymes in the saliva begin to digest starches and fats. With the help of the tongue, the resulting bolus is moved into the esophagus by swallowing. (credit: modification of work by the National Cancer Institute)

Saliva: The extensive chemical process of digestion begins in the mouth. As food is being chewed, saliva, produced by the salivary glands, mixes with the food. Saliva is a watery substance produced in the mouths of many animals. Saliva contains mucus that moistens food and buffers the pH of the food. Saliva also contains lysozymes, which have antibacterial action to reduce tooth decay by inhibiting growth of some bacteria. Saliva also contains an enzyme called **salivary amylase** that begins the process of converting starches in the food into a disaccharide called maltose. Another enzyme called **lipase** is produced by the cells in the tongue. Lipases are a class of enzymes that can break down triglycerides- fats. The chewing and wetting action provided by the teeth and saliva prepare the food into a mass called the **bolus** for swallowing. The tongue helps in swallowing—moving the bolus from the mouth into the pharynx.

HOMEOSTATIC IMBALANCES : MUMPS

The Parotid Glands: Mumps

Infections of the nasal passages and pharynx can attack any salivary gland. The parotid glands are the usual site of infection with the virus that causes mumps (paramyxovirus). Mumps manifests by enlargement and inflammation of the parotid glands, causing a characteristic swelling between the ears and the jaw. Symptoms include fever and throat pain, which can be severe when swallowing acidic substances such as orange juice.

In about one-third of men who are past puberty, mumps also causes testicular inflammation, typically affecting only one testis and rarely resulting in sterility. With the increasing use and effectiveness of mumps vaccines, the incidence of mumps has decreased dramatically.

The pharynx:

The **pharynx** (throat) is involved in both digestion and respiration. It receives food and air from the mouth, and air from the nasal cavities. The pharynx located in throat, connects mouth to the esophagus. Pharynx has three subdivisions. The most superior, the nasopharynx, is involved only in breathing and speech. The other two subdivisions, the **oropharynx** and the **laryngopharynx**, are used for both breathing and digestion.

Pharynx opens to two passageways:

- **the trachea,** which leads to the **lungs, (air way)** and
- **the esophagus,** which leads to the **stomach.**

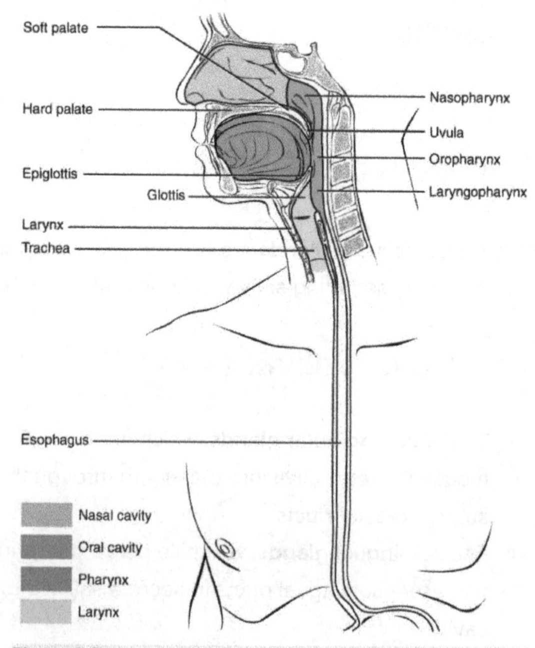

Figure 6: The pharynx, which runs from the nostrils to the esophagus and the larynx.

The trachea has an opening called the **glottis**, which is covered by a cartilaginous flap called the **epiglottis.** When swallowing, the epiglottis closes the glottis and food passes into the esophagus and not the trachea. This arrangement allows food to be kept out of the trachea. When the food "goes down the wrong way," it goes into the trachea. When food enters the trachea, the reaction is to cough, which usually forces the food up and out of the trachea, and back into the pharynx.

Esophagus

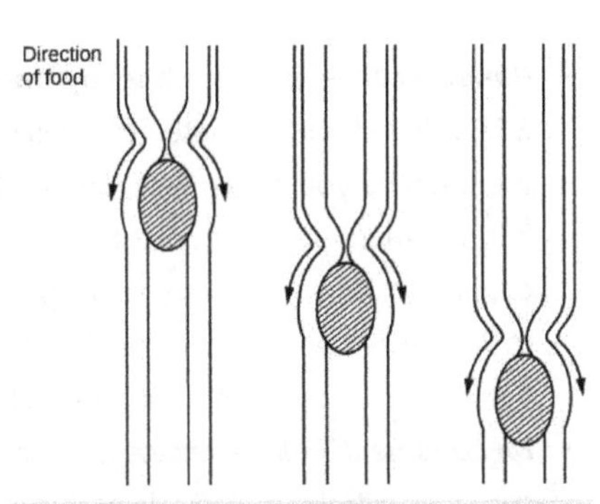

Figure 7. The esophagus transfers food from the mouth to the stomach through peristaltic movements.

The **esophagus** is a tubular organ that connects the mouth to the stomach. The chewed and softened food passes through the esophagus after being swallowed. The smooth muscles of the esophagus undergo a series of wave like movements called **peristalsis** that push the food toward the stomach. The peristalsis wave is unidirectional—it moves food from the mouth to the stomach.

A ring-like muscle called a **sphincter** forms valves in the digestive system. The gastro-esophageal sphincter is located at the stomach end of the esophagus. In response to swallowing and the pressure exerted by the bolus of food, this sphincter opens, and the bolus enters the stomach. When there is no swallowing action, this sphincter is shut and prevents the contents of the stomach from traveling up the esophagus. When the lower esophageal sphincter does not completely close, the stomach's contents can reflux (that is, back up into the esophagus), causing **heartburn or gastroesophageal reflux disease (GERD)**..

The Stomach

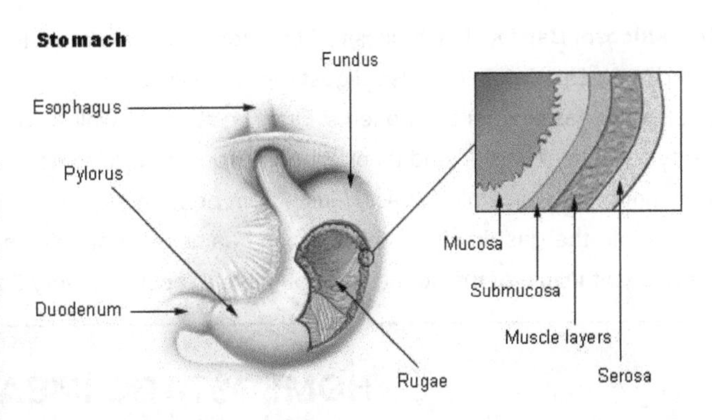

Figure 8: Layers of stomach lining

The stomach is a J-shaped saclike organ in the upper abdomen. The stomach is much like a bag which **store food and secretes gastric digestive juices.** The wall of the stomach is made of the same four layers as most of the rest of the alimentary canal. In addition to the typical circular and longitudinal smooth muscle layers, the muscularis has an inner oblique smooth muscle layer. As a result, in addition to moving food through the canal, the stomach can vigorously churn food, mechanically breaking it down into smaller particles. In the absence of food, the stomach deflates inward, and its mucosa and submucosa fall into a large fold called a **ruga.**

The **pH** in the stomach is between **1.5 and 2.5.** This highly acidic environment is required for the chemical breakdown of food and the extraction of nutrients.A large part of digestion occurs in the stomach.

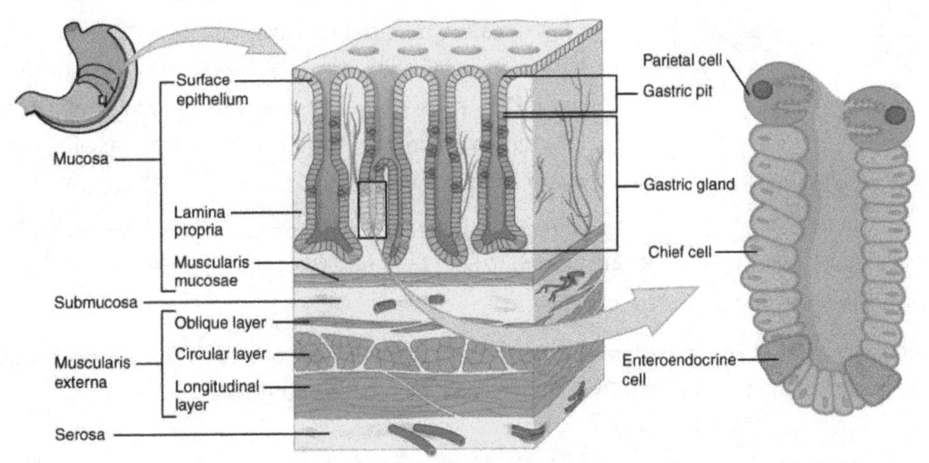

Figure 9: The stomach wall showing wall layers, gastric pit and gastric glands.(one gland is shown enlarged on the right)

The stomach wall is adapted for the functions of the stomach. In the epithelium, **gastric pits** lead to **gastric glands** that secrete **gastric juice**. The gastric glands contain different types of cells that secrete a variety of enzymes, including hydrochloride acid (HCl), which activates the protein-digesting enzyme pepsin.

- ***Parietal cells***— produce both **hydrochloric acid (HCl)** and **intrinsic factor.** HCl is responsible for the high acidity (pH 1.5 to 3.5) of the stomach contents and is needed to activate the protein-digesting enzyme, pepsin. The acidity also kills much of the bacteria you ingest with food and helps to denature proteins, making them more available for enzymatic digestion. Intrinsic factor is a glycoprotein necessary for the absorption of vitamin B_{12} in the small intestine. The parietal cells are capable of creating a very acidic pH of 2.0 within the stomach.
- ***Chief cells***—Located primarily in the basal regions of gastric glands are **chief cells**, which secrete **pepsinogen**, the inactive proenzyme form of pepsin. HCl is necessary for the conversion of pepsinogen to pepsin.
- ***Mucous neck cells***—secrete thin, acidic mucus that is much different from the mucus secreted by the goblet cells of the surface epithelium. The role of this mucus is not currently known.
- ***Enteroendocrine cells***— secrete various hormones – include gastrin,

The combination of all of the stomach secretions is known as **gastric juice.** Food enters the stomach and combines with gastric juice to form a pasty substance called **chyme**. Chyme then leaves the stomach by way of the pyloric sphincter and enters the duodenum.

The Mucosal Barrier: The mucosa of the stomach is exposed to the highly corrosive acidity of gastric juice. Gastric enzymes that can digest protein can also digest the stomach itself. The stomach is protected from self-digestion by the **mucosal barrier**. This barrier has several components. First, the stomach wall is covered by a thick coating of bicarbonate-rich mucus. This mucus forms a physical barrier, and its bicarbonate ions neutralize acid. Second, the epithelial cells of the stomach's mucosa meet at tight junctions, which block gastric juice from penetrating the underlying tissue layers. Finally, stem cells located where gastric glands join the gastric pits quickly replace damaged epithelial mucosal cells, when the epithelial cells are shed. In fact, the surface epithelium of the stomach is completely replaced every 3 to 6 days.

HOMEOSTATIC IMBALANCES: ULCERS

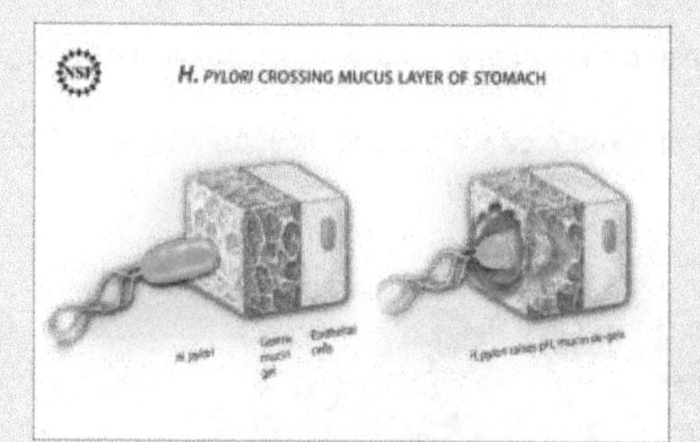

Figure 10: Ulcer-causing Bacterium (H.Pylori) Crossing Mucus Layer of Stomach

As effective as the mucosal barrier is, it is not a "fail-safe" mechanism. Sometimes, gastric juice eats away at the superficial lining of the stomach mucosa, creating erosions, which mostly heal on their own. Deeper and larger erosions are called ulcers.

Why does the mucosal barrier break down? A number of factors can interfere with its ability to protect the stomach lining. The majority of all ulcers are caused by either excessive intake of non-steroidal anti-inflammatory drugs (NSAIDs), including aspirin, or ***Helicobacter pylori*** infection.

Antacids help relieve symptoms of ulcers such as "burning" pain and indigestion. When ulcers are caused by NSAID use, switching to other classes of pain relievers allows healing. When caused by *H. pylori* infection, antibiotics are effective.

A potential complication of ulcers is perforation: Perforated ulcers create a hole in the stomach wall, resulting in peritonitis (inflammation of the peritoneum). These ulcers must be repaired surgically.

Small Intestine

The **small intestine** is a long muscular tube of about 20 feet long and about an inch in diameter. Chyme moves from the stomach to the small intestine. The **small intestine** is the organ where the digestion of protein, fats, and carbohydrates is completed and absorb most of the nutrients from what we eat and drink.

Small Intestine – Structure

The coiled tube of the small intestine is subdivided into three regions. From proximal (at the stomach) to distal, these are the **Duodenum, Jejunum, and Ileum.**

- The shortest region is the 25.4-cm (10-in) **duodenum**, which begins at the pyloric sphincter and makes a C-shaped curve and join the jejunum.. Digestive juices from the pancreas, liver, and gallbladder enter the duodenum. **Bile** is produced in the liver and stored and concentrated in the gallbladder. Bile contains bile salts which emulsify lipids while the pancreas produces enzymes that catabolize starches, disaccharides, proteins, and fats. These digestive juices break down the food particles in the chyme into glucose, triglycerides, and amino acids. Some chemical digestion of food takes place in the duodenum. Absorption of fatty acids also takes place in the duodenum.

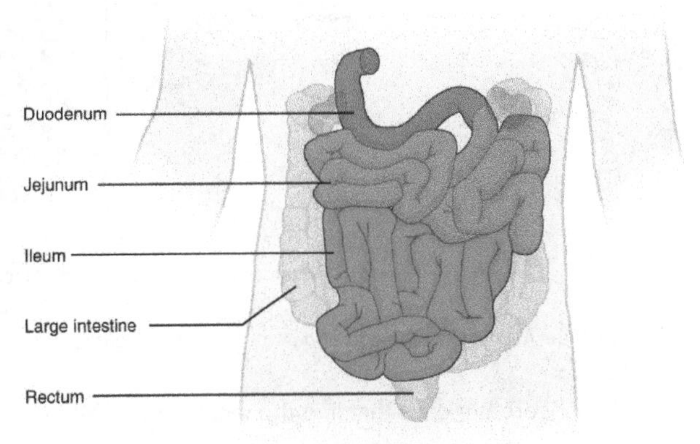

Figure 11. The three regions of the small intestine are the duodenum, jejunum, and ileum

- The **jejunum** is about 0.9 meters (3 feet) long (in life) and runs from the duodenum to the ileum. Here, hydrolysis of nutrients is continued while most of the carbohydrates and amino acids are absorbed through the intestinal lining. The bulk of chemical digestion and nutrient absorption occurs in the jejunum.
- The **ileum** is the longest part of the small intestine, measuring about 1.8 meters (6 feet) in length. The ileum joins the cecum, the first portion of the large intestine, at the **ileocecal sphincter** (or valve). The undigested food is sent to the colon from the ileum via peristaltic movements of the muscle. The vermiform, "worm-like," **appendix** is located at the ileocecal valve. The appendix of humans secretes no enzymes and has an insignificant role in immunity.

Histology

The wall of the small intestine is composed of the same four layers typically present in the alimentary system. The small intestine has a highly folded inner surface containing finger-like projections called the **villi**. The apical surface of each villus has many microscopic projections called **microvilli.** Although their small size makes it difficult to see each microvillus, their combined microscopic appearance suggests a mass of bristles, which is termed the **brush border.** These structures are lined with epithelial cells on the luminal side and allow for the nutrients to be absorbed from the digested food and absorbed into the blood stream on the other side. **The villi and microvilli, with their many folds, increase the surface area of the intestine and increase absorption efficiency of the nutrients.** Absorbed nutrients in the blood are carried into the blood in hepatic portal vein, which leads to the liver. There, the liver regulates the distribution of nutrients to the rest of the body and removes toxic substances, including drugs, alcohol, and some pathogens.

ART CONNECTION

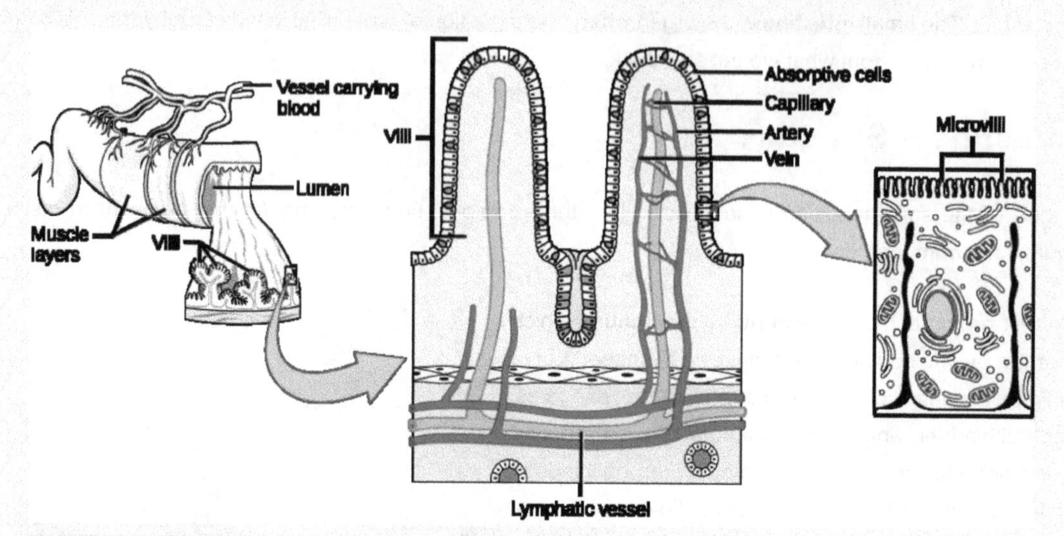

Figure 12. Villi are folds on the small intestine lining that increase the surface area to facilitate the absorption of nutrients.

Which of the following statements about the small intestine is false?

1. Absorptive cells that line the small intestine have microvilli, small projections that increase surface area and aid in the absorption of food.
2. The inside of the small intestine has many folds, called villi.
3. Microvilli are lined with blood vessels as well as lymphatic vessels.
4. The inside of the small intestine is called the lumen.

Large Intestine

The **large intestine**, illustrated in Figure 13, reabsorbs the water from the undigested food material and processes the waste material. The human large intestine is much smaller in length compared to the small intestine but larger in diameter. **It has three parts: the cecum, the colon, and the rectum.** The cecum joins the ileum to the colon and is the receiving pouch for the waste matter. The colon can be divided into four regions, **the ascending colon, the transverse colon, the descending colon and the sigmoid colon**. The main functions of the colon are to extract the water and mineral salts from undigested food, and to store waste material. The colon is home to many **bacteria or "intestinal flora"** that aid in the digestive processes.

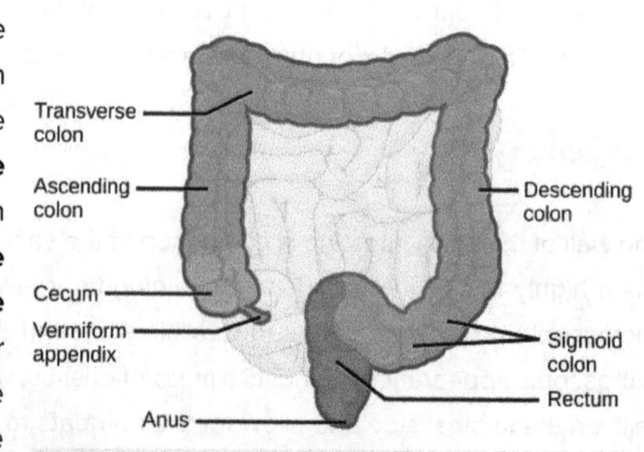

Figure 13. Structure of the large intestine

Bacterial Flora

Most bacteria that enter the alimentary canal are killed by lysozyme, defensins, HCl, or protein-digesting enzymes. However, trillions of bacteria live within the large intestine and are referred to as the **bacterial flora**. Most of the more than 700 species of these bacteria are nonpathogenic commensal organisms that cause no harm as long as they stay in the gut lumen. In fact, many facilitate chemical digestion and absorption, and some synthesize certain vitamins, mainly biotin, pantothenic acid, and vitamin K. **Fecal transplantation (or bacteriotherapy)** is the transfer of stool from a healthy donor into the gastrointestinal tract for the purpose of treating recurrent *C. difficile colitis*. When antibiotics kill off too many "good" bacteria in the digestive tract, fecal transplants can help replenish bacterial balance.

HOMEOSTATIC IMBALANCES: COLORECTAL CANCER

Each year, approximately 140,000 Americans are diagnosed with colorectal cancer, and another 49,000 die from it, making it one of the most deadly malignancies. People with a family history of colorectal cancer are at increased risk. Smoking, excessive alcohol consumption, and a diet high in animal fat and protein also increase the risk. Despite popular opinion to the contrary, studies support the conclusion that dietary fiber and calcium do not reduce the risk of colorectal cancer.

Colorectal cancer may be signaled by constipation or diarrhea, cramping, abdominal pain, and rectal bleeding. Bleeding from the rectum may be either obvious or occult (hidden in feces). Since most colon cancers arise from benign mucosal growths called polyps, cancer prevention is focused on identifying these polyps. The colonoscopy is both diagnostic and therapeutic. Colonoscopy not only allows identification of precancerous polyps, the procedure also enables them to be removed before they become malignant. Screening for fecal occult blood tests and colonoscopy is recommended for those over 50 years of age.

Rectum and Anus

The **rectum** is the terminal end of the large intestine, as shown in Figure 12. The primary role of the rectum is to store the feces until defecation. The feces are propelled using peristaltic movements during elimination. The **anus** is an opening at the far-end of the digestive tract and is the exit point for the waste material.

The small intestine absorbs about 90 percent of the water you ingest (either as liquid or within solid food). The large intestine absorbs most of the remaining water, a process that converts the liquid chyme residue into semisolid **feces** ("stool"). Feces is composed of undigested food residues, unabsorbed digested substances, millions of bacteria, old epithelial cells from the GI mucosa, inorganic salts, and enough water to let it pass smoothly out of the body.

Feces are eliminated through contractions of the rectal muscles. You help this process by a voluntary procedure called **Valsalva's maneuver**, in which you increase intra-abdominal pressure by contracting your diaphragm and abdominal wall muscles, and closing your glottis.

IRRITABLE BOWEL SYNDROME – IBS

Irritable bowel syndrome (IBS) is characterized by abdominal discomfort associated with altered bowel movements and change in your bowel habits. IBS cause abdominal discomfort in different ways, such as sharp pain, cramping, bloating, distention, fullness or even burning. IBS may be triggered by eating specific foods, following a meal, emotional stress, constipation or diarrhea. Many symptoms are related to **hypersensitivity of the nerves** found in the wall of the gastrointestinal tract.

Accessory Organs

Each **accessory digestive organ** aids in the breakdown of food. Within the mouth, the teeth and tongue begin mechanical digestion. Other accessory organs add secretions (enzymes) that catabolize food into nutrients.

Accessory organs include **salivary glands, the liver, the pancreas, and the gallbladder.**

1. Salivary glands: There are three major salivary glands that secrete saliva into the oral cavity —the parotid, the submandibular, and the sublingual. Saliva contains mucus that moistens food and buffers the pH of the food.

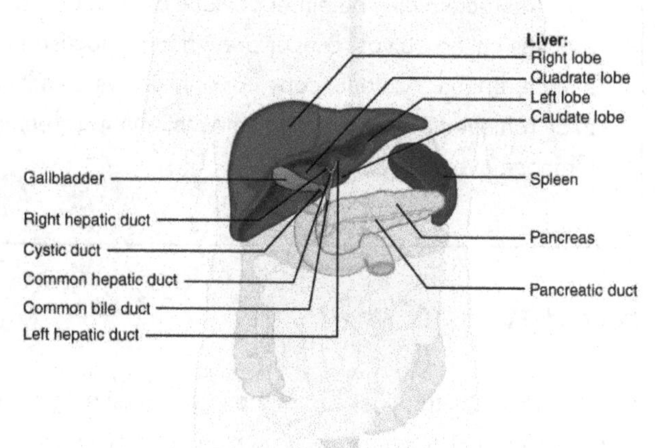

Figure:14 Accessory organs of digestive system

2. The Liver: The liver is the largest gland in the body, weighing about three pounds in an adult. It is the largest internal organ in humans and it plays a very important role in digestion of fats and detoxifying blood. The liver lies inferior to the diaphragm in the right upper quadrant of the abdominal cavity and receives protection from the surrounding ribs. The liver is divided into two primary lobes: a large right lobe and a much smaller left lobe. The lobes are further divided into lobules by blood vessels and connective tissue. The liver performs many functions and is considered a vital organ.

Liver functions include:

Detoxifying the blood, producing **bile,** (a digestive juice that is required for the breakdown of fatty components of the food in the duodenum.), metabolism of carbohydrates, fats and proteins, storing iron, blood and vitamins, recycling red blood cells and producing plasma proteins.

DISORDERS OF THE LIVER: CIRRHOSIS

Cirrhosis is when scar tissue replaces healthy liver tissue. This stops the liver from working normally.Cirrhosis is a long-term (chronic) liver disease. The damage to your liver builds up over time.The most common causes of cirrhosis are:

- **Hepatitis:** inflammation of the liver. Liver inflammation can be caused by several viruses (viral hepatitis), chemicals, drugs, alcohol, certain genetic disorders or by an overactive immune system that mistakenly attacks the liver, called autoimmune hepatitis. Depending on its course, hepatitis can be acute, which flares up suddenly and then goes away, or chronic, which is a long-term condition usually producing more subtle symptoms and progressive liver damage.

- Nonalcoholic fatty liver disease (this happens from metabolic syndrome and is caused by conditions such as obesity, high cholesterol and triglycerides, and high blood pressure)
- What is Hepatitis C and Why Should You Care?
    - https://www.youtube.com/watch?v=IxCelFhuhQo

Bile : composition and function.

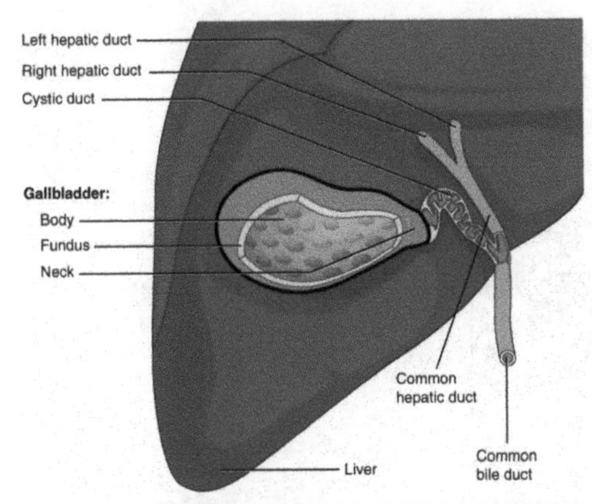

Figure 15. The gallbladder stores and concentrates bile, and releases it into the two-way cystic duct when it is needed by the small intestine.

Bile is a mixture secreted by the liver to accomplish the emulsification of lipids in the small intestine. Recall that lipids are hydrophobic, that is, they do not dissolve in water. Thus, before they can be digested in the watery environment of the small intestine, large lipid globules must be broken down into smaller lipid globules, a process called emulsification.

Hepatocytes (cells of liver) secrete about one liter of bile each day. A yellow-brown or yellow-green alkaline solution (pH 7.6 to 8.6), bile is a mixture of water, bile salts, bile pigments, phospholipids (such as lecithin), electrolytes, cholesterol, and triglycerides. The components most critical to emulsification are bile salts and phospholipids, which have a nonpolar (hydrophobic) region as well as a polar (hydrophilic) region. The hydrophobic region interacts with the large lipid molecules, whereas the hydrophilic region interacts with the watery chyme in the intestine. This results in the large lipid globules being pulled apart into many tiny lipid fragments of about 1 μm in diameter. This change dramatically increases the surface area available for lipid-digesting enzyme activity. This is the same way dish soap works on fats mixed with water. Bile salts act as emulsifying agents, so they are also important for the absorption of digested lipids.

Bilirubin, the main bile pigment, is a waste product produced when the spleen removes old or damaged red blood cells from the circulation. Bilirubin is eventually transformed by intestinal bacteria into stercobilin, a brown pigment that gives your stool its characteristic color!

Gallbladder stores and releases bile: The **gallbladder** is 8–10 cm (~3–4 in) long and is nested in the right lobe of the liver. The liver secretes **bile** which is stored in the **gallbladder.** This muscular sac stores, concentrates, and, when stimulated, propels the bile into the duodenum via the common bile duct. When chyme containing fatty acids enters the duodenum, the bile is secreted from the gallbladder into the duodenum.

3. Pancreas

The **pancreas** is another important gland that secretes digestive juices. The pancreas produces over a liter of pancreatic juice each day. Unlike bile, it is clear and composed mostly of water along with some salts, sodium bicarbonate, and several digestive enzymes.

- Sodium bicarbonate is responsible for the slight alkalinity of pancreatic juice (pH 7.1 to 8.2), which serves to buffer the acidic gastric juice in chyme, inactivate pepsin from the stomach, and create an optimal environment for the activity of pH-sensitive digestive enzymes in the small intestine.

- Pancreatic enzymes are active in the digestion of sugars, proteins, and fats.
    - Pancreatic amylase digesting starch
    - Pancreatic lipase digesting fat
    - Pancreatic trypsin digesting proteins
    - Pancreatic nuclease digesting nucleic acids

Pancreas also perform endocrine functions releasing hormones into the blood.

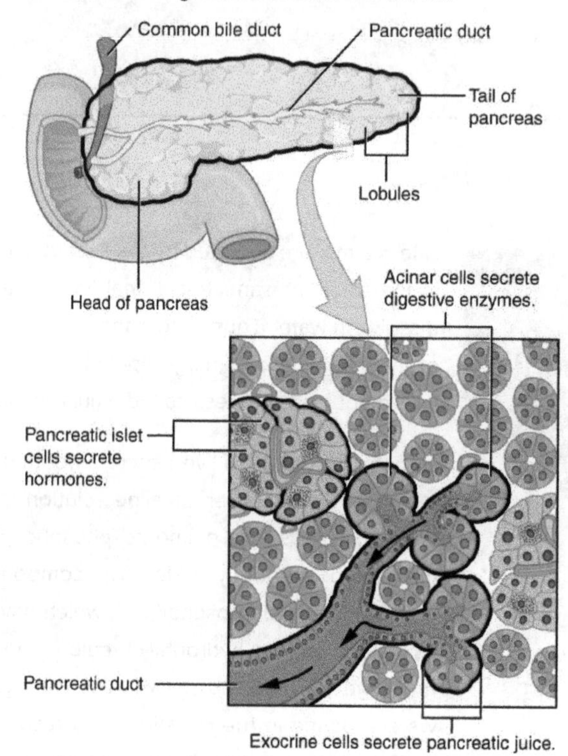

Figure 16. The pancreas has a head, a body, and a tail. It delivers pancreatic juice to the duodenum through the pancreatic duct.

Table 1. Functions of the Digestive Organs (summary)

Organ	Major functions	Other functions
Mouth	• Ingests food • Chews and mixes food • Begins chemical breakdown of carbohydrates • Moves food into the pharynx • Begins breakdown of lipids via lingual lipase	• Moistens and dissolves food, allowing you to taste it • Cleans and lubricates the teeth and oral cavity • Has some antimicrobial activity
Pharynx	• Propels food from the oral cavity to the esophagus	• Lubricates food and passageways
Esophagus	• Propels food to the stomach	• Lubricates food and passageways

Organ	Major functions	Other functions
Stomach	• Mixes and churns food with gastric juices to form chyme • Begins chemical breakdown of proteins • Releases food into the duodenum as chyme • Absorbs some fat-soluble substances (for example, alcohol, aspirin) • Possesses antimicrobial functions	• Stimulates protein-digesting enzymes • Secretes intrinsic factor required for vitamin B_{12}absorption in small intestine
Small intestine	• Mixes chyme with digestive juices • Propels food at a rate slow enough for digestion and absorption • Absorbs breakdown products of carbohydrates, proteins, lipids, and nucleic acids, along with vitamins, minerals, and water • Performs physical digestion via segmentation	• Provides optimal medium for enzymatic activity
Accessory organs	• Liver: produces bile salts, which emulsify lipids, aiding their digestion and absorption • Gallbladder: stores, concentrates, and releases bile • Pancreas: produces digestive enzymes and bicarbonate	• Bicarbonate-rich pancreatic juices help neutralize acidic chyme and provide optimal environment for enzymatic activity
Large intestine	• Further breaks down food residues • Absorbs most residual water, electrolytes, and vitamins produced by enteric bacteria • Propels feces toward rectum • Eliminates feces	• Food residue is concentrated and temporarily stored prior to defecation • Mucus eases passage of feces through colon
Rectum	• receive stool from the colon, • to let the person know that there is stool to be evacuated, and • to hold the stool until evacuation happens.	
Anus	• the opening at the end of the alimentary canal through which solid waste matter leaves the body.	

Processes of Digestion and Absorption

Food is the body's source of fuel. The nutrients in food give the body's cells the energy they need to operate. Before food can be used it has to be mechanically broken down into tiny pieces, then chemically broken down so nutrients can be absorbed. In humans, proteins need to be broken down into amino acids, starches into sugars, and fats into fatty acids and glycerol. This **mechanical and chemical breakdown** encompasses the process of digestion.

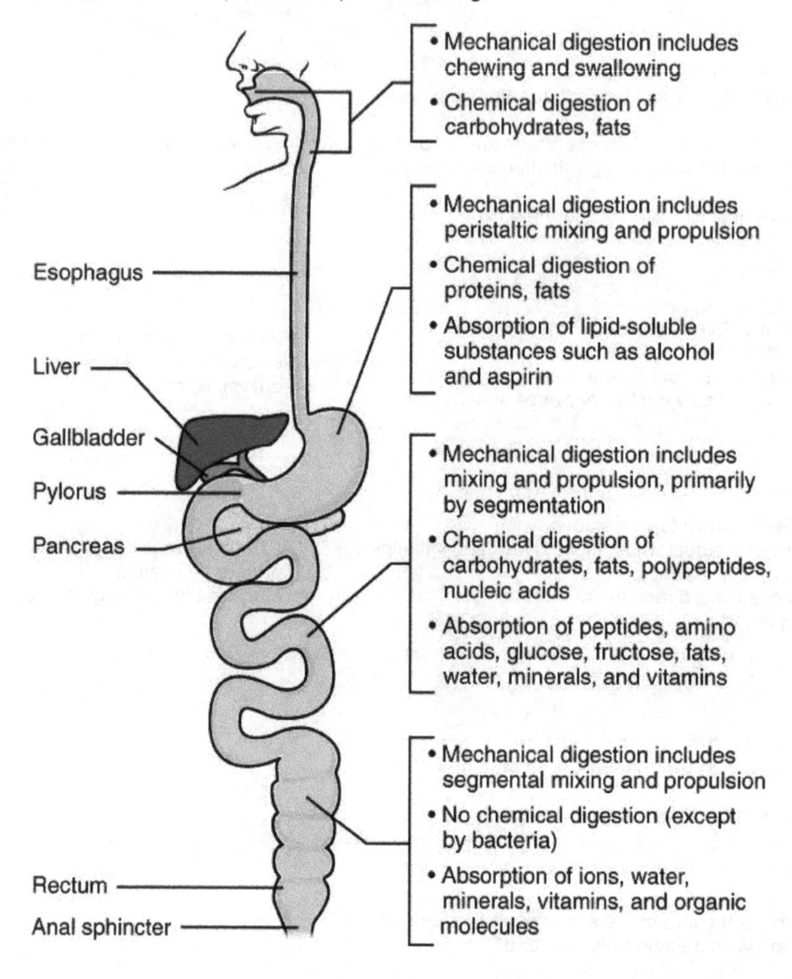

Figure 1. Digestion begins in the mouth and continues as food travels through the small intestine. Most absorption occurs in the small intestine.

Mechanical digestion

Mechanical digestion is a purely physical process that does not change the chemical nature of the food. Instead, it makes the food smaller to increase both surface area and mobility. It includes **mastication, or chewing,** as well as tongue movements that help break food into smaller bits and mix food with saliva.

Although there may be a tendency to think that mechanical digestion is limited to the first steps of the digestive process, it occurs after the food leaves the mouth, as well in the stomach and intestine by peristalsis and segmentation. **Peristalsis** is a series of wave-like muscle contractions that moves food to different processing stations in the digestive tract. Peristalsis in the esophagus propel the food bolus down to the stomach. When food enters the stomach, a highly muscular organ, powerful peristaltic contractions help mash, pulverize, and churn food into chyme. **Chyme** is a semiliquid mass of partially digested food that also contains gastric juices secreted by cells in the stomach. The mechanical **churning of food in the stomach** serves to further break it apart and expose more of its surface area to digestive juices. **Vomiting** is a reverse peristaltic action of the stomach. The vomiting center in the brain (medulla oblongata) is sensitive to stimuli such as toxins and rapid body movements. The vomiting center also receives cortical input so certain thoughts can cause vomiting. **Segmentation** occurs mainly in the small intestine, it is caused by localized contractions of circular muscle of the muscularis layer of the alimentary canal. These contractions isolate small sections of the intestine, moving their contents back and forth while continuously subdividing, breaking up, and mixing the contents. By moving food back and forth in the intestinal lumen, segmentation mixes food with digestive juices and facilitates absorption.

Chemical Digestion

Large food molecules (for example, proteins, lipids, nucleic acids, and starches) must be broken down into subunits that are small enough to be absorbed by the lining of the alimentary canal. This is accomplished by **enzymes** through **hydrolysis.** Chemical digestion breaks large food molecules down into their chemical building blocks, which can then be absorbed through the intestinal wall and into the general circulation

Carbohydrate Digestion

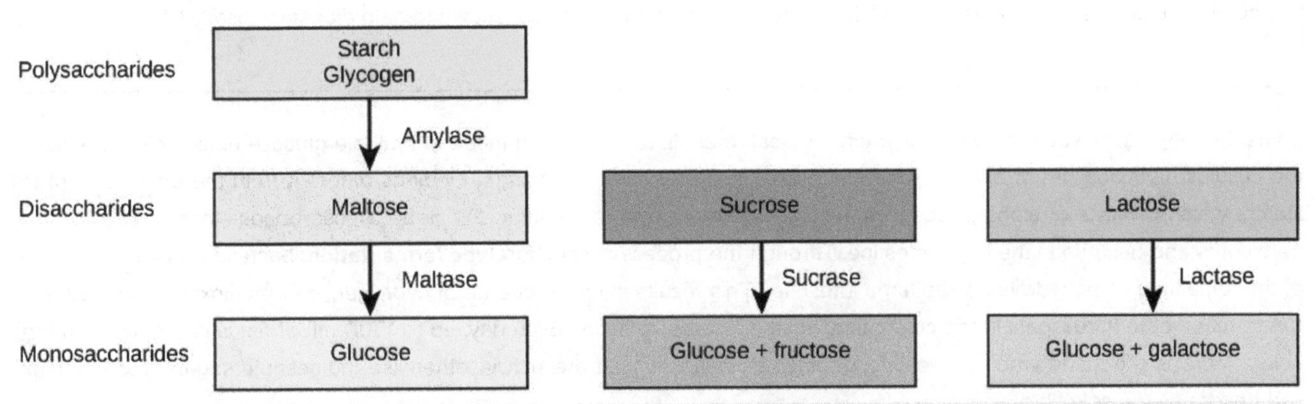

Figure 2: The steps in carbohydrate digestion Digestion of carbohydrates is performed by several enzymes. Starch and glycogen are broken down into glucose by amylase and maltase. Sucrose (table sugar) and lactose (milk sugar) are broken down by sucrase and lactase, respectively.

The average American diet is about 50 percent carbohydrates, which may be classified according to the number of monomers they contain of simple sugars (monosaccharides and disaccharides) and/or complex sugars (polysaccharides). Glucose, galactose, and fructose are the three monosaccharides that are commonly consumed and are readily absorbed. Your digestive system is also able to break down the disaccharide sucrose (regular table sugar: glucose + fructose), lactose (milk sugar: glucose + galactose), and maltose (grain sugar: glucose + glucose), and the polysaccharides glycogen and starch (chains of monosaccharides). Your bodies do not produce enzymes that can break down most fibrous polysaccharides, such as cellulose. While indigestible polysaccharides do not provide any nutritional value, they do provide **dietary fiber, which helps propel food through the alimentary canal.**

The chemical digestion of starches begins in the mouth. In the small intestine, **pancreatic amylase** does the 'heavy lifting' for starch and carbohydrate digestion (Figure 2). Three brush border enzymes hydrolyze sucrose, lactose, and maltose into monosaccharides. **Sucrase** splits sucrose into one molecule of fructose and one molecule of glucose; **maltase** breaks down maltose and maltotriose into two and three glucose molecules, respectively; and **lactase** breaks down lactose into one molecule of glucose and one molecule of galactose. Insufficient lactase can lead to **lactose intolerance.**

DISORDERS OF THE SMALL INTESTINE: LACTOSE INTOLERANCE

Lactose intolerance is a condition characterized by indigestion caused by dairy products. It occurs when the absorptive cells of the small intestine do not produce enough lactase, the enzyme that digests the milk sugar lactose. In most mammals, lactose intolerance increases with age. In contrast, some human populations, most notably Caucasians, are able to maintain the ability to produce lactase as adults.

In people with lactose intolerance, the lactose in chyme is not digested. Bacteria in the large intestine ferment the undigested lactose, a process that produces gas. In addition to gas, symptoms include abdominal cramps, bloating, and diarrhea. Symptom severity ranges from mild discomfort to severe pain; however, symptoms resolve once the lactose is eliminated in feces.

The hydrogen breath test is used to help diagnose lactose intolerance. Lactose-tolerant people have very little hydrogen in their breath. Those with lactose intolerance exhale hydrogen, which is one of the gases produced by the bacterial fermentation of lactose in the colon. After the hydrogen is absorbed from the intestine, it is transported through blood vessels into the lungs. There are a number of lactose-free dairy products available in grocery stores. In addition, dietary supplements are available. Taken with food, they provide lactase to help digest lactose.

Some carbohydrates, such as **cellulose**, are not digested at all, despite being made of multiple glucose units. This is because the cellulose is made out of beta-glucose that makes the inter-monosaccharidal bindings different from the ones present in starch, which consists of alpha-glucose. Humans lack the enzyme for splitting the beta-glucose-bonds—that is reserved for herbivores and bacteria in the large intestine. Through the process of **saccharolytic fermentation**, bacteria break down some of the remaining carbohydrates in the large intestine . This results in the discharge of hydrogen, carbon dioxide, and methane gases that create **flatus** (gas) in the colon; flatulence is excessive flatus. Each day, up to 1500 mL of flatus is produced in the colon. More is produced when you eat foods such as beans, which are rich in otherwise indigestible sugars and complex carbohydrates like soluble dietary fiber.

Protein Digestion

Proteins are polymers composed of amino acids linked by peptide bonds to form long chains. Digestion reduces them to their constituent amino acids. The digestion of protein starts in the stomach, where **HCl and pepsin** break proteins into smaller polypeptides, which then travel to the small intestine. Chemical digestion in the small intestine is continued by pancreatic enzymes, including **chymotrypsin and trypsin,** each of which act on specific bonds in amino acid sequences.. This results in molecules small enough to enter the bloodstream.

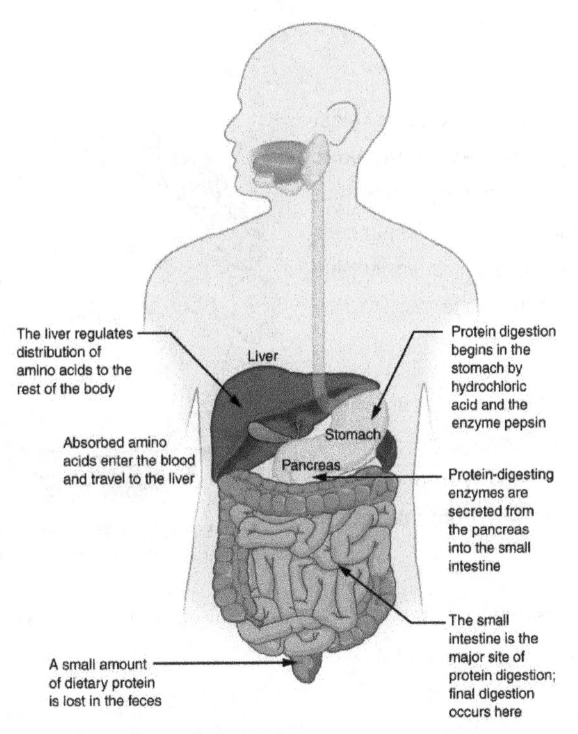

Figure 3. The digestion of protein begins in the stomach and is completed in the small intestine.

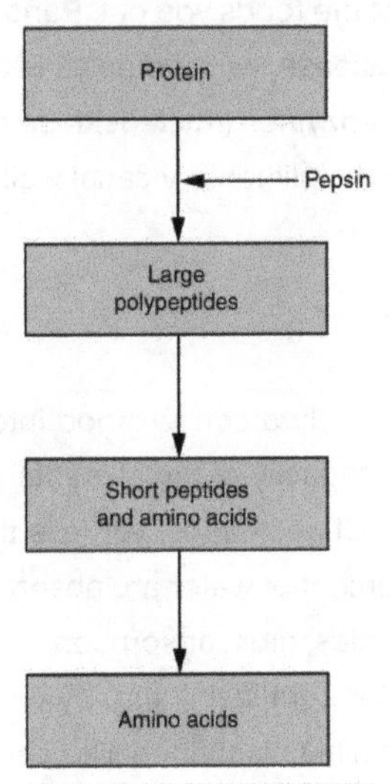

Figure 4. Proteins are successively broken down into their amino acid components.

Lipid Digestion

Lipids (fats) are degraded into fatty acids and glycerol.Most fat-digesting enzymes produced by **pancreas . Pancreatic lipase** breaks down triglycerides into free fatty acids and monoglycerides. The fatty acids include both short-chain (less than 10 to 12 carbons) and long-chain fatty acids. Pancreatic lipase works with the help of the salts from bile secreted by the liver and the gallbladder.

Bile salts attach to triglycerides and help to emulsify them; this aids access by pancreatic lipase because the lipase is water-soluble, but the fatty triglycerides are hydrophobic and tend to orient toward each other and away from the watery intestinal surroundings.

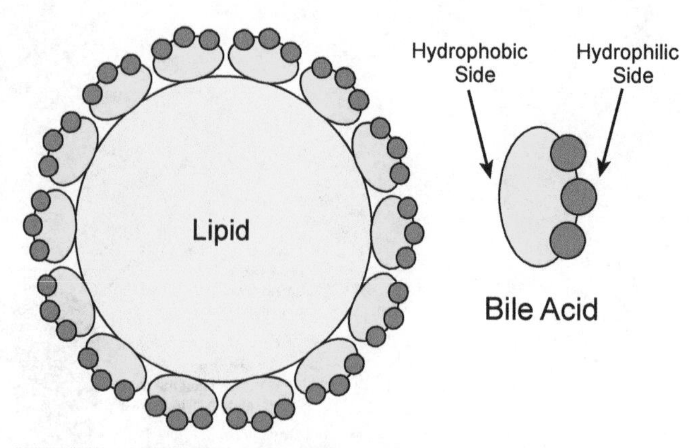

Figure 5: Bile salt action on lipids: Bile salts congregate around fat and separate them into small droplets called micelles.

The bile salts act to hold the triglycerides in their watery surroundings until the lipase can break them into the smaller components that are able to enter the villi for absorption.

Nucleic Acid Digestion

The nucleic acids DNA and RNA are found in most of the foods you eat. **Pancreatic nuclease** are responsible for their digestion: **deoxyribonuclease**, which digests DNA, and **ribonuclease**, which digests RNA. The nucleotides produced by this digestion are further broken down by two intestinal brush border enzymes (**nucleosidase** and **phosphatase**) into pentoses, phosphates, and nitrogenous bases, which can be absorbed through the alimentary canal wall.

Absorption

The mechanical and digestive processes have one goal: to convert food into molecules small enough to be **absorbed by the epithelial cells of the intestinal villi.** The absorptive capacity of the alimentary canal is almost endless. Each day, the alimentary canal processes up to 10 liters of food, liquids, and GI secretions, yet less than one liter enters the large intestine. Almost all ingested food, 80 percent of electrolytes, and 90 percent of water are absorbed in the small intestine. Although the entire small intestine is involved in the absorption of water and lipids, most absorption of carbohydrates and proteins occurs in the jejunum. Notably, bile salts and vitamin B_{12} are absorbed in the terminal ileum. By the time chyme passes from the ileum into the large intestine, it is essentially indigestible food residue (mainly plant fibers like cellulose), some water, and millions of bacteria.

Mineral Absorption

In general, all minerals that enter the intestine are absorbed, whether you need them or not. Iron and calcium are exceptions; they are absorbed in the duodenum in amounts that meet the body's current requirements. ,

Vitamin Absorption

The small intestine absorbs the vitamins that occur naturally in food and supplements. Fat-soluble vitamins (A, D, E, and K) are absorbed along with dietary lipids in micelles via simple diffusion. This is why you are advised to eat some fatty foods when you take fat-soluble vitamin supplements. Most water-soluble vitamins (including most B vitamins and vitamin C) also are absorbed by simple diffusion.

Water Absorption

Each day, about nine liters of fluid enter the small intestine. About 2.3 liters are ingested in foods and beverages, and the rest is from GI secretions. About 90 percent of this water is absorbed in the small intestine. Water absorption is driven by the concentration gradient of the water: The concentration of water is higher in chyme than it is in epithelial cells. Thus, water moves down its concentration gradient from the chyme into cells. As noted earlier, much of the remaining water is then absorbed in the colon.

The Endocrine Pancreas - Glucose Homeostasis

The **pancreas** is a long, slender organ, most of which is located posterior to the bottom half of the stomach. Although it is primarily an exocrine gland, secreting a variety of digestive enzymes, the pancreas has an endocrine function. Its **pancreatic islets**—clusters of cells formerly known as the islets of Langerhans—secrete two major **hormones glucagon and insulin.** These two hormones regulate the rate of glucose metabolism / homeostasis in the body.

The pancreatic islets each contain two major types of cells:

- The **alpha cell** produces the **hormone glucagon** and makes up approximately 20 percent of each islet. Glucagon plays an important role in blood glucose regulation; low blood glucose levels stimulate its release.
- The **beta cell** produces the **hormone insulin** and makes up approximately 75 percent of each islet. Elevated blood glucose levels stimulate the release of insulin.

Regulation of Blood Glucose Levels by Insulin and Glucagon

Glucose is required for cellular respiration and is the preferred fuel for all body cells. The body derives glucose from the breakdown of the carbohydrate-containing foods and drinks we consume. Glucose not immediately taken up by cells for fuel can be stored by the liver and muscles as glycogen, or converted to triglycerides and stored in the adipose tissue. Hormones regulate both the storage and the utilization of glucose as required. Receptors located in the pancreas sense blood glucose levels, and subsequently the pancreatic cells secrete glucagon or insulin to maintain normal levels.

Glucagon

Receptors in the pancreas can sense the decline in blood glucose levels, such as during periods of fasting or during prolonged labor or exercise (Figure 2). In response, the alpha cells of the pancreas secrete the hormone **glucagon**, which has several effects:

- It stimulates the liver to convert its stores of glycogen back into glucose. This response is known as glycogenolysis. The glucose is then released into the circulation for use by body cells.
- It stimulates the liver to take up amino acids from the blood and convert them into glucose. This response is known as gluconeogenesis.
- It stimulates lipolysis, the breakdown of stored triglycerides into free fatty acids and glycerol. Some of the free glycerol released into the bloodstream travels to the liver, which converts it into glucose. This is also a form of gluconeogenesis.

Taken together, these actions increase blood glucose levels. The activity of glucagon is regulated through a negative feedback mechanism; rising blood glucose levels inhibit further glucagon production and secretion.

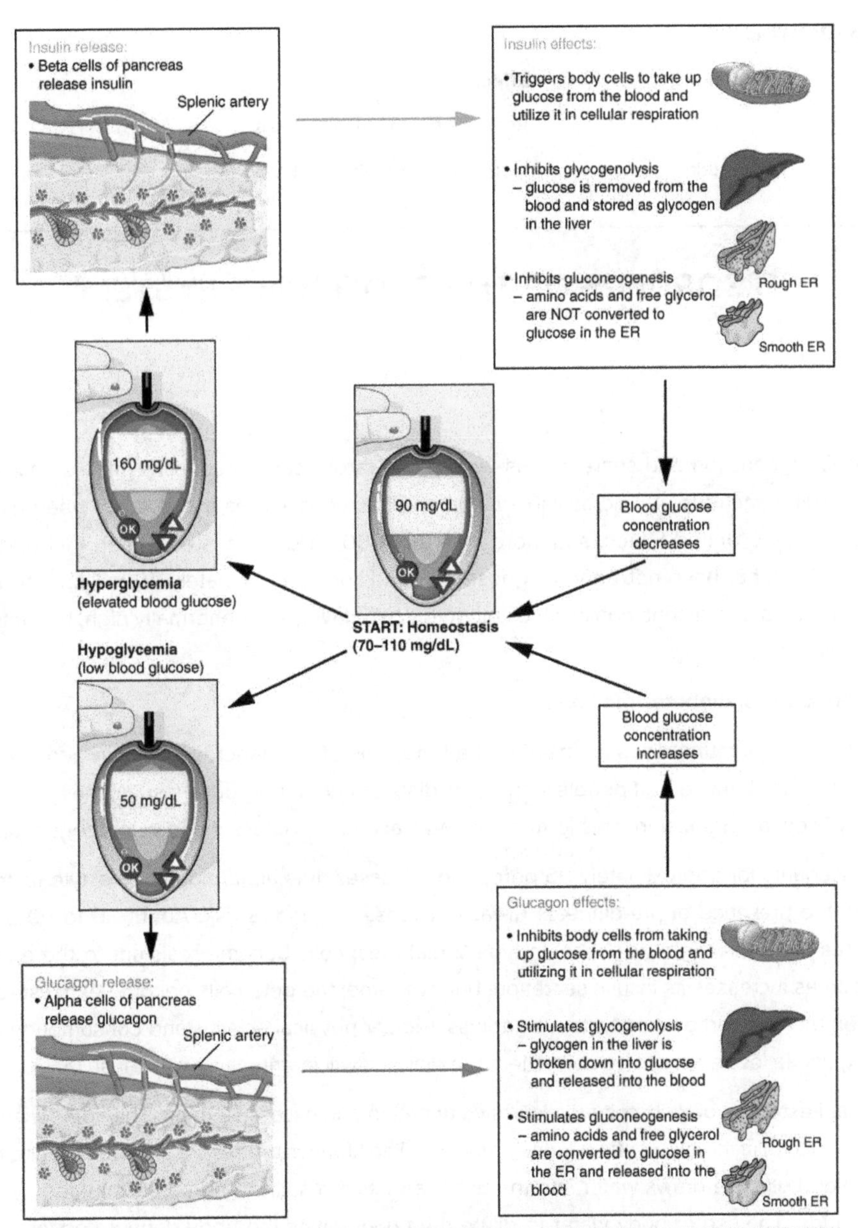

Figure 2. Blood glucose concentration is tightly maintained between 70 mg/dL and 110 mg/dL. If blood glucose concentration rises above this range, insulin is released, which stimulates body cells to remove glucose from the blood. If blood glucose concentration drops below this range, glucagon is released, which stimulates body cells to release glucose into the blood.

Insulin

The primary function of **insulin** is to facilitate the uptake of glucose into body cells. Red blood cells, as well as cells of the brain, liver, kidneys, and the lining of the small intestine, do not have insulin receptors on their cell membranes do not require insulin for glucose uptake. Although all other body cells do require insulin if they are to take glucose from the bloodstream, skeletal muscle cells and adipose cells are the primary targets of insulin.

Insulin also reduces blood glucose levels by stimulating glycolysis, the metabolism of glucose for generation of ATP. Moreover, it stimulates the liver to convert excess glucose into glycogen for storage, and it inhibits enzymes involved in glycogenolysis and gluconeogenesis. Finally, insulin promotes triglyceride and protein synthesis. The secretion of insulin is regulated through a negative feedback mechanism. As blood glucose levels decrease, further insulin release is inhibited.

Table 1. Hormones of the Pancreas

Associated hormones	Chemical class	Effect
Insulin (beta cells)	Protein	Reduces blood glucose levels
Glucagon (alpha cells)	Protein	Increases blood glucose levels

DISORDERS OF THE ENDOCRINE SYSTEM

Diabetes Mellitus

Dysfunction of insulin production and secretion, as well as the target cells' responsiveness to insulin, can lead to a condition called **diabetes mellitus**. An increasingly common disease, diabetes mellitus has been diagnosed in more than 18 million adults in the United States, and more than 200,000 children. It is estimated that up to 7 million more adults have the condition but have not been diagnosed. In addition, approximately 79 million people in the US are estimated to have pre-diabetes, a condition in which blood glucose levels are abnormally high, but not yet high enough to be classified as diabetes.

There are two main forms of diabetes mellitus.

Type 1 diabetes is an autoimmune disease affecting the beta cells of the pancreas. Certain genes are recognized to increase susceptibility. The beta cells of people with type 1 diabetes do not produce insulin; thus, synthetic insulin must be administered by injection or infusion. This form of diabetes accounts for less than five percent of all diabetes cases.

Type 2 diabetes accounts for approximately 95 percent of all cases. It is acquired, and lifestyle factors such as poor diet, inactivity, and the presence of pre-diabetes greatly increase a person's risk. About 80 to 90 percent of people with type 2 diabetes are overweight or obese. In type 2 diabetes, cells become resistant to the effects of insulin. In response, the pancreas increases its insulin secretion, but over time, the beta cells become exhausted. In many cases, type 2 diabetes can be reversed by moderate weight loss, regular physical activity, and consumption of a healthy diet; however, if blood glucose levels cannot be controlled, the diabetic will eventually require insulin.

Two of the early manifestations of diabetes are excessive urination and excessive thirst. They demonstrate how the out-of-control levels of glucose in the blood affect kidney function. The kidneys are responsible for filtering glucose from the blood. Excessive blood glucose draws water into the urine, and as a result the person eliminates an abnormally large quantity of sweet urine. The use of body water to dilute the urine leaves the body dehydrated, and so the person is unusually and continually thirsty. The person may also experience persistent hunger because the body cells are unable to access the glucose in the bloodstream.

Over time, persistently high levels of glucose in the blood injure tissues throughout the body, especially those of the blood vessels and nerves. Inflammation and injury of the lining of arteries lead to atherosclerosis and an **increased risk of heart attack and stroke.** Damage to the microscopic blood vessels of the kidney impairs kidney function and can lead to **kidney failure**. Damage to blood vessels that serve the eyes can lead to blindness. Blood vessel damage also reduces circulation to the limbs, whereas nerve damage leads to a loss of sensation, called **neuropathy**, particularly in the hands and feet. Together, these changes increase the risk of injury, infection, and **tissue death (necrosis),** contributing to a high rate of toe, foot, and lower leg amputations in people with diabetes. Uncontrolled diabetes can also lead to a dangerous form of metabolic acidosis called **ketoacidosis.** Deprived of glucose, cells increasingly rely on fat stores for fuel. However, in a glucose-deficient state, the liver is forced to use an alternative lipid metabolism pathway that results in the increased production of ketone bodies (or ketones), which are acidic. The build-up of ketones in the blood causes ketoacidosis, which—if left untreated—may lead to a life-threatening **"diabetic coma."** Together, these complications make diabetes the seventh leading cause of death in the United States.

Diabetes is diagnosed when lab tests reveal that blood glucose levels are higher than normal, a condition called **hyperglycemia**. The treatment of diabetes depends on the type, the severity of the condition, and the ability of the patient to make lifestyle changes. As noted earlier, moderate weight loss, regular physical activity, and consumption of a healthful diet can reduce blood glucose levels. Some patients with type 2 diabetes may be unable to control their disease with these lifestyle changes, and will require medication. Historically, the first-line treatment of type 2 diabetes was insulin. Research advances have resulted in alternative options, including medications that enhance pancreatic function.

Concept check

1. **Why** are directional terms important to the study of anatomy?
2. **What** are the three body planes?
3. **How** is the dorsal cavity different from the ventral cavity?
4. **What** is the important of serous membranes and **where** are they found?
5. **List and define** the four different types of tissues.
6. **Where** is epithelial tissue located?
7. **List** the different types of connective tissue.
8. **What** are the differences between the three types of muscle tissue?
9. **What** are the two major types of cells of the nervous tissue?
10. **List** the steps in tissue repair.
11. **List** the levels of organization in in the body.
12. **What** are the body systems and **what** are their functions?
13. **What** prevents swallowed food from entering the airways?
14. **Explain** the mechanism responsible for gastroesophageal reflux.
15. **Explain** how the stomach is protected from self-digestion and why this is necessary.
16. **Describe** unique anatomical features that enable the stomach to perform digestive functions.
17. **How** nutrients absorbed in the small intestine pass into the general circulation.
18. **Why** is it important that chyme from the stomach is delivered to the small intestine slowly and in small amounts?
19. **Describe** three of the differences between the walls of the large and small intestines.
20. **Why** does the pancreas secrete some enzymes in their inactive forms, and where are these enzymes activated?
21. **Explain** the role of bile salts in the emulsification of lipids (fats).
22. By clicking on this link, you can **watch a short** video of what happens to the food you eat as it passes from your mouth to your intestine. Along the way, note how the food changes consistency and form. How does this change in consistency facilitate your gaining nutrients from food?
23. Visit this site for an overview of digestion of food in different regions of the digestive tract. Note the route of non-fat nutrients from the small intestine to their release as nutrients to the body.
24. **Watch this** animation that depicts the structure of the small intestine, and, in particular, the villi. Epithelial cells continue the digestion and absorption of nutrients and transport these nutrients to the lymphatic and circulatory systems. In the small intestine, the products of food digestion are absorbed by different structures in the villi. Which structure absorbs and transports fats?
25. By **watching this animation,** you will see that for the various food groups—proteins, fats, and carbohydrates—digestion begins in different parts of the digestion system, though all end in the same place. Of the three major food classes (carbohydrates, fats, and proteins), which is digested in the mouth, the stomach, and the small intestine?

CHAPTER 5 — REPRODUCTIVE SYSTEM

Introduction to the Reproductive System

LEARNING OBJECTIVES

- Describe the anatomy of the male and female reproductive systems, including their accessory structures
- Describe the development and maturation of the sex organs and the emergence of secondary sex characteristics during puberty
- Explain the role of hypothalamic and pituitary hormones in male and female reproductive function
- Trace the path of a sperm cell from its initial production through fertilization of an oocyte
- Explain the events in the ovary prior to ovulation
- Trace the path of an oocyte from ovary to fertilization
- List the different types of birth control methods
- Identify the different types of STDs/Is

Small, uncoordinated, and slick with amniotic fluid, a newborn encounters the world outside of her mother's womb. We do not often consider that a child's birth is proof of the healthy functioning of both her mother's and father's reproductive systems. Moreover, her parents' endocrine systems had to secrete the appropriate regulating hormones to induce the production and release of unique male and female gametes, reproductive cells containing the parents' genetic material (one set of 23 chromosomes). Her parent's reproductive behavior had to facilitate the transfer of male gametes—the sperm—to the female reproductive tract at just the right time to encounter the female gamete, an oocyte (egg). Finally, combination of the gametes (fertilization) had to occur, followed by implantation and development. In this chapter, you will explore the male and female reproductive systems, whose healthy functioning can culminate in the powerful sound of a newborn's first cry.

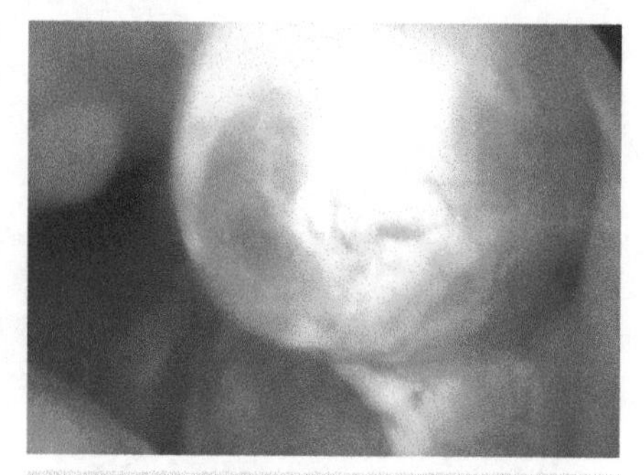

Following a surge of luteinizing hormone (LH), an oocyte (immature egg cell) will be released into the uterine tube, where it will then be available to be fertilized by a male's sperm. Ovulation marks the end of the follicular phase of the ovarian cycle and the start of the luteal phase.

The development of the reproductive systems begins soon after fertilization of the egg, with primordial gonads beginning to develop approximately one month after conception. Reproductive development continues in utero, but there is little change in the reproductive system between infancy and puberty.

Development of the Sexual Organs in the Embryo and Fetus

In both male and female embryos, the same group of cells has the potential to develop into either the male or female gonads; this tissue is considered bipotential. The *SRY* (Sex-determining *R*egion of the *Y* chromosome) gene actively recruits other genes that begin to develop the testes, and suppresses genes that are important in female development. As part of this *SRY*-prompted cascade, germ cells in the bipotential gonads differentiate into spermatogonia. Without *SRY*, different genes are expressed, oogonia form, and primordial follicles develop in the primitive ovary.

Soon after the formation of the testis, the Leydig cells begin to secrete testosterone. Testosterone can influence tissues that are bipotential to become male reproductive structures. For example, with exposure to testosterone, cells that could become either the glans penis or the glans clitoris form the glans penis. Without testosterone, these same cells differentiate into the clitoris.

Not all tissues in the reproductive tract are bipotential. The internal reproductive structures (for example the uterus, uterine tubes, and part of the vagina in females; and the epididymis, ductus deferens, and seminal vesicles in males) form from one of two rudimentary duct systems in the embryo. For proper reproductive function in the adult, one set of these ducts must develop properly, and the other must degrade. In males, secretions from sustentacular cells trigger a degradation of the female duct, called the **Müllerian duct**. At the same time, testosterone secretion stimulates growth of the male tract, the **Wolffian duct**. Without such sustentacular cell secretion, the Müllerian duct will develop; without testosterone, the Wolffian duct will degrade. Thus, the developing offspring will be female. For more information and a figure of differentiation of the gonads, seek additional content on fetal development.

PRACTICE QUESTIONS

A baby's gender is determined at conception, and the different genitalia of male and female fetuses develop from the same tissues in the embryo. View this animation to see a comparison of the development of structures of the female and male reproductive systems in a growing fetus. Where are the testes located for most of gestational time?

Further Sexual Development Occurs at Puberty

Puberty is the stage of development at which individuals become sexually mature. Though the outcomes of puberty for boys and girls are very different, the hormonal control of the process is very similar. In addition, though the timing of these events varies between individuals, the sequence of changes that occur is predictable for male and female adolescents. As shown in the image below, a concerted release of hormones from the hypothalamus (GnRH), the anterior pituitary (LH and FSH), and the gonads (either testosterone or estrogen) is responsible for the maturation of the reproductive systems and the development of **secondary sex characteristics**, which are physical changes that serve auxiliary roles in reproduction.

The first changes begin around the age of eight or nine when the production of LH becomes detectable. The release of LH occurs primarily at night during sleep and precedes the physical changes of puberty by several years. In pre-pubertal children, the sensitivity of the negative feedback system in the hypothalamus and pituitary is very high. This means that very low concentrations of androgens or estrogens will negatively feed back onto the hypothalamus and pituitary, keeping the production of GnRH, LH, and FSH low.

As an individual approaches puberty, two changes in sensitivity occur. The first is a decrease of sensitivity in the hypothalamus and pituitary to negative feedback, meaning that it takes increasingly larger concentrations of sex steroid hormones to stop the production of LH and FSH. The second change in sensitivity is an increase in sensitivity of the gonads to the FSH and LH signals, meaning the gonads of adults are more responsive to gonadotropins than are the gonads of children. As a result of these two changes, the levels of LH and FSH slowly increase and lead to the enlargement and maturation of the gonads, which in turn leads to secretion of higher levels of sex hormones and the initiation of spermatogenesis and folliculogenesis.

In addition to age, multiple factors can affect the age of onset of puberty, including genetics, environment, and psychological stress. One of the more important influences may be nutrition; historical data demonstrate the effect of better and more consistent nutrition on the age of menarche in girls in the United States, which decreased from an average age of approximately 17 years of age in 1860 to the current age of approximately 12.75 years in 1960, as it remains today. Some studies indicate a link between puberty onset and the amount of stored fat in an individual. This effect is more pronounced in girls, but has been documented in both sexes. Body fat, corresponding with secretion of the hormone leptin by adipose cells, appears to have a strong role in determining menarche. This may reflect to some extent the high metabolic costs of gestation and lactation. In girls who are lean and highly active, such as gymnasts, there is often a delay in the onset of puberty.

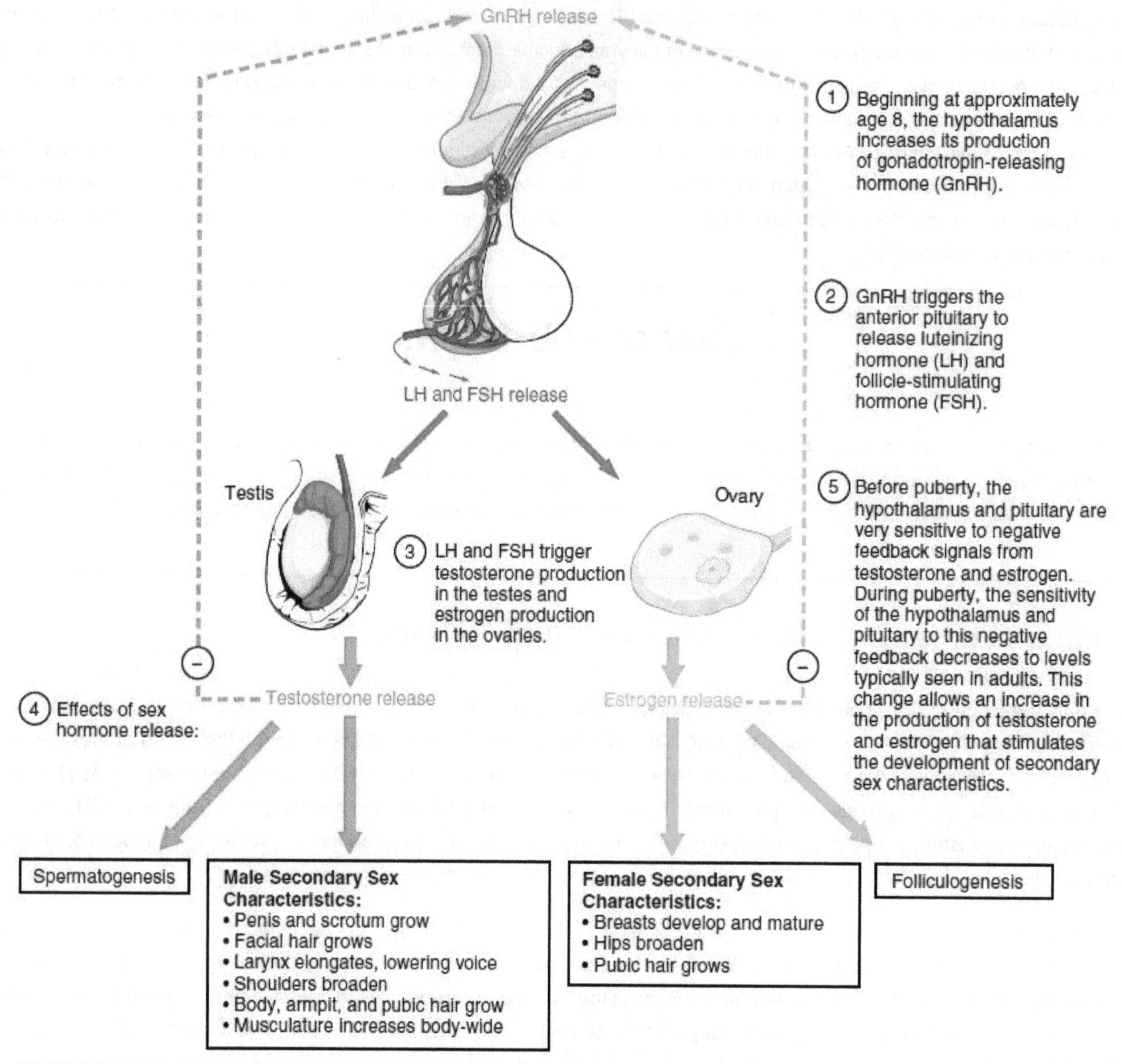

Click to view a larger image. During puberty, the release of LH and FSH from the anterior pituitary stimulates the gonads to produce sex hormones in both male and female adolescents.

Signs of Puberty

Different sex steroid hormone concentrations between the sexes also contribute to the development and function of secondary sexual characteristics. Examples of secondary sexual characteristics are listed in Table 1.

Table 1. Development of the Secondary Sexual Characteristics

Male	Female
Increased larynx size and deepening of the voice	Deposition of fat, predominantly in breasts and hips
Increased muscular development	Breast development
Growth of facial, axillary, and pubic hair, and increased growth of body hair	Broadening of the pelvis and growth of axillary and pubic hair

As a girl reaches puberty, typically the first change that is visible is the development of the breast tissue. This is followed by the growth of axillary and pubic hair. A growth spurt normally starts at approximately age 9 to 11, and may last two years or more. During this time, a girl's height can increase 3 inches a year. The next step in puberty is menarche, the start of menstruation.

In boys, the growth of the testes is typically the first physical sign of the beginning of puberty, which is followed by growth and pigmentation of the scrotum and growth of the penis. The next step is the growth of hair, including armpit, pubic, chest, and facial hair. Testosterone stimulates the growth of the larynx and thickening and lengthening of the vocal folds, which causes the voice to drop in pitch. The first fertile ejaculations typically appear at approximately 15 years of age, but this age can vary widely across individual boys. Unlike the early growth spurt observed in females, the male growth spurt occurs toward the end of puberty, at approximately age 11 to 13, and a boy's height can increase as much as 4 inches a year. In some males, pubertal development can continue through the early 20s.

Anatomy and Physiology of the Male Reproductive System

Unique for its role in human reproduction, a **gamete** is a specialized sex cell carrying 23 chromosomes—one half the number in body cells. At fertilization, the chromosomes in one male gamete, called a **sperm** (or spermatozoon), combine with the chromosomes in one female gamete, called an oocyte. The function of the male reproductive system is to produce sperm and transfer them to the female reproductive tract. The paired testes are a crucial component in this process, as they produce both sperm and androgens, the hormones that support male reproductive physiology.

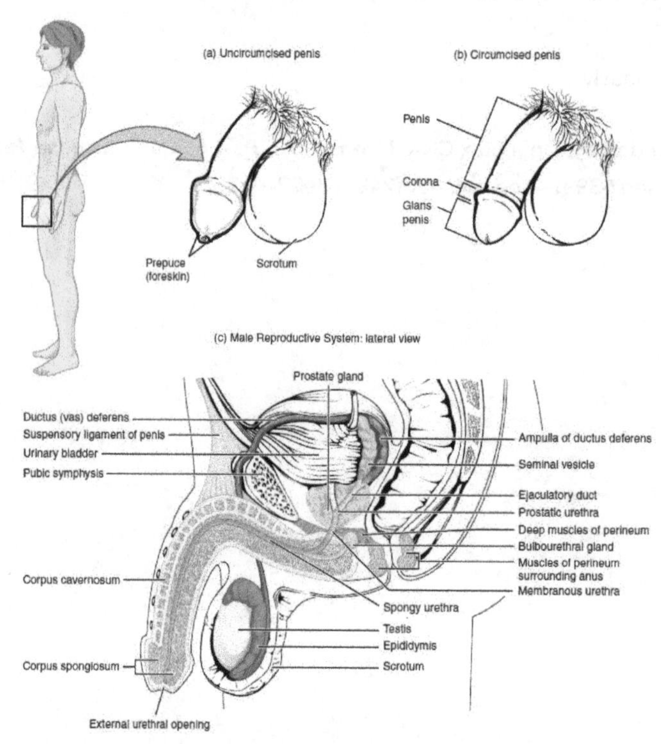

Figure 1. Click for a larger image. The structures of the male reproductive system include the testes, the epididymides, the penis, and the ducts and glands that produce and carry semen. Sperm exit the scrotum through the ductus deferens, which is bundled in the spermatic cord. The seminal vesicles and prostate gland add fluids to the sperm to create semen.

Scrotum

The testes are located in a skin-covered, highly pigmented, muscular sack called the **scrotum** that extends from the body behind the penis. This location is important in sperm production, which occurs within the testes, and proceeds more efficiently when the testes are kept 2 to 4°C below core body temperature.

The dartos muscle makes up the subcutaneous muscle layer of the scrotum. It continues internally to make up the scrotal septum, a wall that divides the scrotum into two compartments, each housing one testis. Descending from the internal oblique muscle of the abdominal wall are the two cremaster muscles, which cover each testis like a muscular net. By contracting simultaneously, the dartos and cremaster muscles can elevate the testes in cold weather (or water), moving the testes closer to the body and decreasing the surface area of the scrotum to retain heat. Alternatively, as the environmental temperature increases, the scrotum relaxes, moving the testes farther from the body core and increasing scrotal surface area, which promotes heat loss. Externally, the scrotum has a raised medial thickening on the surface called the raphae.

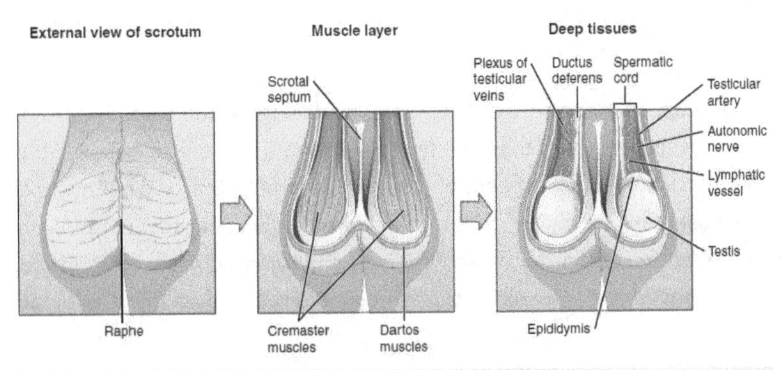

Figure 2. This anterior view shows the structures of the scrotum and testes.

Testes

The **testes** (singular = testis) are the male **gonads**—that is, the male reproductive organs. They produce both sperm and androgens, such as testosterone, and are active throughout the reproductive lifespan of the male.

Paired ovals, the testes are each approximately 4 to 5 cm in length and are housed within the scrotum. They are surrounded by two distinct layers of protective connective tissue. The outer tunica vaginalis is a serous membrane that has both a parietal and a thin visceral layer. Beneath the tunica vaginalis is the tunica albuginea, a tough, white, dense connective tissue layer covering the testis itself. Not only does the tunica albuginea cover the outside of the testis, it also invaginates to form septa that divide the testis into 300 to 400 structures called lobules. Within the lobules, sperm develop in structures called seminiferous tubules. During the seventh month of the developmental period of a male fetus, each testis moves through the abdominal musculature to descend into the scrotal cavity. This is called the "descent of the testis." Cryptorchidism is the clinical term used when one or both of the testes fail to descend into the scrotum prior to birth.

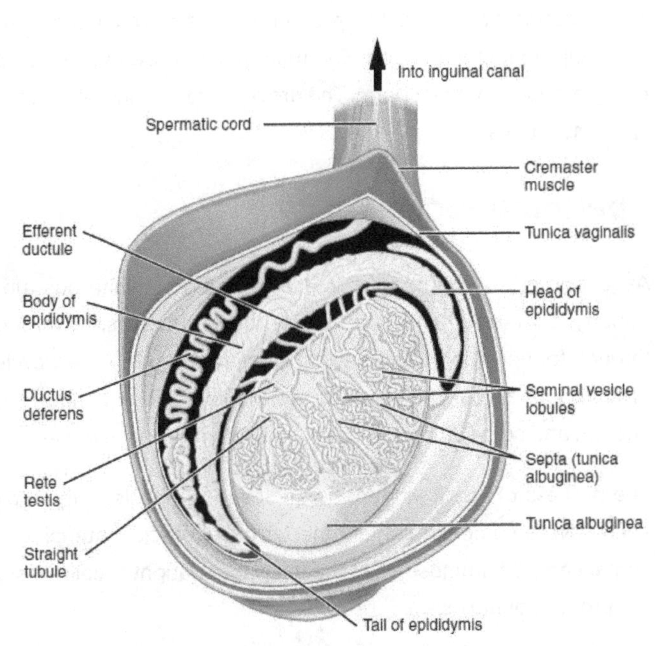

Figure 3. This sagittal view shows the seminiferous tubules, the site of sperm production. Formed sperm are transferred to the epididymis, where they mature. They leave the epididymis during an ejaculation via the ductus deferens.

The tightly coiled **seminiferous tubules** form the bulk of each testis. They are composed of developing sperm cells surrounding a lumen, the hollow center of the tubule, where formed sperm are released into the duct system of the testis. Specifically, from the lumens of the seminiferous tubules, sperm move into the straight tubules (or tubuli recti), and from there into a fine meshwork of tubules called the rete testes. Sperm leave the rete testes, and the testis itself, through the 15 to 20 efferent ductules that cross the tunica albuginea.

Inside the seminiferous tubules are six different cell types. These include supporting cells called sustentacular cells, as well as five types of developing sperm cells called germ cells. Germ cell development progresses from the basement membrane—at the perimeter of the tubule—toward the lumen. Let's look more closely at these cell types.

Sertoli Cells

Surrounding all stages of the developing sperm cells are elongate, branching **Sertoli cells**. Sertoli cells are a type of supporting cell called a sustentacular cell, or sustenocyte, that are typically found in epithelial tissue. Sertoli cells secrete signaling molecules that promote sperm production and can control whether germ cells live or die. They extend physically around the germ cells from the peripheral basement membrane of the seminiferous tubules to the lumen. Tight junctions between these sustentacular cells create the **blood–testis barrier**, which keeps bloodborne substances from reaching the germ cells and, at the same time, keeps surface antigens on developing germ cells from escaping into the bloodstream and prompting an autoimmune response.

Germ Cells

The least mature cells, the **spermatogonia** (singular = spermatogonium), line the basement membrane inside the tubule. Spermatogonia are the stem cells of the testis, which means that they are still able to differentiate into a variety of different cell types throughout adulthood. Spermatogonia divide to produce primary and secondary spermatocytes, then spermatids, which finally produce formed sperm. The process that begins with spermatogonia and concludes with the production of sperm is called **spermatogenesis**.

Spermatogenesis

As just noted, spermatogenesis occurs in the seminiferous tubules that form the bulk of each testis. The process begins at puberty, after which time sperm are produced constantly throughout a man's life. One production cycle, from spermatogonia through formed sperm, takes approximately 64 days. A new cycle starts approximately every 16 days, although this timing is not synchronous across the seminiferous tubules. Sperm counts—the total number of sperm a man produces—slowly decline after age 35, and some studies suggest that smoking can lower sperm counts irrespective of age.

The process of spermatogenesis begins with mitosis of the diploid spermatogonia. Because these cells are diploid ($2n$), they each have a complete copy of the father's genetic material, or 46 chromosomes. However, mature gametes are haploid ($1n$), containing 23 chromosomes—meaning that daughter cells of spermatogonia must undergo a second cellular division through the process of meiosis.

Two identical diploid cells result from spermatogonia mitosis. One of these cells remains a spermatogonium, and the other becomes a primary **spermatocyte**, the next stage in the process of spermatogenesis. As in mitosis, DNA is replicated in a primary spermatocyte, and the cell undergoes cell division to produce two cells with identical chromosomes. Each of these is a secondary spermatocyte. Now a second round of cell division occurs in both of the secondary spermatocytes, separating the chromosome pairs. This second meiotic division results in a total of four cells with only half of the number of chromosomes. Each of these new cells is a **spermatid**. Although haploid, early spermatids look very similar to cells in the earlier stages of spermatogenesis, with a round shape, central nucleus, and large amount of cytoplasm. A process called **spermiogenesis** transforms these early spermatids, reducing the cytoplasm, and beginning the formation of the parts of a true sperm. The fifth stage of germ cell formation—spermatozoa, or formed sperm—is the end result of this process, which occurs in the portion of the tubule nearest the lumen. Eventually, the sperm are released into the lumen and are moved along a series of ducts in the testis toward a structure called the epididymis for the next step of sperm maturation.

Structure of Formed Sperm

Sperm are smaller than most cells in the body; in fact, the volume of a sperm cell is 85,000 times less than that of the female gamete. Approximately 100 to 300 million sperm are produced each day, whereas women typically ovulate only one oocyte per month as is true for most cells in the body, the structure of sperm cells speaks to their function. Sperm have a distinctive head, mid-piece, and tail region. The head of the sperm contains the extremely compact haploid nucleus with very little cytoplasm. These qualities contribute to the overall small size of the sperm (the head is only 5 μm long). A structure called the acrosome covers most of the head of the sperm cell as a "cap" that is filled with lysosomal enzymes important for preparing sperm to participate in fertilization. Tightly packed mitochondria fill the mid-piece of the sperm. ATP produced by these mitochondria will power the flagellum, which extends from the neck and the mid-piece through the tail of the sperm, enabling it to move the entire sperm cell. The central strand of the flagellum, the axial filament, is formed from one centriole inside the maturing sperm cell during the final stages of spermatogenesis.

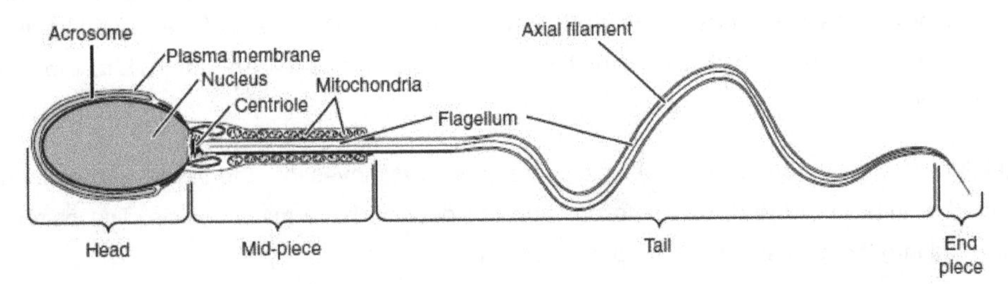

Figure 5. Sperm cells are divided into a head, containing DNA; a mid-piece, containing mitochondria; and a tail, providing motility. The acrosome is oval and somewhat flattened.

Sperm Transport

To fertilize an egg, sperm must be moved from the seminiferous tubules in the testes, through the epididymis, and—later during ejaculation—along the length of the penis and out into the female reproductive tract.

Role of the Epididymis

From the lumen of the seminiferous tubules, the immotile sperm are surrounded by testicular fluid and moved to the **epididymis** (plural = epididymides), a coiled tube attached to the testis where newly formed sperm continue to mature. Though the epididymis does not take up much room in its tightly coiled state, it would be approximately 6 m (20 feet) long if straightened. It takes an average of 12 days for sperm to move through the coils of the epididymis, with the shortest recorded transit time in humans being one day. Sperm enter the head of the epididymis and are moved along predominantly by the contraction of smooth muscles lining the epididymal tubes. As they are moved along the length of the epididymis, the sperm further mature and acquire the ability to move under their own power. Once inside the female reproductive tract, they will use this ability to move independently toward the unfertilized egg. The more mature sperm are then stored in the tail of the epididymis (the final section) until ejaculation occurs.

Duct System

During ejaculation, sperm exit the tail of the epididymis and are pushed by smooth muscle contraction to the **ductus deferens** (also called the vas deferens). The ductus deferens is a thick, muscular tube that is bundled together inside the scrotum with connective tissue, blood vessels, and nerves into a structure called the **spermatic cord**. Because the ductus deferens is physically accessible within the scrotum, surgical sterilization to interrupt sperm delivery can be performed by cutting and sealing a small section of the ductus (vas) deferens. This procedure is called a vasectomy, and it is an effective form of male birth control. Although it may be possible to reverse a vasectomy, clinicians consider the procedure permanent, and advise men to undergo it only if they are certain they no longer wish to father children. From each epididymis, each ductus deferens extends superiorly into the abdominal cavity through the **inguinal canal** in the abdominal wall. From here, the ductus deferens continues posteriorly to the pelvic cavity, ending posterior to the bladder where it dilates in a region called the ampulla (meaning "flask").

Sperm make up only 5 percent of the final volume of **semen**, the thick, milky fluid that the male ejaculates. The bulk of semen is produced by three critical accessory glands of the male reproductive system: the seminal vesicles, the prostate, and the bulbourethral glands.

Seminal Vesicles

As sperm pass through the ampulla of the ductus deferens at ejaculation, they mix with fluid from the associated **seminal vesicle**. The paired seminal vesicles are glands that contribute approximately 60 percent of the semen volume. Seminal vesicle fluid contains large amounts of fructose, which is used by the sperm mitochondria to generate ATP to allow movement through the female reproductive tract.

The fluid, now containing both sperm and seminal vesicle secretions, next moves into the associated **ejaculatory duct**, a short structure formed from the ampulla of the ductus deferens and the duct of the seminal vesicle. The paired ejaculatory ducts transport the seminal fluid into the next structure, the prostate gland.

Prostate Gland

As shown in Figure 1, the centrally located **prostate gland** sits anterior to the rectum at the base of the bladder surrounding the prostatic urethra (the portion of the urethra that runs within the prostate). About the size of a walnut, the prostate is formed of both muscular and glandular tissues. It excretes an alkaline, milky fluid to the passing seminal fluid—now called semen—that is critical to first coagulate and then decoagulate the semen following ejaculation. The temporary thickening of semen helps retain it within the female reproductive tract, providing time for sperm to utilize the fructose provided by seminal vesicle secretions. When the semen regains its fluid state, sperm can then pass farther into the female reproductive tract.

The prostate normally doubles in size during puberty. At approximately age 25, it gradually begins to enlarge again. This enlargement does not usually cause problems; however, abnormal growth of the prostate, or benign prostatic hyperplasia (BPH), can cause constriction of the urethra as it passes through the middle of the prostate gland, leading to a number of lower urinary tract symptoms, such as a frequent and intense urge to urinate, a weak stream, and a sensation that the bladder has not emptied completely. By age 60, approximately 40 percent of men have some degree of BPH. By age 80, the number of affected individuals has jumped to as many as 80 percent. Treatments for BPH attempt to relieve the pressure on the urethra so that urine can flow more normally. Mild to moderate symptoms are treated with medication, whereas severe enlargement of the prostate is treated by surgery in which a portion of the prostate tissue is removed.

Another common disorder involving the prostate is prostate cancer. According to the Centers for Disease Control and Prevention (CDC), prostate cancer is the second most common cancer in men. However, some forms of prostate cancer grow very slowly and thus may not ever require treatment. Aggressive forms of prostate cancer, in contrast, involve metastasis to vulnerable organs like the lungs and brain. There is no link between BPH and prostate cancer, but the symptoms are similar. Prostate cancer is detected by a medical history, a blood test, and a rectal exam that allows physicians to palpate the prostate and check for unusual masses. If a mass is detected, the cancer diagnosis is confirmed by biopsy of the cells.

Bulbourethral Glands

The final addition to semen is made by two **bulbourethral glands** (or Cowper's glands) that release a thick, salty fluid that lubricates the end of the urethra and the vagina, and helps to clean urine residues from the penile urethra. The fluid from these accessory glands is released after the male becomes sexually aroused, and shortly before the release of the semen. It is therefore sometimes called pre-ejaculate. It is important to note that, in addition to the lubricating proteins, it is possible for bulbourethral fluid to pick up sperm already present in the urethra, and therefore it may be able to cause pregnancy.

The Penis

The **penis** is the male organ of copulation (sexual intercourse). It is flaccid for non-sexual actions, such as urination, and turgid and rod-like with sexual arousal. When erect, the stiffness of the organ allows it to penetrate into the vagina and deposit semen into the female reproductive tract.

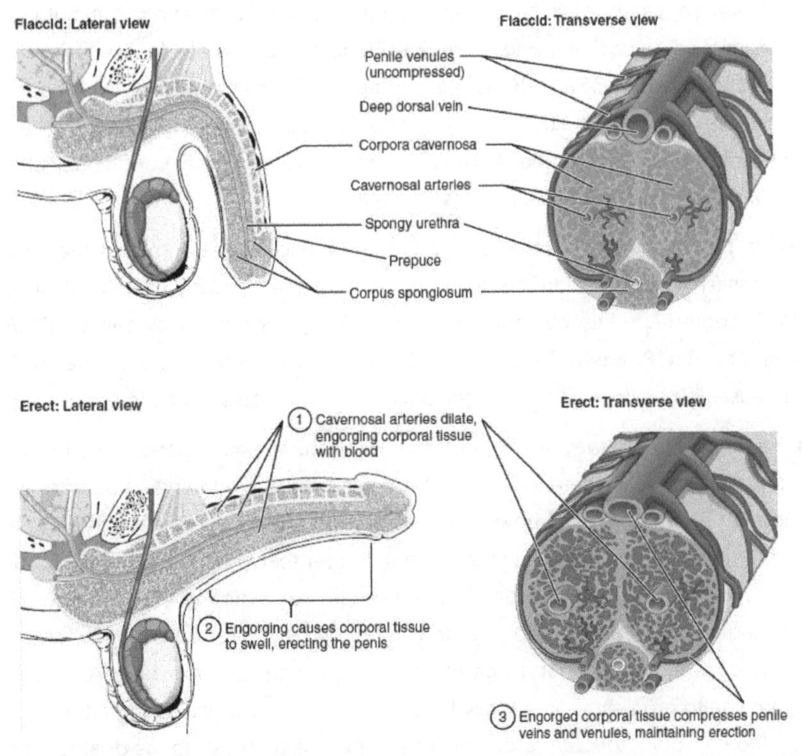

Figure 6. Three columns of erectile tissue make up most of the volume of the penis.

The shaft of the penis surrounds the urethra. The shaft is composed of three column-like chambers of erectile tissue that span the length of the shaft. Each of the two larger lateral chambers is called a **corpus cavernosum** (plural = corpora cavernosa). Together, these make up the bulk of the penis. The **corpus spongiosum**, which can be felt as a raised ridge on the erect penis, is a smaller chamber that surrounds the spongy, or penile, urethra. The end of the penis, called the **glans penis**, has a high concentration of nerve endings, resulting in very sensitive skin that influences the likelihood of ejaculation. The skin from the shaft extends down over the glans and forms a collar called the **prepuce** (or foreskin). The foreskin also contains a dense concentration of nerve endings, and both lubricate and protect the sensitive skin of the glans penis. A surgical procedure called circumcision, often performed for religious or social reasons, removes the prepuce, typically within days of birth.

Both sexual arousal and REM sleep (during which dreaming occurs) can induce an erection. Penile erections are the result of vasocongestion, or engorgement of the tissues because of more arterial blood flowing into the penis than is leaving in the veins. During sexual arousal, nitric oxide (NO) is released from nerve endings near blood vessels within the corpora cavernosa and spongiosum. Release of NO activates a signaling pathway that results in relaxation of the smooth muscles that surround the penile arteries, causing them to dilate. This dilation increases the amount of blood that can enter the penis and induces the endothelial cells in the penile arterial walls to also secrete NO and perpetuate the vasodilation. The rapid increase in blood volume fills the erectile chambers, and the increased pressure of the filled chambers compresses the thin-walled penile venules, preventing venous drainage of the penis. The result of this increased blood flow to the penis and reduced blood return from the penis is erection. Depending on the flaccid dimensions of a penis, it can increase in size slightly or greatly during erection, with the average length of an erect penis measuring approximately 15 cm.

DISORDERS OF THE MALE REPRODUCTIVE SYSTEM: ERECTILE DYSFUNCTION (ED)

Erectile dysfunction (ED) is a condition in which a man has difficulty either initiating or maintaining an erection. The combined prevalence of minimal, moderate, and complete ED is approximately 40 percent in men at age 40, and reaches nearly 70 percent by 70 years of age. In addition to aging, ED is associated with diabetes, vascular disease, psychiatric disorders, prostate disorders, the use of some drugs such as certain antidepressants, and problems with the testes resulting in low testosterone concentrations. These physical and emotional conditions can lead to interruptions in the vasodilation pathway and result in an inability to achieve an erection.

Recall that the release of NO induces relaxation of the smooth muscles that surround the penile arteries, leading to the vasodilation necessary to achieve an erection. To reverse the process of vasodilation, an enzyme called phosphodiesterase (PDE) degrades a key component of the NO signaling pathway called cGMP. There are several different forms of this enzyme, and PDE type 5 is the type of PDE found in the tissues of the penis. Scientists discovered that inhibiting PDE5 increases blood flow, and allows vasodilation of the penis to occur.

PDEs and the vasodilation signaling pathway are found in the vasculature in other parts of the body. In the 1990s, clinical trials of a PDE5 inhibitor called sildenafil were initiated to treat hypertension and angina pectoris (chest pain caused by poor blood flow through the heart). The trial showed that the drug was not effective at treating heart conditions, but many men experienced erection and priapism (erection lasting longer than 4 hours). Because of this, a clinical trial was started to investigate the ability of sildenafil to promote erections in men suffering from ED. In 1998, the FDA approved the drug, marketed as Viagra®. Since approval of the drug, sildenafil and similar PDE inhibitors now generate over a billion dollars a year in sales, and are reported to be effective in treating approximately 70 to 85 percent of cases of ED. Importantly, men with health problems—especially those with cardiac disease taking nitrates—should avoid Viagra or talk to their physician to find out if they are a candidate for the use of this drug, as deaths have been reported for at-risk users.

Testosterone

Testosterone, an androgen, is a steroid hormone produced by **Leydig cells**. The alternate term for Leydig cells, interstitial cells, reflects their location between the seminiferous tubules in the testes. In male embryos, testosterone is secreted by Leydig cells by the seventh week of development, with peak concentrations reached in the second trimester. This early release of testosterone results in the anatomical differentiation of the male sexual organs. In childhood, testosterone concentrations are low. They increase during puberty, activating characteristic physical changes and initiating spermatogenesis.

Functions of Testosterone

The continued presence of testosterone is necessary to keep the male reproductive system working properly, and Leydig cells produce approximately 6 to 7 mg of testosterone per day. Testicular steroidogenesis (the manufacture of androgens, including testosterone) results in testosterone concentrations that are 100 times higher in the testes than in the circulation. Maintaining these normal concentrations of testosterone promotes spermatogenesis, whereas low levels of testosterone can lead to infertility. In addition to intratesticular secretion, testosterone is also released into the systemic circulation and plays an important role in muscle development, bone growth, the development of secondary sex characteristics, and maintaining libido (sex drive) in both males and females. In females, the ovaries secrete small amounts of testosterone, although most is converted to estradiol. A small amount of testosterone is also secreted by the adrenal glands in both sexes.

Control of Testosterone

The regulation of testosterone concentrations throughout the body is critical for male reproductive function. The intricate interplay between the endocrine system and the reproductive system is shown in Figure 7.

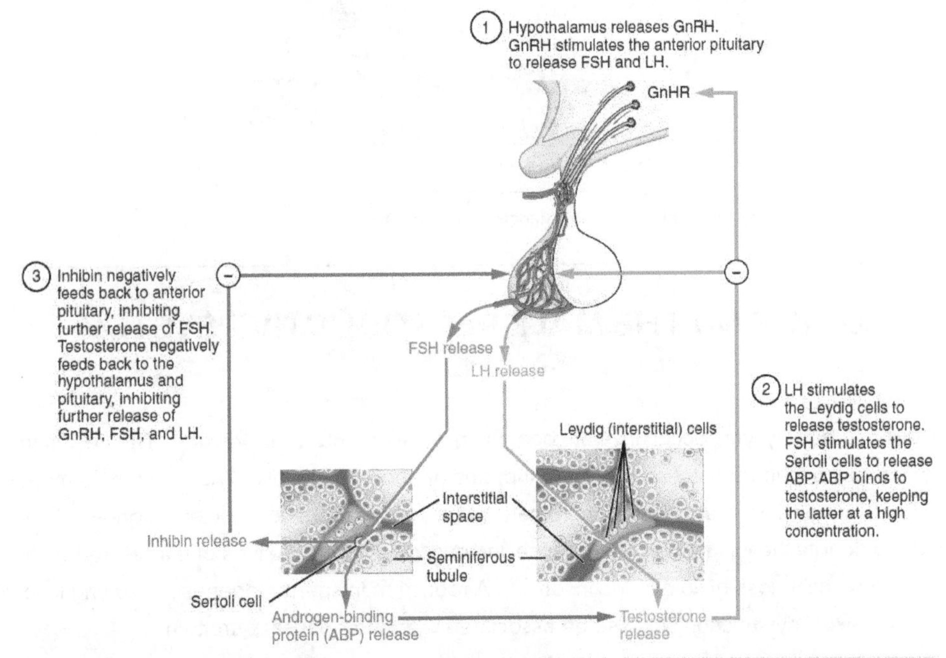

Figure 7. The hypothalamus and pituitary gland regulate the production of testosterone and the cells that assist in spermatogenesis. GnRH activates the anterior pituitary to produce LH and FSH, which in turn stimulate Leydig cells and Sertoli cells, respectively. The system is a negative feedback loop because the end products of the pathway, testosterone and inhibin, interact with the activity of GnRH to inhibit their own production.

The regulation of Leydig cell production of testosterone begins outside of the testes. The hypothalamus and the pituitary gland in the brain integrate external and internal signals to control testosterone synthesis and secretion. The regulation begins in the hypothalamus. Pulsatile release of a hormone called **gonadotropin-releasing hormone (GnRH)** from the hypothalamus stimulates the endocrine release of hormones from the pituitary gland. Binding of GnRH to its receptors on the anterior pituitary gland stimulates release of the two gonadotropins: luteinizing hormone (LH) and follicle-stimulating hormone (FSH). These two hormones are critical for reproductive function in both men and women. In men, FSH binds predominantly to the Sertoli cells within the seminiferous tubules to promote spermatogenesis. FSH also stimulates the Sertoli cells to produce hormones called inhibins, which function to inhibit FSH release from the pituitary, thus reducing testosterone secretion. These polypeptide hormones correlate directly with Sertoli cell function and sperm number; inhibin B can be used as a marker of spermatogenic activity. In men, LH binds to receptors on Leydig cells in the testes and upregulates the production of testosterone.

A negative feedback loop predominantly controls the synthesis and secretion of both FSH and LH. Low blood concentrations of testosterone stimulate the hypothalamic release of GnRH. GnRH then stimulates the anterior pituitary to secrete LH into the bloodstream. In the testis, LH binds to LH receptors on Leydig cells and stimulates the release of testosterone. When concentrations of testosterone in the blood reach a critical threshold, testosterone itself will bind to androgen receptors on both the hypothalamus and the anterior pituitary, inhibiting the synthesis and secretion of GnRH and LH, respectively. When the blood concentrations of testosterone once again decline, testosterone no longer interacts with the receptors to the same degree and GnRH and LH are once again secreted, stimulating more testosterone production. This same process occurs with FSH and inhibin to control spermatogenesis.

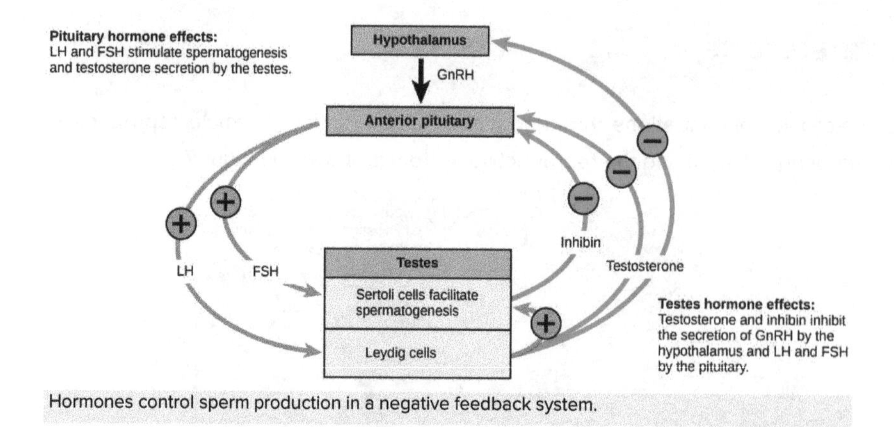

Hormones control sperm production in a negative feedback system.

AGING AND THE MALE REPRODUCTIVE SYSTEM

Declines in Leydig cell activity can occur in men beginning at 40 to 50 years of age. The resulting reduction in circulating testosterone concentrations can lead to symptoms of andropause, also known as male menopause. While the reduction in sex steroids in men is akin to female menopause, there is no clear sign—such as a lack of a menstrual period—to denote the initiation of andropause. Instead, men report feelings of fatigue, reduced muscle mass, depression, anxiety, irritability, loss of libido, and insomnia. A reduction in spermatogenesis resulting in lowered fertility is also reported, and sexual dysfunction can also be associated with andropausal symptoms.

Whereas some researchers believe that certain aspects of andropause are difficult to tease apart from aging in general, testosterone replacement is sometimes prescribed to alleviate some symptoms. Recent studies have shown a benefit from androgen replacement therapy on the new onset of depression in elderly men; however, other studies caution against testosterone replacement for long-term treatment of andropause symptoms, showing that high doses can sharply increase the risk of both heart disease and prostate cancer.

Anatomy and Physiology of the Female Reproductive System

The female reproductive system functions to produce gametes and reproductive hormones, just like the male reproductive system; however, it also has the additional task of supporting the developing fetus and delivering it to the outside world. Unlike its male counterpart, the female reproductive system is located primarily inside the pelvic cavity. Recall that the ovaries are the female gonads. The gamete they produce is called an **oocyte**. We'll discuss the production of oocytes in detail shortly. First, let's look at some of the structures of the female reproductive system.

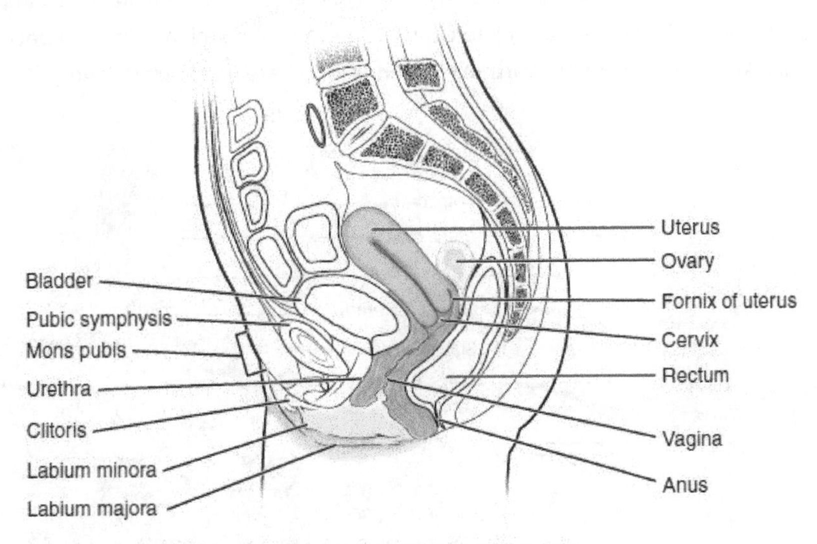

(a) Human female reproductive system: lateral view

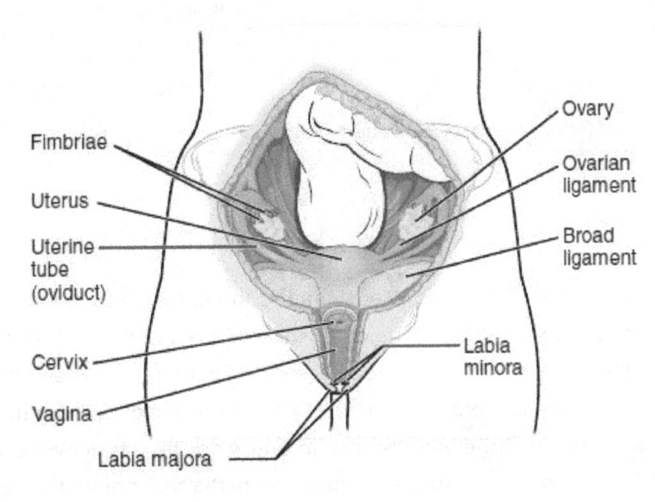

(b) Human female reproductive system: anterior view

Figure 1. The major organs of the female reproductive system are located inside the pelvic cavity.

External Female Genitals

The external female reproductive structures are referred to collectively as the **vulva**. The **mons pubis** is a pad of fat that is located at the anterior, over the pubic bone. After puberty, it becomes covered in pubic hair. The **labia majora** (labia = "lips"; majora = "larger") are folds of hair-covered skin that begin just posterior to the mons pubis. The thinner and more pigmented **labia minora** (labia = "lips"; minora = "smaller") extend medial to the labia majora. Although they naturally vary in shape and size from woman to woman, the labia minora serve to protect the female urethra and the entrance to the female reproductive tract.

The superior, anterior portions of the labia minora come together to encircle the **clitoris** (or glans clitoris), an organ that originates from the same cells as the glans penis and has abundant nerves that make it important in sexual sensation and orgasm. The **hymen** is a thin membrane that sometimes partially covers the entrance to the vagina. An intact hymen cannot be used as an indication of "virginity"; even at birth, this is only a partial membrane, as menstrual fluid and other secretions must be able to exit the body, regardless of penile–vaginal intercourse. The vaginal opening is located between the opening of the urethra and the anus. It is flanked by outlets to the **Bartholin's glands** (or greater vestibular glands).

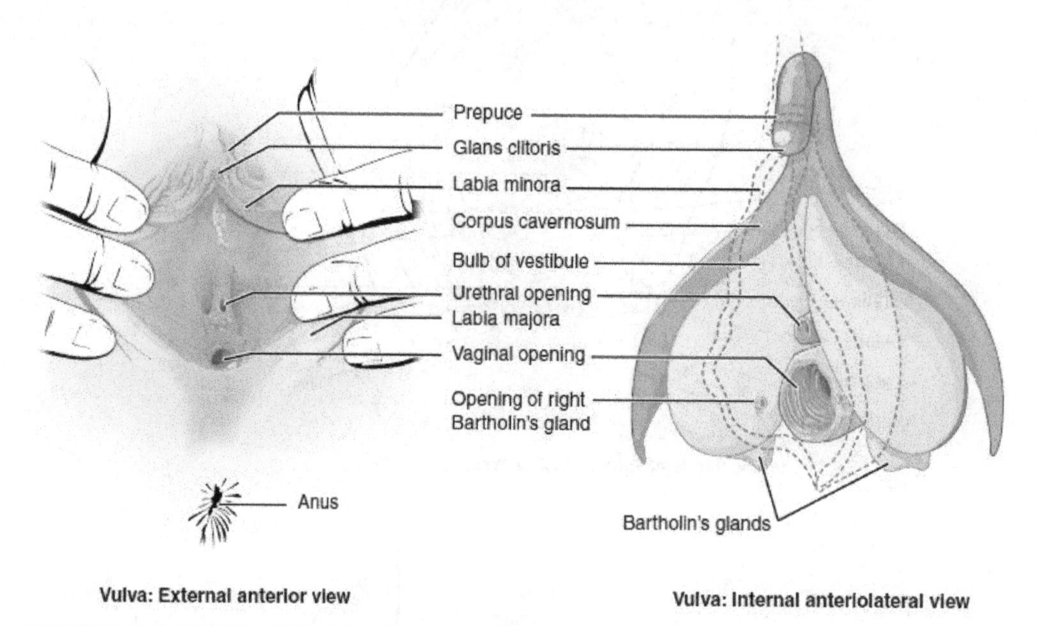

Figure 2. The external female genitalia are referred to collectively as the vulva.

Vagina

The **vagina** is a muscular canal (approximately 10 cm long) that serves as the entrance to the reproductive tract. It also serves as the exit from the uterus during menses and childbirth. The outer walls of the anterior and posterior vagina are formed into longitudinal columns, or ridges, and the superior portion of the vagina—called the fornix—meets the protruding uterine cervix. The walls of the vagina are lined with an outer, fibrous adventitia; a middle layer of smooth muscle; and an inner mucous membrane with transverse folds called **rugae**. Together, the middle and inner layers allow the expansion of the vagina to accommodate intercourse and childbirth. The thin, perforated hymen can partially surround the opening to the vaginal orifice. The hymen can be ruptured with strenuous physical exercise, penile–vaginal intercourse, and childbirth. The Bartholin's glands and the lesser vestibular glands (located near the clitoris) secrete mucus, which keeps the vestibular area moist.

The vagina is home to a normal population of microorganisms that help to protect against infection by pathogenic bacteria, yeast, or other organisms that can enter the vagina. In a healthy woman, the most predominant type of vaginal bacteria is from the genus *Lactobacillus*. This family of beneficial bacterial flora secretes lactic acid, and thus protects the vagina by maintaining an acidic pH (below 4.5). Potential pathogens are less likely to survive in these acidic conditions. Lactic acid, in combination with other vaginal secretions, makes the vagina a self-cleansing organ. However, douching—or washing out the vagina with fluid—can disrupt the normal balance of healthy microorganisms, and actually increase a woman's risk for infections and irritation. Indeed, the American College of Obstetricians and Gynecologists recommend that women do not douche, and that they allow the vagina to maintain its normal healthy population of protective microbial flora.

Ovaries

The **ovaries** are the female gonads. Paired ovals, they are each about 2 to 3 cm in length, about the size of an almond. The ovaries are located within the pelvic cavity, and are supported by the mesovarium, an extension of the peritoneum that connects the ovaries to the **broad ligament**. Extending from the mesovarium itself is the suspensory ligament that contains the ovarian blood and lymph vessels. Finally, the ovary itself is attached to the uterus via the ovarian ligament.

The ovary comprises an outer covering of cuboidal epithelium called the ovarian surface epithelium that is superficial to a dense connective tissue covering called the tunica albuginea. Beneath the tunica albuginea is the cortex, or outer portion, of the organ. The cortex is composed of a tissue framework called the ovarian stroma that forms the bulk of the adult ovary. Oocytes develop within the outer layer of this stroma, each surrounded by supporting cells. This grouping of an oocyte and its supporting cells is called a **follicle**. The growth and development of ovarian follicles will be described shortly. Beneath the cortex lies the inner ovarian medulla, the site of blood vessels, lymph vessels, and the nerves of the ovary. You will learn more about the overall anatomy of the female reproductive system at the end of this section.

The Ovarian Cycle

The **ovarian cycle** is a set of predictable changes in a female's oocytes and ovarian follicles. During a woman's reproductive years, it is a roughly 28-day cycle that can be correlated with, but is not the same as, the menstrual cycle (discussed shortly). The cycle includes two interrelated processes: oogenesis (the production of female gametes) and folliculogenesis (the growth and development of ovarian follicles).

Oogenesis

Gametogenesis in females is called **oogenesis**. The process begins with the ovarian stem cells, or **oogonia**. Oogonia are formed during fetal development, and divide via mitosis, much like spermatogonia in the testis. Unlike spermatogonia, however, oogonia form primary oocytes in the fetal ovary prior to birth. These primary oocytes are then arrested in this stage of meiosis I, only to resume it years later, beginning at puberty and continuing until the woman is near menopause (the cessation of a woman's reproductive functions). The number of primary oocytes present in the ovaries declines from one to two million in an infant, to approximately 400,000 at puberty, to zero by the end of menopause.

The initiation of **ovulation**—the release of an oocyte from the ovary—marks the transition from puberty into reproductive maturity for women. From then on, throughout a woman's reproductive years, ovulation occurs approximately once every 28 days. Just prior to ovulation, a surge of luteinizing hormone triggers the resumption of meiosis in a primary oocyte. This initiates the transition from primary to secondary oocyte. However, as you can see in Figure 3, this cell division does not result in two identical cells. Instead, the cytoplasm is divided unequally, and one daughter cell is much larger than the other. This larger cell, the secondary oocyte, eventually leaves the ovary during ovulation. The smaller cell, called the first **polar body**, may or may not complete meiosis and produce second polar bodies; in either case, it eventually disintegrates. Therefore, even though oogenesis produces up to four cells, only one survives.

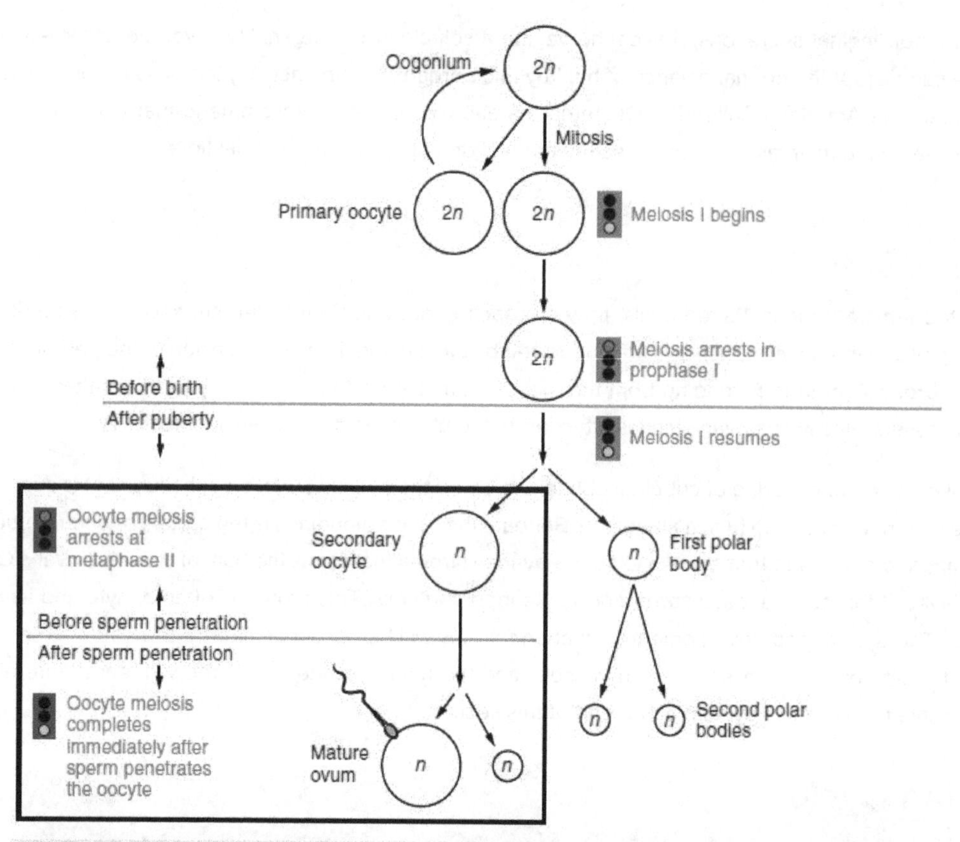

Figure 3. Click for a larger image. The unequal cell division of oogenesis produces one to three polar bodies that later degrade, as well as a single haploid ovum, which is produced only if there is penetration of the secondary oocyte by a sperm cell.

How does the diploid secondary oocyte become an **ovum**—the haploid female gamete? Meiosis of a secondary oocyte is completed only if a sperm succeeds in penetrating its barriers. Meiosis II then resumes, producing one haploid ovum that, at the instant of fertilization by a (haploid) sperm, becomes the first diploid cell of the new offspring (a zygote). Thus, the ovum can be thought of as a brief, transitional, haploid stage between the diploid oocyte and diploid zygote.

The larger amount of cytoplasm contained in the female gamete is used to supply the developing zygote with nutrients during the period between fertilization and implantation into the uterus. Interestingly, sperm contribute only DNA at fertilization —not cytoplasm. Therefore, the cytoplasm and all of the cytoplasmic organelles in the developing embryo are of maternal origin. This includes mitochondria, which contain their own DNA. Scientific research in the 1980s determined that mitochondrial DNA was maternally inherited, meaning that you can trace your mitochondrial DNA directly to your mother, her mother, and so on back through your female ancestors.

EVERYDAY CONNECTIONS: MAPPING HUMAN HISTORY WITH MITOCHONDRIAL DNA

When we talk about human DNA, we're usually referring to nuclear DNA; that is, the DNA coiled into chromosomal bundles in the nucleus of our cells. We inherit half of our nuclear DNA from our father, and half from our mother. However, mitochondrial DNA (mtDNA) comes only from the mitochondria in the cytoplasm of the fat ovum we inherit from our mother. She received her mtDNA from her mother, who got it from her mother, and so on. Each of our cells contains approximately 1700 mitochondria, with each mitochondrion packed with mtDNA containing approximately 37 genes.

Mutations (changes) in mtDNA occur spontaneously in a somewhat organized pattern at regular intervals in human history. By analyzing these mutational relationships, researchers have been able to determine that we can all trace our ancestry back to one woman who lived in Africa about 200,000 years ago. Scientists have given this woman the biblical name Eve, although she is not, of course, the first *Homo sapiens* female. More precisely, she is our most recent common ancestor through matrilineal descent.

This doesn't mean that everyone's mtDNA today looks exactly like that of our ancestral Eve. Because of the spontaneous mutations in mtDNA that have occurred over the centuries, researchers can map different "branches" off of the "main trunk" of our mtDNA family tree. Your mtDNA might have a pattern of mutations that aligns more closely with one branch, and your neighbor's may align with another branch. Still, all branches eventually lead back to Eve.

But what happened to the mtDNA of all of the other *Homo sapiens* females who were living at the time of Eve? Researchers explain that, over the centuries, their female descendants died childless or with only male children, and thus, their maternal line—and its mtDNA—ended.

Folliculogenesis

Again, ovarian follicles are oocytes and their supporting cells. They grow and develop in a process called **folliculogenesis**, which typically leads to ovulation of one follicle approximately every 28 days, along with death to multiple other follicles. The death of ovarian follicles is called atresia, and can occur at any point during follicular development. Recall that, a female infant at birth will have one to two million oocytes within her ovarian follicles, and that this number declines throughout life until menopause, when no follicles remain. As you'll see next, follicles progress from primordial, to primary, to secondary and tertiary stages prior to ovulation—with the oocyte inside the follicle remaining as a primary oocyte until right before ovulation.

Folliculogenesis begins with follicles in a resting state. These small **primordial follicles** are present in newborn females and are the prevailing follicle type in the adult ovary. Primordial follicles have only a single flat layer of support cells, called **granulosa cells**, that surround the oocyte, and they can stay in this resting state for years—some until right before menopause.

After puberty, a few primordial follicles will respond to a recruitment signal each day, and will join a pool of immature growing follicles called **primary follicles**. Primary follicles start with a single layer of granulosa cells, but the granulosa cells then become active and transition from a flat or squamous shape to a rounded, cuboidal shape as they increase in size and proliferate. As the granulosa cells divide, the follicles—now called **secondary follicles**—increase in diameter, adding a new outer layer of connective tissue, blood vessels, and **theca cells**—cells that work with the granulosa cells to produce estrogens.

Within the growing secondary follicle, the primary oocyte now secretes a thin acellular membrane called the zona pellucida that will play a critical role in fertilization. A thick fluid, called follicular fluid, that has formed between the granulosa cells also begins to collect into one large pool, or **antrum**. Follicles in which the antrum has become large and fully formed are considered **tertiary follicles** (or antral follicles). Several follicles reach the tertiary stage at the same time, and most of these will undergo atresia. The one that does not die will continue to grow and develop until ovulation, when it will expel its secondary oocyte surrounded by several layers of granulosa cells from the ovary. Keep in mind that most follicles don't make it to this point. In fact, roughly 99 percent of the follicles in the ovary will undergo atresia, which can occur at any stage of folliculogenesis.

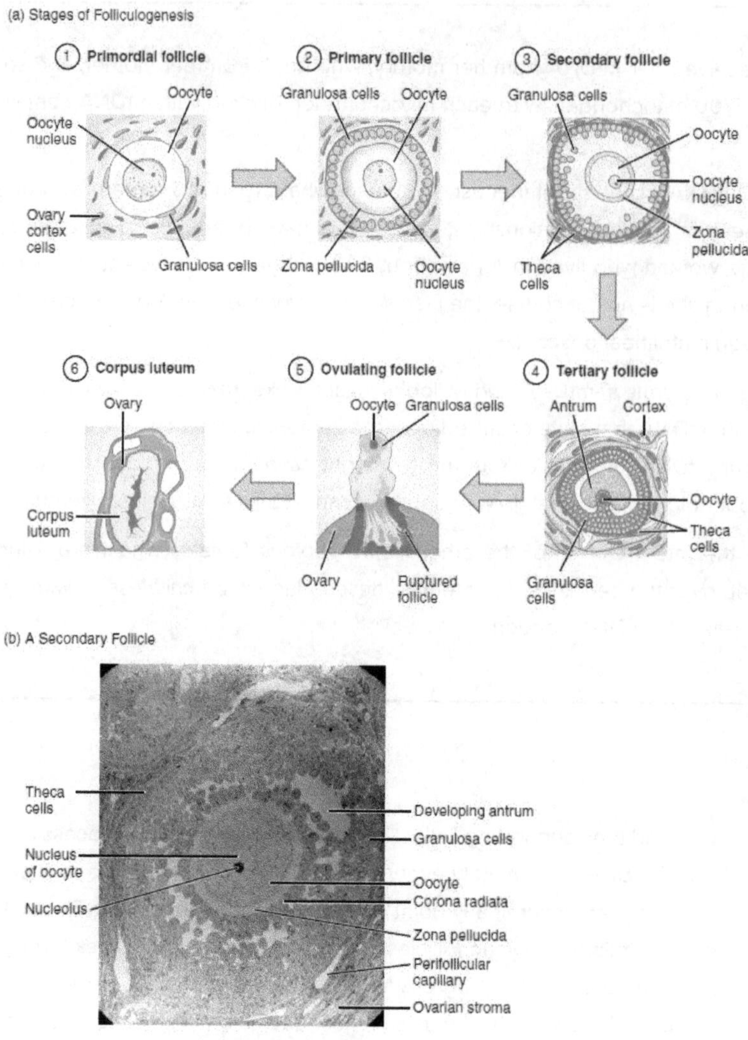

Figure 4. Click for a larger image. (a) The maturation of a follicle is shown in a clockwise direction proceeding from the primordial follicles. FSH stimulates the growth of a tertiary follicle, and LH stimulates the production of estrogen by granulosa and theca cells. Once the follicle is mature, it ruptures and releases the oocyte. Cells remaining in the follicle then develop into the corpus luteum. (b) In this electron micrograph of a secondary follicle, the oocyte, theca cells (thecae folliculi), and developing antrum are clearly visible. EM × 1100. (Micrograph provided by the Regents of University of Michigan Medical School © 2012)

Hormonal Control of the Ovarian Cycle

The process of development that we have just described, from primordial follicle to early tertiary follicle, takes approximately two months in humans. The final stages of development of a small cohort of tertiary follicles, ending with ovulation of a secondary oocyte, occur over a course of approximately 28 days. These changes are regulated by many of the same hormones that regulate the male reproductive system, including GnRH, LH, and FSH.

As in men, the hypothalamus produces GnRH, a hormone that signals the anterior pituitary gland to produce the gonadotropins FSH and LH. These gonadotropins leave the pituitary and travel through the bloodstream to the ovaries, where they bind to receptors on the granulosa and theca cells of the follicles. FSH stimulates the follicles to grow (hence its name of follicle-stimulating hormone), and the five or six tertiary follicles expand in diameter. The release of LH also stimulates the granulosa and theca cells of the follicles to produce the sex steroid hormone estradiol, a type of estrogen. This phase of the ovarian cycle, when the tertiary follicles are growing and secreting estrogen, is known as the follicular phase.

The more granulosa and theca cells a follicle has (that is, the larger and more developed it is), the more estrogen it will produce in response to LH stimulation. As a result of these large follicles producing large amounts of estrogen, systemic plasma estrogen concentrations increase. Following a classic negative feedback loop, the high concentrations of estrogen will stimulate the hypothalamus and pituitary to reduce the production of GnRH, LH, and FSH. Because the large tertiary follicles require FSH to grow and survive at this point, this decline in FSH caused by negative feedback leads most of them to die (atresia). Typically only one follicle, now called the dominant follicle, will survive this reduction in FSH, and this follicle will be the one that releases an oocyte. Scientists have studied many factors that lead to a particular follicle becoming dominant: size, the number of granulosa cells, and the number of FSH receptors on those granulosa cells all contribute to a follicle becoming the one surviving dominant follicle.

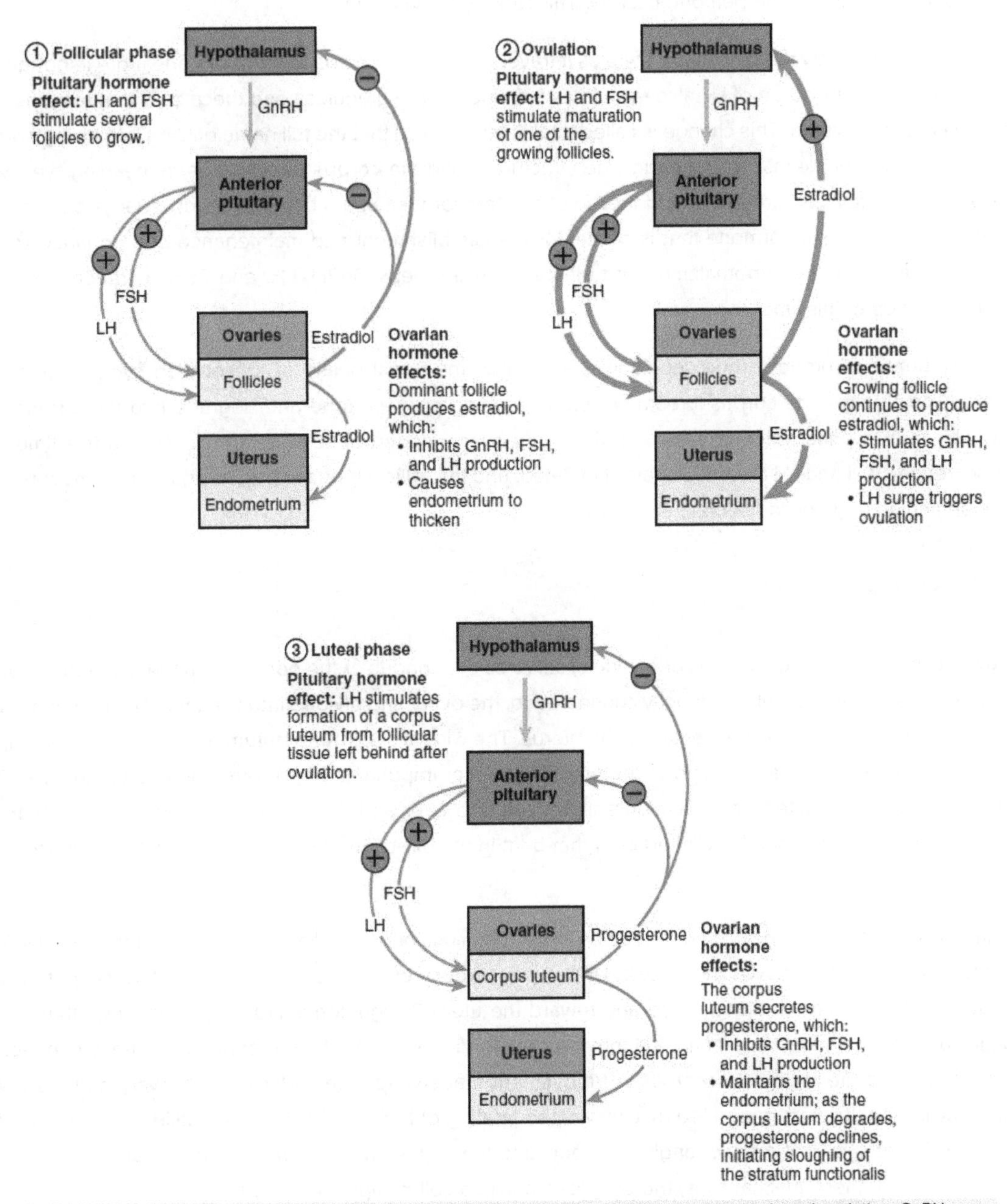

Figure 5. Click for a larger image. The hypothalamus and pituitary gland regulate the ovarian cycle and ovulation. GnRH activates the anterior pituitary to produce LH and FSH, which stimulate the production of estrogen and progesterone by the ovaries.

When only the one dominant follicle remains in the ovary, it again begins to secrete estrogen. It produces more estrogen than all of the developing follicles did together before the negative feedback occurred. It produces so much estrogen that the normal negative feedback doesn't occur. Instead, these extremely high concentrations of systemic plasma estrogen trigger a regulatory switch in the anterior pituitary that responds by secreting large amounts of LH and FSH into the bloodstream. The positive feedback loop by which more estrogen triggers release of more LH and FSH only occurs at this point in the cycle.

It is this large burst of LH (called the LH surge) that leads to ovulation of the dominant follicle. The LH surge induces many changes in the dominant follicle, including stimulating the resumption of meiosis of the primary oocyte to a secondary oocyte. As noted earlier, the polar body that results from unequal cell division simply degrades. The LH surge also triggers proteases (enzymes that cleave proteins) to break down structural proteins in the ovary wall on the surface of the bulging dominant follicle. This degradation of the wall, combined with pressure from the large, fluid-filled antrum, results in the expulsion of the oocyte surrounded by granulosa cells into the peritoneal cavity. This release is ovulation.

In the next section, you will follow the ovulated oocyte as it travels toward the uterus, but there is one more important event that occurs in the ovarian cycle. The surge of LH also stimulates a change in the granulosa and theca cells that remain in the follicle after the oocyte has been ovulated. This change is called luteinization (recall that the full name of LH is luteinizing hormone), and it transforms the collapsed follicle into a new endocrine structure called the **corpus luteum**, a term meaning "yellowish body". Instead of estrogen, the luteinized granulosa and theca cells of the corpus luteum begin to produce large amounts of the sex steroid hormone progesterone, a hormone that is critical for the establishment and maintenance of pregnancy. Progesterone triggers negative feedback at the hypothalamus and pituitary, which keeps GnRH, LH, and FSH secretions low, so no new dominant follicles develop at this time.

The post-ovulatory phase of progesterone secretion is known as the luteal phase of the ovarian cycle. If pregnancy does not occur within 10 to 12 days, the corpus luteum will stop secreting progesterone and degrade into the **corpus albicans**, a nonfunctional "whitish body" that will disintegrate in the ovary over a period of several months. During this time of reduced progesterone secretion, FSH and LH are once again stimulated, and the follicular phase begins again with a new cohort of early tertiary follicles beginning to grow and secrete estrogen.

The Uterine Tubes

The **uterine tubes** (also called fallopian tubes or oviducts) serve as the conduit of the oocyte from the ovary to the uterus. Each of the two uterine tubes is close to, but not directly connected to, the ovary and divided into sections. The **isthmus** is the narrow medial end of each uterine tube that is connected to the uterus. The wide distal **infundibulum** flares out with slender, finger-like projections called **fimbriae**. The middle region of the tube, called the **ampulla**, is where fertilization often occurs. The uterine tubes also have three layers: an outer serosa, a middle smooth muscle layer, and an inner mucosal layer. In addition to its mucus-secreting cells, the inner mucosa contains ciliated cells that beat in the direction of the uterus, producing a current that will be critical to move the oocyte.

Following ovulation, the secondary oocyte surrounded by a few granulosa cells is released into the peritoneal cavity. The nearby uterine tube, either left or right, receives the oocyte. Unlike sperm, oocytes lack flagella, and therefore cannot move on their own. So how do they travel into the uterine tube and toward the uterus? High concentrations of estrogen that occur around the time of ovulation induce contractions of the smooth muscle along the length of the uterine tube. These contractions occur every 4 to 8 seconds, and the result is a coordinated movement that sweeps the surface of the ovary and the pelvic cavity. Current flowing toward the uterus is generated by coordinated beating of the cilia that line the outside and lumen of the length of the uterine tube. These cilia beat more strongly in response to the high estrogen concentrations that occur around the time of ovulation. As a result of these mechanisms, the oocyte–granulosa cell complex is pulled into the interior of the tube. Once inside, the muscular contractions and beating cilia move the oocyte slowly toward the uterus. When fertilization does occur, sperm typically meet the egg while it is still moving through the ampulla.

If the oocyte is successfully fertilized, the resulting zygote will begin to divide into two cells, then four, and so on, as it makes its way through the uterine tube and into the uterus. There, it will implant and continue to grow. If the egg is not fertilized, it will simply degrade—either in the uterine tube or in the uterus, where it may be shed with the next menstrual period.

The open-ended structure of the uterine tubes can have significant health consequences if bacteria or other contagions enter through the vagina and move through the uterus, into the tubes, and then into the pelvic cavity. If this is left unchecked, a bacterial infection (sepsis) could quickly become life-threatening. The spread of an infection in this manner is of special concern when unskilled practitioners perform abortions in non-sterile conditions. Sepsis is also associated with sexually transmitted bacterial infections, especially gonorrhea and chlamydia. These increase a woman's risk for pelvic inflammatory disease (PID), infection of the uterine tubes or other reproductive organs. Even when resolved, PID can leave scar tissue in the tubes, leading to infertility.

The Uterus and Cervix

The **uterus** is the muscular organ that nourishes and supports the growing embryo. Its average size is approximately 5 cm wide by 7 cm long (approximately 2 in by 3 in) when a female is not pregnant. It has three sections. The portion of the uterus superior to the opening of the uterine tubes is called the **fundus**. The middle section of the uterus is called the **body of uterus** (or corpus). The **cervix** is the narrow inferior portion of the uterus that projects into the vagina. The cervix produces mucus secretions that become thin and stringy under the influence of high systemic plasma estrogen concentrations, and these secretions can facilitate sperm movement through the reproductive tract.

Several ligaments maintain the position of the uterus within the abdominopelvic cavity. The broad ligament is a fold of peritoneum that serves as a primary support for the uterus, extending laterally from both sides of the uterus and attaching it to the pelvic wall. The round ligament attaches to the uterus near the uterine tubes, and extends to the labia majora. Finally, the uterosacral ligament stabilizes the uterus posteriorly by its connection from the cervix to the pelvic wall.

The wall of the uterus is made up of three layers. The most superficial layer is the serous membrane, or **perimetrium**, which consists of epithelial tissue that covers the exterior portion of the uterus. The middle layer, or **myometrium**, is a thick layer of smooth muscle responsible for uterine contractions. Most of the uterus is myometrial tissue, and the muscle fibers run horizontally, vertically, and diagonally, allowing the powerful contractions that occur during labor and the less powerful contractions (or cramps) that help to expel menstrual blood during a woman's period. Anteriorly directed myometrial contractions also occur near the time of ovulation, and are thought to possibly facilitate the transport of sperm through the female reproductive tract.

The innermost layer of the uterus is called the **endometrium**. The endometrium contains a connective tissue lining, the lamina propria, which is covered by epithelial tissue that lines the lumen. Structurally, the endometrium consists of two layers: the stratum basalis and the stratum functionalis (the basal and functional layers). The stratum basalis layer is part of the lamina propria and is adjacent to the myometrium; this layer does not shed during menses. In contrast, the thicker stratum functionalis layer contains the glandular portion of the lamina propria and the endothelial tissue that lines the uterine lumen. It is the stratum functionalis that grows and thickens in response to increased levels of estrogen and progesterone. In the luteal phase of the menstrual cycle, special branches off of the uterine artery called spiral arteries supply the thickened stratum functionalis. This inner functional layer provides the proper site of implantation for the fertilized egg, and—should fertilization not occur—it is only the stratum functionalis layer of the endometrium that sheds during menstruation.

Recall that during the follicular phase of the ovarian cycle, the tertiary follicles are growing and secreting estrogen. At the same time, the stratum functionalis of the endometrium is thickening to prepare for a potential implantation. The post-ovulatory increase in progesterone, which characterizes the luteal phase, is key for maintaining a thick stratum functionalis. As long as a functional corpus luteum is present in the ovary, the endometrial lining is prepared for implantation. Indeed, if an embryo implants, signals are sent to the corpus luteum to continue secreting progesterone to maintain the endometrium, and thus maintain the pregnancy. If an embryo does not implant, no signal is sent to the corpus luteum and it degrades, ceasing progesterone production and ending the luteal phase. Without progesterone, the endometrium thins and, under the influence of prostaglandins, the spiral arteries of the endometrium constrict and rupture, preventing oxygenated blood from reaching the endometrial tissue. As a result, endometrial tissue dies and blood, pieces of the endometrial tissue, and white blood cells are shed through the vagina during menstruation, or the **menses**. The first menses after puberty, called **menarche**, can occur either before or after the first ovulation.

The Menstrual Cycle

Now that we have discussed the maturation of the cohort of tertiary follicles in the ovary, the build-up and then shedding of the endometrial lining in the uterus, and the function of the uterine tubes and vagina, we can put everything together to talk about the three phases of the **menstrual cycle**—the series of changes in which the uterine lining is shed, rebuilds, and prepares for implantation.

The timing of the menstrual cycle starts with the first day of menses, referred to as day one of a woman's period. Cycle length is determined by counting the days between the onset of bleeding in two subsequent cycles. Because the average length of a woman's menstrual cycle is 28 days, this is the time period used to identify the timing of events in the cycle. However, the length of the menstrual cycle varies among women, and even in the same woman from one cycle to the next, typically from 21 to 32 days.

Just as the hormones produced by the granulosa and theca cells of the ovary "drive" the follicular and luteal phases of the ovarian cycle, they also control the three distinct phases of the menstrual cycle. These are the menses phase, the proliferative phase, and the secretory phase.

Menses Phase

The **menses phase** of the menstrual cycle is the phase during which the lining is shed; that is, the days that the woman menstruates. Although it averages approximately five days, the menses phase can last from 2 to 7 days, or longer. As shown in Figure 7, the menses phase occurs during the early days of the follicular phase of the ovarian cycle, when progesterone, FSH, and LH levels are low. Recall that progesterone concentrations decline as a result of the degradation of the corpus luteum, marking the end of the luteal phase. This decline in progesterone triggers the shedding of the stratum functionalis of the endometrium.

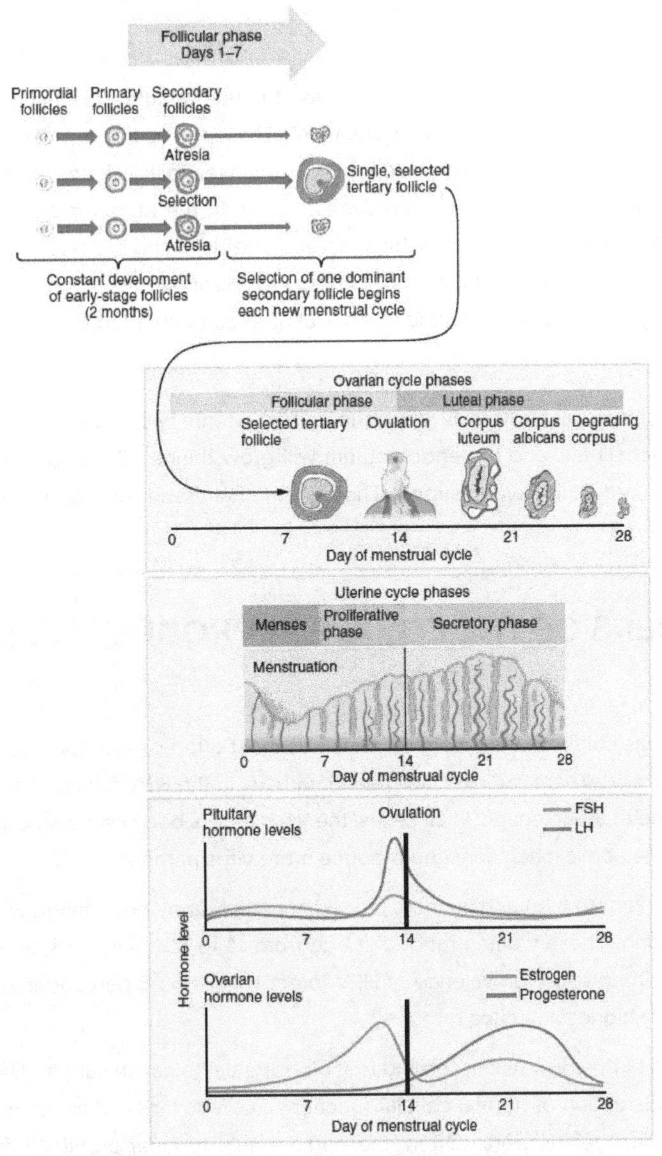

Figure 7. Click for a larger image. The correlation of the hormone levels and their effects on the female reproductive system is shown in this timeline of the ovarian and menstrual cycles. The menstrual cycle begins at day one with the start of menses. Ovulation occurs around day 14 of a 28-day cycle, triggered by the LH surge.

Proliferative Phase

Once menstrual flow ceases, the endometrium begins to proliferate again, marking the beginning of the **proliferative phase** of the menstrual cycle. It occurs when the granulosa and theca cells of the tertiary follicles begin to produce increased amounts of estrogen. These rising estrogen concentrations stimulate the endometrial lining to rebuild.

Recall that the high estrogen concentrations will eventually lead to a decrease in FSH as a result of negative feedback, resulting in atresia of all but one of the developing tertiary follicles. The switch to positive feedback—which occurs with the elevated estrogen production from the dominant follicle—then stimulates the LH surge that will trigger ovulation. In a typical 28-day menstrual cycle, ovulation occurs on day 14. Ovulation marks the end of the proliferative phase as well as the end of the follicular phase.

Secretory Phase

In addition to prompting the LH surge, high estrogen levels increase the uterine tube contractions that facilitate the pick-up and transfer of the ovulated oocyte. High estrogen levels also slightly decrease the acidity of the vagina, making it more hospitable to sperm. In the ovary, the luteinization of the granulosa cells of the collapsed follicle forms the progesterone-producing corpus luteum, marking the beginning of the luteal phase of the ovarian cycle. In the uterus, progesterone from the corpus luteum begins the **secretory phase** of the menstrual cycle, in which the endometrial lining prepares for implantation. Over the next 10 to 12 days, the endometrial glands secrete a fluid rich in glycogen. If fertilization has occurred, this fluid will nourish the ball of cells now developing from the zygote. At the same time, the spiral arteries develop to provide blood to the thickened stratum functionalis.

If no pregnancy occurs within approximately 10 to 12 days, the corpus luteum will degrade into the corpus albicans. Levels of both estrogen and progesterone will fall, and the endometrium will grow thinner. Prostaglandins will be secreted that cause constriction of the spiral arteries, reducing oxygen supply. The endometrial tissue will die, resulting in menses—or the first day of the next cycle.

DISORDERS OF THE FEMALE REPRODUCTIVE SYSTEM

Research over many years has confirmed that cervical cancer is most often caused by a sexually transmitted infection with human papillomavirus (HPV). There are over 100 related viruses in the HPV family, and the characteristics of each strain determine the outcome of the infection. In all cases, the virus enters body cells and uses its own genetic material to take over the host cell's metabolic machinery and produce more virus particles.

HPV infections are common in both men and women. Indeed, a recent study determined that 42.5 percent of females had HPV at the time of testing. These women ranged in age from 14 to 59 years and differed in race, ethnicity, and number of sexual partners. Of note, the prevalence of HPV infection was 53.8 percent among women aged 20 to 24 years, the age group with the highest infection rate.

HPV strains are classified as high or low risk according to their potential to cause cancer. Though most HPV infections do not cause disease, the disruption of normal cellular functions in the low-risk forms of HPV can cause the male or female human host to develop genital warts. Often, the body is able to clear an HPV infection by normal immune responses within 2 years. However, the more serious, high-risk infection by certain types of HPV can result in cancer of the cervix. Infection with either of the cancer-causing variants HPV 16 or HPV 18 has been linked to more than 70 percent of all cervical cancer diagnoses. Although even these high-risk HPV strains can be cleared from the body over time, infections persist in some individuals. If this happens, the HPV infection can influence the cells of the cervix to develop precancerous changes.

Risk factors for cervical cancer include having unprotected sex; having multiple sexual partners; a first sexual experience at a younger age, when the cells of the cervix are not fully mature; failure to receive the HPV vaccine; a compromised immune system; and smoking. The risk of developing cervical cancer is doubled with cigarette smoking.

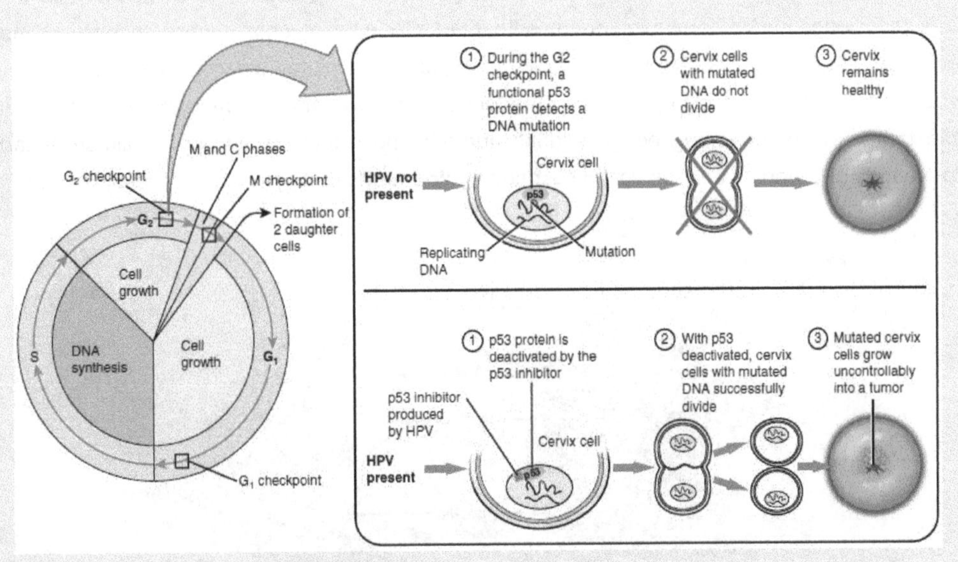

Figure 8. In most cases, cells infected with the HPV virus heal on their own. In some cases, however, the virus continues to spread and becomes an invasive cancer.

When the high-risk types of HPV enter a cell, two viral proteins are used to neutralize proteins that the host cells use as checkpoints in the cell cycle. The best studied of these proteins is p53. In a normal cell, p53 detects DNA damage in the cell's genome and either halts the progression of the cell cycle—allowing time for DNA repair to occur—or initiates apoptosis. Both of these processes prevent the accumulation of mutations in a cell's genome. High-risk HPV can neutralize p53, keeping the cell in a state in which fast growth is possible and impairing apoptosis, allowing mutations to accumulate in the cellular DNA.

The prevalence of cervical cancer in the United States is very low because of regular screening exams called pap smears. Pap smears sample cells of the cervix, allowing the detection of abnormal cells. If pre-cancerous cells are detected, there are several highly effective techniques that are currently in use to remove them before they pose a danger. However, women in developing countries often do not have access to regular pap smears. As a result, these women account for as many as 80 percent of the cases of cervical cancer worldwide.

In 2006, the first vaccine against the high-risk types of HPV was approved. There are now two HPV vaccines available: Gardasil® and Cervarix®. Whereas these vaccines were initially only targeted for women, because HPV is sexually transmitted, both men and women require vaccination for this approach to achieve its maximum efficacy. A recent study suggests that the HPV vaccine has cut the rates of HPV infection by the four targeted strains at least in half. Unfortunately, the high cost of manufacturing the vaccine is currently limiting access to many women worldwide.

The Breasts

Whereas the breasts are located far from the other female reproductive organs, they are considered accessory organs of the female reproductive system. The function of the breasts is to supply milk to an infant in a process called lactation. The external features of the breast include a nipple surrounded by a pigmented **areola**, whose coloration may deepen during pregnancy. The areola is typically circular and can vary in size from 25 to 100 mm in diameter. The areolar region is characterized by small, raised areolar glands that secrete lubricating fluid during lactation to protect the nipple from chafing. When a baby nurses, or draws milk from the breast, the entire areolar region is taken into the mouth.

Breast milk is produced by the **mammary glands**, which are modified sweat glands. The milk itself exits the breast through the nipple via 15 to 20 **lactiferous ducts** that open on the surface of the nipple. These lactiferous ducts each extend to a **lactiferous**

sinus that connects to a glandular lobe within the breast itself that contains groups of milk-secreting cells in clusters called **alveoli**. The clusters can change in size depending on the amount of milk in the alveolar lumen. Once milk is made in the alveoli, stimulated myoepithelial cells that surround the alveoli contract to push the milk to the lactiferous sinuses. From here, the baby can draw milk through the lactiferous ducts by suckling. The lobes themselves are surrounded by fat tissue, which determines the size of the breast; breast size differs between individuals and does not affect the amount of milk produced. Supporting the breasts are multiple bands of connective tissue called **suspensory ligaments** that connect the breast tissue to the dermis of the overlying skin.

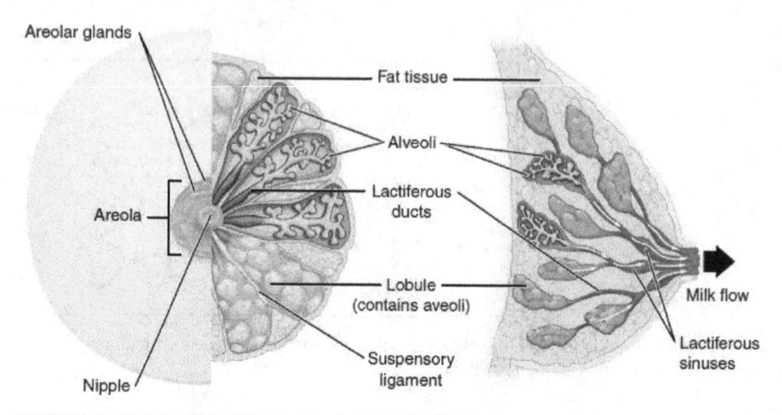

Figure 9. During lactation, milk moves from the alveoli through the lactiferous ducts to the nipple.

During the normal hormonal fluctuations in the menstrual cycle, breast tissue responds to changing levels of estrogen and progesterone, which can lead to swelling and breast tenderness in some individuals, especially during the secretory phase. If pregnancy occurs, the increase in hormones leads to further development of the mammary tissue and enlargement of the breasts.

Hormonal Birth Control

Birth control pills take advantage of the negative feedback system that regulates the ovarian and menstrual cycles to stop ovulation and prevent pregnancy. Typically they work by providing a constant level of both estrogen and progesterone, which negatively feeds back onto the hypothalamus and pituitary, thus preventing the release of FSH and LH. Without FSH, the follicles do not mature, and without the LH surge, ovulation does not occur. Although the estrogen in birth control pills does stimulate some thickening of the endometrial wall, it is reduced compared with a normal cycle and is less likely to support implantation.

Some birth control pills contain 21 active pills containing hormones, and 7 inactive pills (placebos). The decline in hormones during the week that the woman takes the placebo pills triggers menses, although it is typically lighter than a normal menstrual flow because of the reduced endometrial thickening. Newer types of birth control pills have been developed that deliver low-dose estrogens and progesterone for the entire cycle (these are meant to be taken 365 days a year), and menses never occurs. While some women prefer to have the proof of a lack of pregnancy that a monthly period provides, menstruation every 28 days is not required for health reasons, and there are no reported adverse effects of not having a menstrual period in an otherwise healthy individual.

Because birth control pills function by providing constant estrogen and progesterone levels and disrupting negative feedback, skipping even just one or two pills at certain points of the cycle (or even being several hours late taking the pill) can lead to an increase in FSH and LH and result in ovulation. It is important, therefore, that the woman follow the directions on the birth control pill package to successfully prevent pregnancy.

AGING AND THE FEMALE REPRODUCTIVE SYSTEM

Female fertility (the ability to conceive) peaks when women are in their twenties, and is slowly reduced until a women reaches 35 years of age. After that time, fertility declines more rapidly, until it ends completely at the end of menopause. Menopause is the cessation of the menstrual cycle that occurs as a result of the loss of ovarian follicles and the hormones that they produce. A woman is considered to have completed menopause if she has not menstruated in a full year. After that point, she is considered postmenopausal. The average age for this change is consistent worldwide at between 50 and 52 years of age, but it can normally occur in a woman's forties, or later in her fifties. Poor health, including smoking, can lead to earlier loss of fertility and earlier menopause.

As a woman reaches the age of menopause, depletion of the number of viable follicles in the ovaries due to atresia affects the hormonal regulation of the menstrual cycle. During the years leading up to menopause, there is a decrease in the levels of the hormone inhibin, which normally participates in a negative feedback loop to the pituitary to control the production of FSH. The menopausal decrease in inhibin leads to an increase in FSH. The presence of FSH stimulates more follicles to grow and secrete estrogen. Because small, secondary follicles also respond to increases in FSH levels, larger numbers of follicles are stimulated to grow; however, most undergo atresia and die. Eventually, this process leads to the depletion of all follicles in the ovaries, and the production of estrogen falls off dramatically. It is primarily the lack of estrogens that leads to the symptoms of menopause.

The earliest changes occur during the menopausal transition, often referred to as peri-menopause, when a women's cycle becomes irregular but does not stop entirely. Although the levels of estrogen are still nearly the same as before the transition, the level of progesterone produced by the corpus luteum is reduced. This decline in progesterone can lead to abnormal growth, or hyperplasia, of the endometrium. This condition is a concern because it increases the risk of developing endometrial cancer. Two harmless conditions that can develop during the transition are uterine fibroids, which are benign masses of cells, and irregular bleeding. As estrogen levels change, other symptoms that occur are hot flashes and night sweats, trouble sleeping, vaginal dryness, mood swings, difficulty focusing, and thinning of hair on the head along with the growth of more hair on the face. Depending on the individual, these symptoms can be entirely absent, moderate, or severe.

After menopause, lower amounts of estrogens can lead to other changes. Cardiovascular disease becomes as prevalent in women as in men, possibly because estrogens reduce the amount of cholesterol in the blood vessels. When estrogen is lacking, many women find that they suddenly have problems with high cholesterol and the cardiovascular issues that accompany it. Osteoporosis is another problem because bone density decreases rapidly in the first years after menopause. The reduction in bone density leads to a higher incidence of fractures.

Hormone therapy (HT), which employs medication (synthetic estrogens and progestins) to increase estrogen and progestin levels, can alleviate some of the symptoms of menopause. In 2002, the Women's Health Initiative began a study to observe women for the long-term outcomes of hormone replacement therapy over 8.5 years. However, the study was prematurely terminated after 5.2 years because of evidence of a higher than normal risk of breast cancer in patients taking estrogen-only HT. The potential positive effects on cardiovascular disease were also not realized in the estrogen-only patients. The results of other hormone replacement studies over the last 50 years, including a 2012 study that followed over 1,000 menopausal women for 10 years, have shown cardiovascular benefits from estrogen and no increased risk for cancer. Some researchers believe that the age group tested in the 2002 trial may have been too old to benefit from the therapy, thus skewing the results. In the meantime, intense debate and study of the benefits and risks of replacement therapy is ongoing. Current guidelines approve HT for the reduction of hot flashes or flushes, but this treatment is generally only considered when women first start showing signs of menopausal changes, is used in the lowest dose possible for the shortest time possible (5 years or less), and it is suggested that women on HT have regular pelvic and breast exams.

Types of Contraceptive Methods

Birth Control Methods

Many elements need to be considered by women, men, or couples at any given point in their lifetimes when choosing the most appropriate contraceptive method. These elements include safety, effectiveness, availability (including accessibility and affordability), and acceptability. Voluntary informed choice of contraceptive methods is an essential guiding principle, and contraceptive counseling, when applicable, might be an important contributor to the successful use of contraceptive methods.

In choosing a method of contraception, dual protection from the simultaneous risk for HIV and other STDs also should be considered. Although hormonal contraceptives and IUDs are highly effective at preventing pregnancy, they do not protect against STDs, including HIV. Consistent and correct use of the male latex condom reduces the risk for HIV infection and other STDs, including chlamydial infection, gonococcal infection, and trichomoniasis.

Reversible Methods of Birth Control

Intrauterine Contraception

- **Copper T intrauterine device (IUD)** —This IUD is a small device that is shaped in the form of a "T." Your doctor places it inside the uterus to prevent pregnancy. It can stay in your uterus for up to 10 years. Typical use failure rate: 0.8%.[1]
- **Levonorgestrel intrauterine system (LNG IUD)**—The LNG IUD is a small T-shaped device like the Copper T IUD. It is placed inside the uterus by a doctor. It releases a small amount of progestin each day to keep you from getting pregnant. The LNG IUD stays in your uterus for up to 5 years. Typical use failure rate: 0.1-0.4%.[1]

Hormonal Methods

- **Implant**—The implant is a single, thin rod that is inserted under the skin of a women's upper arm. The rod contains a progestin that is released into the body over 3 years. Typical use failure rate: 0.01%.[1]
- **Injection or "shot"**—Women get shots of the hormone progestin in the buttocks or arm every three months from their doctor. Typical use failure rate: 4%.[1]
- **Combined oral contraceptives**—Also called "the pill," combined oral contraceptives contain the hormones estrogen and progestin. It is prescribed by a doctor. A pill is taken at the same time each day. If you are older than 35 years and smoke, have a history of blood clots or breast cancer, your doctor may advise you not to take the pill. Typical use failure rate: 7%.[1]
- **Progestin only pill**—Unlike the combined pill, the progestin-only pill (sometimes called the mini-pill) only has one hormone, progestin, instead of both estrogen and progestin. It is prescribed by a doctor. It is taken at the same time each day. It may be a good option for women who can't take estrogen. Typical use failure rate: 7%.[1]
- **Patch**—This skin patch is worn on the lower abdomen, buttocks, or upper body (but not on the breasts). This method is prescribed by a doctor. It releases hormones progestin and estrogen into the bloodstream. You put on a new patch once a week for three weeks. During the fourth week, you do not wear a patch, so you can have a menstrual period. Typical use failure rate: 7%.[1]

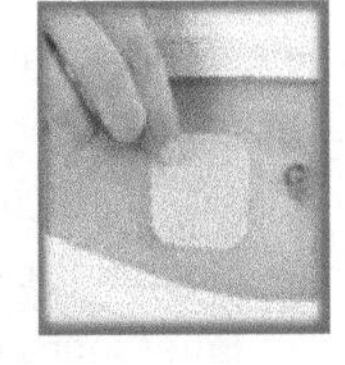

- **Hormonal vaginal contraceptive ring**—The ring releases the hormones progestin and estrogen. You place the ring inside your vagina. You wear the ring for three weeks, take it out for the week you have your period, and then put in a new ring. Typical use failure rate: 7%.[1]

Barrier Methods

- **Diaphragm or cervical cap**—Each of these barrier methods are placed inside the vagina to cover the cervix to block sperm. The diaphragm is shaped like a shallow cup. The cervical cap is a thimble-shaped cup. Before sexual intercourse, you insert them with spermicide to block or kill sperm. Visit your doctor for a proper fitting because diaphragms and cervical caps come in different sizes. Typical use failure rate for the diaphragm: 17%.[1]
- **Sponge**—The contraceptive sponge contains spermicide and is placed in the vagina where it fits over the cervix. The sponge works for up to 24 hours, and must be left in the vagina for at least 6 hours after the last act of intercourse, at which time it is removed and discarded. Typical use failure rate: 14% for women who have never had a baby and 27% for women who have had a baby.[1]
- **Male condom**—Worn by the man, a male condom keeps sperm from getting into a woman's body. Latex condoms, the most common type, help prevent pregnancy, and HIV and other STDs, as do the newer synthetic condoms. "Natural" or "lambskin" condoms also help prevent pregnancy, but may not provide protection against STDs, including HIV. Typical use failure rate: 13%.[1] Condoms can only be used once. You can buy condoms, KY jelly, or water-based lubricants at a drug store. Do not use oil-based lubricants such as massage oils, baby oil, lotions, or petroleum jelly with latex condoms. They will weaken the condom, causing it to tear or break.
- **Female condom**—Worn by the woman, the female condom helps keeps sperm from getting into her body. It is packaged with a lubricant and is available at drug stores. It can be inserted up to eight hours before sexual intercourse. Typical use failure rate: 21%,[1] and also may help prevent STDs.
- **Spermicides**—These products work by killing sperm and come in several forms—foam, gel, cream, film, suppository, or tablet. They are placed in the vagina no more than one hour before intercourse. You leave them in place at least six to eight hours after intercourse. You can use a spermicide in addition to a male condom, diaphragm, or cervical cap. They can be purchased at drug stores. Typical use failure rate: 21%.[1]

Fertility Awareness-Based Methods

- **Fertility awareness-based methods**—Understanding your monthly fertility patternExternal can help you plan to get pregnant or avoid getting pregnant. Your fertility pattern is the number of days in the month when you are fertile (able to get pregnant), days when you are infertile, and days when fertility is unlikely, but possible. If you have a regular menstrual cycle, you have about nine or more fertile days each month. If you do not want to get pregnant, you do not have sex on the days you are fertile, or you use a barrier method of birth control on those days. Failure rates vary across these methods.[1-2] Range of typical use failure rates: 2-23%.[1]

Lactational Amenorrhea Method

For women who have recently had a baby and are breastfeeding, the Lactational Amenorrhea Method (LAM) can be used as birth control when three conditions are met: 1) amenorrhea (not having any menstrual periods after delivering a baby), 2) fully or nearly fully breastfeeding, and 3) less than 6 months after delivering a baby. LAM is a temporary method of birth control, and another birth control method must be used when any of the three conditions are not met.

Emergency Contraception

Emergency contraception is NOT a regular method of birth control. Emergency contraception can be used after no birth control was used during sex, or if the birth control method failed, such as if a condom broke.

- Copper IUD—Women can have the copper T IUD inserted within five days of unprotected sex.
- Emergency contraceptive pills—Women can take emergency contraceptive pills up to 5 days after unprotected sex, but the sooner the pills are taken, the better they will work. There are three different types of emergency contraceptive pills

available in the United States. Some emergency contraceptive pills are available over the counter.

Permanent Methods of Birth Control

- **Female Sterilization—Tubal ligation or "tying tubes"**— A woman can have her fallopian tubes tied (or closed) so that sperm and eggs cannot meet for fertilization. The procedure can be done in a hospital or in an outpatient surgical center. You can go home the same day of the surgery and resume your normal activities within a few days. This method is effective immediately. Typical use failure rate: 0.5%.[1]
- **Male Sterilization–Vasectomy**—This operation is done to keep a man's sperm from going to his penis, so his ejaculate never has any sperm in it that can fertilize an egg. The procedure is typically done at an outpatient surgical center. The man can go home the same day. Recovery time is less than one week. After the operation, a man visits his doctor for tests to count his sperm and to make sure the sperm count has dropped to zero; this takes about 12 weeks. Another form of birth control should be used until the man's sperm count has dropped to zero. Typical use failure rate: 0.15%.

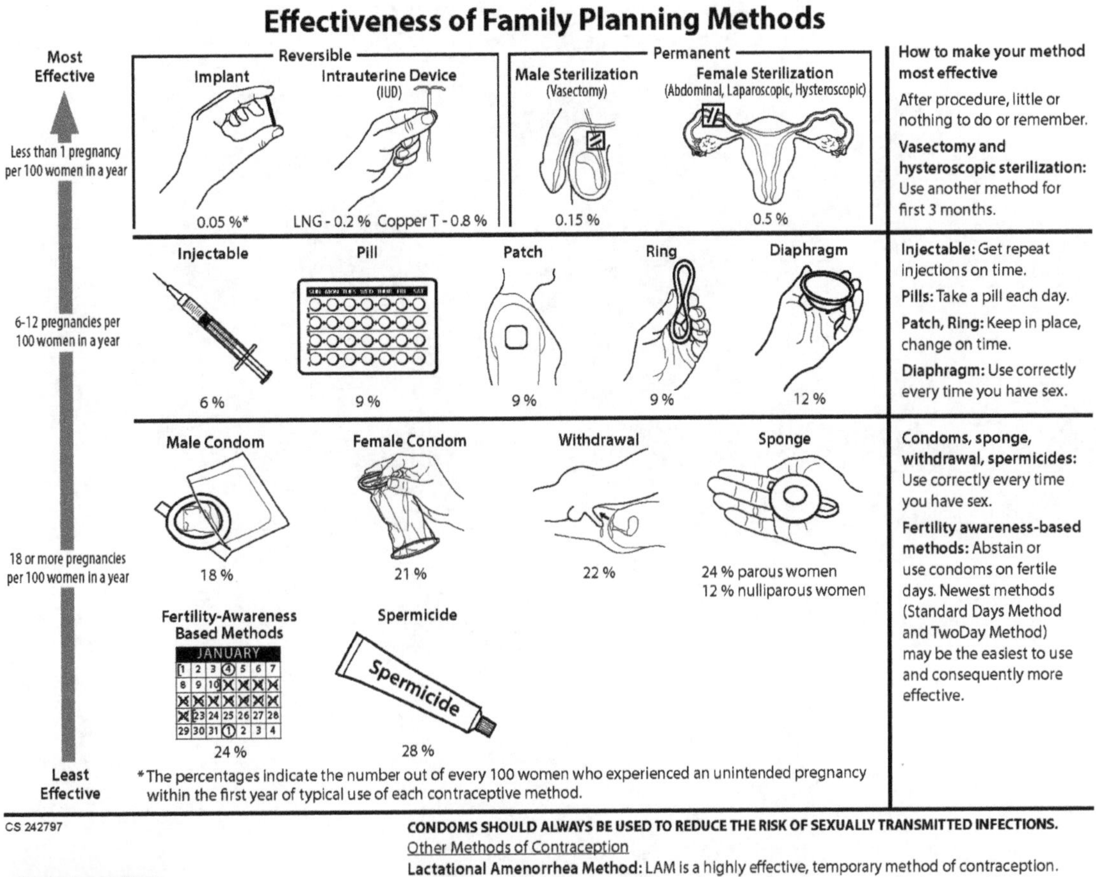

References

1. Trussell J, Aiken ARA, Micks E, Guthrie KA. Efficacy, safety, and personal considerations. In: Hatcher RA, Nelson AL, Trussell J, Cwiak C, Cason P, Policar MS, Edelman A, Aiken ARA, Marrazzo J, Kowal D, eds. Contraceptive technology. 21st ed. New York, NY: Ayer Company Publishers, Inc., 2018.

2. Peragallo Urrutia R, Polis CB, Jensen ET, Greene ME, Kennedy E, Stanford JB. Effectiveness of fertility awareness-based methods for pregnancy prevention: A systematic reviewExternal. Obstet Gynecol 2018;132:591-604.

Sexually Transmitted Infections and Diseases

While sexually transmitted diseases (STDs) affect individuals of all ages, STDs take a particularly heavy toll on young people. CDC estimates that youth ages 15-24 make up just over one quarter of the sexually active population, but account for half of the 20 million new sexually transmitted infections that occur in the United States each year. Many infections are asymptomatic. When a person is infected with a sexually transmitted pathogen and does not show signs or symptoms that person is said to have a sexually transmitted infection (STI). Although STIs and STDs are caused by the same pathogen, they differ as whether symptoms are present.

STDs/STIs can be transmitted by the following activities (depending upon the type of pathogen)

- Vaginal Sex
- Oral Sex
- Rectal Sex
- Rubbing of infected area

COMMON TYPES OF STD/I

Bacterial vaginosis (BV), a common cause of vaginal discharge in women of childbearing age, is a polymicrobial clinical syndrome resulting from a change in the vaginal community of bacteria. Although BV is often not considered an STD, it has been linked to sexual activity. Women may have no symptoms or may complain of a foul-smelling, fishy, vaginal discharge.

Chlamydia is a common sexually transmitted disease (STD) caused by infection with *Chlamydia trachomatis*. It can cause cervicitis in women and urethritis and proctitis in both men and women. Chlamydial infections in women can lead to serious consequences including pelvic inflammatory disease (PID), tubal factor infertility, ectopic pregnancy, and chronic pelvic pain. Lymphogranuloma venereum (LGV), another type of STD caused by different serovars of the same bacterium, occurs commonly in the developing world, and has more recently emerged as a cause of outbreaks of proctitis among men who have sex with men (MSM) worldwide.[1,2]

Gonorrhea is a sexually transmitted disease (STD) that can infect both men and women. It can cause infections in the genitals, rectum, and throat. It is a very common infection, especially among young people ages 15-24 years.

Genital herpes is an STD caused by two types of viruses. The viruses are called herpes simplex virus type 1 (HSV-1) and herpes simplex virus type 2 (HSV-2).Oral herpes is usually caused by HSV-1 and can result in cold sores or fever blisters on or around the mouth. However, most people do not have any symptoms. Most people with oral herpes were infected during childhood or young adulthood from non-sexual contact with saliva. Oral herpes caused by HSV-1 can be spread from the mouth to the genitals through oral sex. This is why some cases of genital herpes are caused by HSV-1.Genital herpes is common in the United States. More than one out of every six people aged 14 to 49 years have genital herpes.

Human Papillomavirus (HPV) is the most common sexually transmitted infection (STI). HPV is a different virus than HIV and HSV (herpes). 79 million Americans, most in their late teens and early 20s, are infected with HPV. There are many different types of HPV. Some types can cause health problems including genital warts and cancers. But there are vaccines that can stop these health problems from happening.

Syphilis is a sexually transmitted infection that can cause serious health problems if it is not treated. Syphilis is divided into stages (primary, secondary, latent, and tertiary), with different signs and symptoms associated with each stage. A person with **primary syphilis** generally has a sore or sores at the original site of infection. These sores usually occur on or around the genitals, around the anus or in the rectum, or in or around the mouth. These sores are usually (but not always) firm, round, and painless. Symptoms of **secondary syphilis** include skin rash, swollen lymph nodes, and fever. The signs and symptoms of primary and secondary syphilis can be mild, and they might not be noticed. During the **latent stage**, there are no signs or symptoms. **Tertiary syphilis** is associated with severe medical problems. A doctor can usually diagnose tertiary syphilis with the help of multiple tests. It can affect the heart, brain, and other organs of the body.

Trichomoniasis (or "trich") is a very common sexually transmitted disease (STD). It is caused by infection with a protozoan parasite called Trichomonas vaginalis. Although symptoms of the disease vary, most people who have the parasite cannot tell they are infected.T richomoniasis is the most common curable STD. In the United States, an estimated 3.7 million people have the infection. However, only about 30% develop any symptoms of trichomoniasis. Infection is more common in women than in men. Older women are more likely than younger women to have been infected with trichomoniasis.

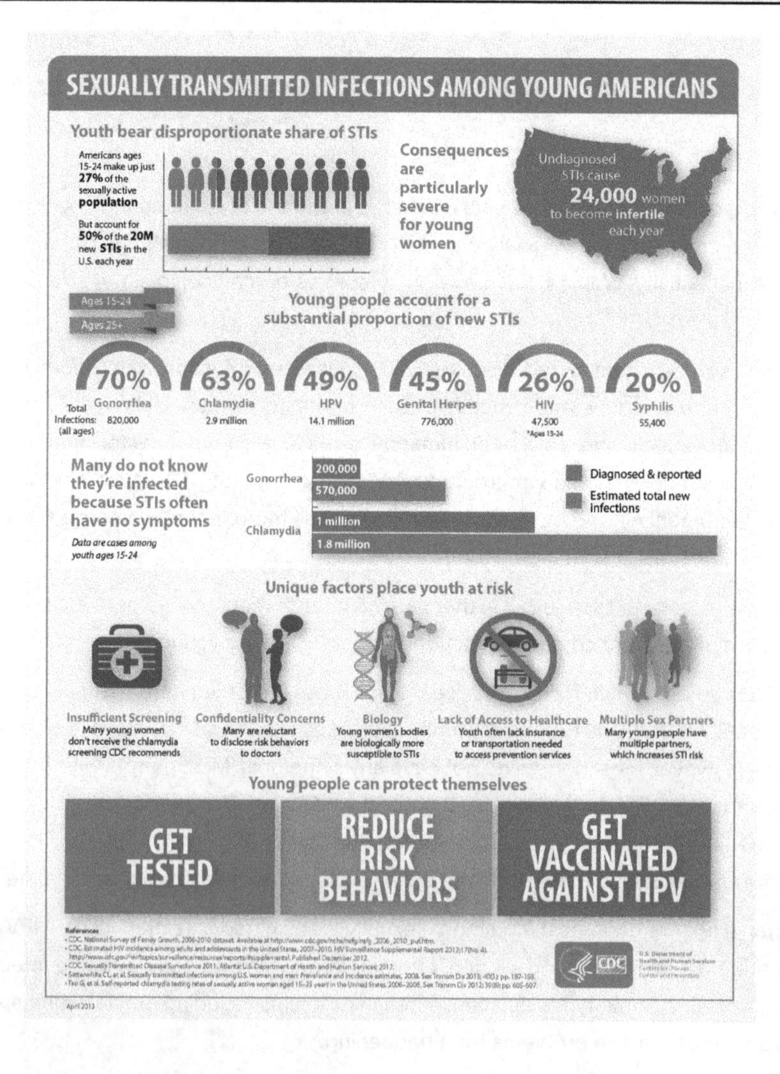

Public domain content

- **Provided by**: CDC. **Located at**: https://www.cdc.gov/std/general/default.htm. **License**: *Public Domain: No Known Copyright*

Concept check

1. **Which** gene causes maleness to develop?
2. **What** are secondary sex characteristics and **which** hormones cause them?
3. **What** are signs of puberty in males and females?
4. **List and describe** the major structures involved in the male reproductive system.
5. **What** is the process of making sperm called? **Where** does it happen? **How** many sperm are formed?
6. **List and describe** the major structures involved in the female reproductive system.
7. **What** is the process of making eggs called? **Where** does it happen? **How** many eggs are formed?
8. **What** is a follicle?
9. **What** are major difference between the ovarian cycle and the menstrual cycle?
10. **What** are the three layers of the uterus?
11. **How** does hormonal birth control work?
12. **What** is the different between reversible vs permanent birth control?
13. **What** are some barriers that can prevent STDs/STIs?
14. **What** are the most common STDs?

CHAPTER 6 — DEVELOPMENT AND INHERITANCE

Fertilization

LEARNING OBJECTIVES

- List and explain the steps involved in fertilization
- Briefly describe the process of sexual differentiation
- Trace the development of a fetus from the end of the embryonic period to birth
- Explain how estrogen, progesterone, and hCG are involved in maintaining pregnancy
- Summarize the events leading to labor
- Identify and describe each of the three stages of childbirth
- Discuss the adaptations of a woman's body to pregnancy
- Describe labor and delivery
- Summarize the physiology of lactation
- Classify and describe the different patterns of inheritance

In approximately nine months, a single cell—a fertilized egg—develops into a fully formed infant consisting of trillions of cells with myriad specialized functions. The dramatic changes of fertilization, embryonic development, and fetal development are followed by remarkable adaptations of the newborn to life outside the womb.

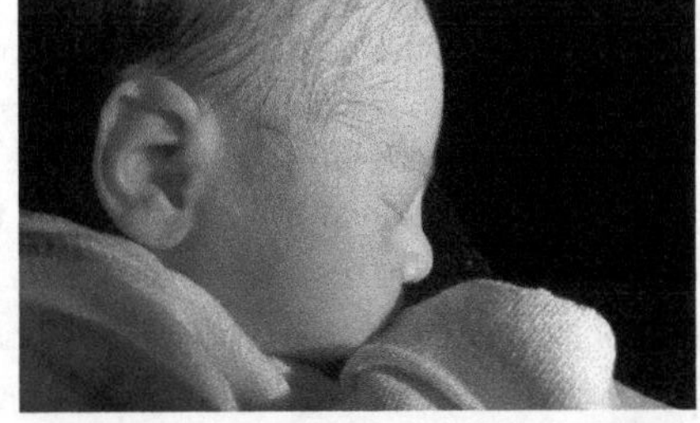

A single fertilized egg develops over the span of nine months into an infant consisting of trillions of cells and capable of surviving outside the womb. (credit: "Seattleye"/flickr.com)

An offspring's normal development depends upon the appropriate synthesis of structural and functional proteins. This, in turn, is governed by the genetic material inherited from the parental egg and sperm, as well as environmental factors.

Fertilization occurs when a sperm and an oocyte (egg) combine and their nuclei fuse. Because each of these reproductive cells is a haploid cell containing half of the genetic material needed to form a human being, their combination forms a diploid cell. This new single cell, called a **zygote**, contains all of the genetic material needed to form a human—half from the mother and half from the father.

Transit of Sperm

Fertilization is a numbers game. During ejaculation, hundreds of millions of sperm (spermatozoa) are released into the vagina. Almost immediately, millions of these sperm are overcome by the acidity of the vagina (approximately pH 3.8), and millions more may be blocked from entering the uterus by thick cervical mucus. Of those that do enter, thousands are destroyed by phagocytic uterine leukocytes. Thus, the race into the uterine tubes, which is the most typical site for sperm to encounter the oocyte, is reduced to a few thousand contenders. Their journey—thought to be facilitated by uterine contractions—usually takes from 30 minutes to 2 hours. If the sperm do not encounter an oocyte immediately, they can survive in the uterine tubes for another 3–5 days. Thus, fertilization can still occur if intercourse takes place a few days before ovulation. In comparison, an oocyte can survive independently for only approximately 24 hours following ovulation. Intercourse more than a day after ovulation will therefore usually not result in fertilization.

During the journey, fluids in the female reproductive tract prepare the sperm for fertilization through a process called **capacitation**, or priming. The fluids improve the motility of the spermatozoa. They also deplete cholesterol molecules embedded in the membrane of the head of the sperm, thinning the membrane in such a way that will help facilitate the release of the lysosomal (digestive) enzymes needed for the sperm to penetrate the oocyte's exterior once contact is made. Sperm must undergo the process of capacitation in order to have the "capacity" to fertilize an oocyte. If they reach the oocyte before capacitation is complete, they will be unable to penetrate the oocyte's thick outer layer of cells.

Contact Between Sperm and Oocyte

Upon ovulation, the oocyte released by the ovary is swept into—and along—the uterine tube. Fertilization must occur in the distal uterine tube because an unfertilized oocyte cannot survive the 72-hour journey to the uterus. As you will recall from your study of the oogenesis, this oocyte (specifically a secondary oocyte) is surrounded by two protective layers. The **corona radiata** is an outer layer of follicular (granulosa) cells that form around a developing oocyte in the ovary and remain with it upon ovulation. The underlying **zona pellucida** (pellucid = "transparent") is a transparent, but thick, glycoprotein membrane that surrounds the cell's plasma membrane.

As it is swept along the distal uterine tube, the oocyte encounters the surviving capacitated sperm, which stream toward it in response to chemical attractants released by the cells of the corona radiata. To reach the oocyte itself, the sperm must penetrate the two protective layers. The sperm first burrow through the cells of the corona radiata. Then, upon contact with the zona pellucida, the sperm bind to receptors in the zona pellucida. This initiates a process called the **acrosomal reaction** in which the enzyme-filled "cap" of the sperm, called the **acrosome**, releases its stored digestive enzymes. These enzymes clear a path through the zona pellucida that allows sperm to reach the oocyte. Finally, a single sperm makes contact with sperm-binding receptors on the oocyte's plasma membrane. The plasma membrane of that sperm then fuses with the oocyte's plasma membrane, and the head and mid-piece of the "winning" sperm enter the oocyte interior.

How do sperm penetrate the corona radiata? Some sperm undergo a spontaneous acrosomal reaction, which is an acrosomal reaction not triggered by contact with the zona pellucida. The digestive enzymes released by this reaction digest the extracellular matrix of the corona radiata. As you can see, the first sperm to reach the oocyte is never the one to fertilize it. Rather, hundreds of sperm cells must undergo the acrosomal reaction, each helping to degrade the corona radiata and zona pellucida until a path is created to allow one sperm to contact and fuse with the plasma membrane of the oocyte. If you consider the loss of millions of sperm between entry into the vagina and degradation of the zona pellucida, you can understand why a low sperm count can cause male infertility.

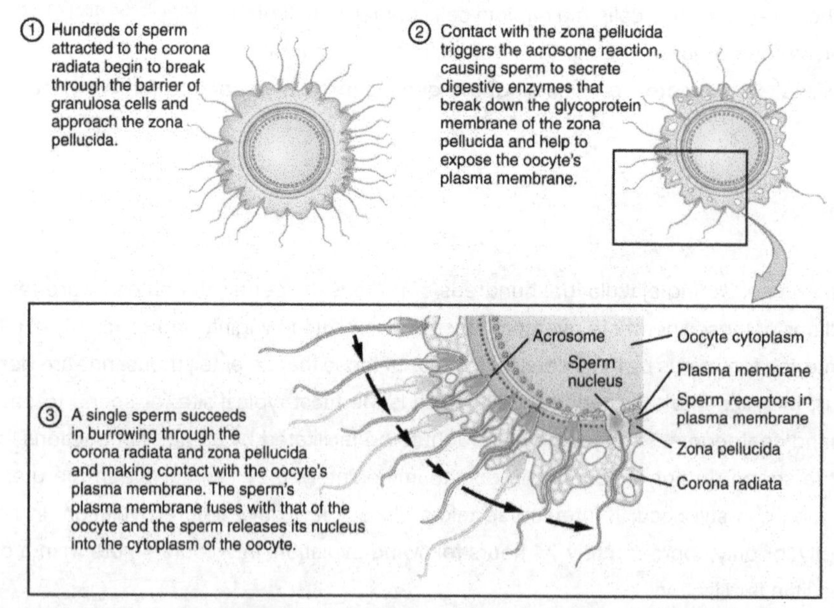

Figure 1. Before fertilization, hundreds of capacitated sperm must break through the surrounding corona radiata and zona pellucida so that one can contact and fuse with the oocyte plasma membrane.

When the first sperm fuses with the oocyte, the oocyte deploys two mechanisms to prevent **polyspermy**, which is penetration by more than one sperm. This is critical because if more than one sperm were to fertilize the oocyte, the resulting zygote would be a triploid organism with three sets of chromosomes. This is incompatible with life.

The first mechanism is the fast block, which involves a near instantaneous change in sodium ion permeability upon binding of the first sperm, depolarizing the oocyte plasma membrane and preventing the fusion of additional sperm cells. The fast block sets in almost immediately and lasts for about a minute, during which time an influx of calcium ions following sperm penetration triggers the second mechanism, the slow block. In this process, referred to as the **cortical reaction**, cortical granules sitting immediately below the oocyte plasma membrane fuse with the membrane and release zonal inhibiting proteins and mucopolysaccharides into the space between the plasma membrane and the zona pellucida. Zonal inhibiting proteins cause the release of any other attached sperm and destroy the oocyte's sperm receptors, thus preventing any more sperm from binding. The mucopolysaccharides then coat the nascent zygote in an impenetrable barrier that, together with hardened zona pellucida, is called a **fertilization membrane**.

The Zygote

Recall that at the point of fertilization, the oocyte has not yet completed meiosis; all secondary oocytes remain arrested in metaphase of meiosis II until fertilization. Only upon fertilization does the oocyte complete meiosis. The unneeded complement of genetic material that results is stored in a second polar body that is eventually ejected. At this moment, the oocyte has become an ovum, the female haploid gamete. The two haploid nuclei derived from the sperm and oocyte and contained within the egg are referred to as pronuclei. They decondense, expand, and replicate their DNA in preparation for mitosis. The pronuclei then migrate toward each other, their nuclear envelopes disintegrate, and the male- and female-derived genetic material intermingles. This step completes the process of fertilization and results in a single-celled diploid zygote with all the genetic instructions it needs to develop into a human.

Most of the time, a woman releases a single egg during an ovulation cycle. However, in approximately 1 percent of ovulation cycles, two eggs are released and both are fertilized. Two zygotes form, implant, and develop, resulting in the birth of dizygotic (or fraternal) twins. Because dizygotic twins develop from two eggs fertilized by two sperm, they are no more identical than siblings born at different times.

Much less commonly, a zygote can divide into two separate offspring during early development. This results in the birth of monozygotic (or identical) twins. Although the zygote can split as early as the two-cell stage, splitting occurs most commonly during the early blastocyst stage, with roughly 70–100 cells present. These two scenarios are distinct from each other, in that the twin embryos that separated at the two-cell stage will have individual placentas, whereas twin embryos that form from separation at the blastocyst stage will share a placenta and a chorionic cavity.

EVERYDAY CONNECTIONS: IN VITRO FERTILIZATION

IVF, which stands for in vitro fertilization, is an assisted reproductive technology. In vitro, which in Latin translates to "in glass," refers to a procedure that takes place outside of the body. There are many different indications for IVF. For example, a woman may produce normal eggs, but the eggs cannot reach the uterus because the uterine tubes are blocked or otherwise compromised. A man may have a low sperm count, low sperm motility, sperm with an unusually high percentage of morphological abnormalities, or sperm that are incapable of penetrating the zona pellucida of an egg.

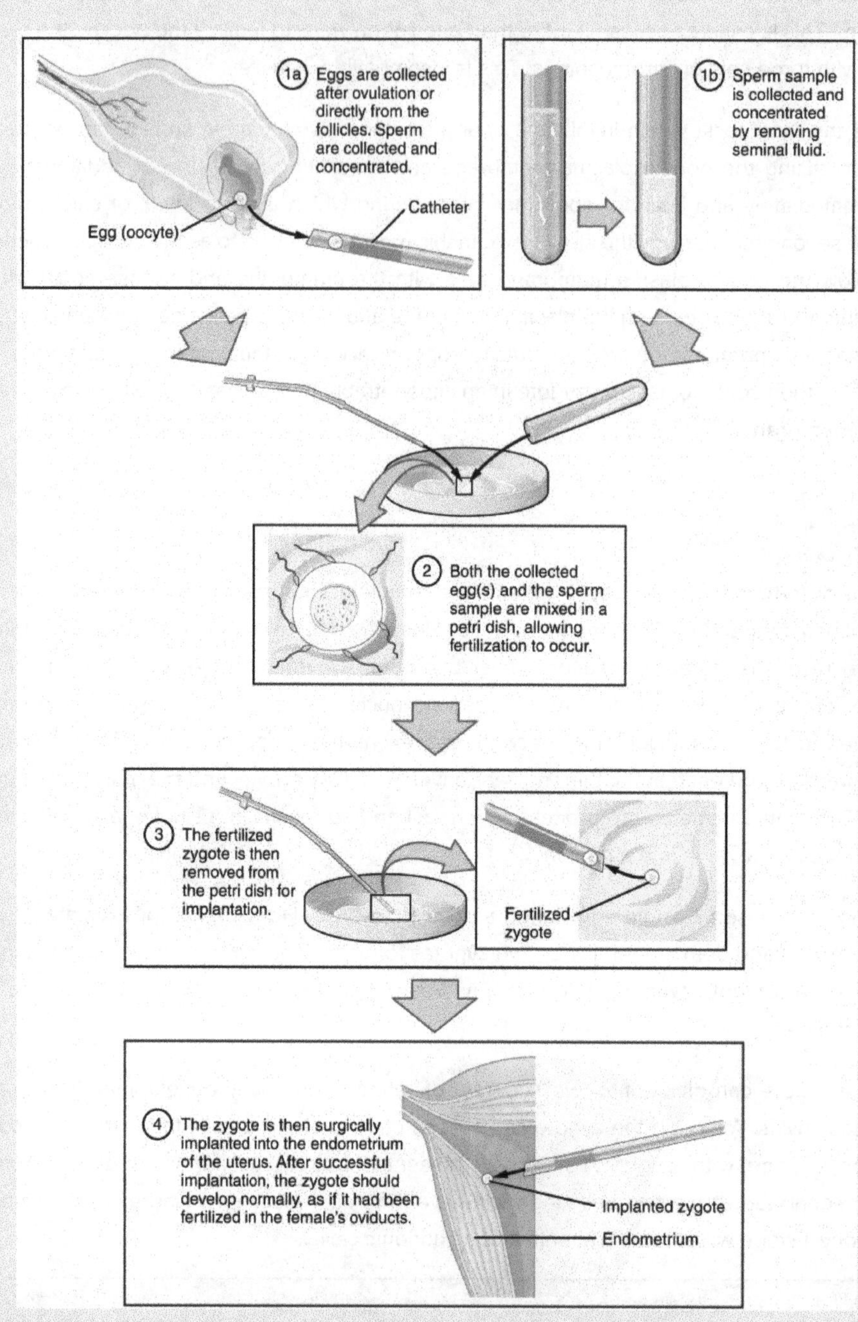

Figure 2. Click for a larger image. In vitro fertilization involves egg collection from the ovaries, fertilization in a petri dish, and the transfer of embryos into the uterus.

A typical IVF procedure begins with egg collection. A normal ovulation cycle produces only one oocyte, but the number can be boosted significantly (to 10–20 oocytes) by administering a short course of gonadotropins. The course begins with follicle-stimulating hormone (FSH) analogs, which support the development of multiple follicles, and ends with a luteinizing hormone (LH) analog that triggers ovulation. Right before the ova would be released from the ovary, they are harvested using ultrasound-guided oocyte retrieval. In this procedure, ultrasound allows a physician to visualize mature follicles. The ova are aspirated (sucked out) using a syringe.

In parallel, sperm are obtained from the male partner or from a sperm bank. The sperm are prepared by washing to remove seminal fluid because seminal fluid contains a peptide, FPP (or, fertilization promoting peptide), that—in high concentrations—prevents capacitation of the sperm. The sperm sample is also concentrated, to increase the sperm count per milliliter.

Next, the eggs and sperm are mixed in a petri dish. The ideal ratio is 75,000 sperm to one egg. If there are severe problems with the sperm—for example, the count is exceedingly low, or the sperm are completely nonmotile, or incapable of binding to or penetrating the zona pellucida—a sperm can be injected into an egg. This is called intracytoplasmic sperm injection (ICSI).

The embryos are then incubated until they either reach the eight-cell stage or the blastocyst stage. In the United States, fertilized eggs are typically cultured to the blastocyst stage because this results in a higher pregnancy rate. Finally, the embryos are transferred to a woman's uterus using a plastic catheter (tube).The diagram below illustrates the steps involved in IVF.

IVF is a relatively new and still evolving technology, and until recently it was necessary to transfer multiple embryos to achieve a good chance of a pregnancy. Today, however, transferred embryos are much more likely to implant successfully, so countries that regulate the IVF industry cap the number of embryos that can be transferred per cycle at two. This reduces the risk of multiple-birth pregnancies.

The rate of success for IVF is correlated with a woman's age. More than 40 percent of women under 35 succeed in giving birth following IVF, but the rate drops to a little over 10 percent in women over 40.

Embryonic Development

Throughout this chapter, we will express embryonic and fetal ages in terms of weeks from fertilization, commonly called conception. The period of time required for full development of a fetus in utero is referred to as **gestation** (gestare = "to carry" or "to bear"). It can be subdivided into distinct gestational periods. The first 2 weeks of prenatal development are referred to as the pre-embryonic stage. A developing human is referred to as an **embryo** during weeks 3–8, and a **fetus** from the ninth week of gestation until birth. In this section, we'll cover the pre-embryonic and embryonic stages of development, which are characterized by cell division, migration, and differentiation. By the end of the embryonic period, all of the organ systems are structured in rudimentary form, although the organs themselves are either nonfunctional or only semi-functional.

Pre-Implantation Embryonic Development

Following fertilization, the zygote and its associated membranes, together referred to as the **conceptus**, continue to be projected toward the uterus by peristalsis and beating cilia. During its journey to the uterus, the zygote undergoes five or six rapid mitotic cell divisions. Although each **cleavage** results in more cells, it does not increase the total volume of the conceptus. Each daughter cell produced by cleavage is called a **blastomere** (blastos = "germ," in the sense of a seed or sprout).

Approximately 3 days after fertilization, a 16-cell conceptus reaches the uterus. The cells that had been loosely grouped are now compacted and look more like a solid mass. The name given to this structure is the **morula** (morula = "little mulberry"). Once inside the uterus, the conceptus floats freely for several more days. It continues to divide, creating a ball of approximately 100 cells, and consuming nutritive endometrial secretions called uterine milk while the uterine lining thickens. The ball of now tightly bound cells starts to secrete fluid and organize themselves around a fluid-filled cavity, the **blastocoel**. At this developmental stage, the conceptus is referred to as a **blastocyst**. Within this structure, a group of cells forms into an **inner cell mass**, which is fated to become the embryo. The cells that form the outer shell are called **trophoblasts** (trophe = "to feed" or "to nourish"). These cells will develop into the chorionic sac and the fetal portion of the **placenta** (the organ of nutrient, waste, and gas exchange between mother and the developing offspring).

The inner mass of embryonic cells is totipotent during this stage, meaning that each cell has the potential to differentiate into any cell type in the human body. Totipotency lasts for only a few days before the cells' fates are set as being the precursors to a specific lineage of cells.

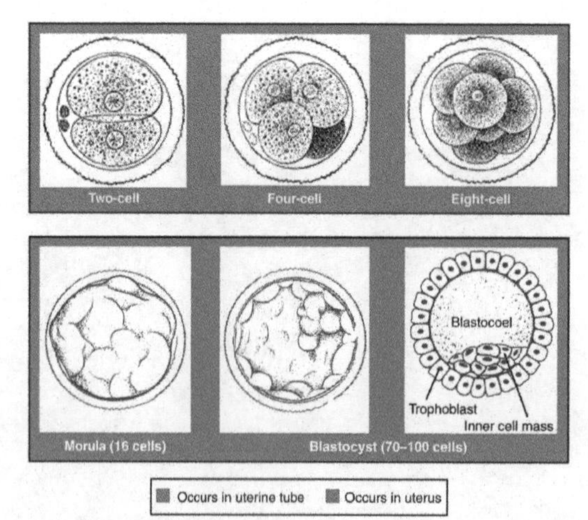

Figure 1. Pre-embryonic cleavages make use of the abundant cytoplasm of the conceptus as the cells rapidly divide without changing the total volume.

As the blastocyst forms, the trophoblast excretes enzymes that begin to degrade the zona pellucida. In a process called "hatching," the conceptus breaks free of the zona pellucida in preparation for implantation.

Implantation

At the end of the first week, the blastocyst comes in contact with the uterine wall and adheres to it, embedding itself in the uterine lining via the trophoblast cells. Thus begins the process of **implantation**, which signals the end of the pre-embryonic stage of development. Implantation can be accompanied by minor bleeding. The blastocyst typically implants in the fundus of the uterus or on the posterior wall. However, if the endometrium is not fully developed and ready to receive the blastocyst, the blastocyst will detach and find a better spot. A significant percentage (50–75 percent) of blastocysts fail to implant; when this occurs, the blastocyst is shed with the endometrium during menses. The high rate of implantation failure is one reason why pregnancy typically requires several ovulation cycles to achieve.

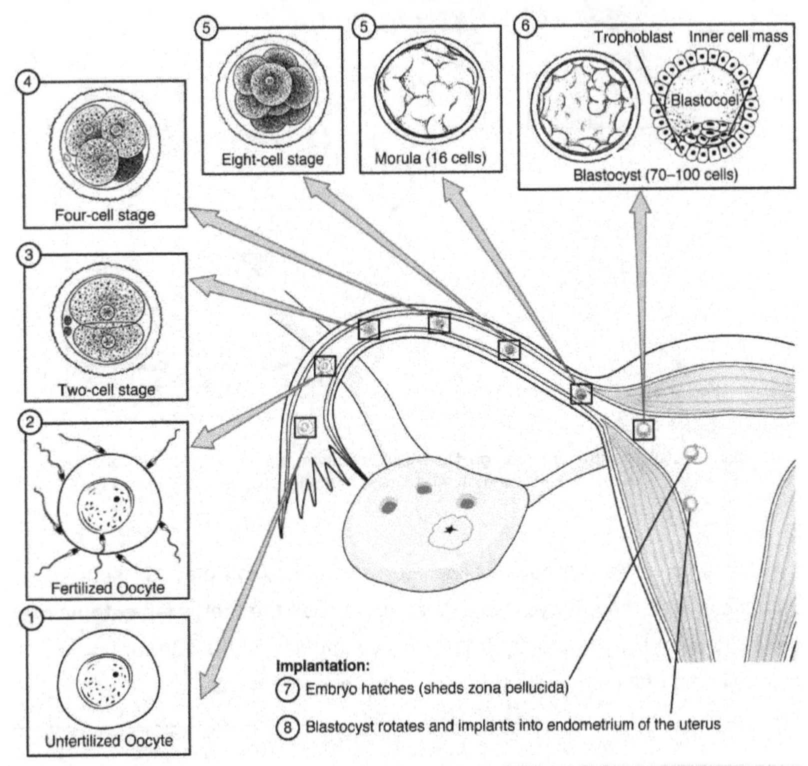

Figure 2. Click for a larger image. Ovulation, fertilization, pre-embryonic development, and implantation occur at specific locations within the female reproductive system in a time span of approximately 1 week.

When implantation succeeds and the blastocyst adheres to the endometrium, the superficial cells of the trophoblast fuse with each other, forming the **syncytiotrophoblast**, a multinucleated body that digests endometrial cells to firmly secure the blastocyst to the uterine wall. In response, the uterine mucosa rebuilds itself and envelops the blastocyst. The trophoblast secretes **human chorionic gonadotropin (hCG)**, a hormone that directs the corpus luteum to survive, enlarge, and continue producing progesterone and estrogen to suppress menses. These functions of hCG are necessary for creating an environment suitable for the developing embryo. As a result of this increased production, hCG accumulates in the maternal bloodstream and is excreted in the urine. Implantation is complete by the middle of the second week. Just a few days after implantation, the trophoblast has secreted enough hCG for an at-home urine pregnancy test to give a positive result.

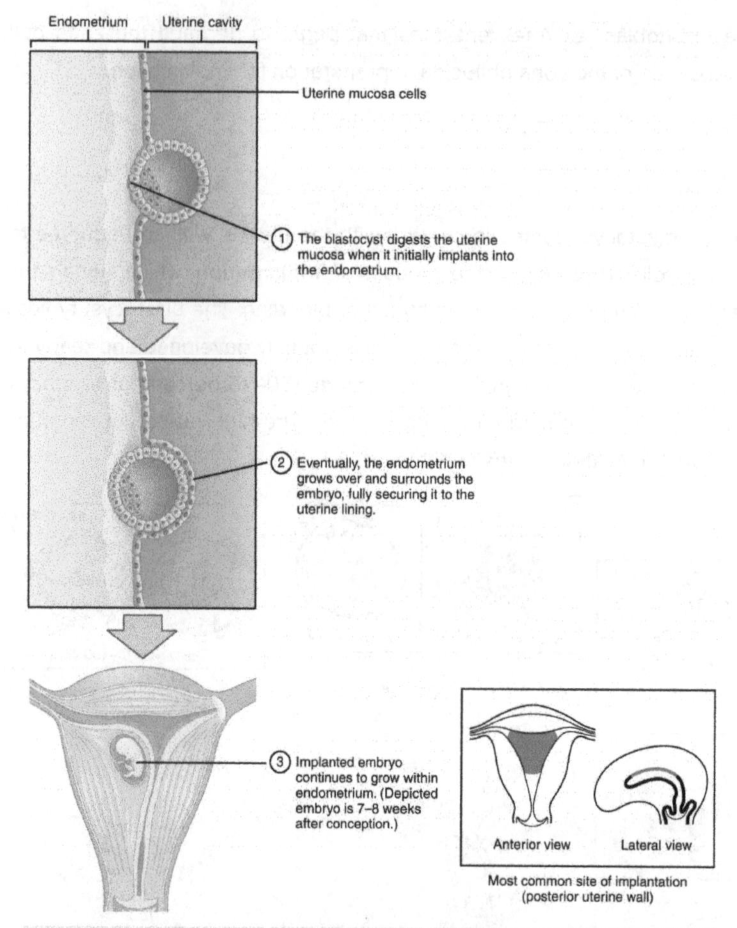

Figure 3. Click to show larger image. During implantation, the trophoblast cells of the blastocyst adhere to the endometrium and digest endometrial cells until it is attached securely.

Most of the time an embryo implants within the body of the uterus in a location that can support growth and development. However, in one to two percent of cases, the embryo implants either outside the uterus (an **ectopic pregnancy**) or in a region of uterus that can create complications for the pregnancy. If the embryo implants in the inferior portion of the uterus, the placenta can potentially grow over the opening of the cervix, a condition call **placenta previa**.

DISORDERS OF THE DEVELOPMENT OF THE EMBRYO

In the vast majority of ectopic pregnancies, the embryo does not complete its journey to the uterus and implants in the uterine tube, referred to as a tubal pregnancy. However, there are also ovarian ectopic pregnancies (in which the egg never left the ovary) and abdominal ectopic pregnancies (in which an egg was "lost" to the abdominal cavity during the transfer from ovary to uterine tube, or in which an embryo from a tubal pregnancy re-implanted in the abdomen). Once in the abdominal cavity, an embryo can implant into any well-vascularized structure—the rectouterine cavity (Douglas' pouch), the mesentery of the intestines, and the greater omentum are some common sites.

Tubal pregnancies can be caused by scar tissue within the tube following a sexually transmitted bacterial infection. The scar tissue impedes the progress of the embryo into the uterus—in some cases "snagging" the embryo and, in other cases, blocking the tube completely. Approximately one half of tubal pregnancies resolve spontaneously. Implantation in a uterine tube causes bleeding, which appears to stimulate smooth muscle contractions and expulsion of the embryo.

In the remaining cases, medical or surgical intervention is necessary. If an ectopic pregnancy is detected early, the embryo's development can be arrested by the administration of the cytotoxic drug methotrexate, which inhibits the metabolism of folic acid. If diagnosis is late and the uterine tube is already ruptured, surgical repair is essential.

Even if the embryo has successfully found its way to the uterus, it does not always implant in an optimal location (the fundus or the posterior wall of the uterus). Placenta previa can result if an embryo implants close to the internal os of the uterus (the internal opening of the cervix). As the fetus grows, the placenta can partially or completely cover the opening of the cervix. Although it occurs in only 0.5 percent of pregnancies, placenta previa is the leading cause of antepartum hemorrhage (profuse vaginal bleeding after week 24 of pregnancy but prior to childbirth).

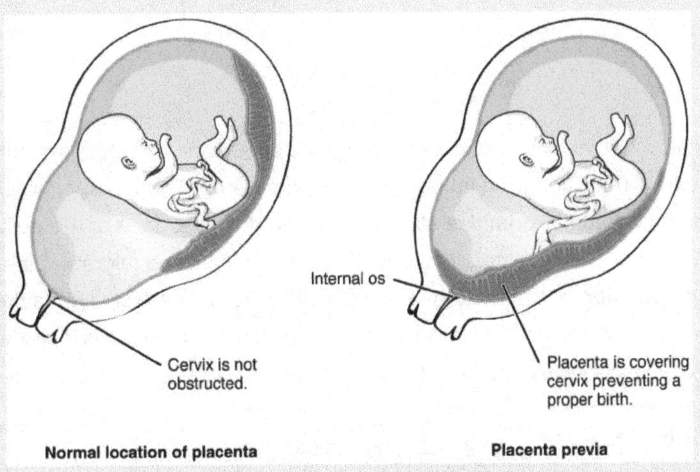

Figure 4. An embryo that implants too close to the opening of the cervix can lead to placenta previa, a condition in which the placenta partially or completely covers the cervix.

Embryonic Membranes

During the second week of development, with the embryo implanted in the uterus, cells within the blastocyst start to organize into layers. Some grow to form the extra-embryonic membranes needed to support and protect the growing embryo: the amnion, the yolk sac, the allantois, and the chorion.

At the beginning of the second week, the cells of the inner cell mass form into a two-layered disc of embryonic cells, and a space—the **amniotic cavity**—opens up between it and the trophoblast (Figure 5). Cells from the upper layer of the disc (the **epiblast**) extend around the amniotic cavity, creating a membranous sac that forms into the **amnion** by the end of the second week. The amnion fills with amniotic fluid and eventually grows to surround the embryo. Early in development, amniotic fluid consists almost entirely of a filtrate of maternal plasma, but as the kidneys of the fetus begin to function at approximately the eighth week, they add urine to the volume of amniotic fluid. Floating within the amniotic fluid, the embryo—and later, the fetus—is protected from trauma and rapid temperature changes. It can move freely within the fluid and can prepare for swallowing and breathing out of the uterus.

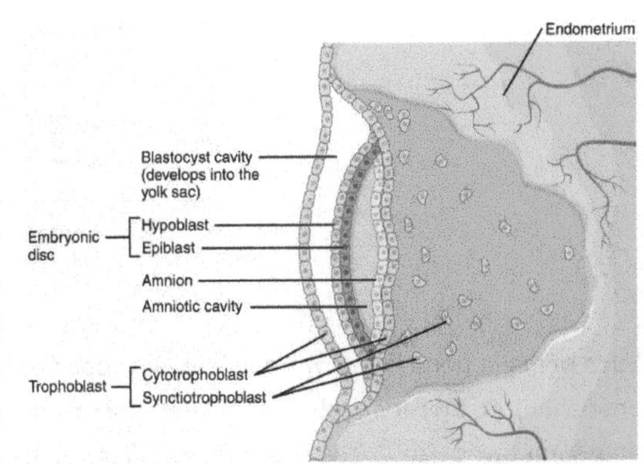

Figure 5. Formation of the embryonic disc leaves spaces on either side that develop into the amniotic cavity and the yolk sac.

On the ventral side of the embryonic disc, opposite the amnion, cells in the lower layer of the embryonic disk (the **hypoblast**) extend into the blastocyst cavity and form a **yolk sac**. The yolk sac supplies some nutrients absorbed from the trophoblast and also provides primitive blood circulation to the developing embryo for the second and third week of development. When the placenta takes over nourishing the embryo at approximately week 4, the yolk sac has been greatly reduced in size and its main function is to serve as the source of blood cells and germ cells (cells that will give rise to gametes). During week 3, a finger-like outpocketing of the yolk sac develops into the **allantois**, a primitive excretory duct of the embryo that will become part of the urinary bladder. Together, the stalks of the yolk sac and allantois establish the outer structure of the umbilical cord.

The last of the extra-embryonic membranes is the **chorion**, which is the one membrane that surrounds all others. The development of the chorion will be discussed in more detail shortly, as it relates to the growth and development of the placenta.

Embryogenesis

As the third week of development begins, the two-layered disc of cells becomes a three-layered disc through the process of **gastrulation**, during which the cells transition from totipotency to multipotency. The embryo, which takes the shape of an oval-shaped disc, forms an indentation called the **primitive streak** along the dorsal surface of the epiblast. A node at the caudal or "tail" end of the primitive streak emits growth factors that direct cells to multiply and migrate. Cells migrate toward and through the primitive streak and then move laterally to create two new layers of cells. The first layer is the **endoderm**, a sheet of cells that displaces the hypoblast and lies adjacent to the yolk sac. The second layer of cells fills in as the middle layer, or **mesoderm**. The cells of the epiblast that remain (not having migrated through the primitive streak) become the **ectoderm**.

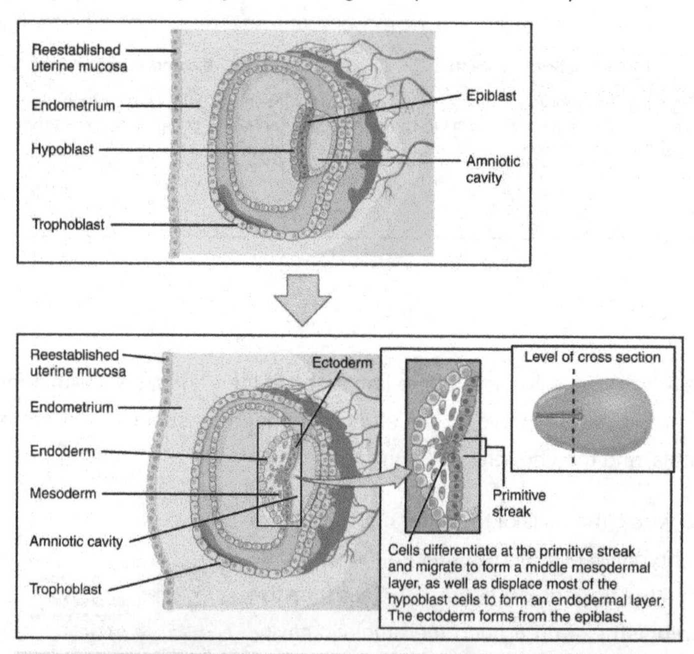

Figure 6. Click for a larger image. Formation of the three primary germ layers occurs during the first 2 weeks of development. The embryo at this stage is only a few millimeters in length.

Each of these germ layers will develop into specific structures in the embryo. Whereas the ectoderm and endoderm form tightly connected epithelial sheets, the mesodermal cells are less organized and exist as a loosely connected cell community. The ectoderm gives rise to cell lineages that differentiate to become the central and peripheral nervous systems, sensory organs, epidermis, hair, and nails. Mesodermal cells ultimately become the skeleton, muscles, connective tissue, heart, blood vessels, and kidneys. The endoderm goes on to form the epithelial lining of the gastrointestinal tract, liver, and pancreas, as well as the lungs (Figure 7).

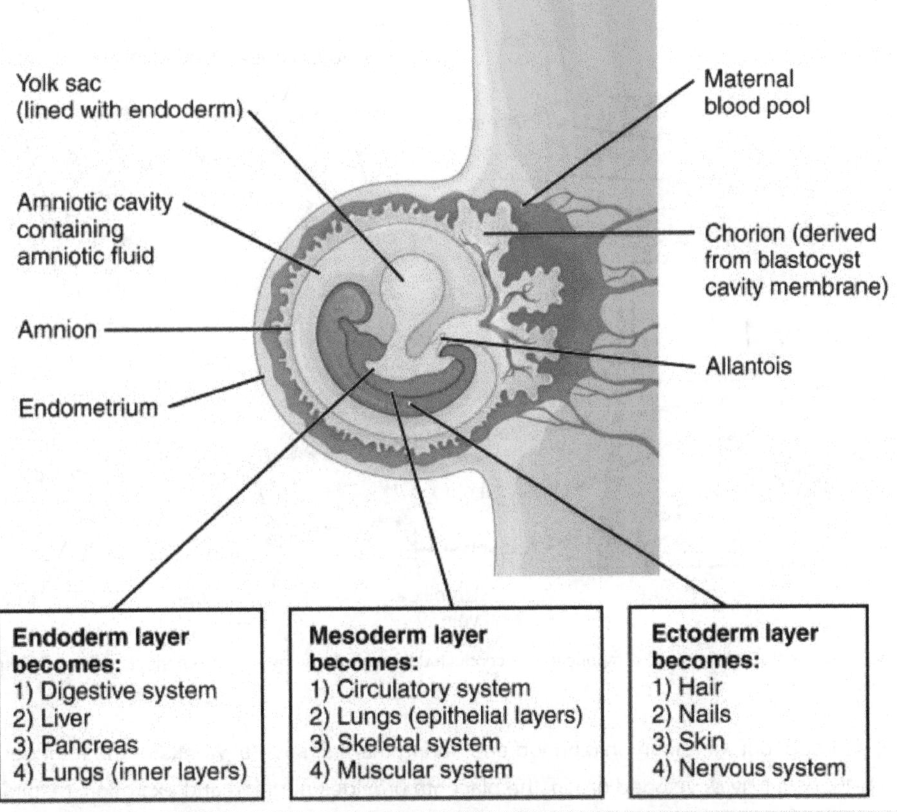

Figure 7. Following gastrulation of the embryo in the third week, embryonic cells of the ectoderm, mesoderm, and endoderm begin to migrate and differentiate into the cell lineages that will give rise to mature organs and organ systems in the infant.

Development of the Placenta

During the first several weeks of development, the cells of the endometrium—referred to as decidual cells—nourish the nascent embryo. During prenatal weeks 4–12, the developing placenta gradually takes over the role of feeding the embryo, and the decidual cells are no longer needed. The mature placenta is composed of tissues derived from the embryo, as well as maternal tissues of the endometrium. The placenta connects to the conceptus via the **umbilical cord**, which carries deoxygenated blood and wastes from the fetus through two umbilical arteries; nutrients and oxygen are carried from the mother to the fetus through the single umbilical vein. The umbilical cord is surrounded by the amnion, and the spaces within the cord around the blood vessels are filled with Wharton's jelly, a mucous connective tissue.

The maternal portion of the placenta develops from the deepest layer of the endometrium, the decidua basalis. To form the embryonic portion of the placenta, the syncytiotrophoblast and the underlying cells of the trophoblast (cytotrophoblast cells) begin to proliferate along with a layer of extraembryonic mesoderm cells. These form the **chorionic membrane**, which envelops the entire conceptus as the chorion. The chorionic membrane forms finger-like structures called **chorionic villi** that burrow into the endometrium like tree roots, making up the fetal portion of the placenta. The cytotrophoblast cells perforate the chorionic villi, burrow farther into the endometrium, and remodel maternal blood vessels to augment maternal blood flow surrounding the villi. Meanwhile, fetal mesenchymal cells derived from the mesoderm fill the villi and differentiate into blood vessels, including the three umbilical blood vessels that connect the embryo to the developing placenta.

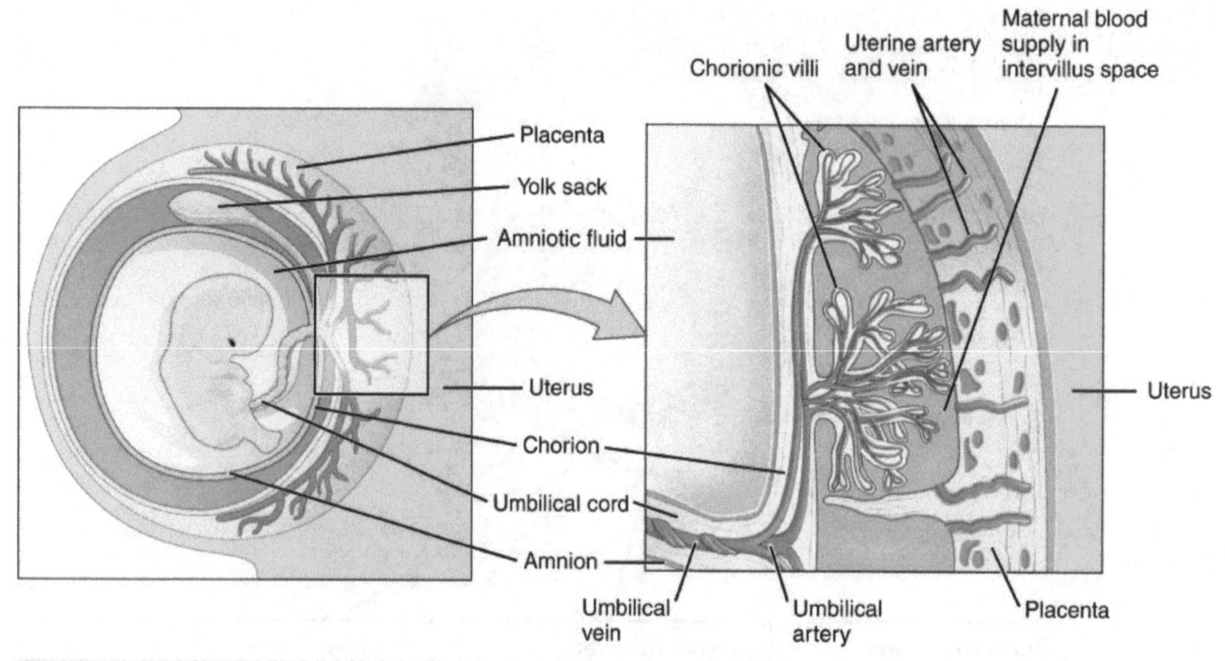

Figure 8. In the placenta, maternal and fetal blood components are conducted through the surface of the chorionic villi, but maternal and fetal bloodstreams never mix directly.

The placenta develops throughout the embryonic period and during the first several weeks of the fetal period; **placentation** is complete by weeks 14–16. As a fully developed organ, the placenta provides nutrition and excretion, respiration, and endocrine function. It receives blood from the fetus through the umbilical arteries. Capillaries in the chorionic villi filter fetal wastes out of the blood and return clean, oxygenated blood to the fetus through the umbilical vein. Nutrients and oxygen are transferred from maternal blood surrounding the villi through the capillaries and into the fetal bloodstream. Some substances move across the placenta by simple diffusion. Oxygen, carbon dioxide, and any other lipid-soluble substances take this route. Other substances move across by facilitated diffusion. This includes water-soluble glucose. The fetus has a high demand for amino acids and iron, and those substances are moved across the placenta by active transport.

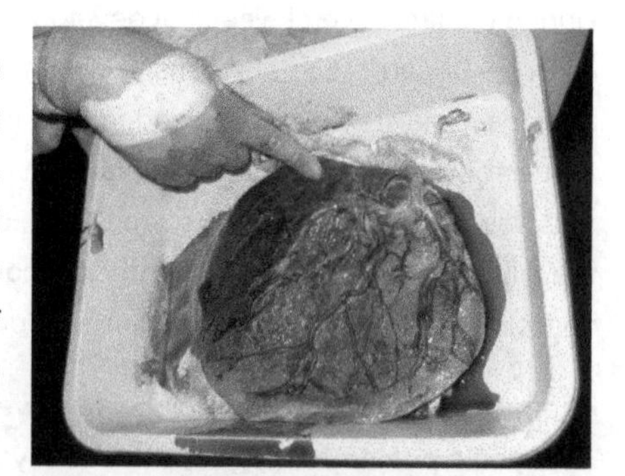

Figure 9. This post-expulsion placenta and umbilical cord (white) are viewed from the fetal side.

Maternal and fetal blood does not commingle because blood cells cannot move across the placenta. This separation prevents the mother's cytotoxic T cells from reaching and subsequently destroying the fetus, which bears "non-self" antigens. Further, it ensures the fetal red blood cells do not enter the mother's circulation and trigger antibody development (if they carry "non-self" antigens)—at least until the final stages of pregnancy or birth. This is the reason that, even in the absence of preventive treatment, an Rh^- mother doesn't develop antibodies that could cause hemolytic disease in her first Rh^+ fetus.

Although blood cells are not exchanged, the chorionic villi provide ample surface area for the two-way exchange of substances between maternal and fetal blood. The rate of exchange increases throughout gestation as the villi become thinner and increasingly branched. The placenta is permeable to lipid-soluble fetotoxic substances: alcohol, nicotine, barbiturates, antibiotics, certain pathogens, and many other substances that can be dangerous or fatal to the developing embryo or fetus. For these reasons, pregnant women should avoid fetotoxic substances. Alcohol consumption by pregnant women, for example, can result in a range of abnormalities referred to as fetal alcohol spectrum disorders (FASD). These include organ and facial malformations, as well as cognitive and behavioral disorders.

Table 1. Functions of the Placenta

Nutrition and digestion	Respiration	Endocrine function
• Mediates diffusion of maternal glucose, amino acids, fatty acids, vitamins, and minerals • Stores nutrients during early pregnancy to accommodate increased fetal demand later in pregnancy • Excretes and filters fetal nitrogenous wastes into maternal blood	• Mediates maternal-to-fetal oxygen transport and fetal-to-maternal carbon dioxide transport	• Secretes several hormones, including hCG, estrogens, and progesterone, to maintain the pregnancy and stimulate maternal and fetal development • Mediates the transmission of maternal hormones into fetal blood and vice versa

Fetal Development

As you will recall, a developing human is called a fetus from the ninth week of gestation until birth. This 30-week period of development is marked by continued cell growth and differentiation, which fully develop the structures and functions of the immature organ systems formed during the embryonic period. The completion of fetal development results in a newborn who, although still immature in many ways, is capable of survival outside the womb.

Sexual Differentiation

Sexual differentiation does not begin until the fetal period, during weeks 9–12. Embryonic males and females, though genetically distinguishable, are morphologically identical. Bipotential gonads, or gonads that can develop into male or female sexual organs, are connected to a central cavity called the cloaca via Müllerian ducts and Wolffian ducts. (The cloaca is an extension of the primitive gut.) Several events lead to sexual differentiation during this period.

During male fetal development, the bipotential gonads become the testes and associated epididymis. The Müllerian ducts degenerate. The Wolffian ducts become the vas deferens, and the cloaca becomes the urethra and rectum.

During female fetal development, the bipotential gonads develop into ovaries. The Wolffian ducts degenerate. The Müllerian ducts become the uterine tubes and uterus, and the cloaca divides and develops into a vagina, a urethra, and a rectum.

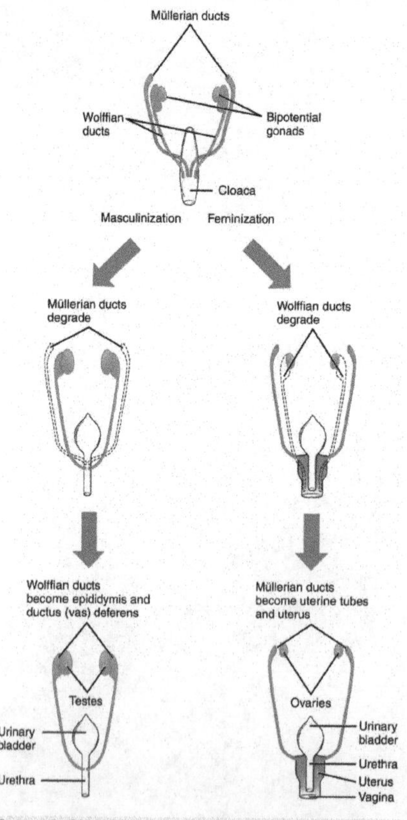

Figure 1. Click for a larger image. Differentiation of the male and female reproductive systems does not occur until the fetal period of development.

The Fetal Circulatory System

During prenatal development, the fetal circulatory system is integrated with the placenta via the umbilical cord so that the fetus receives both oxygen and nutrients from the placenta. However, after childbirth, the umbilical cord is severed, and the newborn's circulatory system must be reconfigured. When the heart first forms in the embryo, it exists as two parallel tubes derived from mesoderm and lined with endothelium, which then fuse together. As the embryo develops into a fetus, the tube-shaped heart folds and further differentiates into the four chambers present in a mature heart. Unlike a mature cardiovascular system, however, the fetal cardiovascular system also includes circulatory shortcuts, or shunts. A **shunt** is an anatomical (or sometimes surgical) diversion that allows blood flow to bypass immature organs such as the lungs and liver until childbirth.

The placenta provides the fetus with necessary oxygen and nutrients via the umbilical vein. (Remember that veins carry blood toward the heart. In this case, the blood flowing to the fetal heart is oxygenated because it comes from the placenta. The respiratory system is immature and cannot yet oxygenate blood on its own.) From the umbilical vein, the oxygenated blood flows toward the inferior vena cava, all but bypassing the immature liver, via the **ductus venosus** shunt. The liver receives just a trickle of blood, which is all that it needs in its immature, semifunctional state. Blood flows from the inferior vena cava to the right atrium, mixing with fetal venous blood along the way.

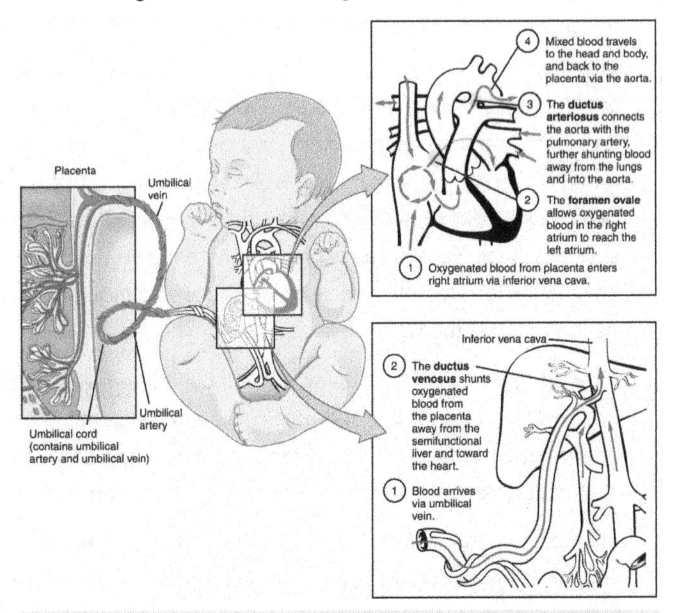

Figure 2. Click for a larger image. The fetal circulatory system includes three shunts to divert blood from undeveloped and partially functioning organs, as well as blood supply to and from the placenta.

Although the fetal liver is semifunctional, the fetal lungs are nonfunctional. The fetal circulation therefore bypasses the lungs by shifting some of the blood through the **foramen ovale**, a shunt that directly connects the right and left atria and avoids the pulmonary trunk altogether. Most of the rest of the blood is pumped to the right ventricle, and from there, into the pulmonary trunk, which splits into pulmonary arteries. However, a shunt within the pulmonary artery, the **ductus arteriosus**, diverts a portion of this blood into the aorta. This ensures that only a small volume of oxygenated blood passes through the immature pulmonary circuit, which has only minor metabolic requirements. Blood vessels of uninflated lungs have high resistance to flow, a condition that encourages blood to flow to the aorta, which presents much lower resistance. The oxygenated blood moves through the foramen ovale into the left atrium, where it mixes with the now deoxygenated blood returning from the pulmonary circuit. This blood then moves into the left ventricle, where it is pumped into the aorta. Some of this blood moves through the coronary arteries into the myocardium, and some moves through the carotid arteries to the brain.

The descending aorta carries partially oxygenated and partially deoxygenated blood into the lower regions of the body. It eventually passes into the umbilical arteries through branches of the internal iliac arteries. The deoxygenated blood collects waste as it circulates through the fetal body and returns to the umbilical cord. Thus, the two umbilical arteries carry blood low in oxygen and high in carbon dioxide and fetal wastes. This blood is filtered through the placenta, where wastes diffuse into the maternal circulation. Oxygen and nutrients from the mother diffuse into the placenta and from there into the fetal blood, and the process repeats.

DISORDERS OF THE DEVELOPING FETUS

Throughout the second half of gestation, the fetal intestines accumulate a tarry, greenish black meconium. The newborn's first stools consist almost entirely of meconium; they later transition to seedy yellow stools or slightly formed tan stools as meconium is cleared and replaced with digested breast milk or formula, respectively. Unlike these later stools, meconium is sterile; it is devoid of bacteria because the fetus is in a sterile environment and has not consumed any breast milk or formula. Typically, an infant does not pass meconium until after birth. However, in 5–20 percent of births, the fetus has a bowel movement in utero, which can cause major complications in the newborn.

The passage of meconium in the uterus signals fetal distress, particularly fetal hypoxia (i.e., oxygen deprivation). This may be caused by maternal drug abuse (especially tobacco or cocaine), maternal hypertension, depletion of amniotic fluid, long labor or difficult birth, or a defect in the placenta that prevents it from delivering adequate oxygen to the fetus. Meconium passage is typically a complication of full-term or post-term newborns because it is rarely passed before 34 weeks of gestation, when the gastrointestinal system has matured and is appropriately controlled by nervous system stimuli. Fetal distress can stimulate the vagus nerve to trigger gastrointestinal peristalsis and relaxation of the anal sphincter. Notably, fetal hypoxic stress also induces a gasping reflex, increasing the likelihood that meconium will be inhaled into the fetal lungs.

Although meconium is a sterile substance, it interferes with the antibiotic properties of the amniotic fluid and makes the newborn and mother more vulnerable to bacterial infections at birth and during the perinatal period. Specifically, inflammation of the fetal membranes, inflammation of the uterine lining, or neonatal sepsis (infection in the newborn) may occur. Meconium also irritates delicate fetal skin and can cause a rash.

The first sign that a fetus has passed meconium usually does not come until childbirth, when the amniotic sac ruptures. Normal amniotic fluid is clear and watery, but amniotic fluid in which meconium has been passed is stained greenish or yellowish. Antibiotics given to the mother may reduce the incidence of maternal bacterial infections, but it is critical that meconium is aspirated from the newborn before the first breath. Under these conditions, an obstetrician will extensively aspirate the infant's airways as soon as the head is delivered, while the rest of the infant's body is still inside the birth canal.

Aspiration of meconium with the first breath can result in labored breathing, a barrel-shaped chest, or a low Apgar score. An obstetrician can identify meconium aspiration by listening to the lungs with a stethoscope for a coarse rattling sound. Blood gas tests and chest X-rays of the infant can confirm meconium aspiration. Inhaled meconium after birth could obstruct a newborn's airways leading to alveolar collapse, interfere with surfactant function by stripping it from the lungs, or cause pulmonary inflammation or hypertension. Any of these complications will make the newborn much more vulnerable to pulmonary infection, including pneumonia.

Human Pregnancy and Birth

Pregnancy begins with the fertilization of an egg and continues through to the birth of the individual. The length of time of gestation varies among animals, but is very similar among the great apes: human gestation is 266 days, while chimpanzee gestation is 237 days, a gorilla's is 257 days, and orangutan gestation is 260 days long. The fox has a 57-day gestation. Dogs and cats have similar gestations averaging 60 days. The longest gestation for a land mammal is an African elephant at 640 days. The longest gestations among marine mammals are the beluga and sperm whales at 460 days.

Human Gestation

Twenty-four hours before fertilization, the egg has finished meiosis and becomes a mature oocyte. When fertilized (at conception) the egg becomes known as a zygote. The zygote travels through the oviduct to the uterus (Figure 1). The developing embryo must implant into the wall of the uterus within seven days, or it will deteriorate and die. The outer layers of the zygote (blastocyst) grow into the endometrium by digesting the endometrial cells, and wound healing of the endometrium closes up the blastocyst into the tissue. Another layer of the blastocyst, the chorion, begins releasing a hormone called human beta chorionic gonadotropin (β-HCG) which makes its way to the corpus luteum and keeps that structure active. This ensures adequate levels of progesterone that will maintain the endometrium of the uterus for the support of the developing embryo. Pregnancy tests determine the level of β-HCG in urine or serum. If the hormone is present, the test is positive.

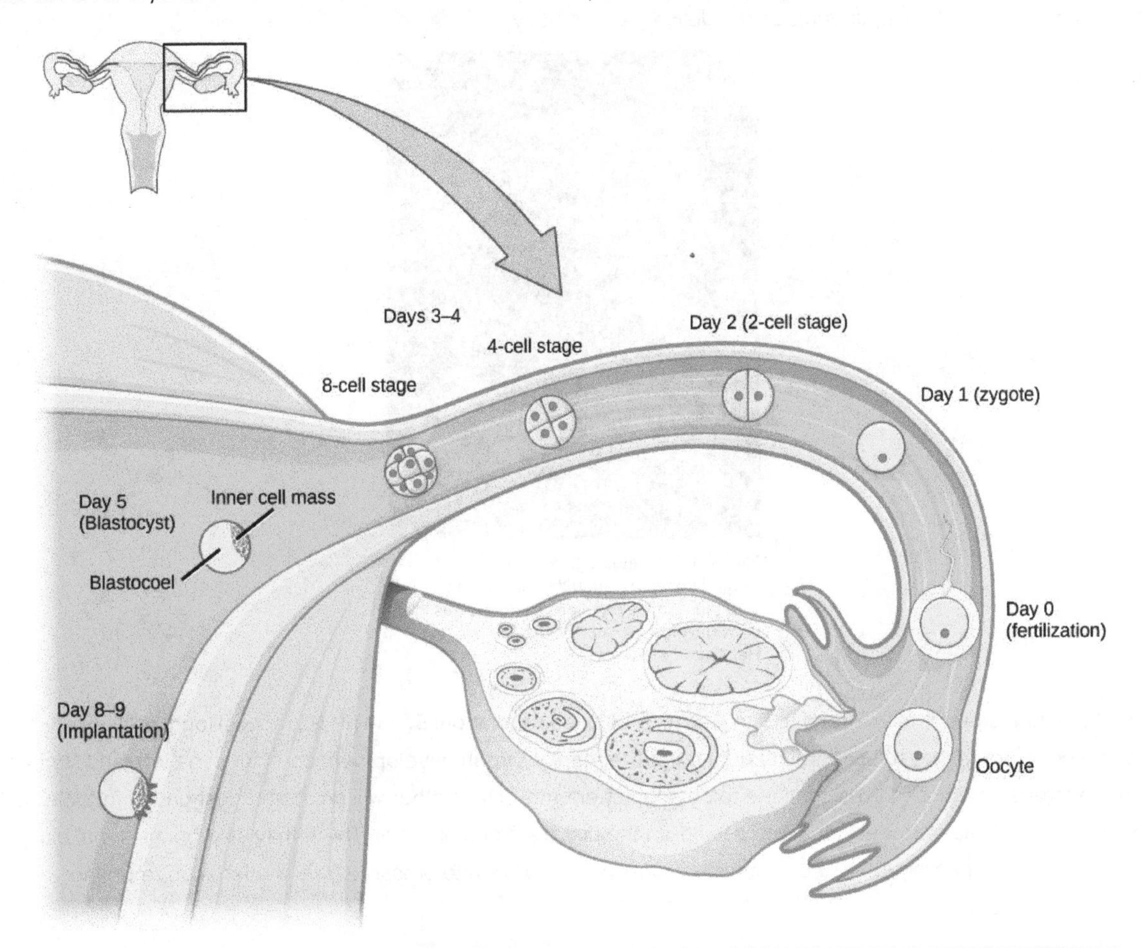

Figure 1. In humans, fertilization occurs soon after the oocyte leaves the ovary. Implantation occurs eight or nine days later.(credit: Ed Uthman)

The gestation period is divided into three equal periods or trimesters. During the first two to four weeks of the first trimester, nutrition and waste are handled by the endometrial lining through diffusion. As the trimester progresses, the outer layer of the embryo begins to merge with the endometrium, and the placenta forms. This organ takes over the nutrient and waste requirements of the embryo and fetus, with the mother's blood passing nutrients to the placenta and removing waste from it. Chemicals from the fetus, such as bilirubin, are processed by the mother's liver for elimination. Some of the mother's immunoglobulins will pass through the placenta, providing passive immunity against some potential infections.

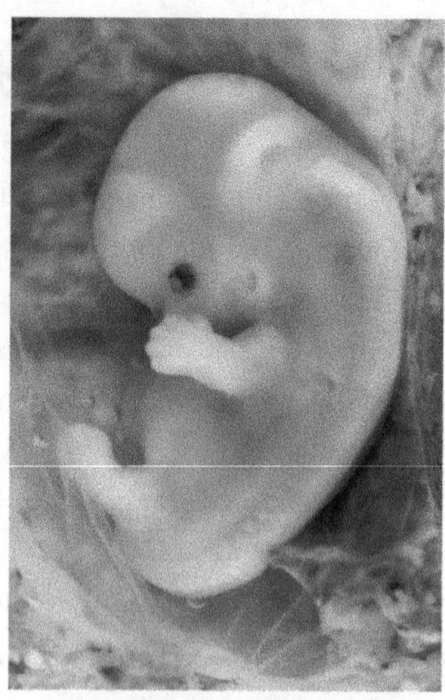

Figure 2. Fetal development is shown at nine weeks gestation. (credit: Ed Uthman)

Internal organs and body structures begin to develop during the first trimester. By five weeks, limb buds, eyes, the heart, and liver have been basically formed. By eight weeks, the term fetus applies, and the body is essentially formed, as shown in (Figure 2). The individual is about five centimeters (two inches) in length and many of the organs, such as the lungs and liver, are not yet functioning. Exposure to any toxins is especially dangerous during the first trimester, as all of the body's organs and structures are going through initial development. Anything that affects that development can have a severe effect on the fetus' survival.

During the second trimester, the fetus grows to about 30 cm (12 inches), as shown in (Figure 3). It becomes active and the mother usually feels the first movements. All organs and structures continue to develop. The placenta has taken over the functions of nutrition and waste and the production of estrogen and progesterone from the corpus luteum, which has degenerated. The placenta will continue functioning up through the delivery of the baby.

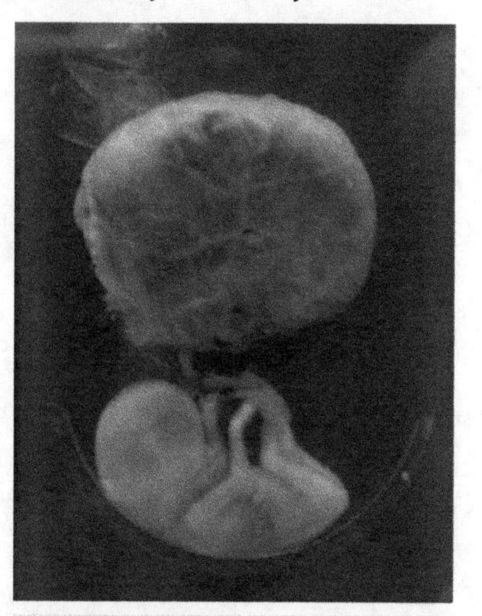

Figure 3. This fetus is just entering the second trimester, when the placenta takes over more of the functions performed as the baby develops. (credit: National Museum of Health and Medicine)

During the third trimester, the fetus grows to 3 to 4 kg (6 ½ -8 ½ lbs.) and about 50 cm (19-20 inches) long, as illustrated in (Figure 4). This is the period of the most rapid growth during the pregnancy. Organ development continues to birth (and some systems, such as the nervous system and liver, continue to develop after birth). The mother will be at her most uncomfortable during this trimester. She may urinate frequently due to pressure on the bladder from the fetus. There may also be intestinal blockage and circulatory problems, especially in her legs. Clots may form in her legs due to pressure from the fetus on returning veins as they enter the abdominal cavity.

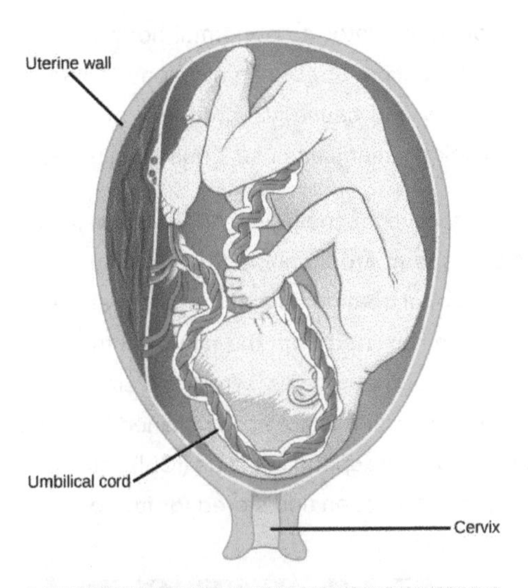

Figure 4. There is rapid fetal growth during the third trimester. (credit: modification of work by Gray's Anatomy)

Labor and Birth

Labor is the physical efforts of expulsion of the fetus and the placenta from the uterus during birth (parturition). Toward the end of the third trimester, estrogen causes receptors on the uterine wall to develop and bind the hormone oxytocin. At this time, the baby reorients, facing forward and down with the back or crown of the head engaging the cervix (uterine opening). This causes the cervix to stretch and nerve impulses are sent to the hypothalamus, which signals for the release of oxytocin from the posterior pituitary. The oxytocin causes the smooth muscle in the uterine wall to contract. At the same time, the placenta releases prostaglandins into the uterus, increasing the contractions. A positive feedback relay occurs between the uterus, hypothalamus, and the posterior pituitary to assure an adequate supply of oxytocin. As more smooth muscle cells are recruited, the contractions increase in intensity and force.

There are three stages to labor. During stage one, the cervix thins and dilates. This is necessary for the baby and placenta to be expelled during birth. The cervix will eventually dilate to about 10 cm. During stage two, the baby is expelled from the uterus. The uterus contracts and the mother pushes as she compresses her abdominal muscles to aid the delivery. The last stage is the passage of the placenta after the baby has been born and the organ has completely disengaged from the uterine wall. If labor should stop before stage two is reached, synthetic oxytocin, known as Pitocin, can be administered to restart and maintain labor.

An alternative to labor and delivery is the surgical delivery of the baby through a procedure called a Caesarian section. This is major abdominal surgery and can lead to post-surgical complications for the mother, but in some cases it may be the only way to safely deliver the baby.

The mother's mammary glands go through changes during the third trimester to prepare for lactation and breastfeeding. When the baby begins suckling at the breast, signals are sent to the hypothalamus causing the release of prolactin from the anterior pituitary. Prolactin causes the mammary glands to produce milk. Oxytocin is also released, promoting the release of the milk. The milk contains nutrients for the baby's development and growth as well as immunoglobulins to protect the child from bacterial and viral infections.

Infertility

Infertility is the inability to conceive a child or carry a child to birth. About 75 percent of causes of infertility can be identified; these include diseases, such as sexually transmitted diseases that can cause scarring of the reproductive tubes in either men or women, or developmental problems frequently related to abnormal hormone levels in one of the individuals. Inadequate nutrition, especially starvation, can delay menstruation. Stress can also lead to infertility. Short-term stress can affect hormone levels, while long-term stress can delay puberty and cause less frequent menstrual cycles. Other factors that affect fertility include toxins (such as cadmium), tobacco smoking, marijuana use, gonadal injuries, and aging.

If infertility is identified, several assisted reproductive technologies (ART) are available to aid conception. A common type of ART is *in vitro* fertilization (IVF) where an egg and sperm are combined outside the body and then placed in the uterus. Eggs are obtained from the woman after extensive hormonal treatments that prepare mature eggs for fertilization and prepare the uterus for implantation of the fertilized egg. Sperm are obtained from the man and they are combined with the eggs and supported through several cell divisions to ensure viability of the zygotes. When the embryos have reached the eight-cell stage, one or more is implanted into the woman's uterus. If fertilization is not accomplished by simple IVF, a procedure that injects the sperm into an egg can be used. This is called intracytoplasmic sperm injection (ICSI) and is shown in (Figure 5). IVF procedures produce a surplus of fertilized eggs and embryos that can be frozen and stored for future use. The procedures can also result in multiple births.

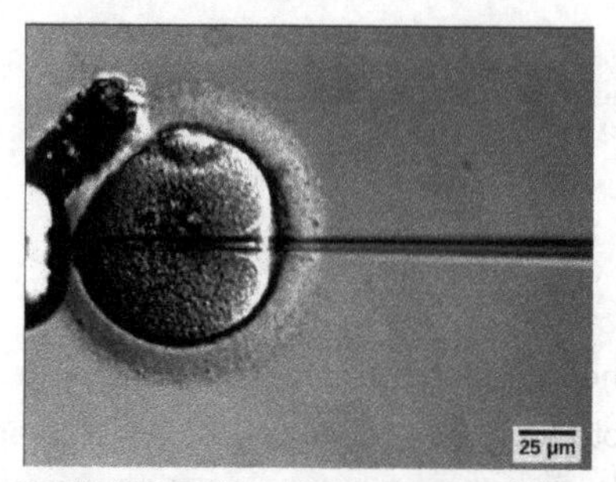

Figure 5. A sperm is inserted into an egg for fertilization during intracytoplasmic sperm injection (ICSI). (credit: scale-bar data from Matt Russell)

Maternal Changes During Pregnancy, Labor, Birth, and Lactation

A full-term pregnancy lasts approximately 270 days (approximately 38.5 weeks) from conception to birth. Because it is easier to remember the first day of the last menstrual period (LMP) than to estimate the date of conception, obstetricians set the due date as 284 days (approximately 40.5 weeks) from the LMP. This assumes that conception occurred on day 14 of the woman's cycle, which is usually a good approximation. The 40 weeks of an average pregnancy are usually discussed in terms of three **trimesters**, each approximately 13 weeks. During the second and third trimesters, the pre-pregnancy uterus—about the size of a fist—grows dramatically to contain the fetus, causing a number of anatomical changes in the mother.

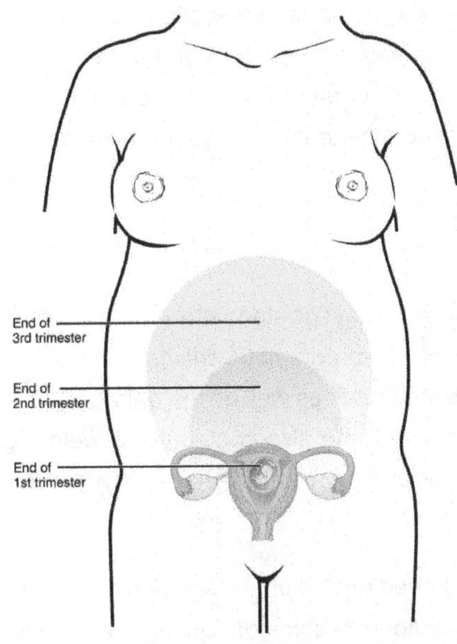

Figure 1. The uterus grows throughout pregnancy to accommodate the fetus.

Effects of Hormones

Virtually all of the effects of pregnancy can be attributed in some way to the influence of hormones—particularly estrogens, progesterone, and hCG. During weeks 7–12 from the LMP, the pregnancy hormones are primarily generated by the corpus luteum. Progesterone secreted by the corpus luteum stimulates the production of decidual cells of the endometrium that nourish the blastocyst before placentation. As the placenta develops and the corpus luteum degenerates during weeks 12–17, the placenta gradually takes over as the endocrine organ of pregnancy.

The placenta converts weak androgens secreted by the maternal and fetal adrenal glands to estrogens, which are necessary for pregnancy to progress. Estrogen levels climb throughout the pregnancy, increasing 30-fold by childbirth. Estrogens have the following actions:

- They suppress FSH and LH production, effectively preventing ovulation. (This function is the biological basis of hormonal birth control pills.)
- They induce the growth of fetal tissues and are necessary for the maturation of the fetal lungs and liver.
- They promote fetal viability by regulating progesterone production and triggering fetal synthesis of cortisol, which helps with the maturation of the lungs, liver, and endocrine organs such as the thyroid gland and adrenal gland.
- They stimulate maternal tissue growth, leading to uterine enlargement and mammary duct expansion and branching.

Relaxin, another hormone secreted by the corpus luteum and then by the placenta, helps prepare the mother's body for childbirth. It increases the elasticity of the symphysis pubis joint and pelvic ligaments, making room for the growing fetus and allowing expansion of the pelvic outlet for childbirth. Relaxin also helps dilate the cervix during labor.

The placenta takes over the synthesis and secretion of progesterone throughout pregnancy as the corpus luteum degenerates. Like estrogen, progesterone suppresses FSH and LH. It also inhibits uterine contractions, protecting the fetus from preterm birth. This hormone decreases in late gestation, allowing uterine contractions to intensify and eventually progress to true labor. The placenta also produces hCG. In addition to promoting survival of the corpus luteum, hCG stimulates the male fetal gonads to secrete testosterone, which is essential for the development of the male reproductive system.

The anterior pituitary enlarges and ramps up its hormone production during pregnancy, raising the levels of thyrotropin, prolactin, and adrenocorticotropic hormone (ACTH). Thyrotropin, in conjunction with placental hormones, increases the production of thyroid hormone, which raises the maternal metabolic rate. This can markedly augment a pregnant woman's appetite and cause hot flashes. Prolactin stimulates enlargement of the mammary glands in preparation for milk production. ACTH stimulates maternal cortisol secretion, which contributes to fetal protein synthesis. In addition to the pituitary hormones, increased parathyroid levels mobilize calcium from maternal bones for fetal use.

Weight Gain

The second and third trimesters of pregnancy are associated with dramatic changes in maternal anatomy and physiology. The most obvious anatomical sign of pregnancy is the dramatic enlargement of the abdominal region, coupled with maternal weight gain. This weight results from the growing fetus as well as the enlarged uterus, amniotic fluid, and placenta. Additional breast tissue and dramatically increased blood volume also contribute to weight gain. Surprisingly, fat storage accounts for only approximately 2.3 kg (5 lbs) in a normal pregnancy and serves as a reserve for the increased metabolic demand of breastfeeding.

During the first trimester, the mother does not need to consume additional calories to maintain a healthy pregnancy. However, a weight gain of approximately 0.45 kg (1 lb) per month is common. During the second and third trimesters, the mother's appetite increases, but it is only necessary for her to consume an additional 300 calories per day to support the growing fetus. Most women gain approximately 0.45 kg (1 lb) per week.

Table 1. Contributors to Weight Gain During Pregnancy

Component	Weight (kg)	Weight (lb)
Fetus	3.2–3.6	7–8
Placenta and fetal membranes	0.9–1.8	2–4
Amniotic fluid	0.9–1.4	2–3
Breast tissue	0.9–1.4	2–3
Blood	1.4	4
Fat	0.9–4.1	3–9
Uterus	0.9–2.3	2–5
Total	10–16.3	22–36

Physiology of Labor

Childbirth, or **parturition**, typically occurs within a week of a woman's due date, unless the woman is pregnant with more than one fetus, which usually causes her to go into labor early. As a pregnancy progresses into its final weeks, several physiological changes occur in response to hormones that trigger labor.

First, recall that progesterone inhibits uterine contractions throughout the first several months of pregnancy. As the pregnancy enters its seventh month, progesterone levels plateau and then drop. Estrogen levels, however, continue to rise in the maternal circulation. The increasing ratio of estrogen to progesterone makes the myometrium (the uterine smooth muscle) more sensitive to stimuli that promote contractions (because progesterone no longer inhibits them). Moreover, in the eighth month of pregnancy, fetal cortisol rises, which boosts estrogen secretion by the placenta and further overpowers the uterine-calming effects of progesterone. Some women may feel the result of the decreasing levels of progesterone in late pregnancy as weak and irregular peristaltic **Braxton Hicks contractions**, also called false labor. These contractions can often be relieved with rest or hydration.

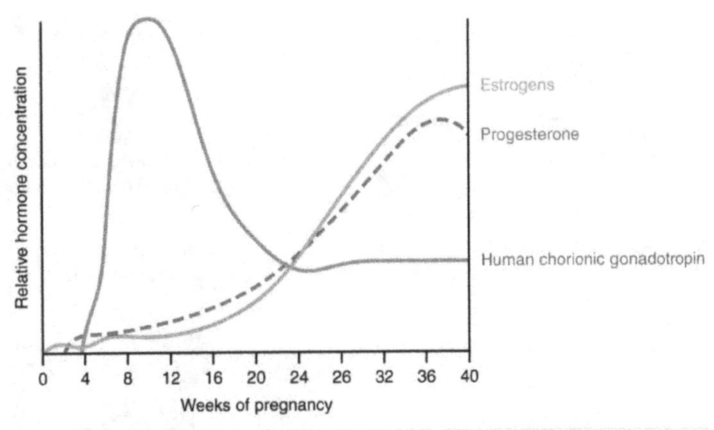

Figure 3. A positive feedback loop of hormones works to initiate labor.

A common sign that labor will be short is the so-called "bloody show." During pregnancy, a plug of mucus accumulates in the cervical canal, blocking the entrance to the uterus. Approximately 1–2 days prior to the onset of true labor, this plug loosens and is expelled, along with a small amount of blood.

Meanwhile, the posterior pituitary has been boosting its secretion of oxytocin, a hormone that stimulates the contractions of labor. At the same time, the myometrium increases its sensitivity to oxytocin by expressing more receptors for this hormone. As labor nears, oxytocin begins to stimulate stronger, more painful uterine contractions, which—in a positive feedback loop—stimulate the secretion of prostaglandins from fetal membranes. Like oxytocin, prostaglandins also enhance uterine contractile strength. The fetal pituitary also secretes oxytocin, which increases prostaglandins even further. Given the importance of oxytocin and prostaglandins to the initiation and maintenance of labor, it is not surprising that, when a pregnancy is not progressing to labor and needs to be induced, a pharmaceutical version of these compounds (called pitocin) is administered by intravenous drip.

Finally, stretching of the myometrium and cervix by a full-term fetus in the vertex (head-down) position is regarded as a stimulant to uterine contractions. The sum of these changes initiates the regular contractions known as **true labor**, which become more powerful and more frequent with time. The pain of labor is attributed to myometrial hypoxia during uterine contractions.

Stages of Childbirth

The process of childbirth can be divided into three stages: cervical dilation, expulsion of the newborn, and afterbirth.

Cervical Dilation

For vaginal birth to occur, the cervix must dilate fully to 10 cm in diameter—wide enough to deliver the newborn's head. The **dilation** stage is the longest stage of labor and typically takes 6–12 hours. However, it varies widely and may take minutes, hours, or days, depending in part on whether the mother has given birth before; in each subsequent labor, this stage tends to be shorter.

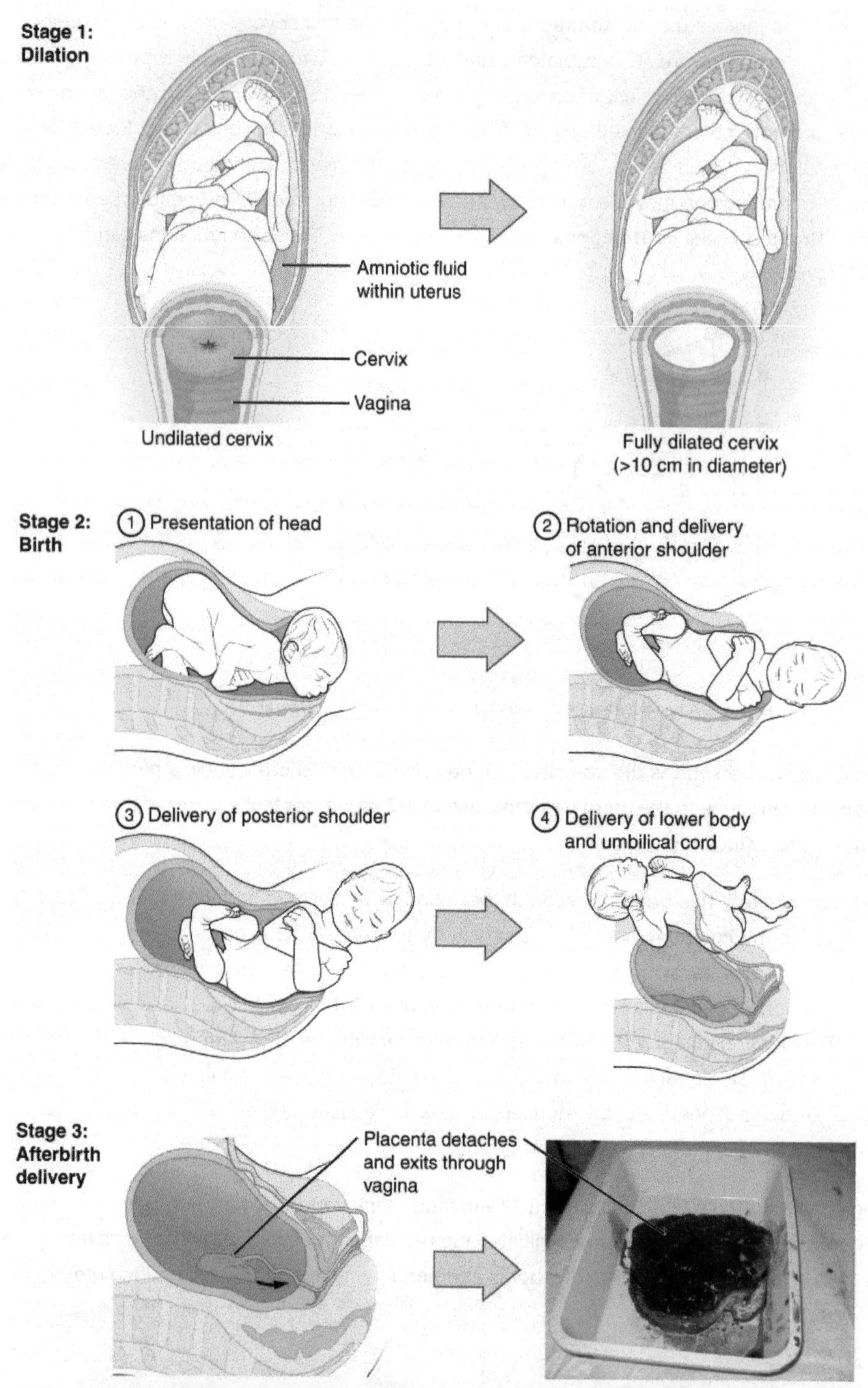

Figure 4. Click for a larger image. The stages of childbirth include Stage 1, early cervical dilation; Stage 2, full dilation and expulsion of the newborn; and Stage 3, delivery of the placenta and associated fetal membranes. (The position of the newborn's shoulder is described relative to the mother.)

True labor progresses in a positive feedback loop in which uterine contractions stretch the cervix, causing it to dilate and efface, or become thinner. Cervical stretching induces reflexive uterine contractions that dilate and efface the cervix further. In addition, cervical dilation boosts oxytocin secretion from the pituitary, which in turn triggers more powerful uterine contractions. When labor begins, uterine contractions may occur only every 3–30 minutes and last only 20–40 seconds; however, by the end of this stage, contractions may occur as frequently as every 1.5–2 minutes and last for a full minute.

Each contraction sharply reduces oxygenated blood flow to the fetus. For this reason, it is critical that a period of relaxation occur after each contraction. Fetal distress, measured as a sustained decrease or increase in the fetal heart rate, can result from severe contractions that are too powerful or lengthy for oxygenated blood to be restored to the fetus. Such a situation can be cause for an emergency birth with vacuum, forceps, or surgically by Caesarian section.

The amniotic membranes rupture before the onset of labor in about 12 percent of women; they typically rupture at the end of the dilation stage in response to excessive pressure from the fetal head entering the birth canal.

Expulsion Stage

The **expulsion** stage begins when the fetal head enters the birth canal and ends with birth of the newborn. It typically takes up to 2 hours, but it can last longer or be completed in minutes, depending in part on the orientation of the fetus. The vertex presentation known as the occiput anterior vertex is the most common presentation and is associated with the greatest ease of vaginal birth. The fetus faces the maternal spinal cord and the smallest part of the head (the posterior aspect called the occiput) exits the birth canal first.

In fewer than 5 percent of births, the infant is oriented in the breech presentation, or buttocks down. In a complete breech, both legs are crossed and oriented downward. In a frank breech presentation, the legs are oriented upward. Before the 1960s, it was common for breech presentations to be delivered vaginally. Today, most breech births are accomplished by Caesarian section.

Vaginal birth is associated with significant stretching of the vaginal canal, the cervix, and the perineum. Until recent decades, it was routine procedure for an obstetrician to numb the perineum and perform an **episiotomy**, an incision in the posterior vaginal wall and perineum. The perineum is now more commonly allowed to tear on its own during birth. Both an episiotomy and a perineal tear need to be sutured shortly after birth to ensure optimal healing. Although suturing the jagged edges of a perineal tear may be more difficult than suturing an episiotomy, tears heal more quickly, are less painful, and are associated with less damage to the muscles around the vagina and rectum.

Upon birth of the newborn's head, an obstetrician will aspirate mucus from the mouth and nose before the newborn's first breath. Once the head is birthed, the rest of the body usually follows quickly. The umbilical cord is then double-clamped, and a cut is made between the clamps. This completes the second stage of childbirth.

Afterbirth

The delivery of the placenta and associated membranes, commonly referred to as the **afterbirth**, marks the final stage of childbirth. After expulsion of the newborn, the myometrium continues to contract. This movement shears the placenta from the back of the uterine wall. It is then easily delivered through the vagina. Continued uterine contractions then reduce blood loss from the site of the placenta. Delivery of the placenta marks the beginning of the postpartum period—the period of approximately 6 weeks immediately following childbirth during which the mother's body gradually returns to a non-pregnant state. If the placenta does not birth spontaneously within approximately 30 minutes, it is considered retained, and the obstetrician may attempt manual removal. If this is not successful, surgery may be required.

It is important that the obstetrician examines the expelled placenta and fetal membranes to ensure that they are intact. If fragments of the placenta remain in the uterus, they can cause postpartum hemorrhage. Uterine contractions continue for several hours after birth to return the uterus to its pre-pregnancy size in a process called **involution**, which also allows the mother's abdominal organs to return to their pre-pregnancy locations. Breastfeeding facilitates this process.

Although postpartum uterine contractions limit blood loss from the detachment of the placenta, the mother does experience a postpartum vaginal discharge called **lochia**. This is made up of uterine lining cells, erythrocytes, leukocytes, and other debris.

Thick, dark, lochia rubra (red lochia) typically continues for 2–3 days, and is replaced by lochia serosa, a thinner, pinkish form that continues until about the tenth postpartum day. After this period, a scant, creamy, or watery discharge called lochia alba (white lochia) may continue for another 1–2 weeks.

Lactation is the process by which milk is synthesized and secreted from the mammary glands of the postpartum female breast in response to an infant sucking at the nipple. Breast milk provides ideal nutrition and passive immunity for the infant, encourages mild uterine contractions to return the uterus to its pre-pregnancy size (i.e., involution), and induces a substantial metabolic increase in the mother, consuming the fat reserves stored during pregnancy.

Structure of the Lactating Breast

Mammary glands are modified sweat glands. The non-pregnant and non-lactating female breast is composed primarily of adipose and collagenous tissue, with mammary glands making up a very minor proportion of breast volume. The mammary gland is composed of milk-transporting lactiferous ducts, which expand and branch extensively during pregnancy in response to estrogen, growth hormone, cortisol, and prolactin. Moreover, in response to progesterone, clusters of breast alveoli bud from the ducts and expand outward toward the chest wall. Breast alveoli are balloon-like structures lined with milk-secreting cuboidal cells, or lactocytes, that are surrounded by a net of contractile myoepithelial cells. Milk is secreted from the lactocytes, fills the alveoli, and is squeezed into the ducts. Clusters of alveoli that drain to a common duct are called lobules; the lactating female has 12–20 lobules organized radially around the nipple. Milk drains from lactiferous ducts into lactiferous sinuses that meet at 4 to 18 perforations in the nipple, called nipple pores. The small bumps of the areola (the darkened skin around the nipple) are called Montgomery glands. They secrete oil to cleanse the nipple opening and prevent chapping and cracking of the nipple during breastfeeding.

The Process of Lactation

The pituitary hormone **prolactin** is instrumental in the establishment and maintenance of breast milk supply. It also is important for the mobilization of maternal micronutrients for breast milk.

Near the fifth week of pregnancy, the level of circulating prolactin begins to increase, eventually rising to approximately 10–20 times the pre-pregnancy concentration. We noted earlier that, during pregnancy, prolactin and other hormones prepare the breasts anatomically for the secretion of milk. The level of prolactin plateaus in late pregnancy, at a level high enough to initiate milk production. However, estrogen, progesterone, and other placental hormones inhibit prolactin-mediated milk synthesis during pregnancy. It is not until the placenta is expelled that this inhibition is lifted and milk production commences.

After childbirth, the baseline prolactin level drops sharply, but it is restored for a 1-hour spike during each feeding to stimulate the production of milk for the next feeding. With each prolactin spike, estrogen and progesterone also increase slightly.

When the infant suckles, sensory nerve fibers in the areola trigger a neuroendocrine reflex that results in milk secretion from lactocytes into the alveoli. The posterior pituitary releases oxytocin, which stimulates myoepithelial cells to squeeze milk from the alveoli so it can drain into the lactiferous ducts, collect in the lactiferous sinuses, and discharge through the nipple pores. It takes less than 1 minute from the time when an infant begins suckling (the latent period) until milk is secreted (the let-down). The image below summarizes the positive feedback loop of the **let-down reflex.**

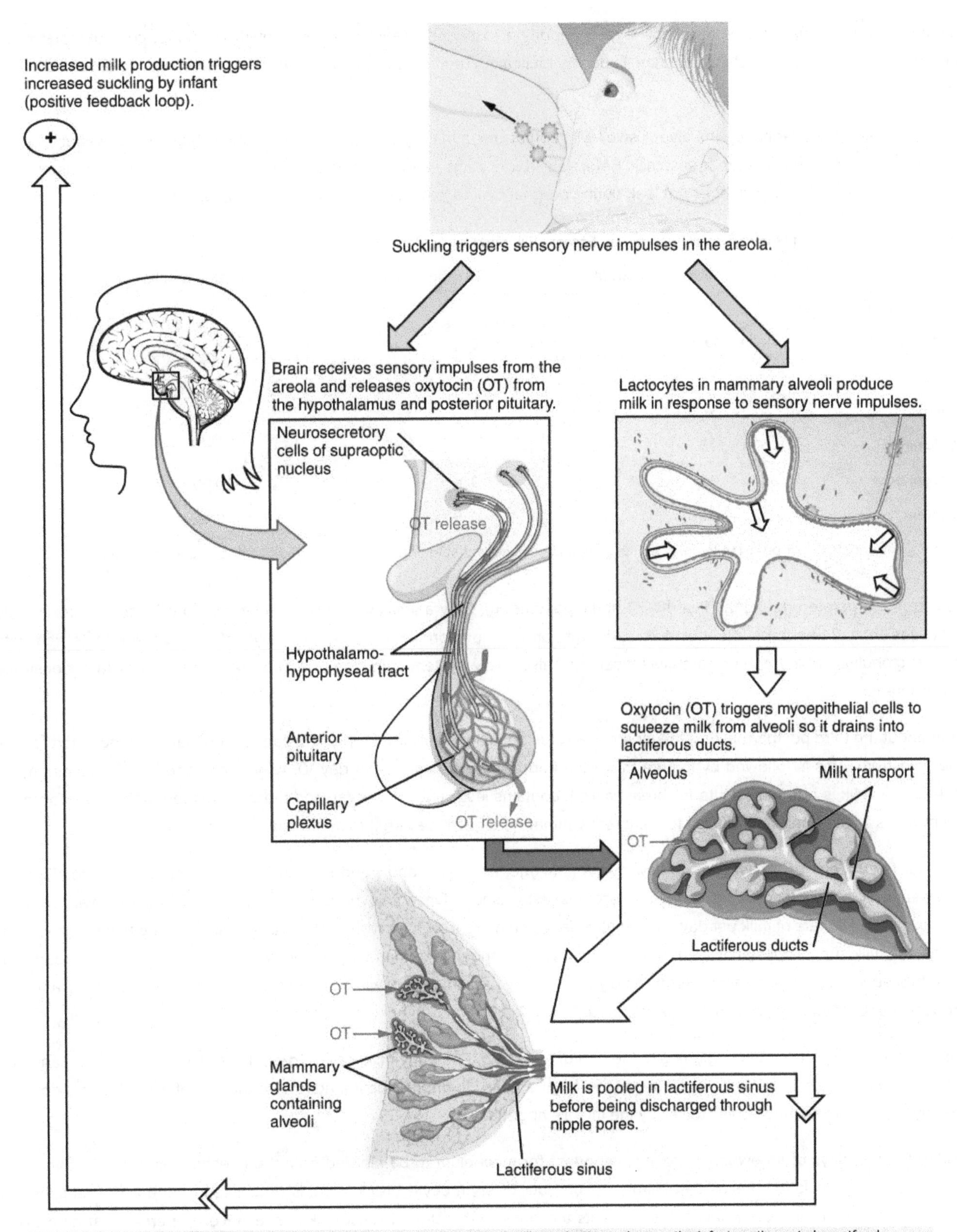

Figure 5. Click for a larger image. A positive feedback loop ensures continued milk production as long as the infant continues to breastfeed.

The prolactin-mediated synthesis of milk changes with time. Frequent milk removal by breastfeeding (or pumping) will maintain high circulating prolactin levels for several months. However, even with continued breastfeeding, baseline prolactin will decrease over time to its pre-pregnancy level. In addition to prolactin and oxytocin, growth hormone, cortisol, parathyroid hormone, and insulin contribute to lactation, in part by facilitating the transport of maternal amino acids, fatty acids, glucose, and calcium to breast milk.

In the final weeks of pregnancy, the alveoli swell with **colostrum**, a thick, yellowish substance that is high in protein but contains less fat and glucose than mature breast milk. Before childbirth, some women experience leakage of colostrum from the nipples. In contrast, mature breast milk does not leak during pregnancy and is not secreted until several days after childbirth.

Table 1. Compositions of Human Colostrum, Mature Breast Milk, and Cow's Milk (g/L)

	Human colostrum	Human breast milk	Cow's milk*
Total protein	23	11	31
Immunoglobulins	19	0.1	1
Fat	30	45	38
Lactose	57	71	47
Calcium	0.5	0.3	1.4
Phosphorus	0.16	0.14	0.90
Sodium	0.50	0.15	0.41
*Cow's milk should never be given to an infant. Its composition is not suitable and its proteins are difficult for the infant to digest.			

Colostrum is secreted during the first 48–72 hours postpartum. Only a small volume of colostrum is produced—approximately 3 ounces in a 24-hour period—but it is sufficient for the newborn in the first few days of life. Colostrum is rich with immunoglobulins, which confer gastrointestinal, and also likely systemic, immunity as the newborn adjusts to a nonsterile environment.

After about the third postpartum day, the mother secretes transitional milk that represents an intermediate between mature milk and colostrum. This is followed by mature milk from approximately postpartum day 10. As you can see in the accompanying table, cow's milk is not a substitute for breast milk. It contains less lactose, less fat, and more protein and minerals. Moreover, the proteins in cow's milk are difficult for an infant's immature digestive system to metabolize and absorb.

The first few weeks of breastfeeding may involve leakage, soreness, and periods of milk engorgement as the relationship between milk supply and infant demand becomes established. Once this period is complete, the mother will produce approximately 1.5 liters of milk per day for a single infant, and more if she has twins or triplets. As the infant goes through growth spurts, the milk supply constantly adjusts to accommodate changes in demand. A woman can continue to lactate for years, but once breastfeeding is stopped for approximately 1 week, any remaining milk will be reabsorbed; in most cases, no more will be produced, even if suckling or pumping is resumed.

Mature milk changes from the beginning to the end of a feeding. The early milk, called **foremilk**, is watery, translucent, and rich in lactose and protein. Its purpose is to quench the infant's thirst. **Hindmilk** is delivered toward the end of a feeding. It is opaque, creamy, and rich in fat, and serves to satisfy the infant's appetite.

During the first days of a newborn's life, it is important for meconium to be cleared from the intestines and for bilirubin to be kept low in the circulation. Recall that bilirubin, a product of erythrocyte breakdown, is processed by the liver and secreted in bile. It enters the gastrointestinal tract and exits the body in the stool. Breast milk has laxative properties that help expel meconium from the intestines and clear bilirubin through the excretion of bile. A high concentration of bilirubin in the blood causes jaundice. Some degree of jaundice is normal in newborns, but a high level of bilirubin—which is neurotoxic—can cause brain damage. Newborns, who do not yet have a fully functional blood–brain barrier, are highly vulnerable to the bilirubin circulating in the blood. Indeed, hyperbilirubinemia, a high level of circulating bilirubin, is the most common condition requiring medical attention in newborns. Newborns with hyperbilirubinemia are treated with phototherapy because UV light helps to break down the bilirubin quickly.

Patterns of Inheritance

We have discussed the events that lead to the development of a newborn. But what makes each newborn unique? The answer lies, of course, in the DNA in the sperm and oocyte that combined to produce that first diploid cell, the human zygote.

From Genotype to Phenotype

Each human body cell has a full complement of DNA stored in 23 pairs of chromosomes. The image below shows the pairs in a systematic arrangement called a **karyotype**. Among these is one pair of chromosomes, called the **sex chromosomes**, that determines the sex of the individual (XX in females, XY in males). The remaining 22 chromosome pairs are called **autosomal chromosomes**. Each of these chromosomes carries hundreds or even thousands of genes, each of which codes for the assembly of a particular protein—that is, genes are "expressed" as proteins. An individual's complete genetic makeup is referred to as his or her **genotype**. The characteristics that the genes express, whether they are physical, behavioral, or biochemical, are a person's **phenotype**.

You inherit one chromosome in each pair—a full complement of 23—from each parent. This occurs when the sperm and oocyte combine at the moment of your conception. Homologous chromosomes—those that make up a complementary pair—have genes for the same characteristics in the same location on the chromosome. Because one copy of a gene, an **allele**, is inherited from each parent, the alleles in these complementary pairs may vary. Take for example an allele that encodes for dimples. A child may inherit the allele encoding for dimples on the chromosome from the father and the allele that encodes for smooth skin (no dimples) on the chromosome from the mother.

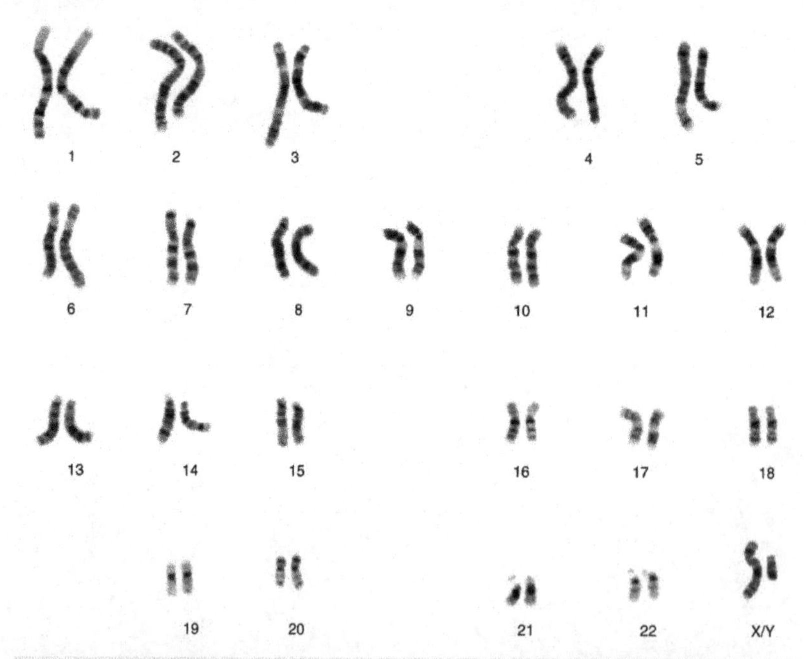

Figure 1. Each pair of chromosomes contains hundreds to thousands of genes. The banding patterns are nearly identical for the two chromosomes within each pair, indicating the same organization of genes. As is visible in this karyotype, the only exception to this is the XY sex chromosome pair in males. (credit: National Human Genome Research Institute)

Although a person can have two identical alleles for a single gene (a **homozygous** state), it is also possible for a person to have two different alleles (a **heterozygous** state). The two alleles can interact in several different ways. The expression of an allele can be dominant, for which the activity of this gene will mask the expression of a nondominant, or recessive, allele. Sometimes dominance is complete; at other times, it is incomplete. In some cases, both alleles are expressed at the same time in a form of expression known as codominance.

In the simplest scenario, a single pair of genes will determine a single heritable characteristic. However, it is quite common for multiple genes to interact to confer a feature. For instance, eight or more genes—each with their own alleles—determine eye color in humans. Moreover, although any one person can only have two alleles corresponding to a given gene, more than two alleles commonly exist in a population. This phenomenon is called multiple alleles. For example, there are three different alleles that encode ABO blood type; these are designated I^A, I^B, and *i*.

Over 100 years of theoretical and experimental genetics studies, and the more recent sequencing and annotation of the human genome, have helped scientists to develop a better understanding of how an individual's genotype is expressed as their phenotype. This body of knowledge can help scientists and medical professionals to predict, or at least estimate, some of the features that an offspring will inherit by examining the genotypes or phenotypes of the parents. One important application of this knowledge is to identify an individual's risk for certain heritable genetic disorders. However, most diseases have a multigenic pattern of inheritance and can also be affected by the environment, so examining the genotypes or phenotypes of a person's parents will provide only limited information about the risk of inheriting a disease. Only for a handful of single-gene disorders can genetic testing allow clinicians to calculate the probability with which a child born to the two parents tested may inherit a specific disease.

Mendel's Theory of Inheritance

Our contemporary understanding of genetics rests on the work of a nineteenth-century monk. Working in the mid-1800s, long before anyone knew about genes or chromosomes, Gregor Mendel discovered that garden peas transmit their physical characteristics to subsequent generations in a discrete and predictable fashion. When he mated, or crossed, two pure-breeding pea plants that differed by a certain characteristic, the first-generation offspring all looked like one of the parents. For instance, when he crossed tall and dwarf pure-breeding pea plants, all of the offspring were tall. Mendel called tallness **dominant** because it was expressed in offspring when it was present in a purebred parent. He called dwarfism **recessive** because it was masked in the offspring if one of the purebred parents possessed the dominant characteristic. Note that tallness and dwarfism are variations on the characteristic of height. Mendel called such a variation a **trait**. We now know that these traits are the expression of different alleles of the gene encoding height.

Mendel performed thousands of crosses in pea plants with differing traits for a variety of characteristics. And he repeatedly came up with the same results—among the traits he studied, one was always dominant, and the other was always recessive. (Remember, however, that this dominant–recessive relationship between alleles is not always the case; some alleles are codominant, and sometimes dominance is incomplete.)

Using his understanding of dominant and recessive traits, Mendel tested whether a recessive trait could be lost altogether in a pea lineage or whether it would resurface in a later generation. By crossing the second-generation offspring of purebred parents with each other, he showed that the latter was true: recessive traits reappeared in third-generation plants in a ratio of 3:1 (three offspring having the dominant trait and one having the recessive trait). Mendel then proposed that characteristics such as height were determined by heritable "factors" that were transmitted, one from each parent, and inherited in pairs by offspring.

In the language of genetics, Mendel's theory applied to humans says that if an individual receives two dominant alleles, one from each parent, the individual's phenotype will express the dominant trait. If an individual receives two recessive alleles, then the recessive trait will be expressed in the phenotype. Individuals who have two identical alleles for a given gene, whether dominant or recessive, are said to be homozygous for that gene (homo- = "same"). Conversely, an individual who has one dominant allele and one recessive allele is said to be heterozygous for that gene (hetero- = "different" or "other"). In this case, the dominant trait will be expressed, and the individual will be phenotypically identical to an individual who possesses two dominant alleles for the trait.

It is common practice in genetics to use capital and lowercase letters to represent dominant and recessive alleles. Using Mendel's pea plants as an example, if a tall pea plant is homozygous, it will possess two tall alleles (*TT*). A dwarf pea plant must be homozygous because its dwarfism can only be expressed when two recessive alleles are present (*tt*). A heterozygous pea plant (*Tt*) would be tall and phenotypically indistinguishable from a tall homozygous pea plant because of the dominant tall allele. Mendel deduced that a 3:1 ratio of dominant to recessive would be produced by the random segregation of heritable factors (genes) when crossing two heterozygous pea plants. In other words, for any given gene, parents are equally likely to pass down either one of their alleles to their offspring in a haploid gamete, and the result will be expressed in a dominant–recessive pattern if both parents are heterozygous for the trait.

Because of the random segregation of gametes, the laws of chance and probability come into play when predicting the likelihood of a given phenotype. Consider a cross between an individual with two dominant alleles for a trait (*AA*) and an individual with two recessive alleles for the same trait (*aa*). All of the parental gametes from the dominant individual would be *A*, and all of the parental gametes from the recessive individual would be *a*. All of the offspring of that second generation, inheriting one allele from each parent, would have the genotype *Aa*, and the probability of expressing the phenotype of the dominant allele would be 4 out of 4, or 100 percent.

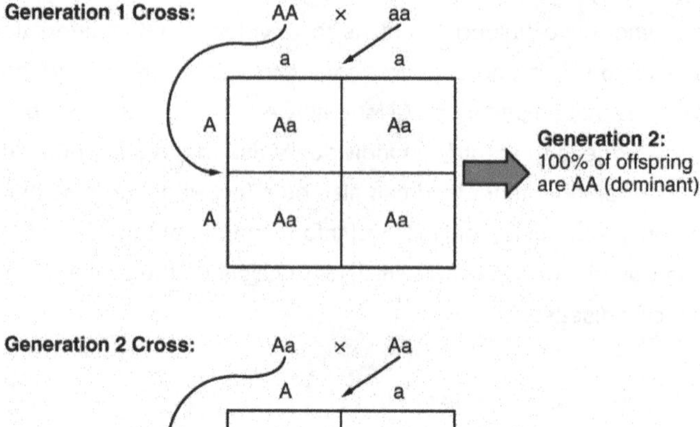

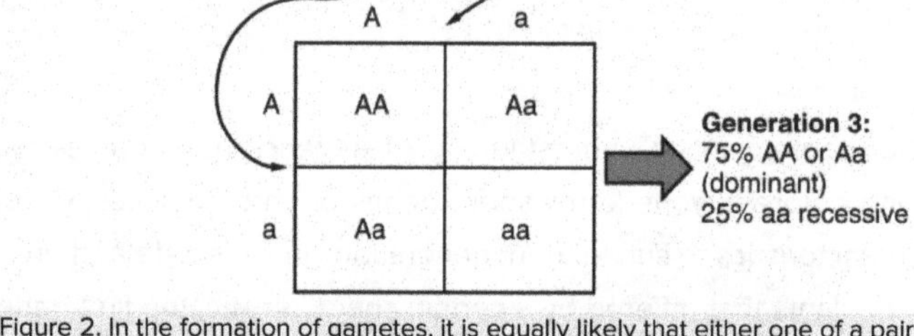

Figure 2. In the formation of gametes, it is equally likely that either one of a pair alleles from one parent will be passed on to the offspring. This figure follows the possible combinations of alleles through two generations following a first-generation cross of homozygous dominant and homozygous recessive parents. The recessive phenotype, which is masked in the second generation, has a 1 in 4, or 25 percent, chance of reappearing in the third generation.

This seems simple enough, but the inheritance pattern gets interesting when the second-generation *Aa* individuals are crossed. In this generation, 50 percent of each parent's gametes are *A* and the other 50 percent are *a*. By Mendel's principle of random segregation, the possible combinations of gametes that the offspring can receive are *AA*, *Aa*, *aA* (which is the same as *Aa*), and *aa*. Because segregation and fertilization are random, each offspring has a 25 percent chance of receiving any of these combinations. Therefore, if an *Aa* × *Aa* cross were performed 1000 times, approximately 250 (25 percent) of the offspring would be *AA*; 500 (50 percent) would be *Aa* (that is, *Aa* plus *aA*); and 250 (25 percent) would be *aa*. The genotypic ratio for this inheritance pattern is 1:2:1. However, we have already established that *AA* and *Aa* (and *aA*) individuals all express the dominant trait (i.e., share the same phenotype), and can therefore be combined into one group. The result is Mendel's third-generation phenotype ratio of 3:1.

Mendel's observation of pea plants also included many crosses that involved multiple traits, which prompted him to formulate the principle of independent assortment. The law states that the members of one pair of genes (alleles) from a parent will sort independently from other pairs of genes during the formation of gametes. Applied to pea plants, that means that the alleles associated with the different traits of the plant, such as color, height, or seed type, will sort independently of one another. This holds true except when two alleles happen to be located close to one other on the same chromosome. Independent assortment provides for a great degree of diversity in offspring.

Mendelian genetics represent the fundamentals of inheritance, but there are two important qualifiers to consider when applying Mendel's findings to inheritance studies in humans. First, as we've already noted, not all genes are inherited in a dominant–recessive pattern. Although all diploid individuals have two alleles for every gene, allele pairs may interact to create several types of inheritance patterns, including incomplete dominance and codominance.

Secondly, Mendel performed his studies using thousands of pea plants. He was able to identify a 3:1 phenotypic ratio in second-generation offspring because his large sample size overcame the influence of variability resulting from chance. In contrast, no human couple has ever had thousands of children. If we know that a man and woman are both heterozygous for a recessive genetic disorder, we would predict that one in every four of their children would be affected by the disease. In real life, however, the influence of chance could change that ratio significantly. For example, if a man and a woman are both heterozygous for cystic fibrosis, a recessive genetic disorder that is expressed only when the individual has two defective alleles, we would expect one in four of their children to have cystic fibrosis. However, it is entirely possible for them to have seven children, none of whom is affected, or for them to have two children, both of whom are affected. For each individual child, the presence or absence of a single gene disorder depends on which alleles that child inherits from his or her parents.

Autosomal Dominant Inheritance

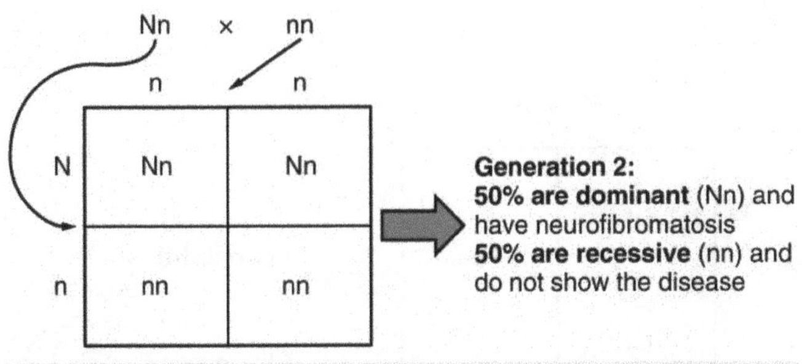

Figure 3. Inheritance pattern of an autosomal dominant disorder, such as neurofibromatosis, is shown in a Punnett square.

In the case of cystic fibrosis, the disorder is recessive to the normal phenotype. However, a genetic abnormality may be dominant to the normal phenotype. When the dominant allele is located on one of the 22 pairs of autosomes (non-sex chromosomes), we refer to its inheritance pattern as **autosomal dominant**. An example of an autosomal dominant disorder is neurofibromatosis type I, a disease that induces tumor formation within the nervous system that leads to skin and skeletal deformities. Consider a couple in which one parent is heterozygous for this disorder (and who therefore has neurofibromatosis), *Nn*, and one parent is homozygous for the normal gene, *nn*. The heterozygous parent would have a 50 percent chance of passing the dominant allele for this disorder to his or her offspring, and the homozygous parent would always pass the normal allele. Therefore, four possible offspring genotypes are equally likely to occur: *Nn*, *Nn*, *nn*, and *nn*. That is, every child of this couple would have a 50 percent chance of inheriting neurofibromatosis. This inheritance pattern is shown in the table below, in a form called a **Punnett square**, named after its creator, the British geneticist Reginald Punnett.

Other genetic diseases that are inherited in this pattern are achondroplastic dwarfism, Marfan syndrome, and Huntington's disease. Because autosomal dominant disorders are expressed by the presence of just one gene, an individual with the disorder will know that he or she has at least one faulty gene. The expression of the disease may manifest later in life, after the childbearing years, which is the case in Huntington's disease (discussed in more detail later in this section).

Autosomal Recessive Inheritance

When a genetic disorder is inherited in an **autosomal recessive** pattern, the disorder corresponds to the recessive phenotype. Heterozygous individuals will not display symptoms of this disorder, because their unaffected gene will compensate. Such an individual is called a **carrier**. Carriers for an autosomal recessive disorder may never know their genotype unless they have a child with the disorder.

An example of an autosomal recessive disorder is cystic fibrosis (CF), which we introduced earlier. CF is characterized by the chronic accumulation of a thick, tenacious mucus in the lungs and digestive tract. Decades ago, children with CF rarely lived to adulthood. With advances in medical technology, the average lifespan in developed countries has increased into middle adulthood. CF is a relatively common disorder that occurs in approximately 1 in 2000 Caucasians. A child born to two CF carriers would have a 25 percent chance of inheriting the disease. This is the same 3:1 dominant:recessive ratio that Mendel observed in his pea plants would apply here. The pattern is shown in the image below, using a diagram that tracks the likely incidence of an autosomal recessive disorder on the basis of parental genotypes.

On the other hand, a child born to a CF carrier and someone with two unaffected alleles would have a 0 percent probability of inheriting CF, but would have a 50 percent chance of being a carrier. Other examples of autosome recessive genetic illnesses include the blood disorder sickle-cell anemia, the fatal neurological disorder Tay–Sachs disease, and the metabolic disorder phenylketonuria.

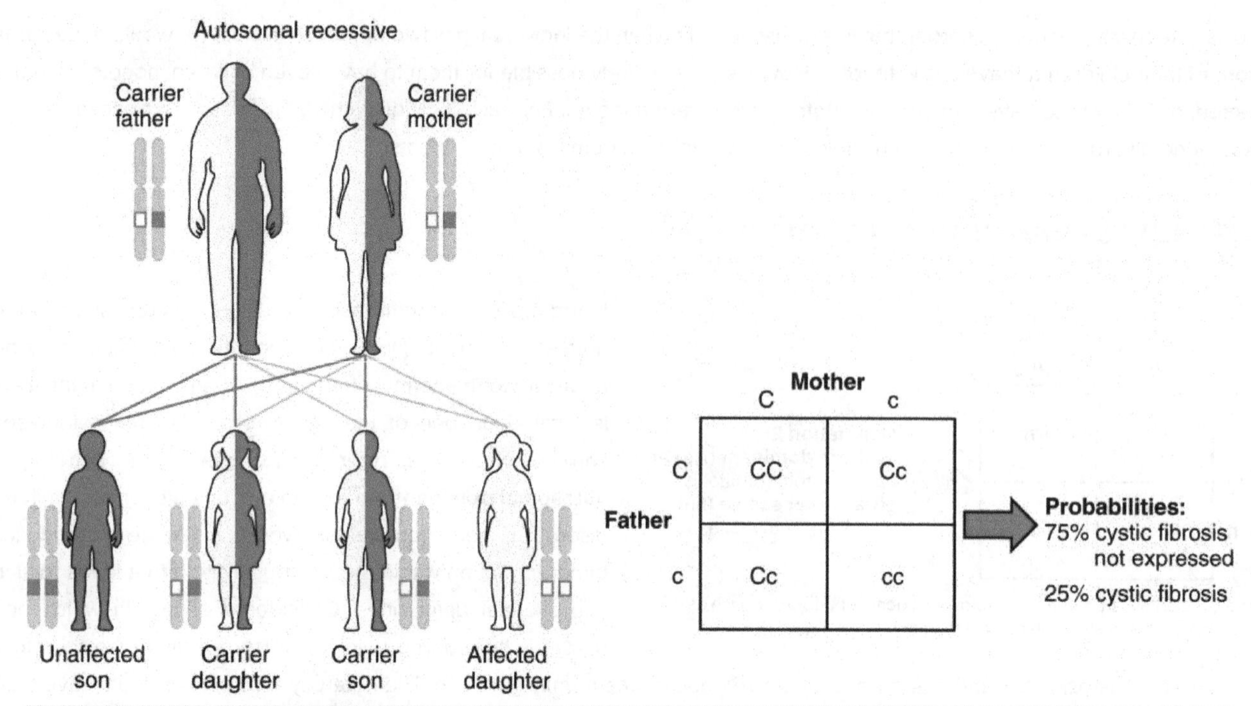

Figure 4. The inheritance pattern of an autosomal recessive disorder with two carrier parents reflects a 3:1 probability of expression among offspring. (credit: U.S. National Library of Medicine)

X-linked Dominant or Recessive Inheritance

An **X-linked** transmission pattern involves genes located on the X chromosome of the 23rd pair. Recall that a male has one X and one Y chromosome. When a father transmits a Y chromosome, the child is male, and when he transmits an X chromosome, the child is female. A mother can transmit only an X chromosome, as both her sex chromosomes are X chromosomes.

X-linked Dominant Inheritance

When an abnormal allele for a gene that occurs on the X chromosome is dominant over the normal allele, the pattern is described as **X-linked dominant**. This is the case with vitamin D–resistant rickets: an affected father would pass the disease gene to all of his daughters, but none of his sons, because he donates only the Y chromosome to his sons. If it is the mother who is affected, all of her children—male or female—would have a 50 percent chance of inheriting the disorder because she can only pass an X chromosome on to her children. For an affected female, the inheritance pattern would be identical to that of an autosomal dominant inheritance pattern in which one parent is heterozygous and the other is homozygous for the normal gene.

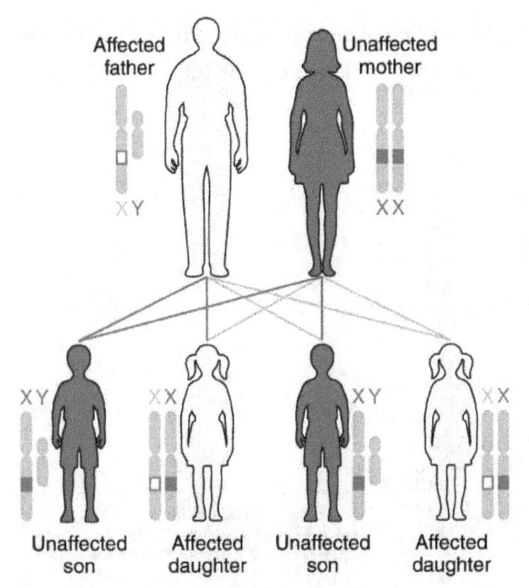

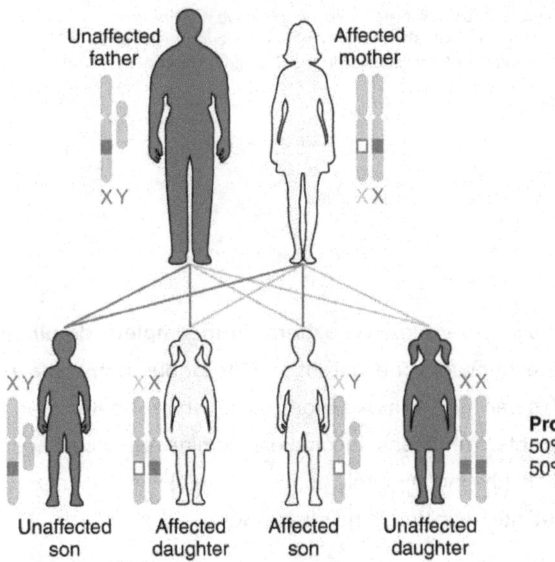

Figure 5. Click for a larger image. A chart of X-linked dominant inheritance patterns differs depending on whether (a) the father or (b) the mother is affected with the disease. (credit: U.S. National Library of Medicine)

X-linked Recessive Inheritance

X-linked recessive inheritance is much more common because females can be carriers of the disease yet still have a normal phenotype. Diseases transmitted by X-linked recessive inheritance include color blindness, the blood-clotting disorder hemophilia, and some forms of muscular dystrophy. For an example of X-linked recessive inheritance, consider parents in which the mother is an unaffected carrier and the father is normal. None of the daughters would have the disease because they receive a normal gene from their father. However, they have a 50 percent chance of receiving the disease gene from their mother and becoming a carrier. In contrast, 50 percent of the sons would be affected.

With X-linked recessive diseases, males either have the disease or are genotypically normal—they cannot be carriers. Females, however, can be genotypically normal, a carrier who is phenotypically normal, or affected with the disease. A daughter can inherit the gene for an X-linked recessive illness when her mother is a carrier or affected, or her father is affected. The daughter will be affected by the disease only if she inherits an X-linked recessive gene from both parents. As you can imagine, X-linked recessive disorders affect many more males than females. For example, color blindness affects at least 1 in 20 males, but only about 1 in 400 females.

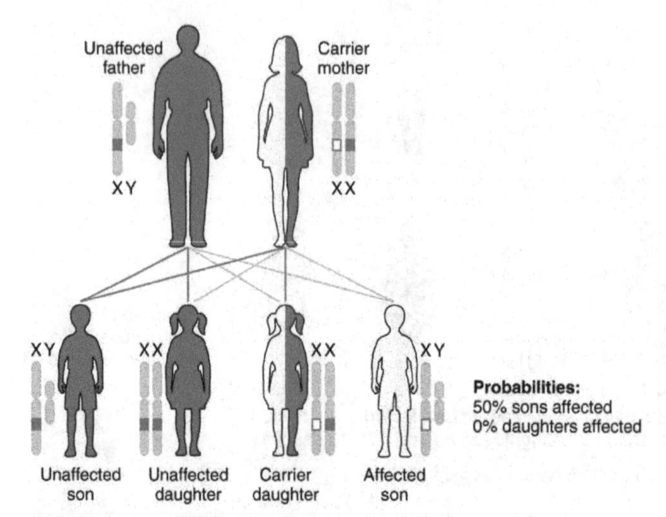

Figure 6. Given two parents in which the father is normal and the mother is a carrier of an X-linked recessive disorder, a son would have a 50 percent probability of being affected with the disorder, whereas daughters would either be carriers or entirely unaffected. (credit: U.S. National Library of Medicine)

Other Inheritance Patterns

Incomplete Dominance

Not all genetic disorders are inherited in a dominant–recessive pattern. In **incomplete dominance**, the offspring express a heterozygous phenotype that is intermediate between one parent's homozygous dominant trait and the other parent's homozygous recessive trait. An example of this can be seen in snapdragons when red-flowered plants and white-flowered plants are crossed to produce pink-flowered plants. In humans, incomplete dominance occurs with one of the genes for hair texture. When one parent passes a curly hair allele (the incompletely dominant allele) and the other parent passes a straight-hair allele, the effect on the offspring will be intermediate, resulting in hair that is wavy.

Codominance

Codominance is characterized by the equal, distinct, and simultaneous expression of both parents' different alleles. This pattern differs from the intermediate, blended features seen in incomplete dominance. A classic example of codominance in humans is ABO blood type. People are blood type A if they have an allele for an enzyme that facilitates the production of surface antigen A on their erythrocytes. This allele is designated I^A. In the same manner, people are blood type B if they express an enzyme for the production of surface antigen B. People who have alleles for both enzymes (I^A and I^B) produce both surface antigens A and B. As a result, they are blood type AB. Because the effect of both alleles (or enzymes) is observed, we say that the I^A and I^B alleles are codominant. There is also a third allele that determines blood type. This allele (*i*) produces a nonfunctional enzyme. People who have two *i* alleles do not produce either A or B surface antigens: they have type O blood. If a person has I^A and *i* alleles, the person will have blood type A. Notice that it does not make any difference whether a person has two I^A alleles or one I^A and one *i* allele. In both cases, the person is blood type A. Because I^A masks *i*, we say that I^A is dominant to *i*. The following table summarizes the expression of blood type.

Table 1. Expression of Blood Types

Blood type	Genotype	Pattern of inheritance
A	I^AI^A or I^Ai	$I^A$is dominant to i
B	I^BI^B orI^Bi	I^B is dominant to i
AB	I^AI^B	I^A is co-dominant to I^B
O	ii	Two recessive alleles

Lethal Alleles

Certain combinations of alleles can be lethal, meaning they prevent the individual from developing in utero, or cause a shortened life span. In **recessive lethal** inheritance patterns, a child who is born to two heterozygous (carrier) parents and who inherited the faulty allele from both would not survive. An example of this is Tay–Sachs, a fatal disorder of the nervous system. In this disorder, parents with one copy of the allele for the disorder are carriers. If they both transmit their abnormal allele, their offspring will develop the disease and will die in childhood, usually before age 5.

Dominant lethal inheritance patterns are much more rare because neither heterozygotes nor homozygotes survive. Of course, dominant lethal alleles that arise naturally through mutation and cause miscarriages or stillbirths are never transmitted to subsequent generations. However, some dominant lethal alleles, such as the allele for Huntington's disease, cause a shortened life span but may not be identified until after the person reaches reproductive age and has children. Huntington's disease causes irreversible nerve cell degeneration and death in 100 percent of affected individuals, but it may not be expressed until the individual reaches middle age. In this way, dominant lethal alleles can be maintained in the human population. Individuals with a family history of Huntington's disease are typically offered genetic counseling, which can help them decide whether or not they wish to be tested for the faulty gene.

Mutations

A **mutation** is a change in the sequence of DNA nucleotides that may or may not affect a person's phenotype. Mutations can arise spontaneously from errors during DNA replication, or they can result from environmental insults such as radiation, certain viruses, or exposure to tobacco smoke or other toxic chemicals. Because genes encode for the assembly of proteins, a mutation in the nucleotide sequence of a gene can change amino acid sequence and, consequently, a protein's structure and function. Spontaneous mutations occurring during meiosis are thought to account for many spontaneous abortions (miscarriages).

Chromosomal Disorders

Sometimes a genetic disease is not caused by a mutation in a gene, but by the presence of an incorrect number of chromosomes. For example, Down syndrome is caused by having three copies of chromosome 21. This is known as trisomy 21. The most common cause of trisomy 21 is chromosomal nondisjunction during meiosis. The frequency of nondisjunction events appears to increase with age, so the frequency of bearing a child with Down syndrome increases in women over 36. The age of the father matters less because nondisjunction is much less likely to occur in a sperm than in an egg.

Whereas Down syndrome is caused by having three copies of a chromosome, Turner syndrome is caused by having just one copy of the X chromosome. This is known as monosomy. The affected child is always female. Women with Turner syndrome are sterile because their sexual organs do not mature.

CAREER CONNECTIONS: GENETIC COUNSELOR

Given the intricate orchestration of gene expression, cell migration, and cell differentiation during prenatal development, it is amazing that the vast majority of newborns are healthy and free of major birth defects. When a woman over 35 is pregnant or intends to become pregnant, or her partner is over 55, or if there is a family history of a genetic disorder, she and her partner may want to speak to a genetic counselor to discuss the likelihood that their child may be affected by a genetic or chromosomal disorder. A genetic counselor can interpret a couple's family history and estimate the risks to their future offspring.

For many genetic diseases, a DNA test can determine whether a person is a carrier. For instance, carrier status for Fragile X, an X-linked disorder associated with mental retardation, or for cystic fibrosis can be determined with a simple blood draw to obtain DNA for testing. A genetic counselor can educate a couple about the implications of such a test and help them decide whether to undergo testing. For chromosomal disorders, the available testing options include a blood test, amniocentesis (in which amniotic fluid is tested), and chorionic villus sampling (in which tissue from the placenta is tested). Each of these has advantages and drawbacks. A genetic counselor can also help a couple cope with the news that either one or both partners is a carrier of a genetic illness, or that their unborn child has been diagnosed with a chromosomal disorder or other birth defect.

To become a genetic counselor, one needs to complete a 4-year undergraduate program and then obtain a Master of Science in Genetic Counseling from an accredited university. Board certification is attained after passing examinations by the American Board of Genetic Counseling. Genetic counselors are essential professionals in many branches of medicine, but there is a particular demand for preconception and prenatal genetic counselors.

Concept check

1. **What** is fertilization and **what** is the resulting cell called?
2. **What** are the different stages a blastocyst formation?
3. **How** does implantation occur?
4. **What** are the major tissues present during gastrulation?
5. **How** does the placenta develop?
6. **What** is organogenesis?
7. **How** does sex differentiation occur?
8. **What** are some effects of hormones during pregnancy?
9. **What** are some body system changed that occur during pregnancy?
10. **Explain** Braxton Hicks contractions and **why** they occur.
11. **Explain** the process of lactation.
12. **How** are genotype and phenotype connected?
13. **Define** homozygous and heterozygous.
14. **Explain** dominant and recessive alleles.
15. **Define** the following: sex-linked trait, multiple alleles, codominance, and incomplete dominance.
16. **What** is a mutation?

CHAPTER 7 — THE SKELETAL SYSTEM

Introduction to the Skeletal System

LEARNING OBJECTIVES

- Describe the functions of the skeletal system.
- Distinguish between long bones, short bones, flat bones, and irregular bones and provide an example of each.
- Identify the anatomical features of a long bone
- Describe the microscopic structure of compact bone, and compare it with that of spongy bone.
- Identify all the bones of the axial and appendicular skeleton.
- Describe various skeletal joints and the movements possible
- Describe the steps involved in bone development and bone repair
- Describe the effect exercise has on bone tissue
- Describe the disorders of the skeletal system
- Describe the effects of hormones on bone tissue, the process of calcium homeostasis

Bone, or **osseous tissue**, is a hard, dense connective tissue that forms most of the adult skeleton, the support structure of the body. In the areas of the skeleton where bones move (for example, the ribcage and joints), **cartilage**, a semi-rigid form of connective tissue, provides flexibility and smooth surfaces for movement.

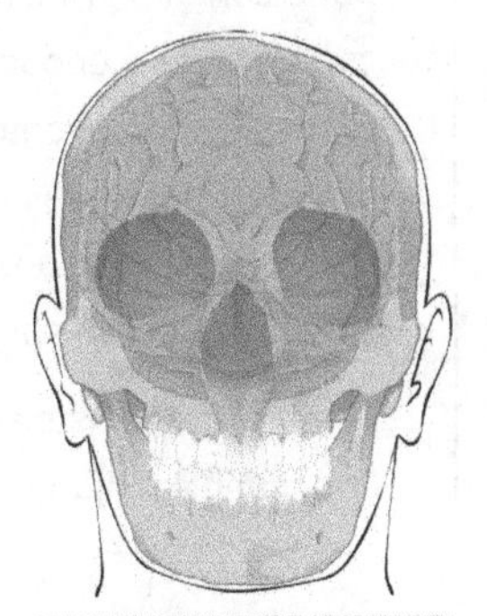

Figure 1. Bones Protect Brain. The cranium completely surrounds and protects the brain from non-traumatic injury.

The **skeletal system** is the body system composed of **bones and cartilage** and performs the following critical functions for the human body:

- protection of vital structures, such as the spinal cord, brain, heart, and lungs.
- support of body structures.
- body locomotion through coordination with the muscular system.
- hematopoiesis, or generation of blood cells, within the red marrow spaces of bones.
- storage and release of the inorganic minerals calcium and phosphorous, which are needed for functions such as muscle contraction and neural signal conduction.

The most apparent functions of the skeletal system are the gross functions—those visible by observation. Simply by looking at a person, you can see how the bones support, facilitate movement, and protect the human body.

Just as the steel beams of a building provide a scaffold to support its weight, the bones and cartilage of your skeletal system compose the scaffold that supports the rest of your body. Without the skeletal system, you would be a limp mass of organs, muscle, and skin.

Bones also facilitate movement by serving as points of attachment for your muscles. While some bones only serve as a support for the muscles, others also transmit the forces produced when your muscles contract. From a mechanical point of view, bones act as levers and joints serve as fulcrums.

Unless a muscle spans a joint and contracts, a bone is not going to move. For information on the interaction of the skeletal and muscular systems, that is, the musculoskeletal system, seek additional content.

Bones also protect internal organs from injury by covering or surrounding them. For example, your ribs protect your lungs and heart, the bones of your vertebral column (spine) protect your spinal cord, and the bones of your cranium (skull) protect your brain (Figure 1).

CAREER CONNECTION: ORTHOPEDIST

An **orthopedist** is a doctor who specializes in diagnosing and treating disorders and injuries related to the musculoskeletal system. Some orthopedic problems can be treated with medications, exercises, braces, and other devices, but others may be best treated with surgery (Figure 2).

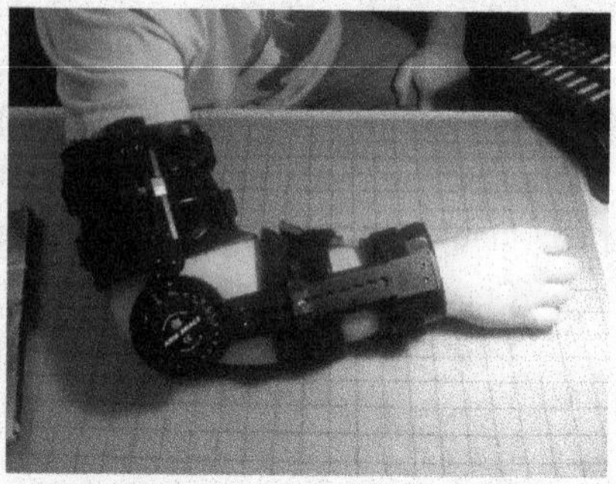

Figure 2. Arm Brace. An orthopedist will sometimes prescribe the use of a brace that reinforces the underlying bone structure it is being used to support. (credit: Juhan Sonin)

While the origin of the word "orthopedics" (ortho- = "straight"; paed- = "child"), literally means "straightening of the child," orthopedists can have patients who range from pediatric to geriatric. In recent years, orthopedists have even performed prenatal surgery to correct spina bifida, a congenital defect in which the neural canal in the spine of the fetus fails to close completely during embryologic development.

Orthopedists commonly treat bone and joint injuries but they also treat other bone conditions including curvature of the spine. Lateral curvatures (scoliosis) can be severe enough to slip under the shoulder blade (scapula) forcing it up as a hump. Spinal curvatures can also be excessive dorsoventrally (kyphosis) causing a hunch back and thoracic compression. These curvatures often appear in preteens as the result of poor posture, abnormal growth, or indeterminate causes. Mostly, they are readily treated by orthopedists. As people age, accumulated spinal column injuries and diseases like osteoporosis can also lead to curvatures of the spine, hence the stooping you sometimes see in the elderly.

Some orthopedists sub-specialize in sports medicine, which addresses both simple injuries, such as a sprained ankle, and complex injuries, such as a torn rotator cuff in the shoulder. Treatment can range from exercise to surgery.

Mineral Storage, Energy Storage, and Hematopoiesis

On a metabolic level, bone tissue performs several critical functions. For one, the bone matrix acts as a reservoir for a number of minerals important to the functioning of the body, especially calcium, and potassium. These minerals, incorporated into bone tissue, can be released back into the bloodstream to maintain levels needed to support physiological processes. Calcium ions, for example, are essential for muscle contractions and controlling the flow of other ions involved in the transmission of nerve impulses.

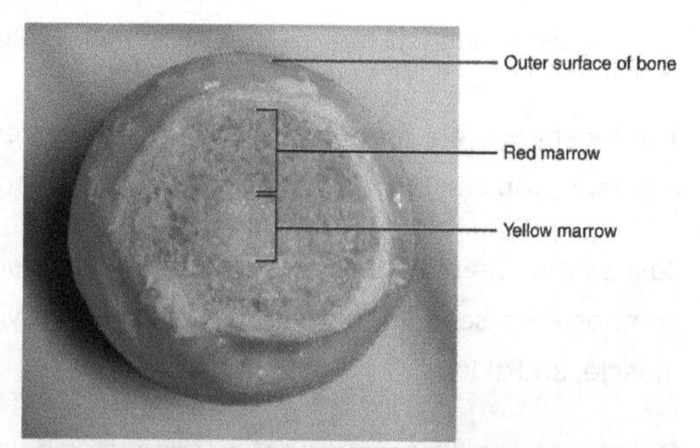

Figure 3. Head of Femur Showing Red and Yellow Marrow. The head of the femur contains both yellow and red marrow. Yellow marrow stores fat. Red marrow is responsible for hematopoiesis. (credit: modification of work by "stevenfruitsmaak"/Wikimedia Commons)

Bone also serves as a site for fat storage and blood cell production. The softer connective tissue that fills the interior of most bone is referred to as bone marrow (Figure 3). There are two types of bone marrow: yellow marrow and red marrow. **Yellow marrow** contains adipose tissue; the triglycerides stored in the adipocytes of the tissue can serve as a source of energy. **Red marrow** is where **hematopoiesis**—the production of blood cells—takes place. Red blood cells, white blood cells, and platelets are all produced in the red marrow.

Classify bones according to their shapes

The 206 bones that compose the adult skeleton are divided into five categories based on their shapes (Figure 4). Their shapes and their functions are related such that each categorical shape of bone has a distinct function.

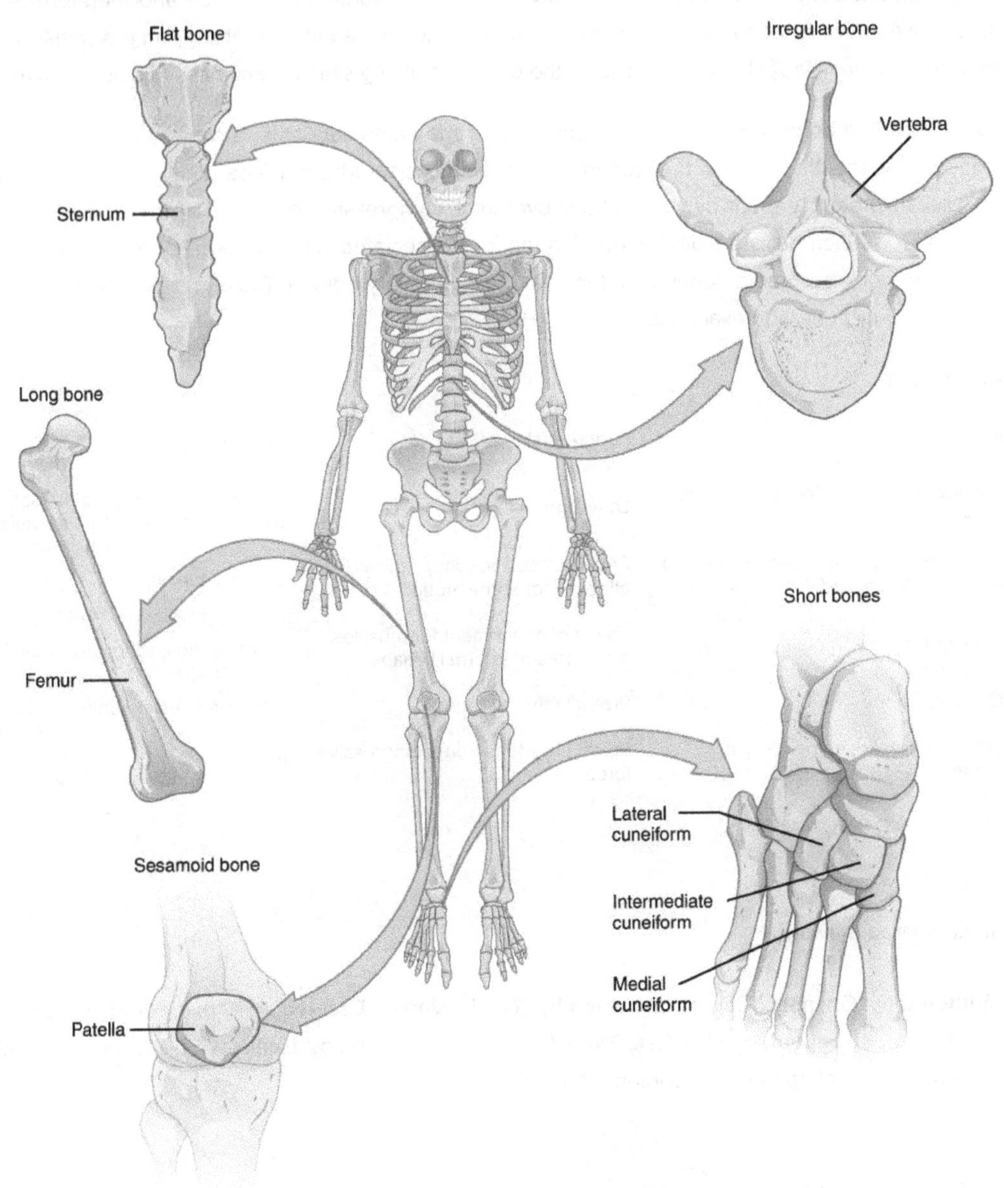

Figure 4. Classifications of Bones. Bones are classified according to their shape.

Long Bones: A **long bone** is one that is cylindrical in shape, being longer than it is wide. Keep in mind, however, that the term describes the shape of a bone, not its size. Long bones are found in the arms (humerus, ulna, radius) and legs (femur, tibia, fibula), as well as in the fingers (metacarpals, phalanges) and toes (metatarsals, phalanges). Long bones function as levers; they move when muscles contract.

Short Bones: A **short bone** is one that is cube-like in shape, being approximately equal in length, width, and thickness. The only short bones in the human skeleton are in the carpals of the wrists and the tarsals of the ankles. Short bones provide stability and support as well as some limited motion.

Flat Bones: The term ***flat bone*** is somewhat of a misnomer because, although a flat bone is typically thin, it is also often curved. Examples include the cranial (skull) bones, the scapulae (shoulder blades), the sternum (breastbone), and the ribs. Flat bones serve as points of attachment for muscles and often protect internal organs.

Irregular Bones: An **irregular bone** is one that does not have any easily characterized shape and therefore does not fit any other classification. These bones tend to have more complex shapes, like the vertebrae that support the spinal cord and protect it from compressive forces. Many facial bones, particularly the ones containing sinuses, are classified as irregular bones.

Sesamoid Bones: A **sesamoid bone** is a small, round bone that, as the name suggests, is shaped like a sesame seed. These bones form in tendons (the sheaths of tissue that connect bones to muscles) where a great deal of pressure is generated in a joint. The sesamoid bones protect tendons by helping them overcome compressive forces. Sesamoid bones vary in number and placement from person to person but are typically found in tendons associated with the feet, hands, and knees. The patellae (singular = patella) are the only sesamoid bones found in common with every person. Table 1 reviews bone classifications with their associated features, functions, and examples.

Table 1. Bone Classifications

Bone classification	Features	Function(s)	Examples
Long	Cylinder-like shape, longer than it is wide	Leverage	Femur, tibia, fibula, metatarsals, humerus, ulna, radius, metacarpals, phalanges
Short	Cube-like shape, approximately equal in length, width, and thickness	Provide stability, support, while allowing for some motion	Carpals, tarsals
Flat	Thin and curved	Points of attachment for muscles; protectors of internal organs	Sternum, ribs, scapulae, cranial bones
Irregular	Complex shape	Protect internal organs	Vertebrae, facial bones
Sesamoid	Small and round; embedded in tendons	Protect tendons from compressive forces	Patellae

Human Skeletal System

The skeletal system includes all of the **bones, cartilages, and ligaments** of the body that support and give shape to the body and body structures. The **skeleton** consists of the bones of the body. For adults, there are **206 bones** in the skeleton. Younger individuals have higher numbers of bones because some bones fuse together during childhood and adolescence to form an adult bone. The primary functions of the skeleton are to provide a rigid, internal structure that can support the weight of the body against the force of gravity, and to provide a structure upon which muscles can act to produce movements of the body. The lower portion of the skeleton is specialized for stability during walking or running. In contrast, the upper skeleton has greater mobility and ranges of motion, features that allow you to lift and carry objects or turn your head and trunk.

In addition to providing for support and movements of the body, the skeleton has protective and storage functions. It protects the internal organs, including the brain, spinal cord, heart, lungs, and pelvic organs. The bones of the skeleton serve as the primary storage site for important minerals such as calcium and phosphate. The bone marrow found within bones stores fat and houses the blood-cell producing tissue of the body.

The skeleton is subdivided into two major divisions—the axial and appendicular.

The Axial Skeleton

The skeleton is subdivided into two major divisions—the axial and appendicular. The **axial skeleton** forms the vertical, central axis of the body and includes all bones of the head, neck, chest, and back. It serves to protect the brain, spinal cord, heart, and lungs. It also serves as the attachment site for muscles that move the head, neck, and back, and for muscles that act across the shoulder and hip joints to move their corresponding limbs.

The axial skeleton of the adult consists of 80 bones, including the **skull**, the **vertebral column**, and the **thoracic cage**. The skull is formed by 22 bones. Also associated with the head are an additional seven bones, including the **hyoid bone** and the **ear ossicles** (three small bones found in each middle ear). The vertebral column consists of 24 bones, each called a **vertebra**, plus the **sacrum** and **coccyx**. The thoracic cage includes the 12 pairs of **ribs**, and the **sternum**, the flattened bone of the anterior chest.

The Appendicular Skeleton

The **appendicular skeleton** includes all bones of the upper and lower limbs, plus the bones that attach each limb to the axial skeleton. There are 126 bones in the appendicular skeleton of an adult. The bones of the appendicular skeleton are covered in a separate section.

The Axial Skeleton - General Features And Functions Of The Skull

The **cranium** (skull) is the skeletal structure of the head that supports the face and protects the brain. It is subdivided into the **facial bones** and the **brain case**. The facial bones underlie the facial structures, form the nasal cavity, enclose the eyeballs, and support the teeth of the upper and lower jaws. The rounded brain case surrounds and protects the brain and houses the middle and inner ear structures. In the adult, the skull consists of 22 individual bones, 21 of which are immobile and united into a single unit. The 22nd bone is the **mandible** (lower jaw), which is the only moveable bone of the skull.

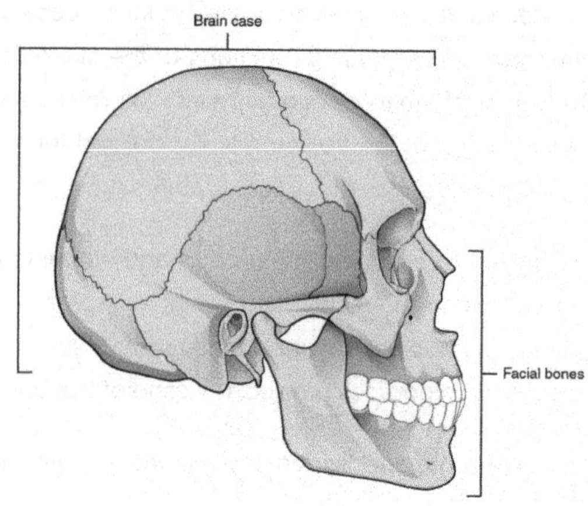

Parts of the Skull

Cranial Bones of the Skull:

Cranial bones form the cranial cavity that surrounds and protects the brain and brainstem. There are eight cranial bones. **frontal, sphenoid, ethmoid, occipital, 2 temporal, and 2 parietal bones.**These bones are important as they provide an articulation point for the 1st cervical vertebra (atlas), as well as the facial bones and the mandible (jaw bone). The bones that form the top and sides of the brain case are usually referred to as the **"flat" bones** of the skull.

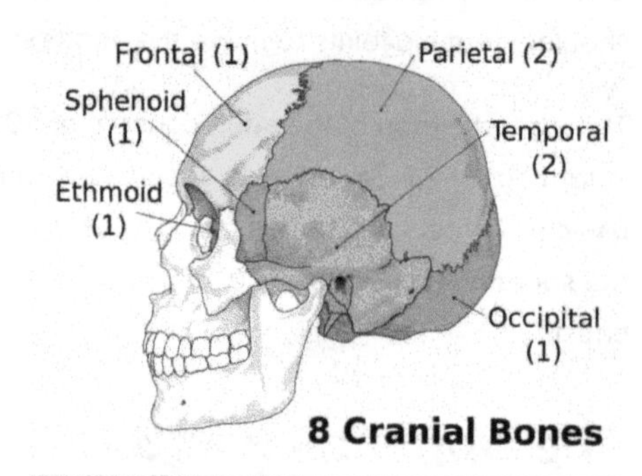

Cranial Bones of Human Skull

- **Frontal bone (1) –** The **frontal bone** is the single bone that forms the forehead.
- **Parietal bone (2) –** The **parietal bone** forms most of the upper lateral side of the skull . These are paired bones, with the right and left parietal bones joining together at the top of the skull. They are examples of flat bones.
- Temporal bone **(2)** – The temporal bone forms the lower lateral side of the skull. Common wisdom has it that the temporal bone (temporal = "time") is so named because this area of the head (the temple) is where hair typically first turns gray, indicating the passage of time. Two major structures if the temporal bone are the external auditory meatus and the mastoid process. The external auditory meatus (the ear canal) which is a prominent canal in the tympanic part of the temporal bone. It transmits sounds to the tympanic membrane (eardrum) in the middle ear. The mastoid process is a large prominence behind your earlobe which serves as a muscle attachment site
- **Occipital bone** (1) – The occipital bone is the single bone that forms the posterior skull and posterior base of the cranial cavity. On the base of the skull, the occipital bone contains the large opening of the foramen magnum, which allows for passage of the spinal cord as it exits the skull. On either side of the foramen magnum is an oval-shaped occipital condyle. These condyles form joints with the first cervical vertebra and thus support the skull on top of the vertebral column.
- Sphenoid bone (1) – The sphenoid bone is a single, complex bone of the central skull. It serves as a "keystone" bone, because it joins with almost every other bone of the skull. The sphenoid forms much of the base of the central skull and also extends laterally to contribute to the sides of the skull. Inside the cranial cavity, it has a pair of lesser wings and a pair of greater wings. The right and left lesser wings of the sphenoid bone, resemble the wings of a flying bird. The greater wings of the sphenoid bone extend laterally to either side. The greater wing is best seen on the outside of the lateral skull, where it forms a rectangular area immediately anterior to the temporal bone.
- Ethmoid bone: The ethmoid bone is a single, midline bone that forms the roof and lateral walls of the upper nasal cavity,

the upper portion of the nasal septum, and contributes to the medial wall of the orbit . On the interior of the skull, the ethmoid also forms a portion of the floor of the anterior cranial cavity.

Sutures of the Skull

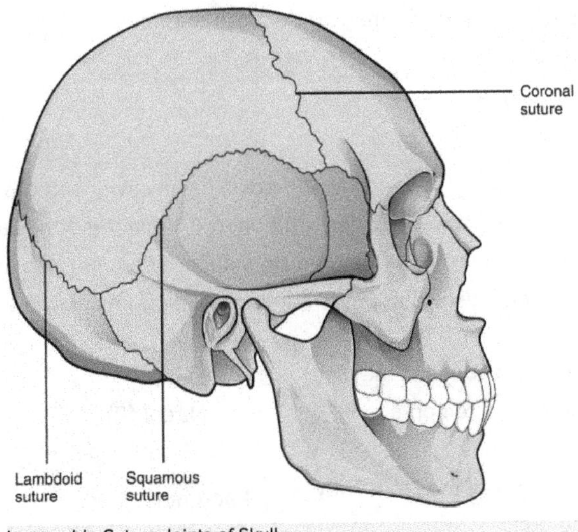

Immovable Suture Joints of Skull

A ***suture*** is an immobile joint between adjacent bones of the skull. The narrow gap between the bones is filled with dense, fibrous connective tissue that unites the bones. The long sutures located between the bones of the brain case are not straight, but instead follow irregular, tightly twisting paths. These twisting lines serve to tightly interlock the adjacent bones, thus adding strength to the skull for brain protection.

Fontanelles

A fontanelle is an anatomical feature on an infant's skull that allows its plates to be flexible to pass through the birth canal. Fontanelles are soft spots on a baby's head that, during birth, enable the bony plates of the skull to flex and allow the child's head to pass through the birth canal. The ossification of the bones of the skull causes the fontanelles to close over a period of 18 to 24 months; they eventually form the sutures of the neurocranium. The cranium of a newborn consists of five main bones: two frontal bones, two parietal bones, and one occipital bone. These are joined by fibrous sutures that allow movement that facilitates childbirth and brain growth. At birth, the skull features a small posterior fontanelle (an open area covered by a tough membrane) where the two parietal bones adjoin the occipital bone (at the lambda). This fontanelle usually closes during the first two to three months of an infant's life.

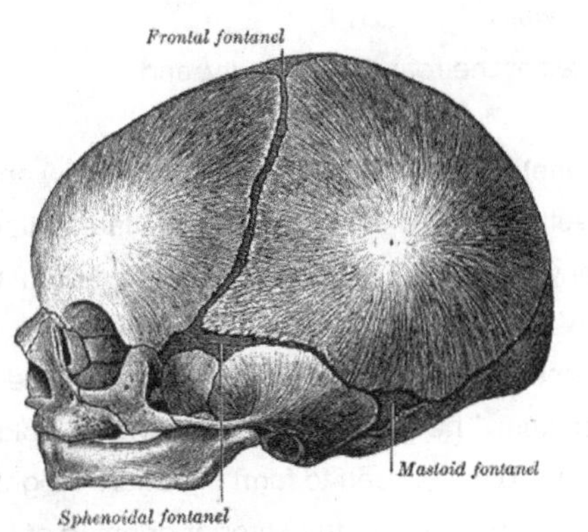

Lateral view of infant skull:

Frontal fontanel
Occipital fontanel

Superior view of infant skull.

The much larger, diamond-shaped anterior fontanelle—where the two frontal and two parietal bones join—generally remains open until a child is about two years old. The anterior fontanelle is useful clinically, as examination of an infant includes palpating the anterior fontanelle. Two smaller fontanelles are located on each side of the head. The more anterior one is the sphenoidal (between the sphenoid, parietal, temporal, and frontal bones), while the more posterior one is the mastoid (between the temporal, occipital, and parietal bones).

The fontanelle may pulsate. Although the precise cause of this is not known, it is perfectly normal and seems to echo the heartbeat, perhaps via the arterial pulse within the brain vasculature, or in the meninges. This pulsating action is how the soft spot got its name: fontanelle means little fountain. Parents may worry that their infant may be more prone to injury at the fontanelles. In fact, although they may colloquially be called soft spots, the membrane covering the fontanelles is extremely tough and difficult to penetrate. The fontanelles allow the infant brain to be imaged using ultrasonography. Once they are closed, most of the brain is inaccessible to ultrasound imaging because the bony skull presents an acoustic barrier.

Facial Bones of the Skull

The facial bones of the skull form the upper and lower jaws, the nose, nasal cavity and nasal septum, and the orbit. The facial bones include 14 bones, with six paired bones and two unpaired bones. The paired bones are the maxilla, palatine, zygomatic, nasal, lacrimal, and inferior nasal conchae bones. The unpaired bones are the vomer and mandible bones. Although classified with the brain-case bones, the ethmoid bone also contributes to the nasal septum and the walls of the nasal cavity and orbit.

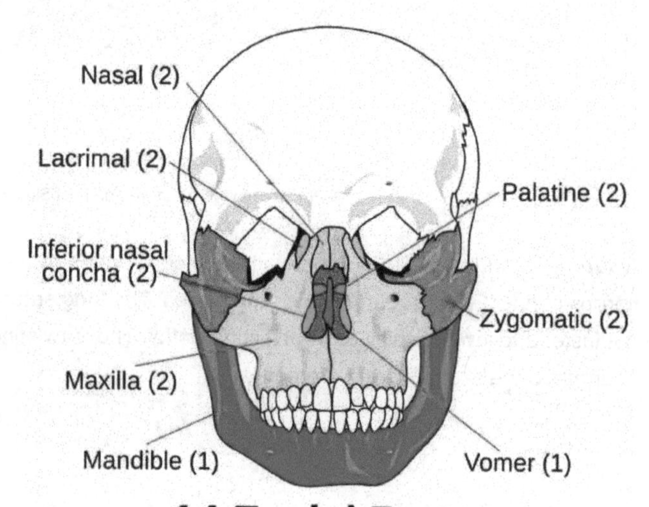

Facial Bones of Human Skull

- **Zygomatic Bone (2) –** Forms the cheek bones of the face, and articulates with the frontal, sphenoid, temporal and maxilla bones.
- **Lacrimal Bone (2)** – The smallest bones of the face. They form part of the medial wall of the orbit.
- **Nasal Bone (2)** – The ***nasal bone*** is one of two small bones that articulate (join) with each other to form the bony base (bridge) of the nose. They also support the cartilages that form the lateral walls of the nose. These are the bones that are damaged when the nose is broken.
- **Inferior nasal conchae (2)** – Located within the nasal cavity, these bones increase the surface area of the nasal cavity, thus increasing the amount of inspired air that can come into contact with the cavity walls.
- **Palatine Bone (2)** – Situated at the rear of oral cavity, and forms part of the hard palate.
- **Maxilla (2) (upper jaw bone) –** Comprises part of the upper jaw and hard palate. The maxillary bone forms the upper jaw and supports the upper teeth. Each maxilla also forms the lateral floor of each orbit and the majority of the hard palate. The ***hard palate*** is the bony plate that forms the roof of the mouth and floor of the nasal cavity, separating the oral and nasal cavities.
- **Vomer –** The unpaired vomer bone, often referred to simply as the vomer, is triangular-shaped and forms the posterior-inferior part of the nasal septum. The vomer is best seen when looking from behind into the posterior openings of the nasal cavity . In this view, the vomer is seen to form the entire height of the nasal septum. A much smaller portion of the vomer can also be seen when looking into the anterior opening of the nasal cavity.
- **Mandible (lower jaw bone)** – The ***mandible*** forms the lower jaw and is the only moveable bone of the skull. The posterior part of mandible is topped by the oval-shaped ***condyle***. The condyle of the mandible articulates (joins) with the temporal bone. Together these articulations form the temporomandibular joint, which allows for opening and closing of the mouth. The broad U-shaped curve located between the coronoid and condylar processes is the ***mandibular notch***.

HOMEOSTATIC IMBALANCES: CLEFT LIP AND CLEFT PALATE

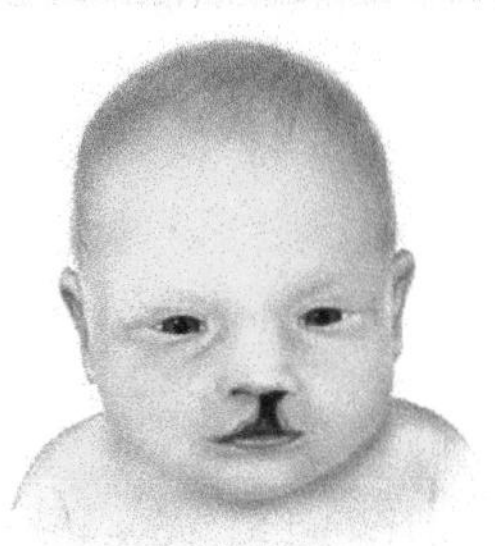

Baby with cleft lip

During embryonic development, the right and left maxilla bones come together at the midline to form the upper jaw. At the same time, the muscle and skin overlying these bones join together to form the upper lip. Inside the mouth, the palatine processes of the maxilla bones, along with the horizontal plates of the right and left palatine bones, join together to form the hard palate. If an error occurs in these developmental processes, a birth defect of cleft lip or cleft palate may result.

Cleft lip is a common development defect that affects approximately 1:1000 births, most of which are male. This defect involves a partial or complete failure of the right and left portions of the upper lip to fuse together, leaving a cleft (gap).

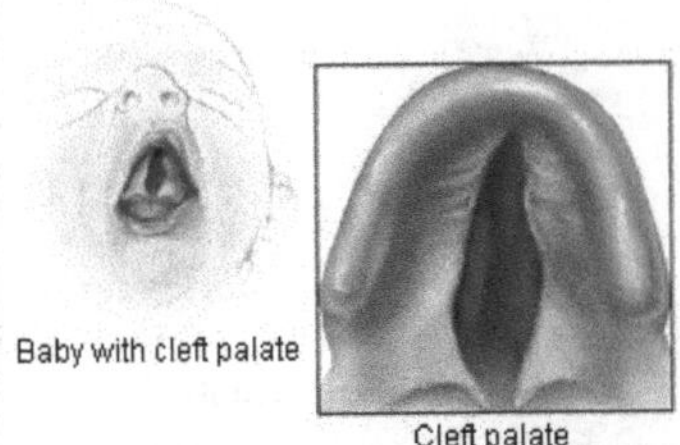

Baby with cleft palate

Cleft palate

A more severe developmental defect is cleft palate, which affects the hard palate. The hard palate is the bony structure that separates the nasal cavity from the oral cavity. It is formed during embryonic development by the midline fusion of the horizontal plates from the right and left palatine bones and the palatine processes of the maxilla bones. Cleft palate affects approximately 1:2500 births and is more common in females. It results from a failure of the two halves of the hard palate to completely come together and fuse at the midline, thus leaving a gap between them. This gap allows for communication between the nasal and oral cavities. In severe cases, the bony gap continues into the anterior upper jaw where the alveolar processes of the maxilla bones also do not properly join together above the front teeth. If this occurs, a cleft lip will also be seen. Because of the communication between the oral and nasal cavities, a cleft palate makes it very difficult for an infant to generate the suckling needed for nursing, thus leaving the infant at risk for malnutrition. Surgical repair is required to correct cleft palate defects.

Paranasal Sinuses

The ***paranasal sinuses*** are hollow, air-filled spaces located within certain bones of the skull. All of the sinuses communicate with the nasal cavity (paranasal = “next to nasal cavity”) and are lined with nasal mucosa. Their function is to reduce bone mass and thus lighten the skull, and they also add resonance to the voice. This second feature is most obvious when you have a cold or sinus congestion. These produce swelling of the mucosa and excess mucus production, which can obstruct the narrow passageways between the sinuses and the nasal cavity, causing your voice to sound different to yourself and others. This blockage can also allow the sinuses to fill with fluid, with the resulting pressure producing pain and discomfort.

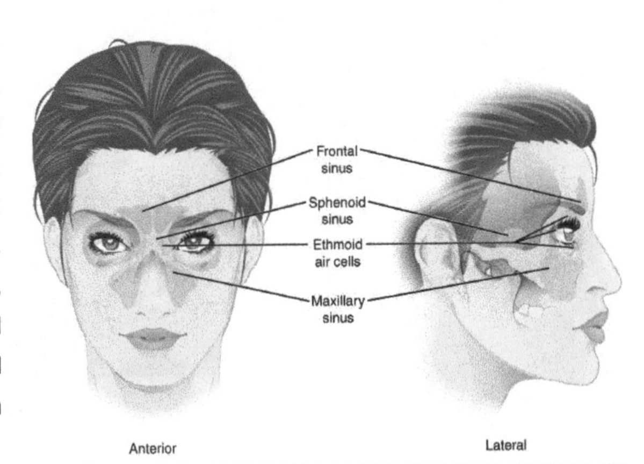

Paranasal Sinuses

The paranasal sinuses are named for the skull bone that each occupies. The ***frontal sinus*** is located just above the eyebrows, within the frontal bone. This irregular space may be divided at the midline into bilateral spaces, or these may be fused into a single sinus space. The frontal sinus is the most anterior of the paranasal sinuses. The largest sinus is the ***maxillary sinus***. These are paired and located within the right and left maxillary bones, where they occupy the area just below the orbits. The maxillary sinuses are most commonly involved during sinus infections. Because their connection to the nasal cavity is located high on their medial wall, they are difficult to drain. The ***sphenoid sinus*** is a single, midline sinus. It is located within the body of the sphenoid bone, just anterior and inferior to the sella turcica, thus making it the most posterior of the paranasal sinuses. The lateral aspects of the ethmoid bone contain multiple small spaces separated by very thin bony walls. Each of these spaces is called an ***ethmoid air cell***. These are located on both sides of the ethmoid bone, between the upper nasal cavity and medial orbit, just behind the superior nasal conchae.

Hyoid Bone

The hyoid bone is an independent bone that does not contact any other bone and thus is not part of the skull. It is a small U-shaped bone located in the upper neck near the level of the inferior mandible, with the tips of the "U" pointing posteriorly. The hyoid serves as the base for the tongue above, and is attached to the larynx below and the pharynx posteriorly. The hyoid is held in position by a series of small muscles that attach to it either from above or below. These muscles act to move the hyoid up/down or forward/back. Movements of the hyoid are coordinated with movements of the tongue, larynx, and pharynx during swallowing and speaking.

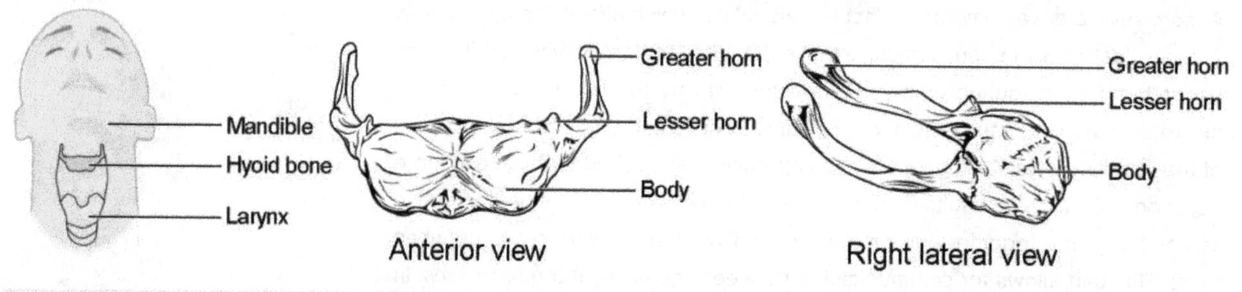

Figure. Hyoid Bone.

Axial Skeleton - The Vertebral Column and The Thoracic Cage

The vertebral column is also known as the spinal column or spine (Figure 1). It consists of a sequence of vertebrae (singular = vertebra), each of which is separated and united by an **intervertebral disc**. Together, the vertebrae and intervertebral discs form the vertebral column. It is a flexible column that supports the head, neck, and body and allows for their movements. It also protects the spinal cord, which passes down the back through openings in the vertebrae.

Regions of the Vertebral Column

The vertebral column originally develops as a series of 33 vertebrae, but this number is eventually reduced to 24 vertebrae, plus the sacrum and coccyx. The vertebral column is subdivided into five regions, with the vertebrae in each area named for that region and numbered in descending order. In the neck, there are seven cervical vertebrae, each designated with the letter "C" followed by its number. Superiorly, the C1 vertebra articulates (forms a joint) with the occipital condyles of the skull. Inferiorly, C1 articulates with the C2 vertebra, and so on. Below these are the 12 thoracic vertebrae, designated T1–T12. The lower back contains the L1–L5 lumbar vertebrae. The single sacrum, which is also part of the pelvis, is formed by the fusion of five sacral vertebrae. Similarly, the coccyx, or tailbone, results from the fusion of four small coccygeal vertebrae. However, the sacral and coccygeal fusions do not start until age 20 and are not completed until middle age.

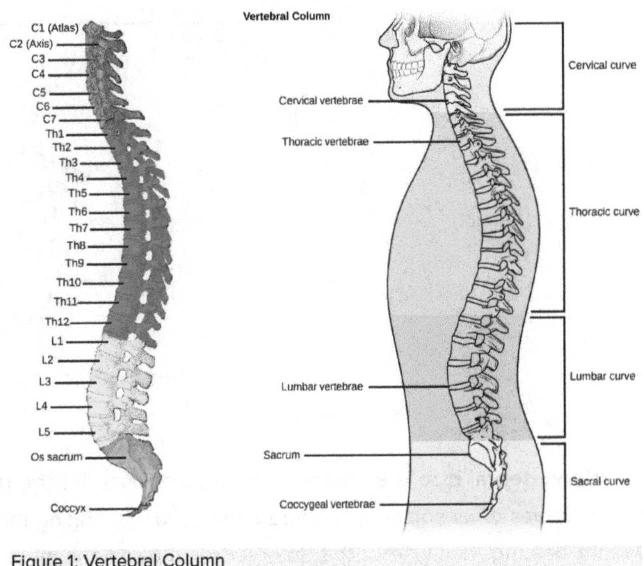

Figure 1: Vertebral Column

An interesting anatomical fact is that almost all mammals have seven cervical vertebrae, regardless of body size. This means that there are large variations in the size of cervical vertebrae, ranging from the very small cervical vertebrae of a shrew to the greatly elongated vertebrae in the neck of a giraffe. In a full-grown giraffe, each cervical vertebra is 11 inches tall.

Curvatures of the Vertebral Column

The adult vertebral column does not form a straight line, but instead has four curvatures along its length (see Figure 1). These curves increase the vertebral column's strength, flexibility, and ability to absorb shock. When the load on the spine is increased, by carrying a heavy backpack for example, the curvatures increase in depth (become more curved) to accommodate the extra weight. They then spring back when the weight is removed. The four adult curvatures are classified as either primary or secondary curvatures. Primary curves are retained from the original fetal curvature, while secondary curvatures develop after birth.

During fetal development, the body is flexed anteriorly into the fetal position, giving the entire vertebral column a single curvature that is concave anteriorly. In the adult, this fetal curvature is retained in two regions of the vertebral column as the **thoracic curve**, which involves the thoracic vertebrae, and the **sacrococcygeal curve**, formed by the sacrum and coccyx. Each of these is thus called a **primary curve** because they are retained from the original fetal curvature of the vertebral column.

A **secondary curve** develops gradually after birth as the child learns to sit upright, stand, and walk. Secondary curves are concave posteriorly, opposite in direction to the original fetal curvature. The **cervical curve** of the neck region develops as the infant begins to hold their head upright when sitting. Later, as the child begins to stand and then to walk, the **lumbar curve** of the lower back develops. In adults, the lumbar curve is generally deeper in females.

Intervertebral Disc

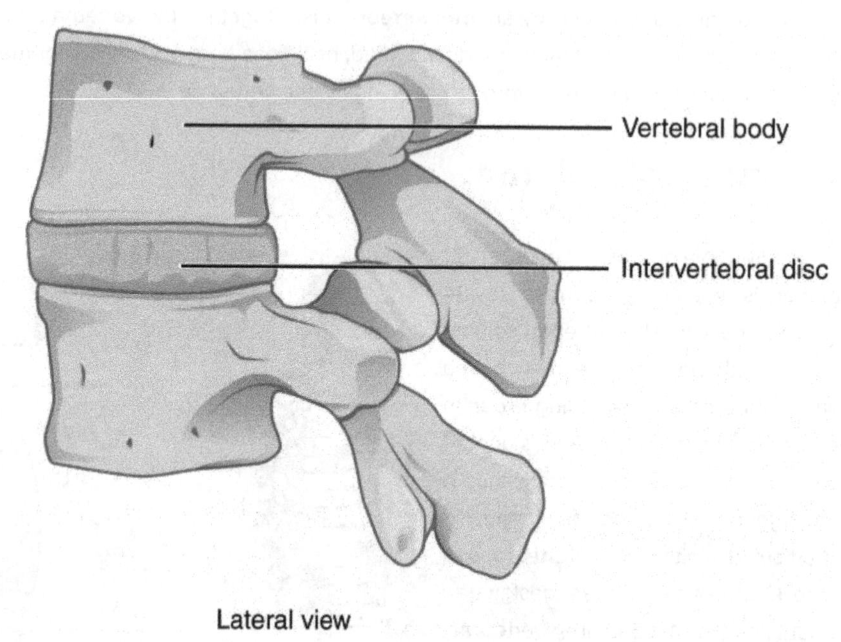

Figure 2: Intervertebral Disc

An **intervertebral disc** is a fibrocartilaginous pad that fills the gap between adjacent vertebral bodies. Each disc is anchored to the bodies of its adjacent vertebrae, thus strongly uniting these. The discs also provide padding between vertebrae during weight bearing. Because of this, intervertebral discs are thin in the cervical region and thickest in the lumbar region, which carries the most body weight. In total, the intervertebral discs account for approximately 25 percent of your body height between the top of the pelvis and the base of the skull. Intervertebral discs are also flexible and can change shape to allow for movements of the vertebral column.

Each intervertebral disc consists of two parts. The **anulus fibrosus** is the tough, fibrous outer layer of the disc. It forms a circle (anulus = "ring" or "circle") and is firmly anchored to the outer margins of the adjacent vertebral bodies. Inside is the **nucleus pulposus**, consisting of a softer, more gel-like material. It has a high water content that serves to resist compression and thus is important for weight bearing. With increasing age, the water content of the nucleus pulposus gradually declines. This causes the disc to become thinner, decreasing total body height somewhat, and reduces the flexibility and range of motion of the disc, making bending more difficult.

The gel-like nature of the nucleus pulposus also allows the intervertebral disc to change shape as one vertebra rocks side to side or forward and back in relation to its neighbors during movements of the vertebral column. Thus, bending forward causes compression of the anterior portion of the disc but expansion of the posterior disc. If the posterior anulus fibrosus is weakened due to injury or increasing age, the pressure exerted on the disc when bending forward and lifting a heavy object can cause the nucleus pulposus to protrude posteriorly through the anulus fibrosus, resulting in a herniated disc ("ruptured" or "slipped" disc) (Figure 3). The posterior bulging of the nucleus pulposus can cause compression of a spinal nerve at the point where it exits through the intervertebral foramen, with resulting pain and/or muscle weakness in those body regions supplied by that nerve. The most common sites for disc herniation are the L4/L5 or L5/S1 intervertebral discs, which can cause sciatica, a widespread pain that radiates from the lower back down the thigh and into the leg. Similar injuries of the C5/C6 or C6/C7 intervertebral discs, following forcible hyperflexion of the neck from a collision accident or football injury, can produce pain in the neck, shoulder, and upper limb.

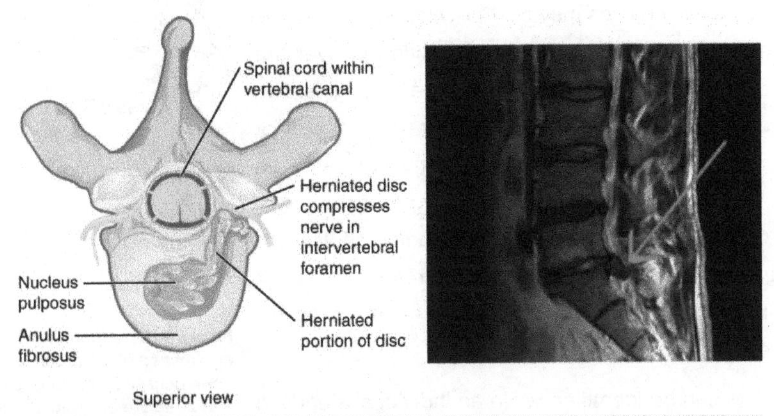

Figure 3. Herniated Intervertebral Disc. Weakening of the anulus fibrosus can result in herniation (protrusion) of the nucleus pulposus and compression of a spinal nerve, resulting in pain and/or muscle weakness in the body regions supplied by that nerve.

DISORDERS OF THE VERTEBRAL COLUMN

Developmental anomalies, pathological changes, or obesity can enhance the normal vertebral column curves, resulting in the development of abnormal or excessive curvatures (Figure 4). Kyphosis, also referred to as humpback or hunchback, is an excessive posterior curvature of the thoracic region. This can develop when osteoporosis causes weakening and erosion of the anterior portions of the upper thoracic vertebrae, resulting in their gradual collapse (Figure 5). Lordosis, or swayback, is an excessive anterior curvature of the lumbar region and is most commonly associated with obesity or late pregnancy. The accumulation of body weight in the abdominal region results an anterior shift in the line of gravity that carries the weight of the body. This causes in an anterior tilt of the pelvis and a pronounced enhancement of the lumbar curve.

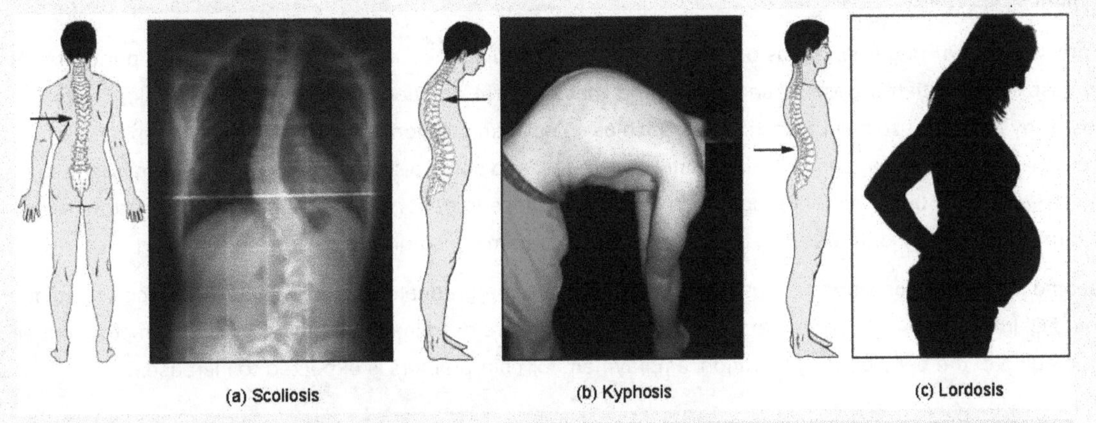

Figure 4. Abnormal Curvatures of the Vertebral Column. (a) Scoliosis is an abnormal lateral bending of the vertebral column. (b) An excessive curvature of the upper thoracic vertebral column is called kyphosis. (c) Lordosis is an excessive curvature in the lumbar region of the vertebral column.

Scoliosis is an abnormal, lateral curvature, accompanied by twisting of the vertebral column. Compensatory curves may also develop in other areas of the vertebral column to help maintain the head positioned over the feet. Scoliosis is the most common vertebral abnormality among girls. The cause is usually unknown, but it may result from weakness of the back muscles, defects such as differential growth rates in the right and left sides of the vertebral column, or differences in the length of the lower limbs. When present, scoliosis tends to get worse during adolescent growth spurts. Although most individuals do not require treatment, a back brace may be recommended for growing children. In extreme cases, surgery may be required.

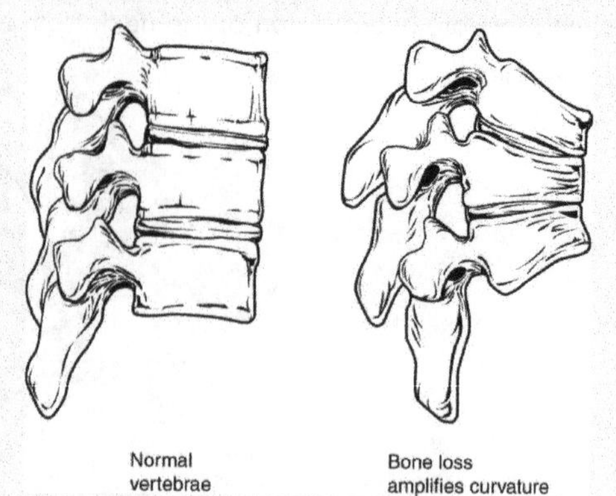

Figure 5. Osteoporosis. Osteoporosis is an age-related disorder that causes the gradual loss of bone density and strength. When the thoracic vertebrae are affected, there can be a gradual collapse of the vertebrae. This results in kyphosis, an excessive curvature of the thoracic region.

Excessive vertebral curves can be identified while an individual stands in the anatomical position. Observe the vertebral profile from the side and then from behind to check for kyphosis or lordosis. Then have the person bend forward. If scoliosis is present, an individual will have difficulty in bending directly forward, and the right and left sides of the back will not be level with each other in the bent position.

CAREER CONNECTIONS: CHIROPRACTOR

Chiropractors are health professionals who use nonsurgical techniques to help patients with musculoskeletal system problems that involve the bones, muscles, ligaments, tendons, or nervous system. They treat problems such as neck pain, back pain, joint pain, or headaches. Chiropractors focus on the patient's overall health and can also provide counseling related to lifestyle issues, such as diet, exercise, or sleep problems. If needed, they will refer the patient to other medical specialists.

Chiropractors use a drug-free, hands-on approach for patient diagnosis and treatment. They will perform a physical exam, assess the patient's posture and spine, and may perform additional diagnostic tests, including taking X-ray images. They primarily use manual techniques, such as spinal manipulation, to adjust the patient's spine or other joints. They can recommend therapeutic or rehabilitative exercises, and some also include acupuncture, massage therapy, or ultrasound as part of the treatment program. In addition to those in general practice, some chiropractors specialize in sport injuries, neurology, orthopaedics, pediatrics, nutrition, internal disorders, or diagnostic imaging.

To become a chiropractor, students must have 3–4 years of undergraduate education, attend an accredited, four-year Doctor of Chiropractic (D.C.) degree program, and pass a licensure examination to be licensed for practice in their state. With the aging of the baby-boom generation, employment for chiropractors is expected to increase.

The Thoracic Cage

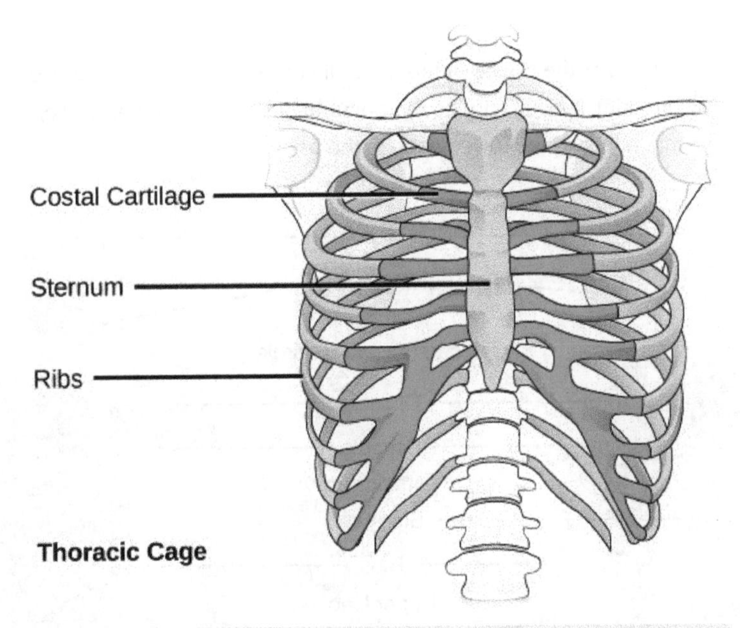

Figure 6: The thoracic cage, or rib cage, protects the heart and the lungs. (credit: modification of work by NCI, NIH)

The thoracic cage (rib cage) forms the thorax (chest) portion of the body. It consists of the **12 pairs of ribs with their costal cartilages and the sternum** . The ribs are anchored posteriorly to the 12 thoracic vertebrae (T1–T12). The thoracic cage protects the heart and lungs.

The sternum is the elongated bony structure that anchors the anterior thoracic cage. Each rib is a curved, flattened bone that contributes to the wall of the thorax. The ribs articulate posteriorly with the T1–T12 thoracic vertebrae, and most attach anteriorly via their costal cartilages to the sternum. There are 12 pairs of ribs. The ribs are numbered 1–12 in accordance with the thoracic vertebrae. The bony ribs do not extend anteriorly completely around to the sternum. Instead, each rib ends in a **costal cartilage**. These cartilages are made of hyaline cartilage and can extend for several inches. Most ribs are then attached, either directly or indirectly, to the sternum via their costal cartilage. The ribs are classified into three groups based on their relationship to the sternum. Ribs 1–7 are classified as **true ribs** (vertebrosternal ribs). The costal cartilage from each of these ribs attaches directly to the sternum. Ribs 8–12 are called **false ribs** (vertebrochondral ribs). The costal cartilages from these ribs do not attach directly to the sternum. For ribs 8–10, the costal cartilages are attached to the cartilage of the next higher rib. Thus, the cartilage of rib 10 attaches to the cartilage of rib 9, rib 9 then attaches to rib 8, and rib 8 is attached to rib 7. The last two false ribs (11–12) are also called **floating ribs** (vertebral ribs). These are short ribs that do not attach to the sternum at all. Instead, their small costal cartilages terminate within the musculature of the lateral abdominal wall.

Human Appendicular Skeleton

The **appendicular skeleton** is composed of the bones of the upper limbs (which function to grasp and manipulate objects) and the lower limbs (which permit locomotion). It also includes the pectoral girdle, or shoulder girdle, that attaches the upper limbs to the body, and the pelvic girdle that attaches the lower limbs to the body.

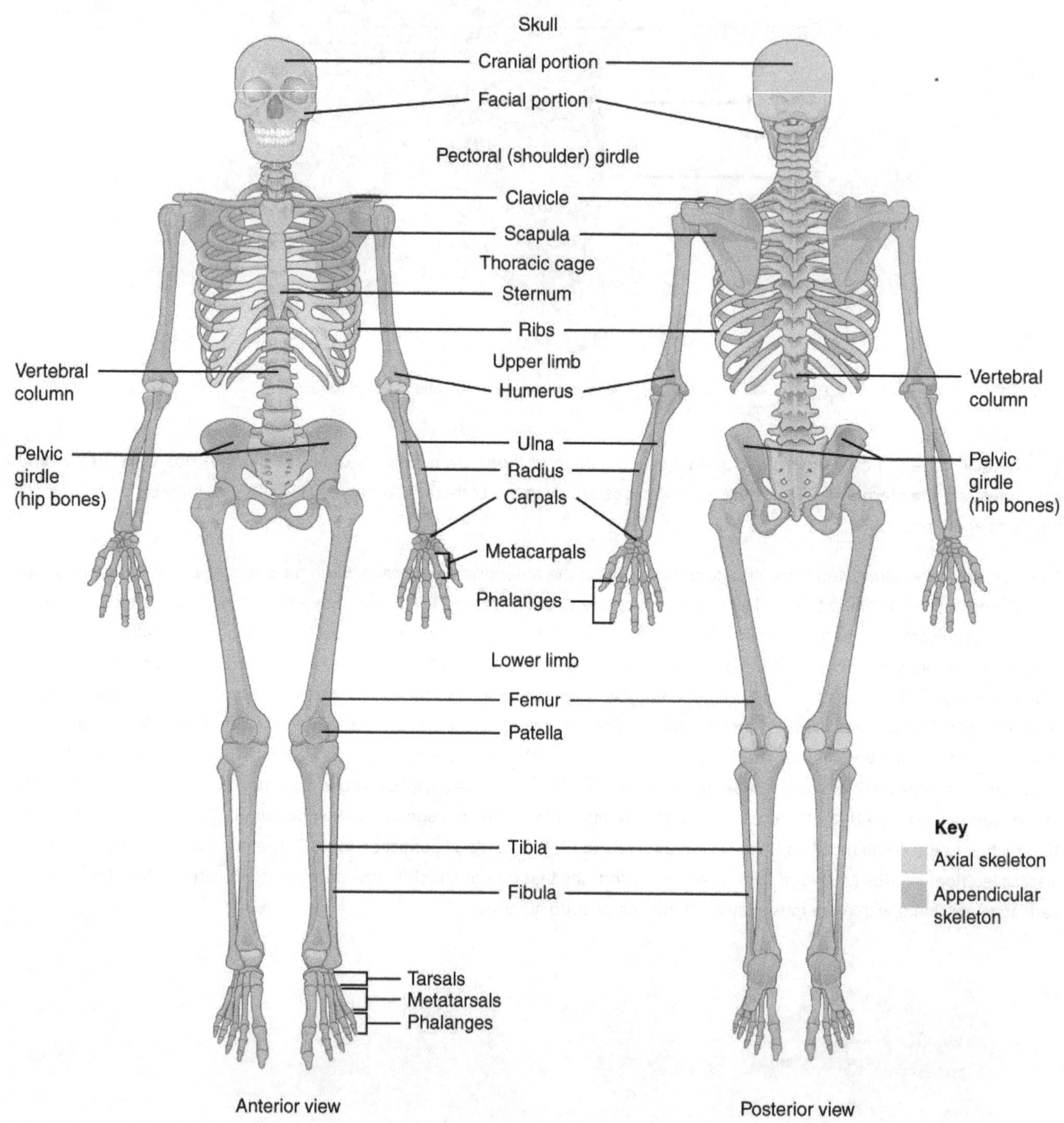

Figure 1: Appendicular Skeletons: The appendicular skeleton consists of the pectoral and pelvic girdles, the limb bones, and the bones of the hands and feet.

The Pectoral Girdle

The **pectoral girdle** bones provide the points of attachment of the upper limbs to the axial skeleton. The human pectoral girdle consists of the clavicle (or collarbone) in the anterior, and the scapula (or shoulder blades) in the posterior.

The **clavicles** are S-shaped bones that position the arms on the body. The clavicles lie horizontally across the front of the thorax (chest) just above the first rib. These bones are fairly fragile and are susceptible to fractures. For example, a fall with the arms outstretched causes the force to be transmitted to the clavicles, which can break if the force is excessive. The clavicle articulates with the sternum and the scapula.

The **scapulae** are flat, triangular bones that are located at the back of the pectoral girdle. They support the muscles crossing the shoulder joint. A ridge, called the spine, runs across the back of the scapula and can easily be felt through the skin (Figure 2). The spine of the scapula is a good example of a bony protrusion that facilitates a broad area of attachment for muscles to bone.

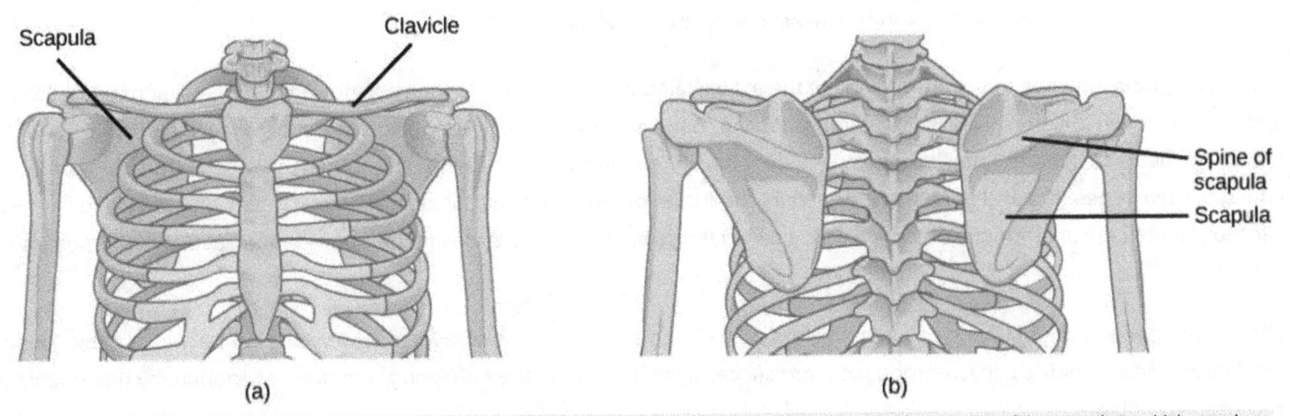

Figure 2: (a) The pectoral girdle in primates consists of the clavicles and scapulae. (b) The posterior view reveals the spine of the scapula to which muscle attaches.

The Upper Limb

The upper limb contains 30 bones in three regions: the arm (shoulder to elbow), the forearm (ulna and radius), and the wrist and hand (Figure 3).

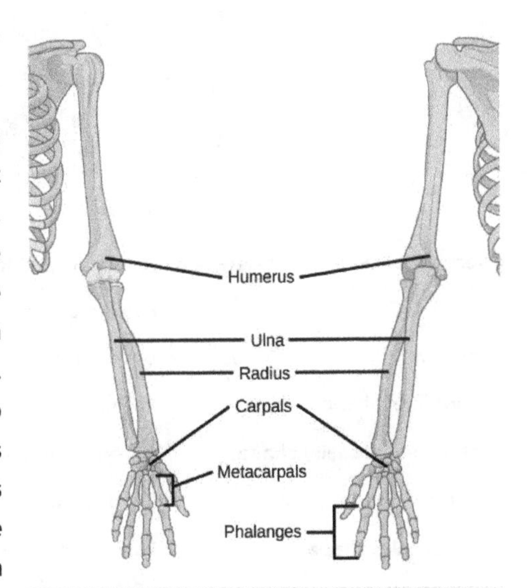

Figure 3: Bones of The Upper Limb.

An **articulation** is any place at which two bones are joined. The **humerus**is the largest and longest bone of the upper limb and the only bone of the arm. It articulates with the scapula at the shoulder and with the forearm at the elbow. The **forearm** extends from the elbow to the wrist and consists of two bones: the ulna and the radius. The **radius** is located along the lateral (thumb) side of the forearm and articulates with the humerus at the elbow. The **ulna** is located on the medial aspect (pinky-finger side) of the forearm. It is longer than the radius. The ulna articulates with the humerus at the elbow. The radius and ulna also articulate with the carpal bones and with each other, which in vertebrates enables a variable degree of rotation of the carpus with respect to the long axis of the limb. The hand includes the eight bones of the **carpus** (wrist), the five bones of the **metacarpus**(palm), and the 14 bones of the **phalanges**(digits). Each digit consists of three phalanges, except for the thumb, when present, which has only two.

The Pelvic Girdle

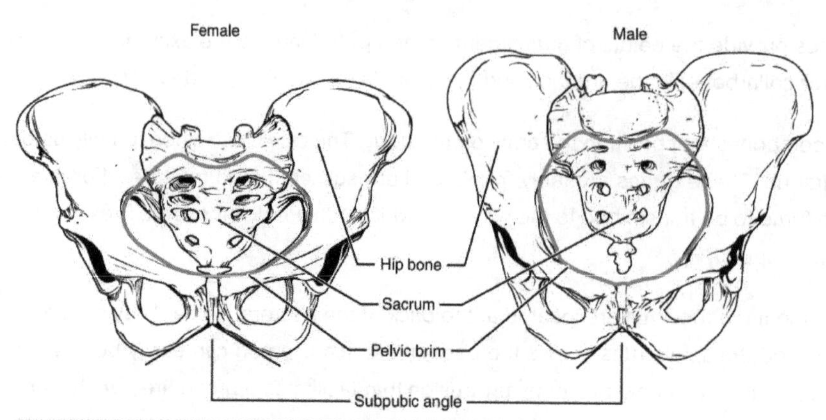

Figure 4: Overview of Differences between the Female and Male Pelvis

The **pelvic girdle** attaches to the lower limbs of the axial skeleton. Because it is responsible for bearing the weight of the body and for locomotion, the pelvic girdle is securely attached to the axial skeleton by strong ligaments. It also has deep sockets with robust ligaments to securely attach the femur to the body. The pelvic girdle is further strengthened by two large hip bones. In adults, the hip bones, or **coxal bones**, are formed by the fusion of three pairs of bones: the ilium, ischium, and pubis. The pelvis joins together in the anterior of the body at a joint called the pubic symphysis and with the bones of the sacrum at the posterior of the body.

The female pelvis is slightly different from the male pelvis. Over generations of evolution, females with a wider pubic angle and larger diameter pelvic canal reproduced more successfully. Therefore, their offspring also had pelvic anatomy that enabled successful childbirth.

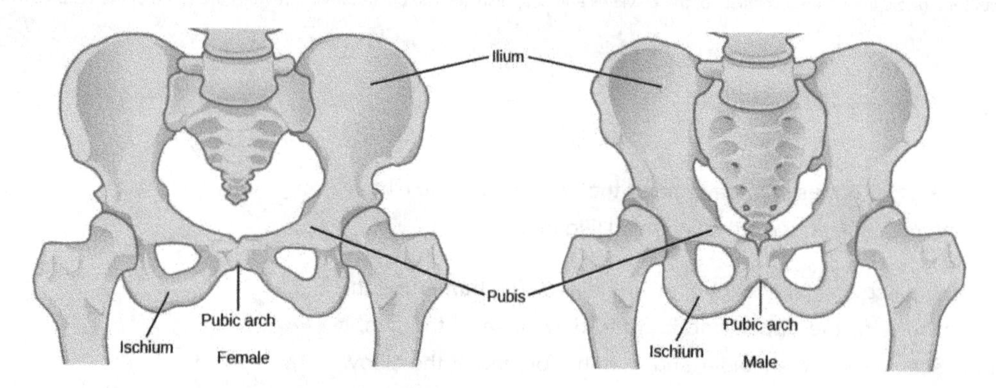

Overview of Differences between the Female and Male Pelvis

	Female pelvis	**Male pelvis**
Pelvic weight	Bones of the pelvis are lighter and thinner	Bones of the pelvis are thicker and heavier
Pelvic inlet shape	Pelvic inlet has a round or oval shape	Pelvic inlet is heart-shaped
Lesser pelvic cavity shape	Lesser pelvic cavity is shorter and wider	Lesser pelvic cavity is longer and narrower
Subpubic angle	Subpubic angle is greater than 80 degrees	Subpubic angle is less than 70 degrees
Pelvic outlet shape	Pelvic outlet is rounded and larger	Pelvic outlet is smaller

The Lower Limb

The **femur**, or thighbone, is the longest, heaviest, and strongest bone in the body. The femur and pelvis form the hip joint at the proximal end. At the distal end, the femur, tibia, and patella form the knee joint. The **patella**, or kneecap, is a triangular bone that lies anterior to the knee joint. The patella is embedded in the tendon of the femoral extensors (quadriceps). It improves knee extension by reducing friction. The **tibia**, or shinbone, is a large bone of the leg that is located directly below the knee. The tibia articulates with the femur at its proximal end, with the fibula and the tarsal bones at its distal end. It is the second largest bone in the human body and is responsible for transmitting the weight of the body from the femur to the foot. The **fibula**, or calf bone, parallels and articulates with the tibia. It does not articulate with the femur and does not bear weight. The fibula acts as a site for muscle attachment and forms the lateral part of the ankle joint.The **lower limb** consists of the thigh, the leg, and the foot. The bones of the lower limb are the femur (thigh bone), patella (kneecap), tibia and fibula (bones of the leg), tarsals (bones of the ankle), and metatarsals and phalanges (bones of the foot) (Figure 5). The bones of the lower limbs are thicker and stronger than the bones of the upper limbs because of the need to support the entire weight of the body and the resulting forces from locomotion. In addition to evolutionary fitness, the bones of an individual will respond to forces exerted upon them.

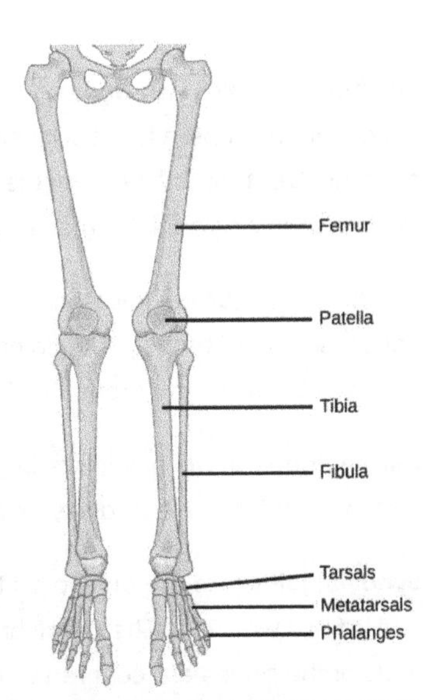

Figure 5. The lower limb consists of the thigh (femur), kneecap (patella), leg (tibia and fibula), ankle (tarsals), and foot (metatarsals and phalanges) bones.

The **tarsals** are the seven bones of the ankle. The ankle transmits the weight of the body from the tibia and the fibula to the foot.

The **metatarsals** are the five bones of the foot. The phalanges are the 14 bones of the toes. Each toe consists of three phalanges, except for the big toe that has only two (Figure 5).

Joints and Skeletal Movements

The point at which two or more bones meet is called a **joint, or articulation.** Joints are responsible for movement, such as the movement of limbs, and stability, such as the stability found in the bones of the skull. There are two ways to classify joints: **based on their structure or based on their function.** The structural classification divides joints into **fibrous, cartilaginous, and synovial joints** depending on the material composing the joint and the presence or absence of a cavity in the joint.

The bones of **fibrous joints** are held together by fibrous connective tissue. There is no cavity, or space, present between the bones, so most fibrous joints do not move at all, or are only capable of minor movements. The joints between the bones in the skull and between the teeth and the bone of their sockets are examples of fibrous joints.

Cartilaginous joints are joints in which the bones are connected by cartilage. An example is found at the joints between vertebrae, the so-called "disks" of the backbone. Cartilaginous joints allow for very little movement.

Synovial joints are the only joints that have a space between the adjoining bones. This space is referred to as the joint cavity and is filled with fluid. The fluid lubricates the joint, reducing friction between the bones and allowing for greater movement. The ends of the bones are covered with cartilage and the entire joint is surrounded by a capsule. Synovial joints are capable of the greatest movement of the joint types. Knees, elbows, and shoulders are examples of synovial joints.

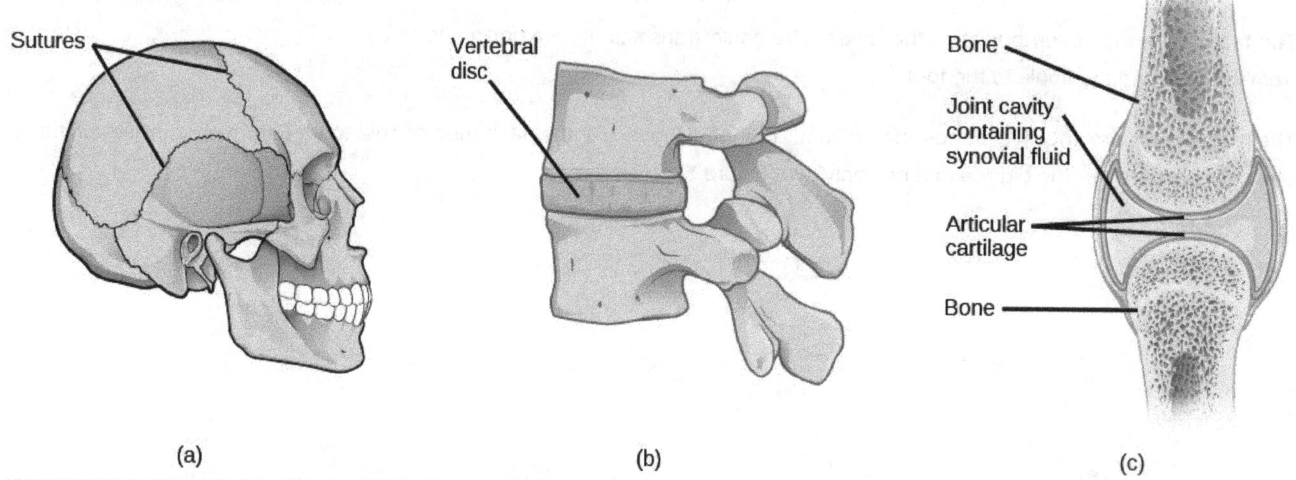

Figure 1(a) Sutures are fibrous joints found only in the skull. (b) Cartilaginous joints are bones connected by cartilage, such as between vertebrae. (c) Synovial joints are the only joints that have a space or "synovial cavity" in the joint.

Types of Synovial Joints

Synovial joints are further classified into six different categories on the basis of the shape and structure of the joint. The shape of the joint affects the type of movement permitted by the joint (Figure 3). These joints can be described as **planar, hinge, pivot, condyloid, saddle, or ball-and-socket joints.**

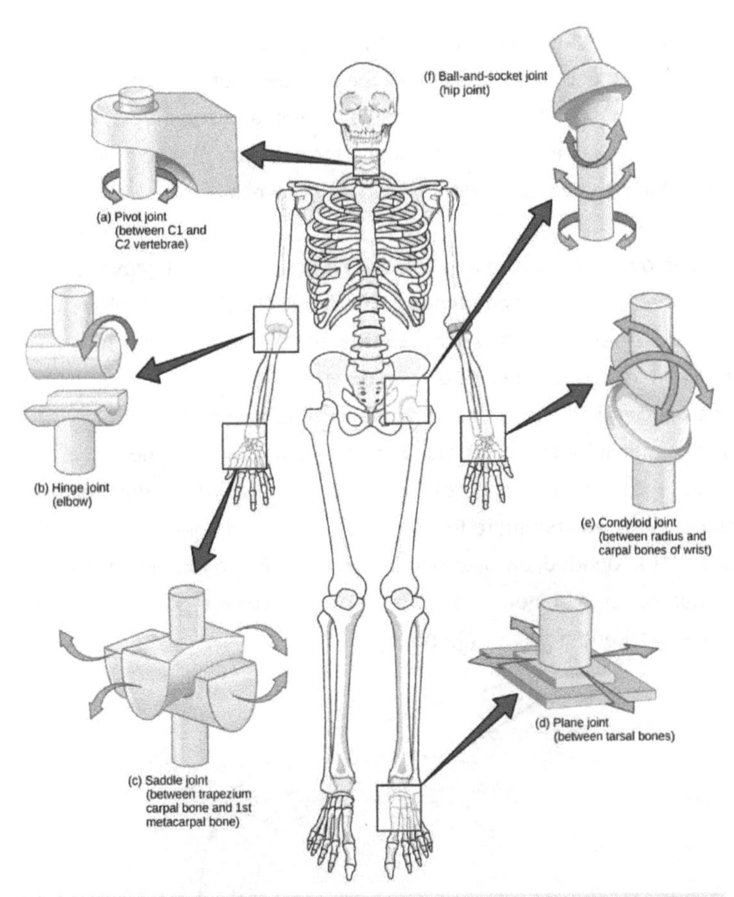

Figure 2. Different types of joints allow different types of movement. Planar, hinge, pivot, condyloid, saddle, and ball-and-socket are all types of synovial joints.

- **Planar joints** have bones with articulating surfaces that are flat or slightly curved faces. These joints allow for gliding movements, and so the joints are sometimes referred to as gliding joints. The range of motion is limited in these joints and does not involve rotation. Planar joints are found in the carpal bones in the hand and the tarsal bones of the foot, as well as between vertebrae (Figure 3).

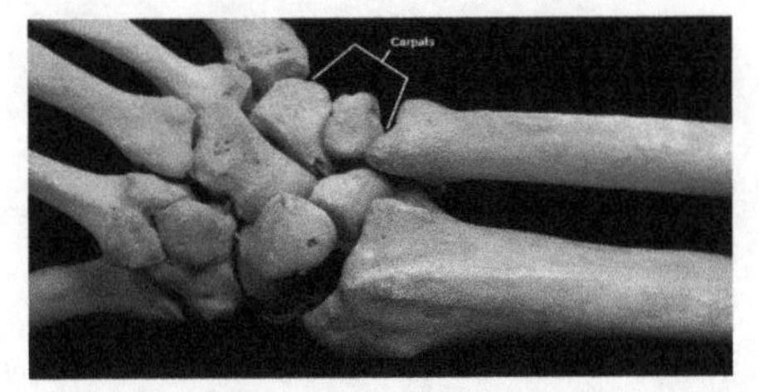

Figure 3. The joints of the carpal bones in the wrist are examples of planar joints. (credit: modification of work by Brian C. Goss)

- **Hinge Joints:** In **hinge joints**, the slightly rounded end of one bone fits into the slightly hollow end of the other bone. In this way, one bone moves while the other remains stationary, like the hinge of a door. The elbow is an example of a hinge joint. The knee is sometimes classified as a modified hinge joint (Figure 4).

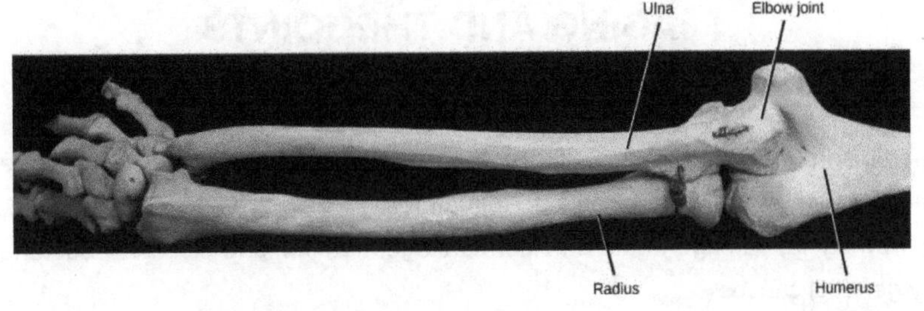

Figure 4. The elbow joint, where the radius articulates with the humerus, is an example of a hinge joint. (credit: modification of work by Brian C. Goss)

- **Pivot joints:** consist of the rounded end of one bone fitting into a ring formed by the other bone. This structure allows rotational movement, as the rounded bone moves around its own axis. An example of a pivot joint is the joint of the first and second vertebrae of the neck that allows the head to move back and forth (Figure 5). The joint of the wrist that allows the palm of the hand to be turned up and down is also a pivot joint.
- **Condyloid joints** consist of an oval-shaped end of one bone fitting into a similarly oval-shaped hollow of another bone. This is also sometimes called an ellipsoidal joint. This type of joint allows angular movement along two axes, as seen in the joints of the wrist and fingers, which can move both side to side and up and down.
- **Saddle joints** are so named because the ends of each bone resemble a saddle, with concave and convex portions that fit together. Saddle joints allow angular movements similar to condyloid joints but with a greater range of motion. An example of a saddle joint is the thumb joint, which can move back and forth and up and down, but more freely than the wrist or fingers.
- **Ball-and-socket joints** possess a rounded, ball-like end of one bone fitting into a cuplike socket of another bone. This organization allows the greatest range of motion, as all movement types are possible in all directions. Examples of ball-and-socket joints are the shoulder and hip joints (Figure 6).

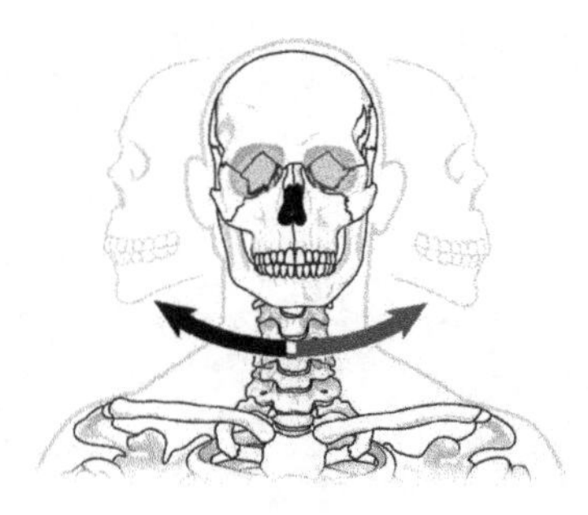

Figure 5. The joint in the neck that allows the head to move back and forth is an example of a pivot joint.

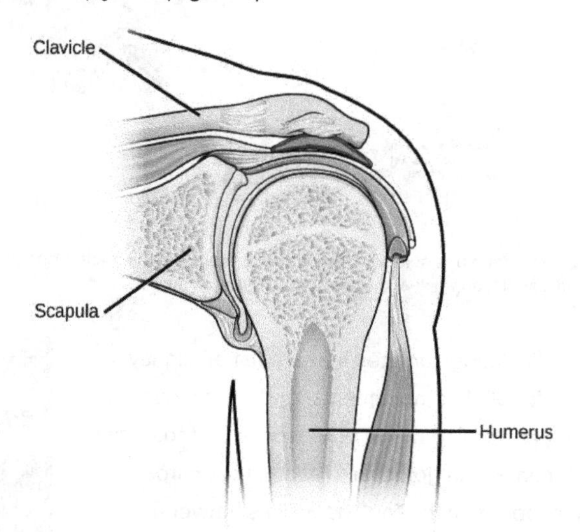

Figure 6. The shoulder joint is an example of a ball-and-socket joint.Movement at Synovial Joints

The wide range of movement allowed by synovial joints produces different types of movements. The movement of synovial joints can be classified as one of four different types: gliding, angular, rotational, or special movement.

AGING AND THE JOINTS

Arthritis is a common disorder of synovial joints that involves inflammation of the joint. This often results in significant joint pain, along with swelling, stiffness, and reduced joint mobility. There are more than 100 different forms of arthritis. Arthritis may arise from aging, damage to the articular cartilage, autoimmune diseases, bacterial or viral infections, or unknown (probably genetic) causes.

The most common type of arthritis is **osteoarthritis**, which is associated with aging and **"wear and tear"** of the articular cartilage. Risk factors that may lead to osteoarthritis later in life include injury to a joint; jobs that involve physical labor; sports with running, twisting, or throwing actions; and being overweight. These factors put stress on the articular cartilage that covers the surfaces of bones at synovial joints, causing the cartilage to gradually become thinner. As the articular cartilage layer wears down, more pressure is placed on the bones. The joint responds by increasing production of the lubricating synovial fluid, but this can lead to swelling of the joint cavity, causing pain and joint stiffness as the articular capsule is stretched. The bone tissue underlying the damaged articular cartilage also responds by thickening, producing irregularities and causing the articulating surface of the bone to become rough or bumpy. Joint movement then results in pain and inflammation. In its early stages, symptoms of osteoarthritis may be reduced by mild activity that "warms up" the joint, but the symptoms may worsen following exercise. In individuals with more advanced osteoarthritis, the affected joints can become more painful and therefore are difficult to use effectively, resulting in increased immobility. There is no cure for osteoarthritis, but several treatments can help alleviate the pain. Treatments may include lifestyle changes, such as weight loss and low-impact exercise, and over-the-counter or prescription medications that help to alleviate the pain and inflammation. For severe cases, joint replacement surgery (arthroplasty) may be required.

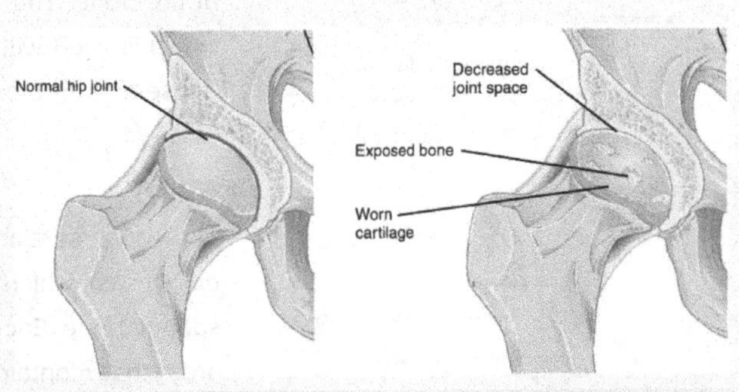

Figure 7: Osteoarthritis of a synovial joint results from aging or prolonged joint wear and tear. These cause erosion and loss of the articular cartilage covering the surfaces of the bones, resulting in inflammation that causes joint stiffness and pain.

Joint replacement is a very invasive procedure, so other treatments are always tried before surgery. However arthroplasty can provide relief from chronic pain and can enhance mobility within a few months following the surgery. This type of surgery involves replacing the articular surfaces of the bones with prosthesis (artificial components). For example, in hip arthroplasty, the worn or damaged parts of the hip joint, including the head and neck of the femur and the acetabulum of the pelvis, are removed and replaced with artificial joint components. The replacement head for the femur consists of a rounded ball attached to the end of a shaft that is inserted inside the diaphysis of the femur. The acetabulum of the pelvis is reshaped and a replacement socket is fitted into its place. The parts, which are always built in advance of the surgery, are sometimes custom made to produce the best possible fit for a patient.

Gout is a form of arthritis that results from the deposition of uric acid crystals within a body joint. Usually only one or a few joints are affected, such as the big toe, knee, or ankle. The attack may only last a few days, but may return to the same or another joint. Gout occurs when the body makes too much uric acid or the kidneys do not properly excrete it. A diet with excessive fructose has been implicated in raising the chances of a susceptible individual developing gout.

Other forms of arthritis are associated with various autoimmune diseases, bacterial infections of the joint, or unknown genetic causes. Autoimmune diseases, including rheumatoid arthritis, scleroderma, or systemic lupus erythematosus, produce arthritis because the immune system of the body attacks the body joints. In **rheumatoid arthritis**, the joint capsule and synovial membrane become inflamed. As the disease progresses, the articular cartilage is severely damaged or destroyed, resulting in joint deformation, loss of movement, and severe disability. The most commonly involved joints are the hands, feet, and cervical spine, with corresponding joints on both sides of the body usually affected, though not always to the same extent. Rheumatoid arthritis is also associated with lung fibrosis, vasculitis (inflammation of blood vessels), coronary heart disease, and premature mortality. With no known cure, treatments are aimed at alleviating symptoms. Exercise, anti-inflammatory and pain medications, various specific disease-modifying anti-rheumatic drugs, or surgery are used to treat rheumatoid arthritis.

Bone Structure, Formation, Development And Bone Repair

Bone tissue (osseous tissue) differs greatly from other tissues in the body. Bone is hard and many of its functions depend on that characteristic hardness.

Gross Anatomy of Bone

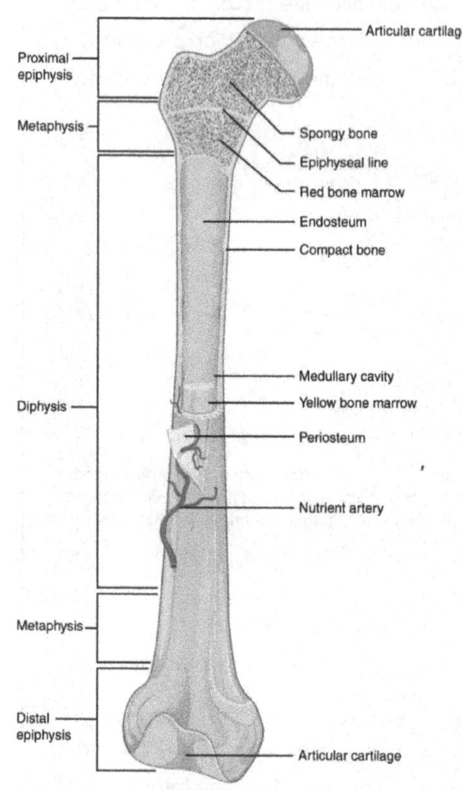

Figure 1. Anatomy of a Long Bone. A typical long bone shows the gross anatomical characteristics of bone.

The structure of a long bone allows for the best visualization of all of the parts of a bone (Figure 1). A long bone has two parts: the **diaphysis** and the **epiphysis**. The diaphysis is the tubular shaft that runs between the proximal and distal ends of the bone. The hollow region in the diaphysis is called the **medullary cavity**, which is filled with yellow marrow. The walls of the diaphysis are composed of dense and hard **compact bone**.

The wider section at each end of the bone is called the *epiphysis* (plural = *epiphyses*), which is filled with spongy bone. Red marrow fills the spaces in the spongy bone. Each epiphysis meets the diaphysis at the metaphysis, the narrow area that contains the **epiphyseal plate** (growth plate), a layer of hyaline (transparent) cartilage in a growing bone. When the bone stops growing in early adulthood (approximately 18–21 years), the cartilage is replaced by osseous tissue and the epiphyseal plate becomes an epiphyseal line.

The medullary cavity has a delicate membranous lining called the **endosteum** (*end–* = "inside"; *oste–* = "bone"), where bone growth, repair, and remodeling occur. The outer surface of the bone is covered with a fibrous membrane called the **periosteum** (*peri–* = "around" or "surrounding"). The periosteum contains blood vessels, nerves, and lymphatic vessels that nourish compact bone. Tendons and ligaments also attach to bones at the periosteum. The periosteum covers the entire outer surface except where the epiphyses meet other bones to form joints (Figure 2). In this region, the epiphyses are covered with **articular cartilage**, a thin layer of cartilage that reduces friction and acts as a shock absorber.

Flat bones, like those of the cranium, consist of a layer of **diploë** (spongy bone), lined on either side by a layer of compact bone (Figure 2). The two layers of compact bone and the interior spongy bone work together to protect the internal organs. If the outer layer of a cranial bone fractures, the brain is still protected by the intact inner layer.

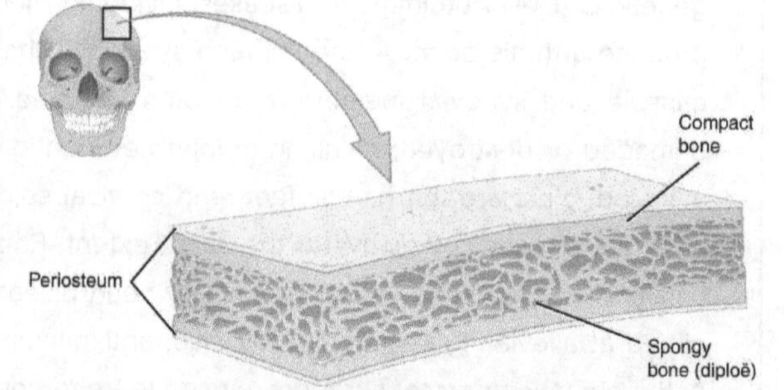

Figure 2. Anatomy of a Flat Bone. This cross-section of a flat bone shows the spongy bone (diploë) lined on either side by a layer of compact bone.

Bone Cells and Tissue

Bone contains a relatively small number of cells entrenched in a matrix of collagen fibers that provide a surface for inorganic salt crystals to adhere. These salt crystals form when calcium phosphate and calcium carbonate combine to create **hydroxyapatite,** which incorporates other inorganic salts like magnesium hydroxide, fluoride, and sulfate as it crystallizes, or calcifies, on the collagen fibers. The hydroxyapatite crystals give bones their hardness and strength, while the collagen fibers give them flexibility so that they are not brittle.

Although bone cells compose a small amount of the bone volume, they are crucial to the function of bones. Four types of cells are found within bone tissue: **osteoblasts, osteocytes, osteogenic cells, and osteoclasts** (Figure 3).

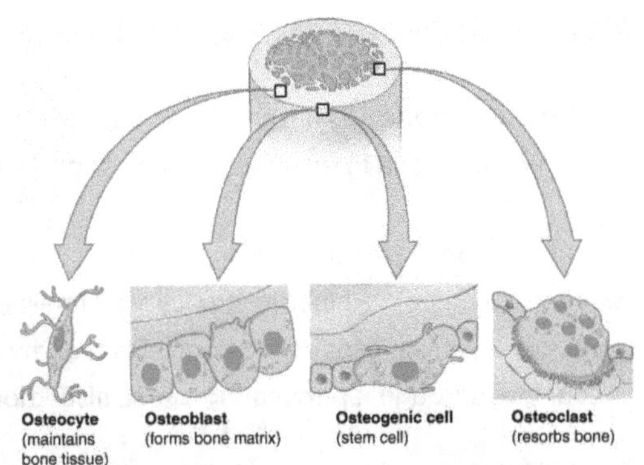

Figure 3. Bone Cells. Four types of cells are found within bone tissue. Osteogenic cells are undifferentiated and develop into osteoblasts. When osteoblasts get trapped within the calcified matrix, their structure and function changes, and they become osteocytes. Osteoclasts develop from monocytes and macrophages and differ in appearance from other bone cells.

The **osteoblast**. is the bone cell responsible for forming new bone and is found in the growing portions of bone, including the periosteum and endosteum. Osteoblasts, which do not divide, synthesize and secrete the collagen matrix and calcium salts. As the secreted matrix surrounding the osteoblast calcifies, the osteoblast become trapped within it; as a result, it changes in structure and becomes an **osteocyte**, the primary cell of mature bone and the most common type of bone cell. Each osteocyte is located in a space called a **lacuna** and is surrounded by bone tissue. Osteocytes maintain the mineral concentration of the matrix via the secretion of enzymes. Like osteoblasts, osteocytes lack mitotic activity. They can communicate with each other and receive nutrients via long cytoplasmic processes that extend through **canaliculi** (singular = *canaliculus*), channels within the bone matrix.

If osteoblasts and osteocytes are incapable of mitosis, then how are they replenished when old ones die? The answer lies in the properties of a third category of bone cells—the **osteogenic cell**. These osteogenic cells are undifferentiated with high mitotic activity and they are the only bone cells that divide. Immature osteogenic cells are found in the deep layers of the periosteum and the marrow. They differentiate and develop into osteoblasts.

The dynamic nature of bone means that new tissue is constantly formed, and old, injured, or unnecessary bone is dissolved for repair or for calcium release. The cell responsible for bone resorption, or breakdown, is the **osteoclast**. They are found on bone surfaces, are multinucleated, and originate from monocytes and macrophages, two types of white blood cells, not from osteogenic cells. Osteoclasts are continually breaking down old bone while osteoblasts are continually forming new bone. The ongoing balance between osteoblasts and osteoclasts is responsible for the constant but subtle reshaping of bone. Table 2 reviews the bone cells, their functions, and locations.

Table 2. Bone Cells

Cell type	Function	Location
Osteogenic cells	Develop into osteoblasts	Deep layers of the periosteum and the marrow
Osteoblasts	Bone formation	Growing portions of bone, including periosteum and endosteum
Osteocytes	Maintain mineral concentration of matrix	Entrapped in matrix
Osteoclasts	Bone resorption	Bone surfaces and at sites of old, injured, or unneeded bone

Bone Tissue Types

Bones are considered organs because they contain various types of tissue, such as blood, connective tissue, nerves, and bone tissue. Osteocytes, the living cells of bone tissue, form the mineral matrix of bones. There are two types of bone tissue: **compact and spongy.**

The differences between compact and spongy bone are best explored via their histology. Most bones contain compact and spongy osseous tissue, but their distribution and concentration vary based on the bone's overall function. Compact bone is dense so that it can withstand compressive forces, while spongy (cancellous) bone has open spaces and supports shifts in weight distribution.

- **Compact Bone:** Compact bone is the denser, stronger of the two types of bone tissue (Figure 4). It can be found under the periosteum and in the diaphyses of long bones, where it provides support and protection.

The microscopic structural unit of compact bone is called an **osteon**, or Haversian system. Each osteon is composed of concentric rings of calcified matrix called lamellae (singular = lamella). Running down the center of each osteon is the **central canal**, or Haversian canal, which contains blood vessels, nerves, and lymphatic vessels. These vessels and nerves branch off at right angles through a **perforating canal**, also known as Volkmann's canals, to extend to the periosteum and endosteum.

The osteocytes are located inside spaces called lacunae (singular = lacuna), found at the borders of adjacent lamellae. As described earlier, canaliculi connect with the canaliculi of other lacunae and eventually with the central canal. This system allows nutrients to be transported to the osteocytes and wastes to be removed from them.

- **Spongy (Cancellous) Bone:** Like compact bone, spongy bone, also known as cancellous bone, contains osteocytes housed in lacunae, but they are not arranged in concentric circles. Instead, the lacunae and osteocytes are found in a lattice-like network of matrix spikes called trabeculae (singular = *trabecula*) (Figure 5). The trabeculae may appear to be a random network, but each trabecula forms along lines of stress to provide strength to the bone. The spaces of the trabeculated network provide balance to the dense and heavy compact bone by making bones lighter so that muscles can move them more easily. In addition, the spaces in some spongy bones contain red marrow, protected by the trabeculae, where hematopoiesis occurs.

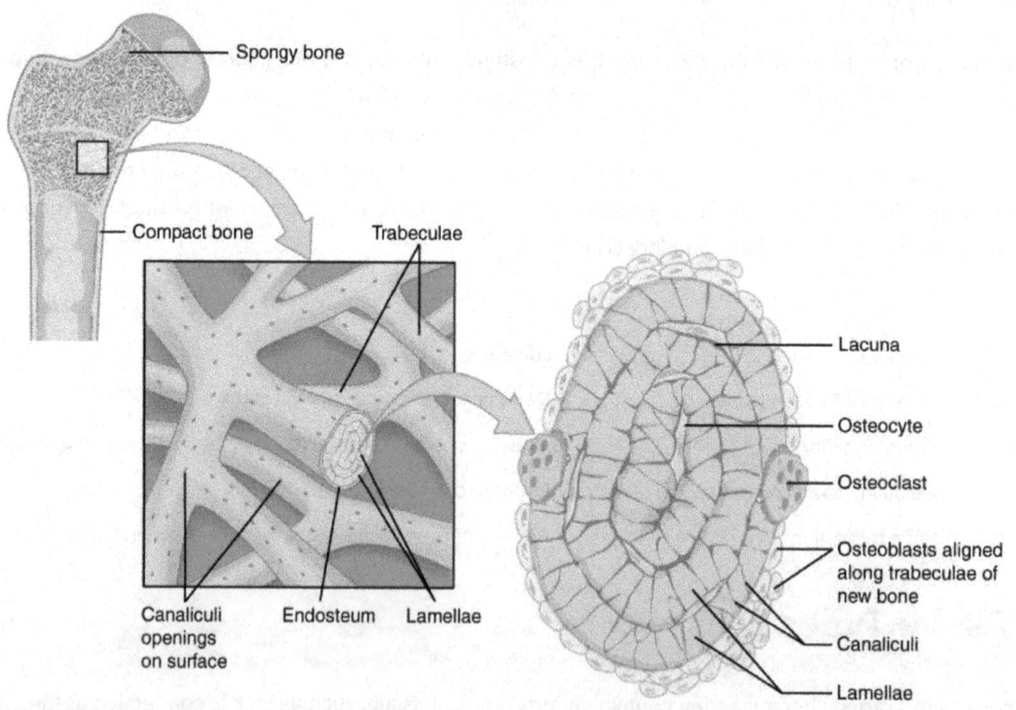

Figure 5. Diagram of Spongy Bone. Spongy bone is composed of trabeculae that contain the osteocytes. Red marrow fills the spaces in some bones.

AGING AND THE SKELETAL SYSTEM: PAGET'S DISEASE

Paget's disease usually occurs in adults over age 40. It is a disorder of the bone remodeling process that begins with overactive osteoclasts. This means more bone is resorbed than is laid down. The osteoblasts try to compensate but the new bone they lay down is weak and brittle and therefore prone to fracture.

While some people with Paget's disease have no symptoms, others experience pain, bone fractures, and bone deformities (Figure 6). Bones of the pelvis, skull, spine, and legs are the most commonly affected. When occurring in the skull, Paget's disease can cause headaches and hearing loss.

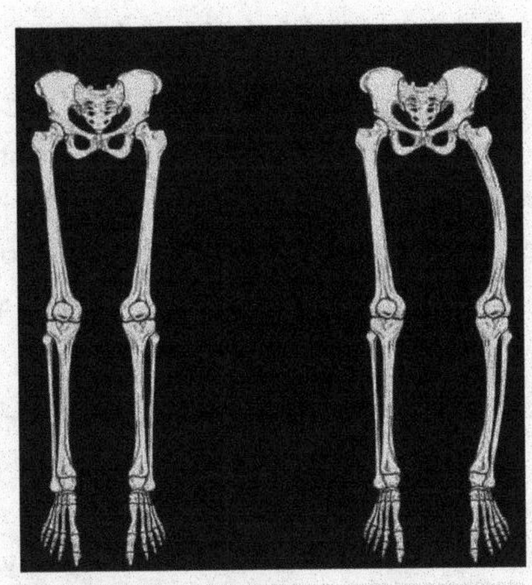

Figure 6. Paget's Disease. Normal leg bones are relatively straight, but those affected by Paget's disease are porous and curved.

What causes the osteoclasts to become overactive? The answer is still unknown, but hereditary factors seem to play a role. Some scientists believe Paget's disease is due to an as-yet-unidentified virus.

Paget's disease is diagnosed via imaging studies and lab tests. X-rays may show bone deformities or areas of bone resorption. Bone scans are also useful. In these studies, a dye containing a radioactive ion is injected into the body. Areas of bone resorption have an affinity for the ion, so they will light up on the scan if the ions are absorbed. In addition, blood levels of an enzyme called alkaline phosphatase are typically elevated in people with Paget's disease.

Bisphosphonates, drugs that decrease the activity of osteoclasts, are often used in the treatment of Paget's disease. However, in a small percentage of cases, bisphosphonates themselves have been linked to an increased risk of fractures because the old bone that is left after bisphosphonates are administered becomes worn out and brittle. Still, most doctors feel that the benefits of bisphosphonates more than outweigh the risk; the medical professional has to weigh the benefits and risks on a case-by-case basis. Bisphosphonate treatment can reduce the overall risk of deformities or fractures, which in turn reduces the risk of surgical repair and its associated risks and complications.

Blood and Nerve Supply

The spongy bone and medullary cavity receive nourishment from arteries that pass through the compact bone. The arteries enter through the **nutrient foramen** (plural = *foramina*), small openings in the diaphysis (Figure 7). The osteocytes in spongy bone are nourished by blood vessels of the periosteum that penetrate spongy bone and blood that circulates in the marrow cavities. As the blood passes through the marrow cavities, it is collected by veins, which then pass out of the bone through the foramina.

In addition to the blood vessels, nerves follow the same paths into the bone where they tend to concentrate in the more metabolically active regions of the bone. The nerves sense pain, and it appears the nerves also play roles in regulating blood supplies and in bone growth, hence their concentrations in metabolically active sites of the bone.

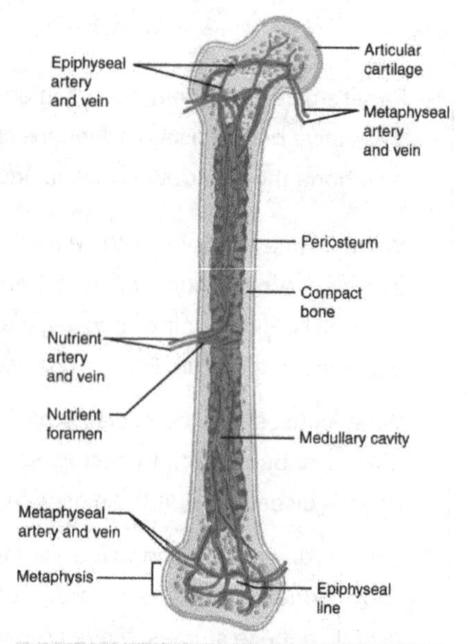

Figure 7. Diagram of Blood and Nerve Supply to Bone. Blood vessels and nerves enter the bone through the nutrient foramen.

Bone Formation and Development

In the early stages of embryonic development, the embryo's skeleton consists of fibrous membranes and hyaline cartilage. By the sixth or seventh week of embryonic life, the actual process of bone development, **ossification** (osteogenesis), begins. There are two osteogenic pathways—intramembranous ossification and endochondral ossification—but bone is the same regardless of the pathway that produces it.

Cartilage Templates

Bone is a replacement tissue; that is, it uses a model tissue on which to lay down its mineral matrix. For skeletal development, the most common template is cartilage. During fetal development, a framework is laid down that determines where bones will form. This framework is a flexible, semi-solid matrix produced by chondroblasts and consists of hyaluronic acid, chondroitin sulfate, collagen fibers, and water. As the matrix surrounds and isolates chondroblasts, they are called chondrocytes. Unlike most connective tissues, cartilage is avascular, meaning that it has no blood vessels supplying nutrients and removing metabolic wastes. All of these functions are carried on by diffusion through the matrix. This is why damaged cartilage does not repair itself as readily as most tissues do.

Throughout fetal development and into childhood growth and development, bone forms on the cartilaginous matrix. By the time a fetus is born, most of the cartilage has been replaced with bone. Some additional cartilage will be replaced throughout childhood, and some cartilage remains in the adult skeleton.

Intramembranous Ossification

Intramembranous ossification is the process of bone development from fibrous membranes. It is involved in the formation of the flat bones of the skull, the mandible, and the clavicles. Ossification begins as mesenchymal cells form a template of the future bone. They then differentiate into osteoblasts at the ossification center. Osteoblasts secrete the extracellular matrix and deposit calcium, which hardens the matrix. The non-mineralized portion of the bone or osteoid continues to form around blood vessels, forming spongy bone. Connective tissue in the matrix differentiates into red bone marrow in the fetus. The spongy bone is remodeled into a thin layer of compact bone on the surface of the spongy bone.

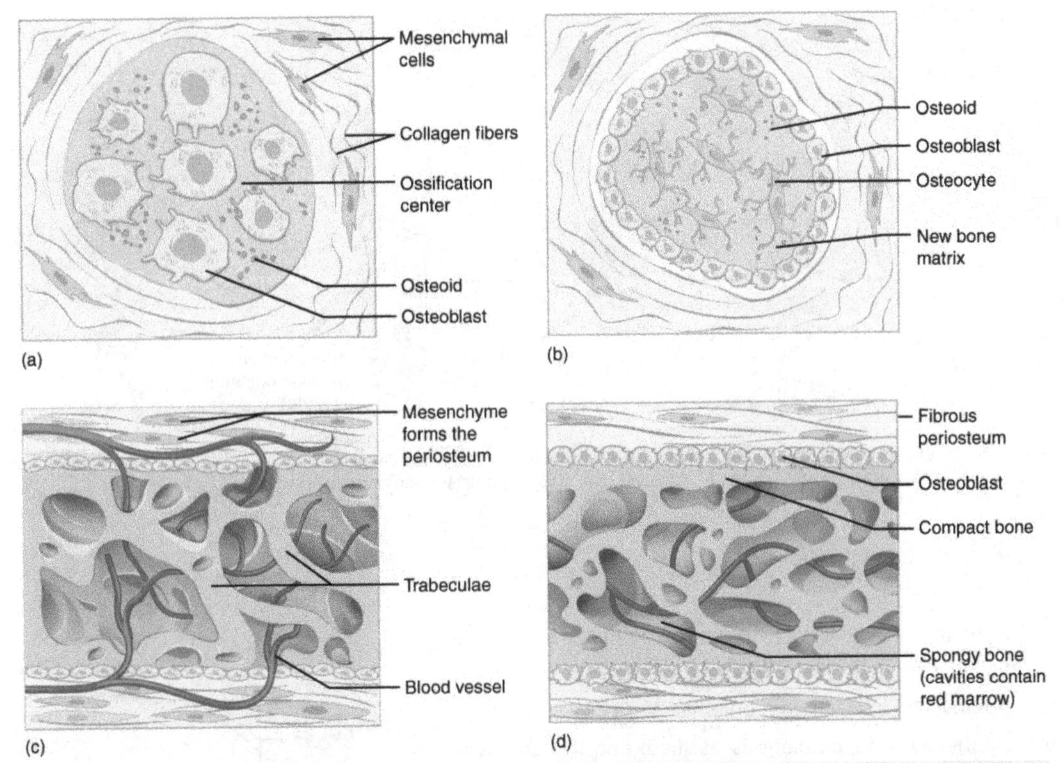

Figure 8. Intramembranous Ossification. Intramembranous ossification follows four steps. (a) Mesenchymal cells group into clusters, and ossification centers form. (b) Secreted osteoid traps osteoblasts, which then become osteocytes. (c) Trabecular matrix and periosteum form. (d) Compact bone develops superficial to the trabecular bone, and crowded blood vessels condense into red marrow.

Intramembranous ossification begins *in utero* during fetal development and continues on into adolescence. At birth, the skull and clavicles are not fully ossified nor are the sutures of the skull closed. This allows the skull and shoulders to deform during passage through the birth canal. The last bones to ossify via intramembranous ossification are the flat bones of the face, which reach their adult size at the end of the adolescent growth spurt.

Endochondral Ossification

Endochondral ossification is the process of bone development from hyaline cartilage. All of the bones of the body, except for the flat bones of the skull, mandible, and clavicles, are formed through endochondral ossification.

In long bones, chondrocytes form a template of the hyaline cartilage diaphysis. Responding to complex developmental signals, the matrix begins to calcify. This calcification prevents diffusion of nutrients into the matrix, resulting in chondrocytes dying and the opening up of cavities in the diaphysis cartilage. Blood vessels invade the cavities, and osteoblasts and osteoclasts modify the calcified cartilage matrix into spongy bone. Osteoclasts then break down some of the spongy bone to create a marrow, or medullary, cavity in the center of the diaphysis. Dense, irregular connective tissue forms a sheath (periosteum) around the bones. The periosteum assists in attaching the bone to surrounding tissues, tendons, and ligaments. The bone continues to grow and elongate as the cartilage cells at the epiphyses divide.

In the last stage of prenatal bone development, the centers of the epiphyses begin to calcify. Secondary ossification centers form in the epiphyses as blood vessels and osteoblasts enter these areas and convert hyaline cartilage into spongy bone. Until adolescence, hyaline cartilage persists at the epiphyseal plate (growth plate), which is the region between the diaphysis and epiphysis that is responsible for the lengthwise growth of long bones.

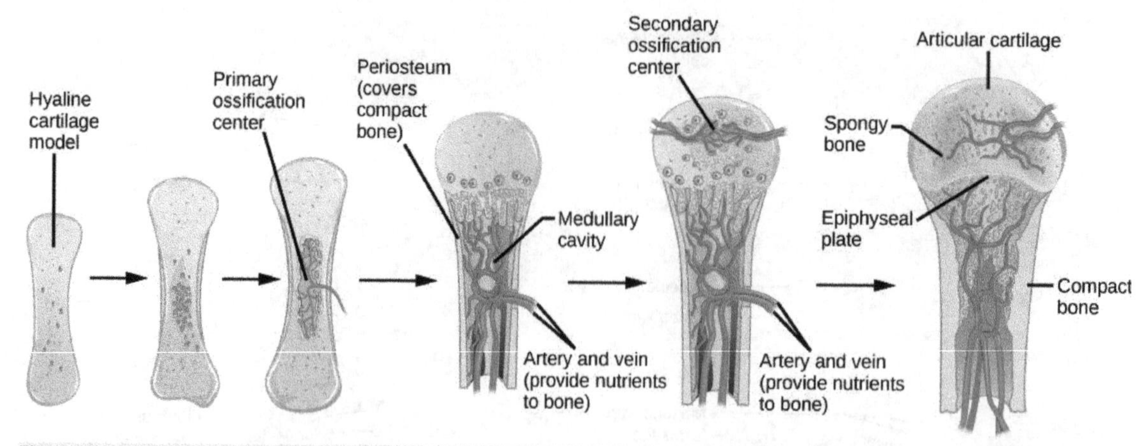

Figure 9. Endochondral ossification is the process of bone development from hyaline cartilage. The periosteum is the connective tissue on the outside of bone that acts as the interface between bone, blood vessels, tendons, and ligaments.

How Bones Grow in Length

The **epiphyseal plate** (growth plate) is the area of growth in a long bone. It is a layer of hyaline cartilage where ossification occurs in immature bones. On the epiphyseal side of the epiphyseal plate, cartilage is formed. On the diaphyseal side, cartilage is ossified, and the diaphysis grows in length.

Bones continue to grow in length until early adulthood. The rate of growth is controlled by hormones, which will be discussed later. When the chondrocytes in the epiphyseal plate cease their proliferation and bone replaces the cartilage, longitudinal growth stops. All that remains of the epiphyseal plate is the **epiphyseal line** (Figure 10).

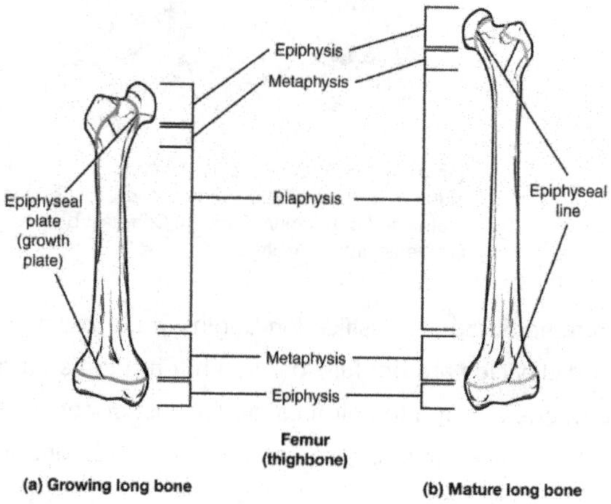

Epiphysis
Metaphysis
Epiphyseal plate (growth plate)
Diaphysis
Epiphyseal line
Metaphysis
Epiphysis
Femur (thighbone)
(a) Growing long bone
(b) Mature long bone

Figure 10. Progression from Epiphyseal Plate to Epiphyseal Line. As a bone matures, the epiphyseal plate progresses to an epiphyseal line. (a) Epiphyseal plates are visible in a growing bone. (b) Epiphyseal lines are the remnants of epiphyseal plates in a mature bone.

How Bones Grow in Diameter

While bones are increasing in length, they are also increasing in diameter; growth in diameter can continue even after longitudinal growth ceases. This is called appositional growth. Osteoclasts resorb old bone that lines the medullary cavity, while osteoblasts, via intramembranous ossification, produce new bone tissue beneath the periosteum. The erosion of old bone along the medullary cavity and the deposition of new bone beneath the periosteum not only increase the diameter of the diaphysis but also increase the diameter of the medullary cavity. This process is called **modeling**.

Bone Remodeling

The process in which matrix is resorbed on one surface of a bone and deposited on another is known as bone modeling. Modeling primarily takes place during a bone's growth. However, in adult life, bone undergoes **remodeling**, in which resorption of old or damaged bone takes place on the same surface where osteoblasts lay new bone to replace that which is resorbed. Injury, exercise, and other activities lead to remodeling. Those influences are discussed later in the chapter, but even without injury or exercise, about 5 to 10 percent of the skeleton is remodeled annually just by destroying old bone and renewing it with fresh bone.

DISEASES OF THE SKELETAL SYSTEM

Osteogenesis imperfecta (OI) is a genetic disease in which bones do not form properly and therefore are fragile and break easily. It is also called brittle bone disease. The disease is present from birth and affects a person throughout life.

The genetic mutation that causes OI affects the body's production of collagen, one of the critical components of bone matrix. The severity of the disease can range from mild to severe. Those with the most severe forms of the disease sustain many more fractures than those with a mild form. Frequent and multiple fractures typically lead to bone deformities and short stature. Bowing of the long bones and curvature of the spine are also common in people afflicted with OI. Curvature of the spine makes breathing difficult because the lungs are compressed.

Because collagen is such an important structural protein in many parts of the body, people with OI may also experience fragile skin, weak muscles, loose joints, easy bruising, frequent nosebleeds, brittle teeth, blue sclera, and hearing loss. There is no known cure for OI. Treatment focuses on helping the person retain as much independence as possible while minimizing fractures and maximizing mobility. Toward that end, safe exercises, like swimming, in which the body is less likely to experience collisions or compressive forces, are recommended. Braces to support legs, ankles, knees, and wrists are used as needed. Canes, walkers, or wheelchairs can also help compensate for weaknesses.

When bones do break, casts, splints, or wraps are used. In some cases, metal rods may be surgically implanted into the long bones of the arms and legs. Research is currently being conducted on using bisphosphonates to treat OI. Smoking and being overweight are especially risky in people with OI, since smoking is known to weaken bones, and extra body weight puts additional stress on the bones.

Bone Remodeling and Repair

Bone renewal continues after birth into adulthood. Bone remodeling is the replacement of old bone tissue by new bone tissue. It involves the processes of bone deposition by osteoblasts and bone resorption by osteoclasts. Normal bone growth requires vitamins D, C, and A, plus minerals such as calcium, phosphorous, and magnesium. Hormones such as parathyroid hormone, growth hormone, and calcitonin are also required for proper bone growth and maintenance.

Bone turnover rates are quite high, with five to seven percent of bone mass being recycled every week. Differences in turnover rate exist in different areas of the skeleton and in different areas of a bone. For example, the bone in the head of the femur may be fully replaced every six months, whereas the bone along the shaft is altered much more slowly.

Bone remodeling allows bones to adapt to stresses by becoming thicker and stronger when subjected to stress. Bones that are not subject to normal stress, for example when a limb is in a cast, will begin to lose mass. A fractured or broken bone undergoes repair through four stages

Fractures & Bone Repair

A **fracture** is a broken bone. It will heal whether or not a physician resets it in its anatomical position. If the bone is not reset correctly, the healing process will keep the bone in its deformed position.

When a broken bone is manipulated and set into its natural position without surgery, the procedure is called a **closed reduction**. **Open reduction** requires surgery to expose the fracture and reset the bone. While some fractures can be minor, others are quite severe and result in grave complications. For example, a fractured diaphysis of the femur has the potential to release fat globules into the bloodstream. These can become lodged in the capillary beds of the lungs, leading to respiratory distress and if not treated quickly, death.

Types of Fractures

Fractures are classified by their complexity, location, and other features. Table 1 outlines common types of fractures. Some fractures may be described using more than one term because it may have the features of more than one type (e.g., an open transverse fracture).

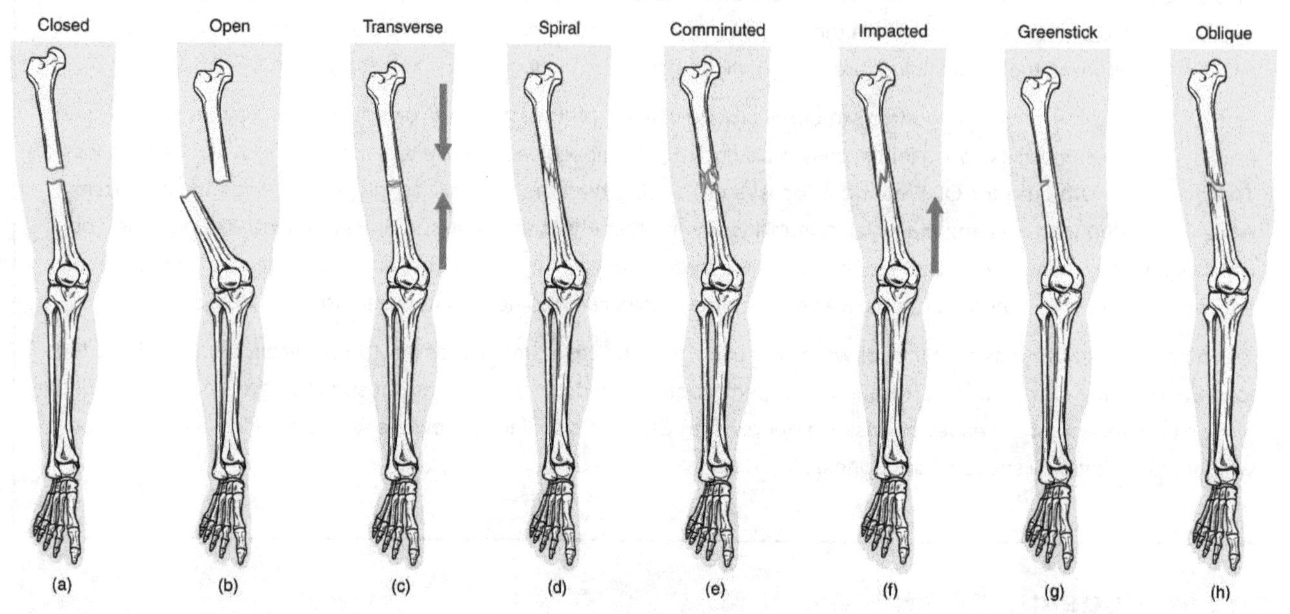

Figure 11. Types of Fractures. Compare healthy bone with different types of fractures: (a) closed fracture, (b) open fracture, (c) transverse fracture, (d) spiral fracture, (e) comminuted fracture, (f) impacted fracture, (g) greenstick fracture, and (h) oblique fracture.

Table 1. Types of Fractures

Type of fracture	Description
Transverse	Occurs straight across the long axis of the bone
Oblique	Occurs at an angle that is not 90 degrees
Spiral	Bone segments are pulled apart as a result of a twisting motion
Comminuted	Several breaks result in many small pieces between two large segments
Impacted	One fragment is driven into the other, usually as a result of compression
Greenstick	A partial fracture in which only one side of the bone is broken
Open (or compound)	A fracture in which at least one end of the broken bone tears through the skin; carries a high risk of infection
Closed (or simple)	A fracture in which the skin remains intact

Bone Repair

When a bone breaks, blood flows from any vessel torn by the fracture. These vessels could be in the periosteum, osteons, and/or medullary cavity. The blood begins to clot, and about six to eight hours after the fracture, the clotting blood has formed a **fracture hematoma** (Figure 12).

The disruption of blood flow to the bone results in the death of bone cells around the fracture.

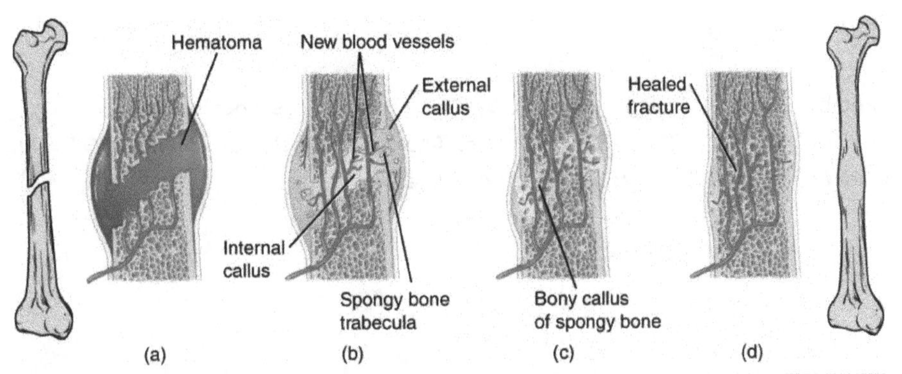

Figure 12. Stages in Fracture Repair. The healing of a bone fracture follows a series of progressive steps: (a) A fracture hematoma forms. (b) Internal and external calli form. (c) Cartilage of the calli is replaced by trabecular bone. (d) Remodeling occurs.

Within about 48 hours after the fracture, chondrocytes from the endosteum have created an **internal callus** (plural = *calli*) by secreting a fibrocartilaginous matrix between the two ends of the broken bone, while the periosteal chondrocytes and osteoblasts create an **external callus** of hyaline cartilage and bone, respectively, around the outside of the break (Figure 12b). This stabilizes the fracture.

Over the next several weeks, osteoclasts resorb the dead bone; osteogenic cells become active, divide, and differentiate into osteoblasts. The cartilage in the calli is replaced by trabecular bone via endochondral ossification (Figure 12c).

Eventually, the internal and external calli unite, compact bone replaces spongy bone at the outer margins of the fracture, and healing is complete. A slight swelling may remain on the outer surface of the bone, but quite often, that region undergoes remodeling (Figure 12d), and no external evidence of the fracture remains.

SCIENTIFIC METHOD CONNECTION

Decalcification of Bones

Question: What effect does the removal of calcium and collagen have on bone structure?

Background: Conduct a literature search on the role of calcium and collagen in maintaining bone structure. Conduct a literature search on diseases in which bone structure is compromised.

Hypothesis: Develop a hypothesis that states predictions of the flexibility, strength, and mass of bones that have had the calcium and collagen components removed. Develop a hypothesis regarding the attempt to add calcium back to decalcified bones.

Test the hypothesis: Test the prediction by removing calcium from chicken bones by placing them in a jar of vinegar for seven days. Test the hypothesis regarding adding calcium back to decalcified bone by placing the decalcified chicken bones into a jar of water with calcium supplements added. Test the prediction by denaturing the collagen from the bones by baking them at 250°C for three hours.

Analyze the data: Create a table showing the changes in bone flexibility, strength, and mass in the three different environments.

Report the results: Under which conditions was the bone most flexible? Under which conditions was the bone the strongest?

Draw a conclusion: Did the results support or refute the hypothesis? How do the results observed in this experiment correspond to diseases that destroy bone tissue?

Exercise, Nutrition, Hormones, and Calcium Homeostasis

All of the organ systems of your body are interdependent, and the skeletal system is no exception. The food you take in via your digestive system and the hormones secreted by your endocrine system affect your bones. Even using your muscles to engage in exercise has an impact on your bones.

Exercise and Bone Tissue

During long space missions, astronauts can lose approximately 1 to 2 percent of their bone mass per month. This loss of bone mass is thought to be caused by the lack of mechanical stress on astronauts' bones due to the low gravitational forces in space. Lack of mechanical stress causes bones to lose mineral salts and collagen fibers, and thus strength. Similarly, mechanical stress stimulates the deposition of mineral salts and collagen fibers. The internal and external structure of a bone will change as stress increases or decreases so that the bone is an ideal size and weight for the amount of activity it endures. That is why people who exercise regularly have thicker bones than people who are more sedentary. It is also why a broken bone in a cast atrophies while its contralateral mate maintains its concentration of mineral salts and collagen fibers. The bones undergo remodeling as a result of forces (or lack of forces) placed on them.

Numerous, controlled studies have demonstrated that people who exercise regularly have greater bone density than those who are more sedentary. Any type of exercise will stimulate the deposition of more bone tissue, but resistance training has a greater effect than cardiovascular activities. Resistance training is especially important to slow down the eventual bone loss due to aging and for preventing osteoporosis.

Nutrition and Bone Tissue

The vitamins and minerals contained in all of the food we consume are important for all of our organ systems. However, there are certain nutrients that affect bone health.

Calcium and Vitamin D

You already know that calcium is a critical component of bone, especially in the form of calcium phosphate and calcium carbonate. Since the body cannot make calcium, it must be obtained from the diet. However, calcium cannot be absorbed from the small intestine without vitamin D. Therefore, intake of vitamin D is also critical to bone health. In addition to vitamin D's role in calcium absorption, it also plays a role, though not as clearly understood, in bone remodeling.

Milk and other dairy foods are not the only sources of calcium. This important nutrient is also found in green leafy vegetables, broccoli, and intact salmon and canned sardines with their soft bones. Nuts, beans, seeds, and shellfish provide calcium in smaller quantities.

Except for fatty fish like salmon and tuna, or fortified milk or cereal, vitamin D is not found naturally in many foods. The action of sunlight on the skin triggers the body to produce its own vitamin D (Figure 1), but many people, especially those of darker complexion and those living in northern latitudes where the sun's rays are not as strong, are deficient in vitamin D. In cases of deficiency, a doctor can prescribe a vitamin D supplement.

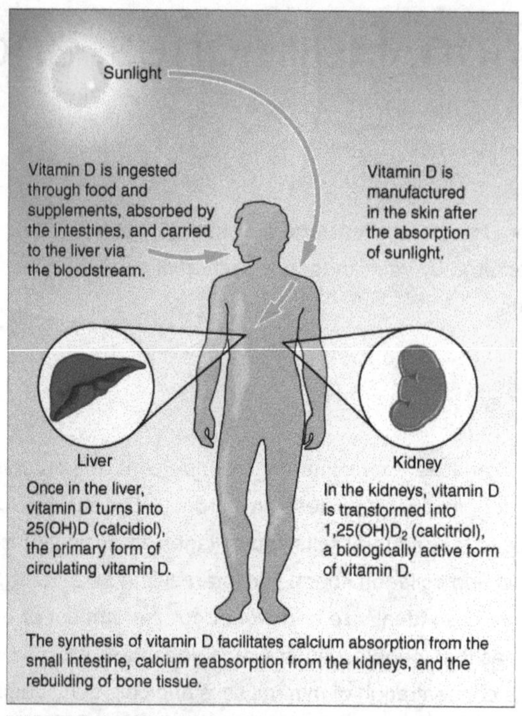

Figure 1. Synthesis of Vitamin D. Sunlight is one source of vitamin D.

Other Nutrients

Vitamin K also supports bone mineralization and may have a synergistic role with vitamin D in the regulation of bone growth. Green leafy vegetables are a good source of vitamin K.

The minerals magnesium and fluoride may also play a role in supporting bone health. While magnesium is only found in trace amounts in the human body, more than 60 percent of it is in the skeleton, suggesting it plays a role in the structure of bone. Fluoride can displace the hydroxyl group in bone's hydroxyapatite crystals and form fluorapatite. Similar to its effect on dental enamel, fluorapatite helps stabilize and strengthen bone mineral. Fluoride can also enter spaces within hydroxyapatite crystals, thus increasing their density.

Omega-3 fatty acids have long been known to reduce inflammation in various parts of the body. Inflammation can interfere with the function of osteoblasts, so consuming omega-3 fatty acids, in the diet or in supplements, may also help enhance production of new osseous tissue. Table 1 summarizes the role of nutrients in bone health.

Table 1. Nutrients and Bone Health

Nutrient	Role in bone health
Calcium	Needed to make calcium phosphate and calcium carbonate, which form the hydroxyapatite crystals that give bone its hardness
Vitamin D	Needed for calcium absorption
Vitamin K	Supports bone mineralization; may have synergistic effect with vitamin D
Magnesium	Structural component of bone
Fluoride	Structural component of bone
Omega-3 fatty acids	Reduces inflammation that may interfere with osteoblast function

RHEUMATOID ARTHRITIS (RA) AND OSTEOARTHRITIS

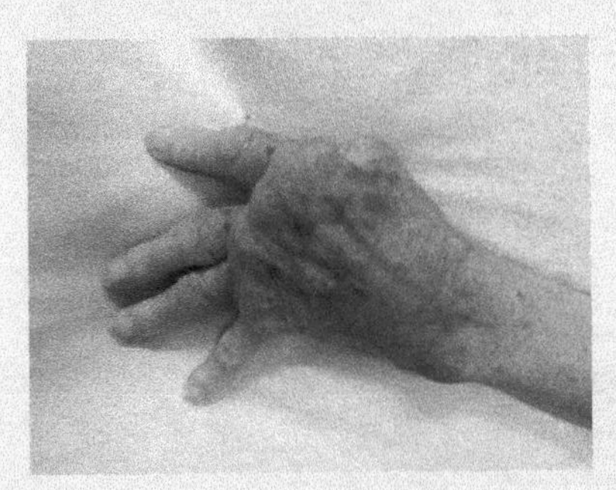

Rheumatoid Arthritis:

Rheumatoid arthritis (RA) is a chronic, systemic inflammatory disorder that may affect many tissues and organs, but principally attacks flexible (synovial) joints. The process involves an inflammatory response of the capsule around the joints (synovium) secondary to swelling (hyperplasia) of synovial cells, excess synovial fluid, and the development of fibrous tissue (pannus) in the synovium. The pathology of the disease process often leads to the destruction of articular cartilage and ankylosis (fusion) of the joints. Rheumatoid arthritis can also produce diffuse inflammation in the lungs, the membrane around the heart (pericardium), the membranes of the lung (pleura), and the white of the eye (sclera), and also nodular lesions, most common in subcutaneous tissue. Although the cause of rheumatoid arthritis is unknown, autoimmunity plays a pivotal role in both its chronicity and progression, and RA is considered a **systemic autoimmune disease.**

There are over 100 different forms of arthritis. The most common form, **osteoarthritis** (degenerative joint disease), is a result of trauma to the joint, infection of the joint, or age. The major complaint by individuals who have arthritis is joint pain. Pain is often a constant and may be localized to the joint affected. The pain from arthritis is due to inflammation that occurs around the joint, damage to the joint from disease, daily wear and tear of joint, muscle strains caused by forceful movements against stiff, painful joints, and fatigue

Arthritis is the most common cause of disability in the United States. More than 20 million individuals with arthritis have severe limitations in function on a daily basis. Absenteeism and frequent visits to the physician are common in individuals who have arthritis. Arthritis makes it very difficult for individuals to be physically active and many become home bound. Treatment options vary depending on the type of arthritis and include physical therapy, lifestyle changes (including exercise and weight control), orthopedic bracing, and medications. Joint replacement surgery may be required in eroding forms of arthritis. Medications can help reduce inflammation in the joint which decreases pain. Moreover, by decreasing inflammation, the joint damage may be slowed

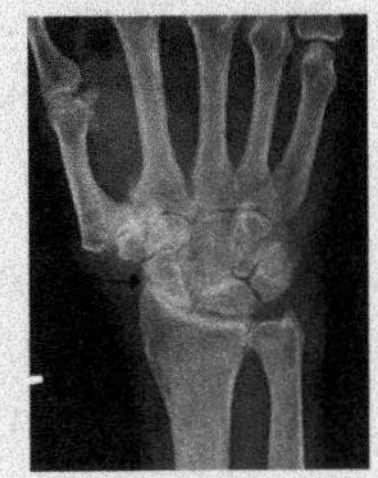

Osteoarthritis

Hormones and Bone Tissue

The endocrine system produces and secretes hormones, many of which interact with the skeletal system. These hormones are involved in controlling bone growth, maintaining bone once it is formed, and remodeling it.

Hormones That Influence Osteoblasts and/or Maintain the Matrix

Several hormones are necessary for controlling bone growth and maintaining the bone matrix. The pituitary gland secretes growth hormone (GH), which, as its name implies, controls bone growth in several ways. It triggers chondrocyte proliferation in epiphyseal plates, resulting in the increasing length of long bones. GH also increases calcium retention, which enhances mineralization, and stimulates osteoblastic activity, which improves bone density.

GH is not alone in stimulating bone growth and maintaining osseous tissue. Thyroxine, a hormone secreted by the thyroid gland promotes osteoblastic activity and the synthesis of bone matrix. During puberty, the sex hormones (estrogen in girls, testosterone in boys) also come into play. They too promote osteoblastic activity and production of bone matrix, and in addition, are responsible for the growth spurt that often occurs during adolescence. They also promote the conversion of the epiphyseal plate to the epiphyseal line (i.e., cartilage to its bony remnant), thus bringing an end to the longitudinal growth of bones. Additionally, calcitriol, the active form of vitamin D, is produced by the kidneys and stimulates the absorption of calcium and phosphate from the digestive tract.

AGING AND THE SKELETAL SYSTEM

Figure 2 shows that women lose bone mass more quickly than men starting at about 50 years of age. This occurs because 50 is the approximate age at which women go through menopause. Not only do their menstrual periods lessen and eventually cease, but their ovaries reduce in size and then cease the production of estrogen, a hormone that promotes osteoblastic activity and production of bone matrix. Thus, osteoporosis is more common in women than in men, but men can develop it, too. Anyone with a family history of osteoporosis has a greater risk of developing the disease, so the best treatment is prevention, which should start with a childhood diet that includes adequate intake of calcium and vitamin D and a lifestyle that includes weight-bearing exercise. These actions, as discussed above, are important in building bone mass. Promoting proper nutrition and weight-bearing exercise early in life can maximize bone mass before the age of 30, thus reducing the risk of osteoporosis.

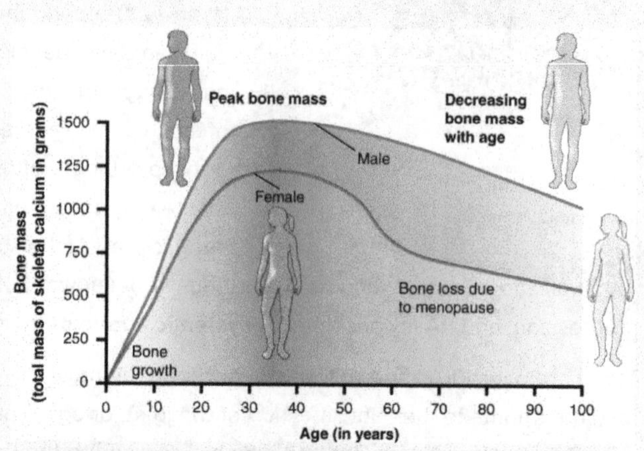

Figure 2. Graph Showing Relationship Between Age and Bone Mass. Bone density peaks at about 30 years of age. Women lose bone mass more rapidly than men.

For many elderly people, a hip fracture can be life threatening. The fracture itself may not be serious, but the immobility that comes during the healing process can lead to the formation of blood clots that can lodge in the capillaries of the lungs, resulting in respiratory failure; pneumonia due to the lack of poor air exchange that accompanies immobility; pressure sores (bed sores) that allow pathogens to enter the body and cause infections; and urinary tract infections from catheterization.

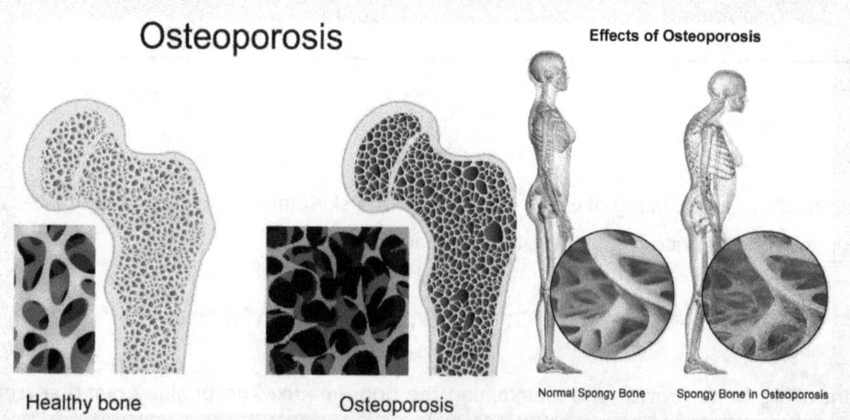

Osteoporosis is the weakening of bones in the body. It is caused by lack of calcium deposited in the bones. This lack of calcium causes the bones to become brittle. They break easily. Side effects include limping. Some symptoms late in the disease include pain in the bones, bones breaking very easily and lower back pain due to spinal bone fractures. It is more likely for a woman to get osteoporosis than a man. Elderly people are more likely to develop osteoporosis than younger people. The amount of calcium in the bones decreases as a person gets older.

Treatment of Osteoporosis

There is no cure for osteoporosis. Osteoporosis risks can be reduced with lifestyle changes and sometimes medication, or treatment may involve both. Lifestyle change includes **diet, exercise, and preventing falls**. Medication includes **calcium, vitamin D,** bisphosphonates, and several others.

Consuming too much alcohol and caffeine can cause an increase in the risk of weak bones. Too much salt increases the amount of calcium lost in urine. This is bad because calcium makes bones strong. People should not take iron and calcium supplements at the same time. This is because calcium competes with the absorption of minerals such as iron.

Hormones That Influence Osteoclasts

Bone modeling and remodeling require osteoclasts to resorb unneeded, damaged, or old bone, and osteoblasts to lay down new bone. Two hormones that affect the osteoclasts are parathyroid hormone (PTH) and calcitonin.

PTH stimulates osteoclast proliferation and activity. As a result, calcium is released from the bones into the circulation, thus increasing the calcium ion concentration in the blood. PTH also promotes the reabsorption of calcium by the kidney tubules, which can affect calcium homeostasis (see below).

The small intestine is also affected by PTH, albeit indirectly. Because another function of PTH is to stimulate the synthesis of vitamin D, and because vitamin D promotes intestinal absorption of calcium, PTH indirectly increases calcium uptake by the small intestine. Calcitonin, a hormone secreted by the thyroid gland, has some effects that counteract those of PTH. Calcitonin inhibits osteoclast activity and stimulates calcium uptake by the bones, thus reducing the concentration of calcium ions in the blood. As evidenced by their opposing functions in maintaining calcium homeostasis, PTH and calcitonin are generally *not* secreted at the same time. Table 2 summarizes the hormones that influence the skeletal system.

Table 2. Hormones That Affect the Skeletal System

Hormone	Role
Growth hormone	Increases length of long bones, enhances mineralization, and improves bone density
Thyroxine	Stimulates bone growth and promotes synthesis of bone matrix
Sex hormones	Promote osteoblastic activity and production of bone matrix; responsible for adolescent growth spurt; promote conversion of epiphyseal plate to epiphyseal line
Calcitriol	Stimulates absorption of calcium and phosphate from digestive tract
Parathyroid hormone	Stimulates osteoclast proliferation and resorption of bone by osteoclasts; promotes reabsorption of calcium by kidney tubules; indirectly increases calcium absorption by small intestine
Calcitonin	Inhibits osteoclast activity and stimulates calcium uptake by bones

Calcium Homeostasis: Interactions of the Skeletal System and Other Organ Systems

Calcium is not only the most abundant mineral in bone, it is also the most abundant mineral in the human body. Calcium ions are needed not only for bone mineralization but for tooth health, regulation of the heart rate and strength of contraction, blood coagulation, contraction of smooth and skeletal muscle cells, and regulation of nerve impulse conduction. The normal level of calcium in the blood is about 10 mg/dL. When the body cannot maintain this level, a person will experience hypo- or hypercalcemia.

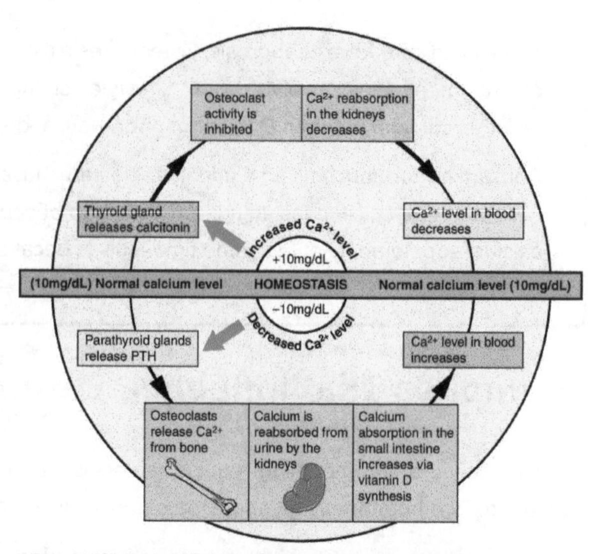

Figure 3. Pathways in Calcium Homeostasis. The body regulates calcium homeostasis with two pathways; one is signaled to turn on when blood calcium levels drop below normal and one is the pathway that is signaled to turn on when blood calcium levels are elevated.

Hypocalcemia, a condition characterized by abnormally low levels of calcium, can have an adverse effect on a number of different body systems including circulation, muscles, nerves, and bone. Without adequate calcium, blood has difficulty coagulating, the heart may skip beats or stop beating altogether, muscles may have difficulty contracting, nerves may have difficulty functioning, and bones may become brittle. The causes of hypocalcemia can range from hormonal imbalances to an improper diet. Treatments vary according to the cause, but prognoses are generally good.

Conversely, in **hypercalcemia**, a condition characterized by abnormally high levels of calcium, the nervous system is underactive, which results in lethargy, sluggish reflexes, constipation and loss of appetite, confusion, and in severe cases, coma.

Obviously, calcium homeostasis is critical. The skeletal, endocrine, and digestive systems play a role in this, but the kidneys do, too. These body systems work together to maintain a normal calcium level in the blood

Calcium is a chemical element that cannot be produced by any biological processes. The only way it can enter the body is through the diet. The bones act as a storage site for calcium: The body deposits calcium in the bones when blood levels get too high, and it releases calcium when blood levels drop too low. This process is regulated by PTH, vitamin D, and calcitonin.

Cells of the parathyroid gland have plasma membrane receptors for calcium. When calcium is not binding to these receptors, the cells release PTH, which stimulates osteoclast proliferation and resorption of bone by osteoclasts. This demineralization process releases calcium into the blood. PTH promotes reabsorption of calcium from the urine by the kidneys, so that the calcium returns to the blood. Finally, PTH stimulates the synthesis of vitamin D, which in turn, stimulates calcium absorption from any digested food in the small intestine.

When all these processes return blood calcium levels to normal, there is enough calcium to bind with the receptors on the surface of the cells of the parathyroid glands, and this cycle of events is turned off

When blood levels of calcium get too high, the thyroid gland is stimulated to release calcitonin, which inhibits osteoclast activity and stimulates calcium uptake by the bones, but also decreases reabsorption of calcium by the kidneys. All of these actions lower blood levels of calcium. When blood calcium levels return to normal, the thyroid gland stops secreting calcitonin.

Concept Check

1. **What** are the primary functions of the skeletal system?
2. **Which** part of the typical long bone includes the red bone marrow and the yellow bone marrow?
3. **What** are the major differences between spongy bone and compact bone?
4. **Define** the two divisions of the skeleton.
5. **Discuss** the functions of the axial skeleton.
6. **Define t**he parts and functions of the thoracic cage.
7. **Discuss** the different types of joints and the movement provided by each.
8. **What** are the major differences between the male pelvis and female pelvis that permit childbirth in females?
9. **What** are the major differences between the pelvic girdle and the pectoral girdle that allow the pelvic girdle to bear the weight of the body?
10. **Define** how joints are classified based on function. **Describe** and give an example for each functional type of joint. Briefly **define** the types of joint movements available at a ball-and-socket joint.
11. **What** are the roles of osteoblasts, osteocytes, and osteoclasts?
12. **How** are bone remodeling and bone repair related?
13. If the articular cartilage at the end of one of your long bones were to degenerate, what symptoms do you think you would experience? **Why?**
14. **Watch this** video to view a rotating and exploded skull with color-coded bones. Which bone (yellow) is centrally located and joins with most of the other bones of the skull?
15. Osteoporosis is a common age-related bone disease in which bone density and strength is decreased. **Watch this** video to get a better understanding of how thoracic vertebrae may become weakened and may fractured due to this disease. How may vertebral osteoporosis contribute to kyphosis?

CHAPTER 8 — MUSCULAR SYSTEM

Introduction to Muscle Tissue

LEARNING OBJECTIVES

- Describe the properties of muscle tissue.
- Compare and contrast the three types of muscle tissue.
- Describe the significant events of a skeletal muscle contraction within a muscle in generating force
- Describe the Nervous System Control of muscle tension
- Describe the types of skeletal muscle fibers
- Identify the major muscles of the human body and their actions.
- Compare and contrast agonist and antagonist muscles
- Explain the structure and function of smooth muscle and cardiac muscle tissue.

Muscle is one of the four primary tissue types of the body. Muscle cells are specialized for contraction. Muscles allow for motions such as walking, and they also facilitate bodily processes such as respiration and digestion.

Properties of muscle tissue

All muscle cells share several properties: contractility, excitability, extensibility, and elasticity:

1. ***Contractility*** is the ability of muscle cells to forcefully shorten. Contractility allows muscle tissue to pull on its attachment points and shorten with force.*(muscles can only pull, never push.)*
2. ***Excitability*** is the ability to respond to a stimulus, which may be delivered from a motor neuron or a hormone.
3. ***Extensibility*** is the ability of a muscle to be stretched or extended.
4. ***Elasticity*** is the ability to a muscle to return to its original length when relaxed

Three types of muscle tissue:

Muscle cells are specialized for contraction. Muscles allow for motions such as walking, and they also facilitate bodily processes such as respiration and digestion. The body contains three types of muscle tissue: **(a) skeletal muscle tissue (b) smooth muscle tissue (c) cardiac muscle tissue.**

- **Skeletal muscle tissue** forms skeletal muscles, which attach to bones or skin and control locomotion and any movement that can be consciously controlled. Because it can be controlled by thought, skeletal muscle is also called voluntary muscle. Skeletal muscles are long and cylindrical in appearance; when viewed under a microscope, skeletal muscle tissue has a striped or striated appearance. The striations are caused by the regular arrangement of contractile proteins (actin and myosin). Actin is a globular contractile protein that interacts with myosin for muscle contraction. Skeletal muscle also has multiple nuclei present in a single cell
- **Smooth muscle tissue** occurs in the walls of hollow organs such as the intestines, stomach, and urinary bladder, and around passages such as the respiratory tract and blood vessels. Smooth muscle has no striations, is not under voluntary control, has only one nucleus per cell, is tapered at both ends, and is called involuntary muscle.
- **Cardiac muscle tissue** is only found in the heart, and cardiac contractions pump blood throughout the body and maintain

blood pressure. Like skeletal muscle, cardiac muscle is striated, but unlike skeletal muscle, cardiac muscle cannot be consciously controlled and is called involuntary muscle. It has one nucleus per cell, is branched, and is distinguished by the presence of intercalated disk.

Table 1: Comparison of Structure and Properties of Muscle Tissue Types

Tissue	Histology	Function	Location
Skeletal	Long cylindrical fiber, striated, many peripherally located nuclei	Voluntary movement, produces heat, protects organs	Attached to bones and around entrance points to body (e.g., mouth, anus)
Cardiac	Short, branched, striated, single central nucleus	Contracts to pump blood	Heart
Smooth	Short, spindle-shaped, no evident striation, single nucleus in each fiber	Involuntary movement, moves food, involuntary control of respiration, moves secretions, regulates flow of blood in arteries by contraction	Walls of major organs and passageways

Skeletal Muscle: Structure and Contraction

Skeletal muscle tissue forms skeletal muscles. The best-known feature of skeletal muscle is its **ability to contract and cause movement.** Skeletal muscles act not only to produce movement but also to stop movement, such as resisting gravity to maintain posture. Small, constant adjustments of the skeletal muscles are needed to hold a body upright or balanced in any position. Muscles also prevent excess movement of the bones and joints, maintaining skeletal stability and preventing skeletal structure damage or deformation. Joints can become misaligned or dislocated entirely by pulling on the associated bones; muscles work to keep joints stable. Skeletal muscles are located throughout the body at the openings of internal tracts to control the movement of various substances. These muscles allow functions, such as swallowing, urination, and defecation, to be under voluntary control. Skeletal muscles also protect internal organs (particularly abdominal and pelvic organs) by acting as an external barrier or shield to external trauma and by supporting the weight of the organs.

Skeletal muscles contribute to the **maintenance of homeostasis in the body by generating heat.** Muscle contraction requires energy, and when ATP is broken down, heat is produced. This heat is very noticeable during exercise, when sustained muscle movement causes body temperature to rise, and in cases of extreme cold, when shivering produces random skeletal muscle contractions to generate heat.

Skeletal Muscle Structure

Each skeletal muscle is an organ that consists of various integrated tissues. These tissues include the skeletal muscle fibers, blood vessels, nerve fibers, and connective tissue. Each skeletal muscle has three layers of connective tissue (called "mysia") that enclose it and provide structure to the muscle as a whole, and also compartmentalize the muscle fibers within the muscle (Figure 1). Each muscle is wrapped in a sheath of dense, irregular connective tissue called the **epimysium**, which allows a muscle to contract and move powerfully while maintaining its structural integrity. The epimysium also separates muscle from other tissues and organs in the area, allowing the muscle to move independently.

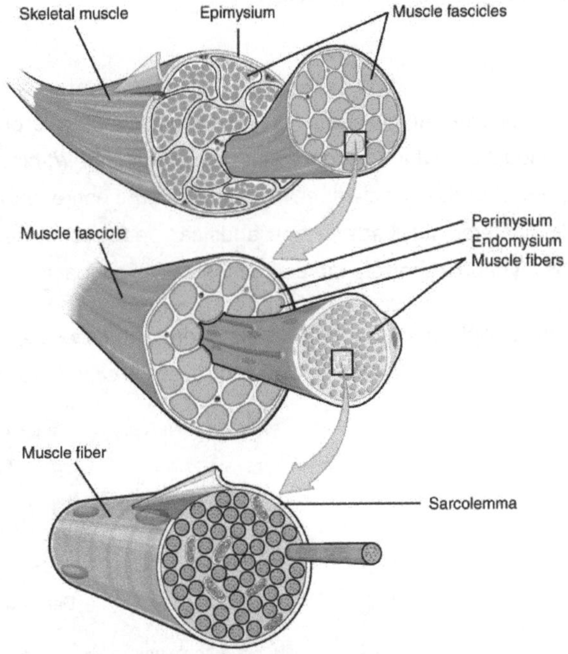

Inside each skeletal muscle, muscle fibers are organized into individual bundles, each called a **fascicle**, by a middle layer of connective tissue called the **perimysium**. This fascicular organization is common in muscles of the limbs; it allows the nervous system to trigger a specific movement of a muscle by activating a subset of muscle fibers within a bundle, or fascicle of the muscle. Inside each fascicle, each muscle fiber is encased in a thin connective tissue layer of collagen and reticular fibers called the **endomysium**. The endomysium contains the extracellular fluid and nutrients to support the muscle fiber. These nutrients are supplied via blood to the muscle tissue.

Figure 1. The Three Connective Tissue Layers. Bundles of muscle fibers, called fascicles, are covered by the perimysium. Muscle fibers are covered by the endomysium.

In skeletal muscles that work with tendons to pull on bones, the collagen in the three tissue layers (the mysia) intertwines with the collagen of a tendon. At the other end of the tendon, it fuses with the periosteum coating the bone. The tension created by contraction of the muscle fibers is then transferred though the mysia, to the tendon, and then to the periosteum to pull on the bone for movement of the skeleton. In other places, the mysia may fuse with a broad, tendon-like sheet called an **aponeurosis**, or to fascia, the connective tissue between skin and bones. The broad sheet of connective tissue in the lower back that the latissimus dorsi muscles (the "lats") fuse into is an example of an aponeurosis.

Every skeletal muscle is also richly supplied by blood vessels for nourishment, oxygen delivery, and waste removal. In addition, every muscle fiber in a skeletal muscle is supplied by the axon branch of a somatic motor neuron, which signals the fiber to contract. Unlike cardiac and smooth muscle, the only way to functionally contract a skeletal muscle is through signaling from the nervous system.

Skeletal Muscle Fiber (Cell) Structure

Because skeletal muscle cells are long and cylindrical, they are commonly referred to as muscle fibers. Skeletal muscle fibers can be quite large for human cells, with diameters up to 100 μm and lengths up to 30 cm (11.8 in) in the Sartorius of the upper leg. During early development, embryonic myoblasts, each with its own nucleus, fuse with up to hundreds of other myoblasts to form the multinucleated skeletal muscle fibers. Multiple nuclei mean multiple copies of genes, permitting the production of the large amounts of proteins and enzymes needed for muscle contraction.

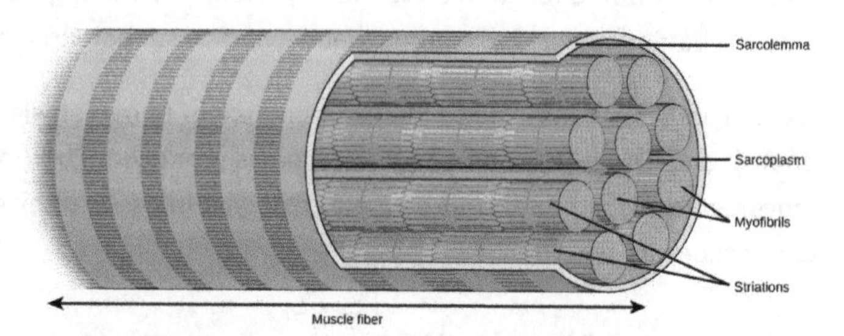

Figure 2. A skeletal muscle cell is surrounded by a plasma membrane called the sarcolemma with a cytoplasm called the sarcoplasm. A muscle fiber is composed of many fibrils, packaged into orderly units.

The plasma membrane of a skeletal muscle fiber is called the **sarcolemma.** The sarcolemma is the site of action potential conduction, which triggers muscle contraction. Within each muscle fiber are **myofibrils**—long cylindrical structures that lie parallel to the muscle fiber. Myofibrils run the entire length of the muscle fiber, and because they are only approximately 1.2 μm in diameter, hundreds to thousands can be found inside one muscle fiber. They attach to the sarcolemma at their ends, so that as myofibrils shorten, the entire muscle cell contracts.

The cytoplasm of muscle fibers is referred to as **sarcoplasm**, and the specialized smooth endoplasmic reticulum, which stores, releases, and retrieves calcium ions (Ca^{++}) is called the **sarcoplasmic reticulum (SR)** (Figure 3).

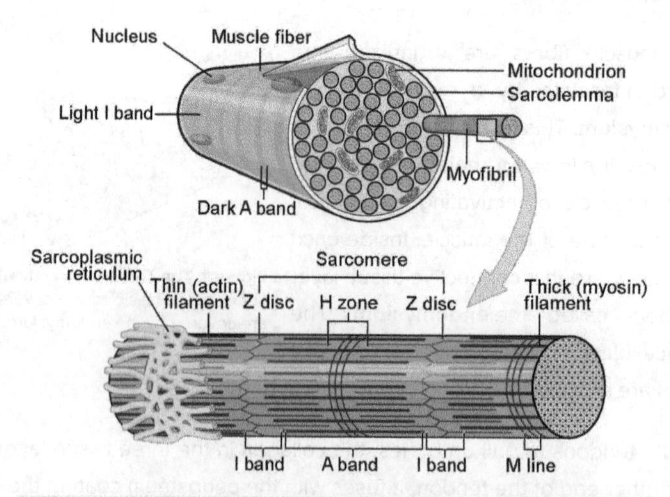

Figure 3. Muscle Fiber. A skeletal muscle fiber is surrounded by a plasma membrane called the sarcolemma, which contains sarcoplasm, the cytoplasm of muscle cells. A muscle fiber is composed of many fibrils, which give the cell its striated appearance.

The Sarcomere

The functional unit of a skeletal muscle fiber is the **sarcomere**, a highly organized arrangement of the contractile myofilaments **actin** (thin filament) and **myosin** (thick filament), along with other support proteins. The striated appearance of skeletal muscle fibers is due to the arrangement of the myofilaments of actin and myosin in sequential order from one end of the muscle fiber to the other.

Microscopically Visible Feature	Composition
A band	Region of thick-filament myosin proteins
H zone	Central region of the A band with no overlapping actin proteins when muscle is relaxed
I band	Region of thin-filament actin proteins, with no myosin
M line	M line accessory proteins in center of myosin thick filament perpendicular to the sarcomere
Z disc	Zig-zag line of Z line proteins and actin binding proteins perpendicular to the sarcomere

The sarcomere itself is bundled within the myofibril that runs the entire length of the muscle fiber and attaches to the sarcolemma at its end. As myofibrils contract, the entire muscle cell contracts. Because myofibrils are only approximately 1.2 μm in diameter, hundreds to thousands (each with thousands of sarcomeres) can be found inside one muscle fiber. Each sarcomere is approximately 2 μm in length with a three-dimensional cylinder-like arrangement and is bordered by structures called Z-discs (also called Z-lines, because pictures are two-dimensional), to which the actin myofilaments are anchored (Figure 4). Because the actin and its troponin-tropomyosin complex (projecting from the Z-discs toward the center of the sarcomere) form strands that are thinner than the myosin, it is called the thin filament of the sarcomere. Likewise, because the myosin strands and their multiple heads (projecting from the center of the sarcomere, toward but not all to way to, the Z-discs) have more mass and are thicker, they are called the thick filament of the sarcomere.

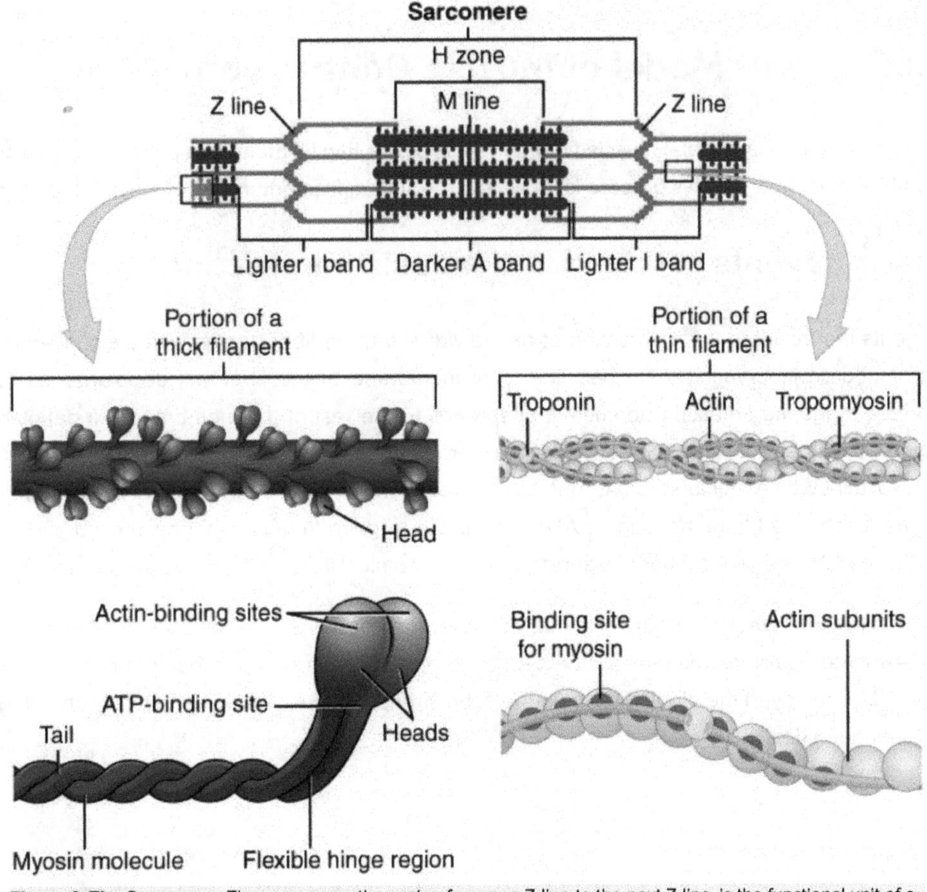

Figure 4. The Sarcomere. The sarcomere, the region from one Z-line to the next Z-line, is the functional unit of a skeletal muscle fiber.

The Neuromuscular Junction

Another specialization of the skeletal muscle is the site where a motor neuron's terminal meets the muscle fiber—called the **neuromuscular junction (NMJ)**. This is where the muscle fiber first responds to signaling by the motor neuron. Every skeletal muscle fiber in every skeletal muscle is innervated by a motor neuron at the NMJ. Excitation signals from the neuron are the only way to functionally activate the fiber to contract.

Signaling begins when a neuronal **action potential** travels along the axon of a motor neuron, and then along the individual branches to terminate at the NMJ. At the NMJ, the axon terminal releases a chemical messenger, or **neurotransmitter, called acetylcholine (ACh).** The ACh molecules diffuse across a minute space called the **synaptic cleft** and bind to ACh receptors located within the **motor end-plate** of the sarcolemma on the other side of the synapse. Once ACh binds, a channel in the ACh receptor opens and positively charged ions can pass through into the muscle fiber, causing it to **depolarize**, meaning that the membrane potential of the muscle fiber becomes less negative (closer to zero.) As the membrane depolarizes, another set of ion channels called **voltage-gated sodium channels** are triggered to open. Sodium ions enter the muscle fiber, and an action potential rapidly spreads (or "fires") along the entire membrane to initiate the muscle to contract.

Things happen very quickly in the world of excitable membranes (just think about how quickly you can snap your fingers as soon as you decide to do it). Immediately following depolarization of the membrane, it repolarizes, re-establishing the negative membrane potential. Meanwhile, the ACh in the synaptic cleft is degraded by the enzyme **acetylcholinesterase (AChE)** so that the ACh cannot rebind to a receptor and reopen its channel, which would cause unwanted extended muscle excitation and contraction.

The Sliding Filament Model of Muscle Contraction

When signaled by a motor neuron, a skeletal muscle fiber contracts as the thin filaments are pulled and then slide past the thick filaments within the fiber's sarcomeres. This process is known as the sliding filament model of muscle contraction (Figure 4).

The sequence of events involved in a muscle contraction:

The sequence of events that result in the contraction of an individual muscle fiber begins with a signal—the neurotransmitter, ACh—from the motor neuron innervating that fiber. The local membrane of the fiber will depolarize as positively charged sodium ions (Na^+) enter, triggering an action potential that spreads to the rest of the membrane will depolarize, including the T-tubules. This triggers the release of calcium ions (Ca^{++}) from storage in the sarcoplasmic reticulum (SR). The Ca^{++} then initiates contraction, which is sustained by ATP (Figure 5). As long as Ca^{++} ions remain in the sarcoplasm to bind to troponin, which keeps the actin-binding sites "unshielded," and as long as ATP is available to drive the cross-bridge cycling and the pulling of actin strands by myosin, the muscle fiber will continue to shorten to an anatomical limit.

Muscle contraction usually stops when signaling from the motor neuron ends, which repolarizes the sarcolemma and T-tubules, and closes the voltage-gated calcium channels in the SR. Ca^{++} ions are then pumped back into the SR, which causes the tropomyosin to reshield (or re-cover) the binding sites on the actin strands. A muscle also can stop contracting when it runs out of ATP and becomes fatigued (Figure 6)

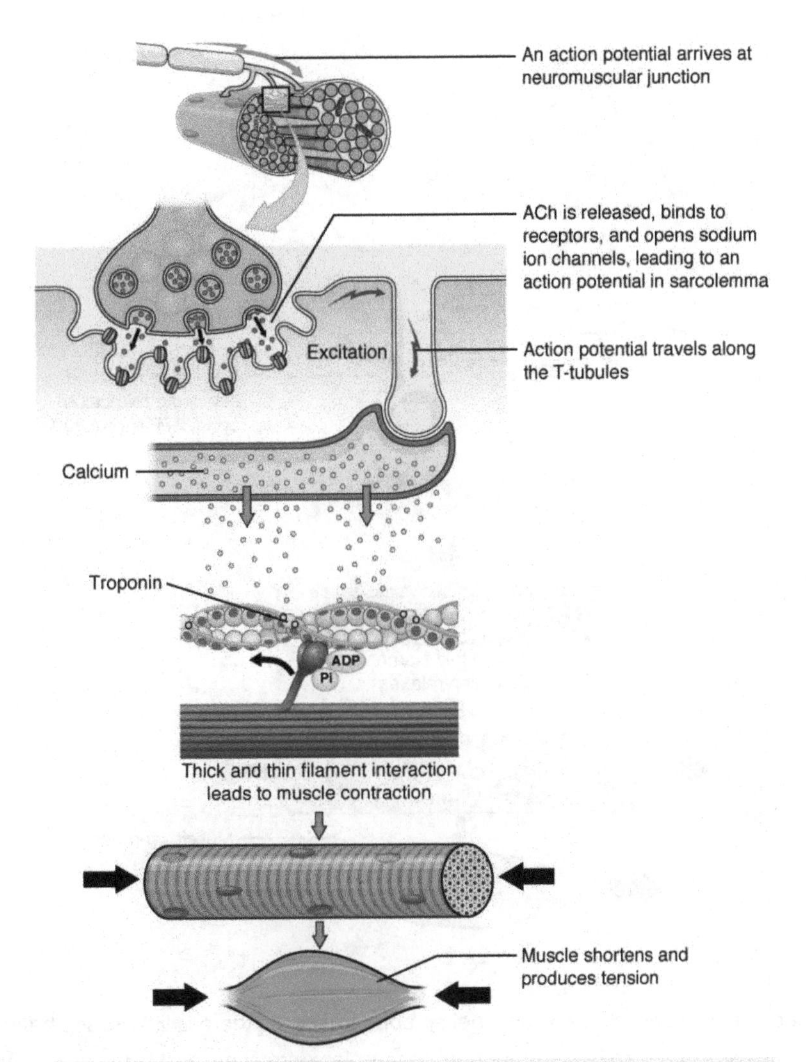

Figure 5. Contraction of a Muscle Fiber.

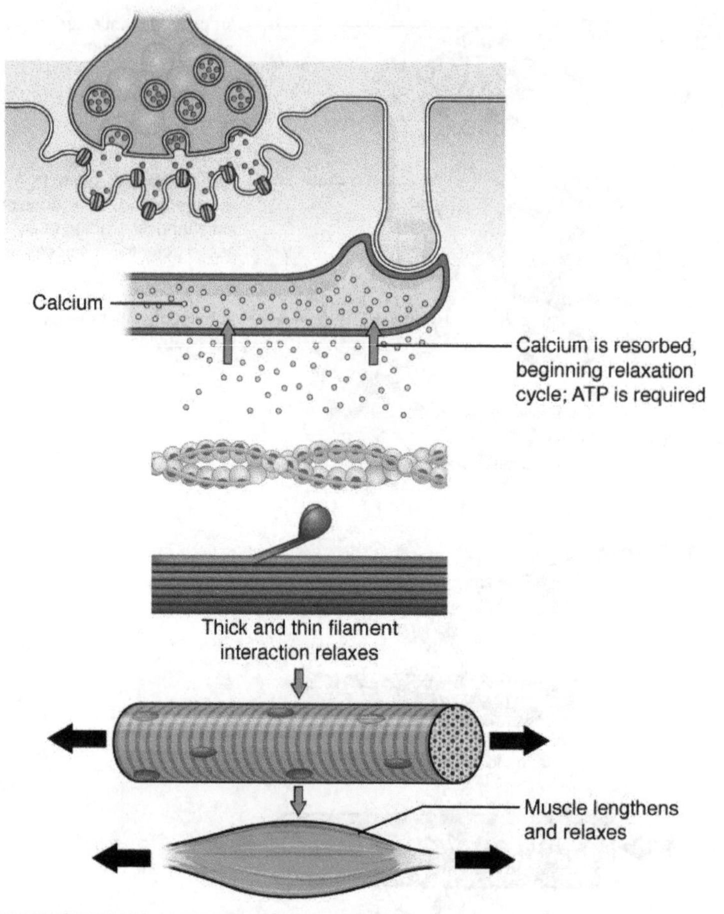

Figure 6. Relaxation of a Muscle Fiber.

The region where thick and thin filaments overlap has a dense appearance, as there is little space between the filaments. This zone where thin and thick filaments overlap is very important to muscle contraction, as it is the site where filament movement starts. Thin filaments, anchored at their ends by the Z-discs, do not extend completely into the central region that only contains thick filaments, anchored at their bases at a spot called the M-line. A myofibril is composed of many sarcomeres running along its length; thus, myofibrils and muscle cells contract as the sarcomeres contract.

The Sliding Filament Model of Contraction

When signaled by a motor neuron, a skeletal muscle fiber contracts as the thin filaments are pulled and then slide past the thick filaments within the fiber's sarcomeres. This process is known as the **sliding filament model of muscle contraction** (Figure 7). The sliding can only occur when myosin-binding sites on the actin filaments are exposed by a series of steps that begins with **Ca^{++}** entry into the sarcoplasm.

Tropomyosin is a protein that winds around the chains of the actin filament and covers the myosin-binding sites to prevent actin from binding to myosin. Tropomyosin binds to troponin to form a troponin-tropomyosin complex. The troponin-tropomyosin complex prevents the myosin "heads" from binding to the active sites on the actin microfilaments. Troponin also has a binding site for Ca^{++} ions.

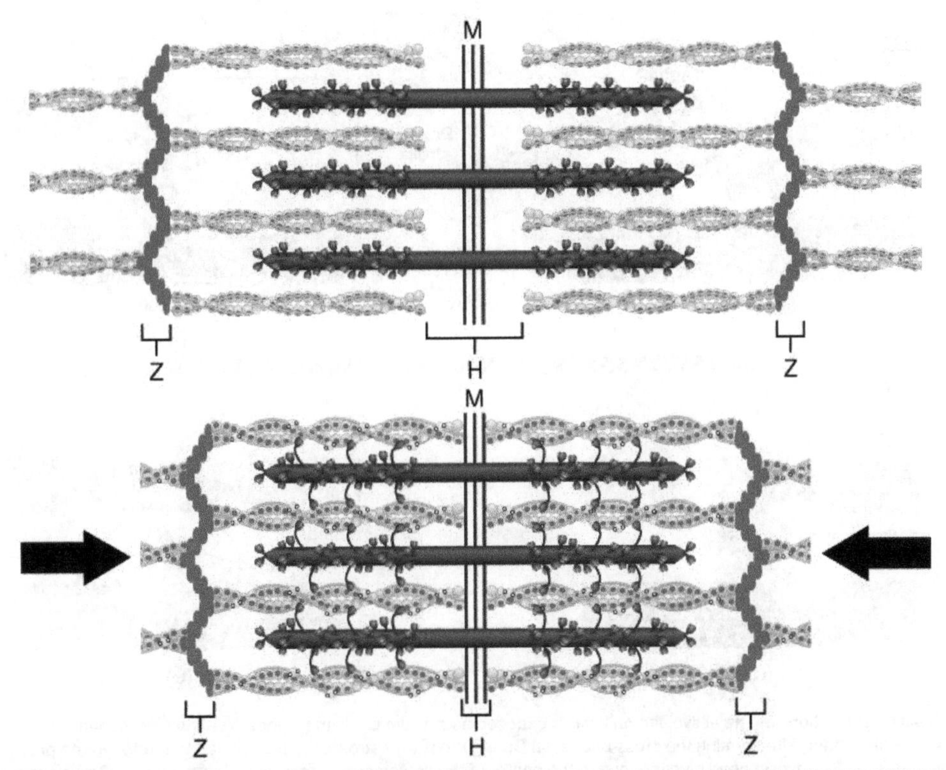

Figure 7. The Sliding Filament Model of Muscle Contraction. When a sarcomere contracts, the Z lines move closer together, and the I band becomes smaller. The A band stays the same width. At full contraction, the thin and thick filaments overlap.

To initiate muscle contraction, tropomyosin has to expose the myosin-binding site on an actin filament to allow cross-bridge formation between the actin and myosin microfilaments. The first step in the process of contraction is for Ca^{++} to bind to troponin so that tropomyosin can slide away from the binding sites on the actin strands. This allows the myosin heads to bind to these exposed binding sites and form cross-bridges. The thin filaments are then pulled by the myosin heads to slide past the thick filaments toward the center of the sarcomere. But each head can only pull a very short distance before it has reached its limit and must be "re-cocked" before it can pull again, a step that requires ATP.

ATP and Muscle Contraction

For thin filaments to continue to slide past thick filaments during muscle contraction, myosin heads must pull the actin at the binding sites, detach, re-cock, attach to more binding sites, pull, detach, re-cock, etc. This repeated movement is known as the cross-bridge cycle. This motion of the myosin heads is similar to the oars when an individual rows a boat: The paddle of the oars (the myosin heads) pull, are lifted from the water (detach), repositioned (re-cocked) and then immersed again to pull (Figure 8). Each cycle requires energy, and the action of the myosin heads in the sarcomeres repetitively pulling on the thin filaments also requires energy, which is provided by ATP.

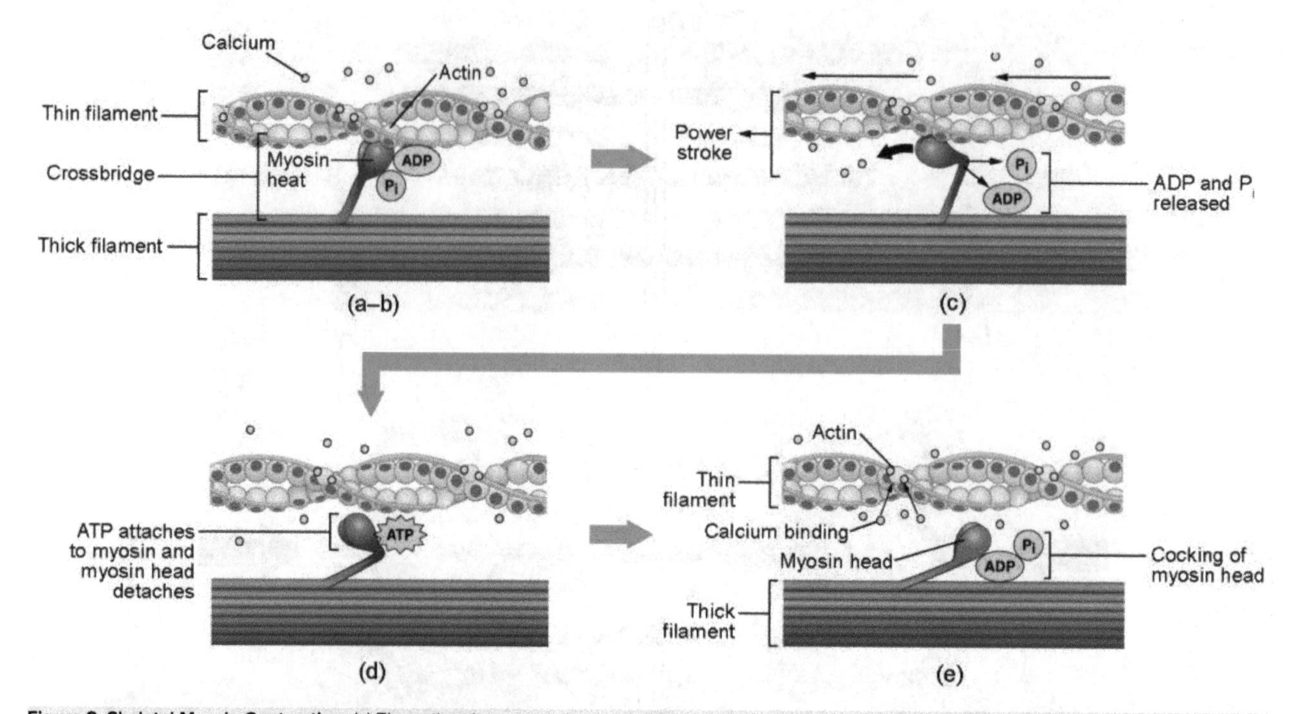

Figure 8. Skeletal Muscle Contraction. (a) The active site on actin is exposed as calcium binds to troponin. (b) The myosin head is attracted to actin, and myosin binds actin at its actin-binding site, forming the cross-bridge. (c) During the power stroke, the phosphate generated in the previous contraction cycle is released. This results in the myosin head pivoting toward the center of the sarcomere, after which the attached ADP and phosphate group are released. (d) A new molecule of ATP attaches to the myosin head, causing the cross-bridge to detach. (e) The myosin head hydrolyzes ATP to ADP and phosphate, which returns the myosin to the cocked position.

Cross-bridge formation occurs when the myosin head attaches to the actin while adenosine diphosphate (ADP) and inorganic phosphate (P_i) are still bound to myosin (Figure 8a,b). P_i is then released, causing myosin to form a stronger attachment to the actin, after which the myosin head moves toward the M-line, pulling the actin along with it. As actin is pulled, the filaments move approximately 10 nm toward the M-line. This movement is called the **power stroke**, as movement of the thin filament occurs at this step (Figure 8c). In the absence of ATP, the myosin head will not detach from actin.

One part of the myosin head attaches to the binding site on the actin, but the head has another binding site for ATP. ATP binding causes the myosin head to detach from the actin (Figure 8d). After this occurs, ATP is converted to ADP and P_i by the intrinsic **ATPase** activity of myosin. The energy released during ATP hydrolysis changes the angle of the myosin head into a cocked position (Figure 8e). The myosin head is now in position for further movement.

When the myosin head is cocked, myosin is in a high-energy configuration. This energy is expended as the myosin head moves through the power stroke, and at the end of the power stroke, the myosin head is in a low-energy position. After the power stroke, ADP is released; however, the formed cross-bridge is still in place, and actin and myosin are bound together. As long as ATP is available, it readily attaches to myosin, the cross-bridge cycle can recur, and muscle contraction can continue.

Note that each thick filament of roughly 300 myosin molecules has multiple myosin heads, and many cross-bridges form and break continuously during muscle contraction. Multiply this by all of the sarcomeres in one myofibril, all the myofibrils in one muscle fiber, and all of the muscle fibers in one skeletal muscle, and you can understand why so much energy (ATP) is needed to keep skeletal muscles working. In fact, it is the loss of ATP that results in the rigor mortis observed soon after someone dies. With no further ATP production possible, there is no ATP available for myosin heads to detach from the actin-binding sites, so the cross-bridges stay in place, causing the rigidity in the skeletal muscles.

Sources of ATP

ATP supplies the energy for muscle contraction to take place. In addition to its direct role in the cross-bridge cycle, ATP also provides the energy for the active-transport Ca^{++} pumps in the SR. Muscle contraction does not occur without sufficient amounts of ATP. The amount of ATP stored in muscle is very low, only sufficient to power a few seconds worth of contractions. As it is broken down, ATP must therefore be regenerated and replaced quickly to allow for sustained contraction. There are three mechanisms by which ATP can be regenerated: creatine phosphate metabolism, anaerobic glycolysis, fermentation and aerobic respiration.

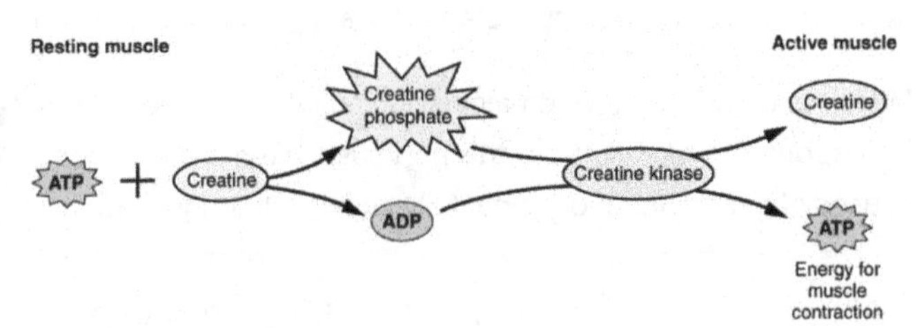

Figure 9. Muscle Metabolism. Some ATP is stored in a resting muscle. As contraction starts, it is used up in seconds. More ATP is generated from creatine phosphate for about 15 seconds.

Creatine phosphate is a molecule that can store energy in its phosphate bonds. In a resting muscle, excess ATP transfers its energy to creatine, producing ADP and creatine phosphate. This acts as an energy reserve that can be used to quickly create more ATP. When the muscle starts to contract and needs energy, creatine phosphate transfers its phosphate back to ADP to form ATP and creatine. This reaction is catalyzed by the enzyme creatine kinase and occurs very quickly; thus, creatine phosphate-derived ATP powers the first few seconds of muscle contraction. However, creatine phosphate can only provide approximately 15 seconds worth of energy, at which point another energy source has to be used (Figure 9).

As the ATP produced by creatine phosphate is depleted, muscles turn to glycolysis as an ATP source. **Glycolysis** is an anaerobic (non-oxygen-dependent) process that breaks down glucose (sugar) to produce ATP; however, glycolysis cannot generate ATP as quickly as creatine phosphate. Thus, the switch to glycolysis results in a slower rate of ATP availability to the muscle. The sugar used in glycolysis can be provided by blood glucose or by metabolizing glycogen that is stored in the muscle. The breakdown of one glucose molecule produces two ATP and two molecules of **pyruvic acid**, which can be used in aerobic respiration or when oxygen levels are low, converted to lactic acid (Figure 10).

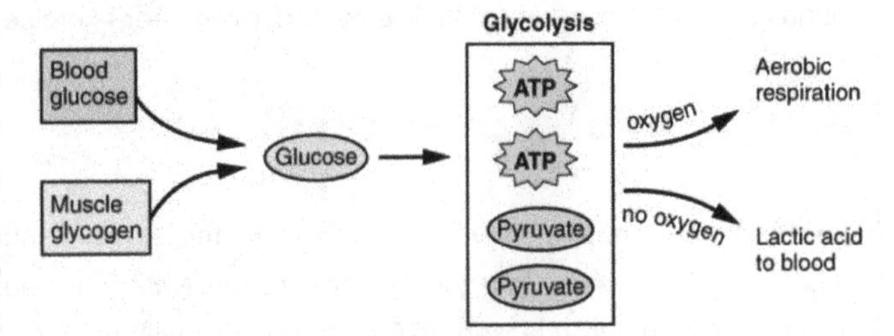

Figure 10. Glycolysis and Aerobic Respiration. Each glucose molecule produces two ATP and two molecules of pyruvic acid, which can be used in aerobic respiration or converted to lactic acid. If oxygen is not available, pyruvic acid is converted to lactic acid, which may contribute to muscle fatigue. This occurs during strenuous exercise when high amounts of energy are needed but oxygen cannot be sufficiently delivered to muscle.

If oxygen is available, pyruvic acid is used in aerobic respiration. However, if oxygen is not available, pyruvic acid is converted to **lactic acid**, which may contribute to muscle fatigue. This conversion allows the recycling of the enzyme NAD^+ from NADH, which is needed for glycolysis to continue. This occurs during strenuous exercise when high amounts of energy are needed but oxygen cannot be sufficiently delivered to muscle. Glycolysis itself cannot be sustained for very long (approximately 1 minute of muscle activity), but it is useful in facilitating short bursts of high-intensity output. This is because glycolysis does not utilize glucose very efficiently, producing a net gain of two ATPs per molecule of glucose, and the end product of lactic acid, which may contribute to muscle fatigue as it accumulates.

Aerobic respiration is the breakdown of glucose or other nutrients in the presence of oxygen (O_2) to produce carbon dioxide, water, and ATP. Approximately 95 percent of the ATP required for resting or moderately active muscles is provided by aerobic respiration, which takes place in mitochondria. The inputs for aerobic respiration include glucose circulating in the bloodstream, pyruvic acid, and fatty acids. Aerobic respiration is much more efficient than anaerobic glycolysis, producing approximately 36 ATPs per molecule of glucose versus four from glycolysis. However, aerobic respiration cannot be sustained without a steady supply of O_2 to the skeletal muscle and is much slower (Figure 11). To compensate, muscles store small amount of excess oxygen in proteins call myoglobin, allowing for more efficient muscle contractions and less fatigue. Aerobic training also increases the efficiency of the circulatory system so that O_2 can be supplied to the muscles for longer periods of time.

Muscle fatigue occurs when a muscle can no longer contract in response to signals from the nervous system. The exact causes of muscle fatigue are not fully known, although certain factors have been correlated with the decreased muscle contraction that occurs during fatigue. ATP is needed for normal muscle contraction, and as ATP reserves are reduced, muscle function may decline. This may be more of a factor in brief, intense muscle output rather than sustained, lower intensity efforts. Lactic acid buildup may lower intracellular pH, affecting enzyme and protein activity. Imbalances in Na^+ and K^+ levels as a result of membrane depolarization may disrupt Ca^{++} flow out of the SR. Long periods of sustained exercise may damage the SR and the sarcolemma, resulting in impaired Ca^{++} regulation.

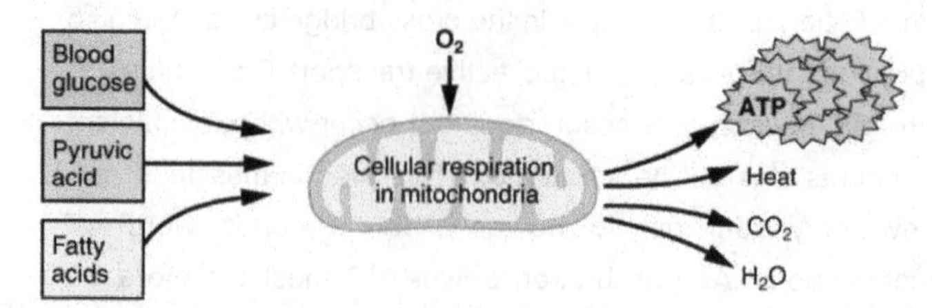

Figure 11. Cellular Respiration. Aerobic respiration is the breakdown of glucose in the presence of oxygen (O_2) to produce carbon dioxide, water, and ATP. Approximately 95 percent of the ATP required for resting or moderately active muscles is provided by aerobic respiration, which takes place in mitochondria.

Intense muscle activity results in an **oxygen debt**, which is the amount of oxygen needed to compensate for ATP produced without oxygen during muscle contraction. Oxygen is required to restore ATP and creatine phosphate levels, convert lactic acid to pyruvic acid, and, in the liver, to convert lactic acid into glucose or glycogen. Other systems used during exercise also require oxygen, and all of these combined processes result in the increased breathing rate that occurs after exercise. Until the oxygen debt has been met, oxygen intake is elevated, even after exercise has stopped.

Relaxation of a Skeletal Muscle

Relaxing skeletal muscle fibers, and ultimately, the skeletal muscle, begins with the motor neuron, which stops releasing its chemical signal, ACh, into the synapse at the NMJ. The muscle fiber will repolarize, which closes the gates in the SR where Ca^{++} was being released. ATP-driven pumps will move Ca^{++} out of the sarcoplasm back into the SR. This results in the "reshielding" of the actin-binding sites on the thin filaments. Without the ability to form cross-bridges between the thin and thick filaments, the muscle fiber loses its tension and relaxes.

Nervous System Control of Muscle Tension

To move an object, referred to as load, the sarcomeres in the muscle fibers of the skeletal muscle must shorten. The force generated by the contraction of the muscle (or shortening of the sarcomeres) is called **muscle tension**. However, muscle tension also is generated when the muscle is contracting against a load that does not move, resulting in two main types of skeletal muscle contractions: isotonic contractions and isometric contractions.

In **isotonic contractions**, where the tension in the muscle stays constant, a load is moved as the length of the muscle changes (shortens). There are two types of isotonic contractions: concentric and eccentric. A **concentric contraction** involves the muscle shortening to move a load. An example of this is the biceps brachii muscle contracting when a hand weight is brought upward with increasing muscle tension. As the biceps brachii contract, the angle of the elbow joint decreases as the forearm is brought toward the body. Here, the biceps brachii contracts as sarcomeres in its muscle fibers are shortening and cross-bridges form; the myosin heads pull the actin. An **eccentric contraction** occurs as the muscle tension diminishes and the muscle lengthens. In this case, the hand weight is lowered in a slow and controlled manner as the amount of cross-bridges being activated by nervous system stimulation decreases. In this case, as tension is released from the biceps brachii, the angle of the elbow joint increases. Eccentric contractions are also used for movement and balance of the body.

An **isometric contraction** occurs as the muscle produces tension without changing the angle of a skeletal joint. Isometric contractions involve sarcomere shortening and increasing muscle tension, but do not move a load, as the force produced cannot overcome the resistance provided by the load. For example, if one attempts to lift a hand weight that is too heavy, there will be sarcomere activation and shortening to a point, and ever-increasing muscle tension, but no change in the angle of the elbow joint. In everyday living, isometric contractions are active in maintaining posture and maintaining bone and joint stability. However, holding your head in an upright position occurs not because the muscles cannot move the head, but because the goal is to remain stationary and not produce movement. Most actions of the body are the result of a combination of isotonic and isometric contractions working together to produce a wide range of outcomes (Figure 1).

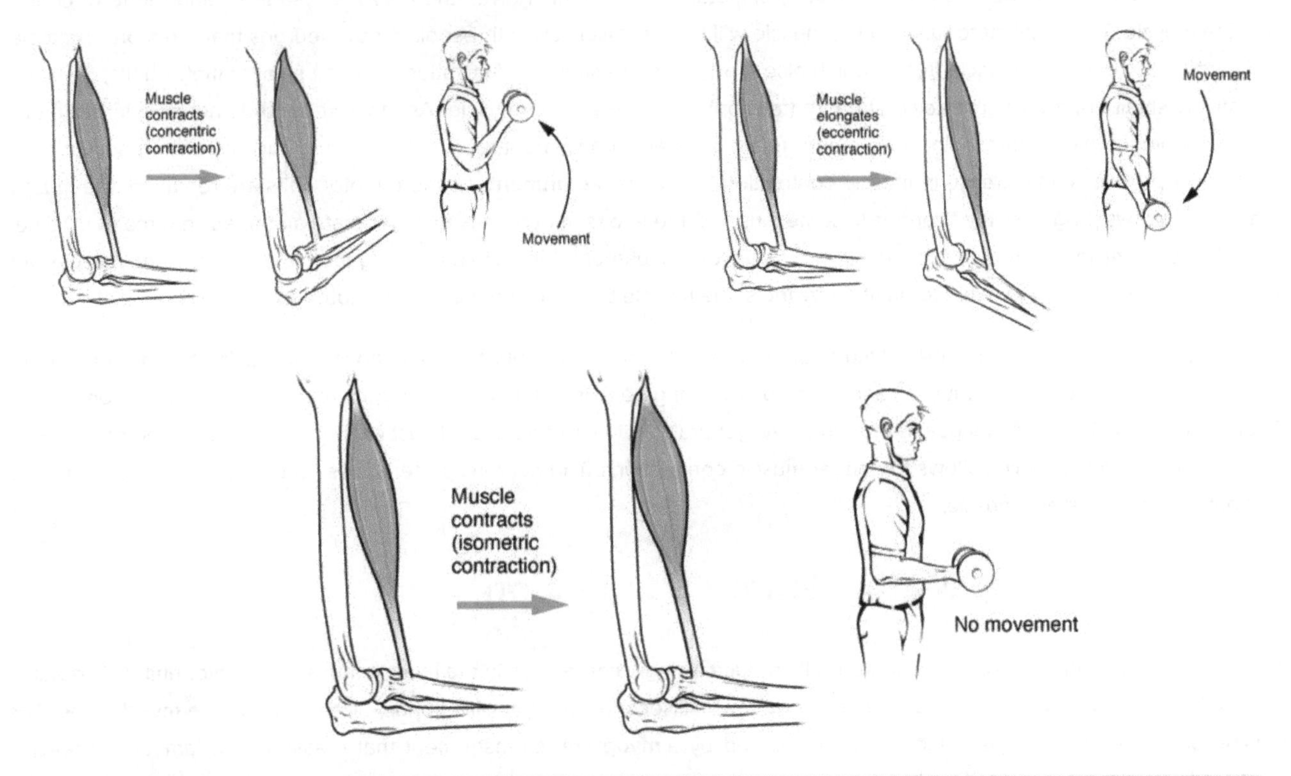

Figure 1. Types of Muscle Contractions. During isotonic contractions, muscle length changes to move a load. During isometric contractions, muscle length does not change because the load exceeds the tension the muscle can generate.

All of these muscle activities are under the exquisite control of the nervous system. Neural control regulates concentric, eccentric and isometric contractions, muscle fiber recruitment, and muscle tone. A crucial aspect of nervous system control of skeletal muscles is the role of motor units.

Motor Units

As you have learned, every skeletal muscle fiber must be innervated by the axon terminal of a motor neuron in order to contract. Each muscle fiber is innervated by only one motor neuron. The actual group of muscle fibers in a muscle innervated by a single motor neuron is called a **motor unit**. The size of a motor unit is variable depending on the nature of the muscle.

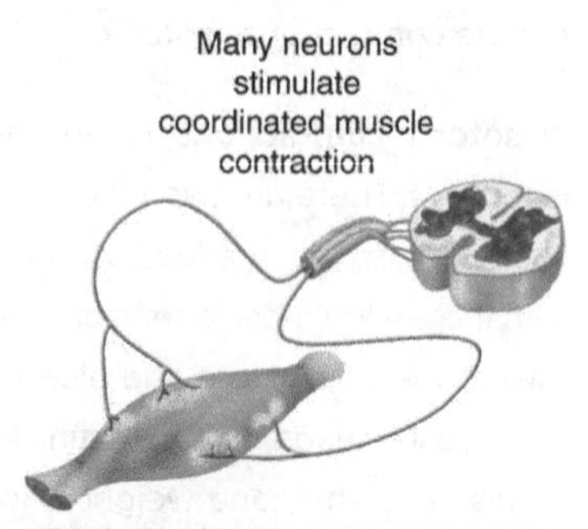

Figure 2 : Motor Neurons

A small motor unit is an arrangement where a single motor neuron supplies a small number of muscle fibers in a muscle. Small motor units permit very fine motor control of the muscle. The best example in humans is the small motor units of the extraocular eye muscles that move the eyeballs. There are thousands of muscle fibers in each muscle, but every six or so fibers are supplied by a single motor neuron, as the axons branch to form synaptic connections at their individual NMJs. This allows for exquisite control of eye movements so that both eyes can quickly focus on the same object. Small motor units are also involved in the many fine movements of the fingers and thumb of the hand for grasping, texting, etc.

A large motor unit is an arrangement where a single motor neuron supplies a large number of muscle fibers in a muscle. Large motor units are concerned with simple, or "gross," movements, such as powerfully extending the knee joint. The best example is the large motor units of the thigh muscles or back muscles, where a single motor neuron will supply thousands of muscle fibers in a muscle, as its axon splits into thousands of branches.

There is a wide range of motor units within many skeletal muscles, which gives the nervous system a wide range of control over the muscle. The small motor units in the muscle will have smaller, lower-threshold motor neurons that are more excitable, firing first to their skeletal muscle fibers, which also tend to be the smallest. Activation of these smaller motor units, results in a relatively small degree of contractile strength (tension) generated in the muscle. As more strength is needed, larger motor units, with bigger, higher-threshold motor neurons are enlisted to activate larger muscle fibers. This increasing activation of motor units produces an increase in muscle contraction known as **recruitment**. As more motor units are recruited, the muscle contraction grows progressively stronger. In some muscles, the largest motor units may generate a contractile force of 50 times more than the smallest motor units in the muscle. This allows a feather to be picked up using the biceps brachii arm muscle with minimal force, and a heavy weight to be lifted by the same muscle by recruiting the largest motor units.

When necessary, the maximal number of motor units in a muscle can be recruited simultaneously, producing the maximum force of contraction for that muscle, but this cannot last for very long because of the energy requirements to sustain the contraction. To prevent complete muscle fatigue, motor units are generally not all simultaneously active, but instead some motor units rest while others are active, which allows for longer muscle contractions. The nervous system uses recruitment as a mechanism to efficiently utilize a skeletal muscle.

The Frequency of Motor Neuron Stimulation

A single action potential from a motor neuron will produce a single contraction in the muscle fibers of its motor unit. This isolated contraction is called a **twitch**. A twitch can last for a few milliseconds or 100 milliseconds, depending on the muscle type. The tension produced by a single twitch can be measured by a **myogram**, an instrument that measures the amount of tension produced over time (Figure 3). Each twitch undergoes three phases.

- The first phase is the **latent period**, during which the action potential is being propagated along the sarcolemma and Ca^{++} ions are released from the SR. This is the phase during which excitation and contraction are being coupled but contraction has yet to occur.
- The **contraction phase** occurs next. The Ca^{++} ions in the sarcoplasm have bound to troponin, tropomyosin has shifted away from actin-binding sites, cross-bridges formed, and sarcomeres are actively shortening to the point of peak tension.
- The last phase is the **relaxation phase**, when tension decreases as contraction stops. Ca^{++} ions are pumped out of the sarcoplasm into the SR, and cross-bridge cycling stops, returning the muscle fibers to their resting state.

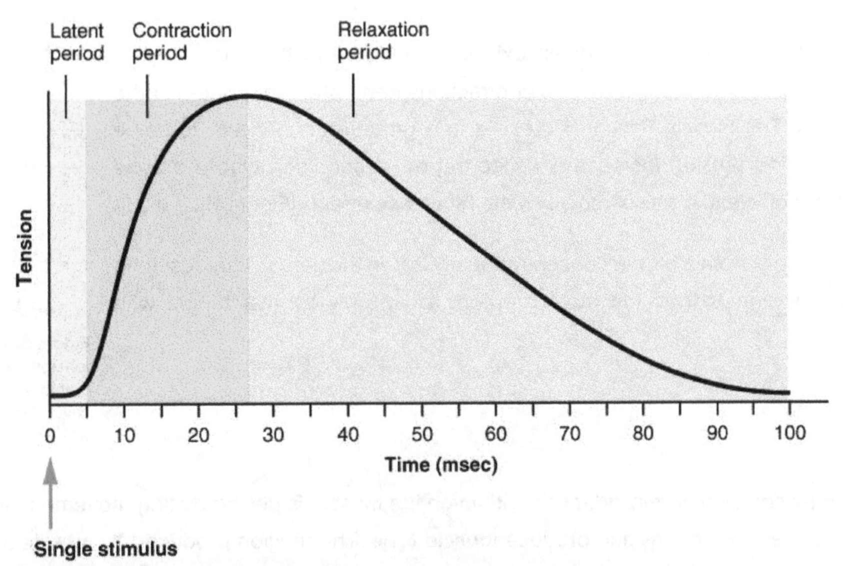

Figure 3. A Myogram of a Muscle Twitch. A single muscle twitch has a latent period, a contraction phase when tension increases, and a relaxation phase when tension decreases. During the latent period, the action potential is being propagated along the sarcolemma. During the contraction phase, Ca^{++} ions in the sarcoplasm bind to troponin, tropomyosin moves from actin-binding sites, cross-bridges form, and sarcomeres shorten. During the relaxation phase, tension decreases as Ca^{++} ions are pumped out of the sarcoplasm and cross-bridge cycling stops.

Although a person can experience a muscle "twitch," a single twitch does not produce any significant muscle activity in a living body. A series of action potentials to the muscle fibers is necessary to produce a muscle contraction that can produce work. Normal muscle contraction is more sustained, and it can be modified by input from the nervous system to produce varying amounts of force; this is called a **graded muscle response**. The frequency of action potentials (nerve impulses) from a motor neuron and the number of motor neurons transmitting action potentials both affect the tension produced in skeletal muscle.

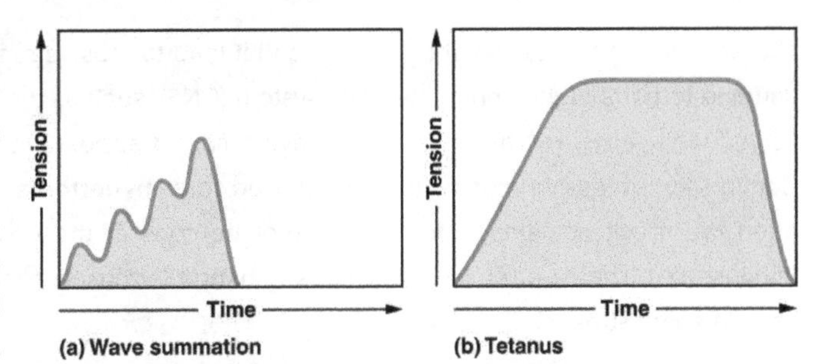

Figure 4. Wave Summation and Tetanus. (a) The excitation-contraction coupling effects of successive motor neuron signaling is added together which is referred to as wave summation. The bottom of each wave, the end of the relaxation phase, represents the point of stimulus. (b) When the stimulus frequency is so high that the relaxation phase disappears completely, the contractions become continuous; this is called tetanus.

The rate at which a motor neuron fires action potentials affects the tension produced in the skeletal muscle. If the fibers are stimulated while a previous twitch is still occurring, the second twitch will be stronger. This response is called **wave summation**, because the excitation-contraction coupling effects of successive motor neuron signaling is summed, or added together (Figure 4a). At the molecular level, summation occurs because the second stimulus triggers the release of more Ca^{++} ions, which become available to activate additional sarcomeres while the muscle is still contracting from the first stimulus. Summation results in greater contraction of the motor unit.

If the frequency of motor neuron signaling increases, summation and subsequent muscle tension in the motor unit continues to rise until it reaches a peak point. The tension at this point is about three to four times greater than the tension of a single twitch, a state referred to as incomplete tetanus. During incomplete tetanus, the muscle goes through quick cycles of contraction with a short relaxation phase for each. If the stimulus frequency is so high that the relaxation phase disappears completely, contractions become continuous in a process called complete **tetanus** (Figure 4b). During tetanus, the concentration of Ca^{++} ions in the sarcoplasm allows virtually all of the sarcomeres to form cross-bridges and shorten, so that a contraction can continue uninterrupted (until the muscle fatigues and can no longer produce tension).

Treppe

When a skeletal muscle has been dormant for an extended period and then activated to contract, with all other things being equal, the initial contractions generate about one-half the force of later contractions. The muscle tension increases in a graded manner that to some looks like a set of stairs. This tension increase is called **treppe,** a condition where muscle contractions become more efficient. It's also known as the "staircase effect" (Figure 5).

It is believed that treppe results from a higher concentration of Ca^{++} in the sarcoplasm resulting from the steady stream of signals from the motor neuron. It can only be maintained with adequate ATP.

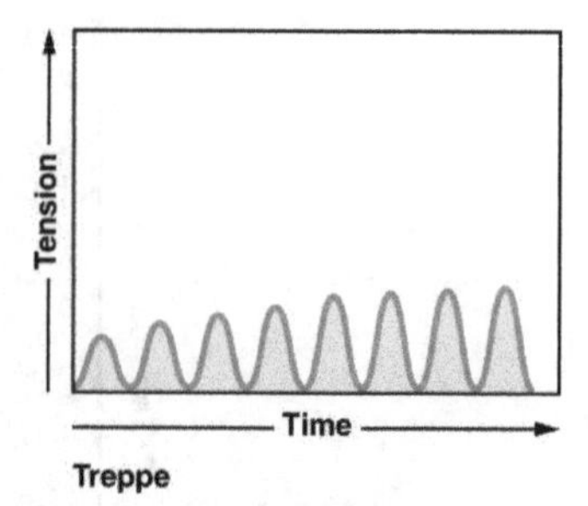

Figure 5. Treppe. When muscle tension increases in a graded manner that looks like a set of stairs, it is called treppe. The bottom of each wave represents the point of stimulus.

Muscle Tone

Skeletal muscles are rarely completely relaxed, or flaccid. Even if a muscle is not producing movement, it is contracted a small amount to maintain its contractile proteins and produce **muscle tone**. The tension produced by muscle tone allows muscles to continually stabilize joints and maintain posture.

Muscle tone is accomplished by a complex interaction between the nervous system and skeletal muscles that results in the activation of a few motor units at a time, most likely in a cyclical manner. In this manner, muscles never fatigue completely, as some motor units can recover while others are active.

The absence of the low-level contractions that lead to muscle tone is referred to as **hypotonia** or atrophy, and can result from damage to parts of the central nervous system (CNS), such as the cerebellum, or from loss of innervations to a skeletal muscle, as in poliomyelitis. Hypotonic muscles have a flaccid appearance and display functional impairments, such as weak reflexes. Conversely, excessive muscle tone is referred to as **hypertonia**, accompanied by hyperreflexia (excessive reflex responses), often the result of damage to upper motor neurons in the CNS. Hypertonia can present with muscle rigidity (as seen in Parkinson's disease) or spasticity, a phasic change in muscle tone, where a limb will "snap" back from passive stretching (as seen in some strokes).

Types of Muscle Fibers and Exercise

Two criteria to consider when classifying the types of muscle fibers are how fast some fibers contract relative to others, and how fibers produce ATP. Using these criteria, there are three main types of skeletal muscle fibers. **Slow oxidative** (SO) fibers contract relatively slowly and use aerobic respiration (oxygen and glucose) to produce ATP. **Fast oxidative** (FO) fibers have fast contractions and primarily use aerobic respiration, but because they may switch to anaerobic respiration (glycolysis), can fatigue more quickly than SO fibers. **Fast glycolytic (FG)** fibers have fast contractions and primarily use anaerobic glycolysis. The FG fibers fatigue more quickly than the others. Most skeletal muscles in a human contain(s) all three types, although in varying proportions.

The speed of contraction is dependent on how quickly myosin's ATPase hydrolyzes ATP to produce cross-bridge action. Fast fibers hydrolyze ATP approximately twice as quickly as slow fibers, resulting in much quicker cross-bridge cycling (which pulls the thin filaments toward the center of the sarcomeres at a faster rate). The primary metabolic pathway used by a muscle fiber determines whether the fiber is classified as oxidative or glycolytic. If a fiber primarily produces ATP through aerobic pathways it is oxidative. More ATP can be produced during each metabolic cycle, making the fiber more resistant to fatigue. Glycolytic fibers primarily create ATP through anaerobic glycolysis, which produces less ATP per cycle. As a result, glycolytic fibers fatigue at a quicker rate.

The oxidative fibers contain many more mitochondria than the glycolytic fibers, because aerobic metabolism, which uses oxygen (O_2) in the metabolic pathway, occurs in the mitochondria. The SO fibers possess a large number of mitochondria and are capable of contracting for longer periods because of the large amount of ATP they can produce, but they have a relatively small diameter and do not produce a large amount of tension. SO fibers are extensively supplied with blood capillaries to supply O_2 from the red blood cells in the bloodstream. The SO fibers also possess myoglobin, an O_2-carrying molecule similar to O_2-carrying hemoglobin in the red blood cells. The myoglobin stores some of the needed O_2 within the fibers themselves (and gives SO fibers their red color). All of these features allow SO fibers to produce large quantities of ATP, which can sustain muscle activity without fatiguing for long periods of time.

The fact that SO fibers can function for long periods without fatiguing makes them useful in maintaining posture, producing isometric contractions, stabilizing bones and joints, and making small movements that happen often but do not require large amounts of energy. They do not produce high tension, and thus they are not used for powerful, fast movements that require high amounts of energy and rapid cross-bridge cycling.

FO fibers are sometimes called intermediate fibers because they possess characteristics that are intermediate between fast fibers and slow fibers. They produce ATP relatively quickly, more quickly than SO fibers, and thus can produce relatively high amounts of tension. They are oxidative because they produce ATP aerobically, possess high amounts of mitochondria, and do not fatigue quickly. However, FO fibers do not possess significant myoglobin, giving them a lighter color than the red SO fibers. FO fibers are used primarily for movements, such as walking, that require more energy than postural control but less energy than an explosive movement, such as sprinting. FO fibers are useful for this type of movement because they produce more tension than SO fibers but they are more fatigue-resistant than FG fibers.

FG fibers primarily use anaerobic glycolysis as their ATP source. They have a large diameter and possess high amounts of glycogen, which is used in glycolysis to generate ATP quickly to produce high levels of tension. Because they do not primarily use aerobic metabolism, they do not possess substantial numbers of mitochondria or significant amounts of myoglobin and therefore have a white color. FG fibers are used to produce rapid, forceful contractions to make quick, powerful movements. These fibers fatigue quickly, permitting them to only be used for short periods. Most muscles possess a mixture of each fiber type. The predominant fiber type in a muscle is determined by the primary function of the muscle.

Exercise and Muscle Performance

Physical training alters the appearance of skeletal muscles and can produce changes in muscle performance. Conversely, a lack of use can result in decreased performance and muscle appearance. Although muscle cells can change in size, new cells are not formed when muscles grow. Instead, structural proteins are added to muscle fibers in a process called **hypertrophy**, so cell diameter increases. The reverse, when structural proteins are lost and muscle mass decreases, is called **atrophy**. Age-related muscle atrophy is called **sarcopenia**. Cellular components of muscles can also undergo changes in response to changes in muscle use.

Endurance Exercise

Slow fibers are predominantly used in endurance exercises that require little force but involve numerous repetitions. The aerobic metabolism used by slow-twitch fibers allows them to maintain contractions over long periods. Endurance training modifies these slow fibers to make them even more efficient by producing more mitochondria to enable more aerobic metabolism and more ATP production. Endurance exercise can also increase the amount of myoglobin in a cell, as increased aerobic respiration increases the need for oxygen. Myoglobin is found in the sarcoplasm and acts as an oxygen storage supply for the mitochondria.

The training can trigger the formation of more extensive capillary networks around the fiber, a process called **angiogenesis**, to supply oxygen and remove metabolic waste. To allow these capillary networks to supply the deep portions of the muscle, muscle mass does not greatly increase in order to maintain a smaller area for the diffusion of nutrients and gases. All of these cellular changes result in the ability to sustain low levels of muscle contractions for greater periods without fatiguing.

Figure 1. Marathoners. Long-distance runners have a large number of SO fibers and relatively few FO and FG fibers. (credit: "Tseo2"/Wikimedia Commons)

The proportion of SO muscle fibers in muscle determines the suitability of that muscle for endurance, and may benefit those participating in endurance activities. Postural muscles have a large number of SO fibers and relatively few FO and FG fibers, to keep the back straight (Figure 1). Endurance athletes, like marathon-runners also would benefit from a larger proportion of SO fibers, but it is unclear if the most-successful marathoners are those with naturally high numbers of SO fibers, or whether the most successful marathon runners develop high numbers of SO fibers with repetitive training. Endurance training can result in overuse injuries such as stress fractures and joint and tendon inflammation.

Resistance Exercise

Resistance exercises, as opposed to endurance exercise, require large amounts of FG fibers to produce short, powerful movements that are not repeated over long periods. The high rates of ATP hydrolysis and cross-bridge formation in FG fibers result in powerful muscle contractions. Muscles used for power have a higher ratio of FG to SO/FO fibers, and trained athletes possess even higher levels of FG fibers in their muscles.

Resistance exercise affects muscles by increasing the formation of myofibrils, thereby increasing the thickness of muscle fibers. This added structure causes hypertrophy, or the enlargement of muscles, exemplified by the large skeletal muscles seen in body builders and other athletes (Figure 2). Because this muscular enlargement is achieved by the addition of structural proteins, athletes trying to build muscle mass often ingest large amounts of protein.

Figure 2. Hypertrophy. Body builders have a large number of FG fibers and relatively few FO and SO fibers. (credit: Lin Mei/flickr)

Except for the hypertrophy that follows an increase in the number of sarcomeres and myofibrils in a skeletal muscle, the cellular changes observed during endurance training do not usually occur with resistance training. There is usually no significant increase in mitochondria or capillary density. However, resistance training does increase the development of connective tissue, which adds to the overall mass of the muscle and helps to contain muscles as they produce increasingly powerful contractions. Tendons also become stronger to prevent tendon damage, as the force produced by muscles is transferred to tendons that attach the muscle to bone.

For effective strength training, the intensity of the exercise must continually be increased. For instance, continued weight lifting without increasing the weight of the load does not increase muscle size. To produce ever-greater results, the weights lifted must become increasingly heavier, making it more difficult for muscles to move the load. The muscle then adapts to this heavier load, and an even heavier load must be used if even greater muscle mass is desired.

If done improperly, resistance training can lead to overuse injuries of the muscle, tendon, or bone. These injuries can occur if the load is too heavy or if the muscles are not given sufficient time between workouts to recover or if joints are not aligned properly during the exercises. Cellular damage to muscle fibers that occurs after intense exercise includes damage to the sarcolemma and myofibrils. This muscle damage contributes to the feeling of soreness after strenuous exercise, but muscles gain mass as this damage is repaired, and additional structural proteins are added to replace the damaged ones. Overworking skeletal muscles can also lead to tendon damage and even skeletal damage if the load is too great for the muscles to bear.

Performance-Enhancing Substances

Some athletes attempt to boost their performance by using various agents that may enhance muscle performance. Anabolic steroids are one of the more widely known agents used to boost muscle mass and increase power output. Anabolic steroids are a form of testosterone, a male sex hormone that stimulates muscle formation, leading to increased muscle mass.

Endurance athletes may also try to boost the availability of oxygen to muscles to increase aerobic respiration by using substances such as erythropoietin (EPO), a hormone normally produced in the kidneys, which triggers the production of red blood cells. The extra oxygen carried by these blood cells can then be used by muscles for aerobic respiration. Human growth hormone (hGH) is another supplement, and although it can facilitate building muscle mass, its main role is to promote the healing of muscle and other tissues after strenuous exercise. Increased hGH may allow for faster recovery after muscle damage, reducing the rest required after exercise, and allowing for more sustained high-level performance.

Although performance-enhancing substances often do improve performance, most are banned by governing bodies in sports and are illegal for nonmedical purposes. Their use to enhance performance raises ethical issues of cheating because they give users an unfair advantage over nonusers. A greater concern, however, is that their use carries serious health risks. The side effects of these substances are often significant, nonreversible, and in some cases fatal. The physiological strain caused by these substances is often greater than what the body can handle, leading to effects that are unpredictable and dangerous. Anabolic steroid use has been linked to infertility, aggressive behavior, cardiovascular disease, and brain cancer.

Similarly, some athletes have used creatine to increase power output. Creatine phosphate provides quick bursts of ATP to muscles in the initial stages of contraction. Increasing the amount of creatine available to cells is thought to produce more ATP and therefore increase explosive power output, although its effectiveness as a supplement has been questioned.

EVERYDAY CONNECTION: AGING AND MUSCLE TISSUE

Although atrophy due to disuse can often be reversed with exercise, muscle atrophy with age, referred to as sarcopenia, is irreversible. This is a primary reason why even highly trained athletes succumb to declining performance with age. This decline is noticeable in athletes whose sports require strength and powerful movements, such as sprinting, whereas the effects of age are less noticeable in endurance athletes such as marathon runners or long-distance cyclists. As muscles age, muscle fibers die, and they are replaced by connective tissue and adipose tissue (Figure 3).

Because those tissues cannot contract and generate force as muscle can, muscles lose the ability to produce powerful contractions. The decline in muscle mass causes a loss of strength, including the strength required for posture and mobility. This may be caused by a reduction in FG fibers that hydrolyze ATP quickly to produce short, powerful contractions. Muscles in older people sometimes possess greater numbers of SO fibers, which are responsible for longer contractions and do not produce powerful movements. There may also be a reduction in the size of motor units, resulting in fewer fibers being stimulated and less muscle tension being produced.

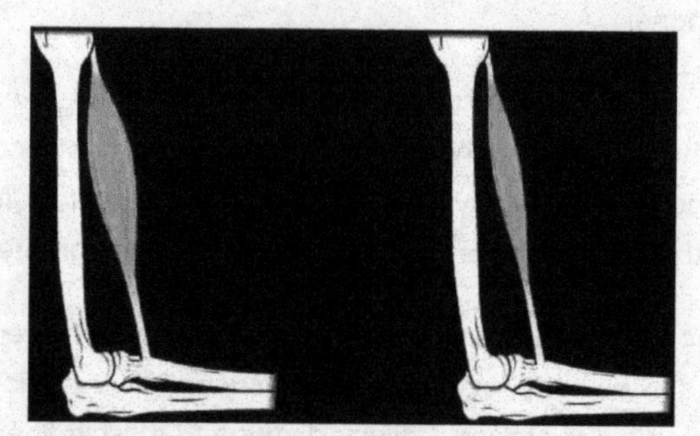

Figure 3. Atrophy. Muscle mass is reduced as muscles atrophy with disuse.

Sarcopenia can be delayed to some extent by exercise, as training adds structural proteins and causes cellular changes that can offset the effects of atrophy. Increased exercise can produce greater numbers of cellular mitochondria, increase capillary density, and increase the mass and strength of connective tissue. The effects of age-related atrophy are especially pronounced in people who are sedentary, as the loss of muscle cells is displayed as functional impairments such as trouble with locomotion, balance, and posture. This can lead to a decrease in quality of life and medical problems, such as joint problems because the muscles that stabilize bones and joints are weakened. Problems with locomotion and balance can also cause various injuries due to falls.

Major Skeletal Muscles of Human Body and Interactions

To move the skeleton, the tension created by the contraction of the fibers in most skeletal muscles is transferred to the tendons. The tendons are strong bands of dense, regular connective tissue that connect muscles to bones. The bone connection is why this muscle tissue is called skeletal muscle.Interactions of Skeletal Muscles in the Body

To pull on a bone, that is, to change the angle at its synovial joint, which essentially moves the skeleton, a skeletal muscle must also be attached to a fixed part of the skeleton. The moveable end of the muscle that attaches to the bone being pulled is called the muscle's **insertion**, and the end of the muscle attached to a fixed (stabilized) bone is called the **origin**. During forearm **flexion**—bending the elbow—the brachioradialis assists the brachialis.

Although a number of muscles may be involved in an action, the principal muscle involved is called the **prime mover**, or **agonist**. To lift a cup, a muscle called the biceps brachii is actually the prime mover; however, because it can be assisted by the brachialis, the brachialis is called a **synergist** in this action. A synergist can also be a **fixator** that stabilizes the bone that is the attachment for the prime mover's origin.

A muscle with the opposite action of the prime mover is called an **antagonist**. Antagonists play two important roles in muscle function:

1. They maintain body or limb position, such as holding the arm out or standing erect
2. They control rapid movement, as in shadow boxing without landing a punch or the ability to check the motion of a limb

For example, to extend the knee, a group of four muscles called the quadriceps femoris in the anterior compartment of the thigh are activated (and would be called the agonists of knee extension). However, to flex the knee joint, an opposite or antagonistic set of muscles called the hamstrings is activated.

As you can see, these terms would also be reversed for the opposing action. If you consider the first action as the knee bending, the hamstrings would be called the agonists and the quadriceps femoris would then be called the antagonists. See Table 1 for a list of some agonists and antagonists.

There are also skeletal muscles that do not pull against the skeleton for movements. For example, there are the muscles that produce facial expressions. The insertions and origins of facial muscles are in the skin, so that certain individual muscles contract to form a smile or frown, form sounds or words, and raise the eyebrows. There also are skeletal muscles in the tongue, and the external urinary and anal sphincters that allow for voluntary regulation of urination and defecation, respectively. In addition, the diaphragm contracts and relaxes to change the volume of the pleural cavities but it does not move the skeleton to do this.

Table 1: Major Muscles of the Human Body and their Actions.

	Muscles	*Actions*
Major Muscles of Upper Limb	**Deltoid**	shoulder abduction, flexion and extension
	Biceps brachii	flexes elbow
	Triceps brachii	extends forearm
Major Muscles of Lower Limb	**Gluteus maximus**	external rotation and extension of the hip joint,
	Sartorius	lateral rotation and abduction of thigh; flexion and medial rotation of leg
	Quadriceps group (4 muscles together)	knee extension; hip flexion

	Hamstrings group (3 muscles together)	flexes knee, extends hip joint, medially rotates leg at knee
	Tibialis anterior	dorsiflex and invert the foot
	Gastrocnemius	plantar flexes foot – extension or flexion of the foot at the ankle, flexes knee
Major Muscles of Trunk – Anterior	**Pectoralis**	control the movement of the arm, create lateral, vertical, or rotational motion.
	Rectus abdominus	pulls the ribs and the pelvis in and curves the back. (doing crunches)
	External oblique	rotate the trunk
Major Muscles of Trunk – Posterior	**Trapezius** (upper back)	tilt and turn the head and neck, shrug, steady the shoulders, and twist the arms. Elevates, depresses, rotates, and retracts the scapula, or shoulder blade
	Latissimus dorsi (lower back)	When flexed, the muscle works at extending, adducting and rotating the arm.

Table 2. Agonist and Antagonist Skeletal Muscle Pairs

Agonist	Antagonist	Movement
Biceps brachii: in the anterior compartment of the arm	Triceps brachii: in the posterior compartment of the arm	The biceps brachii flexes the forearm, whereas the triceps brachii extends it.
Hamstrings: group of three muscles in the posterior compartment of the thigh	Quadriceps femoris: group of four muscles in the anterior compartment of the thigh	The hamstrings flex the leg, whereas the quadriceps femoris extend it.

EVERYDAY CONNECTIONS: EXERCISE AND STRETCHING

When exercising, it is important to first warm up the muscles. Stretching pulls on the muscle fibers and it also results in an increased blood flow to the muscles being worked. Without a proper warm-up, it is possible that you may either damage some of the muscle fibers or pull a tendon. A pulled tendon, regardless of location, results in pain, swelling, and diminished function; if it is moderate to severe, the injury could immobilize you for an extended period.

Recall the discussion about muscles crossing joints to create movement. Most of the joints you use during exercise are synovial joints, which have synovial fluid in the joint space between two bones. Exercise and stretching may also have a beneficial effect on synovial joints. Synovial fluid is a thin, but viscous film with the consistency of egg whites. When you first get up and start moving, your joints feel stiff for a number of reasons. After proper stretching and warm-up, the synovial fluid may become less viscous, allowing for better joint function.

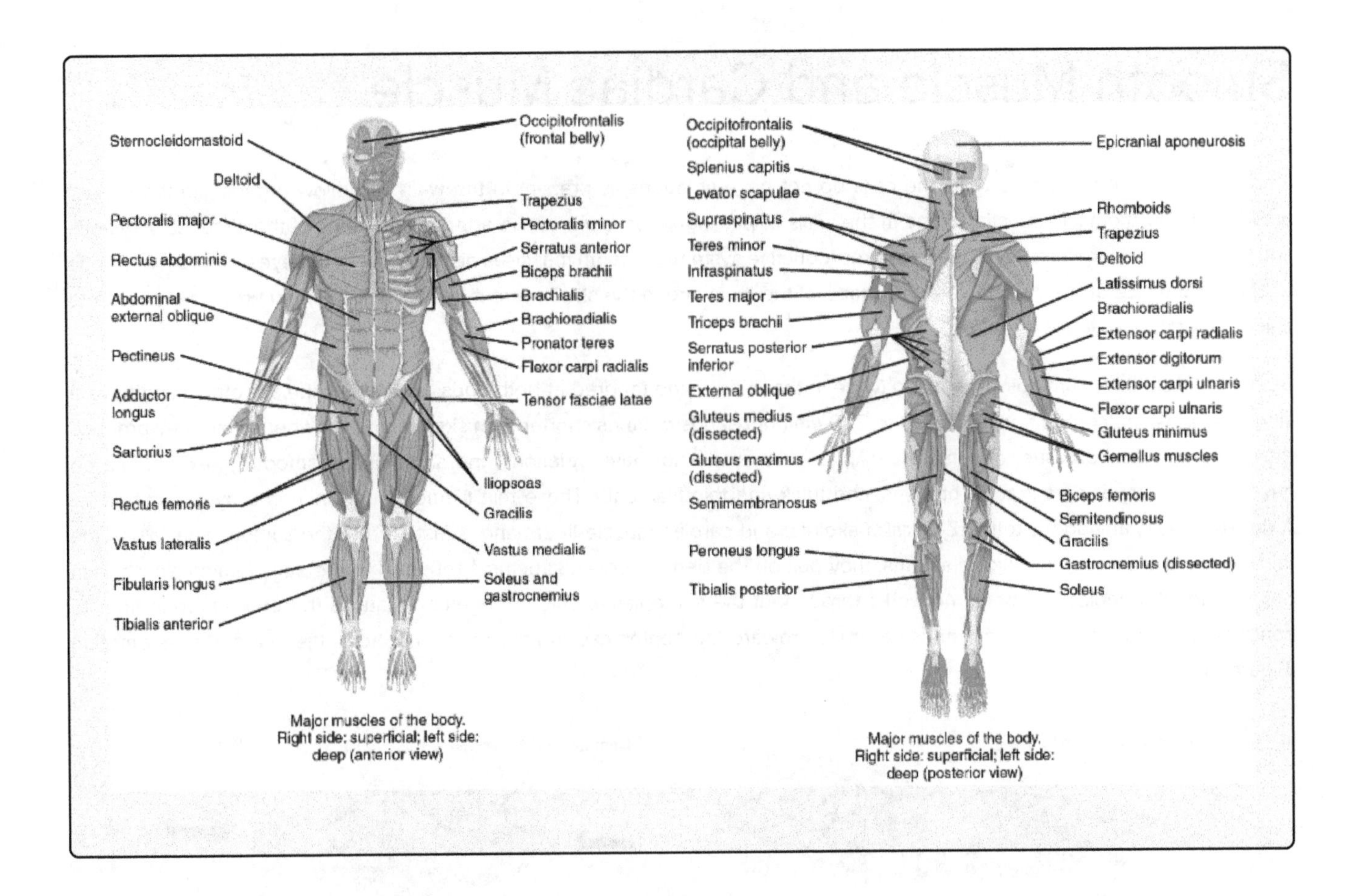

Sternocleidomastoid
Occipitofrontalis (frontal belly)
Deltoid
Trapezius
Pectoralis major
Pectoralis minor
Serratus anterior
Rectus abdominis
Biceps brachii
Brachialis
Abdominal external oblique
Brachioradialis
Pronator teres
Pectineus
Flexor carpi radialis
Adductor longus
Tensor fasciae latae
Sartorius
Iliopsoas
Rectus femoris
Gracilis
Vastus lateralis
Vastus medialis
Fibularis longus
Soleus and gastrocnemius
Tibialis anterior
Major muscles of the body. Right side: superficial; left side: deep (anterior view)
Occipitofrontalis (occipital belly)
Epicranial aponeurosis
Splenius capitis
Levator scapulae
Rhomboids
Supraspinatus
Trapezius
Teres minor
Deltoid
Infraspinatus
Latissimus dorsi
Teres major
Brachioradialis
Triceps brachii
Extensor carpi radialis
Serratus posterior inferior
Extensor digitorum
Extensor carpi ulnaris
External oblique
Flexor carpi ulnaris
Gluteus medius (dissected)
Gluteus minimus
Gemellus muscles
Gluteus maximus (dissected)
Biceps femoris
Semimembranosus
Semitendinosus
Gracilis
Peroneus longus
Gastrocnemius (dissected)
Tibialis posterior
Soleus
Major muscles of the body. Right side: superficial; left side: deep (posterior view)

Smooth Muscle and Cardiac Muscle

Smooth muscle (so-named because the cells do not have striations) is present in the walls of hollow organs like the urinary bladder, uterus, stomach, intestines, and in the walls of passageways, such as the arteries and veins of the circulatory system, and the tracts of the respiratory, urinary, and reproductive systems. Smooth muscle is also present in the eyes, where it functions to change the size of the iris and alter the shape of the lens; and in the skin where it causes hair to stand erect in response to cold temperature or fear.

Smooth muscle fibers are spindle-shaped (wide in the middle and tapered at both ends, somewhat like a football) and have a single nucleus; they range from about 30 to 200 μm (thousands of times shorter than skeletal muscle fibers), and they produce their own connective tissue, endomysium. Although they do not have striations and sarcomeres, smooth muscle fibers do have actin and myosin contractile proteins, and thick and thin filaments. These thin filaments are anchored by dense bodies. A **dense body** is analogous to the Z-discs of skeletal and cardiac muscle fibers and is fastened to the sarcolemma. When the thin filaments slide past the thick filaments, they pull on the dense bodies, structures tethered to the sarcolemma, which then pull on the intermediate filaments networks throughout the sarcoplasm. This arrangement causes the entire muscle fiber to contract in a manner whereby the ends are pulled toward the center, causing the midsection to bulge in a corkscrew motion (Figure 2).

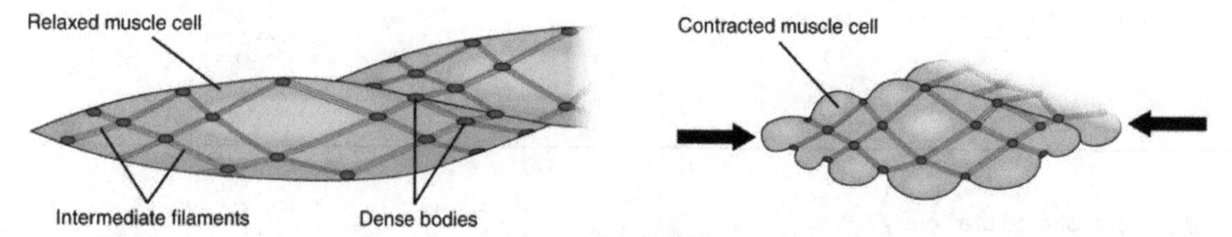

Figure 2. Muscle Contraction. The dense bodies and intermediate filaments are networked through the sarcoplasm, which cause the muscle fiber to contract.

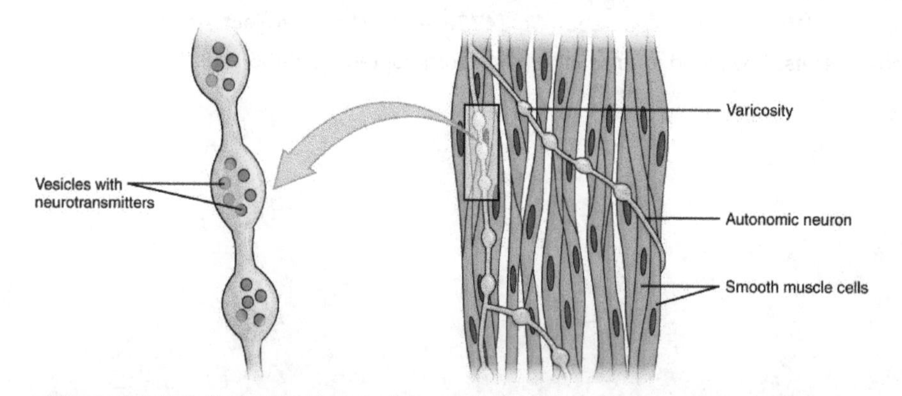

Figure 3. Motor Units. A series of axon-like swelling, called varicosities or "boutons," from autonomic neurons form motor units through the smooth muscle.

Smooth muscle is organized in two ways: as single-unit smooth muscle, which is much more common; and as multiunit smooth muscle. The two types have different locations in the body and have different characteristics.

Single-unit muscle has its muscle fibers joined by gap junctions so that the muscle contracts as a single unit. This type of smooth muscle is found in the walls of all visceral organs except the heart (which has cardiac muscle in its walls), and so it is commonly called **visceral muscle**. Because the muscle fibers are not constrained by the organization and stretchability limits of sarcomeres, visceral smooth muscle has a **stress-relaxation response**. This means that as the muscle of a hollow organ is stretched when it fills, the mechanical stress of the stretching will trigger contraction, but this is immediately followed by relaxation so that the organ does not empty its contents prematurely. This is important for hollow organs, such as the stomach or urinary bladder, which continuously expand as they fill. The smooth muscle around these organs also can maintain a muscle tone when the organ empties and shrinks, a feature that prevents "flabbiness" in the empty organ. In general, visceral smooth muscle produces slow, steady contractions that allow substances, such as food in the digestive tract, to move through the body.

Multiunit smooth muscle cells rarely possess gap junctions, and thus are not electrically coupled. As a result, contraction does not spread from one cell to the next, but is instead confined to the cell that was originally stimulated. Stimuli for multiunit smooth muscles come from autonomic nerves or hormones but not from stretching. This type of tissue is found around large blood vessels, in the respiratory airways, and in the eyes.

Hyperplasia in Smooth Muscle

Similar to skeletal and cardiac muscle cells, smooth muscle can undergo hypertrophy to increase in size. Unlike other muscle, smooth muscle can also divide to produce more cells, a process called **hyperplasia**. This can most evidently be observed in the uterus at puberty, which responds to increased estrogen levels by producing more uterine smooth muscle fibers, and greatly increases the size of the myometrium.

Cardiac muscle tissue

Cardiac muscle tissue is only found in the heart. Highly coordinated contractions of cardiac muscle pump blood into the vessels of the circulatory system. Similar to skeletal muscle, cardiac muscle is striated and organized into sarcomeres, possessing the same banding organization as skeletal muscle.

However, cardiac muscle fibers are shorter than skeletal muscle fibers and usually contain only one nucleus, which is located in the central region of the cell. Cardiac muscle fibers also possess many mitochondria and myoglobin, as ATP is produced primarily through aerobic metabolism. Cardiac muscle fibers cells also are extensively branched and are connected to one another at their ends by intercalated discs. An **intercalated disc** allows the cardiac muscle cells to contract in a wave-like pattern so that the heart can work as a pump.

Intercalated discs are part of the sarcolemma and contain two structures important in cardiac muscle contraction: gap junctions and desmosomes. A gap junction forms channels between adjacent cardiac muscle fibers that allow the depolarizing current produced by cations to flow from one cardiac muscle cell to the next. This joining is called electric coupling, and in cardiac muscle it allows the quick transmission of action potentials and the coordinated contraction of the entire heart. This network of electrically connected cardiac muscle cells creates a functional unit of contraction called a syncytium. The remainder of the intercalated disc is composed of desmosomes. A **desmosome** is a cell structure that anchors the ends of cardiac muscle fibers together so the cells do not pull apart during the stress of individual fibers contracting (Figure 4).

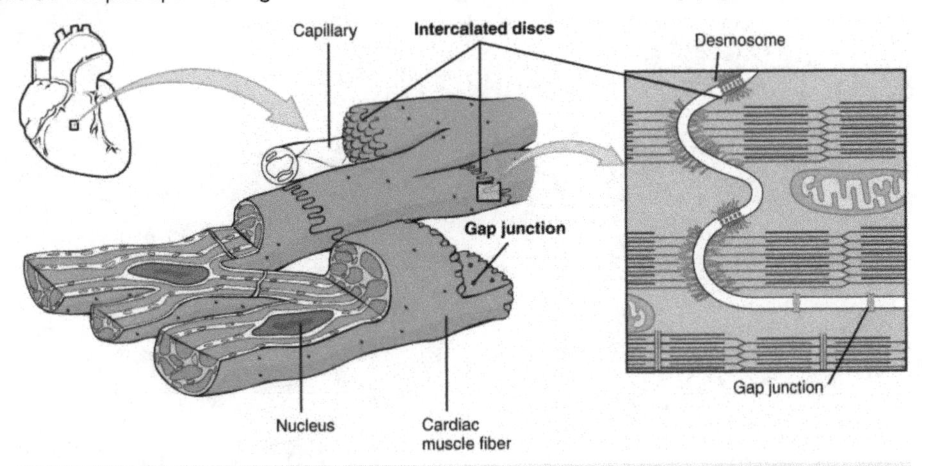

Figure 4. Cardiac Muscle. Intercalated discs are part of the cardiac muscle sarcolemma and they contain gap junctions and desmosomes.

Contractions of the heart (heartbeats) are controlled by specialized cardiac muscle cells called pacemaker cells that directly control heart rate. Although cardiac muscle cannot be consciously controlled, the pacemaker cells respond to signals from the autonomic nervous system (ANS) to speed up or slow down the heart rate. The pacemaker cells can also respond to various hormones that modulate heart rate to control blood pressure.

The wave of contraction that allows the heart to work as a unit, called a functional syncytium, begins with the pacemaker cells. This group of cells is self-excitable and able to depolarize to threshold and fire action potentials on their own, a feature called **autorhythmicity**; they do this at set intervals which determine heart rate. Because they are connected with gap junctions to surrounding muscle fibers and the specialized fibers of the heart's conduction system, the pacemaker cells are able to transfer the depolarization to the other cardiac muscle fibers in a manner that allows the heart to contract in a coordinated manner.

Concept Check

1. **Why** is elasticity an important quality of muscle tissue?
2. **Describe** how tendons facilitate body movement.
3. **What** causes the striated appearance of skeletal muscle tissue?
4. **How** would muscle contractions be affected if ATP was completely depleted in a muscle fiber?
5. **Why** does a motor unit of the eye have few muscle fibers compared to a motor unit of the leg?
6. **How** are cardiac muscle cells similar to and different from skeletal muscle cells?
7. **Watch this** videoto learn more about macro- and microstructures of skeletal muscles. (a) What are the names of the "junction points" between sarcomeres? (b) What are the names of the "subunits" within the myofibrils that run the length of skeletal muscle fibers? (c) What is the "double strand of pearls" described in the video? (d) What gives a skeletal muscle fiber its striated appearance?
8. Every skeletal muscle fiber is supplied by a motor neuron at the NMJ. **Watch this** videoto learn more about what happens at the neuromuscular junction. (a) What is the definition of a motor unit? (b) What is the structural and functional difference between a large motor unit and a small motor unit? Can you give an example of each? (c) Why is the neurotransmitter acetylcholine degraded after binding to its receptor?
9. The release of calcium ions initiates muscle contractions. **Watch this** videoto learn more about the role of calcium. (a) What are "T-tubules" and what is their role? (b) Please also describe how actin-binding sites are made available for cross-bridging with myosin heads during contraction.
10. Movements of the body occur at joints. **Describe** how muscles are arranged around the joints of the body.
11. Which muscles form the hamstrings? How do they function together?
12. Which muscles form the quadriceps? How do they function together?

CHAPTER 9 — THE URINARY SYSTEM

Urinary System - Main Organs and their Functions

LEARNING OBJECTIVES

- Define the functions of the urinary system
- Describe the structure and functions of the organs of the urinary system.Describe how the nephron is the functional unit of the kidney and explain how it actively filters blood and generates urine
- Explain how the urinary system works in maintaining water and electrolyte homeostasis.
- Describe the effect of hormones on urine formation.
- Describe the characteristics of a normal urine sample.
- Major diseases of the urinary system

The Urinary System

The urinary system, is a group of organs in the body that filters out excess fluid and other substances from the bloodstream. The purpose of the renal / urinary system is to eliminate wastes from the body, regulate blood volume and pressure, control levels of electrolytes and metabolites, and regulate blood pH.

Urinary System Functions

The renal system has many functions. Many of these functions are interrelated with the physiological mechanisms in the cardiovascular and respiratory systems.

1. Excretion : Removal of metabolic waste products from the body (mainly urea and uric acid). Metabolic wastes and excess ions are filtered out of the blood, along with water, and leave the body in the form of **urine.**
2. Maintains water balance: adjusts blood volume and blood pressure. The renal system alters water retention and thirst to slowly change blood volume and keep blood pressure in a normal range.(Blood pressure homeostasis)
3. Assist in maintaining electrolyte / salt balance (e.g., sodium, potassium, and calcium). Maintain salt balance in order to control blood volume
4. Assist in maintaining acid base / pH balance of blood (7.35-7.45) by controlling the loss of hydrogen ions and bicarbonate ions in urine.
5. Secretion of hormones. Example : Erythropoietin / EPO.

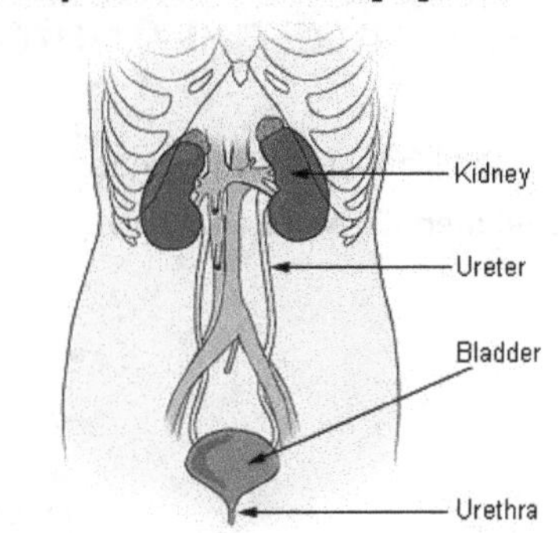

Figure 1.: Components of the Urinary system: Here are the major organs of the renal system.

Organs of the Renal / Urinary System

The renal / urinary system organs include the kidneys, ureters, bladder, and urethra

- **Kidneys:** The kidneys are a pair of bean-shaped organs which filter blood and produce urine.. They remove wastes, control the body's fluid balance, and keep the right levels of electrolytes. Kidneys are the most complex and critical part of the urinary system.

- **Ureter:** Urine passes from the renal tube through tubes called ureters and into the bladder.
- **Bladder:** The bladder is flexible and is used as storage until the urine is allowed to pass through the urethra and out of the body.
- **Urethra:** Duct that transmits urine from the bladder to the exterior of the body during urination. The female and male urinary system are very similar, differing only in the length of the urethra.

Kidneys – External Anatomy

The kidneys lie on either side of the spine in the retroperitoneal space between the parietal peritoneum and the posterior abdominal wall, well protected by muscle, fat, and ribs. They are roughly the size of your fist, and the male kidney is typically a bit larger than the female kidney. The kidneys are well vascularized, receiving about 25 percent of the cardiac output at rest.

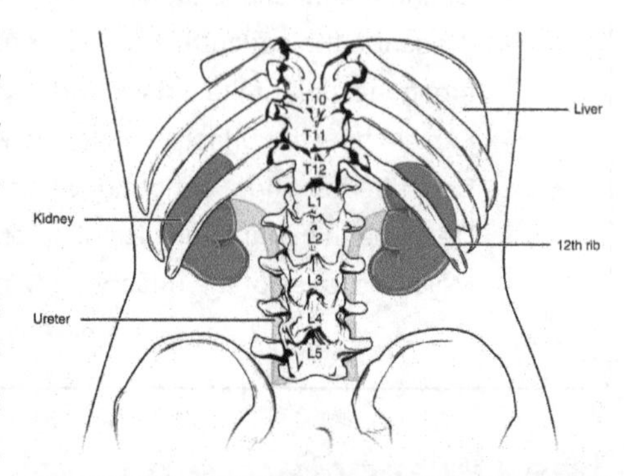

Figure 2. The kidneys are slightly protected by the ribs and are surrounded by fat for protection (not shown).

The left kidney is located at about the T12 to L3 vertebrae, whereas the right is lower due to slight displacement by the liver. Upper portions of the kidneys are somewhat protected by the eleventh and twelfth ribs. Each kidney weighs about 125–175 g in males and 115–155 g in females. They are about 11–14 cm in length, 6 cm wide, and 4 cm thick, and are directly covered by a fibrous capsule composed of dense, irregular connective tissue that helps to hold their shape and protect them. This capsule is covered by a shock-absorbing layer of adipose tissue called the **renal fat pad**, which in turn is encompassed by a tough renal fascia. The fascia and, to a lesser extent, the overlying peritoneum serve to firmly anchor the kidneys to the posterior abdominal wall in a retroperitoneal position.

On the superior aspect (top) of each kidney is the **adrenal gland.** The adrenal cortex directly influences renal function through the production of the hormone aldosterone to stimulate sodium reabsorption.

Kidney – Internal Anatomy

Externally, the kidneys are surrounded by three layers, illustrated in (Figure). The kidney has three regions: **Outer renal cortex, Inner renal medulla and Renal pelvis.**

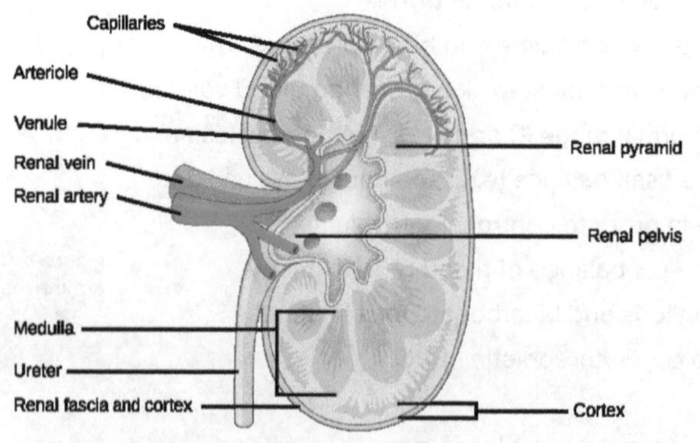

Figure 3. The internal structure of the kidney is shown. (credit: modification of work by NCI)

- **Renal Cortex:** In a dissected kidney, it is easy to identify the cortex; it appears lighter in color compared to the rest of the kidney. The renal cortex is granular due to the presence of **nephrons—the functional unit of the kidney.** Some nephrons have a short **loop of Henle** that does not dip beyond the cortex. These nephrons are called **cortical nephrons**. About 15 percent of nephrons have long loops of Henle that extend deep into the medulla and are called **juxtamedullary nephrons**.

- **Renal Medulla:** The medulla consists of multiple pyramidal tissue masses, called the **renal pyramids.** In between the pyramids are spaces called **renal columns** through which the blood vessels pass. The tips of the pyramids, called renal papillae, point toward the renal pelvis. There are, on average, eight renal pyramids in each kidney.
- **Renal Pelvis:** The renal pelvis leads to the ureter on the outside of the kidney. On the inside of the kidney, the renal pelvis branches out into two or three extensions called the **major calyces,** which further branch into the **minor calyces.** The ureters are urine-bearing tubes that exit the kidney and empty into the urinary bladder.
- The **renal hilum** is the entry and exit site for structures servicing the kidneys: vessels, nerves, lymphatics, and ureters. The medial-facing hila are tucked into the sweeping convex outline of the cortex. Emerging from the hilum is the renal pelvis, which is formed from the major and minor calyxes in the kidney. The smooth muscle in the renal pelvis funnels urine via peristalsis into the ureter. The renal arteries form directly from the descending aorta, whereas the renal veins return cleansed blood directly to the inferior vena cava.

Blood flow through the kidney / Renal circulation:

Blood flows to the kidneys through the right and left renal arteries. The **renal arteries** originate from the abdominal aorta and enter the renal hila to supply the kidneys. Inside each kidney these branch into smaller arterioles and penetrate deep into the renal medulla and renal cortex. Blood coming into the kidney is rich in oxygen and nutrients and waste materials. Eventually, the smaller arterioles divide and become **renal capillaries** where blood gets filtered, and metabolic wastes are removed to form urine. Renal capillaries are located in and around the nephrons. These renal capillaries fuse and form **renal veins** and leave the kidney. Blood leaving the kidney via renal veins will have less waste materials as most of the wastes materials are filtered out of the blood in renal capillaries. They also have less oxygen and nutrients as these materials are used up by the kidney cells.

HEART → AORTA → RENAL ARTERY →KIDNEY → RENAL VEIN → INFERIOR VENACAVA → HEART

Main blood vessels of kidney	Direction of flow of blood	Quality of blood
Renal artery (R and L)	Brings blood to the kidneys. They branch directly from the aorta (the main artery coming off the heart) on either side and extend to each kidney.	Blood coming in via renal artery is rich in oxygen, nutrients and metabolic wastes.
Renal vein (R and L)	The renal veins are blood vessels that drain the kidney. They connect the kidney to the inferior vena cava. They carry the blood filtered by the kidney.	Blood in renal veins have less oxygen, nutrients and metabolic wastes

Nephron - Structure

Nephrons: The Functional Units of the Kidney

Nephrons take a simple filtrate of the blood and modify it into urine. Nephrons are the "functional units" of the kidney; they cleanse the blood and balance the constituents of the circulation. Each kidney is made up of over one million nephrons that dot the renal cortex, giving it a granular appearance when sectioned sagittally. Many changes take place in the different parts of the nephron before urine is created for disposal. The term **forming urine** will be used hereafter to describe the **filtrate** as it is modified into true urine. The principle task of the nephron population is to balance the plasma to homeostatic set points and excrete potential toxins in the urine. They do this by accomplishing three principle functions—filtration, reabsorption, and secretion. They also have additional secondary functions that exert control in three areas: blood pressure (via production of **renin**), red blood cell production (via the hormone EPO), and calcium absorption (via conversion of calcidiol into calcitriol, the active form of vitamin D).

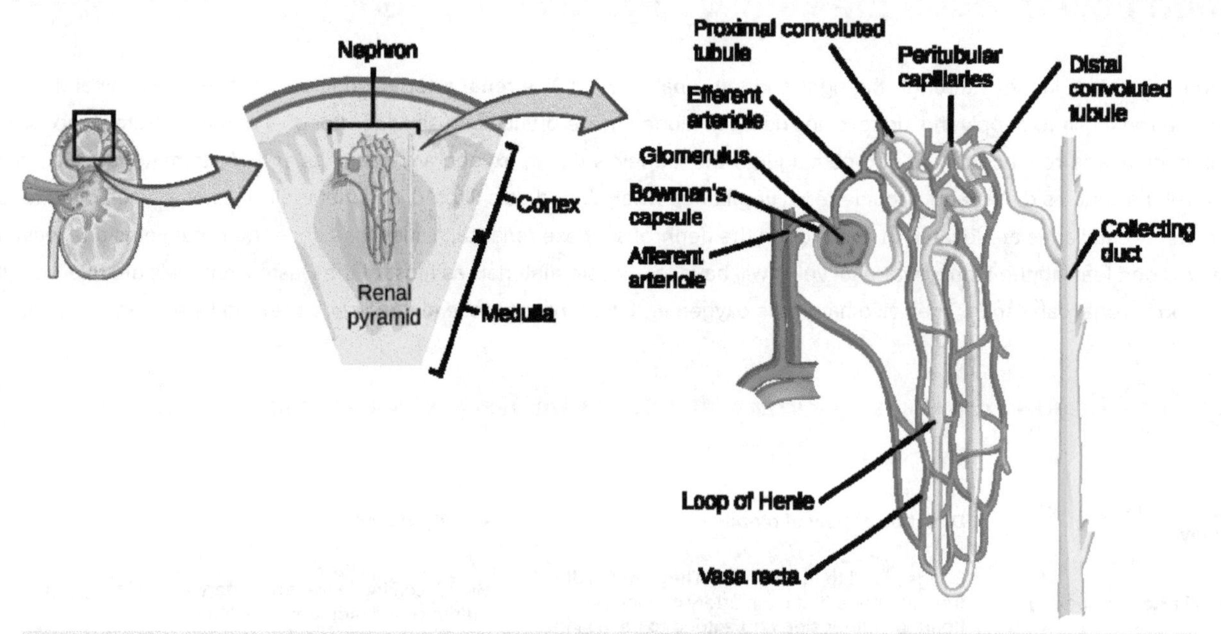

Figure 1. The nephron is the functional unit of the kidney. The glomerulus and convoluted tubules are located in the kidney cortex, while collecting ducts are located in the pyramids of the medulla. (credit: modification of work by NIDDK)

Tubular parts of a Nephron – converts the filtrate into urine

The Bowman's capsule / Glomerular capsule:

The Bowman's capsule (also called the **glomerular capsule**), is the beginning of a nephron. It surrounds the glomerulus. It is a double walled cup, It is composed of inner visceral and outer parietal layers. Parietal (outer) layer of glomerular capsule is simple squamous epithelium. Visceral (inner) layer of glomerular capsule consists of elaborate cells called **podocytes** that wrap around the capillaries of the glomerulus. Capsular space separates the two layers of glomerular capsule.

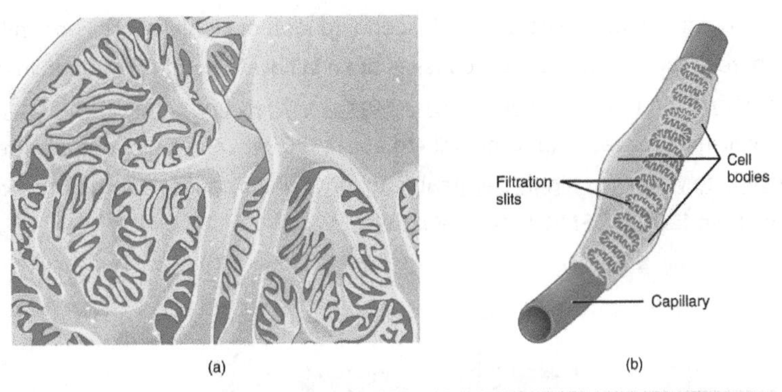

Figure 3: Glomerular capillaries covered by Podocytes .

Renal tubule: long, coiled tube that converts the filtrate into urine

The renal tubule is a long and convoluted structure that emerges from the glomerular capsule and can be divided into **three parts based on function**. The first part is called the **proximal convoluted tubule (PCT)** due to its proximity to the glomerulus; it stays in the renal cortex. The second part is called the **loop of Henle**, because it forms a loop (with descending and ascending limbs) that goes through the renal medulla. The third part of the renal tubule is called the **distal convoluted tubule (DCT)** and this part is also restricted to the renal cortex. The DCT, which is the last part of the nephron, connects and empties its contents into **collecting ducts** that line the medullary pyramids. The collecting ducts amass contents from multiple nephrons and fuse together as they enter the papillae of the renal medulla.

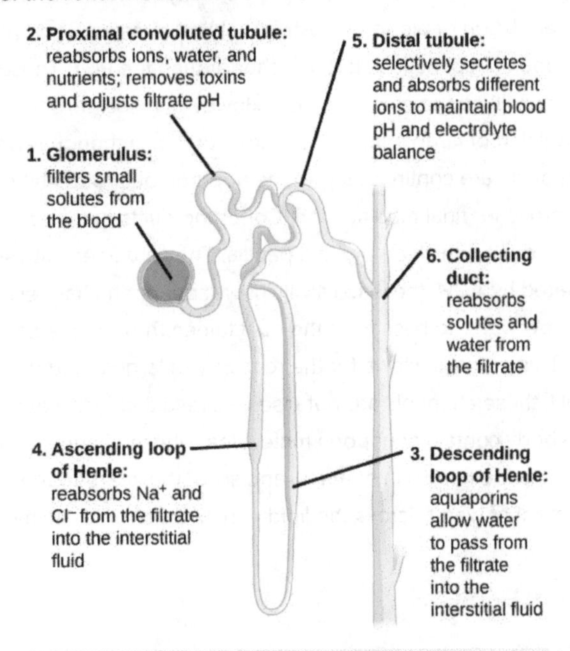

Figure 4. Tubular parts of Nephron and their functions.

- **Proximal Convoluted Tubule:** Filtered fluid collected by Bowman's capsule enters into the PCT. It is called convoluted due to its tortuous path. Simple cuboidal cells form this tubule with prominent microvilli on the luminal surface, forming a brush border. These microvilli create a large surface area to maximize the absorption and secretion of solutes (Na^+, Cl^-, glucose, etc.), the most essential function of this portion of the nephron. These cells actively transport ions across their membranes, so they possess a high concentration of mitochondria in order to produce sufficient ATP.
- **The loop of Henle** is a U-shaped tube that consists of a descending limb and ascending limb. It transfers fluid from the proximal to the distal tubule. The descending and ascending portions of the loop of Henle (sometimes referred to as the nephron loop) are continuations of the same tubule. They run adjacent and parallel to each other after having made a

hairpin turn at the deepest point of their descent. The descending loop of Henle consists of an initial short, thick portion and long, thin portion, whereas the ascending loop consists of an initial short, thin portion followed by a long, thick portion. The descending limb is highly permeable to water but completely impermeable to ions, causing a large amount of water to be reabsorbed, which increases fluid osmolarity to about 1200 mOSm/L. In contrast, the ascending limb of Henle's loop is impermeable to water but highly permeable to ions, which causes a large drop in the osmolarity of fluid passing through the loop, from 1200 mOSM/L to 100 mOSm/L.

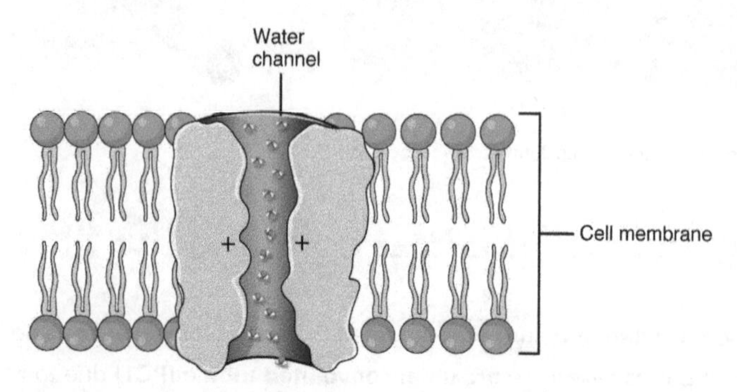

Figure 5. Aquaporin Water Channel : Positive charges inside the channel prevent the leakage of electrolytes across the cell membrane, while allowing water to move due to osmosis.

- **Distal Convoluted Tubule (DCT):** The distal convoluted tubule and collecting duct is the final site of reabsorption in the nephron. Unlike the other components of the nephron, its permeability to water is variable depending on a hormone stimulus to enable the complex regulation of blood osmolarity, volume, pressure, and pH. The DCT, like the PCT, is very tortuous and formed by simple cuboidal epithelium, but it is shorter than the PCT. These cells are not as active as those in the PCT; thus, there are fewer microvilli on the apical surface. However, these cells must also pump ions against their concentration gradient, so you will find of large numbers of mitochondria, although fewer than in the PCT.
- **Collecting Ducts:** The collecting ducts are continuous with the nephron but not technically part of it. In fact, each duct collects filtrate from several nephrons for final modification. Collecting ducts merge as they descend deeper in the medulla to form about 30 terminal ducts, which empty at a papilla. They are lined with simple squamous epithelium with receptors for ADH. When stimulated by ADH, these cells will insert aquaporin channel proteins into their membranes, which as their name suggests, allow water to pass from the duct lumen through the cells and into the interstitial spaces to be recovered by the vasa recta. This process allows for the recovery of large amounts of water from the filtrate back into the blood. In the absence of ADH, these channels are not inserted, resulting in the excretion of water in the form of dilute urine. Most, if not all, cells of the body contain aquaporin molecules, whose channels are so small that only water can pass. At least 10 types of aquaporins are known in humans, and six of those are found in the kidney. The function of all aquaporins is to allow the movement of water across the lipid-rich, hydrophobic cell membrane.

Nephrons and Blood Vessels

Each nephron has its own independent blood supply.

- **Afferent Arteriole**: The renal artery first divides into smaller arteries, followed by further branching and pass through the renal columns to reach the cortex. In the cortex they divide further and form afferent arterioles.The afferent arterioles service about 1.3 million nephrons in each kidney. The branch that enters the glomerulus is called the afferent arteriole. A group of specialized cells known as juxtaglomerular apparatus (JGA) are located around the afferent arteriole where it enters the renal corpuscle. The JGA secretes an enzyme called renin, due to a variety of stimuli, and it is involved in the process of blood volume homeostasis.

- **The Glomerulus:** The glomerulus is a capillary tuft that receives its blood supply from an afferent arteriole of the renal circulation. First step of urine formation "filtration of blood" happens at the glomerulular capillaries. – glomerular filtration. The glomerular capillaries are fenestrated capillaries. *Fenestrated capillaries* have pores in the endothelial cells those provide channels across the capillary wall so we could have small molecules pass through easily. These capillaries are more permeable than continuous capillaries. Fenestrations allow many substances to diffuse from the blood based primarily on size. Substances cross readily if they are less than 4 nm in size and most pass freely up to 8 nm in size. **Water and small molecules like glucose, urea and ions like sodium** cross the glomerular capillaries and get into the glomerular capsule of nephron. The fenestrations (pores) prevent filtration of blood cells or large proteins. **Red blood cells and large proteins,** such as serum albumins, cannot pass through the glomerulus under normal circumstances because they are **too big to pass through glomerular capillaries.** However, in some injuries they may be able to pass through and can cause blood and protein content to enter the urine, which is a sign of problems in the kidney. As blood passes through the glomerular capillaries, 10 to 20 percent of the plasma filters between these sieve-like fingers to be captured by Bowman's capsule and funneled to the PCT. The glomerulus and the double walled cup-shaped chamber that surrounds it, called the glomerular or Bowman's capsule, together is referred to as **Renal corpuscle.**

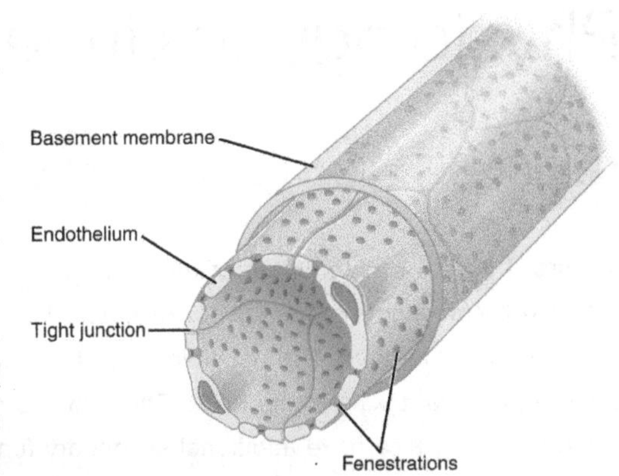

Figure 6: Fenestrated Capillary with fenestrations (pores) : Pores allow many substances to diffuse from the blood based primarily on size.

- **Efferent Arteriole:** After passing through the renal corpuscle, the capillaries form a second arteriole, **the efferent arteriole** (Figure 25.10). The branch that exits the glomerulus is called the **efferent arteriole.** These will next form a capillary network around the more distal portions of the nephron tubule, **the peritubular capillaries** and **vasa recta,** before returning to the venous system. As the glomerular filtrate progresses through the nephron, these capillary networks recover most of the solutes and water, and return them to the circulation. Since a capillary bed (the glomerulus) drains into a vessel that in turn forms a second capillary bed, the definition of **a portal system** is met. This is the only portal system in which an arteriole is found between the first and second capillary beds.

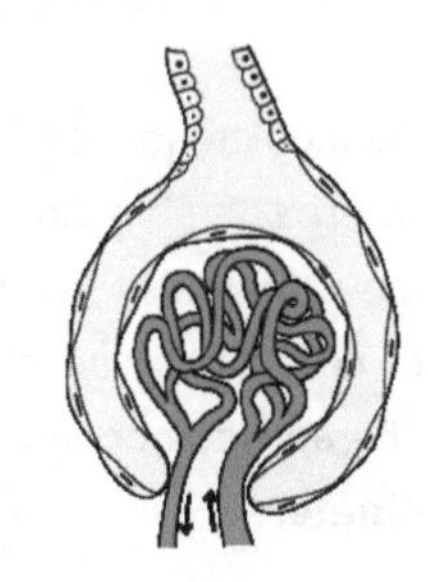

- **Peritubular capillaries: Peritubular capillaries** are tiny blood vessels, supplied by the efferent arteriole, that travel alongside nephrons and surrounds them allowing reabsorption and **secretion** between blood and the inner lumen of the nephron. In cortical nephrons, the peritubular capillary network surrounds the PCT and DCT. In juxtamedullary nephrons, the peritubular capillary network forms a network around the loop of Henle and is called the **vasa recta.** Peritubular capillaries join together and form the **renal veins** and return the filtered blood with less waste materials to the venous system.

Physiology of Urine Formation in the Nephrons

Nephrons: The Functional Unit

Nephrons take a simple filtrate of the blood and modify it into urine. Many changes take place in the different parts of the nephron before urine is created for disposal. The term **forming urine** will be used hereafter to describe the filtrate as it is modified into true urine. The principle task of the nephron population is to balance the plasma to homeostatic set points and excrete potential toxins in the urine. They do this by accomplishing **three principle functions—filtration, reabsorption, and secretion.** They also have additional secondary functions that exert control in three areas: blood pressure (via production of **renin**), red blood cell production (via the hormone EPO), and calcium absorption (via conversion of calcidiol into calcitriol, the active form of vitamin D).

Urine is a waste byproduct formed from excess water and metabolic waste molecules during the process of renal system filtration. The primary function of the renal system is to regulate blood volume and plasma osmolarity, and waste removal via urine is essentially a convenient way that the body performs many functions using one process.

Steps involved in urine formation are **Glomerular filtration, Tubular reabsorption** **and Tubular secretion**

Glomerular Filtration

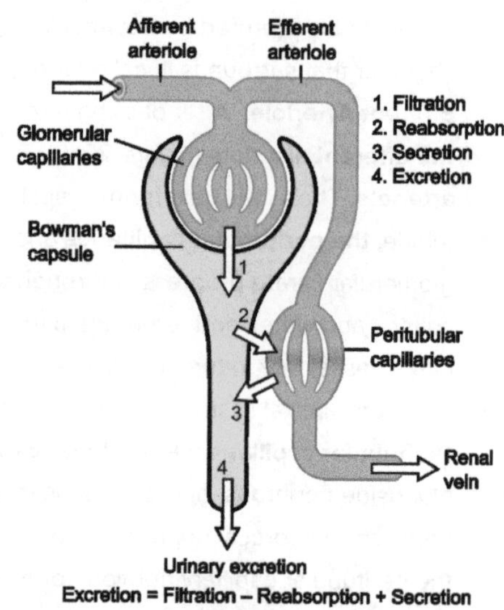

Figure 1: Steps involved in urine formation.

During filtration, blood enters the afferent arteriole and flows into the glomerulus where filterable blood components, such as water and nitrogenous waste, will move towards the inside of the glomerulus, and nonfilterable components, such as cells and serum albumins, proteins will exit via the efferent arteriole. These filterable components accumulate in the glomerulus to form the **glomerular filtrate.**

Glomerular filtration filters out most of the solutes due to high blood pressure and specialized membranes in the afferent arteriole. The blood pressure in the glomerulus is maintained independent of factors that affect systemic blood pressure. The "leaky" connections between the endothelial cells of the glomerular capillary network allow solutes to pass through easily. All solutes in the glomerular capillaries, except for macromolecules like proteins, pass through by passive diffusion. There is no energy requirement at this stage of the filtration process. **Glomerular filtration rate (GFR)** is the volume of glomerular filtrate formed per minute by the kidneys. GFR is regulated by multiple mechanisms and is an important indicator of kidney function.

Tubular Reabsorption and Secretion

Tubular reabsorption occurs in the PCT part of the renal tubule. Almost all nutrients are reabsorbed, and this occurs either by passive or active transport. Reabsorption of water and some key electrolytes are regulated and can be influenced by hormones. Sodium (Na) is the most abundant ion and most of it is reabsorbed by active transport and then transported to the peritubular capillaries. Because Na is actively transported out of the tubule, water follows it to even out the osmotic pressure. Water is also independently reabsorbed into the peritubular capillaries due to the presence of **aquaporins, or water channels**, in the PCT. This occurs due to the low blood pressure and high osmotic pressure in the peritubular capillaries. However, every solute has a **transport maximum** and the excess is not reabsorbed

In the loop of Henle, the permeability of the membrane changes. The descending limb is permeable to water, not solutes; the opposite is true for the ascending limb. Additionally, the loop of Henle invades the renal medulla, which is naturally high in salt concentration and tends to absorb water from the renal tubule and concentrate the filtrate. The osmotic gradient increases as it moves deeper into the medulla. Because two sides of the loop of Henle perform opposing functions, as illustrated in Figure 2, it acts as a **countercurrent multiplier**. The vasa recta around it acts as the **countercurrent exchanger**.

The loop of Henle (seen in Figure 2) acts as a countercurrent multiplier that uses energy to create concentration gradients. The descending limb is water permeable. Water flows from the filtrate to the interstitial fluid, so osmolality inside the limb increases as it descends into the renal medulla. At the bottom, the osmolality is higher inside the loop than in the interstitial fluid. Thus, as filtrate enters the ascending limb, Na and Cl^- ions exit through ion channels present in the cell membrane. Further up, Na is actively transported out of the filtrate and Cl^- follows. Osmolarity is given in units of milliosmoles per liter (mOsm/L)

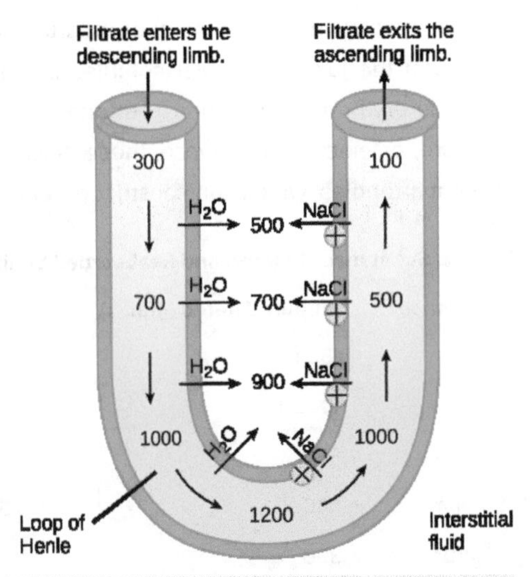

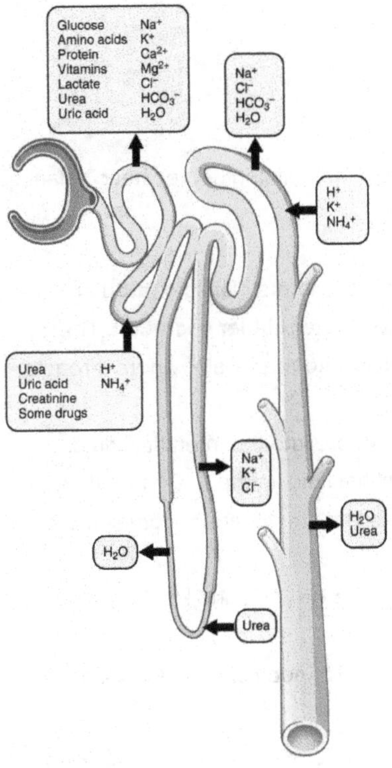

Figure 2: Countercurrent exchange in Loop of Henle

By the time the filtrate reaches the DCT, most of the urine and solutes have been reabsorbed. If the body requires additional water, all of it can be reabsorbed at this point. Further reabsorption is controlled by hormones, which will be discussed in a later section. Excretion of wastes occurs due to lack of reabsorption combined with tubular secretion. Undesirable products like metabolic wastes, urea, uric acid, and certain drugs, are excreted by tubular secretion. Most of the tubular secretion happens in the DCT, but some occurs in the early part of the collecting duct. Kidneys also maintain an acid-base balance by secreting excess H ions.

Figure 3: Substances Filtered and Reabsorbed by the Nephron

Although parts of the renal tubules are named proximal and distal, in a cross-section of the kidney, the tubules are placed close together and in contact with each other and the glomerulus. This allows for exchange of chemical messengers between the different cell types. For example, the DCT ascending limb of the loop of Henle has masses of cells called **macula densa**, which are in contact with cells of the afferent arterioles called **juxtaglomerular cells**. Together, the macula densa and juxtaglomerular cells form the juxtaglomerular complex (JGC). The JGC is an endocrine structure that secretes the enzyme renin and the hormone erythropoietin. When hormones trigger the macula densa cells in the DCT due to variations in blood volume, blood pressure, or electrolyte balance, these cells can immediately communicate the problem to the capillaries in the afferent and efferent arterioles, which can constrict or relax to change the glomerular filtration rate of the kidneys.

By the time the filtrate reaches the DCT, most of the urine and solutes have been reabsorbed. If the body requires additional water, all of it can be reabsorbed at this point. Further reabsorption is controlled by hormones, which will be discussed in a later section. Excretion of wastes occurs due to lack of reabsorption combined with tubular secretion. Undesirable products like metabolic wastes, urea, uric acid, and certain drugs, are excreted by tubular secretion. Most of the tubular secretion happens in the DCT, but some occurs in the early part of the collecting duct. Kidneys also maintain an acid-base balance by secreting excess H ions.

Although parts of the renal tubules are named proximal and distal, in a cross-section of the kidney, the tubules are placed close together and in contact with each other and the glomerulus. This allows for exchange of chemical messengers between the different cell types. For example, the DCT ascending limb of the loop of Henle has masses of cells called macula densa, which are in contact with cells of the afferent arterioles called juxtaglomerular cells. Together, the macula densa and juxtaglomerular cells form the juxtaglomerular complex (JGC). The JGC is an endocrine structure that secretes the enzyme renin and the hormone erythropoietin. When hormones trigger the macula densa cells in the DCT due to variations in blood volume, blood pressure, or electrolyte balance, these cells can immediately communicate the problem to the capillaries in the afferent and efferent arterioles, which can constrict or relax to change the glomerular filtration rate of the kidneys.

Table 2. Substances Filtered and Reabsorbed by the Kidney per 24 Hours

Substance	Amount filtered (grams)	Amount reabsorbed (grams)	Amount in urine (grams)
Water	180 L	179 L	1 L
Proteins	10–20	10–20	0
Chlorine	630	625	5
Sodium	540	537	3
Bicarbonate	300	299.7	0.3
Glucose	180	180	0
Urea	53	28	25
Potassium	28	24	4
Uric acid	8.5	7.7	0.8
Creatinine	1.4	0	1.4

Each part of the nephron seen in this figure performs a different function in filtering waste and maintaining homeostatic balance.

1. The glomerulus forces small solutes out of the blood by pressure.
2. The proximal convoluted tubule reabsorbs ions, water, and nutrients from the filtrate into the interstitial fluid (**tubular reabsorption**), and actively transports toxins and drugs from the interstitial fluid into the filtrate **tubular secretion**. The proximal convoluted tubule also adjusts blood pH by selectively secreting ammonia (NH_3) into the filtrate, where it reacts with H to form NH_4. The more acidic the filtrate, the more ammonia is secreted.
3. The **descending loop of Henle** is lined with cells containing **aquaporins** that from the filtrate into the interstitial fluid.
4. In the thin part of the **ascending loop of Henle, Na and Cl^- ions diffuse** into the interstitial fluid. In the thick part, these same ions are actively transported into the interstitial fluid. Because salt but not water is lost, the filtrate becomes more dilute as it travels up the limb.
5. In the distal convoluted tubule, K and H ions are selectively secreted into the filtrate, while Na , Cl^-, and HCO_3^- ions are reabsorbed to maintain pH and electrolyte balance in the blood.
6. The collecting duct reabsorbs solutes and water from the filtrate, forming dilute urine. (credit: modification of work by NIDDK)

Table 3. Summary of steps involved in urine production.

Process	Purpose	Site	Direction of flow of materials	Materials moved
Glomerular Filtration	removes a large amount of fluid and solutes from the blood.	Glomerulus	Blood → Urineglomerulus → Nephron	Water, glucose, sodium, urea and similar small molecules,
Tubular Reabsorption	returns useful substances to the blood	Tubules of nephron (PCT)	Urine → BloodNephron → peritubular capillaries	• Water – Mostly reabsorbed • Glucose – Totally reabsorbed. • Aminoacid – Mostly reabsorbed. • Salt – Mostly reabsorbed • Urea – 50 % reabsorbed.
Tubular Secretion	•regulating chemical levels in body, •excretion of harmful chemicals	Tubules of nephron (DCT)	Blood → Urineperitubular capillaries → Nephron	• penicillin, cocaine, marijuana, pesticides, preservatives, H , ammonium, K , uric acid, hemoglobin breakdown products, and other wastes

CAREER CONNECTION

A **nephrologist** studies and deals with diseases of the kidneys—both those that cause kidney failure (such as diabetes) and the conditions that are produced by kidney disease (such as hypertension). Blood pressure, blood volume, and changes in electrolyte balance come under the purview of a nephrologist.

Nephrologists usually work with other physicians who refer patients to them or consult with them about specific diagnoses and treatment plans. Patients are usually referred to a nephrologist for symptoms such as blood or protein in the urine, very high blood pressure, kidney stones, or renal failure.

Nephrology is a subspecialty of internal medicine. To become a nephrologist, medical school is followed by additional training to become certified in internal medicine. An additional two or more years is spent specifically studying kidney disorders and their accompanying effects on the body.

Physical Characteristics of Urine

The urinary system's ability to filter the blood resides in about 2 to 3 million tufts of specialized capillaries—the glomeruli—distributed more or less equally between the two kidneys. Because the glomeruli filter the blood based mostly on particle size, large elements like blood cells, platelets, antibodies, and albumen are excluded. The glomerulus is the first part of the nephron, which then continues as a highly specialized tubular structure responsible for creating the final urine composition. All other solutes, such as ions, amino acids, vitamins, and wastes, are filtered to create a filtrate composition very similar to plasma. The glomeruli create about 200 liters (189 quarts) of this filtrate every day, yet you excrete less than two liters of waste you call urine.

Characteristics of the urine change, depending on influences such as water intake, exercise, environmental temperature, nutrient intake, and other factors (See Table 1). Some of the characteristics such as color and odor are rough descriptors of your state of hydration. For example, if you exercise or work outside, and sweat a great deal, your urine will turn darker and produce a slight odor, even if you drink plenty of water. Athletes are often advised to consume water until their urine is clear. This is good advice; however, it takes time for the kidneys to process body fluids and store it in the bladder. Another way of looking at this is that the quality of the urine produced is an average over the time it takes to make that urine. Producing clear urine may take only a few minutes if you are drinking a lot of water or several hours if you are working outside and not drinking much.

Table 1. Normal Urine Characteristics

Characteristic	Normal values
Color	Pale yellow to deep amber
Odor	Odorless
Volume	750–2000 mL/24 hour
pH	4.5–8.0
Specific gravity	1.003–1.032
Osmolarity	40–1350 mOsmol/kg
Urobilinogen	0.2–1.0 mg/100 mL
White blood cells	0–2 HPF (per high-power field of microscope)
Leukocyte esterase	None
Protein	None or trace
Bilirubin	<0.3 mg/100 mL
Ketones	None
Nitrites	None
Blood	None
Glucose	None

Urinalysis (urine analysis) often provides clues to renal disease. Normally, only traces of protein are found in urine, and when higher amounts are found, damage to the glomeruli is the likely basis. Unusually large quantities of urine may point to diseases like diabetes mellitus or hypothalamic tumors that cause diabetes insipidus. The color of urine is determined mostly by the breakdown products of red blood cell destruction (Figure 1).

The "heme" of hemoglobin is converted by the liver into water-soluble forms that can be excreted into the bile and indirectly into the urine. This yellow pigment is **urochrome**. Urine color may also be affected by certain foods like beets, berries, and fava beans. A kidney stone or a cancer of the urinary system may produce sufficient bleeding to manifest as pink or even bright red urine. Diseases of the liver or obstructions of bile drainage from the liver impart a dark "tea" or "cola" hue to the urine. Dehydration produces darker, concentrated urine that may also possess the slight odor of ammonia. Most of the ammonia produced from protein breakdown is converted into urea by the liver, so ammonia is rarely detected in fresh urine. The strong ammonia odor you may detect in bathrooms or alleys is due to the breakdown of urea into ammonia by bacteria in the environment. About one in five people detect a distinctive odor in their urine after consuming asparagus; other foods such as onions, garlic, and fish can impart their own aromas! These food-caused odors are harmless.

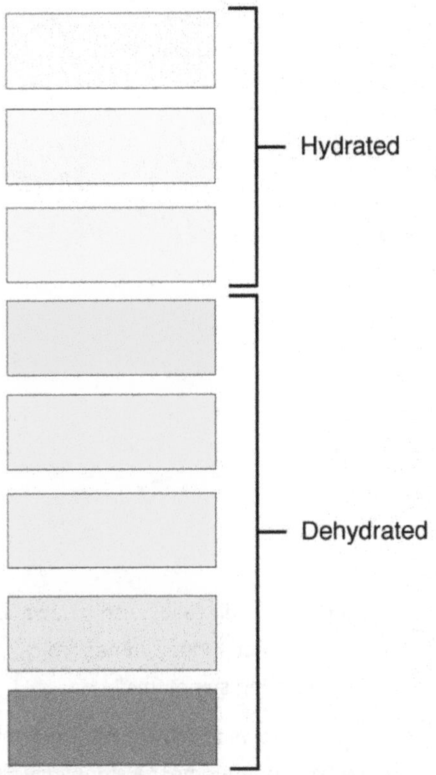

Hydrated

Dehydrated

Figure 1. Urine Color

Urine volume varies considerably. The normal range is one to two liters per day. The kidneys must produce a minimum urine volume of about 500 mL/day to rid the body of wastes. Output below this level may be caused by severe dehydration or renal disease and is termed **oliguria**. The virtual absence of urine production is termed **anuria**. Excessive urine production is **polyuria**, which may be due to diabetes mellitus or diabetes insipidus. In diabetes mellitus, blood glucose levels exceed the number of available sodium-glucose transporters in the kidney, and glucose appears in the urine. The osmotic nature of glucose attracts water, leading to its loss in the urine. In the case of diabetes insipidus, insufficient pituitary antidiuretic hormone (ADH) release or insufficient numbers of ADH receptors in the collecting ducts means that too few water channels are inserted into the cell membranes that line the collecting ducts of the kidney. Insufficient numbers of water channels (aquaporins) reduce water absorption, resulting in high volumes of very dilute urine.

Table 2. Urine Volumes

Volume condition	Volume	Causes
Normal	1–2 L/day	
Polyuria	>2.5 L/day	Diabetes mellitus; diabetes insipidus; excess caffeine or alcohol; kidney disease; certain drugs, such as diuretics; sickle cell anemia; excessive water intake
Oliguria	300–500 mL/day	Dehydration; blood loss; diarrhea; cardiogenic shock; kidney disease; enlarged prostate
Anuria	<50 mL/day	Kidney failure; obstruction, such as kidney stone or tumor; enlarged prostate

The pH (hydrogen ion concentration) of the urine can vary more than 1000-fold, from a normal low of 4.5 to a maximum of 8.0. Diet can influence pH; meats lower the pH, whereas citrus fruits, vegetables, and dairy products raise the pH. Chronically high or low pH can lead to disorders, such as the development of kidney stones or osteomalacia.

KIDNEY STONES

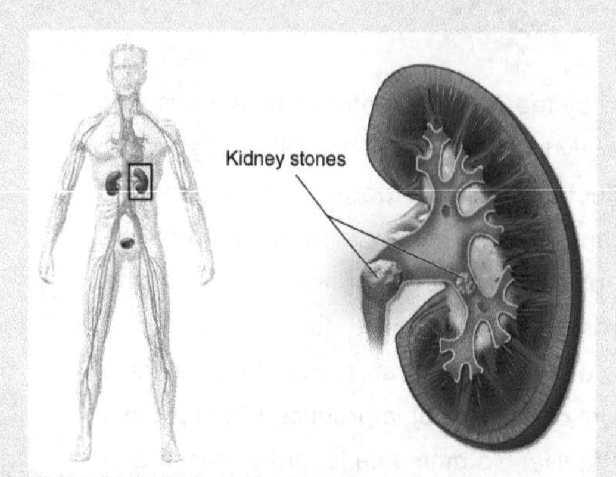

Kidney stone disease, also known as **urolithiasis**, is when a solid piece of material (kidney stone) happens in the urinary tract. Kidney stones typically form in the kidney and leave the body during urination. A small stone may pass without causing symptoms.

If a stone grows to more than 5 millimeters (0.2 in) it can cause blockage of the ureter resulting in severe pain in the lower back or abdomen. A stone may also result in blood in the urine, vomiting, or painful urination. About half of people will have another stone within ten years.

Specific gravity is a measure of the quantity of solutes per unit volume of a solution and is traditionally easier to measure than osmolarity. Urine will always have a specific gravity greater than pure water (water = 1.0) due to the presence of solutes. Laboratories can now measure urine osmolarity directly, which is a more accurate indicator of urinary solutes than **specific gravity**. Remember that osmolarity is the number of osmoles or milliosmoles per liter of fluid (mOsmol/L). Urine osmolarity ranges from a low of 50–100 mOsmol/L to as high as 1200 mOsmol/L H_2O.

Cells are not normally found in the urine. The presence of leukocytes may indicate a urinary tract infection. **Leukocyte esterase** is released by leukocytes; if detected in the urine, it can be taken as indirect evidence of a urinary tract infection (UTI).

Protein does not normally leave the glomerular capillaries, so only trace amounts of protein should be found in the urine, approximately 10 mg/100 mL in a random sample. If excessive protein is detected in the urine, it usually means that the glomerulus is damaged and is allowing protein to "leak" into the filtrate.

Ketones are byproducts of fat metabolism. Finding ketones in the urine suggests that the body is using fat as an energy source in preference to glucose. In diabetes mellitus when there is not enough insulin (type I diabetes mellitus) or because of insulin resistance (type II diabetes mellitus), there is plenty of glucose, but without the action of insulin, the cells cannot take it up, so it remains in the bloodstream. Instead, the cells are forced to use fat as their energy source, and fat consumed at such a level produces excessive ketones as byproducts. These excess ketones will appear in the urine. Ketones may also appear if there is a severe deficiency of proteins or carbohydrates in the diet.

Nitrates (NO_3^-) occur normally in the urine. Gram-negative bacteria metabolize nitrate into nitrite (NO_2^-), and its presence in the urine is indirect evidence of infection.

Blood: There should be no blood found in the urine. It may sometimes appear in urine samples as a result of menstrual contamination, but this is not an abnormal condition. Now that you understand what the normal characteristics of urine are, the next section will introduce you to how you store and dispose of this waste product and how you make it.

Gross Anatomy of Urine Transport

Urine is a fluid of variable composition that requires specialized structures to remove it from the body safely and efficiently. Blood is filtered, and the filtrate is transformed into urine at a relatively constant rate throughout the day. This processed liquid is stored until a convenient time for excretion. All structures involved in the transport and storage of the urine are large enough to be visible to the naked eye. This transport and storage system not only stores the waste, but it protects the tissues from damage due to the wide range of pH and osmolarity of the urine, prevents infection by foreign organisms, and for the male, provides reproductive functions.

Urethra

The **urethra** transports urine from the bladder to the outside of the body for disposal. The urethra is the only urologic organ that shows any significant anatomic difference between males and females; all other urine transport structures are identical.

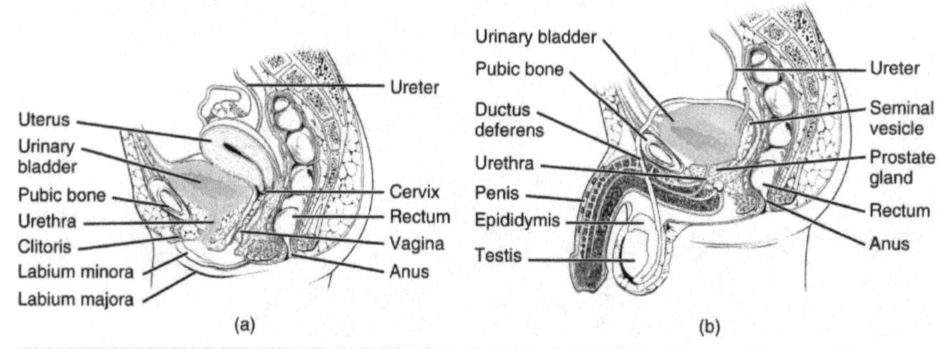

Figure 1. The urethra transports urine from the bladder to the outside of the body. This image shows (a) a female urethra and (b) a male urethra.

The urethra in both males and females begins inferior and central to the two ureteral openings forming the three points of a triangular-shaped area at the base of the bladder called the **trigone** (Greek *tri–* = "triangle" and the root of the word "trigonometry"). The urethra tracks posterior and inferior to the pubic symphysis (see Figure 1). In both males and females, the proximal urethra is lined by transitional epithelium, whereas the terminal portion is a nonkeratinized, stratified squamous epithelium. In the male, pseudostratified columnar epithelium lines the urethra between these two cell types. Voiding is regulated by an involuntary autonomic nervous system-controlled **internal urinary sphincter**, consisting of smooth muscle and voluntary skeletal muscle that forms the **external urinary sphincter** below it.

Female Urethra

The external urethral orifice is embedded in the anterior vaginal wall inferior to the clitoris, superior to the vaginal opening (introitus), and medial to the labia minora. Its short length, about 4 cm, is less of a barrier to fecal bacteria than the longer male urethra and the best explanation for the greater incidence of UTI in women. Voluntary control of the external urethral sphincter is a function of the pudendal nerve. It arises in the sacral region of the spinal cord, traveling via the S2–S4 nerves of the sacral plexus.

Male Urethra

The male urethra passes through the prostate gland immediately inferior to the bladder before passing below the pubic symphysis (see Figure 1b). The length of the male urethra varies between men but averages 20 cm in length. It is divided into four regions: the preprostatic urethra, the prostatic urethra, the membranous urethra, and the spongy or penile urethra. The preprostatic urethra is very short and incorporated into the bladder wall. The prostatic urethra passes through the prostate gland. During sexual intercourse, it receives sperm via the ejaculatory ducts and secretions from the seminal vesicles. Paired Cowper's glands (bulbourethral glands) produce and secrete mucus into the urethra to buffer urethral pH during sexual stimulation. The mucus neutralizes the usually acidic environment and lubricates the urethra, decreasing the resistance to ejaculation. The membranous urethra passes through the deep muscles of the perineum, where it is invested by the overlying urethral sphincters. The spongy urethra exits at the tip (external urethral orifice) of the penis after passing through the corpus spongiosum. Mucous glands are found along much of the length of the urethra and protect the urethra from extremes of urine pH. Innervation is the same in both males and females.

Bladder

The urinary bladder collects urine from both ureters. The bladder lies anterior to the uterus in females, posterior to the pubic bone and anterior to the rectum. During late pregnancy, its capacity is reduced due to compression by the enlarging uterus, resulting in increased frequency of urination. In males, the anatomy is similar, minus the uterus, and with the addition of the prostate inferior to the bladder. The bladder is partially **retroperitoneal** (outside the peritoneal cavity) with its peritoneal-covered "dome" projecting into the abdomen when the bladder is distended with urine.

The bladder is a highly distensible organ comprised of irregular crisscrossing bands of smooth muscle collectively called the **detrusor muscle**. The interior surface is made of transitional cellular epithelium that is structurally suited for the large volume fluctuations of the bladder. When empty, it resembles columnar epithelia, but when stretched, it "transitions" (hence the name) to a squamous appearance. Volumes in adults can range from nearly zero to 500–600 mL.

The detrusor muscle contracts with significant force in the young. The bladder's strength diminishes with age, but voluntary contractions of abdominal skeletal muscles can increase intra-abdominal pressure to promote more forceful bladder emptying. Such voluntary contraction is also used in forceful defecation and childbirth.

Micturition Reflex

Micturition is a less-often used, but proper term for urination or voiding. It results from an interplay of involuntary and voluntary actions by the internal and external urethral sphincters. When bladder volume reaches about 150 mL, an urge to void is sensed but is easily overridden. Voluntary control of urination relies on consciously preventing relaxation of the external urethral sphincter to maintain urinary continence. As the bladder fills, subsequent urges become harder to ignore. Ultimately, voluntary constraint fails with resulting **incontinence**, which will occur as bladder volume approaches 300 to 400 mL.

Normal micturition is a result of stretch receptors in the bladder wall that transmit nerve impulses to the sacral region of the spinal cord to generate a spinal reflex. The resulting parasympathetic neural outflow causes contraction of the detrusor muscle and relaxation of the involuntary internal urethral sphincter. At the same time, the spinal cord inhibits somatic motor neurons, resulting in the relaxation of the skeletal muscle of the external urethral sphincter. The micturition reflex is active in infants but with maturity, children learn to override the reflex by asserting external sphincter control, thereby delaying voiding (potty training). This reflex may be preserved even in the face of spinal cord injury that results in paraplegia or quadriplegia. However, relaxation of the external sphincter may not be possible in all cases, and therefore, periodic catheterization may be necessary for bladder emptying.

Nerves involved in the control of urination include the hypogastric, pelvic, and pudendal. Voluntary micturition requires an intact spinal cord and functional pudendal nerve arising from the **sacral micturition center**. Since the external urinary sphincter is voluntary skeletal muscle, actions by cholinergic neurons maintain contraction (and thereby continence) during filling of the bladder. At the same time, sympathetic nervous activity via the hypogastric nerves suppresses contraction of the detrusor muscle. With further bladder stretch, afferent signals traveling over sacral pelvic nerves activate parasympathetic neurons. This activates efferent neurons to release acetylcholine at the neuromuscular junctions, producing detrusor contraction and bladder emptying.

Ureters

The kidneys and ureters are completely retroperitoneal, and the bladder has a peritoneal covering only over the dome. As urine is formed, it drains into the calyces of the kidney, which merge to form the funnel-shaped renal pelvis in the hilum of each kidney. The hilum narrows to become the ureter of each kidney. As urine passes through the ureter, it does not passively drain into the bladder but rather is propelled by waves of **peristalsis.** As the ureters enter the pelvis, they sweep laterally, hugging the pelvic walls. As they approach the bladder, they turn medially and pierce the bladder wall obliquely. This is important because it creates an one-way valve (a **physiological sphincter** rather than an **anatomical sphincter**) that allows urine into the bladder but prevents reflux of urine from the bladder back into the ureter. Children born lacking this oblique course of the ureter through the bladder wall are susceptible to "vesicoureteral reflux," which dramatically increases their risk of serious UTI. Pregnancy also increases the likelihood of reflux and UTI.

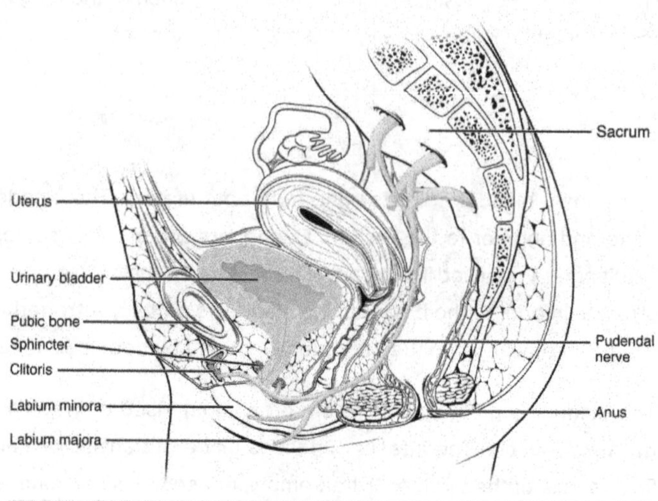

Figure 2. Nerves Innervating the Urinary System

The ureters are approximately 30 cm long. The inner mucosa is lined with transitional epithelium and scattered goblet cells that secrete protective mucus. The muscular layer of the ureter consists of longitudinal and circular smooth muscles that create the peristaltic contractions to move the urine into the bladder without the aid of gravity. Finally, a loose adventitial layer composed of collagen and fat anchors the ureters between the parietal peritoneum and the posterior abdominal wall.

Hormonal Regulation in Urine Production

Several hormones have specific, important roles in regulating kidney function. They act to stimulate or inhibit blood flow. Some of these are endocrine, acting from a distance, whereas others are paracrine, acting locally.

- Water levels in the body are controlled by **antidiuretic hormone (ADH),** which is produced in the hypothalamus and triggers the reabsorption of water by the kidneys. Underproduction of ADH can cause diabetes insipidus.
- **Aldosterone**, a hormone produced by the adrenal cortex of the kidneys, enhances Na reabsorption from the extracellular fluids and subsequent water reabsorption by diffusion.
- The **renin-angiotensin-aldosterone system** is one way that aldosterone release is controlled.

Antidiuretic hormone (ADH)

ADH, a 9-amino acid peptide released by the posterior pituitary – in brain, works to do the exact opposite. **It promotes the recovery of water, decreases urine volume, and maintains plasma osmolarity and blood pressure.** It does so by stimulating the movement of aquaporin proteins into the apical cell membrane of principal cells of the collecting ducts to form water channels.

A proper water balance in the body is important to avoid dehydration or over-hydration. The water concentration of the body is monitored by **osmoreceptors** in the hypothalamus, which detect the concentration of electrolytes in the extracellular fluid. The concentration of electrolytes in the blood rises when there is water loss caused by excessive perspiration, inadequate water intake, or low blood volume due to blood loss. An increase in blood electrolyte levels results in a neuronal signal being sent from the osmoreceptors in hypothalamus.

The **hypothalamus** produces a polypeptide hormone known as **antidiuretic hormone (ADH)**, which is transported to and released from the posterior pituitary gland. The principal action of ADH is to regulate the amount of water excreted by the kidneys. As ADH (which is also known as **vasopressin**) causes direct water reabsorption from the kidney tubules, salts and wastes are concentrated in what will eventually be excreted as urine. The hypothalamus controls the mechanisms of ADH secretion, either by regulating blood volume or the concentration of water in the blood. Dehydration or physiological stress can cause an increase of osmolarity above 300 mOsm/L, which in turn, raises ADH secretion and water will be retained, causing an increase in blood pressure. ADH travels in the bloodstream to the kidneys. Once at the kidneys, ADH changes the kidneys to become more permeable to water by temporarily inserting **water channels, aquaporins,** into the kidney tubules and collecting ducts. . Water moves out of the kidney tubules through the aquaporins, reducing urine volume. The water is reabsorbed into the capillaries lowering blood osmolarity back toward normal. As blood osmolarity decreases, a negative feedback mechanism reduces osmoreceptor activity in the hypothalamus, and ADH secretion is reduced. ADH release can be reduced by certain substances, including alcohol, which can cause increased urine production and dehydration.

DIABETES INSIPIDUS (DI)

Diabetes insipidus (DI) is a rare disease that causes frequent urination. Chronic underproduction of ADH or a mutation in the ADH receptor results in diabetes insipidus. If the posterior pituitary does not release enough ADH, water cannot be retained by the kidneys and is lost as urine. To make up for lost water, a person with diabetes insipidus may feel the need to drink large amounts and is likely to urinate frequently, even at night, which can disrupt sleep and, on occasion, cause bedwetting. Because of the excretion of abnormally large volumes of dilute urine, people with diabetes insipidus may quickly become dehydrated if they do not drink enough water. If the condition is not severe, dehydration may not occur, but severe cases can lead to electrolyte imbalances due to dehydration.

Diabetes insipidus should not be confused with **diabetes mellitus (DM)**, which results from insulin deficiency or resistance leading to high blood glucose, also called blood sugar. **Diabetes mellitus** has two main forms, type 1**diabetes** and type 2 **diabetes. Diabetes insipidus** is a different form of illness altogether.

Diuretics and Fluid Volume

A **diuretic** is a compound that increases urine volume. Three familiar drinks contain diuretic compounds: coffee, tea, and alcohol. The caffeine in coffee and tea works by promoting vasodilation in the nephron, which increases GFR. Alcohol increases GFR by inhibiting ADH release from the posterior pituitary, resulting in less water recovery by the collecting duct. In cases of high blood pressure, diuretics may be prescribed to reduce blood volume and, thereby, reduce blood pressure. The most frequently prescribed anti-hypertensive diuretic is hydrochlorothiazide. It inhibits the Na / Cl^- symporter in the DCT and collecting duct. The result is a loss of Na with water following passively by osmosis.

Osmotic diuretics promote water loss by osmosis. An example is the indigestible sugar mannitol, which is most often administered to reduce brain swelling after head injury. However, it is not the only sugar that can produce a diuretic effect. In cases of poorly controlled diabetes mellitus, glucose levels exceed the capacity of the tubular glucose symporters, resulting in glucose in the urine. The unrecovered glucose becomes a powerful osmotic diuretic. Classically, in the days before glucose could be detected in the blood and urine, clinicians identified diabetes mellitus by the three Ps: polyuria (diuresis), polydipsia (increased thirst), and polyphagia (increased hunger).

Aldosterone:

Another hormone responsible for maintaining electrolyte concentrations in extracellular fluids is **aldosterone**, a steroid hormone that is produced by the **adrenal cortex.** In contrast to ADH, which promotes the reabsorption of water to maintain proper water balance, aldosterone maintains proper water balance by enhancing Na reabsorption and K secretion from extracellular fluid of the cells in kidney tubules. Because it is produced in the cortex of the adrenal gland and affects the concentrations of minerals Na and K, aldosterone is referred to as a **mineralocorticoid**, a corticosteroid that affects ion and water balance. Aldosterone release is stimulated by a decrease in blood sodium levels, blood volume, or blood pressure, or an increase in blood potassium levels. It also prevents the loss of Na from sweat, saliva, and gastric juice. The reabsorption of Na also results in the osmotic reabsorption of water, which alters blood volume and blood pressure.

The renin-angiotensin-aldosterone system (RAAS)

Aldosterone production can be stimulated by low blood pressure, which triggers a sequence of chemical release, as illustrated in Figure 1. When blood pressure drops, the renin-angiotensin-aldosterone system (RAAS) is activated. Cells in the juxtaglomerular apparatus, which regulates the functions of the nephrons of the kidney, detect this and release **renin**. Renin, an enzyme, circulates in the blood and reacts with a plasma protein produced by the liver called angiotensinogen. When angiotensinogen is cleaved by renin, it produces angiotensin I, which is then converted into angiotensin II in the lungs. Angiotensin II functions as a hormone and then causes the release of the hormone aldosterone by the adrenal cortex, resulting in increased Na reabsorption, water retention, and an increase in blood pressure. Angiotensin II in addition to being a potent vasoconstrictor also causes an increase in ADH and increased thirst, both of which help to raise blood pressure.

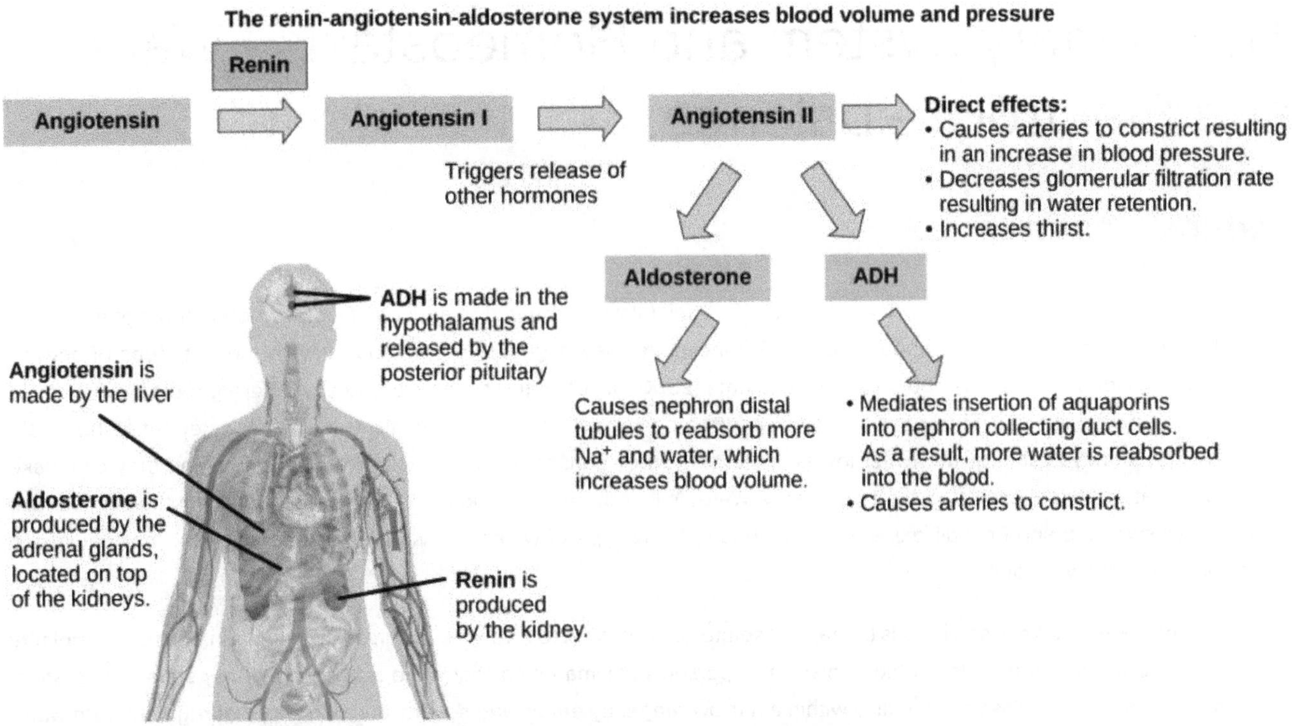

Figure 1. ADH and aldosterone increase blood pressure and volume. Angiotensin II stimulates release of these hormones. Angiotensin II, in turn, is formed when renin cleaves angiotensin. (credit: modification of work by Mikael Häggström)

Hormones That Affect Osmoregulation and kidney function

Hormone	Where produced	Function
Epinephrine and Norepinephrine	Adrenal medulla	Can decrease kidney function temporarily by vasoconstriction
Renin	Kidney nephrons	Increases blood pressure by acting on angiotensinogen
Angiotensin	Liver	Angiotensin II affects multiple processes and increases blood pressure
Aldosterone	Adrenal cortex	Prevents loss of sodium and water
Anti-diuretic hormone – ADH (vasopressin)	Hypothalamus (stored in the posterior pituitary)	Prevents water loss

The Urinary System and Homeostasis - Water and Electrolyte balance.

Water Balance

On a typical day, the average adult will take in about 2500 mL (almost 3 quarts) of aqueous fluids. Although most of the intake comes through the digestive tract, about 230 mL (8 ounces) per day is generated metabolically, in the last steps of aerobic respiration. Additionally, each day about the same volume (2500 mL) of water leaves the body by different routes; most of this lost water is removed as urine. The kidneys also can adjust blood volume though mechanisms that draw water out of the filtrate and urine. The kidneys can regulate water levels in the body; they conserve water if you are dehydrated, and they can make urine more dilute to expel excess water if necessary. Water is lost through the skin through evaporation from the skin surface without overt sweating and from air expelled from the lungs. This type of water loss is called insensible water loss because a person is usually unaware of it.

Regulation of Water Intake: Osmolality is the ratio of solutes in a solution to a volume of solvent in a solution. **Plasma osmolality** is thus the ratio of solutes to water in blood plasma. A person's plasma osmolality value reflects his or her state of hydration. A healthy body maintains plasma osmolality within a narrow range, by employing several mechanisms that regulate both water intake and output.

Drinking water is considered voluntary. So how is water intake regulated by the body? Consider someone who is experiencing **dehydration**, a net loss of water that results in insufficient water in blood and other tissues. The water that leaves the body, as exhaled air, sweat, or urine, is ultimately extracted from blood plasma. As the blood becomes more concentrated, the thirst response—a sequence of physiological processes—is triggered. Osmoreceptors are sensory receptors in the thirst center in the hypothalamus that monitor the concentration of solutes (osmolality) of the blood. If blood osmolality increases above its ideal value, the hypothalamus transmits signals that result in a conscious awareness of thirst. The person should (and normally does) respond by drinking water. The hypothalamus of a dehydrated person also releases antidiuretic hormone (ADH) through the posterior pituitary gland. ADH signals the kidneys to recover water from urine, effectively diluting the blood plasma. To conserve water, the hypothalamus of a dehydrated person also sends signals via the sympathetic nervous system to the salivary glands in the mouth. The signals result in a decrease in watery, serous output (and an increase in stickier, thicker mucus output). These changes in secretions result in a "dry mouth" and the sensation of thirst.

Decreased blood volume resulting from water loss has two additional effects. First, baroreceptors, blood-pressure receptors in the arch of the aorta and the carotid arteries in the neck, detect a decrease in blood pressure that results from decreased blood volume. The heart is ultimately signaled to increase its rate and/or strength of contractions to compensate for the lowered blood pressure.

Second, the kidneys have a renin-angiotensin hormonal system that increases the production of the active form of the hormone angiotensin II, which helps stimulate thirst, but also stimulates the release of the hormone aldosterone from the adrenal glands. Aldosterone increases the reabsorption of sodium in the distal tubules of the nephrons in the kidneys, and water follows this reabsorbed sodium back into the blood.

If adequate fluids are not consumed, dehydration results and a person's body contains too little water to function correctly. A person who repeatedly vomits or who has diarrhea may become dehydrated, and infants, because their body mass is so low, can become dangerously dehydrated very quickly. Endurance athletes such as distance runners often become dehydrated during long races. Dehydration can be a medical emergency, and a dehydrated person may lose consciousness, become comatose, or die, if his or her body is not rehydrated quickly.

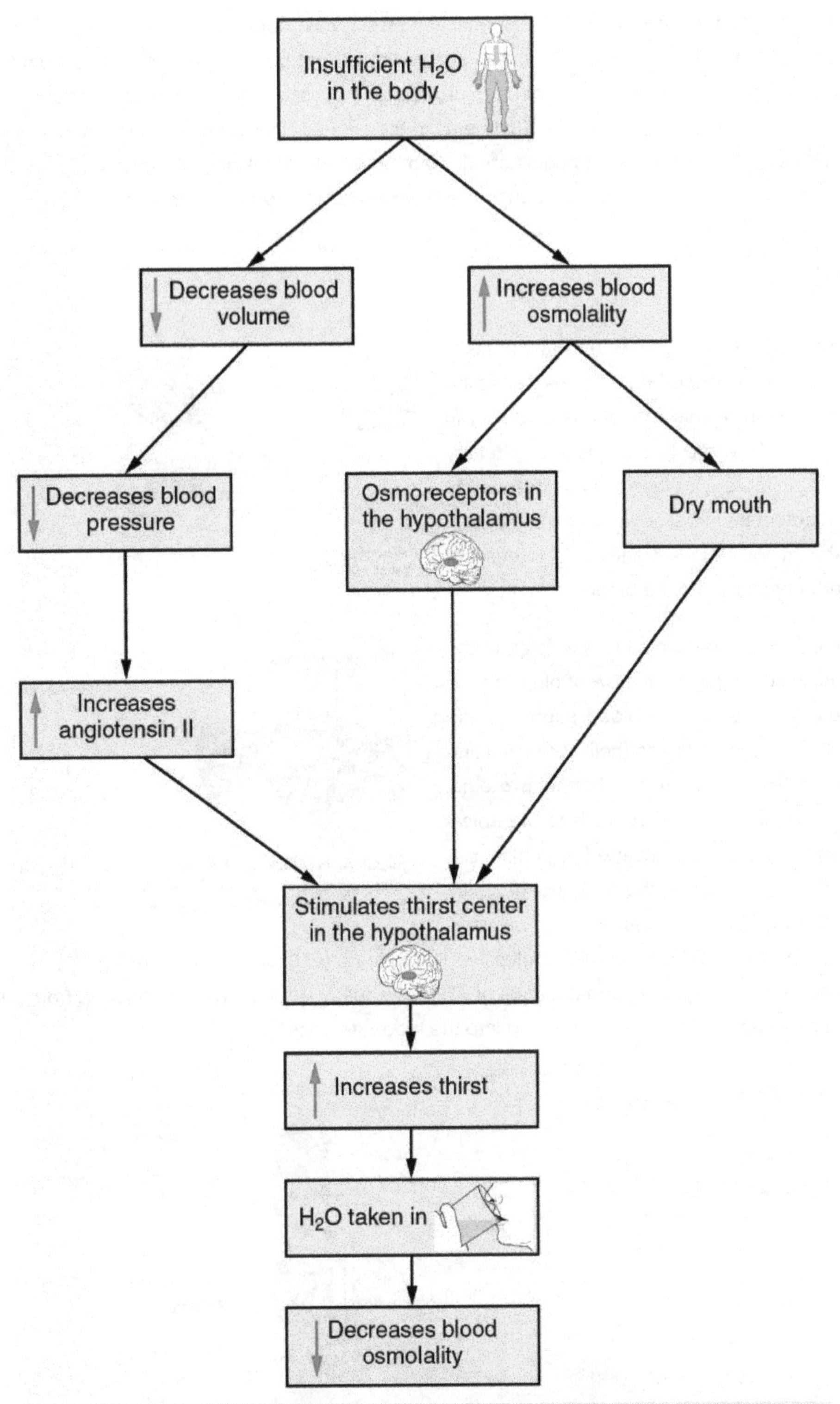

Figure 1. Click to view a larger image. The thirst response begins when osmoreceptors detect a decrease in water levels in the blood.

Regulation of Water Output

Water loss from the body occurs predominantly through the renal system. A person produces an average of 1.5 liters (1.6 quarts) of urine per day. Although the volume of urine varies in response to hydration levels, there is a minimum volume of urine production required for proper bodily functions. The kidney excretes 100 to 1200 milliosmoles of solutes per day to rid the body of a variety of excess salts and other water-soluble chemical wastes, most notably creatinine, urea, and uric acid. Failure to produce the minimum volume of urine means that metabolic wastes cannot be effectively removed from the body, a situation that can impair organ function. The minimum level of urine production necessary to maintain normal function is about 0.47 liters (0.5 quarts) per day. The kidneys also must make adjustments in the event of ingestion of too much fluid. **Diuresis**, which is the production of urine in excess of normal levels, begins about 30 minutes after drinking a large quantity of fluid. Diuresis reaches a peak after about 1 hour, and normal urine production is reestablished after about 3 hours.

Role of ADH

Antidiuretic hormone (ADH), also known as vasopressin, controls the amount of water reabsorbed from the collecting ducts and tubules in the kidney. This hormone is produced in the hypothalamus and is delivered to the posterior pituitary for storage and release (Figure 2.). When the osmoreceptors in the hypothalamus detect an increase in the concentration of blood plasma, the hypothalamus signals the release of ADH from the posterior pituitary into the blood.

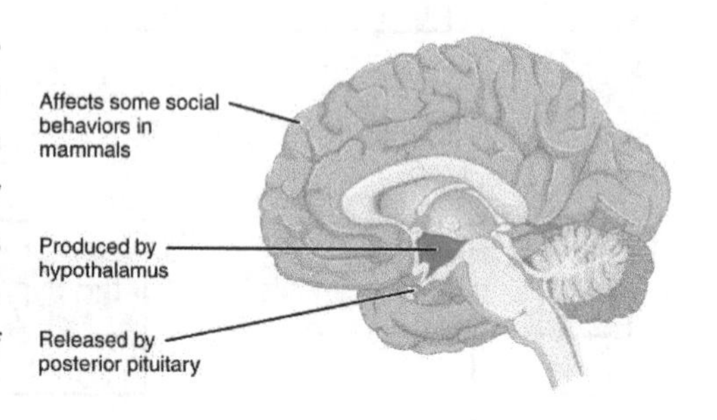

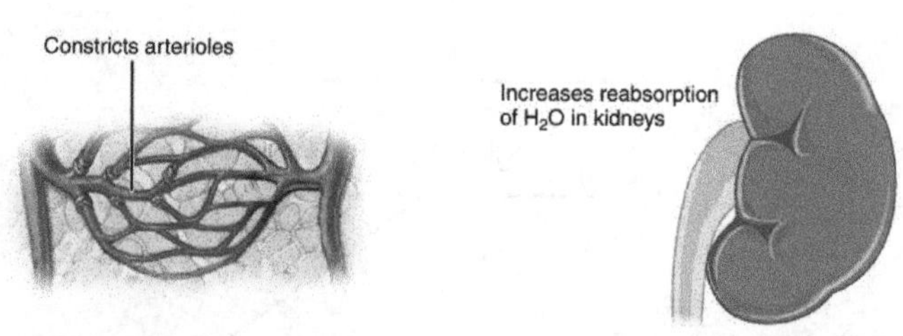

Figure 2. ADH is produced in the hypothalamus and released by the posterior pituitary gland. It causes the kidneys to retain water, constricts arterioles in the peripheral circulation, and affects some social behaviors in mammals.

ADH has two major effects. It constricts the arterioles in the peripheral circulation, which reduces the flow of blood to the extremities and thereby increases the blood supply to the core of the body. ADH also causes the epithelial cells that line the renal collecting tubules to move **water channel proteins, called aquaporins,** from the interior of the cells to the apical surface, where these proteins are inserted into the cell membrane. The result is an increase in the water permeability of these cells and, thus, a large increase in water passage from the urine through the walls of the collecting tubules, leading to more reabsorption of water into the bloodstream. When the blood plasma becomes less concentrated and the level of ADH decreases, aquaporins are removed from collecting tubule cell membranes, and the passage of water out of urine and into the blood decreases.

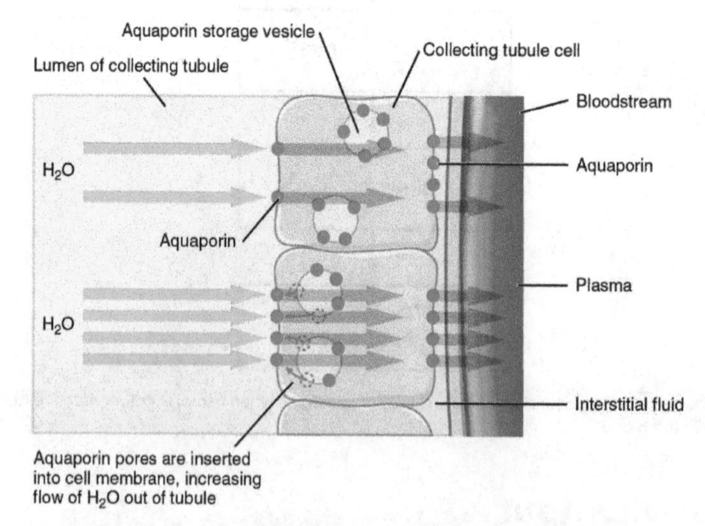

Figure 3. The binding of ADH to receptors on the cells of the collecting tubule results in aquaporins being inserted into the plasma membrane, shown in the lower cell. This dramatically increases the flow of water out of the tubule and into the bloodstream.

A diuretic is a compound that increases urine output and therefore decreases water conservation by the body. Diuretics are used to treat hypertension, congestive heart failure, and fluid retention associated with menstruation. Alcohol acts as a diuretic by inhibiting the release of ADH. Additionally, caffeine, when consumed in high concentrations, acts as a **diuretic.**

Electrolyte balance

The body contains a large variety of ions, or electrolytes, which perform a variety of functions. Some ions assist in the transmission of electrical impulses along cell membranes in neurons and muscles. Other ions help to stabilize protein structures in enzymes. Still others aid in releasing hormones from endocrine glands. All of the ions in plasma contribute to the osmotic balance that controls the movement of water between cells and their environment.

Electrolytes in living systems include sodium, potassium, chloride, bicarbonate, calcium, phosphate, magnesium, copper, zinc, iron, manganese, molybdenum, copper, and chromium. In terms of body functioning, **six electrolytes are most important: sodium, potassium, chloride, bicarbonate, calcium, and phosphate.** These six ions aid in nerve excitability, endocrine secretion, membrane permeability, buffering body fluids, and controlling the movement of fluids between compartments.

Excretion of ions occurs mainly through the kidneys, with lesser amounts lost in sweat and in feces. Excessive sweating may cause a significant loss, especially of sodium and chloride. Severe vomiting or diarrhea will cause a loss of chloride and bicarbonate ions. Adjustments in respiratory and renal functions allow the body to regulate the levels of these ions in the ECF.

The following table lists the reference values for blood plasma, cerebrospinal fluid (CSF), and urine for the six ions addressed in this section. In a clinical setting, sodium, potassium, and chloride are typically analyzed in a routine urine sample. In contrast, calcium and phosphate analysis requires a collection of urine across a 24-hour period, because the output of these ions can vary considerably over the course of a day. Urine values reflect the rates of excretion of these ions. Bicarbonate is the one ion that is not normally excreted in urine; instead, it is conserved by the kidneys for use in the body's buffering systems.

Table 1. Electrolyte and Ion Reference Values

Name	Chemical symbol	Plasma	CSF	Urine
Sodium	Na^+	136.00–146.00 (mM)	138.00–150.00 (mM)	40.00–220.00 (mM)
Potassium	K^+	3.50–5.00 (mM)	0.35–3.5 (mM)	25.00–125.00 (mM)
Chloride	Cl^-	98.00–107.00 (mM)	118.00–132.00 (mM)	110.00–250.00 (mM)
Bicarbonate	HCO_3^-	22.00–29.00 (mM)	—	—
Calcium	Ca^{++}	2.15–2.55 (mmol/day)	—	Up to 7.49 (mmol/day)
Phosphate	PO4	0.81–1.45 (mmol/day)	—	12.90–42.00 (mmol/day)

Imbalances of these ions can result in various problems in the body, and their concentrations are tightly regulated. Aldosterone and angiotensin II control the exchange of sodium and potassium between the renal filtrate and the renal collecting tubule. Calcium and phosphate are regulated by PTH, calcitrol, and calcitonin.

Sodium

Sodium is the major cation of the extracellular fluid. It is responsible for one-half of the osmotic pressure gradient that exists between the interior of cells and their surrounding environment. People eating a typical Western diet, which is very high in NaCl, routinely take in 130 to 160 mmol/day of sodium, but humans require only 1 to 2 mmol/day. This excess sodium appears to be a major factor in hypertension (high blood pressure) in some people. Excretion of sodium is accomplished primarily by the kidneys. Sodium is freely filtered through the glomerular capillaries of the kidneys, and although much of the filtered sodium is reabsorbed in the proximal convoluted tubule, some remains in the filtrate and urine, and is normally excreted.

Hyponatremia is a lower-than-normal concentration of sodium, usually associated with excess water accumulation in the body, which dilutes the sodium. An abnormal loss of sodium from the body can result from several conditions, including excessive sweating, vomiting, or diarrhea; the use of diuretics; excessive production of urine, which can occur in diabetes; and acidosis, either metabolic acidosis or diabetic ketoacidosis. At the cellular level, hyponatremia results in increased entry of water into cells by osmosis, because the concentration of solutes within the cell exceeds the concentration of solutes in the now-diluted ECF. The excess water causes swelling of the cells; the swelling of red blood cells—decreasing their oxygen-carrying efficiency and making them potentially too large to fit through capillaries—along with the swelling of neurons in the brain can result in brain damage or even death. **Hypernatremia** is an abnormal increase of blood sodium. It can result from water loss from the blood. Hormonal imbalances involving ADH and aldosterone may also result in higher-than-normal sodium values.

Potassium

Potassium is the major intracellular cation. It helps establish the resting membrane potential in neurons and muscle fibers after membrane depolarization and action potentials. In contrast to sodium, potassium has very little effect on osmotic pressure.

Hypokalemia is an abnormally low potassium blood level. Similar to the situation with hyponatremia, hypokalemia can occur because of either an absolute reduction of potassium in the body or a relative reduction of potassium in the blood due to the redistribution of potassium. An absolute loss of potassium can arise from decreased intake, frequently related to starvation. It can also come about from vomiting, diarrhea, or alkalosis. **Hyperkalemia**, an elevated potassium blood level, also can impair the function of skeletal muscles, the nervous system, and the heart. Hyperkalemia can result from increased dietary intake of potassium. In such a situation, potassium from the blood ends up in the ECF in abnormally high concentrations. This can result in a partial depolarization (excitation) of the plasma membrane of skeletal muscle fibers, neurons, and cardiac cells of the heart, and can also lead to an inability of cells to repolarize. For the heart, this means that it won't relax after a contraction, and will effectively "seize" and stop pumping blood, which is fatal within minutes. Because of such effects on the nervous system, a person with hyperkalemia may also exhibit mental confusion, numbness, and weakened respiratory muscles.

Chloride

Chloride is the predominant extracellular anion. Chloride is a major contributor to the osmotic pressure gradient between the ICF and ECF, and plays an important role in maintaining proper hydration.

Hypochloremia, or lower-than-normal blood chloride levels, can occur because of defective renal tubular absorption. Vomiting, diarrhea, and metabolic acidosis can also lead to hypochloremia. **Hyperchloremia**, or higher-than-normal blood chloride levels, can occur due to dehydration, excessive intake of dietary salt (NaCl) or swallowing of sea water, aspirin intoxication, congestive heart failure, and the hereditary, chronic lung disease, cystic fibrosis. In people who have cystic fibrosis, chloride levels in sweat are two to five times those of normal levels, and analysis of sweat is often used in the diagnosis of the disease.

Bicarbonate

Bicarbonate is the second most abundant anion in the blood. Its principal function is to maintain your body's acid-base balance by being part of buffer systems.

Calcium

About two pounds of calcium in your body are bound up in bone, which provides hardness to the bone and serves as a mineral reserve for calcium and its salts for the rest of the tissues. Teeth also have a high concentration of calcium within them. A little more than one-half of blood calcium is bound to proteins, leaving the rest in its ionized form. Calcium ions, Ca^{2+}, are necessary for muscle contraction, enzyme activity, and blood coagulation. In addition, calcium helps to stabilize cell membranes and is essential for the release of neurotransmitters from neurons and of hormones from endocrine glands. Calcium is absorbed through the intestines under the influence of activated vitamin D. A deficiency of vitamin D leads to a decrease in absorbed calcium and, eventually, a depletion of calcium stores from the skeletal system, potentially leading to rickets in children and osteomalacia in adults, contributing to osteoporosis.

Hypocalcemia, or abnormally low calcium blood levels, is seen in hypoparathyroidism, which may follow the removal of the thyroid gland, because the four nodules of the parathyroid gland are embedded in it. **Hypercalcemia**, or abnormally high calcium blood levels, is seen in primary hyperparathyroidism. Some malignancies may also result in hypercalcemia.

Phosphate

Bone and teeth bind up 85 percent of the body's phosphate as part of calcium-phosphate salts. Phosphate is found in phospholipids, such as those that make up the cell membrane, and in ATP, nucleotides, and buffers.

Hypophosphatemia, or abnormally low phosphate blood levels, occurs with heavy use of antacids, during alcohol withdrawal, and during malnourishment. In the face of phosphate depletion, the kidneys usually conserve phosphate, but during starvation, this conservation is impaired greatly. **Hyperphosphatemia**, or abnormally increased levels of phosphates in the blood, occurs if there is decreased renal function or in cases of acute lymphocytic leukemia. Additionally, because phosphate is a major constituent of the ICF, any significant destruction of cells can result in dumping of phosphate into the ECF.

Regulation of Sodium and Potassium

Sodium is reabsorbed from the renal filtrate, and potassium is excreted into the filtrate in the renal collecting tubule. The control of this exchange is governed principally by two hormones—aldosterone and angiotensin II.

Aldosterone

Recall that aldosterone increases the excretion of potassium and the reabsorption of sodium in the distal tubule. Aldosterone is released if blood levels of potassium increase, if blood levels of sodium severely decrease, or if blood pressure decreases. Its net effect is to conserve and increase water levels in the plasma by reducing the excretion of sodium, and thus water, from the kidneys. In a negative feedback loop, increased osmolality of the ECF (which follows aldosterone-stimulated sodium absorption) inhibits the release of the hormone.

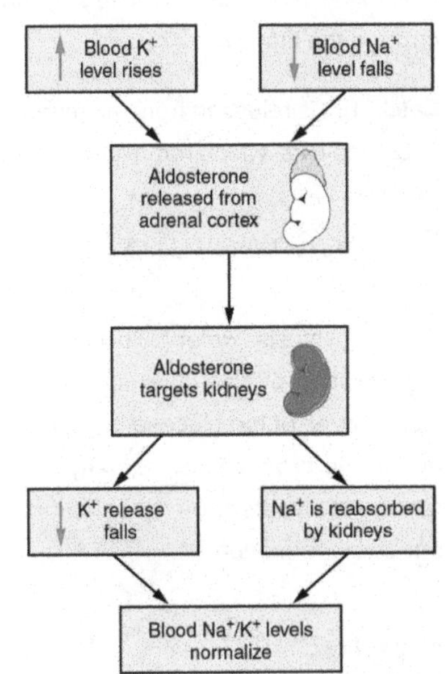

Figure 4. Aldosterone, which is released by the adrenal gland, facilitates reabsorption of Na^+ and thus the reabsorption of water.

Angiotensin II

Angiotensin II causes vasoconstriction and an increase in systemic blood pressure. This action increases the glomerular filtration rate, resulting in more material filtered out of the glomerular capillaries and into Bowman's capsule. Angiotensin II also signals an increase in the release of aldosterone from the adrenal cortex.

In the distal convoluted tubules and collecting ducts of the kidneys, aldosterone stimulates the synthesis and activation of the sodium-potassium pump. Sodium passes from the filtrate, into and through the cells of the tubules and ducts, into the ECF and then into capillaries. Water follows the sodium due to osmosis. Thus, aldosterone causes an increase in blood sodium levels and blood volume. Aldosterone's effect on potassium is the reverse of that of sodium; under its influence, excess potassium is pumped into the renal filtrate for excretion from the body.

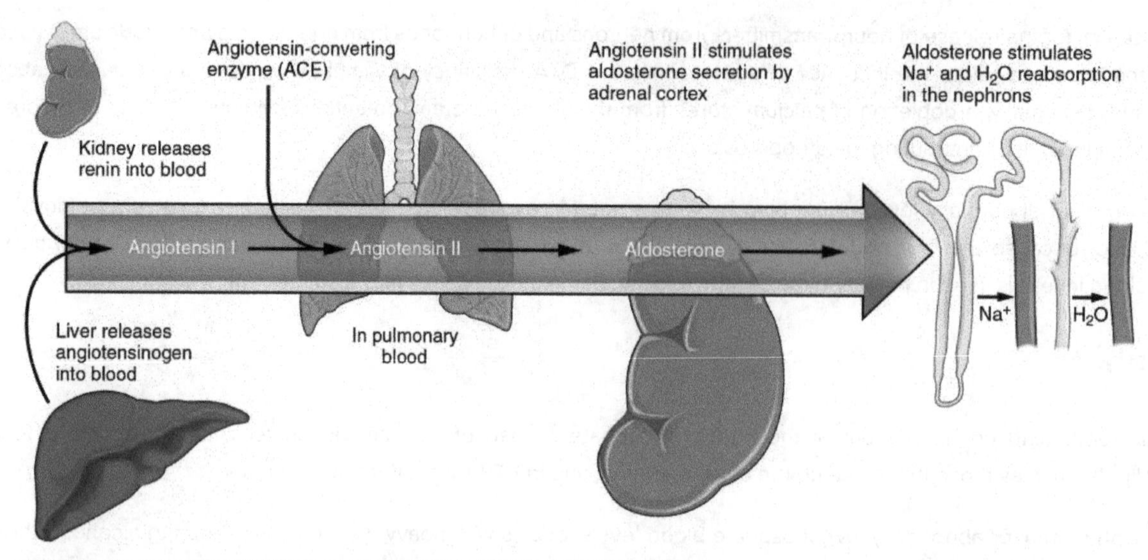

Figure 5. Angiotensin II stimulates the release of aldosterone from the adrenal cortex.

Regulation of Calcium and Phosphate

Calcium and phosphate are both regulated through the actions of three hormones: **parathyroid hormone (PTH), dihydroxyvitamin D (calcitriol), and calcitonin**. All three are released or synthesized in response to the blood levels of calcium.

PTH is released from the parathyroid gland in response to a decrease in the concentration of blood calcium. The hormone activates osteoclasts to break down bone matrix and release inorganic calcium-phosphate salts. PTH also increases the gastrointestinal absorption of dietary calcium by converting vitamin D into **dihydroxyvitamin D** (calcitriol), an active form of vitamin D that intestinal epithelial cells require to absorb calcium.

PTH raises blood calcium levels by inhibiting the loss of calcium through the kidneys. PTH also increases the loss of phosphate through the kidneys.

Calcitonin is released from the thyroid gland in response to elevated blood levels of calcium. The hormone increases the activity of osteoblasts, which remove calcium from the blood and incorporate calcium into the bony matrix.

Blood Pressure Regulation

Due to osmosis, water follows where Na^+ leads. Much of the water the kidneys recover from the forming urine follows the reabsorption of Na^+. ADH stimulation of aquaporin channels allows for regulation of water recovery in the collecting ducts. Normally, all of the glucose is recovered, but loss of glucose control (diabetes mellitus) may result in an osmotic dieresis severe enough to produce severe dehydration and death. A loss of renal function means a loss of effective vascular volume control, leading to hypotension (low blood pressure) or hypertension (high blood pressure), which can lead to stroke, heart attack, and aneurysm formation.

Regulation of Osmolarity

Blood pressure and osmolarity are regulated in a similar fashion. Severe hypo-osmolarity can cause problems like lysis (rupture) of blood cells or widespread edema, which is due to a solute imbalance. Inadequate solute concentration (such as protein) in the plasma results in water moving toward an area of greater solute concentration, in this case, the interstitial space and cell cytoplasm. If the kidney glomeruli are damaged by an autoimmune illness, large quantities of protein may be lost in the urine.

The resultant drop in serum osmolarity leads to widespread edema that, if severe, may lead to damaging or fatal brain swelling. Severe hypertonic conditions may arise with severe dehydration from lack of water intake, severe vomiting, or uncontrolled diarrhea. When the kidney is unable to recover sufficient water from the forming urine, the consequences may be severe (lethargy, confusion, muscle cramps, and finally, death) .

pH Regulation

Proper kidney function is essential for pH homeostasis.

Vitamin D Synthesis

In order for vitamin D to become active, it must undergo a hydroxylation reaction in the kidney, that is, an –OH group must be added to calcidiol to make calcitriol (1,25-dihydroxycholecalciferol). Activated vitamin D is important for absorption of Ca^{++} in the digestive tract, its reabsorption in the kidney, and the maintenance of normal serum concentrations of Ca^{++} and phosphate. Calcium is vitally important in bone health, muscle contraction, hormone secretion, and neurotransmitter release. Inadequate Ca^{++} leads to disorders like osteoporosis and **osteomalacia in adults and rickets in children.** Deficits may also result in problems with cell proliferation, neuromuscular function, blood clotting, and the inflammatory response. Recent research has confirmed that vitamin D receptors are present in most, if not all, cells of the body, reflecting the systemic importance of vitamin D. Many scientists have suggested it be referred to as a hormone rather than a vitamin.

Erythropoiesis: control by EPO hormone.

EPO is a 193-amino acid protein that stimulates the formation of red blood cells in the bone marrow. The kidney produces 85 percent of circulating EPO; the liver, the remainder. If you move to a higher altitude, the partial pressure of oxygen is lower, meaning there is less pressure to push oxygen across the alveolar membrane and into the red blood cell. One way the body compensates is to manufacture more red blood cells by increasing EPO production. If you start an aerobic exercise program, your tissues will need more oxygen to cope, and the kidney will respond with more EPO. If erythrocytes are lost due to severe or prolonged bleeding, or under produced due to disease or severe malnutrition, the kidneys come to the rescue by producing more EPO. Renal failure (loss of EPO production) is associated with anemia, which makes it difficult for the body to cope with increased oxygen demands or to supply oxygen adequately even under normal conditions. Anemia diminishes performance and can be life threatening.

Renal Disease and Failure

Infection in Urinary system

A **urinary tract** infection (**UTI**) is an infection in any part of our urinary system — kidneys, ureters, bladder, and urethra. Most infections involve the lower urinary **tract** — the bladder and the urethra. Women are at higher risk of developing a **UTI** than are men. UTI is treated with antibiotics. Urinary tract infections are *caused* by microbes such as bacteria overcoming the body's defenses in the urinary tract. Researchers have supported the use of cranberry juice to prevent UTIs by preventing the attachment to infection causing bacteria to the surface of urinary tract.

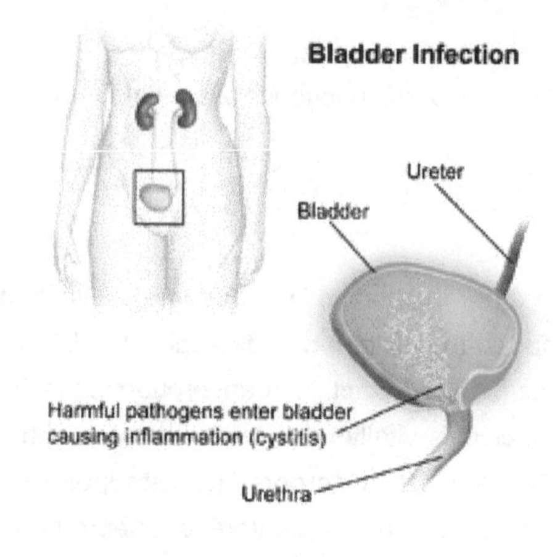

UTI – Bladder infection

- Infection localized in the ureter is called urethritis. **Urethritis** is inflammation of the urethra. The most common symptom is painful or difficult, urination.
- If the infection spreads to urinary bladder, then it is called **cystitis**.
- If the kidneys are infected, it is called **pyelonephritis**

Kidney Stones

The kidney plays a very important role in the body as it filters out waste products. Sometimes kidney stones can form when there is a build up of specific minerals in your urine. Kidney stones are hard, pebble-like pieces of material that form in one or both of your kidneys. They rarely cause permanent damage if treated by a health care professional.

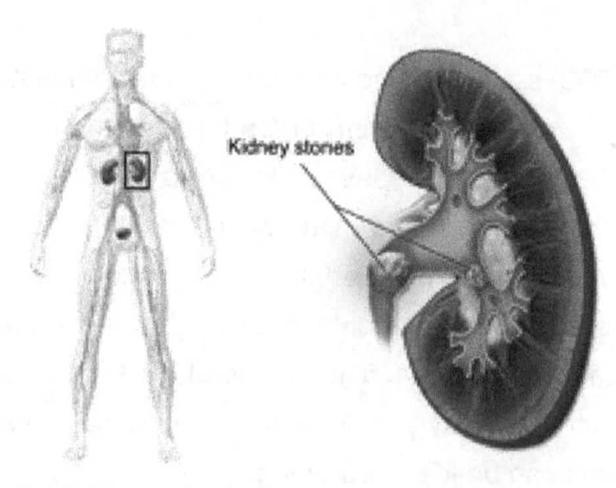

Illustration of kidney stones

Kidney stones vary in size and shape. They may be as small as a grain of sand or as large as a pea. Rarely, some kidney stones are as big as golf balls. Kidney stones may be smooth or jagged and are usually yellow or brown. A small kidney stone may pass through your urinary tract on its own, causing little or no pain. A larger kidney stone may get stuck along the way. A kidney stone that gets stuck can block your flow of urine, causing severe pain or bleeding.

Polycystic Kidney Disease

Polycystic kidney disease (PKD or PCKD, also known as polycystic kidney syndrome) is a cystic genetic disorder of the kidneys. There are two types of PKD: autosomal dominant polycystic kidney disease (ADPKD), and the less-common autosomal recessive polycystic kidney disease (ARPKD). PKD is characterized by the presence of multiple cysts (hence,"polycystic"), typically in both kidneys. The cysts are numerous and are fluid-filled, resulting in massive enlargement of the kidneys. The disease can also damage the liver, pancreas, and, in some rare cases, the heart and brain. The two major forms of polycystic kidney disease are distinguished by their patterns of inheritance. Polycystic kidney disease is one of the most common life-threatening genetic diseases, affecting an estimated 12.5 million people worldwide.

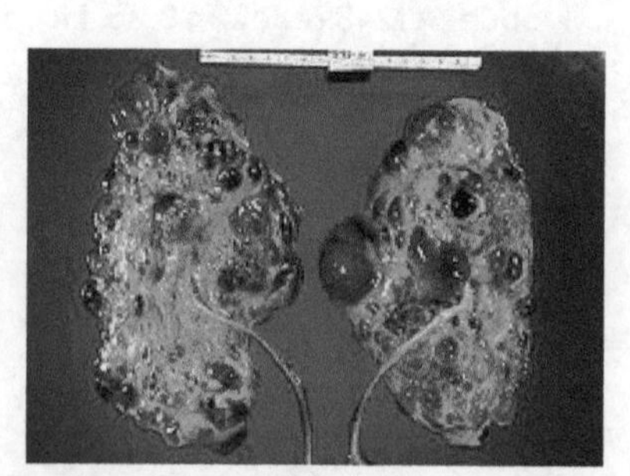
Polycystic Kidney Disease

Benign Prostatic Hyperplasia (BPH)

The **prostate** is a walnut-sized gland of the male reproductive system between the bladder and the penis. The urethra runs through the center of the **prostate. Benign prostatic hyperplasia (BPH)**, also called **prostate enlargement**, is a noncancerous increase in the size of the prostate gland. Symptoms may include frequent urination, trouble starting to urinate, weak stream, inability to urinate, or loss of bladder control. Complications can include urinary tract infections, bladder stones, and chronic kidney problems.

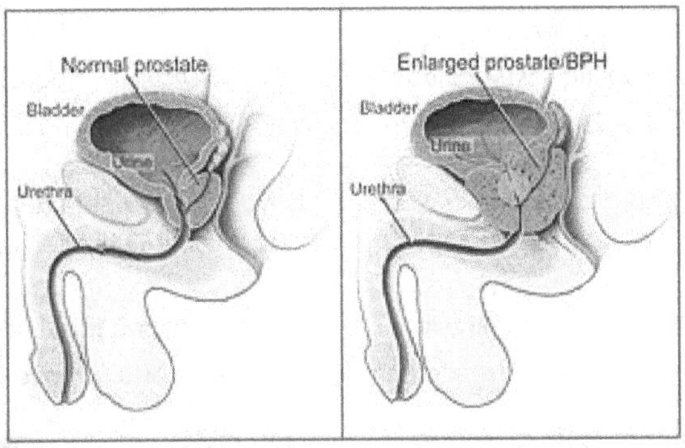

Benign Prostatic Hyperplasia

Renal Failure

Many illness can cause progressive renal diseases and renal failure, especially diabetes and hypertension. **Renal failure** (also kidney failure or renal insufficiency) is a medical condition in which the kidneys fail to adequately filter waste products from the blood. **Diagnosis:** Renal failure is mainly determined by a decrease in the glomerular filtration rate, which is the rate at which blood is filtered in the glomeruli of the kidney. This is detected by a decrease in or absence of urine production or determination of waste products (creatinine or urea) in the blood. Depending on the cause, **hematuria** (blood loss in the urine) and **proteinuria** (protein loss in the urine) may be noted.

In renal failure, there may be problems with increased fluid in the body (leading to swelling), increased acid levels, raised levels of potassium, decreased levels of calcium, increased levels of phosphate, and in later stages, anemia. Bone health may also be affected. Long-term kidney problems are associated with an increased risk of cardiovascular disease.

Categories of Renal Failure

Renal failure can be divided into two categories: **acute kidney injury or chronic kidney disease** .The type of renal failure is determined by the trend in the serum creatinine. Other factors that may help differentiate acute kidney injury from chronic kidney disease include anemia and the kidney size on ultrasound. Chronic kidney disease generally leads to anemia and small kidney size.

Acute kidney injury (AKI), previously called **acute renal failure (ARF),** is a rapidly progressive loss of renal function, generally characterized by oliguria (decreased urine production, quantified as less than 400 mL per day in adults, less than 0.5 mL/kg/h in children or less than 1 mL/kg/h in infants); and fluid and electrolyte imbalance. AKI can result from a variety of causes, generally classified as prerenal, intrinsic, and postrenal. An underlying cause must be identified and treated to arrest the progress, and dialysis may be necessary to bridge the time gap required for treating these fundamental causes.

Chronic kidney disease (CKD) can also develop slowly and, initially, show few symptoms. CKD can be the long-term consequence of irreversible acute disease or part of a disease progression.

Symptoms can vary from person to person. Someone in early stage kidney disease may not feel sick or notice symptoms as they occur. When kidneys fail to filter properly, waste accumulates in the blood and the body, a condition called azotemia. Very low levels of azotaemia may produce few, if any, symptoms. If the disease progresses, symptoms become noticeable (if the failure is of sufficient degree to cause symptoms). **Renal failure uremia** is a syndrome of renal failure characterized by elevated levels of urea and creatinine in the blood.

Renal Failure Uremia

Renal failure uremia is a syndrome of renal failure that includes elevated blood urea and creatinine levels. Acute renal failure can be reversed if diagnosed early. Acute renal failure can be caused by severe hypotension or severe glomerular disease. Diagnostic tests include BUN (Blood Urea Nitrogen) and plasma creatinine level tests. It is considered to be chronic renal failure if the decline of renal function is to less than 25%.

If you catch kidney disease early, you may be able to prevent kidney failure by taking steps to keep your kidneys as healthy as possible for as long as possible. If your kidneys fail, you will need **dialysis** or a **kidney transplant** to survive.

Hemodialysis/dialysis is a process of purifying the blood of a person whose kidneys are not working normally. Dialysis removes excess fluid, correct electrolyte problems, and remove toxins in those with kidney failure, thus maintaining homeostasis.

During hemodialysis, patient's blood is pumped through a filter, called a **dialyzer.** This machine works like an artificial kidney and filters out extra salt, waste, and fluid. Cleaned blood is sent back into your body through the second needle. **Peritoneal dialysis (PD)** uses the peritoneum in a person's abdomen as the membrane through which fluid and dissolved substances are exchanged with the blood.

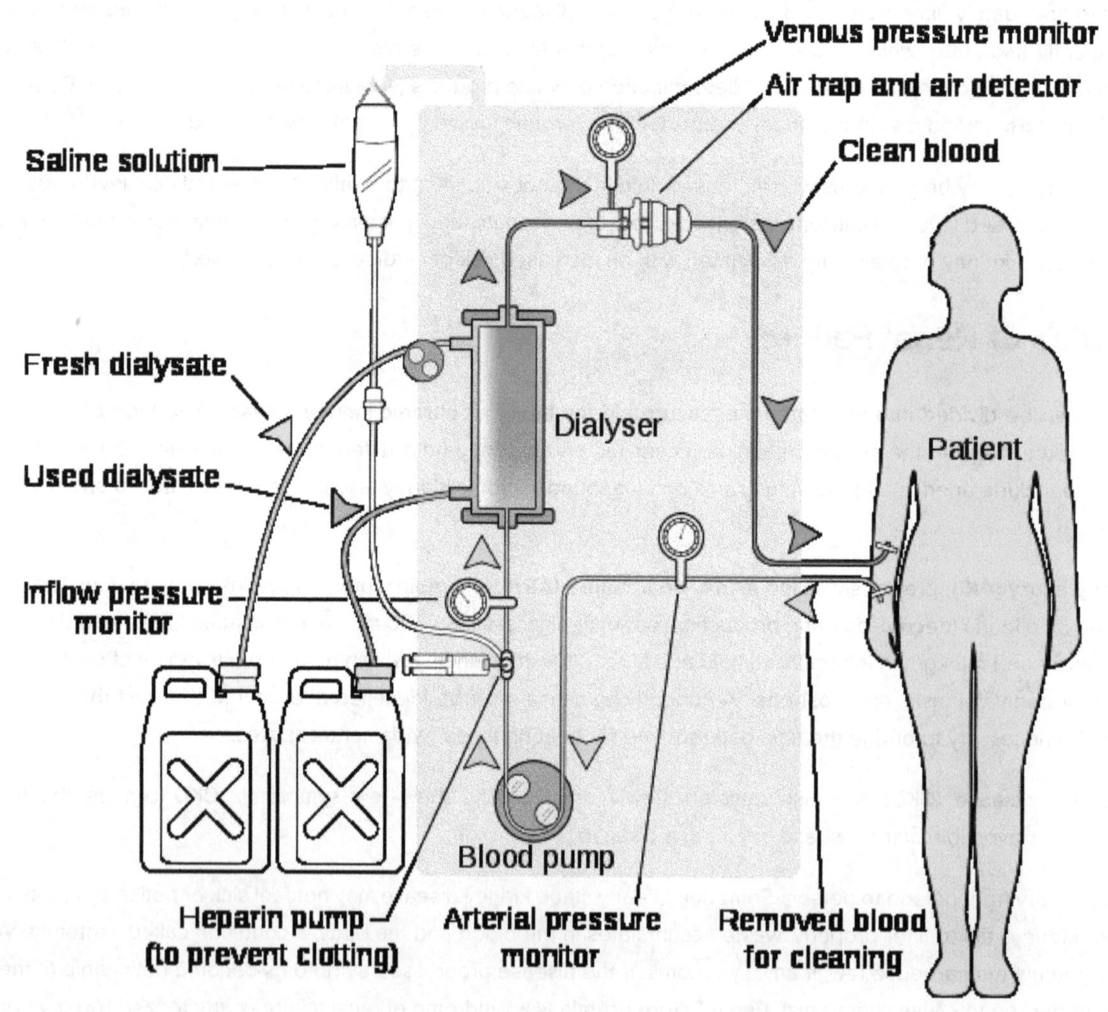

Hemodialysis

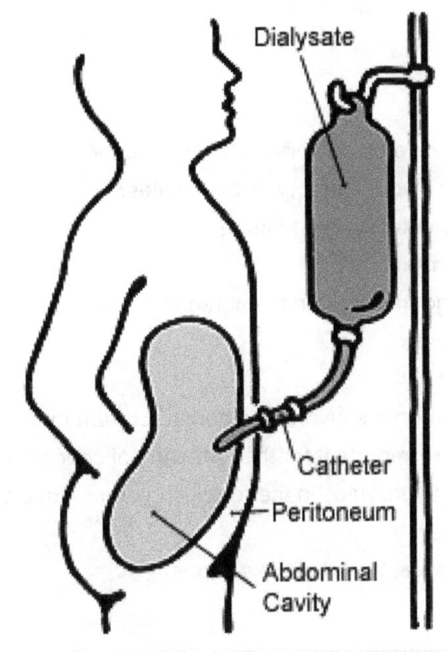

Diagram of peritoneal dialysis

Kidney Transplantation

Kidney transplantation is the organ transplant of a kidney into a patient with end-stage renal disease. The **transplanted kidney** takes over the work of the two **kidneys** that failed, so you no longer need dialysis. Kidney transplantation requires a person to be at the end stage of renal failure.

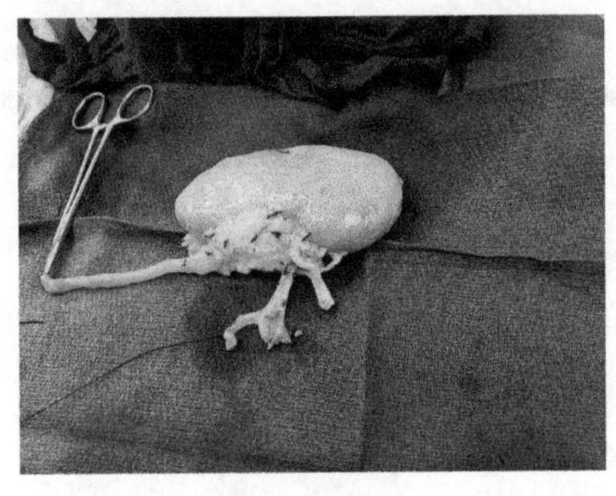

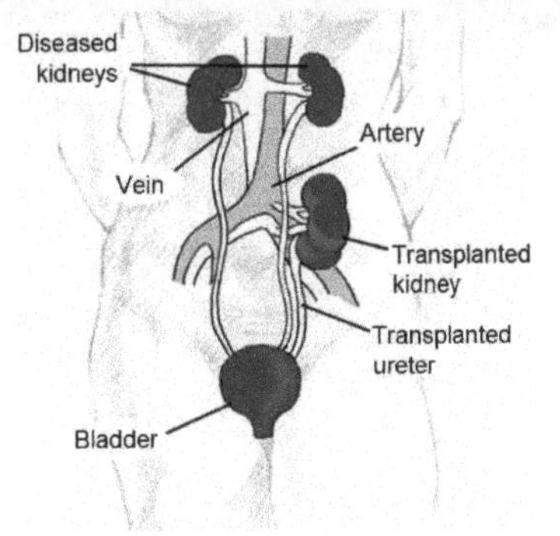

Kidney Transplantation

Concept Check

1. **What** is suggested by the presence of white blood cells found in the urine?
2. **Why** are females more likely to contract bladder infections than males?
3. **What** anatomical structures provide protection to the kidney?
4. **Name** the structures found in the renal hilum.
5. **What** are the major structures comprising the filtration membrane?
6. **Give the formula** for net filtration pressure.
7. **Name** at least five symptoms of kidney failure.
8. **Which** vessels and what part of the nephron are involved in countercurrent multiplication?
9. **Why** are the loop of Henle and vasa recta important for the formation of concentrated urine?
10. **What** organs produce which hormones or enzymes in the renin–angiotensin system?
11. **What** is the role of ADH in water balance?
12. **When** you drink beer,urine output increases. Why does this happen?

CHAPTER 10 — BLOOD

An Overview of Blood

LEARNING OBJECTIVES

- Identify the primary functions of blood, its fluid and cellular components, and its physical characteristics
- Describe the formation of the formed element components of blood
- Discuss the structure and function of red blood cells and hemoglobin
- Classify and characterize white blood cells
- Describe the structure of platelets and explain the process of hemostasis
- Explain the significance of AB and Rh blood groups in blood transfusions
- Discuss a variety of blood disorders

Single-celled organisms do not need blood. They obtain nutrients directly from and excrete wastes directly into their environment. The human organism cannot do that. Our large, complex bodies need blood to deliver nutrients to and remove wastes from our trillions of cells. The heart pumps blood throughout the body in a network of blood vessels. Together, these three components—blood, heart, and vessels—makes up the cardiovascular system. This chapter focuses on the medium of transport: blood.

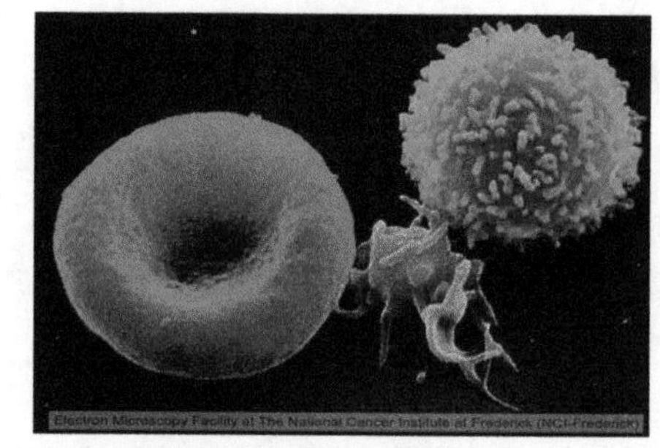

Figure 1: Blood cell types

Functions of Blood

Blood has three main functions: **transport, protection / defense, and regulation.**

Transport: Blood transports the following substances:

- Gases, namely oxygen (O_2) and carbon dioxide (CO_2), between the lungs and rest of the body
- Nutrients from the digestive tract and storage sites to the rest of the body
- Waste products to be detoxified or removed by the liver and kidneys
- Hormones from the glands in which they are produced to their target cells

Protection: Blood has several roles in inflammation:

- Leukocytes, or white blood cells, destroy invading microorganisms and cancer cells and defend our body.
- Antibodies and other proteins destroy pathogenic substances
- Platelet factors initiate blood clotting and protects the body from further blood loss.

Regulation: Blood plays an important role in regulating the body's systems and **maintaining homeostasis.** Blood helps regulate:

- Blood also helps to maintain the chemical balance of the body. Proteins and other compounds in blood act as buffers, which thereby help to regulate the pH of body tissues. Blood also helps to regulate the water content of body cells.
- Water balance by transferring water to and from tissues
- Regulating core body temperature – Our body temperature is regulated via a classic negative-feedback loop. If you were

exercising on a warm day, your rising core body temperature would trigger several homeostatic mechanisms, including increased transport of blood from your core to your body periphery, which is typically cooler. As blood passes through the vessels of the skin, heat would be dissipated to the environment, and the blood returning to your body core would be cooler. In contrast, on a cold day, blood is diverted away from the skin to maintain a warmer body core. In extreme cases, this may result in frostbite.

Composition of Blood

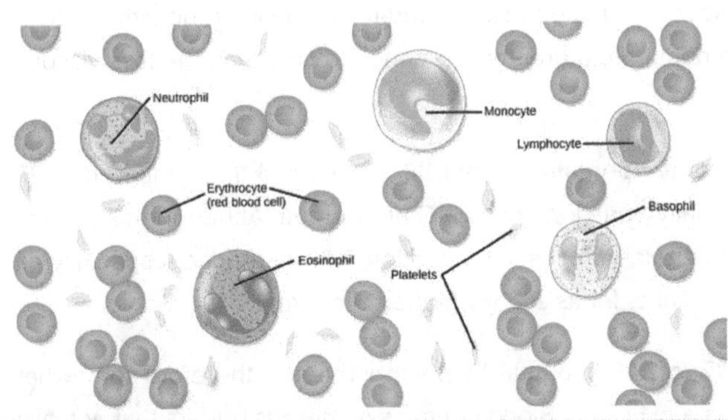

Figure 2: Blood smear showing the components of blood.

Blood is a type of connective tissue. Like all connective tissues, it is made up of cellular elements and an extracellular matrix. The cellular elements—referred to as the formed elements—include red blood cells (RBCs), white blood cells (WBCs), and cell fragments called platelets. The **extracellular matrix, called plasma,** makes blood unique among connective tissues because it is fluid. This fluid, which is mostly water, perpetually suspends the formed elements and enables them to circulate throughout the body within the cardiovascular system.

A hematocrit, measures the percentage of RBCs, clinically known as erythrocytes, in a blood sample. It is performed by spinning the blood sample in a specialized centrifuge, a process that causes the heavier elements suspended within the blood sample to separate from the lightweight, liquid plasma. Because the heaviest elements in blood are the erythrocytes, these settle at the very bottom of the hematocrit tube. Located above the erythrocytes is a pale, thin layer composed of the remaining formed elements of blood. These are the WBCs, clinically known as leukocytes, and the platelets, cell fragments also called thrombocytes. This layer is referred to as the buffy coat because of its color; it normally constitutes less than 1 percent of a blood sample. Above the buffy coat is the blood plasma, normally a pale, straw-colored fluid, which constitutes the remainder of the sample.

The volume of erythrocytes after centrifugation is also commonly referred to as packed cell volume (PCV). In normal blood, about 45 percent of a sample is erythrocytes. The hematocrit of any one sample can vary significantly, however, about 36–50 percent, according to gender and other factors. Normal hematocrit values for females range from 37 to 47, with a mean value of 41; for males, hematocrit ranges from 42 to 52, with a mean of 47. The percentage of other formed elements, the WBCs and platelets, is extremely small so it is not normally considered with the hematocrit. So the mean plasma percentage is the percent of blood that is not erythrocytes: for females, it is approximately 59 (or 100 minus 41), and for males, it is approximately 53 (or 100 minus 47).

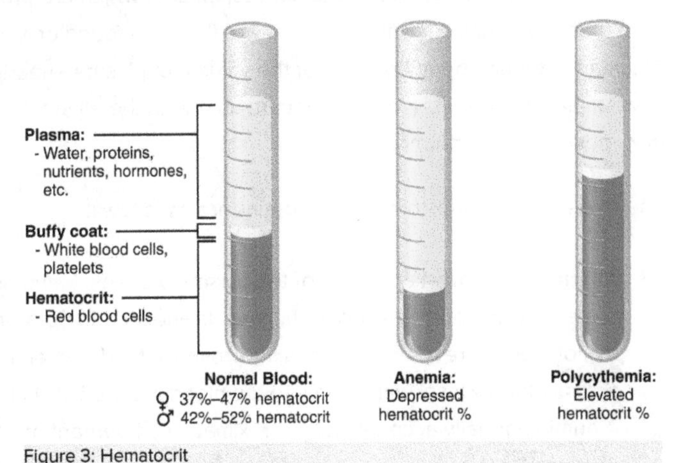

Figure 3: Hematocrit

The cellular elements of blood include a vast number of erythrocytes and comparatively fewer leukocytes and platelets. Plasma is the fluid in which the formed elements are suspended. A sample of blood spun in a centrifuge reveals that plasma is the lightest component. It floats at the top of the tube separated from the heaviest elements, the erythrocytes, by a **buffy coat** of leukocytes and platelets. Hematocrit is the percentage of the total sample that is comprised of erythrocytes. Depressed and elevated hematocrit levels are shown for comparison.

Characteristics of Blood

When you think about blood, the first characteristic that probably comes to mind is its color. Blood that has just taken up oxygen in the lungs is **bright red**, and blood that has released oxygen in the tissues is a more **dusky red.** This is because hemoglobin is a pigment that changes color, depending upon the degree of oxygen saturation.

Blood is viscous and somewhat sticky to the touch. It has a viscosity approximately five times greater than water. Viscosity is a measure of a fluid's thickness or resistance to flow, and is influenced by the presence of the plasma proteins and formed elements within the blood. The viscosity of blood has a dramatic impact on blood pressure and flow. Consider the difference in flow between water and honey. The more viscous honey would demonstrate a greater resistance to flow than the less viscous water. The same principle applies to blood.

The normal temperature of blood is slightly higher than normal body temperature—about 38 °C (or 100.4 °F), compared to 37 °C (or 98.6 °F) for an internal body temperature reading, although daily variations of 0.5 °C are normal. Although the surface of blood vessels is relatively smooth, as blood flows through them, it experiences some friction and resistance, especially as vessels age and lose their elasticity, thereby producing heat. This accounts for its slightly higher temperature.

The pH of blood averages about 7.4; however, it can range from **7.35 to 7.45 in a healthy person**. Blood is therefore somewhat more basic (alkaline) on a chemical scale than pure water, which has a pH of 7.0. Blood contains numerous buffers that actually help to regulate pH.

Blood constitutes approximately 8 percent of adult body weight. Adult males typically average about 5 to 6 liters of blood. Females average 4–5 liters.

Blood Plasma

Like other fluids in the body, plasma is composed primarily of water: In fact, it is about 92 percent water. Dissolved or suspended within this water is a mixture of substances, most of which are proteins. There are literally hundreds of substances dissolved or suspended in the plasma, although many of them are found only in very small quantities.
Plasma Proteins: About 7 percent of the volume of plasma—nearly all that is not water—is made of proteins. These include several plasma proteins (proteins that are unique to the plasma), plus a much smaller number of regulatory proteins, including enzymes and some hormones.

The three major groups of plasma proteins are as follows:

- **Albumin**is the most abundant of the plasma proteins. Manufactured by the liver, albumin molecules serve as binding proteins—transport vehicles for fatty acids and steroid hormones. Albumin is also the most significant contributor to the osmotic pressure of blood; that is, its presence holds water inside the blood vessels and draws water from the tissues, across blood vessel walls, and into the bloodstream. This in turn helps to maintain both blood volume and blood pressure. Albumin normally accounts for approximately 54 percent of the total plasma protein content, in clinical levels of 3.5–5.0 g/dL blood.
- The second most common plasma proteins are the **globulins.** A heterogeneous group, there are three main subgroups known as **alpha, beta, and gamma globulins**. The alpha and beta globulins transport iron, lipids, and the fat-soluble vitamins A, D, E, and K to the cells; like albumin, they also contribute to osmotic pressure. Globulins make up approximately 38 percent of the total plasma protein volume, in clinical levels of 1.0–1.5 g/dL blood.
- The least abundant plasma protein is **fibrinogen.** Like albumin and the alpha and beta globulins, fibrinogen is produced by the liver. It is essential for **blood clotting,** a process described later in this chapter. Fibrinogen accounts for about 7 percent of the total plasma protein volume, in clinical levels of 0.2–0.45 g/dL blood.

CAREER CONNECTION:PHLEBOTOMY AND MEDICAL LAB TECHNOLOGY

Phlebotomists are professionals trained to draw blood (phleb- = "a blood vessel"; -tomy = "to cut"). When more than a few drops of blood are required, phlebotomists perform a venipuncture, typically of a surface vein in the arm. They perform a capillary stick on a finger, an earlobe, or the heel of an infant when only a small quantity of blood is required. An arterial stick is collected from an artery and used to analyze blood gases. After collection, the blood may be analyzed by medical laboratories or perhaps used for transfusions, donations, or research. While many allied health professionals practice phlebotomy, the American Society of Phlebotomy Technicians issues certificates to individuals passing a national examination, and some large labs and hospitals hire individuals expressly for their skill in phlebotomy.

Medical or clinical laboratories employ a variety of individuals in technical positions:

- **Medical technologists (MT)**, also known as clinical laboratory technologists (CLT), typically hold a bachelor's degree and certification from an accredited training program. They perform a wide variety of tests on various body fluids, including blood. The information they provide is essential to the primary care providers in determining a diagnosis and in monitoring the course of a disease and response to treatment.
- **Medical laboratory technicians (MLT)** typically have an associate's degree but may perform duties similar to those of an MT.
- **Medical laboratory assistants (MLA)** spend the majority of their time processing samples and carrying out routine assignments within the lab. Clinical training is required, but a degree may not be essential to obtaining a position.

Production of the Formed Elements from Red Bone Marrow

The lifespan of the formed elements is very brief. Although one type of leukocyte called memory cells can survive for years, most erythrocytes, leukocytes, and platelets normally live only a few hours to a few weeks. Thus, the body must form new blood cells and platelets quickly and continuously. When you donate a unit of blood during a blood drive (approximately 475 mL, or about 1 pint), your body typically replaces the donated plasma within 24 hours, but it takes about 4 to 6 weeks to replace the blood cells. This restricts the frequency with which donors can contribute their blood. The process by which this replacement occurs is called **hemopoiesis, or hematopoiesis** (from the Greek root haima- = "blood"; -poiesis = "production").

Sites of Hemopoiesis

Prior to birth, hemopoiesis occurs in a number of tissues, beginning with the yolk sac of the developing embryo, and continuing in the fetal liver, spleen, lymphatic tissue, and eventually the red bone marrow. Following birth, most hemopoiesis occurs in the red marrow, a connective tissue within the spaces of spongy (cancellous) bone tissue. In children, hemopoiesis can occur in the medullary cavity of long bones; in adults, the process is largely restricted to the cranial and pelvic bones, the vertebrae, the sternum, and the proximal epiphyses of the femur and humerus.

Throughout adulthood, the liver and spleen maintain their ability to generate the formed elements. This process is referred to as extramedullary hemopoiesis (meaning hemopoiesis outside the medullary cavity of adult bones). When a disease such as bone cancer destroys the bone marrow, causing hemopoiesis to fail, extramedullary hemopoiesis may be initiated.

Differentiation of Formed Elements from Stem Cells

All formed elements arise from stem cells of the **red bone marrow.** Hemopoiesis begins when the hemopoietic stem cell is exposed to appropriate chemical stimuli collectively called hemopoietic growth factors, which prompt it to divide and differentiate. One daughter cell remains a hemopoietic stem cell, allowing hemopoiesis to continue. The other daughter cell becomes either of two types of more specialized stem cells:

- **Lymphoid stem cells** give rise to a class of leukocytes known as lymphocytes, which include the various T cells, B cells, and natural killer (NK) cells, all of which function in immunity. However, hemopoiesis of lymphocytes progresses somewhat differently from the process for the other formed elements. In brief, lymphoid stem cells quickly migrate from the bone marrow to lymphatic tissues, including the lymph nodes, spleen, and thymus, where their production and differentiation continues. B cells are so named since they mature in the bone marrow, while T cells mature in the thymus.
- **Myeloid stem cells**give rise to all the other formed elements, including the erythrocytes; megakaryocytes that produce platelets; and a myeloblast lineage that gives rise to monocytes and three forms of granular leukocytes: neutrophils, eosinophils, and basophils.

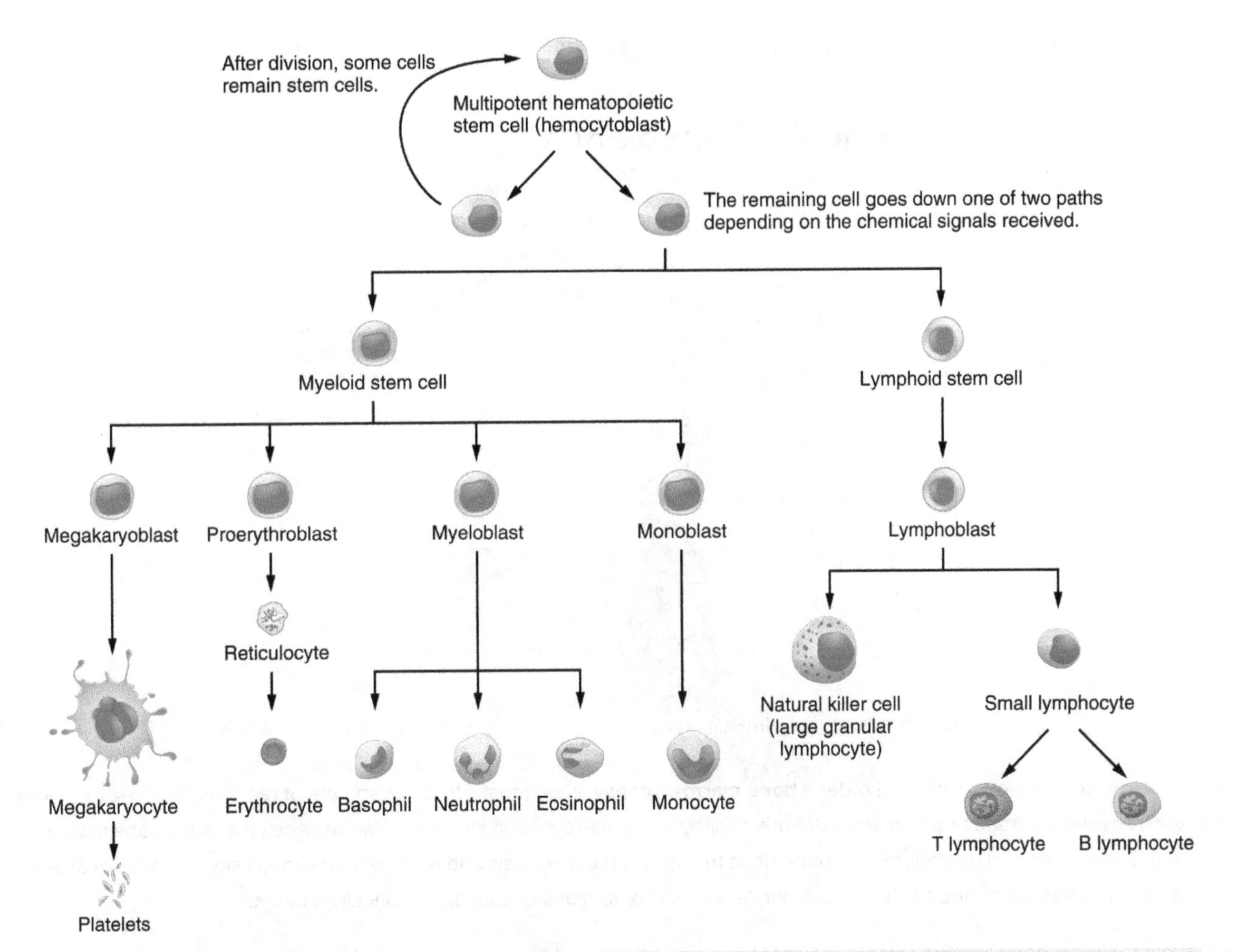

Figure 1:Hematopoietic System of Bone Marrow Hemopoiesis is the proliferation and differentiation of the formed elements of blood.

Hemopoietic Growth Factors

Development from stem cells to precursor cells to mature cells is again initiated by hemopoietic growth factors. These include the following:

- **Erythropoietin (EPO)** is a glycoprotein hormone secreted by the interstitial fibroblast cells of the kidneys in response to low oxygen levels. It prompts the production of erythrocytes. Some athletes use synthetic EPO as a performance-enhancing drug (called blood doping) to increase RBC counts and subsequently increase oxygen delivery to tissues throughout the body. EPO is a banned substance in most organized sports, but it is also used medically in the treatment of certain anemia, specifically those triggered by certain types of cancer, and other disorders in which increased erythrocyte counts and oxygen levels are desirable.
- **Thrombopoietin,** another glycoprotein hormone, is produced by the liver and kidneys. It triggers the development of megakaryocytes into platelets.

Bone Marrow Sampling and Transplants

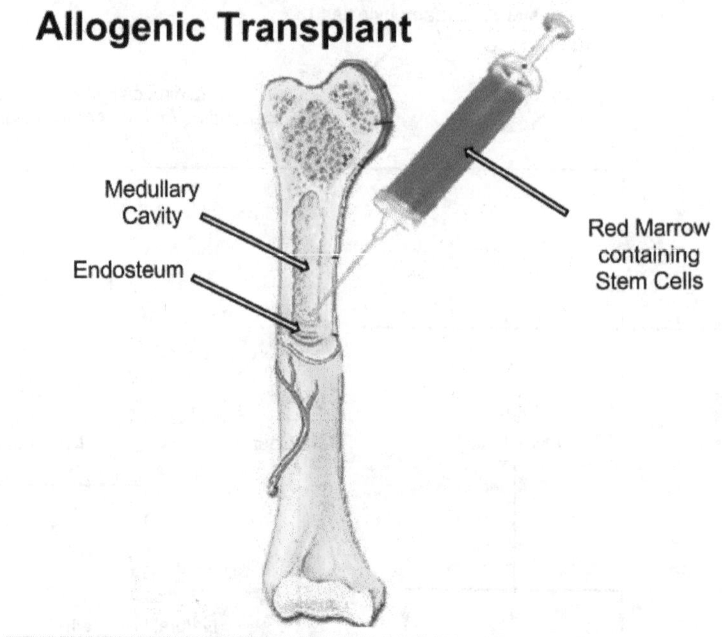

Figure 2: Bone Marrow transplant

Sometimes, a healthcare provider will order a **bone marrow biopsy**, a diagnostic test of a sample of red bone marrow, or a **bone marrow transplant**, a treatment in which a donor's healthy bone marrow—and its stem cells—replaces the faulty bone marrow of a patient. These tests and procedures are often used to assist in the diagnosis and treatment of various severe forms of anemia, such as thalassemia major and sickle cell anemia, as well as some types of cancer, specifically leukemia.

In the past, when a bone marrow sample or transplant was necessary, the procedure would have required inserting a large-bore needle into the region near the iliac crest of the pelvic bones (os coxae). This location was preferred, since its location close to the body surface makes it more accessible, and it is relatively isolated from most vital organs. Unfortunately, the procedure is quite painful.

Now, direct sampling of bone marrow can often be avoided. In many cases, stem cells can be isolated in just a few hours from a sample of a patient's blood. The isolated stem cells are then grown in culture using the appropriate hemopoietic growth factors, and analyzed or sometimes frozen for later use.

For an individual requiring a transplant, a matching donor is essential to prevent the immune system from destroying the donor cells—a phenomenon known as tissue rejection. To treat patients with bone marrow transplants, it is first necessary to destroy the patient's own diseased marrow through radiation and/or chemotherapy. Donor bone marrow stem cells are then intravenously infused. From the bloodstream, they establish themselves in the recipient's bone marrow.

Erythrocytes - RBC

The erythrocyte, commonly known as a **red blood cell (or RBC),** is by far the most common formed element: A single drop of blood contains millions of erythrocytes and just thousands of leukocytes. Specifically, males have about 5.4 million erythrocytes per microliter (µL) of blood, and females have approximately 4.8 million per µL. In fact, erythrocytes are estimated to make up about 25 percent of the total cells in the body. As you can imagine, they are quite small cells, with a mean diameter of only about 7–8 micrometers (µm). **The primary functions of erythrocytes are to pick up inhaled oxygen from the lungs and transport it to the body's tissues, and to pick up some (about 24 percent) carbon dioxide waste at the tissues and transport it to the lungs for exhalation.** Erythrocytes remain within the vascular network. Although leukocytes typically leave the blood vessels to perform their defensive functions, movement of erythrocytes from the blood vessels is abnormal.

Shape and Structure of Erythrocytes

Red Blood Cells

Figure 1: Red Blood Cells (RBCs)

As an erythrocyte matures in the red bone marrow, it extrudes its nucleus and most of its other organelles. Erythrocytes are biconcave disks; that is, they are plump at their periphery and very thin in the center. Since they lack most organelles, there is more interior space for the presence of the hemoglobin molecules that, as you will see shortly, transport gases. The biconcave shape also provides a greater surface area across which gas exchange can occur, relative to its volume; a sphere of a similar diameter would have a lower surface area-to-volume ratio. In the capillaries, the oxygen carried by the erythrocytes can diffuse into the plasma and then through the capillary walls to reach the cells, whereas some of the carbon dioxide produced by the cells as a waste product diffuses into the capillaries to be picked up by the erythrocytes. Capillary beds are extremely narrow, slowing the passage of the erythrocytes and providing an extended opportunity for gas exchange to occur. However, the space within capillaries can be so minute that, despite their own small size, erythrocytes may have to fold in on themselves if they are to make their way through. Fortunately, their structural proteins like spectrin are flexible, allowing them to bend over themselves to a surprising degree, then spring back again when they enter a wider vessel. In wider vessels, erythrocytes may stack up much like a roll of coins, forming a rouleaux, from the French word for "roll."

Shape of Red Blood Cells Erythrocytes are biconcave discs with very shallow centers. This shape optimizes the ratio of surface area to volume, facilitating gas exchange. It also enables them to fold up as they move through narrow blood vessels.

Hemoglobin

Hemoglobin is a large molecule made up of **proteins and iron.** It consists of four folded chains of a protein called **globin,** designated alpha 1 and 2, and beta 1 and 2. Each of these globin molecules is bound to a red pigment molecule called **heme,** which contains an ion of iron (Fe^{2+}).

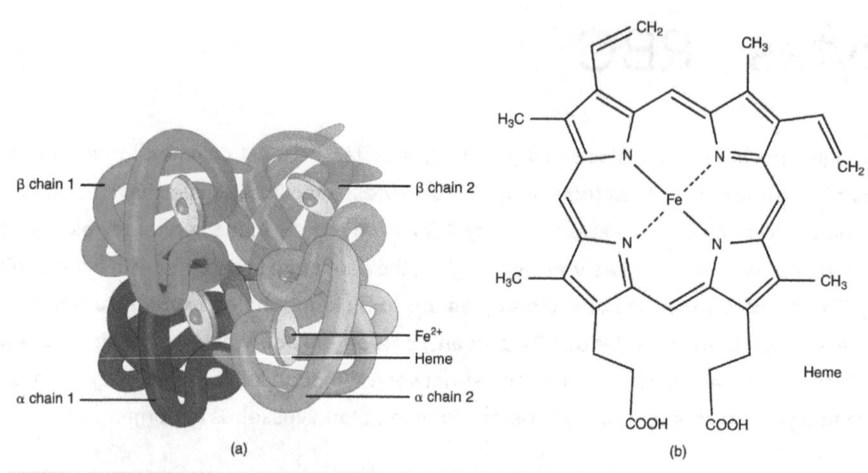

Figure 2: Hemoglobin (a) A molecule of hemoglobin contains four globin proteins, each of which is bound to one molecule of the iron-containing pigment heme. (b) A single erythrocyte can contain 300 million hemoglobin molecules, and thus more than 1 billion oxygen molecules.

Each iron ion in the heme can bind to one oxygen molecule; therefore, each hemoglobin molecule can transport **four** oxygen molecules. An individual erythrocyte may contain about 300 million hemoglobin molecules, and therefore can bind to and transport up to 1.2 billion oxygen molecules.

In the lungs, hemoglobin picks up oxygen, which binds to the iron ions, forming **oxyhemoglobin.** The bright red, oxygenated hemoglobin travels to the body tissues, where it releases some of the oxygen molecules, becoming darker red **deoxyhemoglobin,** sometimes referred to as reduced hemoglobin. Oxygen release depends on the need for oxygen in the surrounding tissues, so hemoglobin rarely if ever leaves all of its oxygen behind. In the capillaries, carbon dioxide enters the bloodstream. About 76 percent dissolves in the plasma, some of it remaining as dissolved CO_2, and the remainder forming bicarbonate ion. About 23–24 percent of it binds to the amino acids in hemoglobin, forming a molecule known as **carbaminohemoglobin.** From the capillaries, the hemoglobin carries carbon dioxide back to the lungs, where it releases it for exchange of oxygen.

Changes in the levels of RBCs can have significant effects on the body's ability to effectively deliver oxygen to the tissues. Ineffective hematopoiesis results in insufficient numbers of RBCs and results in one of several forms of anemia. An overproduction of RBCs produces a condition called **polycythemia.** The primary drawback with polycythemia is not a failure to directly deliver enough oxygen to the tissues, but rather the increased viscosity of the blood, which makes it more difficult for the heart to circulate the blood.

In patients with insufficient hemoglobin, the tissues may not receive sufficient oxygen, resulting in another form of anemia. In determining oxygenation of tissues, the value of greatest interest in healthcare is the percent saturation; that is, the percentage of hemoglobin sites occupied by oxygen in a patient's blood. Clinically this value is commonly referred to simply as "percent sat."

Percent saturation is normally monitored using a device known as a **pulse oximeter**, which is applied to a thin part of the body, typically the tip of the patient's finger. The device works by sending two different wavelengths of light (one red, the other infrared) through the finger and measuring the light with a photodetector as it exits. Hemoglobin absorbs light differentially depending upon its saturation with oxygen. The machine calibrates the amount of light received by the photodetector against the amount absorbed by the partially oxygenated hemoglobin and presents the data as percent saturation. Normal pulse oximeter readings range from 95–100 percent. Lower percentages reflect **hypoxemia, or low blood oxygen**. The term **hypoxia** is more generic and simply refers to low oxygen levels. Oxygen levels are also directly monitored from free oxygen in the plasma typically following an arterial stick. When this method is applied, the amount of oxygen present is expressed in terms of partial pressure of oxygen or simply pO_2 and is typically recorded in units of millimeters of mercury, mm Hg.

The kidneys filter about 180 liters (~380 pints) of blood in an average adult each day, or about 20 percent of the total resting volume, and thus serve as ideal sites for receptors that determine oxygen saturation. In response to hypoxemia, less oxygen will exit the vessels supplying the kidney, resulting in hypoxia (low oxygen concentration) in the tissue fluid of the kidney where oxygen concentration is actually monitored. Interstitial fibroblasts within the kidney secrete **EPO**, thereby increasing erythrocyte production and restoring oxygen levels. In a classic negative-feedback loop, as oxygen saturation rises, EPO secretion falls, and vice versa, thereby maintaining homeostasis. Populations dwelling at high elevations, with inherently lower levels of oxygen in the atmosphere, naturally maintain a hematocrit higher than people living at sea level. Consequently, people traveling to high elevations may experience symptoms of hypoxemia, such as fatigue, headache, and shortness of breath, for a few days after their arrival. In response to the hypoxemia, the kidneys secrete EPO to step up the production of erythrocytes until homeostasis is achieved once again. To avoid the symptoms of hypoxemia, or altitude sickness, mountain climbers typically rest for several days to a week or more at a series of camps situated at increasing elevations to allow EPO levels and, consequently, erythrocyte counts to rise. When climbing the tallest peaks, such as Mt. Everest and K2 in the Himalayas, many mountain climbers rely upon bottled oxygen as they near the summit.

Lifecycle of Erythrocytes

Production of erythrocytes in the marrow occurs at the staggering rate of more than 2 million cells per second. For this production to occur, a number of raw materials must be present in adequate amounts. These include the same nutrients that are essential to the production and maintenance of any cell, such as glucose, lipids, and amino acids. However, erythrocyte production also requires several trace elements:

- **Iron.** We have said that each heme group in a hemoglobin molecule contains an ion of the trace mineral iron. On average, less than 20 percent of the iron we consume is absorbed. Heme iron, from animal foods such as meat, poultry, and fish, is absorbed more efficiently than non-heme iron from plant foods. Upon absorption, iron becomes part of the body's total iron pool. The bone marrow, liver, and spleen can store iron in the protein compounds ferritin and hemosiderin. Ferroportin transports the iron across the intestinal cell plasma membranes and from its storage sites into tissue fluid where it enters the blood. When EPO stimulates the production of erythrocytes, iron is released from storage, bound to transferrin, and carried to the red marrow where it attaches to erythrocyte precursors.
- **Zinc.** The trace mineral zinc functions as a co-enzyme that facilitates the synthesis of the heme portion of hemoglobin.
- **B vitamins.** The B vitamins folate and vitamin B_{12} function as co-enzymes that facilitate DNA synthesis. Thus, both are critical for the synthesis of new cells, including erythrocytes.

Erythrocytes live up to **120 days** in the circulation, after which the worn-out cells are removed by a type of myeloid phagocytic cell called a macrophage, located primarily within the bone marrow, liver, and spleen. The components of the degraded erythrocytes' hemoglobin are further processed as follows:

- **Globin, the protein portion** of hemoglobin, is broken down into amino acids, which can be sent back to the bone marrow to be used in the production of new erythrocytes. Hemoglobin that is not phagocytized is broken down in the circulation, releasing alpha and beta chains that are removed from circulation by the kidneys.
- The iron contained in the **heme portion** of hemoglobin may be stored in the liver or spleen, primarily in the form of ferritin or hemosiderin, or carried through the bloodstream by transferrin to the red bone marrow for recycling into new erythrocytes.
- The non-iron portion of heme is degraded into the waste product **biliverdin,** a green pigment, and then into another waste product,**bilirubin,** a yellow pigment. Bilirubin binds to albumin and travels in the blood to the liver, which uses it in the manufacture of bile, a compound released into the intestines to help emulsify dietary fats. In the large intestine, bacteria breaks the bilirubin apart from the bile and converts it to urobilinogen and then into stercobilin. It is then eliminated from the body in the feces. Broad-spectrum antibiotics typically eliminate these bacteria as well and may alter the color of feces. The kidneys also remove any circulating bilirubin and other related metabolic byproducts such as urobilins and

secrete them into the urine.

The breakdown pigments formed from the destruction of hemoglobin can be seen in a variety of situations. At the site of an injury, biliverdin from damaged RBCs produces some of the dramatic colors associated with bruising. With a failing liver, bilirubin cannot be removed effectively from circulation and causes the body to assume a yellowish tinge associated with jaundice. Stercobilins within the feces produce the typical brown color associated with this waste. And the yellow of urine is associated with the urobilins.

The erythrocyte lifecycle is summarized in Figure 3.

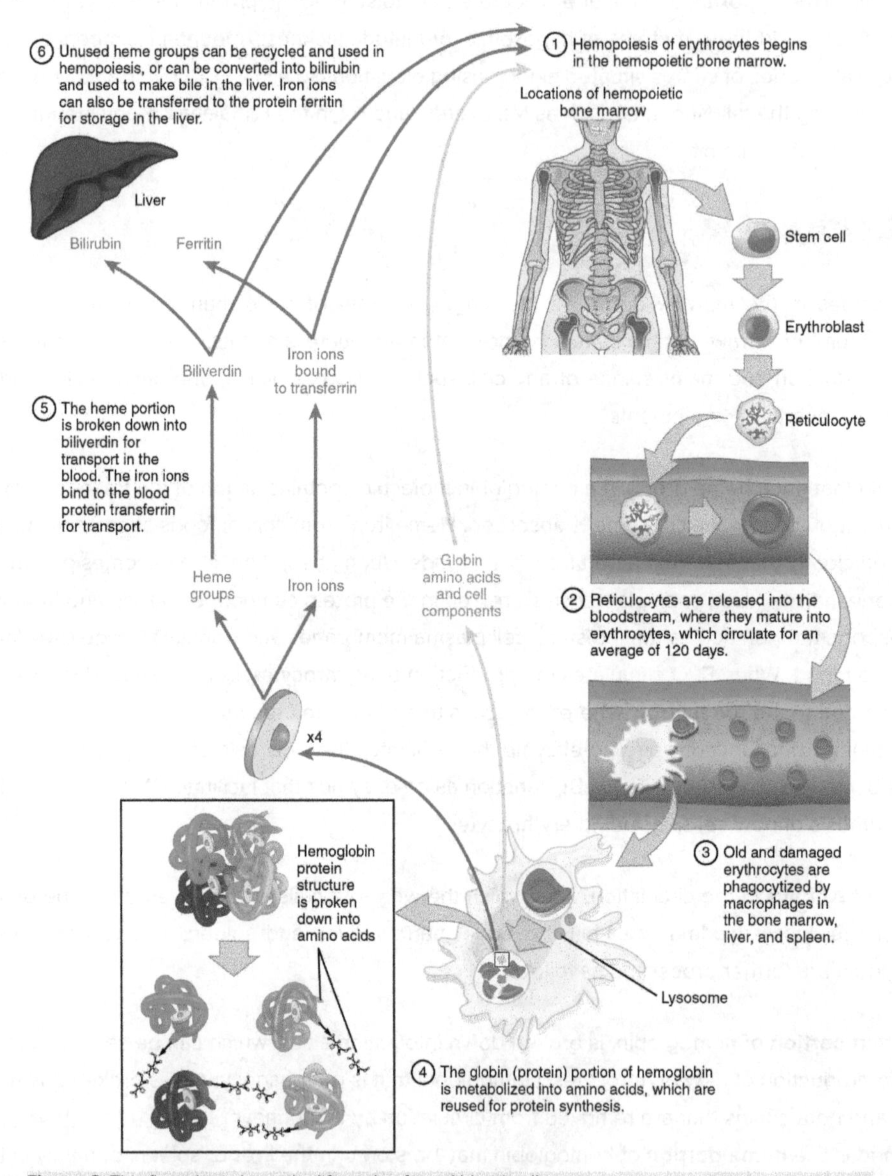

Figure 3: This flow chart shows the life cycle of a red blood cell.

Erythrocyte Lifecycle Erythrocytes are produced in the bone marrow and sent into the circulation. At the end of their lifecycle, they are destroyed by macrophages, and their components are recycled.

Disorders of Erythrocytes

The size, shape, and number of erythrocytes, and the number of hemoglobin molecules can have a major impact on a person's health. When the number of RBCs or hemoglobin is deficient, the general condition is called **anemia.** There are more than 400 types of anemia and more than 3.5 million Americans suffer from this condition. Anemia can be broken down into three major groups:

- those caused by blood loss,
- those caused by faulty or decreased RBC production, and
- caused by excessive destruction of RBCs.

The effects of the anemia are widespread, because reduced numbers of RBCs or hemoglobin will result in lower levels of oxygen being delivered to body tissues. Since oxygen is required for tissue functioning, anemia produces **fatigue, lethargy, and an increased risk for infection**. An oxygen deficit in the brain impairs the ability to think clearly, and may prompt headaches and irritability. Lack of oxygen leaves the patient short of breath, even as the heart and lungs work harder in response to the deficit.

Blood loss anemias are fairly straightforward. In addition to bleeding from wounds or other lesions, these forms of anemia may be due to ulcers, hemorrhoids, inflammation of the stomach (gastritis), and some cancers of the gastrointestinal tract. The excessive use of aspirin or other nonsteroidal anti-inflammatory drugs such as ibuprofen can trigger ulceration and gastritis. Excessive menstruation and loss of blood during childbirth are also potential causes.

Anemias caused by faulty or decreased RBC production include: **sickle cell anemia, iron deficiency anemia, vitamin deficiency anemia, and diseases of the bone marrow and stem cells.**

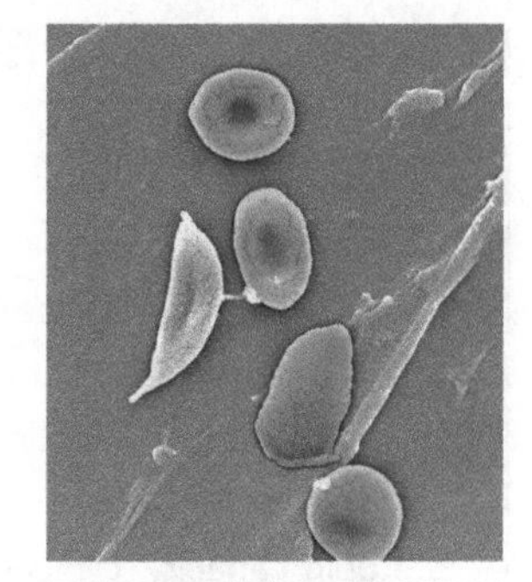

Figure 4: Sickle Cell

- **Sickle cell anemia:** A characteristic change in the shape of erythrocytes is seen in sickle cell disease(also referred to as sickle cell anemia). A genetic disorder, it is caused by production of an abnormal type of hemoglobin, called hemoglobin S, which delivers less oxygen to tissues and causes erythrocytes to assume a sickle (or crescent) shape, especially at low oxygen concentrations. These abnormally shaped cells can then become lodged in narrow capillaries because they are unable to fold in on themselves to squeeze through, blocking blood flow to tissues and causing a variety of serious problems from painful joints to delayed growth and even blindness and cerebrovascular accidents (strokes). Sickle cell anemia is a genetic condition particularly found in individuals of African descent.
- Sickle cell anemia is caused by a mutation in one of the hemoglobin genes. Erythrocytes produce an abnormal type of hemoglobin, which causes the cell to take on a sickle or crescent shape. (credit: Janice Haney Carr)
- **Iron deficiency anemia** is the most common type and results when the amount of available iron is insufficient to allow production of sufficient heme. This condition can occur in individuals with a deficiency of iron in the diet and is especially common in teens and children as well as in vegans and vegetarians. Additionally, iron deficiency anemia may be caused by either an inability to absorb and transport iron or slow, chronic bleeding.
- **Vitamin-deficient anemias** generally involve insufficient vitamin B12 and folate.
- **Megaloblastic anemia** involves a deficiency of vitamin B12 and/or folate, and often involves diets deficient in these essential nutrients. Lack of meat or a viable alternate source, and overcooking or eating insufficient amounts of vegetables may lead to a lack of folate.
- **Pernicious anemia** is caused by poor absorption of vitamin B12 and is often seen in patients with Crohn's disease (a severe intestinal disorder often treated by surgery), surgical removal of the intestines or stomach (common in some weight loss surgeries), intestinal parasites, and AIDS.
- Pregnancies, some medications, excessive alcohol consumption, and some diseases such as celiac disease are also associated with vitamin deficiencies. It is essential to provide sufficient folic acid during the early stages of pregnancy to

reduce the risk of neurological defects, including spina bifida, a failure of the neural tube to close.

- Assorted disease processes can also interfere with the production and formation of RBCs and hemoglobin. If myeloid stem cells are defective or replaced by cancer cells, there will be insufficient quantities of RBCs produced.
- **Aplastic anemia** is the condition in which there are deficient numbers of RBC stem cells. Aplastic anemia is often inherited, or it may be triggered by radiation, medication, chemotherapy, or infection.
- **Thalassemia i**s an inherited condition typically occurring in individuals from the Middle East, the Mediterranean, African, and Southeast Asia, in which maturation of the RBCs does not proceed normally. The most severe form is called Cooley's anemia.
- Lead exposure from industrial sources or even dust from paint chips of iron-containing paints or pottery that has not been properly glazed may also lead to destruction of the red marrow.
- Various disease processes also can lead to anemias. These include chronic kidney diseases often associated with a decreased production of EPO, hypothyroidism, some forms of cancer, lupus, and rheumatoid arthritis.

In contrast to anemia, an elevated RBC count is called **polycythemia** and is detected in a patient's elevated hematocrit. It can occur transiently in a person who is dehydrated; when water intake is inadequate or water losses are excessive, the plasma volume falls. As a result, the hematocrit rises. For reasons mentioned earlier, a mild form of polycythemia is chronic but normal in people living at high altitudes. Some elite athletes train at high elevations specifically to induce this phenomenon. Finally, a type of bone marrow disease called polycythemia vera (from the Greek vera = "true") causes an excessive production of immature erythrocytes. Polycythemia vera can dangerously elevate the viscosity of blood, raising blood pressure and making it more difficult for the heart to pump blood throughout the body. It is a relatively rare disease that occurs more often in men than women, and is more likely to be present in elderly patients those over 60 years of age.

EVERYDAY CONNECTION – BLOOD DOPING

In its original intent, the term blood doping was used to describe the practice of injecting by transfusion supplemental RBCs into an individual, typically to enhance performance in a sport. Additional RBCs would deliver more oxygen to the tissues, providing extra aerobic capacity, clinically referred to as VO2 max. The source of the cells was either from the recipient (autologous) or from a donor with compatible blood (homologous). This practice was aided by the well-developed techniques of harvesting, concentrating, and freezing of the RBCs that could be later thawed and injected, yet still retain their functionality. These practices are considered illegal in virtually all sports and run the risk of infection, significantly increasing the viscosity of the blood and the potential for transmission of blood-borne pathogens if the blood was collected from another individual.

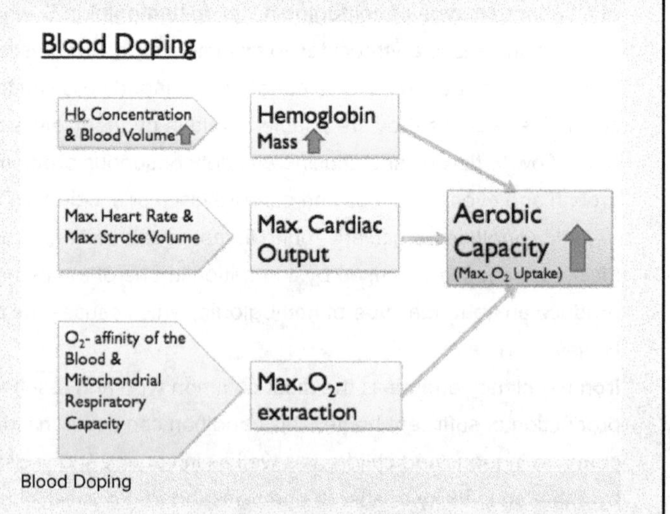

Blood Doping

With the development of synthetic EPO in the 1980s, it became possible to provide additional RBCs by artificially stimulating RBC production in the bone marrow. Originally developed to treat patients suffering from anemia, renal failure, or cancer treatment, large quantities of EPO can be generated by recombinant DNA technology. Synthetic EPO is injected under the skin and can increase hematocrit for many weeks. It may also induce polycythemia and raise hematocrit to 70 or greater. This increased viscosity raises the resistance of the blood and forces the heart to pump more powerfully; in extreme cases, it has resulted in death. Other drugs such as cobalt II chloride have been shown to increase natural EPO gene expression. Blood doping has become problematic in many sports, especially cycling. Lance Armstrong, winner of seven Tour de France and many other cycling titles, was stripped of his victories and admitted to blood doping in 2013.

Leukocytes and Platelets

The leukocyte, commonly known as a white blood cell (or WBC), is a major component of the body's defenses against disease. Leukocytes protect the body against invading microorganisms and body cells with mutated DNA, and they clean up debris. Platelets are essential for the repair of blood vessels when damage to them has occurred; they also provide growth factors for healing and repair.

Classification of Leukocytes

When scientists first began to observe stained blood slides, it quickly became evident that leukocytes could be divided into two groups, according to whether their cytoplasm contained highly visible granules:

- **Granular leukocytes** contain abundant granules within the cytoplasm. They include **neutrophils, eosinophils, and basophils** from myeloid stem cells.
- **Agranular leukocytes:** Granules are not totally lacking in **agranular leukocytes,** they are far fewer and less obvious. Agranular leukocytes include monocytes, which mature into **macrophages** that are phagocytic, and **lymphocytes,** which arise from the lymphoid stem cell line.

Lifespan Characteristics of Leukocytes

Although leukocytes and erythrocytes both originate from **hematopoietic stem cells in the bone marrow,** they are very different from each other in many significant ways. For instance, leukocytes are far less numerous than erythrocytes: Typically there are only 5000 to 10,000 per μL. They are also larger than erythrocytes and are the only formed elements that are complete cells, possessing a nucleus and organelles. And although there is just one type of erythrocyte, there are many types of leukocytes. Most of these types have a much shorter lifespan than that of erythrocytes, some as short as a few days or hours or even a few minutes in the case of acute infection.

One of the most distinctive characteristics of leukocytes is their movement. Whereas erythrocytes spend their days circulating within the blood vessels, leukocytes routinely leave the bloodstream to perform their defensive functions in the body's tissues. For leukocytes, the vascular network is simply a highway they travel and soon exit to reach their true destination. When they arrive, they are often given distinct names, such as macrophage or microglia, depending on their function. They leave the capillaries—the smallest blood vessels—or other small vessels through a process known as emigration (from the Latin for "removal") or diapedesis(dia- = "through"; -pedan = "to leap") in which they squeeze through adjacent cells in a blood vessel wall.

Once they have exited the capillaries, some leukocytes will take up fixed positions in lymphatic tissue, bone marrow, the spleen, the thymus, or other organs. Others will move about through the tissue spaces very much like amoebas, continuously extending their plasma membranes, sometimes wandering freely, and sometimes moving toward the direction in which they are drawn by chemical signals. This attracting of leukocytes occurs because of positive **chemotaxis**(literally "movement in response to chemicals"), a phenomenon in which injured or infected cells and nearby leukocytes emit the equivalent of a chemical "911" call, attracting more leukocytes to the site. In clinical medicine, the differential counts of the types and percentages of leukocytes present are often key indicators in making a diagnosis and selecting a treatment.

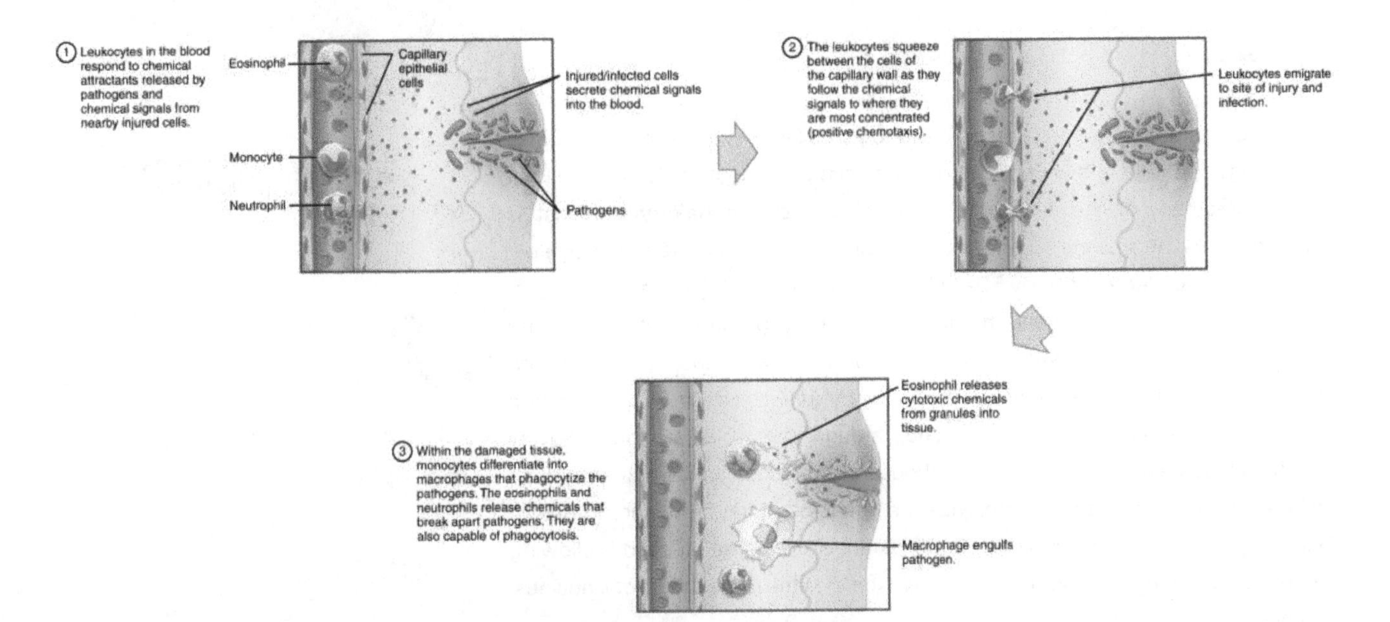

Figure 2: Leukocytes exit the blood vessel and then move through the connective tissue of the dermis toward the site of a wound. Some leukocytes, such as the eosinophil and neutrophil, are characterized as granular leukocytes. They release chemicals from their granules that destroy pathogens; they are also capable of phagocytosis. The monocyte, an agranular leukocyte, differentiates into a macrophage that then phagocytizes the pathogens.

Disorders of Leukocytes

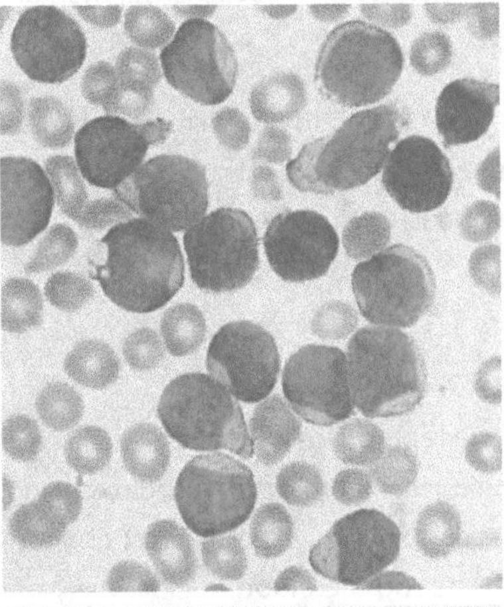

Figure 3: Leukemia blood smear

Leukopenia is a condition in which too few leukocytes are produced. If this condition is pronounced, the individual may be unable to ward off disease. Excessive leukocyte proliferation is known as leukocytosis. Although leukocyte counts are high, the cells themselves are often nonfunctional, leaving the individual at increased risk for disease.

Leukemia is a cancer involving an abundance of leukocytes. It may involve only one specific type of leukocyte from either the myeloid line (myelocytic leukemia) or the lymphoid line (lymphocytic leukemia). In chronic leukemia, mature leukocytes accumulate and fail to die. In acute leukemia, there is an overproduction of young, immature leukocytes. In both conditions the cells do not function properly.

Lymphoma is a form of cancer in which masses of malignant T and/or B lymphocytes collect in lymph nodes, the spleen, the liver, and other tissues. As in leukemia, the malignant leukocytes do not function properly, and the patient is vulnerable to infection. Some forms of lymphoma tend to progress slowly and respond well to treatment. Others tend to progress quickly and require aggressive treatment, without which they are rapidly fatal.

Platelets

You may occasionally see platelets referred to as **thrombocytes**, but because this name suggests they are a type of cell, it is not accurate. A platelet is not a cell but rather a fragment of the cytoplasm of a cell called a **megakaryocyte** that is surrounded by a plasma membrane. Megakaryocytes are descended from myeloid stem cells and are large, typically 50–100 μm in diameter, and contain an enlarged, lobed nucleus. As noted earlier, thrombopoietin, a glycoprotein secreted by the kidneys and liver, stimulates the proliferation of megakaryoblasts, which mature into megakaryocytes. These remain within bone marrow tissue and ultimately form platelet-precursor extensions that extend through the walls of bone marrow capillaries to release into the circulation thousands of cytoplasmic fragments, each enclosed by a bit of plasma membrane. These enclosed fragments are platelets. Each megakarocyte releases 2000–3000 platelets during its lifespan. Following platelet release, megakaryocyte remnants, which are little more than a cell nucleus, are consumed by macrophages.

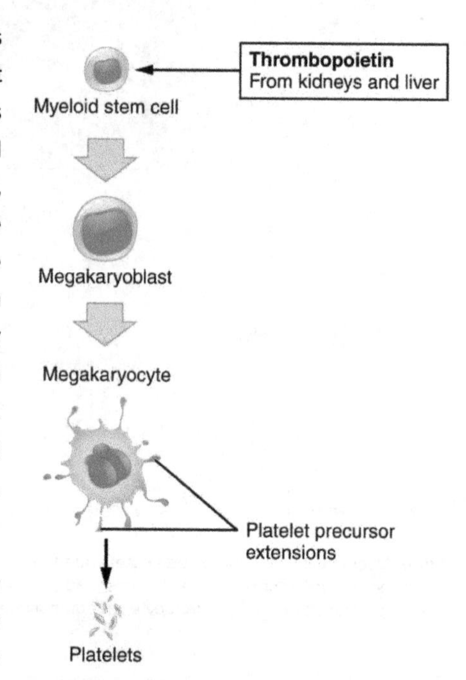

Figure 4: Platelets are derived from cells called megakaryocytes.

Platelets are relatively small, 2–4 μm in diameter, but numerous, with typically 150,000–160,000 per μL of blood. After entering the circulation, approximately one-third migrate to the spleen for storage for later release in response to any rupture in a blood vessel. They then become activated to perform their primary function, which is to limit blood loss. Platelets remain only about 10 days, then are phagocytized by macrophages.

Platelets are critical to hemostasis, the stoppage of blood flow following damage to a vessel. They also secrete a variety of growth factors essential for growth and repair of tissue, particularly connective tissue. Infusions of concentrated platelets are now being used in some therapies to stimulate healing.

Platelets and Coagulation Factors

Blood must clot to heal wounds and prevent excess blood loss. Small cell fragments called platelets (thrombocytes) are attracted to the wound site where they adhere by extending many projections and releasing their contents. These contents activate other platelets and also interact with other coagulation factors, which convert fibrinogen, a water-soluble protein present in blood serum into fibrin (a non-water soluble protein), causing the blood to clot. Many of the clotting factors require vitamin K to work, and vitamin K deficiency can lead to problems with blood clotting. Many platelets converge and stick together at the wound site forming a platelet plug (also called a fibrin clot), as illustrated in Figure 1b. The plug or clot lasts for a number of days and stops the loss of blood. Platelets are formed from the disintegration of larger cells called megakaryocytes, like that shown in Figure 1a. For each megakaryocyte, 2000–3000 platelets are formed with 150,000 to 400,000 platelets present in each cubic millimeter of blood. Each platelet is disc shaped and 2–4 μm in diameter. They contain many small vesicles but do not contain a nucleus.

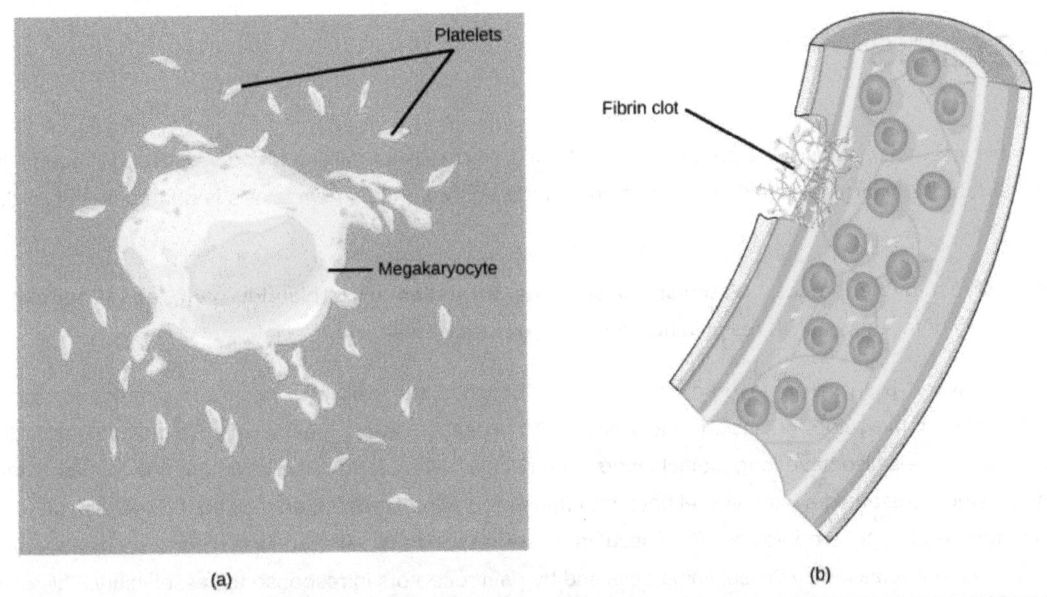

Figure 5: (a) Platelets are formed from large cells called megakaryocytes. The megakaryocyte breaks up into thousands of fragments that become platelets. (b) Platelets are required for clotting of the blood. The platelets collect at a wound site in conjunction with other clotting factors, such as fibrinogen, to form a fibrin clot that prevents blood loss and allows the wound to heal.

Disorders of Platelets

Thrombocytosis is a condition in which there are too many platelets. This may trigger formation of unwanted blood clots (thrombosis), a potentially fatal disorder. If there is an insufficient number of platelets, called **thrombocytopenia,** blood may not clot properly, and excessive bleeding may result.

Hemostasis

Platelets are key players in hemostasis, the process by which the body seals a ruptured blood vessel and prevents further loss of blood. Although rupture of larger vessels usually requires medical intervention, hemostasis is quite effective in dealing with small, simple wounds.

There are three steps to the process: **Vascular spasm, The formation of a platelet plug, and Coagulation (blood clotting).**Failure of any of these steps will result in hemorrhage—excessive bleeding.

1. Vascular Spasm: When a vessel is severed or punctured, or when the wall of a vessel is damaged, vascular spasm occurs. In vascular spasm, the smooth muscle in the walls of the vessel contracts dramatically. This smooth muscle has both circular layers; larger vessels also have longitudinal layers. The circular layers tend to constrict the flow of blood, whereas the longitudinal layers, when present, draw the vessel back into the surrounding tissue, often making it more difficult for a surgeon to locate, clamp, and tie off a severed vessel. The vascular spasm response is believed to be triggered by several chemicals called endothelins that are released by vessel-lining cells and by pain receptors in response to vessel injury. This phenomenon typically lasts for up to 30 minutes, although it can last for hours.

2. Formation of the Platelet Plug: In the second step, platelets, which normally float free in the plasma, encounter the area of vessel rupture with the exposed underlying connective tissue and collagenous fibers. The platelets begin to clump together, become spiked and sticky, and bind to the exposed collagen and endothelial lining. This process is assisted by a glycoprotein in the blood plasma called von Willebrand factor, which helps stabilize the growing platelet plug. As platelets collect, they simultaneously release chemicals from their granules into the plasma that further contribute to hemostasis. A platelet plug can temporarily seal a small opening in a blood vessel. Plug formation, in essence, buys the body time while more sophisticated and durable repairs are being made. In a similar manner, even modern naval warships still carry an assortment of wooden plugs to temporarily repair small breaches in their hulls until permanent repairs can be made.

3. Coagulation: Those more sophisticated and more durable repairs are collectively called **coagulation, the formation of a blood clot**. The process is sometimes characterized as a cascade, because one event prompts the next as in a multi-level waterfall. The result is the production of a gelatinous but robust clot made up of a mesh of fibrin—an insoluble filamentous protein derived from fibrinogen, the plasma protein introduced earlier—in which platelets and blood cells are trapped.

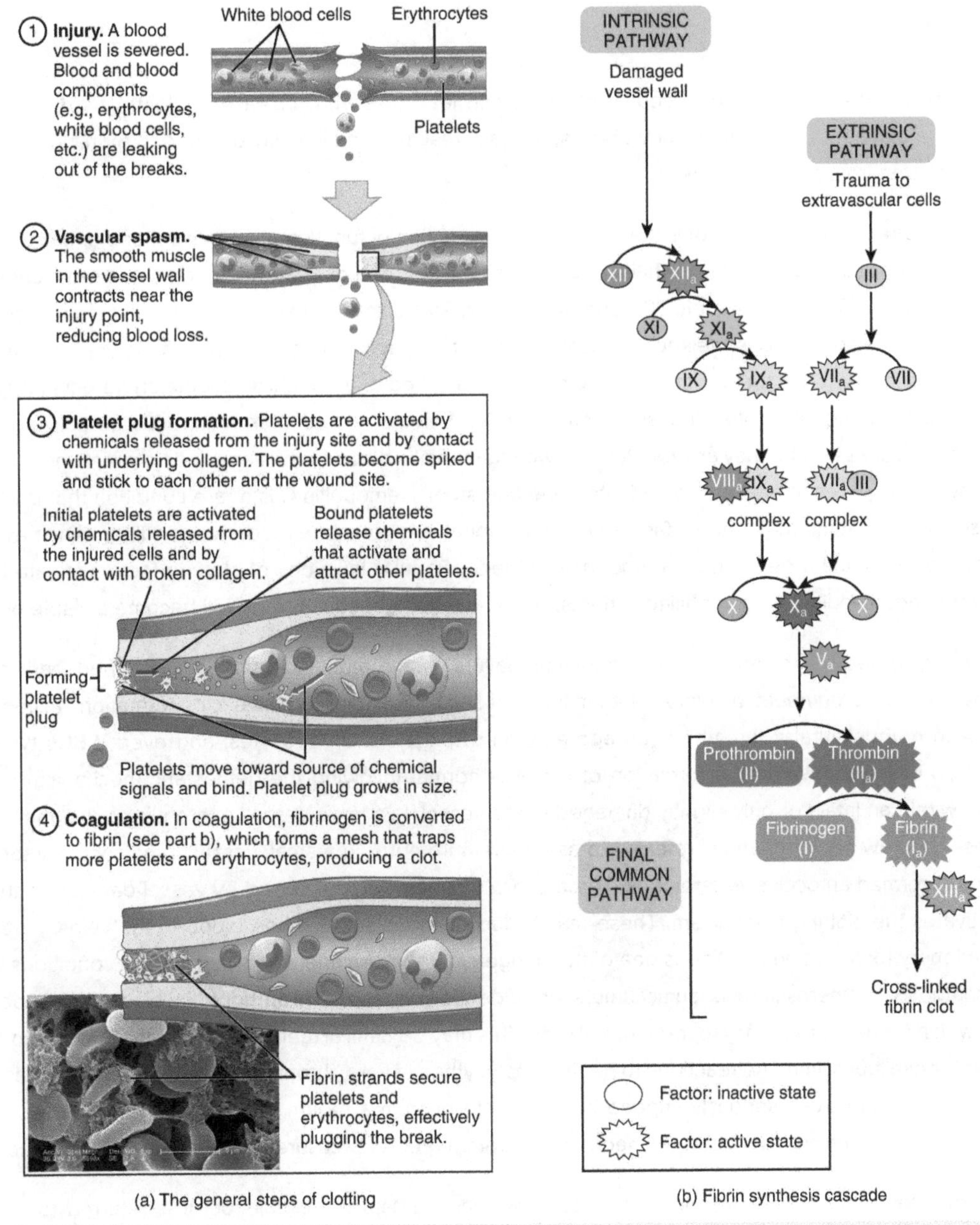

Figure 1: Three steps of hemostasis.

Clotting Factors Involved in Coagulation

In the coagulation cascade, chemicals called clotting factors (or coagulation factors) prompt reactions that activate still more coagulation factors. The process is complex, but is initiated along two basic pathways:

- The extrinsic pathway, which normally is triggered by trauma.
- The intrinsic pathway, which begins in the bloodstream and is triggered by internal damage to the wall of the vessel.

Both of these merge into a third pathway, referred to as the common pathway . All three pathways are dependent upon the 12 known clotting factors, including Ca^{2+} and vitamin K. Clotting factors are secreted primarily by the liver and the platelets. The liver requires the fat-soluble vitamin K to produce many of them. Vitamin K (along with biotin and folate) is somewhat unusual among vitamins in that it is not only consumed in the diet but is also synthesized by bacteria residing in the large intestine. The calcium ion, considered factor IV, is derived from the diet and from the breakdown of bone.

Disorders of Clotting

Either an insufficient or an excessive production of platelets can lead to severe disease or death. As discussed earlier, an insufficient number of platelets, called **thrombocytopenia,** typically results in the inability of blood to form clots. This can lead to excessive bleeding, even from minor wounds.

Another reason for failure of the blood to clot is the inadequate production of functional amounts of one or more clotting factors. This is the case in the **genetic disorder hemophilia,** which is actually a group of related disorders, the most common of which is hemophilia A, accounting for approximately 80 percent of cases. This disorder results in the inability to synthesize sufficient quantities of factor VIII. Hemophilia B is the second most common form, accounting for approximately 20 percent of cases. In this case, there is a deficiency of factor IX. Both of these defects are linked to the X chromosome and are typically passed from a healthy (carrier) mother to her male offspring, since males are XY. Females would need to inherit a defective gene from each parent to manifest the disease, since they are XX. Patients with hemophilia bleed from even minor internal and external wounds, and leak blood into joint spaces after exercise and into urine and stool. Hemophilia C is a rare condition that is triggered by an autosomal (not sex) chromosome that renders factor XI nonfunctional. It is not a true recessive condition, since even individuals with a single copy of the mutant gene show a tendency to bleed. Regular infusions of clotting factors isolated from healthy donors can help prevent bleeding in hemophiliac patients. At some point, genetic therapy will become a viable option.

In contrast to the disorders characterized by coagulation failure is thrombocytosis, also mentioned earlier, a condition characterized by excessive numbers of platelets that increases the risk for excessive clot formation, a condition known as **thrombosis.** A **thrombus**(plural = thrombi) is an aggregation of platelets, erythrocytes, and even WBCs typically trapped within a mass of fibrin strands. While the formation of a clot is normal following the hemostatic mechanism just described, thrombi can form within an intact or only slightly damaged blood vessel. In a large vessel, a thrombus will adhere to the vessel wall and decrease the flow of blood, and is referred to as a mural thrombus. In a small vessel, it may actually totally block the flow of blood and is termed an occlusive thrombus. Thrombi are most commonly caused by vessel damage to the endothelial lining, which activates the clotting mechanism. These may include venous stasis, when blood in the veins, particularly in the legs, remains stationary for long periods. This is one of the dangers of long airplane flights in crowded conditions and may lead to deep vein thrombosis or atherosclerosis, an accumulation of debris in arteries. **Thrombophilia**, also called hypercoagulation, is a condition in which there is a tendency to form thrombosis. This may be familial (genetic) or acquired. Acquired forms include the autoimmune disease lupus, immune reactions to heparin, polycythemia vera, thrombocytosis, sickle cell disease, pregnancy, and even obesity. A thrombus can seriously impede blood flow to or from a region and will cause a local increase in blood pressure. If flow is to be maintained, the heart will need to generate a greater pressure to overcome the resistance.

When a portion of a thrombus breaks free from the vessel wall and enters the circulation, it is referred to as an **embolus.** An embolus that is carried through the bloodstream can be large enough to block a vessel critical to a major organ. When it becomes trapped, an embolus is called an embolism. In the heart, brain, or lungs, an embolism may accordingly cause a heart attack, a stroke, or a **pulmonary embolism**. These are medical emergencies.

Among the many known biochemical activities of aspirin is its role as an **anticoagulant. Aspirin (acetylsalicylic acid)** is very effective at inhibiting the aggregation of platelets. It is routinely administered during a heart attack or stroke to reduce the adverse effects. Physicians sometimes recommend that patients at risk for cardiovascular disease take a low dose of aspirin on a daily basis as a preventive measure. However, aspirin can also lead to serious side effects, including increasing the risk of ulcers. A patient is well advised to consult a physician before beginning any aspirin regimen.

Blood Typing and Blood Transfusion

Blood transfusions in humans were risky procedures until the discovery of the major human blood groups by Karl Landsteiner, an Austrian biologist and physician, in 1900. Until that point, physicians did not understand that death sometimes followed blood transfusions, when the type of donor blood infused into the patient was incompatible with the patient's own blood. Blood groups are determined by the presence or absence of specific marker molecules on the plasma membranes of erythrocytes. With their discovery, it became possible for the first time to match patient-donor blood types and prevent transfusion reactions and deaths.

Antigens, Antibodies, and Transfusion Reactions

Antigens are substances that the body does not recognize as belonging to the "self" and that therefore trigger a defensive response from the leukocytes of the immune system. (Seek more content for additional information on immunity.) Here, we will focus on the role of immunity in blood transfusion reactions. With RBCs in particular, you may see the antigens referred to as isoantigens or agglutinogens (surface antigens) and the antibodies referred to as isoantibodies or agglutinins. In this chapter, we will use the more common terms **antigens and antibodies.**

Antigens are generally large proteins, but may include other classes of organic molecules, including carbohydrates, lipids, and nucleic acids seen on RBCs. Following an infusion of incompatible blood, erythrocytes with foreign antigens appear in the bloodstream and trigger an immune response. Proteins called antibodies (immunoglobulins), which are produced by certain B lymphocytes called plasma cells, attach to the antigens on the plasma membranes of the infused erythrocytes and cause them to adhere to one another.

- Because the arms of the Y-shaped antibodies attach randomly to more than one nonself erythrocyte surface, they form clumps of erythrocytes. This process is called agglutination.
- The clumps of erythrocytes block small blood vessels throughout the body, depriving tissues of oxygen and nutrients.
- As the erythrocyte clumps are degraded, in a process called **hemolysis**, their hemoglobin is released into the bloodstream. This hemoglobin travels to the kidneys, which are responsible for filtration of the blood. However, the load of hemoglobin released can easily overwhelm the kidney's capacity to clear it, and the patient can quickly develop kidney failure.

More than 50 antigens have been identified on erythrocyte membranes, but the most significant in terms of their potential harm to patients are classified in **two groups: the ABO blood group and the Rh blood group.**

1. The ABO Blood Group

Although the ABO blood group name consists of three letters, ABO blood typing designates the presence or absence of just two antigens, A and B. Both are glycoproteins. People whose erythrocytes have A antigens on their erythrocyte membrane surfaces are designated blood type A, and those whose erythrocytes have B antigens are blood type B. People can also have both A and B antigens on their erythrocytes, in which case they are blood type AB. People with neither A nor B antigens are designated blood type O. ABO blood types are genetically determined.
Normally the body must be exposed to a foreign antigen before an antibody can be produced. This is not the case for the ABO blood group. Individuals with type A blood—without any prior exposure to incompatible blood—have preformed antibodies to the B antigen circulating in their blood plasma. These antibodies, referred to as anti-B antibodies, will cause agglutination and hemolysis if they ever encounter erythrocytes with B antigens. Similarly, an individual with type B blood has pre-formed anti-A antibodies. Individuals with type AB blood, which has both antigens, do not have preformed antibodies to either of these. People with type O blood lack antigens A and B on their erythrocytes, but both anti-A and anti-B antibodies circulate in their blood plasma.

	Blood Type			
	A	B	AB	O
Red Blood Cell Type	A	B	AB	O
Antibodies in Plasma	Anti-B	Anti-A	None	Anti-A and Anti-B
Antigens in Red blood Cell	A antigen	B antigen	A and B antigens	None
Blood Types Compatible in an Emergency	A, O	B, O	A, B, AB, O (AB^{+} is the universal recipient)	O (O is the universal donor)

Table 1: ABO Blood Group This chart summarizes the characteristics of the blood types in the ABO blood group. See the text for more on the concept of a universal donor or recipient.

2. Rh Blood Groups – Rhesus Factor

The Rh blood group is classified according to the presence or absence of a second erythrocyte antigen identified as Rh. (It was first discovered in a type of primate known as a ***rhesus macaque***, which is often used in research, because its blood is similar to that of humans.) Although dozens of Rh antigens have been identified, only one, designated D, is clinically important. Those who have the Rh D antigen present on their erythrocytes—about 85 percent of Americans—are described as **Rh positive (Rh^{+})** and those who lack it are **Rh negative (Rh^{-}).** Note that the Rh group is distinct from the ABO group, so any individual, no matter their ABO blood type, may have or lack this Rh antigen. When identifying a patient's blood type, the Rh group is designated by adding the word positive or negative to the ABO type. For example, A positive (A^{+}) means ABO group A blood with the Rh antigen present, and AB negative (AB^{-}) means ABO group AB blood without the Rh antigen.

Besides being a consideration for blood transfusion, parents who differ based on Rh status must be cautious to ensure that maternal antibodies do not destroy their child's red blood cells during fetal development, which can cause hemolytic anemia.

Summary of ABO and Rh Blood Types within the United States

Blood Type	African-Americans	Asian-Americans	Caucasian-Americans	Latino/Latina-Americans
A^{+}	24	27	33	29
A^{-}	2	0.5	7	2
B^{+}	18	25	9	9
B^{-}	1	0.4	2	1
AB^{+}	4	7	3	2
AB^{-}	0.3	0.1	1	0.2
O^{+}	47	39	37	53
O^{-}	4	1	8	4

Determining ABO Blood Types

Clinicians are able to determine a patient's blood type quickly and easily using commercially prepared antibodies. An unknown blood sample is allocated into separate wells. Into one well a small amount of anti-A antibody is added, and to another a small amount of anti-B antibody. If the antigen is present, the antibodies will cause visible agglutination of the cells. The blood should also be tested for Rh antibodies.

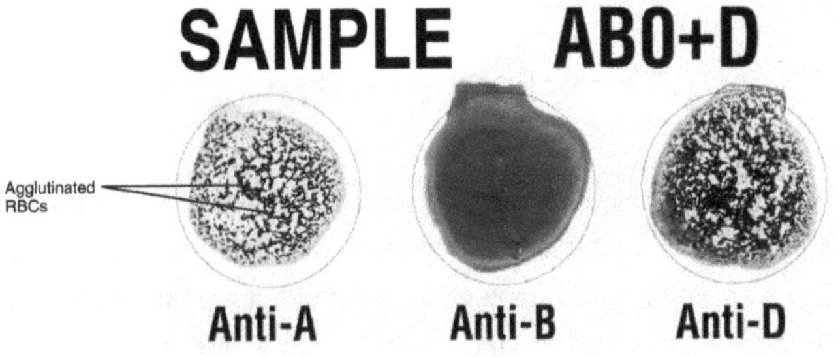

SAMPLE ABO+D

Agglutinated RBCs

Anti-A Anti-B Anti-D

Figure 1: Blood group testing.

Cross Matching Blood Types:

This sample of a commercially produced "bedside" card enables quick typing of both a recipient's and donor's blood before transfusion. The card contains three reaction sites or wells. One is coated with an anti-A antibody, one with an anti-B antibody, and one with an anti-D antibody (tests for the presence of Rh factor D). Mixing a drop of blood and saline into each well enables the blood to interact with a preparation of type-specific antibodies, also called anti-seras. Agglutination of RBCs in a given site indicates a positive identification of the blood antigens, in this case A and Rh antigens for blood type A^+. For the purpose of transfusion, the donor's and recipient's blood types must match.

ABO Transfusion Protocols

To avoid transfusion reactions, it is best to transfuse only matching blood types; that is, a type B^+ recipient should ideally receive blood only from a type B^+ donor and so on. That said, in emergency situations, when acute hemorrhage threatens the patient's life, there may not be time for cross matching to identify blood type. In these cases, blood from a **universal donor—an individual with type O^- blood**—may be transfused. Recall that type O erythrocytes do not display A or B antigens. Thus, anti-A or anti-B antibodies that might be circulating in the patient's blood plasma will not encounter any erythrocyte surface antigens on the donated blood and therefore will not be provoked into a response. One problem with this designation of universal donor is if the O^- individual had prior exposure to Rh antigen, Rh antibodies may be present in the donated blood. Also, introducing type O blood into an individual with type A, B, or AB blood will nevertheless introduce antibodies against both A and B antigens, as these are always circulating in the type O blood plasma. This may cause problems for the recipient, but because the volume of blood transfused is much lower than the volume of the patient's own blood, the adverse effects of the relatively few infused plasma antibodies are typically limited. Rh factor also plays a role. If Rh^-individuals receiving blood have had prior exposure to Rh antigen, antibodies for this antigen may be present in the blood and trigger agglutination to some degree. Although it is always preferable to cross match a patient's blood before transfusing, in a true life-threatening emergency situation, this is not always possible, and these procedures may be implemented.

A patient with **blood type AB^+ is known as the universal recipient.** This patient can theoretically receive any type of blood, because the patient's own blood—having both A and B antigens on the erythrocyte surface—does not produce anti-A or anti-B antibodies. In addition, an Rh^+ patient can receive both Rh^+ and Rh^- blood. However, keep in mind that the donor's blood will contain circulating antibodies, again with possible negative implications.

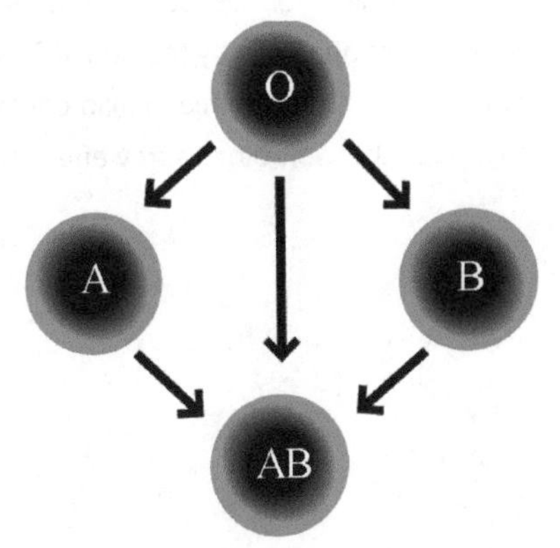

Figure 2: Blood transfusion

At the scene of multiple-vehicle accidents, military engagements, and natural or human-caused disasters, many victims may suffer simultaneously from acute hemorrhage, yet type O blood may not be immediately available. In these circumstances, medics may at least try to replace some of the volume of blood that has been lost. This is done by intravenous administration of a saline solution that provides fluids and electrolytes in proportions equivalent to those of normal blood plasma. Research is ongoing to develop a safe and effective artificial blood that would carry out the oxygen-carrying function of blood without the RBCs, enabling transfusions in the field without concern for incompatibility. These blood substitutes normally contain hemoglobin- as well as perfluorocarbon-based oxygen carriers.

The Cross-Matching Process

Much of the routine work of a blood bank involves testing blood from both donors and recipients to ensure that every recipient is given blood that is compatible and is as safe as possible. Several laboratory tests allow cross-matching of compatible blood between donor and recipient. Patients should ideally receive their own blood or type-specific blood products to minimize the chance of a transfusion reaction. Risks can be further reduced by cross-matching blood, but this process isn't always performed if time is short and the need for transfusion has not been anticipated.

Cross-matching involves mixing a sample of the recipient's serum with a sample of the donor's red blood cells and checking if the mixture agglutinates, or forms clumps. These clumps are the result of antibodies binding the red blood cells together. If agglutination is not obvious by direct vision, blood bank technicians check for agglutination with a microscope. If agglutination occurs, that particular donor's blood cannot be transfused to that particular recipient.

Potential Transfusion Complications

If a patient receives blood during a transfusion that is not compatible with his or her blood type, severe problems can occur. Acute hemolytic transfusion reactions occur if donated blood cells are attacked by matching host antibodies. This can cause shock-like symptoms, such as fever, hypotension, and disseminated intravascular coagulation from immune system mediated endothelial damage. Transfusion reactions are also associated with acute renal failure. Lung injury is common as well, due to pulmonary edema from fluid overload if plasma volume becomes too high or neutrophil activation during a transfusion reaction. If the donated blood is contaminated with bacteria, it may induce septic shock in the patient.

Hemolytic disease of the newborn (HDN) or erythroblastosis fetalis:

In contrast to the ABO group antibodies, which are preformed, antibodies to the Rh antigen are produced only in Rh^- individuals after exposure to the antigen. This process, called sensitization, occurs following a transfusion with Rh-incompatible blood or, more commonly, with the birth of an Rh^+ baby to an Rh^- mother. Problems are rare in a first pregnancy, since the baby's Rh^+ cells rarely cross the placenta (the organ of gas and nutrient exchange between the baby and the mother). However, during or immediately after birth, the Rh^- mother can be exposed to the baby's Rh^+ cells. Research has shown that this occurs in about 13–14 percent of such pregnancies. After exposure, the mother's immune system begins to generate anti-Rh antibodies. If the mother should then conceive another Rh^+ baby, the Rh antibodies she has produced can cross the placenta into the fetal bloodstream and destroy the fetal RBCs. This condition, known as **hemolytic disease of the newborn (HDN) or erythroblastosis fetalis,** may cause anemia in mild cases, but the agglutination and hemolysis can be so severe that without treatment the fetus may die in the womb or shortly after birth.

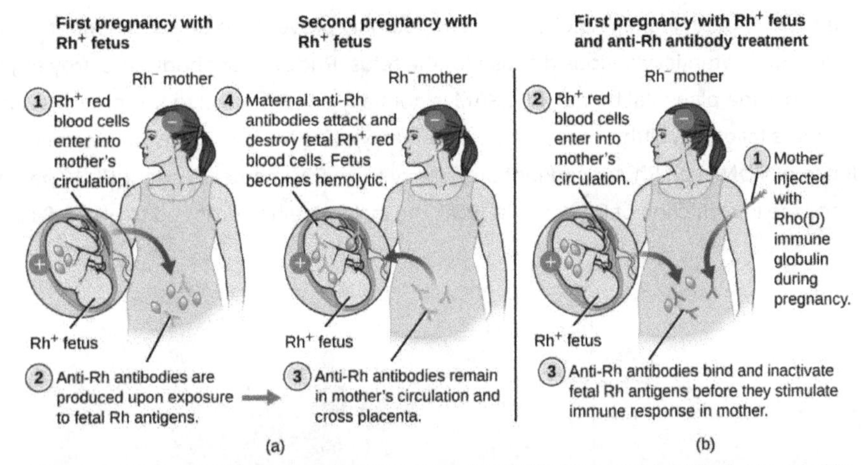

Figure 3: (a) When an Rh− mother has an Rh+ fetus, fetal erythrocytes are introduced into the mother's circulatory system before or during birth, leading to production of anti-Rh IgG antibodies. These antibodies remain in the mother and, if she becomes pregnant with a second Rh+ baby, they can cross the placenta and attach to fetal Rh+ erythrocytes. Complement-mediated hemolysis of fetal erythrocytes results in a lack of sufficient cells for proper oxygenation of the fetus. (b) HDN can be prevented by administering Rho(D) immune globulin during and after each pregnancy with an Rh+ fetus. The immune globulin binds fetal Rh+ RBCs that gain access to the mother's bloodstream, preventing activation of her primary immune response.

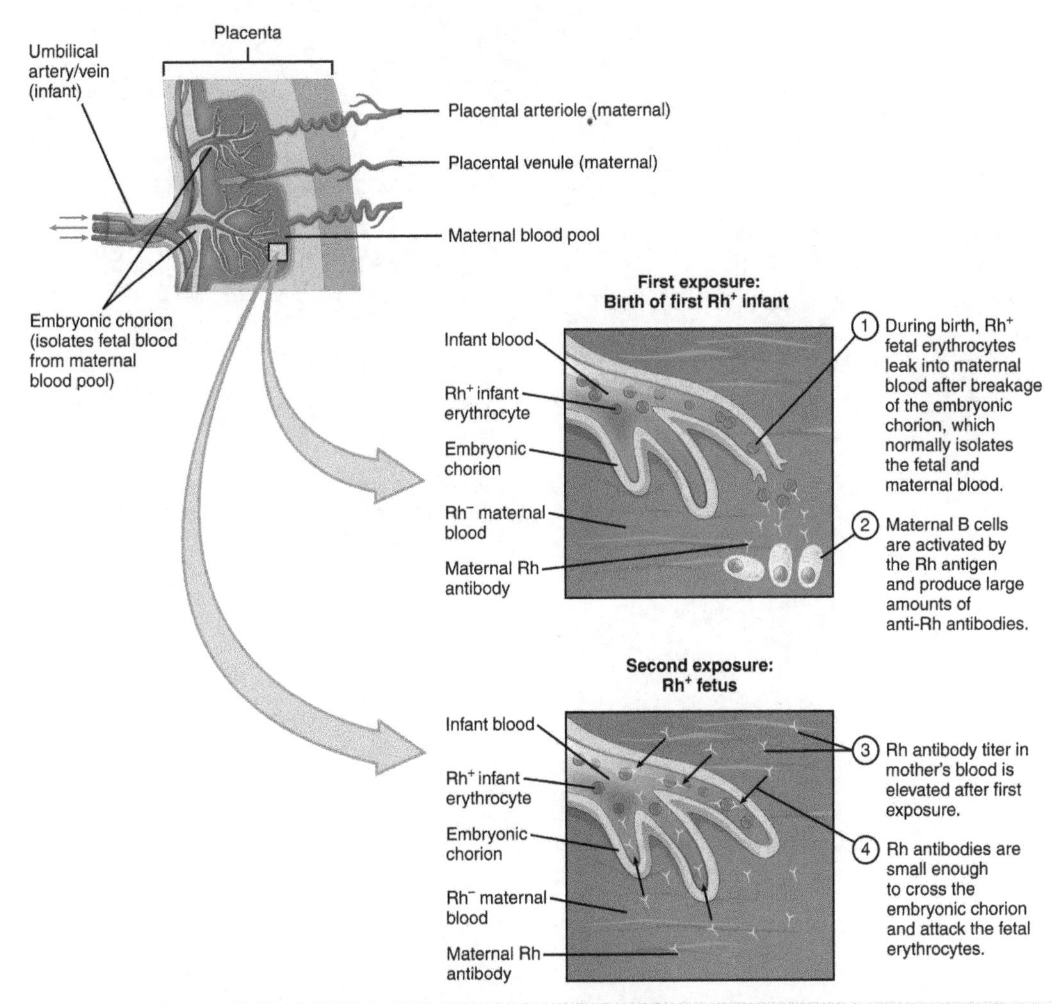

Figure 4: Erythroblastosis Fetalis The first exposure of an Rh− mother to Rh+ erythrocytes during pregnancy induces sensitization. Anti-Rh antibodies begin to circulate in the mother's bloodstream. A second exposure occurs with a subsequent pregnancy with an Rh+ fetus in the uterus. Maternal anti-Rh antibodies may cross the placenta and enter the fetal bloodstream, causing agglutination and hemolysis of fetal erythrocytes.

A drug known as **RhoGAM,** short for Rh immune globulin, can temporarily prevent the development of Rh antibodies in the Rh^- mother, thereby averting this potentially serious disease for the fetus. RhoGAM antibodies destroy any fetal Rh^+ erythrocytes that may cross the placental barrier. RhoGAM is normally administered to Rh^- mothers during weeks 26–28 of pregnancy and within 72 hours following birth. It has proven remarkably effective in decreasing the incidence of HDN. Earlier we noted that the incidence of HDN in an Rh^+subsequent pregnancy to an Rh^- mother is about 13–14 percent without preventive treatment. Since the introduction of RhoGAM in 1968, the incidence has dropped to about 0.1 percent in the United States.

Concept Check

1. A patient's hematocrit is 42 percent. Approximately **what** percentage of the patient's blood is plasma?
2. **Why** would it be incorrect to refer to the formed elements as cells?
3. **True or false:** The buffy coat is the portion of a blood sample that is made up of its proteins.
4. A young woman has been experiencing unusually heavy menstrual bleeding for several years. She follows a strict vegan diet (no animal foods). She is at risk for **what** disorder, and **why**?
5. A patient has thalassemia, a genetic disorder characterized by abnormal synthesis of globin proteins and excessive destruction of erythrocytes. This patient is jaundiced and is found to have an excessive level of bilirubin in his blood. **Explain** the connection.
6. **Explain** why administration of a thrombolytic agent is a first intervention for someone who has suffered a thrombotic stroke.
7. Following a motor vehicle accident, a patient is rushed to the emergency department with multiple traumatic injuries, causing severe bleeding. The patient's condition is critical, and there is no time for determining his blood type. **What** type of blood is transfused, and **why**?
8. In preparation for a scheduled surgery, a patient visits the hospital lab for a blood draw. The technician collects a blood sample and performs a test to determine its type. She places a sample of the patient's blood in two wells. To the first well she adds anti-A antibody. To the second she adds anti-B antibody. Both samples visibly agglutinate. Has the technician made an error, or is this a normal response? If normal, **what** blood type does this indicate?

CHAPTER 11 — THE CARDIOVASCULAR SYSTEM

Heart Anatomy

LEARNING OBJECTIVES

- Relate the structure of the heart to its function as a pump
- Identify and describe the interior and exterior parts of the human heart
- Compare cardiac muscle to skeletal and smooth muscle
- Explain the cardiac cycle
- Compare systemic circulation to pulmonary circulation
- Identify and describe the components of the conducting system that distributes electrical impulses through the heart
- Describe factors affecting heart rate
- Compare and contrast the anatomical structure of arteries, arterioles, capillaries, venules, and veins

The vital importance of the heart is obvious. If one assumes an average rate of 75 contractions per minute, a human heart would contract approximately 108,000 times in one day, more than 39 million times in one year, and nearly 3 billion times during a 75-year lifespan. Each of the major pumping chambers of the heart ejects approximately 70 mL blood per contraction in a resting adult. This would be equal to 5.25 liters of fluid per minute and approximately 14,000 liters per day. Over one year, that would equal 10,000,000 liters or 2.6 million gallons of blood sent through roughly 60,000 miles of vessels. In order to understand how that happens, it is necessary to understand the anatomy and physiology of the heart.

Location of the Heart

The human heart is located within the thoracic cavity, between the lungs in the space known as the mediastinum. Figure 1 shows the position of the heart within the thoracic cavity. Within the mediastinum, the heart is enclosed by a tough membrane known as the pericardium, or pericardial sac, and sits in its own space called the **pericardial cavity**. The dorsal surface of the heart lies near the bodies of the vertebrae, and its anterior surface sits deep to the sternum and costal cartilages. The great veins, the superior and inferior venae cavae, and the great arteries, the aorta and pulmonary trunk, are attached to the superior surface of the heart, called the base. The inferior tip of the heart, the apex, lies just to the left of the sternum.

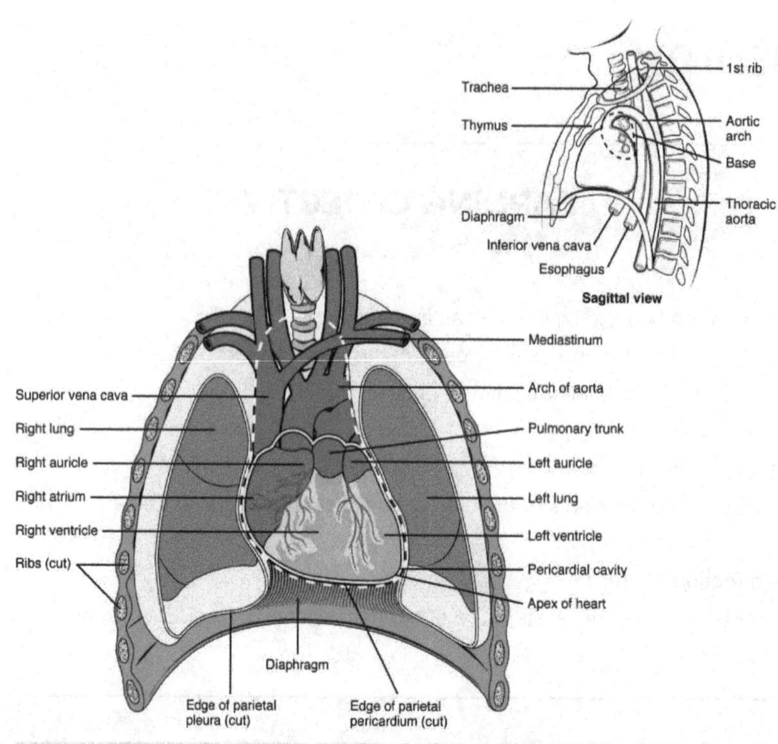

Figure 1. The heart is located within the thoracic cavity, medially between the lungs in the mediastinum. It is about the size of a fist, is broad at the top, and tapers toward the base.

EVERYDAY CONNECTION: CPR

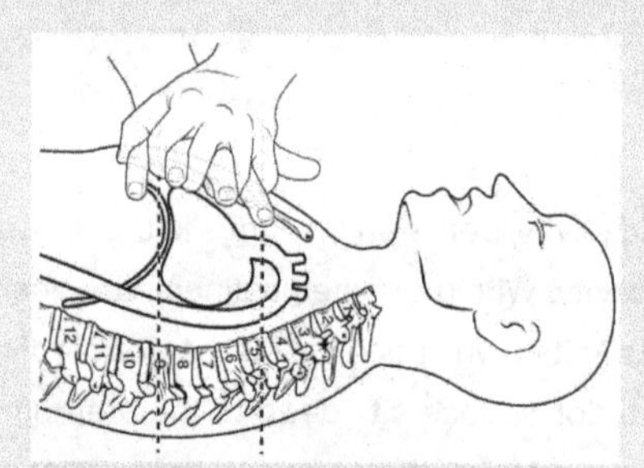

Figure 2. If the heart should stop, CPR can maintain the flow of blood until the heart resumes beating. By applying pressure to the sternum, the blood within the heart will be squeezed out of the heart and into the circulation. Proper positioning of the hands on the sternum to perform CPR would be between the lines at T4 and T9.

The position of the heart in the torso between the vertebrae and sternum (see the image above for the position of the heart within the thorax) allows for individuals to apply an emergency technique known as cardiopulmonary resuscitation (CPR) if the heart of a patient should stop. By applying pressure with the flat portion of one hand on the sternum in the area between the lines in the image below), it is possible to manually compress the blood within the heart enough to push some of the blood within it into the pulmonary and systemic circuits. This is particularly critical for the brain, as irreversible damage and death of neurons occur within minutes of loss of blood flow.

Shape and Size of the Heart

The shape of the heart is similar to a pinecone, rather broad at the superior surface and tapering to the apex. A typical heart is approximately the size of your fist: 12 cm (5 in) in length, 8 cm (3.5 in) wide, and 6 cm (2.5 in) in thickness. The heart of a well-trained athlete, especially one specializing in aerobic sports, can be considerably larger than this. Cardiac muscle responds to exercise in a manner similar to that of skeletal muscle. That is, exercise results in the addition of protein myofilaments that increase the size of the individual cells without increasing their numbers, a concept called hypertrophy. As a result of this hypertrophy athletic hearts can pump blood more effectively at lower rates than nonathletic hearts.

Chambers and Circulation through the Heart

The human heart consists of four chambers: The left side and the right side each have one **atrium** and one **ventricle**. Each of the upper chambers, the right atrium (plural = atria) and the left atrium, acts as a receiving chamber and contracts to push blood into the lower chambers, the right ventricle and the left ventricle. The ventricles serve as the primary pumping chambers of the heart, propelling blood to the lungs or to the rest of the body.

There are two distinct but linked circuits in the human circulation called the pulmonary and systemic circuits. Although both circuits transport blood and everything it carries, we can initially view the circuits from the point of view of gases. The **pulmonary circuit** transports blood to and from the lungs, where it picks up oxygen and delivers carbon dioxide for exhalation. The **systemic circuit** transports oxygenated blood to virtually all of the tissues of the body and returns relatively deoxygenated blood and carbon dioxide to the heart to be sent back to the pulmonary circulation.

The right ventricle pumps deoxygenated blood into the **pulmonary trunk**, which leads toward the lungs and branches into the left and right **pulmonary arteries**. These vessels in turn branch many times before reaching the **pulmonary capillaries**, where gas exchange occurs: Carbon dioxide exits the blood and oxygen enters. The pulmonary trunk arteries and their branches are the only arteries in the post-natal body that carry relatively deoxygenated blood. Highly oxygenated blood returning from the pulmonary capillaries in the lungs passes through a series of vessels that join together to form the **pulmonary veins**—the only post-natal veins in the body that carry highly oxygenated blood. The pulmonary veins direct blood into the left atrium, which pumps the blood into the left ventricle, which in turn pumps oxygenated blood into the aorta and on to the many branches of the systemic circuit. Eventually, these vessels will lead to the systemic capillaries, where exchange with the tissue fluid and cells of the body occurs. In this case, oxygen and nutrients exit the systemic capillaries to be used by the cells in their metabolic processes, and carbon dioxide and waste products will enter the blood.

The blood exiting the systemic capillaries is lower in oxygen concentration than when it entered. The capillaries will ultimately unite to form venules, joining to form ever-larger veins, eventually flowing into the two major systemic veins, the **superior vena cava** and the **inferior vena cava**, which return blood to the right atrium. The blood in the superior and inferior venae cavae flows into the right atrium, which pumps blood into the right ventricle. This process of blood circulation continues as long as the individual remains alive.

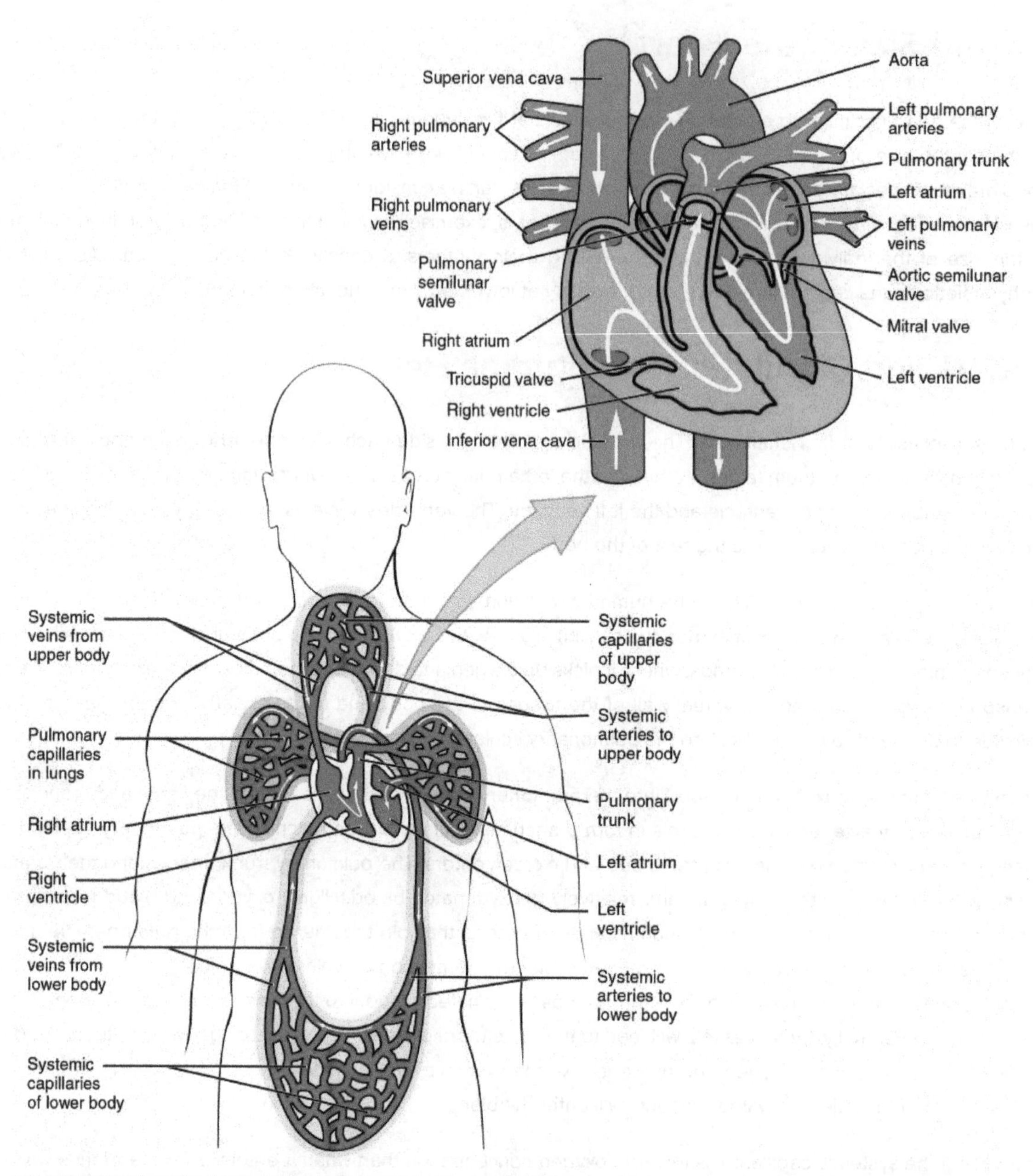

Figure 3. Blood flows from the right atrium to the right ventricle, where it is pumped into the pulmonary circuit. The blood in the pulmonary artery branches is low in oxygen but relatively high in carbon dioxide. Gas exchange occurs in the pulmonary capillaries (oxygen into the blood, carbon dioxide out), and blood high in oxygen and low in carbon dioxide is returned to the left atrium. From here, blood enters the left ventricle, which pumps it into the systemic circuit. Following exchange in the systemic capillaries (oxygen and nutrients out of the capillaries and carbon dioxide and wastes in), blood returns to the right atrium and the cycle is repeated.

Membranes

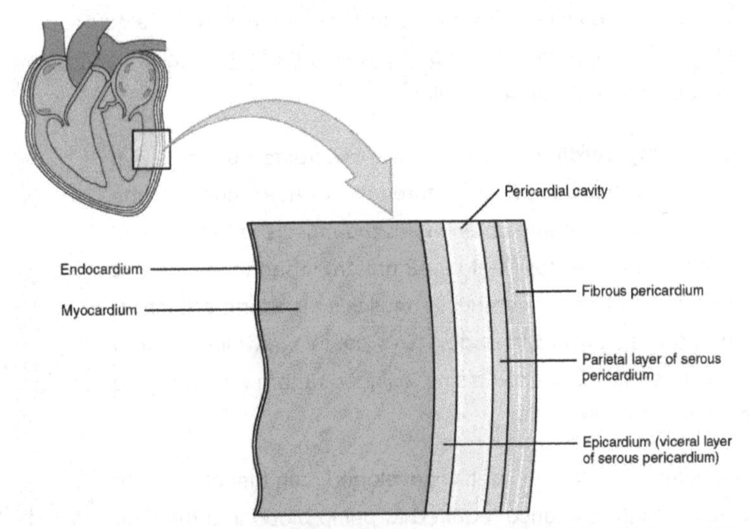

Figure 4. The pericardial membrane that surrounds the heart consists of three layers and the pericardial cavity. The heart wall also consists of three layers. The pericardial membrane and the heart wall share the epicardium.

The membrane that directly surrounds the heart and defines the pericardial cavity is called the **pericardium** or **pericardial sac**. The pericardium, which literally translates as "around the heart," consists of two distinct sublayers: the sturdy outer fibrous pericardium and the inner serous pericardium. The fibrous pericardium is made of tough, dense connective tissue that protects the heart and maintains its position in the thorax. The more delicate serous pericardium consists of two layers: the parietal pericardium, which is fused to the fibrous pericardium, and an inner visceral pericardium; or **epicardium**, which is fused to the heart and is part of the heart wall. The pericardial cavity, filled with lubricating serous fluid, lies between the epicardium and the pericardium.

DISORDERS OF THE HEART

Cardiac Tamponade

If excess fluid builds within the pericardial space, it can lead to a condition called cardiac tamponade, or pericardial tamponade. With each contraction of the heart, more fluid—in most instances, blood—accumulates within the pericardial cavity. In order to fill with blood for the next contraction, the heart must relax. However, the excess fluid in the pericardial cavity puts pressure on the heart and prevents full relaxation, so the chambers within the heart contain slightly less blood as they begin each heart cycle. Over time, less and less blood is ejected from the heart. If the fluid builds up slowly, as in hypothyroidism, the pericardial cavity may be able to expand gradually to accommodate this extra volume. Some cases of fluid in excess of one liter within the pericardial cavity have been reported. Rapid accumulation of as little as 100 mL of fluid following trauma may trigger cardiac tamponade. Other common causes include myocardial rupture, pericarditis, cancer, or even cardiac surgery. Removal of this excess fluid requires insertion of drainage tubes into the pericardial cavity. Premature removal of these drainage tubes, for example, following cardiac surgery, or clot formation within these tubes are causes of this condition. Untreated, cardiac tamponade can lead to death.

Layers

The wall of the heart is composed of three layers of unequal thickness. From superficial to deep, these are the epicardium, the myocardium, and the endocardium. The outermost layer of the wall of the heart is also the innermost layer of the pericardium, the epicardium, or the visceral pericardium discussed earlier.

The middle and thickest layer is the **myocardium**, which consists of cardiac muscle cells. It is the contraction of the myocardium that pumps blood through the heart and into the major arteries. The muscle pattern is elegant and complex, as the muscle cells swirl and spiral around the chambers of the heart. They form a figure 8 pattern around the atria and around the bases of the great vessels. Deeper ventricular muscles also form a figure 8 around the two ventricles and proceed toward the apex. This complex swirling pattern allows the heart to pump blood more effectively than a simple linear pattern would. Figure 6 illustrates the arrangement of muscle cells.

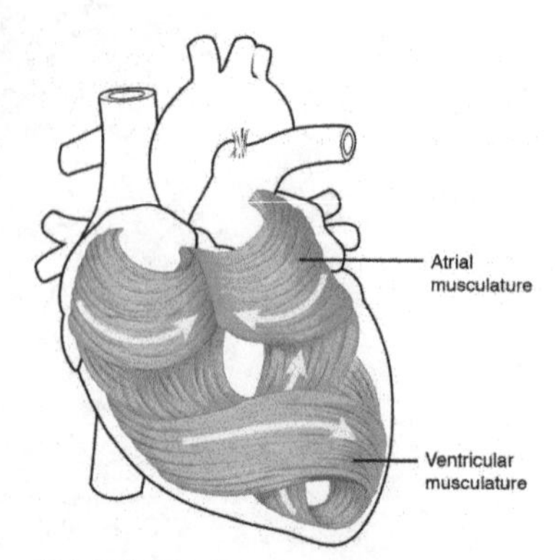

Figure 6. The swirling pattern of cardiac muscle tissue contributes significantly to the heart's ability to pump blood effectively.

The muscle of the left ventricle is much thicker and better developed than that of the right ventricle. In order to overcome the high resistance required to pump blood into the long systemic circuit, the left ventricle must generate a great amount of pressure. The right ventricle does not need to generate as much pressure, since the pulmonary circuit is shorter and provides less resistance. The image below illustrates the differences in muscular thickness needed for each of the ventricles.

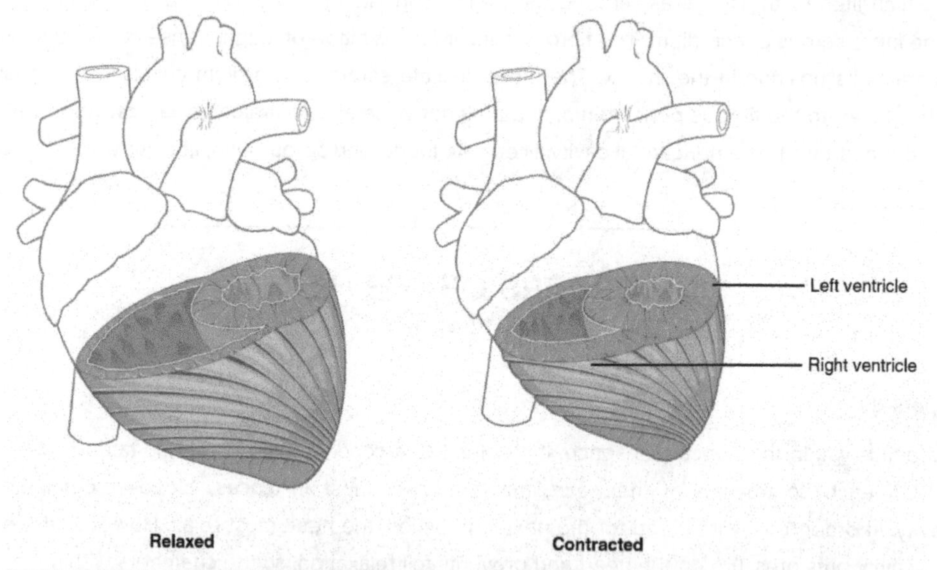

Figure 7. The myocardium in the left ventricle is significantly thicker than that of the right ventricle. The left ventricle must generate a much greater pressure to overcome greater resistance in the systemic circuit. The ventricles are shown in both relaxed and contracting states.

The innermost layer of the heart wall, the **endocardium**, is joined to the myocardium with a thin layer of connective tissue. The endocardium lines the chambers where the blood circulates and covers the heart valves. It is made of simple squamous epithelium called **endothelium**, which is continuous with the endothelial lining of the blood vessels.

Internal Structure of the Heart

Recall that the heart's contraction cycle follows a dual pattern of circulation—the pulmonary and systemic circuits—because of the pairs of chambers that pump blood into the circulation. In order to develop a more precise understanding of cardiac function, it is first necessary to explore the internal anatomical structures in more detail.

Septa of the Heart

The word septum is derived from the Latin for “something that encloses;” in this case, a **septum** (plural = septa) refers to a wall or partition that divides the heart into chambers. The septa are physical extensions of the myocardium lined with endocardium. Located between the two atria is the **interatrial septum**. Between the two ventricles is a second septum known as the **interventricular septum**. It is substantially thicker than the interatrial septum, since the ventricles generate far greater pressure when they contract.

The septum between the atria and ventricles is known as the **atrioventricular septum**. It is marked by the presence of four openings that allow blood to move from the atria into the ventricles and from the ventricles into the pulmonary trunk and aorta. Located in each of these openings between the atria and ventricles is a **valve**, a specialized structure that ensures one-way flow of blood.

The valves between the atria and ventricles are known generically as **atrioventricular valves**. The valves at the openings that lead to the pulmonary trunk and aorta are known generically as **semilunar valves**. The interventricular septum is visible in the image below. In this figure, the atrioventricular septum has been removed to better show the bicuspid and tricuspid valves; the interatrial septum is not visible, since its location is covered by the aorta and pulmonary trunk.

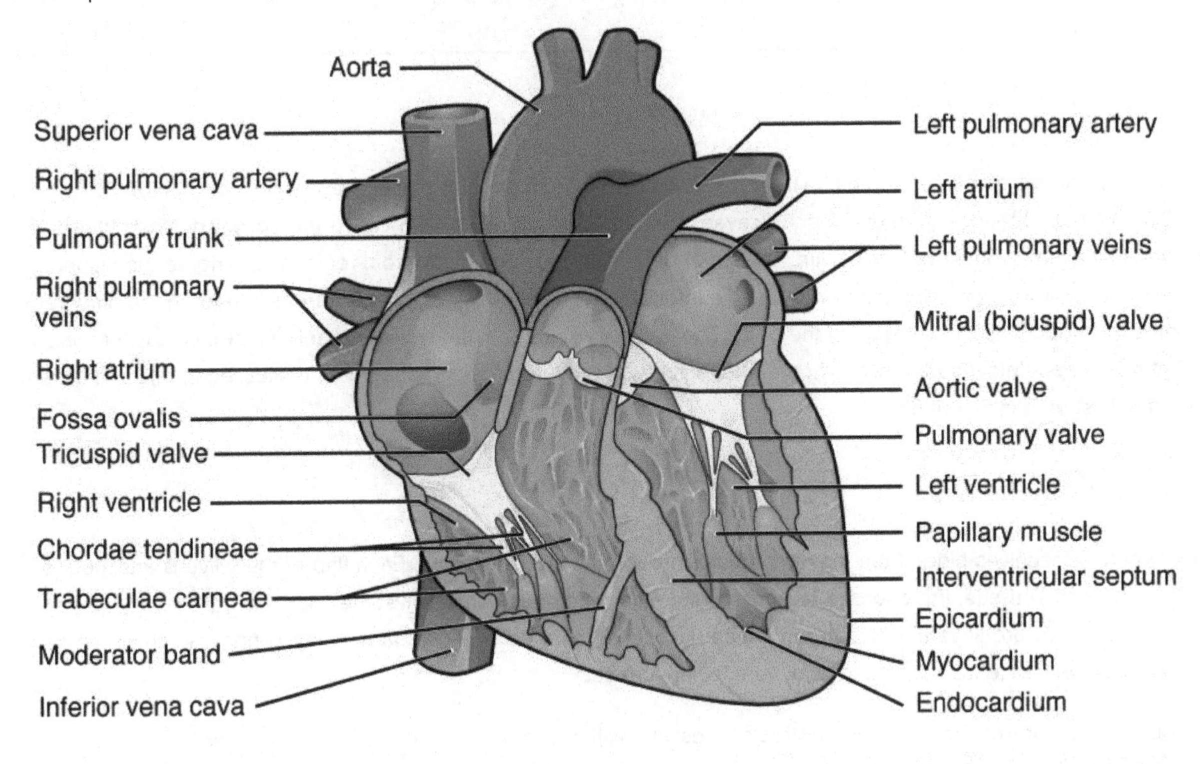

Figure 8. This anterior view of the heart shows the four chambers, the major vessels and their early branches, as well as the valves. The presence of the pulmonary trunk and aorta covers the interatrial septum, and the atrioventricular septum is cut away to show the atrioventricular valves.

DISORDERS OF THE HEART: HEART DEFECTS

One very common form of interatrial septum pathology is patent foramen ovale, which occurs when the septum does not close at birth, and an opening remains between the left and right atria. The word patent is from the Latin root patens for "open." It may be benign or asymptomatic, perhaps never being diagnosed, or in extreme cases, it may require surgical repair to close the opening permanently. As much as 20–25 percent of the general population may have a patent foramen ovale, but fortunately, most have the benign, asymptomatic version. Patent foramen ovale is normally detected by presence of a heart murmur (an abnormal heart sound) and confirmed by imaging with an echocardiogram. Despite its prevalence in the general population, the causes of patent ovale are unknown, and there are no known risk factors. In nonlife-threatening cases, it is better to monitor the condition than to risk heart surgery to repair and seal the opening.

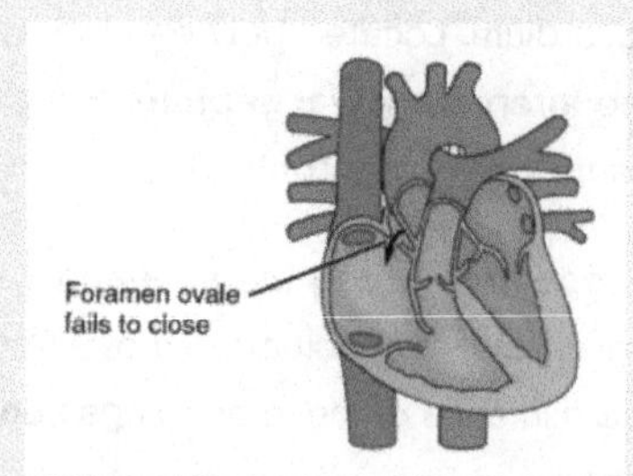

Figure 9. A patent foramen ovale defect is an abnormal opening in the interatrial septum, or more commonly, a failure of the foramen ovale to close.

Right Atrium

The right atrium serves as the receiving chamber for blood returning to the heart from the systemic circulation. The two major systemic veins, the superior and inferior venae cavae, and the large coronary vein called the **coronary sinus** that drains the heart myocardium empty into the right atrium. The superior vena cava drains blood from regions superior to the diaphragm: the head, neck, upper limbs, and the thoracic region. The inferior vena cava drains blood from areas inferior to the diaphragm: the lower limbs and abdominopelvic region of the body. The coronary sinus is a thin-walled vessel that drains most of the coronary veins that return systemic blood from the heart. The majority of the internal heart structures discussed in this and subsequent sections are illustrated in Figure 8.

Right Ventricle

The right ventricle receives blood from the right atrium through the tricuspid valve. Each flap of the valve is attached to strong strands of connective tissue, the **chordae tendineae**, literally "tendinous cords," or sometimes more poetically referred to as "heart strings." There are several chordae tendineae associated with each of the flaps. They connect each of the flaps to a **papillary muscle** that extends from the inferior ventricular surface.

When the myocardium of the ventricle contracts, pressure within the ventricular chamber rises. To prevent any potential backflow, the papillary muscles also contract, generating tension on the chordae tendineae. This prevents the flaps of the valves from being forced into the atria and regurgitation of the blood back into the atria during ventricular contraction.The image below shows papillary muscles and chordae tendineae attached to the tricuspid valve.

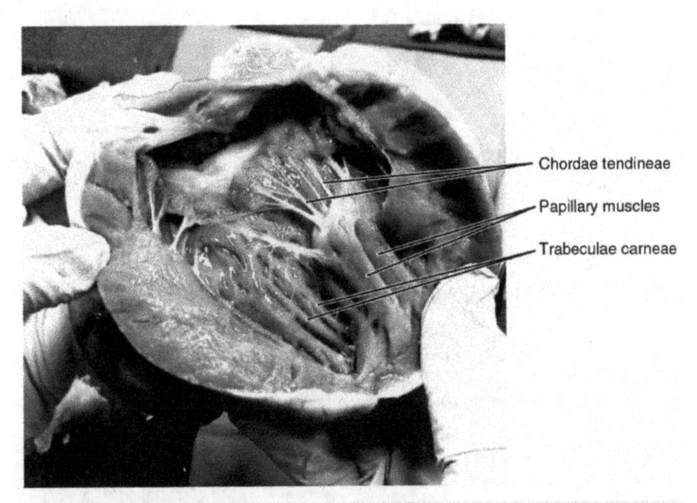

Figure 10. In this frontal section, you can see papillary muscles attached to the tricuspid valve on the right as well as the mitral valve on the left via chordae tendineae. (credit: modification of work by "PV KS"/flickr.com)

When the right ventricle contracts, it ejects blood into the pulmonary trunk, which branches into the left and right pulmonary arteries that carry it to each lung. The superior surface of the right ventricle begins to taper as it approaches the pulmonary trunk. At the base of the pulmonary trunk is the pulmonary semilunar valve that prevents backflow from the pulmonary trunk.

Left Atrium

After exchange of gases in the pulmonary capillaries, blood returns to the left atrium high in oxygen via one of the four pulmonary veins. Blood flows nearly continuously from the pulmonary veins back into the atrium, which acts as the receiving chamber, and from here through an opening into the left ventricle. Most blood flows passively into the heart while both the atria and ventricles are relaxed, but toward the end of the ventricular relaxation period, the left atrium will contract, pumping blood into the ventricle. This atrial contraction accounts for approximately 20 percent of ventricular filling. The opening between the left atrium and ventricle is guarded by the mitral valve.

Left Ventricle

Recall that the muscular layer is much thicker in the left ventricle compared to the right. The mitral valve is connected to papillary muscles via chordae tendineae. There are two papillary muscles on the left—the anterior and posterior—as opposed to three on the right.

The left ventricle is the major pumping chamber for the systemic circuit; it ejects blood into the aorta through the aortic semilunar valve.

Heart Valve Function

In Figure 12 (a), the two atrioventricular valves are open and the two semilunar valves are closed. This occurs when both atria and ventricles are relaxed and when the atria contract to pump blood into the ventricles. Figure 13 (b) shows a frontal view. Although only the left side of the heart is illustrated, the process is virtually identical on the right.

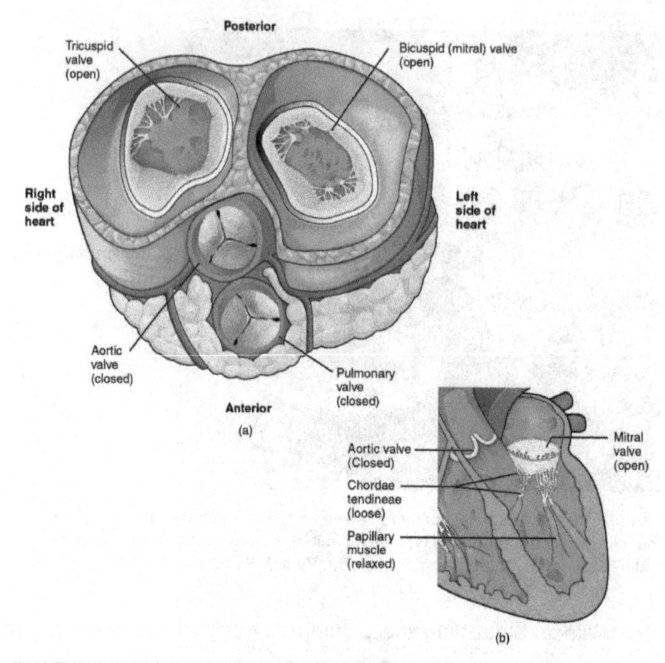

Figure 12. (a) A transverse section through the heart illustrates the four heart valves. The two atrioventricular valves are open; the two semilunar valves are closed. The atria and vessels have been removed. (b) A frontal section through the heart illustrates blood flow through the mitral valve. When the mitral valve is open, it allows blood to move from the left atrium to the left ventricle. The aortic semilunar valve is closed to prevent backflow of blood from the aorta to the left ventricle.

Figure 13 (a) shows the atrioventricular valves closed while the two semilunar valves are open. This occurs when the ventricles contract to eject blood into the pulmonary trunk and aorta. Closure of the two atrioventricular valves prevents blood from being forced back into the atria. This stage can be seen from a frontal view in Figure 13 (b).

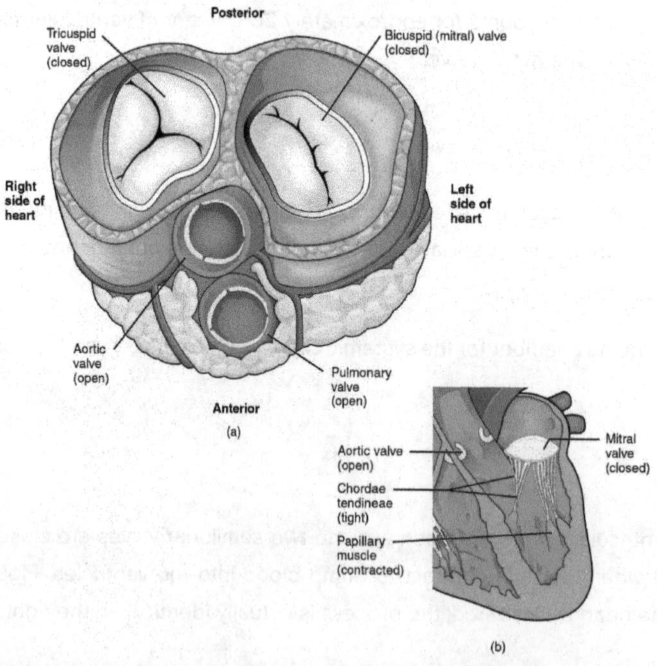

Figure 13. (a) A transverse section through the heart illustrates the four heart valves during ventricular contraction. The two atrioventricular valves are closed, but the two semilunar valves are open. The atria and vessels have been removed. (b) A frontal view shows the closed mitral (bicuspid) valve that prevents backflow of blood into the left atrium. The aortic semilunar valve is open to allow blood to be ejected into the aorta.

Disorders of the Heart Valves

When heart valves do not function properly, they are often described as incompetent and result in valvular heart disease, which can range from benign to lethal. Some of these conditions are congenital, that is, the individual was born with the defect, whereas others may be attributed to disease processes or trauma. Some malfunctions are treated with medications, others require surgery, and still others may be mild enough that the condition is merely monitored since treatment might trigger more serious consequences.

Valvular disorders are often caused by carditis, or inflammation of the heart. One common trigger for this inflammation is rheumatic fever, or scarlet fever, an autoimmune response to the presence of a bacterium, *Streptococcus pyogenes*, normally a disease of childhood.

While any of the heart valves may be involved in valve disorders, mitral regurgitation is the most common, detected in approximately 2 percent of the population, and the pulmonary semilunar valve is the least frequently involved. When a valve malfunctions, the flow of blood to a region will often be disrupted.

If one of the cusps of the valve is forced backward by the force of the blood, the condition is referred to as a prolapsed valve. Prolapse may occur if the chordae tendineae are damaged or broken, causing the closure mechanism to fail. The failure of the valve to close properly disrupts the normal one-way flow of blood and results in regurgitation, when the blood flows backward from its normal path. Using a stethoscope, the disruption to the normal flow of blood produces a heart murmur.

Auscultation, or listening to a patient's heart sounds, is one of the most useful diagnostic tools, since it is proven, safe, and inexpensive. The term auscultation is derived from the Latin for "to listen," and the technique has been used for diagnostic purposes as far back as the ancient Egyptians. Valve and septal disorders will trigger abnormal heart sounds. If a valvular disorder is detected or suspected, a test called an echocardiogram, or simply an "echo," may be ordered. Echocardiograms are sonograms of the heart and can help in the diagnosis of valve disorders as well as a wide variety of heart pathologies.

Coronary Circulation

You will recall that the heart is a remarkable pump composed largely of cardiac muscle cells that are incredibly active throughout life. Like all other cells, a cardiac muscle cell requires a reliable supply of oxygen and nutrients, and a way to remove wastes, so it needs a dedicated, complex, and extensive coronary circulation. And because of the critical and nearly ceaseless activity of the heart throughout life, this need for a blood supply is even greater than for a typical cell.

Coronary Arteries

Coronary arteries supply blood to the myocardium and other components of the heart. The first portion of the aorta after it arises from the left ventricle gives rise to the coronary arteries. The left coronary artery distributes blood to the left side of the heart, the left atrium and ventricle, and the interventricular septum. The right coronary artery distributes blood to the right atrium, portions of both ventricles, and the heart conduction system. Figure 15 presents views of the coronary circulation from both the anterior and posterior views.

Coronary Veins

Coronary veins drain the heart and generally parallel the large surface arteries. The **coronary sinus** is a large, thin-walled vein on the posterior surface of the heart that empties directly into the right atrium. The **anterior cardiac veins** parallel the small cardiac arteries and drain the anterior surface of the right ventricle. Unlike other cardiac veins, it bypasses the coronary sinus and drains directly into the right atrium.

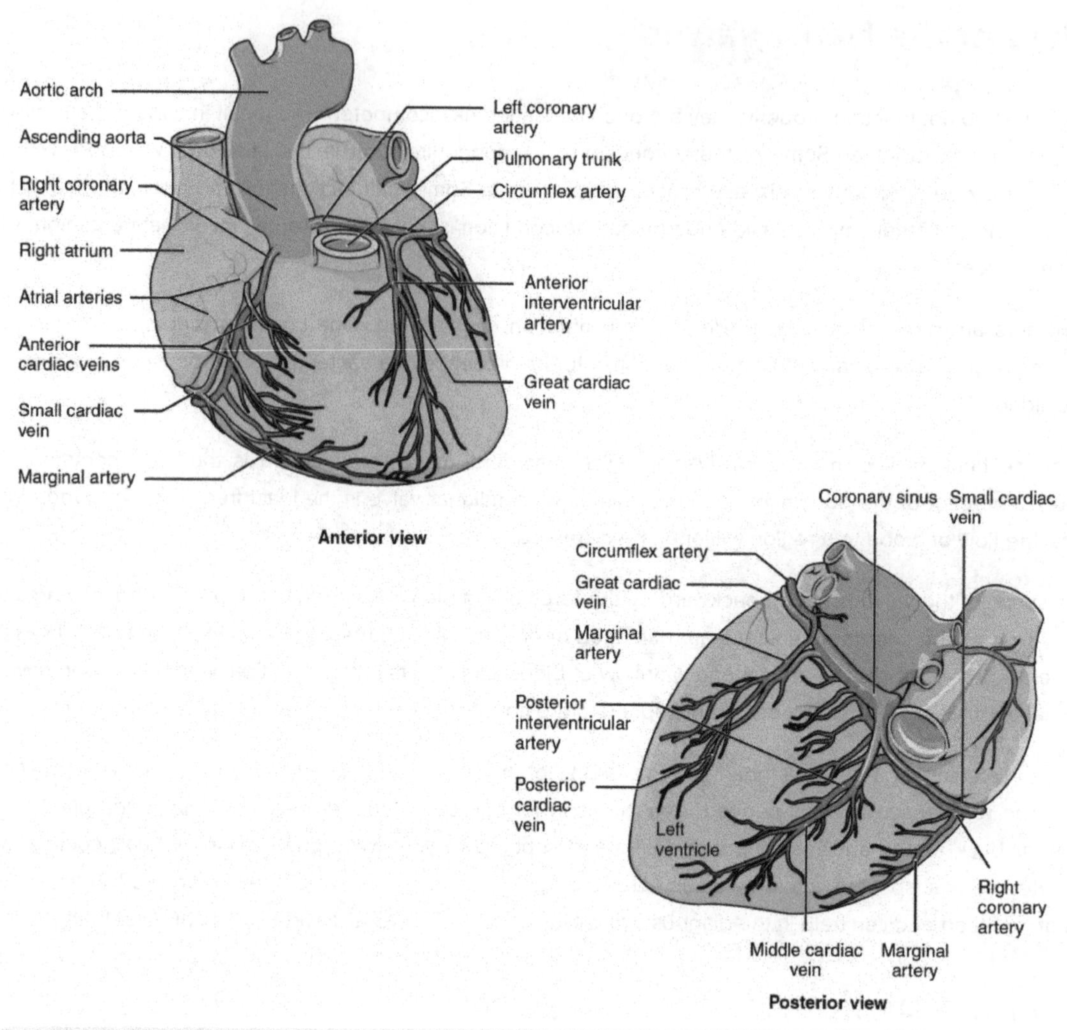

Figure 15. The anterior view of the heart shows the prominent coronary surface vessels. The posterior view of the heart shows the prominent coronary surface vessels.

DISEASES OF THE HEART: MYOCARDIAL INFARCTION

Myocardial infarction (MI) is the formal term for what is commonly referred to as a heart attack. It normally results from a lack of blood flow (ischemia) and oxygen (hypoxia) to a region of the heart, resulting in death of the cardiac muscle cells. A myocardial infarction often occurs when a coronary artery is blocked by the buildup of atherosclerotic plaque consisting of lipids, cholesterol and fatty acids, and white blood cells, primarily macrophages. It can also occur when a portion of an unstable atherosclerotic plaque travels through the coronary arterial system and lodges in one of the smaller vessels. The resulting blockage restricts the flow of blood and oxygen to the myocardium and causes death of the tissue. MIs may be triggered by excessive exercise, in which the partially occluded artery is no longer able to pump sufficient quantities of blood, or severe stress, which may induce spasm of the smooth muscle in the walls of the vessel.

In the case of acute MI, there is often sudden pain beneath the sternum (retrosternal pain) called angina pectoris, often radiating down the left arm in males but not in female patients. Until this anomaly between the sexes was discovered, many female patients suffering MIs were misdiagnosed and sent home. In addition, patients typically present with difficulty breathing and shortness of breath (dyspnea), irregular heartbeat (palpations), nausea and vomiting, sweating (diaphoresis), anxiety, and fainting (syncope), although not all of these symptoms may be present. Many of the symptoms are shared with other medical conditions, including anxiety attacks and simple indigestion, so differential diagnosis is critical. It is estimated that between 22 and 64 percent of MIs present without any symptoms.

An MI can be confirmed by examining the patient's ECG, which frequently reveals alterations in the ST and Q components. Some classification schemes of MI are referred to as ST-elevated MI (STEMI) and non-elevated MI (non-STEMI). In addition, echocardiography or cardiac magnetic resonance imaging may be employed. Common blood tests indicating an MI include elevated levels of creatine kinase MB (an enzyme that catalyzes the conversion of creatine to phosphocreatine, consuming ATP) and cardiac troponin (the regulatory protein for muscle contraction), both of which are released by damaged cardiac muscle cells.

Immediate treatments for MI are essential and include administering supplemental oxygen, aspirin that helps to break up clots, and nitroglycerine administered sublingually (under the tongue) to facilitate its absorption. Despite its unquestioned success in treatments and use since the 1880s, the mechanism of nitroglycerine is still incompletely understood but is believed to involve the release of nitric oxide, a known vasodilator, and endothelium-derived releasing factor, which also relaxes the smooth muscle in the tunica media of coronary vessels. Longer-term treatments include injections of thrombolytic agents such as streptokinase that dissolve the clot, the anticoagulant heparin, balloon angioplasty and stents to open blocked vessels, and bypass surgery to allow blood to pass around the site of blockage. If the damage is extensive, coronary replacement with a donor heart or coronary assist device, a sophisticated mechanical device that supplements the pumping activity of the heart, may be employed. Despite the attention, development of artificial hearts to augment the severely limited supply of heart donors has proven less than satisfactory but will likely improve in the future.

MIs may trigger cardiac arrest, but the two are not synonymous. Important risk factors for MI include cardiovascular disease, age, smoking, high blood levels of the low-density lipoprotein (LDL, often referred to as "bad" cholesterol), low levels of high-density lipoprotein (HDL, or "good" cholesterol), hypertension, diabetes mellitus, obesity, lack of physical exercise, chronic kidney disease, excessive alcohol consumption, and use of illegal drugs.

DISEASES OF THE HEART: CORONARY ARTERY DISEASE

Coronary artery disease is the leading cause of death worldwide. It occurs when the buildup of plaque—a fatty material including cholesterol, connective tissue, white blood cells, and some smooth muscle cells—within the walls of the arteries obstructs the flow of blood and decreases the flexibility or compliance of the vessels. This condition is called atherosclerosis, a hardening of the arteries that involves the accumulation of plaque. As the coronary blood vessels become occluded, the flow of blood to the tissues will be restricted, a condition called ischemia that causes the cells to receive insufficient amounts of oxygen, called hypoxia. The image below shows the blockage of coronary arteries highlighted by the injection of dye. Some individuals with coronary artery disease report pain radiating from the chest called angina pectoris, but others remain asymptomatic. If untreated, coronary artery disease can lead to MI or a heart attack.

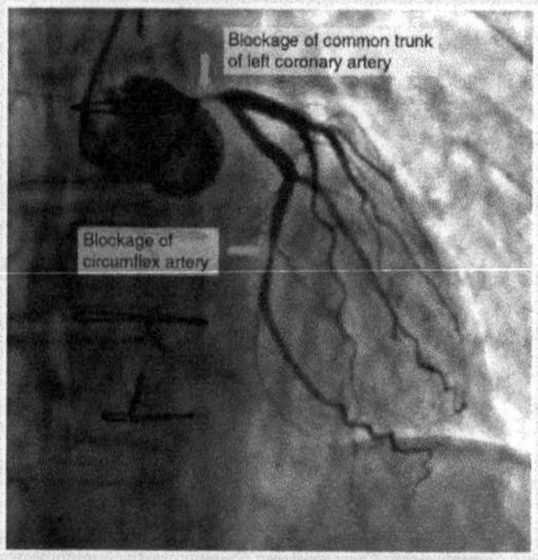

Figure 16. In this coronary angiogram (X-ray), the dye makes visible two occluded coronary arteries. Such blockages can lead to decreased blood flow (ischemia) and insufficient oxygen (hypoxia) delivered to the cardiac tissues. If uncorrected, this can lead to cardiac muscle death (myocardial infarction).

The disease progresses slowly and often begins in children and can be seen as fatty "streaks" in the vessels. It then gradually progresses throughout life. Well-documented risk factors include smoking, family history, hypertension, obesity, diabetes, high alcohol consumption, lack of exercise, stress, and hyperlipidemia or high circulating levels of lipids in the blood. Treatments may include medication, changes to diet and exercise, angioplasty with a balloon catheter, insertion of a stent, or coronary bypass procedure.

Angioplasty is a procedure in which the occlusion is mechanically widened with a balloon. A specialized catheter with an expandable tip is inserted into a superficial vessel, normally in the leg, and then directed to the site of the occlusion. At this point, the balloon is inflated to compress the plaque material and to open the vessel to increase blood flow. Then, the balloon is deflated and retracted. A stent consisting of a specialized mesh is typically inserted at the site of occlusion to reinforce the weakened and damaged walls. Stent insertions have been routine in cardiology for more than 40 years.

Coronary bypass surgery may also be performed. This surgical procedure grafts a replacement vessel obtained from another, less vital portion of the body to bypass the occluded area. This procedure is clearly effective in treating patients experiencing a MI, but overall does not increase longevity. Nor does it seem advisable in patients with stable although diminished cardiac capacity since frequently loss of mental acuity occurs following the procedure. Long-term changes to behavior, emphasizing diet and exercise plus a medicine regime tailored to lower blood pressure, lower cholesterol and lipids, and reduce clotting are equally as effective.

Cardiac Muscle and Electrical Activity

Recall that cardiac muscle shares a few characteristics with both skeletal muscle and smooth muscle, but it has some unique properties of its own. Not the least of these exceptional properties is its ability to initiate an electrical potential at a fixed rate that spreads rapidly from cell to cell to trigger the contractile mechanism. This property is known as **autorhythmicity**. Neither smooth nor skeletal muscle can do this. Even though cardiac muscle has autorhythmicity, heart rate is modulated by the endocrine and nervous systems.

There are two major types of cardiac muscle cells: myocardial contractile cells and myocardial conducting cells. The **myocardial contractile cells** constitute the bulk (99 percent) of the cells in the atria and ventricles. Contractile cells conduct impulses and are responsible for contractions that pump blood through the body. The **myocardial conducting cells** (1 percent of the cells) form the conduction system of the heart. They are generally much smaller than the contractile cells and have few of the myofibrils or filaments needed for contraction. Their function is similar in many respects to neurons, although they are specialized muscle cells. Myocardial conduction cells initiate and propagate the action potential (the electrical impulse) that travels throughout the heart and triggers the contractions that propel the blood.

Structure of Cardiac Muscle

Compared to the giant cylinders of skeletal muscle, cardiac muscle cells, or cardiomyocytes, are considerably shorter with much smaller diameters. Cardiac muscle also demonstrates striations, the alternating pattern of dark A bands and light I bands attributed to the precise arrangement of the myofilaments and fibrils that are organized in sarcomeres along the length of the cell. These contractile elements are virtually identical to skeletal muscle. Typically, cardiomyocytes have a single, central nucleus, but two or more nuclei may be found in some cells.

Cardiac muscle cells branch freely. A junction between two adjoining cells is marked by a critical structure called an **intercalated disc**, which helps support the synchronized contraction of the muscle. The sarcolemmas from adjacent cells bind together at the intercalated discs. They consist of desmosomes, specialized linking proteoglycans, tight junctions, and large numbers of gap junctions that allow the passage of ions between the cells and help to synchronize the contraction.

EVERYDAY CONNECTION: REPAIR AND REPLACEMENT

Damaged cardiac muscle cells have extremely limited abilities to repair themselves or to replace dead cells via mitosis. Recent evidence indicates that at least some stem cells remain within the heart that continue to divide and at least potentially replace these dead cells. However, newly formed or repaired cells are rarely as functional as the original cells, and cardiac function is reduced. In the event of a heart attack or MI, dead cells are often replaced by patches of scar tissue. Autopsies performed on individuals who had successfully received heart transplants show some proliferation of original cells. If researchers can unlock the mechanism that generates new cells and restore full mitotic capabilities to heart muscle, the prognosis for heart attack survivors will be greatly enhanced. To date, myocardial cells produced within the patient (*in situ*) by cardiac stem cells seem to be nonfunctional, although those grown in Petri dishes (*in vitro*) do beat. Perhaps soon this mystery will be solved, and new advances in treatment will be commonplace.

Conduction System of the Heart

If embryonic heart cells are separated into a Petri dish and kept alive, each is capable of generating its own electrical impulse followed by contraction. When two independently beating embryonic cardiac muscle cells are placed together, the cell with the higher inherent rate sets the pace, and the impulse spreads from the faster to the slower cell to trigger a contraction. As more cells are joined together, the fastest cell continues to assume control of the rate. A fully developed adult heart maintains the capability of generating its own electrical impulse, triggered by the fastest cells, as part of the cardiac conduction system. The components of the cardiac conduction system include the sinoatrial node, the atrioventricular node, the atrioventricular bundle, the atrioventricular bundle branches, and the Purkinje cells.

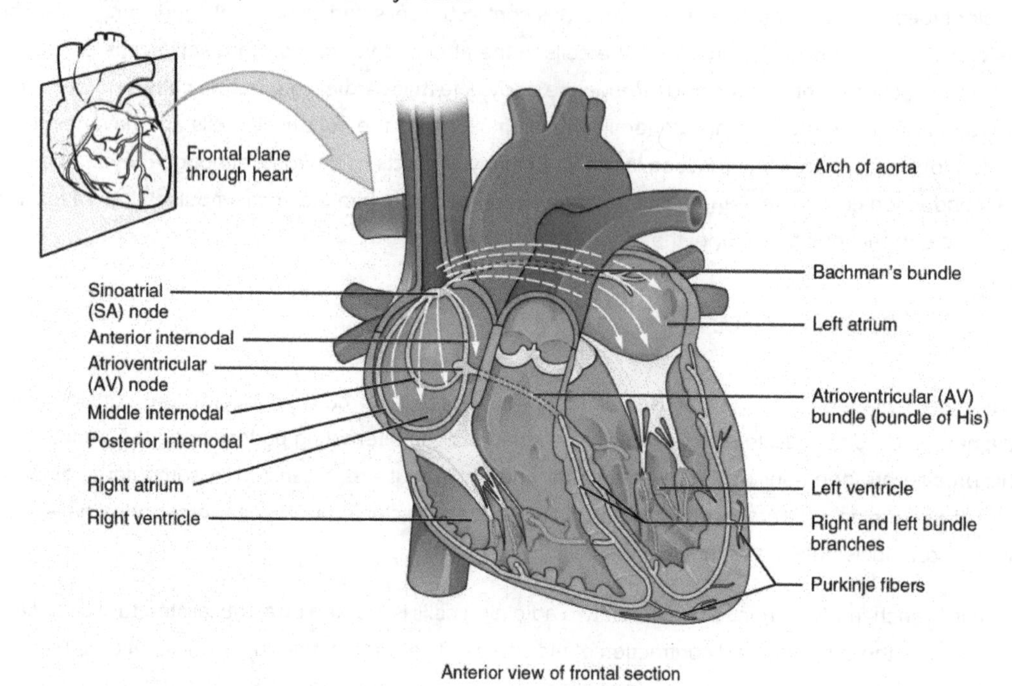

Figure 2. Specialized conducting components of the heart include the sinoatrial node, the internodal pathways, the atrioventricular node, the atrioventricular bundle, the right and left bundle branches, and the Purkinje fibers.

Sinoatrial (SA) Node

Normal cardiac rhythm is established by the **sinoatrial (SA) node**, a specialized clump of myocardial conducting cells located in the superior and posterior walls of the right atrium in close proximity to the superior vena cava. The SA node has the highest rate of depolarization and is known as the **pacemaker** of the heart. It initiates the **sinus rhythm**, or normal electrical pattern followed by contraction of the heart.

This impulse spreads from its initiation in the SA node throughout the atria through specialized **internodal pathways**, to the atrial myocardial contractile cells and the atrioventricular node. The impulse takes approximately 50 ms (milliseconds) to travel between these two nodes. Figure 3 illustrates the initiation of the impulse in the SA node that then spreads the impulse throughout the atria to the atrioventricular node. The electrical event, the wave of depolarization, is the trigger for muscular contraction. The wave of depolarization begins in the right atrium, and the impulse spreads across the superior portions of both atria and then down through the contractile cells. The contractile cells then begin contraction from the superior to the inferior portions of the atria, efficiently pumping blood into the ventricles.

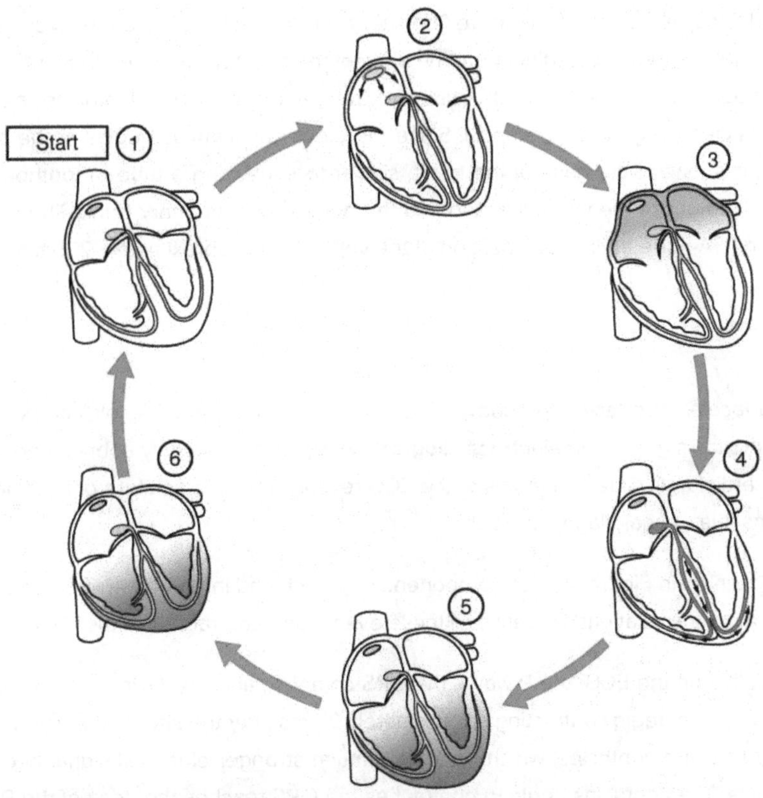

Figure 3. (1) The sinoatrial (SA) node and the remainder of the conduction system are at rest. (2) The SA node initiates the action potential, which sweeps across the atria. (3) After reaching the atrioventricular node, there is a delay of approximately 100 ms that allows the atria to complete pumping blood before the impulse is transmitted to the atrioventricular bundle. (4) Following the delay, the impulse travels through the atrioventricular bundle and bundle branches to the Purkinje fibers, and also reaches the right papillary muscle via the moderator band. (5) The impulse spreads to the contractile fibers of the ventricle. (6) Ventricular contraction begins.

Atrioventricular (AV) Node

The **atrioventricular (AV) node** is a second clump of specialized myocardial conductive cells, located in the inferior portion of the right atrium within the atrioventricular septum. The septum prevents the impulse from spreading directly to the ventricles without passing through the AV node. There is a critical pause before the AV node transmits the impulse to the atrioventricular bundle (see image above, step 3). This pause is critical to heart function, as it allows the atrial cardiomyocytes to complete their contraction that pumps blood into the ventricles before the impulse is transmitted to the cells of the ventricle itself. With extreme stimulation by the SA node, the AV node can transmit impulses maximally at 220 per minute. This establishes the typical maximum heart rate in a healthy young individual. Damaged hearts or those stimulated by drugs can contract at higher rates, but at these rates, the heart can no longer effectively pump blood.

Atrioventricular Bundle (Bundle of His), Bundle Branches, and Purkinje Fibers

Arising from the AV node, the **atrioventricular bundle**, or **bundle of His**, proceeds through the interventricular septum before dividing into two **atrioventricular bundle branches**, commonly called the left and right bundle branches. The left bundle branch supplies the left ventricle, and the right bundle branch the right ventricle. Since the left ventricle is much larger than the right, the left bundle branch is also considerably larger than the right. Both bundle branches descend and reach the apex of the heart where they connect with the Purkinje fibers (see image above, step 4). This passage takes approximately 25 ms.

The **Purkinje fibers** are additional myocardial conductive fibers that spread the impulse to the myocardial contractile cells in the ventricles. They extend throughout the myocardium from the apex of the heart toward the atrioventricular septum and the base of the heart. The Purkinje fibers have a fast inherent conduction rate, and the electrical impulse reaches all of the ventricular muscle cells in about 75 ms (see image above, step 5). Since the electrical stimulus begins at the apex, the contraction also begins at the apex and travels toward the base of the heart, similar to squeezing a tube of toothpaste from the bottom. This allows the blood to be pumped out of the ventricles and into the aorta and pulmonary trunk. The total time elapsed from the initiation of the impulse in the SA node until depolarization of the ventricles is approximately 225 ms.

Electrocardiogram

By careful placement of surface electrodes on the body, it is possible to record the complex, compound electrical signal of the heart. This tracing of the electrical signal is the **electrocardiogram (ECG)**, also commonly abbreviated EKG (K coming kardiology, from the German term for cardiology). Careful analysis of the ECG reveals a detailed picture of both normal and abnormal heart function, and is an indispensable clinical diagnostic tool.

A normal ECG tracing is presented in Figure 4. Each component, segment, and interval is labeled and corresponds to important electrical events, demonstrating the relationship between these events and contraction in the heart.

There are five prominent points on the ECG: the P wave, the QRS complex, and the T wave. The small **P wave** represents the depolarization of the atria. The atria begin contracting approximately 25 ms after the start of the P wave. The large **QRS complex** represents the depolarization of the ventricles, which requires a much stronger electrical signal because of the larger size of the ventricular cardiac muscle. The ventricles begin to contract as the QRS reaches the peak of the R wave. Lastly, the **T wave** represents the repolarization of the ventricles. The repolarization of the atria occurs during the QRS complex, which masks it on an ECG.

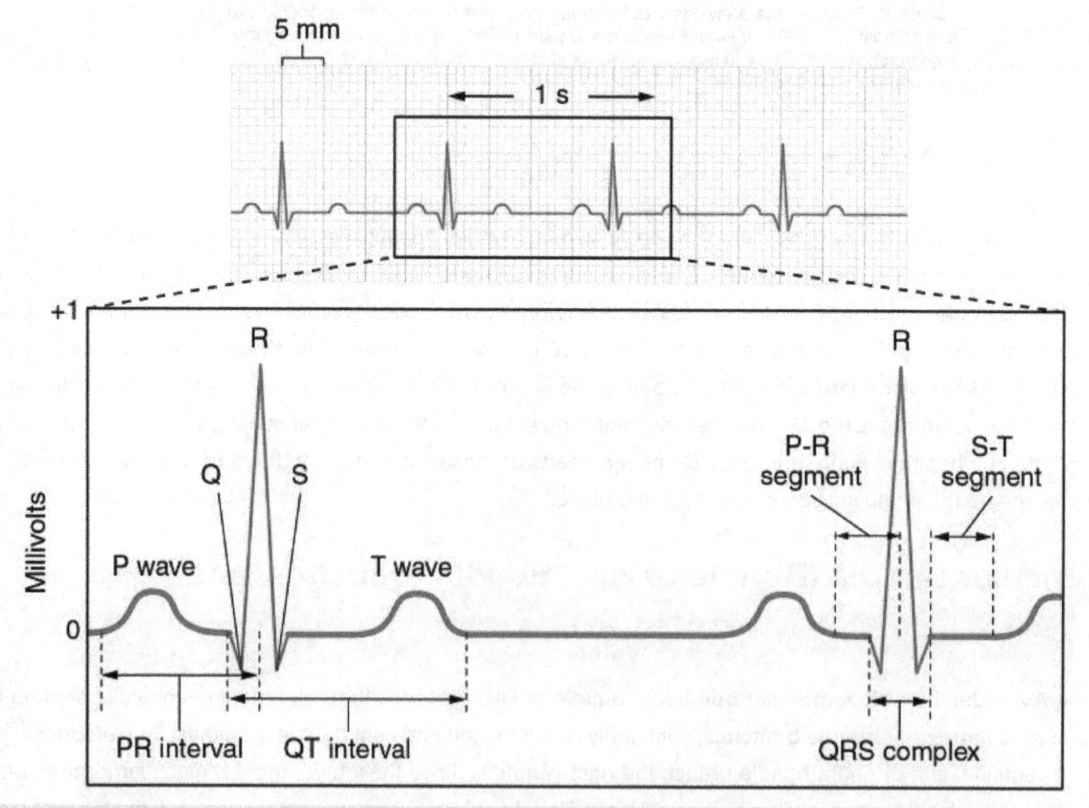

Figure 4. A normal tracing shows the P wave, QRS complex, and T wave. Also indicated are the PR, QT, QRS, and ST intervals, plus the P-R and S-T segments.

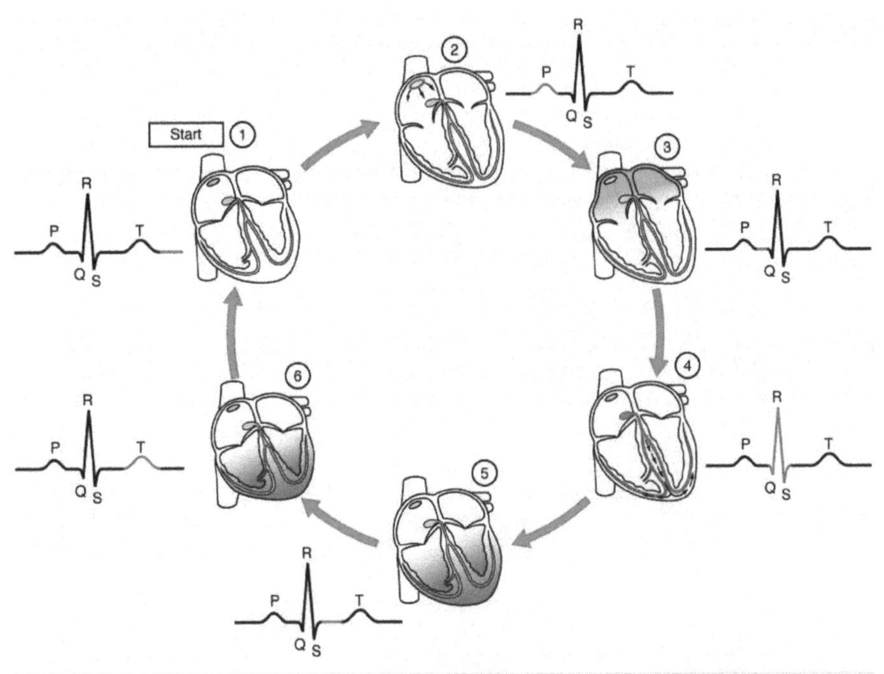

Figure 5. This diagram correlates an ECG tracing with the electrical and mechanical events of a heart contraction. Each segment of an ECG tracing corresponds to one event in the cardiac cycle.

EVERYDAY CONNECTION: EXTERNAL AUTOMATED DEFIBRILLATORS

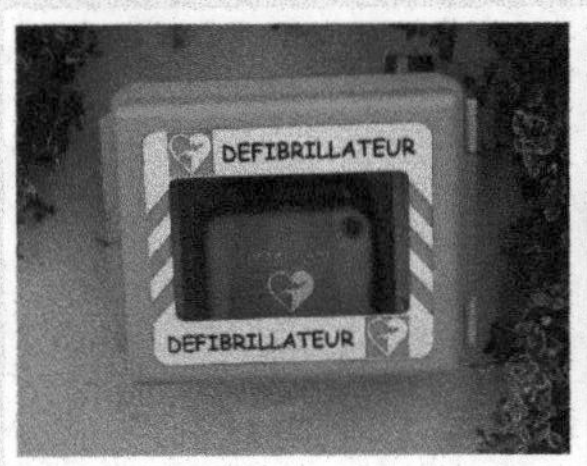

(a) (b)

Figure 10. (a) An external automatic defibrillator can be used by nonmedical personnel to reestablish a normal sinus rhythm in a person with fibrillation. (b) Defibrillator paddles are more commonly used in hospital settings. (credit b: "widerider107"/flickr.com)

In the event that the electrical activity of the heart is severely disrupted, cessation of electrical activity or fibrillation may occur. In fibrillation, the heart beats in a wild, uncontrolled manner, which prevents it from being able to pump effectively. Atrial fibrillation is a serious condition, but as long as the ventricles continue to pump blood, the patient's life may not be in immediate danger. Ventricular fibrillation is a medical emergency that requires life support, because the ventricles are not effectively pumping blood. In a hospital setting, it is often described as "code blue." If untreated for as little as a few minutes, ventricular fibrillation may lead to brain death. The most common treatment is defibrillation, which uses special paddles to apply a charge to the heart from an external electrical source in an attempt to establish a normal sinus rhythm. A defibrillator effectively stops the heart so that the SA node can trigger a normal conduction cycle. Because of their effectiveness in reestablishing a normal sinus rhythm, external automated defibrillators (EADs) are being placed in areas frequented by large numbers of people, such as schools, restaurants, and airports. These devices contain simple and direct verbal instructions that can be followed by nonmedical personnel in an attempt to save a life.

14fb4ad7-39a1-4eee-ab6e-3ef2482e3e22@8.25

Cardiac Cycle

The period of time that begins with contraction of the atria and ends with ventricular relaxation is known as the **cardiac cycle**. The period of contraction that the heart undergoes while it pumps blood into circulation is called **systole**. The period of relaxation that occurs as the chambers fill with blood is called **diastole**. Both the atria and ventricles undergo systole and diastole, and it is essential that these components be carefully regulated and coordinated to ensure blood is pumped efficiently to the body.

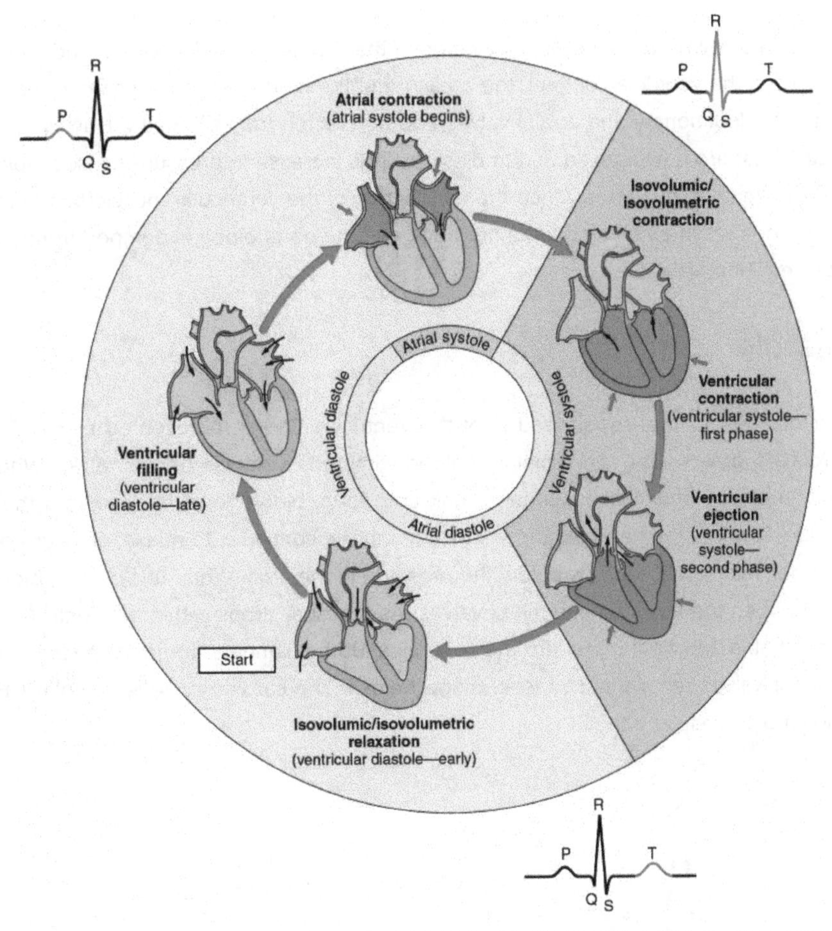

Figure 1. The cardiac cycle begins with atrial systole and progresses to ventricular systole, atrial diastole, and ventricular diastole, when the cycle begins again. Correlations to the ECG are highlighted.

Phases of the Cardiac Cycle

At the beginning of the cardiac cycle, both the atria and ventricles are relaxed (diastole). Blood is flowing into the right atrium from the superior and inferior venae cavae and the coronary sinus. Blood flows into the left atrium from the four pulmonary veins. The two atrioventricular valves, the tricuspid and mitral valves, are both open, so blood flows unimpeded from the atria and into the ventricles. Approximately 70–80 percent of ventricular filling occurs by this method. The two semilunar valves, the pulmonary and aortic valves, are closed, preventing backflow of blood into the right and left ventricles from the pulmonary trunk on the right and the aorta on the left.

Atrial Systole and Diastole

Contraction of the atria follows depolarization, represented by the P wave of the ECG. As the atrial muscles contract from the superior portion of the atria toward the atrioventricular septum, pressure rises within the atria and blood is pumped into the ventricles through the open atrioventricular (tricuspid, and mitral or bicuspid) valves. Atrial systole lasts approximately 100 ms and ends prior to ventricular systole, as the atrial muscle returns to diastole.

Ventricular Systole

Ventricular systole (see image below) follows the depolarization of the ventricles and is represented by the QRS complex in the ECG. Initially, as the muscles in the ventricle contract, the pressure of the blood within the chamber rises, but it is not yet high enough to open the semilunar (pulmonary and aortic) valves and be ejected from the heart. However, blood pressure quickly rises above that of the atria that are now relaxed and in diastole. This increase in pressure causes blood to flow back toward the atria, closing the tricuspid and mitral valves. Once the contraction of the ventricular muscle has raised the pressure to the point that it is greater than the pressures in the pulmonary trunk and the aorta blood is pumped from the heart, pushing open the pulmonary and aortic semilunar valves.

Ventricular Diastole

Ventricular relaxation, or diastole, follows repolarization of the ventricles and is represented by the T wave of the ECG. As the ventricular muscle relaxes, pressure on the remaining blood within the ventricle begins to fall. When pressure within the ventricles drops below pressure in both the pulmonary trunk and aorta, blood flows back toward the heart. The semilunar valves close to prevent backflow into the heart. As the ventricular muscle continues to relaxe, pressure on the blood within the ventricles drops even further. Eventually, it drops below the pressure in the atria. When this occurs, blood flows from the atria into the ventricles, pushing open the tricuspid and mitral valves. As pressure drops within the ventricles, blood flows from the major veins into the relaxed atria and from there into the ventricles. Both chambers are in diastole, the atrioventricular valves are open, and the semilunar valves remain closed (see image below). The cardiac cycle is complete. Figure 2 illustrates the relationship between the cardiac cycle and the ECG.

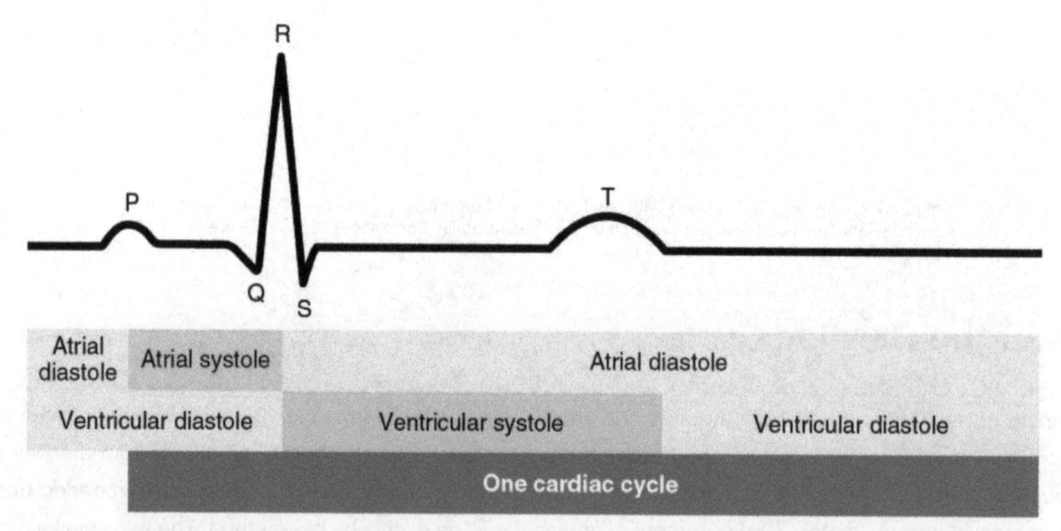

Figure 2. Initially, both the atria and ventricles are relaxed (diastole). The P wave represents depolarization of the atria and is followed by atrial contraction (systole). Atrial systole extends until the QRS complex, at which point, the atria relax. The QRS complex represents depolarization of the ventricles and is followed by ventricular contraction. The T wave represents the repolarization of the ventricles and marks the beginning of ventricular relaxation.

Heart Sounds

One of the simplest, yet effective, diagnostic techniques applied to assess the state of a patient's heart is auscultation using a stethoscope.

In a normal, healthy heart, there are only two audible **heart sounds**: S_1 and S_2. S_1 is the sound created by the closing of the atrioventricular valves during ventricular contraction and is normally described as a "lub," or first heart sound. The second heart sound, S_2, is the sound of the closing of the semilunar valves during ventricular diastole and is described as a "dub" (Figure 3). There is a third heart sound, S_3, but it is rarely heard in healthy individuals. It may be the sound of blood flowing into the atria, or blood sloshing back and forth in the ventricle, or even tensing of the chordae tendineae. S_3 may be heard in youth, some athletes, and pregnant women. If the sound is heard later in life, it may indicate congestive heart failure, warranting further tests.

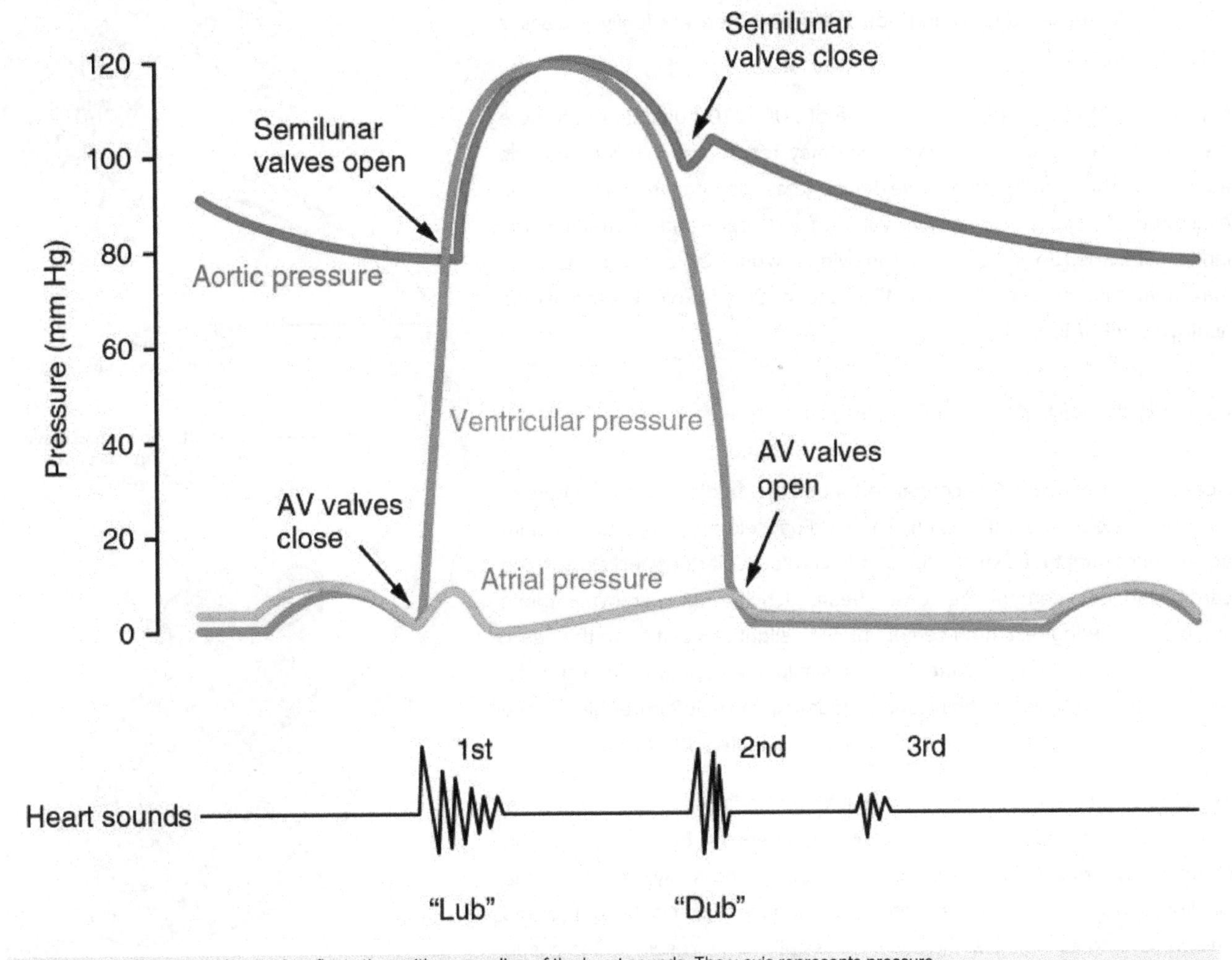

Figure 3. In this illustration, the x-axis reflects time with a recording of the heart sounds. The y-axis represents pressure.

Cardiac Physiology

The autorhythmicity inherent in cardiac cells keeps the heart beating at a regular pace; however, the heart is regulated by and responds to outside influences as well. Neural and endocrine controls are vital to the regulation of cardiac function. In addition, the heart is sensitive to several environmental factors, including electrolytes.

Heart Rates

Heart rates (HRs) vary considerably, not only with exercise and fitness levels, but also with age. Newborn resting HRs may be 120 bpm. HR gradually decreases until young adulthood and then gradually increases again with age.

Maximum HRs are normally in the range of 200–220 bpm, although there are some extreme cases in which they may reach higher levels. As one ages, the ability to generate maximum rates decreases. This may be estimated by taking the maximal value of 220 bpm and subtracting the individual's age. So a 40-year-old individual would be expected to hit a maximum rate of approximately 180, and a 60-year-old person would achieve a HR of 160.

Cardiovascular Centers

Nervous control over HR is centralized within the cardiovascular centers of the medulla oblongata (Figure 1). The cardioaccelerator regions stimulate activity via sympathetic stimulation of the cardioaccelerator nerves, and the cardioinhibitory centers decrease heart activity via parasympathetic stimulation. During rest, both centers provide slight stimulation to the heart, contributing to **autonomic tone**. This is a similar concept to tone in skeletal muscles. Normally, vagal stimulation predominates as, left unregulated, the SA node would initiate a sinus rhythm of approximately 100 bpm.

The cardiovascular center receives input from a series of visceral receptors. Among these receptors are various proprioreceptors, baroreceptors, and chemoreceptors. Collectively, these inputs normally enable the cardiovascular centers to regulate heart function precisely, a process known as **cardiac reflexes**. Increased physical activity results in increased rates of firing by various proprioreceptors located in muscles, joint capsules, and tendons. Any such increase in physical activity would logically warrant increased blood flow. The cardiac centers monitor these increased rates of firing, and suppress parasympathetic stimulation and increase sympathetic stimulation as needed in order to increase blood flow.

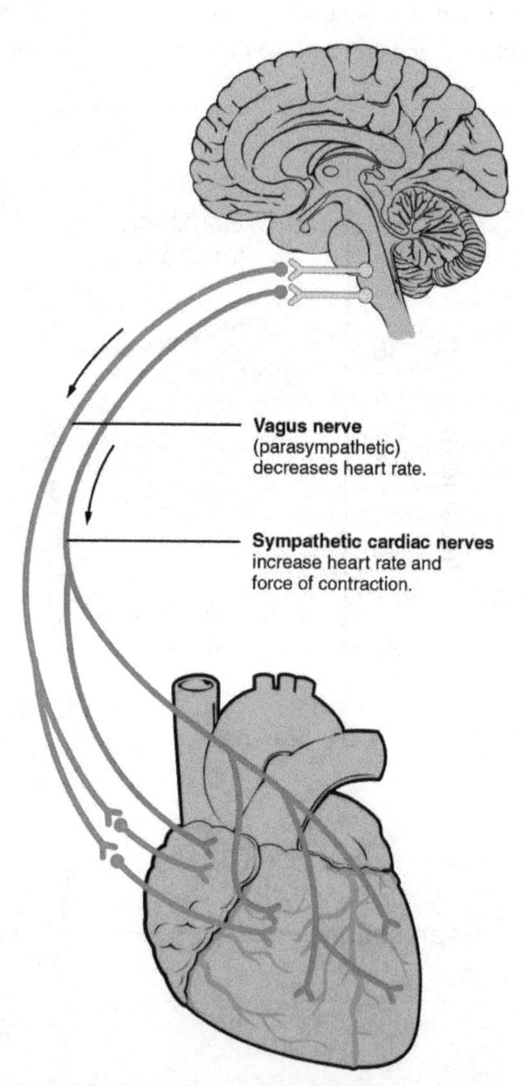

Figure 1. Cardioaccelerator and cardioinhibitory areas are components of the paired cardiac centers located in the medulla oblongata of the brain. They innervate the heart via sympathetic cardiac nerves that increase cardiac activity and vagus (parasympathetic) nerves that slow cardiac activity.

Other Factors Influencing Heart Rate

Table 1. Major Factors Increasing Heart Rate and Force of Contraction

Factor	Effect
Cardioaccelerator nerves	Release of norepinephrine
Proprioreceptors	Increased rates of firing during exercise
Chemoreceptors	Decreased levels of O_2; increased levels of H^+, CO_2, and lactic acid
Baroreceptors	Decreased rates of firing, indicating falling blood volume/pressure
Limbic system	Anticipation of physical exercise or strong emotions
Catecholamines	Increased epinephrine and norepinephrine
Thyroid hormones	Increased T_3 and T_4
Calcium	Increased Ca^{2+}
Potassium	Decreased K^+
Sodium	Decreased Na^+
Body temperature	Increased body temperature
Nicotine and caffeine	Stimulants, increasing heart rate

Table 2. Factors Decreasing Heart Rate and Force of Contraction

Factor	Effect
Cardioinhibitor nerves (vagus)	Release of acetylcholine
Proprioreceptors	Decreased rates of firing following exercise
Chemoreceptors	Increased levels of O_2; decreased levels of H^+ and CO_2
Baroreceptors	Increased rates of firing, indicating higher blood volume/pressure
Limbic system	Anticipation of relaxation
Catecholamines	Decreased epinephrine and norepinephrine
Thyroid hormones	Decreased T_3 and T_4
Calcium	Decreased Ca^{2+}
Potassium	Increased K^+
Sodium	Increased Na^+
Body temperature	Decrease in body temperature

Structure and Function of Blood Vessels

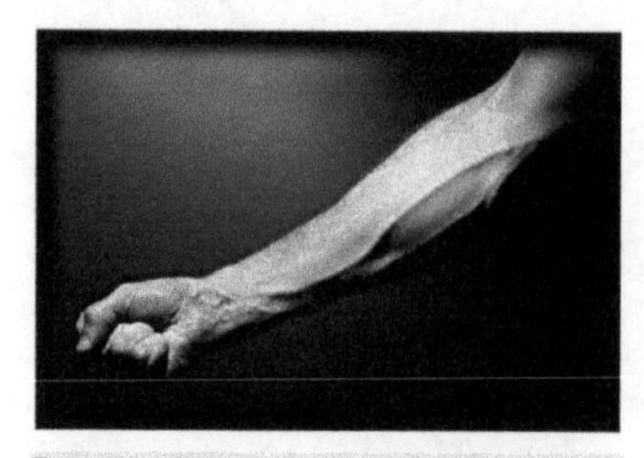

Figure 1. While most blood vessels are located deep from the surface and are not visible, the superficial veins of the upper limb provide an indication of the extent, prominence, and importance of these structures to the body. (credit: Colin Davis)

In this chapter, you will learn about the vascular part of the cardiovascular system, that is, the vessels that transport blood throughout the body and provide the physical site where gases, nutrients, and other substances are exchanged with body cells. When vessel functioning is reduced, blood-borne substances do not circulate effectively throughout the body. As a result, tissue injury occurs, metabolism is impaired, and the functions of every bodily system are threatened.

Blood is carried through the body via blood vessels. An **artery** is a blood vessel that carries blood away from the heart, where it branches into ever-smaller vessels. Eventually, the smallest arteries, vessels called **arterioles**, further branch into tiny **capillaries**, where nutrients and wastes are exchanged, and then combine with other vessels that exit capillaries to form **venules**, small blood vessels that carry blood to a **vein**, a larger blood vessel that returns blood to the heart.

Arteries and veins transport blood in two distinct circuits: the systemic circuit and the pulmonary circuit. Systemic arteries provide blood rich in oxygen to the body's tissues. The blood returned to the heart through systemic veins has less oxygen, since much of the oxygen carried by the arteries has been delivered to the cells. In contrast, in the pulmonary circuit, arteries carry blood low in oxygen exclusively to the lungs for gas exchange. Pulmonary veins then return freshly oxygenated blood from the lungs to the heart to be pumped back out into systemic circulation. Although arteries and veins differ structurally and functionally, they share certain features.

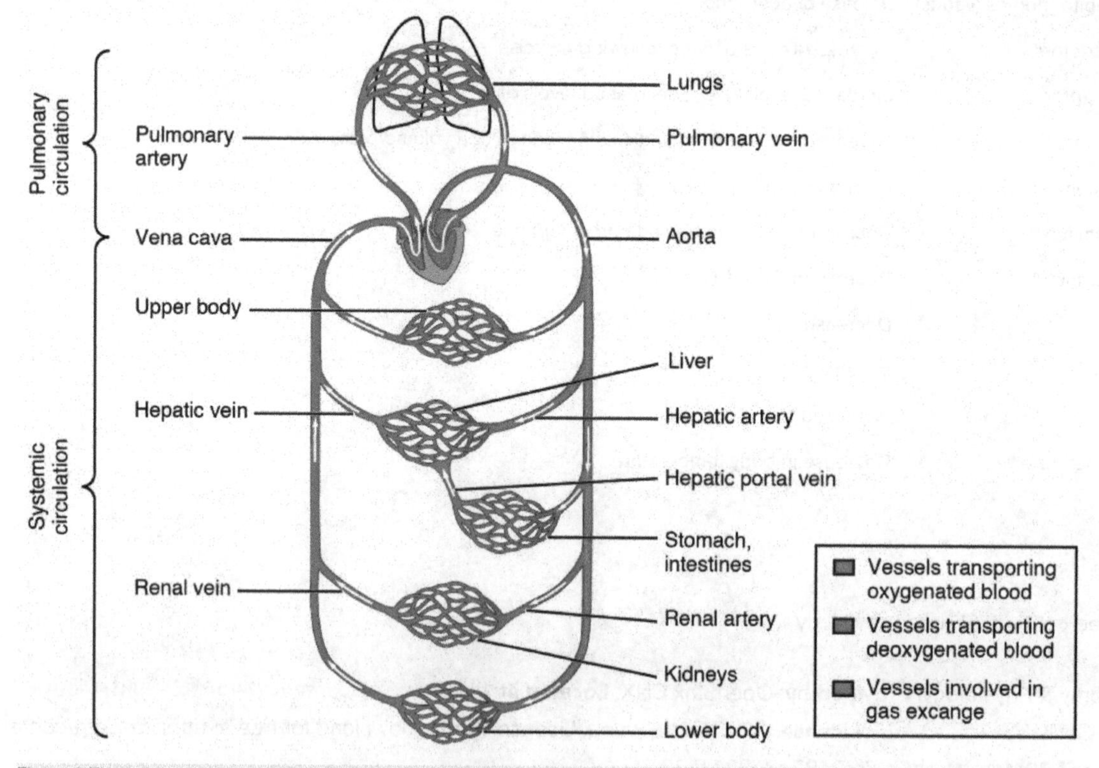

Figure 1. The pulmonary circuit moves blood from the right side of the heart to the lungs and back to the heart. The systemic circuit moves blood from the left side of the heart to the head and body and returns it to the right side of the heart to repeat the cycle. The arrows indicate the direction of blood flow, and the colors show the relative levels of oxygen concentration.

Shared Structures

Different types of blood vessels vary slightly in their structures, but they share the same general features. Arteries and arterioles have thicker walls than veins and venules because they are closer to the heart and receive blood that is surging at a far greater pressure (Figure 2). Each type of vessel has a **lumen**—a hollow passageway through which blood flows. Arteries have smaller lumens than veins, a characteristic that helps to maintain the pressure of blood moving through the system. Together, their thicker walls and smaller diameters give arterial lumens a more rounded appearance in cross section than the lumens of veins.

Both arteries and veins have the same three distinct tissue layers, called tunics (from the Latin term tunica), for the garments first worn by ancient Romans; the term tunic is also used for some modern garments. From the most interior layer to the outer, these tunics are the tunica intima, the tunica media, and the tunica externa. Table 1 compares and contrasts the tunics of the arteries and veins.

Table 1. Comparison of Tunics in Arteries and Veins

	Arteries	Veins
General appearance	Thick walls with small lumens; Generally appear rounded	Thin walls with large lumens; Generally appear flattened
Tunica intima	Endothelium usually appears wavy due to constriction of smooth muscle; Internal elastic membrane present in larger vessels	Endothelium appears smooth; Internal elastic membrane absent
Tunica media	Normally the thickest layer in arteries; Smooth muscle cells and elastic fibers predominate (the proportions of these vary with distance from the heart); External elastic membrane present in larger vessels	Normally thinner than the tunica externa; Smooth muscle cells and collagenous fibers predominate; Nervi vasorum and vasa vasorum present; External elastic membrane absent
Tunica externa	Normally thinner than the tunica media in all but the largest arteries; Collagenous and elastic fibers; Nervi vasorum and vasa vasorum present	Normally the thickest layer in veins; Collagenous and smooth fibers predominate; Some smooth muscle fibers; Nervi vasorum and vasa vasorum present

Tunica Intima

The **tunica intima** (also called the tunica interna) is composed of epithelial and connective tissue layers. Lining the tunica intima is the specialized simple squamous epithelium called the endothelium, which is continuous throughout the entire vascular system, including the lining of the chambers of the heart. Damage to this endothelial lining and exposure of blood to the collagenous fibers beneath is one of the primary causes of clot formation.

Next to the endothelium is the basement membrane that effectively binds the endothelium to the connective tissue. The basement membrane provides strength while maintaining flexibility, and it is permeable, allowing materials to pass through it. The thin outer layer of the tunica intima contains a small amount of connective tissue that consists primarily of elastic fibers to provide the vessel with additional flexibility; it also contains some collagenous fibers to provide additional strength.

Tunica Media

The **tunica media** is the substantial middle layer of the vessel wall (see Figure 2). It is generally the thickest layer in arteries, and it is much thicker in arteries than it is in veins. The tunica media consists of layers of smooth muscle supported by connective tissue that is primarily made up of elastic fibers, most of which are arranged in circular sheets. Contraction and relaxation of the circular muscles decrease and increase the diameter of the vessel lumen, respectively. Specifically in arteries, **vasoconstriction** decreases blood flow as the smooth muscle in the walls of the tunica media contracts, making the lumen narrower and increasing blood pressure. Similarly, **vasodilation** increases blood flow as the smooth muscle relaxes, allowing the lumen to widen and blood pressure to drop. Separating the tunica media from the outer tunica externa in larger arteries is the **external elastic membrane** (also called the external elastic lamina), which also appears wavy in slides. This structure is not usually seen in smaller arteries, nor is it seen in veins.

Tunica Externa

The outer tunic, the **tunica externa** (also called the tunica adventitia), is a substantial sheath of connective tissue composed primarily of collagenous fibers. Some bands of elastic fibers are found here as well. This is normally the thickest tunic in veins and may be thicker than the tunica media in some larger arteries. The outer layers of the tunica externa are not distinct but rather blend with the surrounding connective tissue outside the vessel, helping to hold the vessel in relative position. If you are able to palpate some of the superficial veins on your upper limbs and try to move them, you will find that the tunica externa prevents this. If the tunica externa did not hold the vessel in place, any movement would likely result in disruption of blood flow.

Arteries

An **artery** is a blood vessel that conducts blood away from the heart. All arteries have relatively thick walls that can withstand the high pressure of blood ejected from the heart. However, those close to the heart have the thickest walls, containing a high percentage of elastic fibers in all three of their tunics. This type of artery is known as an **elastic artery** (see Figure 3). Their abundant elastic fibers allow them to expand, as blood pumped from the ventricles passes through them, and then to recoil after the surge has passed.

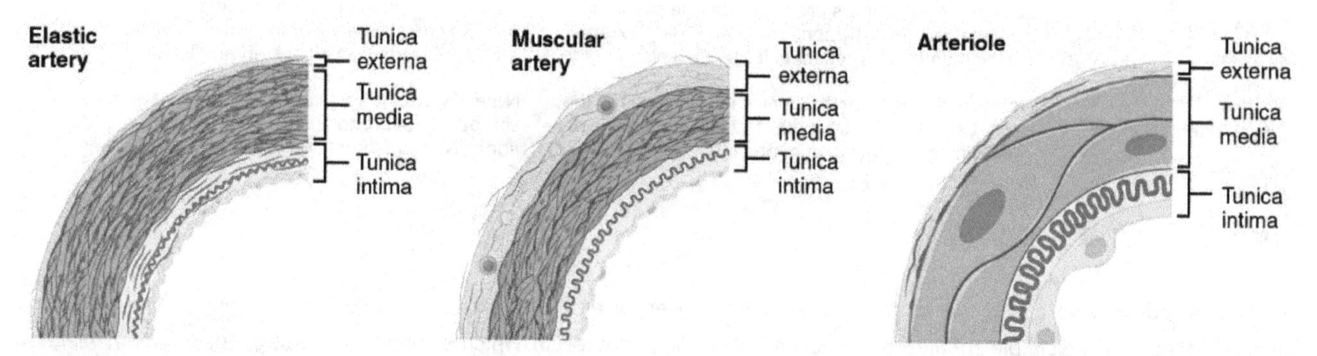

Figure 3. Comparison of the walls of an elastic artery, a muscular artery, and an arteriole is shown. In terms of scale, the diameter of an arteriole is measured in micrometers compared to millimeters for elastic and muscular arteries.

Farther from the heart, where the surge of blood has dampened, the percentage of elastic fibers in an artery's tunica intima decreases and the amount of smooth muscle in its tunica media increases. The artery at this point is described as a **muscular artery.** Their thick tunica media allows muscular arteries to play a leading role in vasoconstriction. In contrast, their decreased quantity of elastic fibers limits their ability to expand. Fortunately, because the blood pressure has eased by the time it reaches these more distant vessels, elasticity has become less important.

Notice that although the distinctions between elastic and muscular arteries are important, there is no "line of demarcation" where an elastic artery suddenly becomes muscular. Rather, there is a gradual transition as the vascular tree repeatedly branches. In turn, muscular arteries branch to distribute blood to the vast network of arterioles. For this reason, a muscular artery is also known as a distributing artery.

Arterioles

An **arteriole** is a very small artery that leads to a capillary. Arterioles have the same three tunics as the larger vessels, but the thickness of each is greatly diminished. The critical endothelial lining of the tunica intima is intact. The tunica media is restricted to one or two smooth muscle cell layers in thickness. The tunica externa remains but is very thin (see Figure 3).

Capillaries

A **capillary** is a microscopic channel that supplies blood to the tissues themselves. Exchange of gases and other substances occurs in the capillaries between the blood and the surrounding cells and their tissue fluid (interstitial fluid). The diameter of a capillary lumen is just barely wide enough for an erythrocyte to squeeze through.

The wall of a capillary consists of the endothelial layer surrounded by a basement membrane with occasional smooth muscle fibers. For capillaries to function, their walls must be leaky, allowing substances to pass through. There are three major types of capillaries, which differ according to their degree of "leakiness:" continuous, fenestrated, and sinusoid capillaries.

Continuous Capillaries

The most common type of capillary, the **continuous capillary**, is found in almost all vascularized tissues. Continuous capillaries are characterized by a complete endothelial lining with tight junctions between endothelial cells. Substances that can pass between cells include metabolic products, such as glucose, water, and small hydrophobic molecules like gases and hormones, as well as various leukocytes.

Fenestrated Capillaries

A **fenestrated capillary** is one that has pores (or fenestrations) that make the capillary permeable to larger molecules. Fenestrated capillaries are common in the small intestine, which is the primary site of nutrient absorption, as well as in the kidneys, which filter the blood. They are also found in the choroid plexus of the brain and many endocrine structures, including the hypothalamus, pituitary, pineal, and thyroid glands.

Sinusoid Capillaries

A **sinusoid capillary** (or sinusoid) is the least common type of capillary. Sinusoid capillaries are flattened, and they have extensive intercellular gaps and incomplete basement membranes, in addition to intercellular clefts and fenestrations. These very large openings allow for the passage of the largest molecules, including plasma proteins and even cells. Blood flow through sinusoids is very slow, allowing more time for exchange of gases, nutrients, and wastes. Sinusoids are found in the liver and spleen, bone marrow, lymph nodes (where they carry lymph, not blood), and many endocrine glands including the pituitary and adrenal glands.

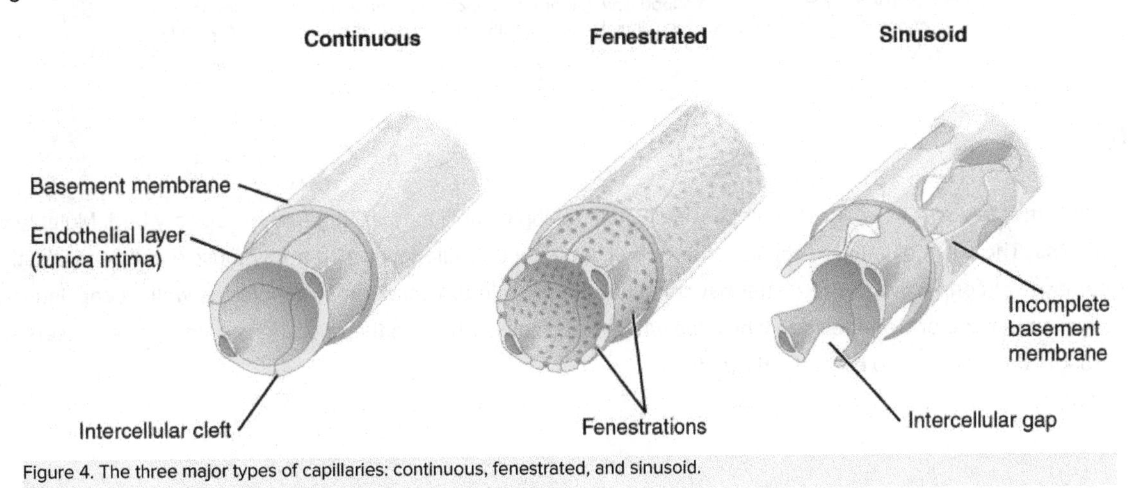

Figure 4. The three major types of capillaries: continuous, fenestrated, and sinusoid.

Metarterioles and Capillary Beds

A **metarteriole** is a type of vessel that has structural characteristics of both an arteriole and a capillary. Slightly larger than the typical capillary, the smooth muscle of the tunica media of the metarteriole is not continuous but forms rings of smooth muscle (sphincters) prior to the entrance to the capillaries. Each metarteriole arises from a terminal arteriole and branches to supply blood to a **capillary bed** that may consist of 10–100 capillaries.

The **precapillary sphincters**, circular smooth muscle cells that surround the capillary at its origin with the metarteriole, tightly regulate the flow of blood from a metarteriole to the capillaries it supplies. Their function is critical: If all of the capillary beds in the body were to open simultaneously, they would collectively hold every drop of blood in the body and there would be none in the arteries, arterioles, venules, veins, or the heart itself. Normally, the precapillary sphincters are closed. When the surrounding tissues need oxygen and have excess waste products, the precapillary sphincters open, allowing blood to flow through and exchange to occur before closing once more (see Figure 5). If all of the precapillary sphincters in a capillary bed are closed, blood will flow from the metarteriole directly into a **thoroughfare channel** and then into the venous circulation, bypassing the capillary bed entirely. This creates what is known as a **vascular shunt**. In addition, an **arteriovenous anastomosis** may bypass the capillary bed and lead directly to the venous system.

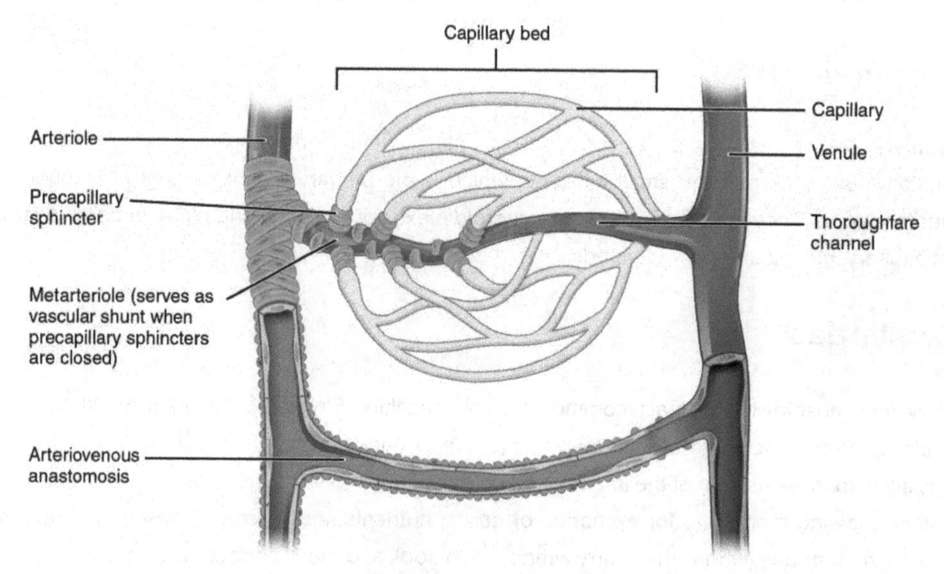

Figure 5. In a capillary bed, arterioles give rise to metarterioles. Precapillary sphincters located at the junction of a metarteriole with a capillary regulate blood flow. A thoroughfare channel connects the metarteriole to a venule. An arteriovenous anastomosis, which directly connects the arteriole with the venule, is shown at the bottom.

Venules

A **venule** is an extremely small vein. Postcapillary venules join multiple capillaries exiting from a capillary bed. Multiple venules join to form veins. The walls of venules consist of endothelium, a thin middle layer with a few muscle cells and elastic fibers, plus an outer layer of connective tissue fibers that constitute a very thin tunica externa. Venules as well as capillaries are the primary sites of emigration or diapedesis, in which the white blood cells adhere to the endothelial lining of the vessels and then squeeze through adjacent cells to enter the tissue fluid.

Veins

A **vein** is a blood vessel that conducts blood toward the heart. Compared to arteries, veins are thin-walled vessels with large and irregular lumens (see Figure 6).

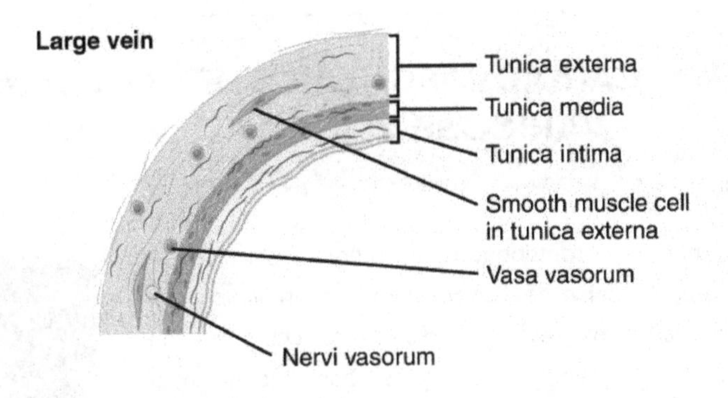

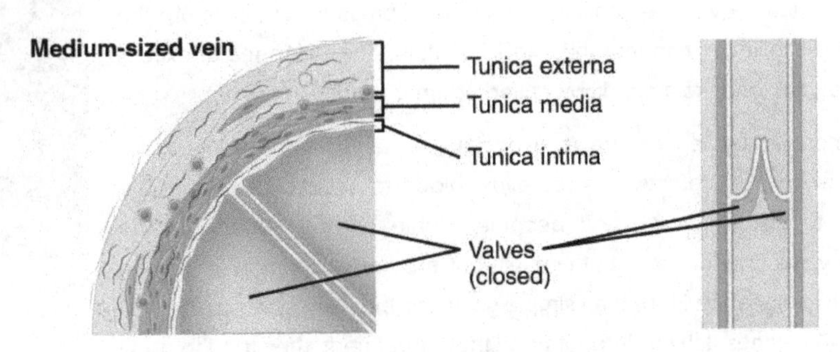

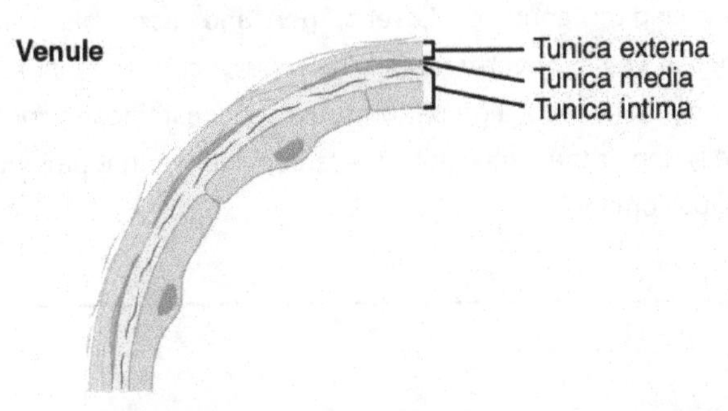

Figure 6. Many veins have valves to prevent back flow of blood, whereas venules do not. In terms of scale, the diameter of a venule is measured in micrometers compared to millimeters for veins.

Because they are low-pressure vessels, larger veins are commonly equipped with valves that promote the unidirectional flow of blood toward the heart and prevent backflow toward the capillaries caused by the inherent low blood pressure in veins as well as the pull of gravity. Table 2 compares the features of arteries and veins.

Table 2. Comparison of Arteries and Veins

	Arteries	Veins
Direction of blood flow	Conducts blood away from the heart	Conducts blood toward the heart
General appearance	Rounded	Irregular, often collapsed
Pressure	High	Low
Wall thickness	Thick	Thin
Relative oxygen concentration	Higher in systemic arteries Lower in pulmonary arteries	Lower in systemic veins Higher in pulmonary veins
Valves	Not present	Present most commonly in limbs and in veins inferior to the heart

DISORDERS OF THE CARDIOVASCULAR SYSTEM: EDEMA AND VARICOSE VEINS

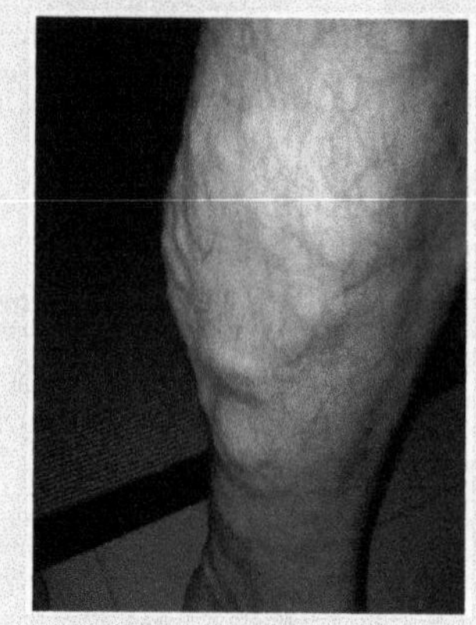

Figure 7. Varicose veins are commonly found in the lower limbs. (credit: Thomas Kriese)

Despite the presence of valves and the contributions of other anatomical and physiological adaptations, over the course of a day, some blood will inevitably pool, especially in the lower limbs, due to the pull of gravity. Any blood that accumulates in a vein will increase the pressure within it, which can then be reflected back into the smaller veins, venules, and eventually even the capillaries. Increased pressure will promote the flow of fluids out of the capillaries and into the interstitial fluid. The presence of excess tissue fluid around the cells leads to a condition called edema.

Edema may be accompanied by varicose veins, especially in the superficial veins of the legs. This disorder arises when defective valves allow blood to accumulate within the veins, causing them to distend, twist, and become visible on the surface of the integument. Varicose veins may occur in both sexes, but are more common in women and are often related to pregnancy. More than simple cosmetic blemishes, varicose veins are often painful and sometimes itchy or throbbing. Without treatment, they tend to grow worse over time. The use of support hose, as well as elevating the feet and legs whenever possible, may be helpful in alleviating this condition. Laser surgery and interventional radiologic procedures can reduce the size and severity of varicose veins. Severe cases may require conventional surgery to remove the damaged vessels. As there are typically redundant circulation patterns, that is, anastomoses, for the smaller and more superficial veins, removal does not typically impair the circulation. There is evidence that patients with varicose veins suffer a greater risk of developing a thrombus or clot.

Blood Flow, Blood Pressure, and Resistance

Components of Arterial Blood Pressure

Arterial blood pressure in the larger vessels consists of several distinct components: systolic and diastolic pressures, pulse pressure, and mean arterial pressure.

Systolic and Diastolic Pressures

When systemic arterial blood pressure is measured, it is recorded as a ratio of two numbers (e.g., 120/80 is a normal adult blood pressure), expressed as systolic pressure over diastolic pressure. The systolic pressure is the higher value (typically around 120 mm Hg) and reflects the arterial pressure resulting from the ejection of blood during ventricular contraction, or systole. The diastolic pressure is the lower value (usually about 80 mm Hg) and represents the arterial pressure of blood during ventricular relaxation, or diastole.

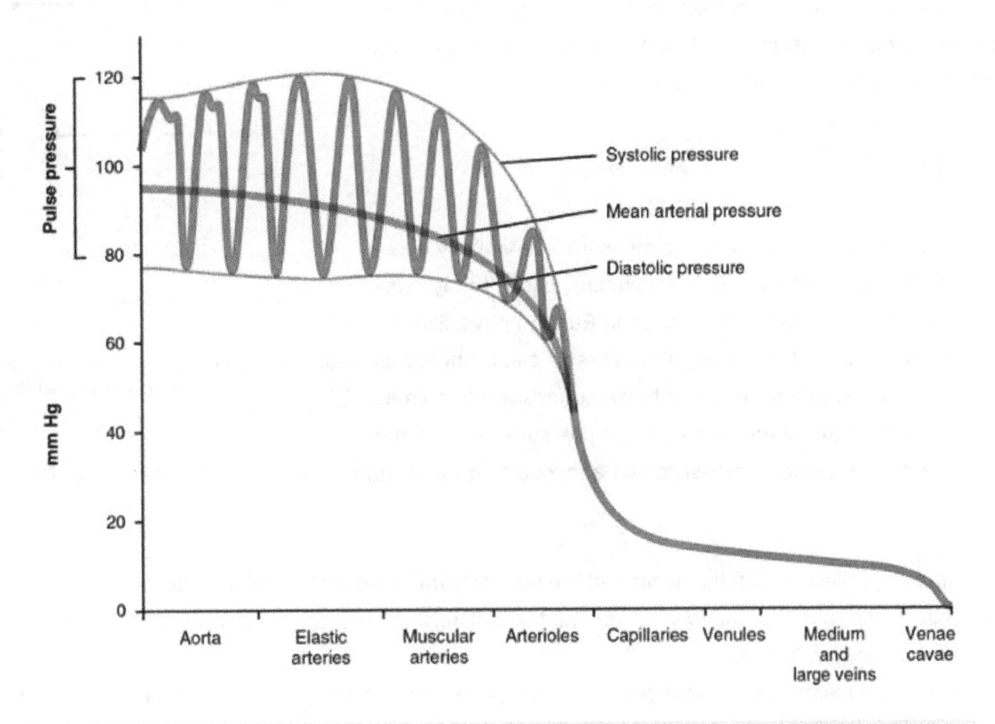

Figure 1. The graph shows the components of blood pressure throughout the blood vessels, including systolic, diastolic, mean arterial, and pulse pressures.

Pulse Pressure

As shown in Figure 1, the difference between the systolic pressure and the diastolic pressure is the pulse pressure. For example, an individual with a systolic pressure of 120 mm Hg and a diastolic pressure of 80 mm Hg would have a pulse pressure of 40 mmHg.

Mean Arterial Pressure

Mean arterial pressure (MAP) represents the "average" pressure of blood in the arteries, that is, the average force driving blood into vessels that serve the tissues. Normally, the MAP falls within the range of 70–110 mm Hg. If the value falls below 60 mm

Hg for an extended time, blood pressure will not be high enough to ensure circulation to and through the tissues, which results in ischemia, or insufficient blood flow. A condition called hypoxia, inadequate oxygenation of tissues, commonly accompanies ischemia. The term hypoxemia refers to low levels of oxygen in systemic arterial blood. Neurons are especially sensitive to hypoxia and may die or be damaged if blood flow and oxygen supplies are not quickly restored.

Pulse

After blood is ejected from the heart, elastic fibers in the arteries help maintain a high-pressure gradient as they expand to accommodate the blood, then recoil. This expansion and recoiling effect, known as the pulse, can be palpated manually or measured electronically.

Because pulse indicates heart rate, it is measured clinically to provide clues to a patient's state of health. It is recorded as beats per minute. A high or irregular pulse rate can be caused by physical activity or other temporary factors, but it may also indicate a heart condition. The pulse strength indicates the strength of ventricular contraction and cardiac output. If the pulse is strong, then systolic pressure is high. If it is weak, systolic pressure has fallen, and medical intervention may be warranted.

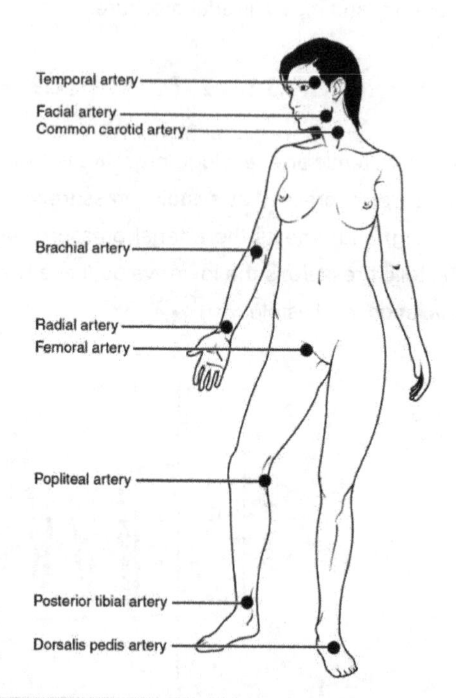

Figure 2. The pulse is most readily measured at the radial artery, but can be measured at any of the pulse points shown.

Measurement of Blood Pressure

Blood pressure is one of the critical parameters measured on virtually every patient in every healthcare setting. The technique used today was developed more than 100 years ago by a pioneering Russian physician, Dr. Nikolai Korotkoff. Turbulent blood flow through the vessels can be heard as a soft ticking while measuring blood pressure; these sounds are known as Korotkoff sounds. The technique of measuring blood pressure requires the use of a sphygmomanometer (a blood pressure cuff attached to a measuring device) and a stethoscope. The technique is as follows:

- The clinician wraps an inflatable cuff tightly around the patient's arm at about the level of the heart.
- The clinician squeezes a rubber pump to inject air into the cuff, raising pressure around the artery and temporarily cutting off blood flow into the patient's arm.
- The clinician places the stethoscope on the patient's antecubital region and, while gradually allowing air within the cuff to escape, listens for the Korotkoff sounds.

Although there are five recognized Korotkoff sounds, only two are normally recorded. Initially, no sounds are heard since there is no blood flow through the vessels, but as air pressure drops, the cuff relaxes, and blood flow returns to the arm. As shown in Figure 3, the first sound heard through the stethoscope—the first Korotkoff sound—indicates systolic pressure. As more air is released from the cuff, blood is able to flow freely through the brachial artery and all sounds disappear. The point at which the last sound is heard is recorded as the patient's diastolic pressure.

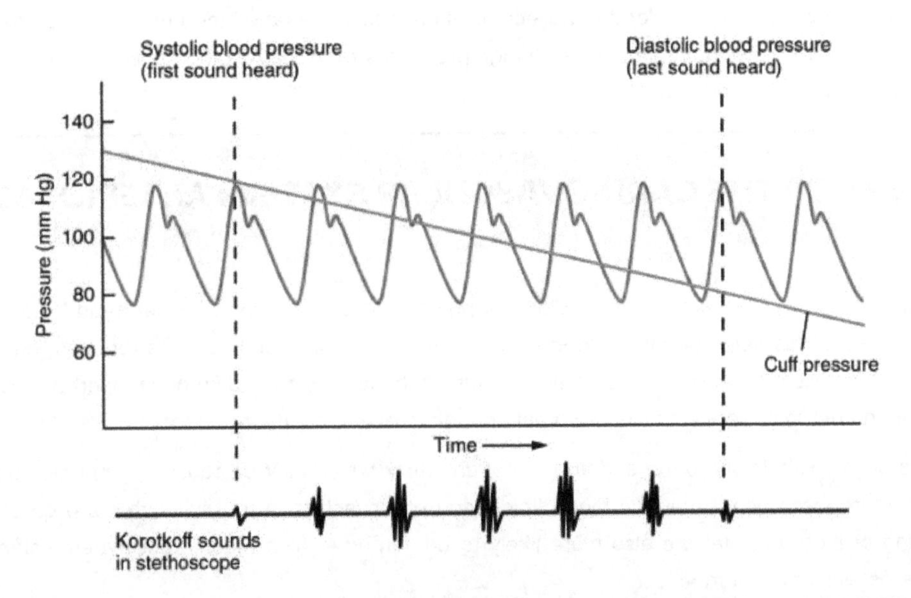

Figure 3. When pressure in a sphygmomanometer cuff is released, a clinician can hear the Korotkoff sounds. In this graph, a blood pressure tracing is aligned to a measurement of systolic and diastolic pressures.

Figure 4 compares vessel diameter, total cross-sectional area, average blood pressure, and blood velocity through the systemic vessels. Notice in parts (a) and (b) that the total cross-sectional area of the body's capillary beds is far greater than any other type of vessel. Although the diameter of an individual capillary is significantly smaller than the diameter of an arteriole, there are vastly more capillaries in the body than there are other types of blood vessels. Part (c) shows that blood pressure drops unevenly as blood travels from arteries to arterioles, capillaries, venules, and veins, and encounters greater resistance. However, the site of the most precipitous drop, and the site of greatest resistance, is the arterioles. This explains why vasodilation and vasoconstriction of arterioles play more significant roles in regulating blood pressure than do the vasodilation and vasoconstriction of other vessels.

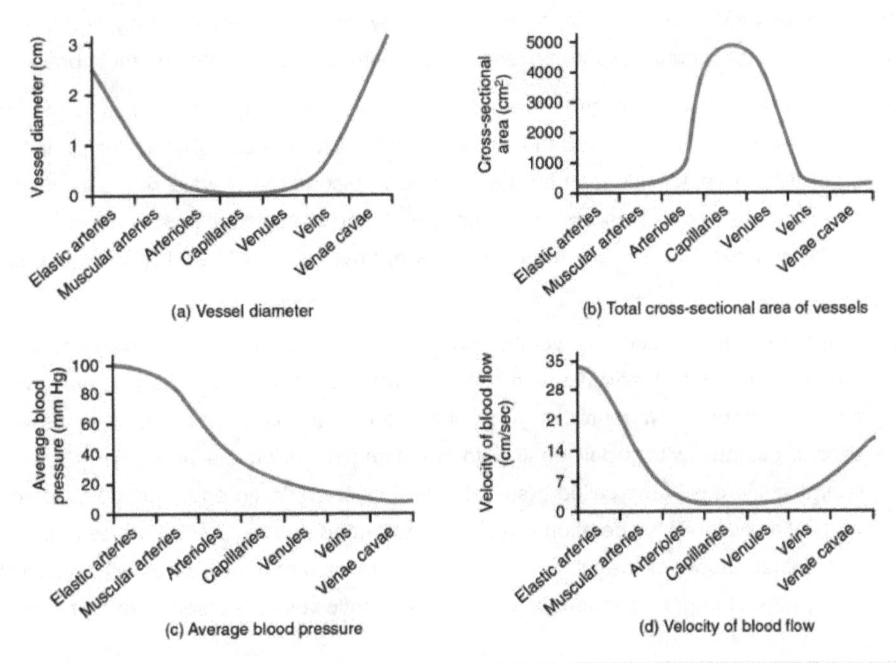

Figure 4. The relationships among blood vessels that can be compared include (a) vessel diameter, (b) total cross-sectional area, (c) average blood pressure, and (d) velocity of blood flow.

Part (d) shows that the velocity (speed) of blood flow decreases dramatically as the blood moves from arteries to arterioles to capillaries. This slow flow rate allows more time for exchange processes to occur. As blood flows through the veins, the rate of velocity increases, as blood is returned to the heart.

DISORDERS OF THE CARDIOVASCULAR SYSTEM: ARTERIOSCLEROSIS

Compliance allows an artery to expand when blood is pumped through it from the heart, and then to recoil after the surge has passed. This helps promote blood flow. In arteriosclerosis, compliance is reduced, and pressure and resistance within the vessel increase. This is a leading cause of hypertension and coronary heart disease, as it causes the heart to work harder to generate a pressure great enough to overcome the resistance.

Arteriosclerosis begins with injury to the endothelium of an artery, which may be caused by irritation from high blood glucose, infection, tobacco use, excessive blood lipids, and other factors. Artery walls that are constantly stressed by blood flowing at high pressure are also more likely to be injured—which means that hypertension can promote arteriosclerosis, as well as result from it.

Recall that tissue injury causes inflammation. As inflammation spreads into the artery wall, it weakens and scars it, leaving it stiff (sclerotic). As a result, compliance is reduced. Moreover, circulating triglycerides and cholesterol can seep between the damaged lining cells and become trapped within the artery wall, where they are frequently joined by leukocytes, calcium, and cellular debris. Eventually, this buildup, called plaque, can narrow arteries enough to impair blood flow. The term for this condition, atherosclerosis (athero- = "porridge") describes the mealy deposits.

Sometimes a plaque can rupture, causing microscopic tears in the artery wall that allow blood to leak into the tissue on the other side. When this happens, platelets rush to the site to clot the blood. This clot can further obstruct the artery and—if it occurs in a coronary or cerebral artery—cause a sudden heart attack or stroke. Alternatively, plaque can break off and travel through the bloodstream as an embolus until it blocks a more distant, smaller artery.

Even without total blockage, vessel narrowing leads to ischemia—reduced blood flow—to the tissue region "downstream" of the narrowed vessel. Ischemia in turn leads to hypoxia—decreased supply of oxygen to the tissues. Hypoxia involving cardiac muscle or brain tissue can lead to cell death and severe impairment of brain or heart function.

A major risk factor for both arteriosclerosis and atherosclerosis is advanced age, as the conditions tend to progress over time. Arteriosclerosis is normally defined as the more generalized loss of compliance, "hardening of the arteries," whereas atherosclerosis is a more specific term for the build-up of plaque in the walls of the vessel and is a specific type of arteriosclerosis. There is also a distinct genetic component, and pre-existing hypertension and/or diabetes also greatly increase the risk. However, obesity, poor nutrition, lack of physical activity, and tobacco use all are major risk factors.

Treatment includes lifestyle changes, such as weight loss, smoking cessation, regular exercise, and adoption of a diet low in sodium and saturated fats. Medications to reduce cholesterol and blood pressure may be prescribed. For blocked coronary arteries, surgery is warranted. In angioplasty, a catheter is inserted into the vessel at the point of narrowing, and a second catheter with a balloon-like tip is inflated to widen the opening. To prevent subsequent collapse of the vessel, a small mesh tube called a stent is often inserted. In an endarterectomy, plaque is surgically removed from the walls of a vessel. This operation is typically performed on the carotid arteries of the neck, which are a prime source of oxygenated blood for the brain. In a coronary bypass procedure, a non-vital superficial vessel from another part of the body (often the great saphenous vein) or a synthetic vessel is inserted to create a path around the blocked area of a coronary artery.

Capillary Exchange

The primary purpose of the cardiovascular system is to circulate gases, nutrients, wastes, and other substances to and from the cells of the body. Small molecules, such as gases, lipids, and lipid-soluble molecules, can diffuse directly through the membranes of the endothelial cells of the capillary wall. Glucose, amino acids, and ions—including sodium, potassium, calcium, and chloride—use transporters to move through specific channels in the membrane by facilitated diffusion. Glucose, ions, and larger molecules may also leave the blood through intercellular clefts. Larger molecules can pass through the pores of fenestrated capillaries, and even large plasma proteins can pass through the great gaps in the sinusoids. Some large proteins in blood plasma can move into and out of the endothelial cells packaged within vesicles by endocytosis and exocytosis. Water moves by osmosis.

Bulk Flow

The mass movement of fluids into and out of capillary beds requires a transport mechanism far more efficient than mere diffusion. This movement, often referred to as bulk flow, involves two pressure-driven mechanisms: Volumes of fluid move from an area of higher pressure in a capillary bed to an area of lower pressure in the tissues via **filtration**. In contrast, the movement of fluid from an area of higher pressure in the tissues into an area of lower pressure in the capillaries is **reabsorption**. Two types of pressure interact to drive each of these movements: hydrostatic pressure and osmotic pressure.

Hydrostatic Pressure

The primary force driving fluid transport between the capillaries and tissues is hydrostatic pressure, which can be defined as the pressure of any fluid enclosed in a space. **Blood hydrostatic pressure** is the force exerted by the blood confined within blood vessels or heart chambers. Hydrostatic pressure is the force that drives fluid out of capillaries and into the tissues, a process called filtration.

Osmotic Pressure

The net pressure that drives reabsorption—the movement of fluid from the interstitial fluid back into the capillaries—is called osmotic pressure. Whereas hydrostatic pressure forces fluid out of the capillary, osmotic pressure draws fluid back in. Osmotic pressure is determined by osmotic concentration gradients, that is, the difference in the solute-to-water concentrations in the blood and tissue fluid. A region higher in solute concentration (and lower in water concentration) draws water across a semipermeable membrane from a region higher in water concentration (and lower in solute concentration).

As we discuss osmotic pressure in blood and tissue fluid, it is important to recognize that the formed elements of blood do not contribute to osmotic concentration gradients. Rather, it is the plasma proteins that play the key role. Solutes also move across the capillary wall according to their concentration gradient, but overall, the concentrations should be similar and not have a significant impact on osmosis. Because of their large size and chemical structure, plasma proteins are not truly solutes, that is, they do not dissolve but are dispersed or suspended in their fluid medium, forming a colloid rather than a solution. The plasma proteins suspended in blood cannot move across the semipermeable capillary cell membrane, and so they remain in the plasma. As a result, blood has a higher colloidal concentration and lower water concentration than tissue fluid. It therefore attracts water.

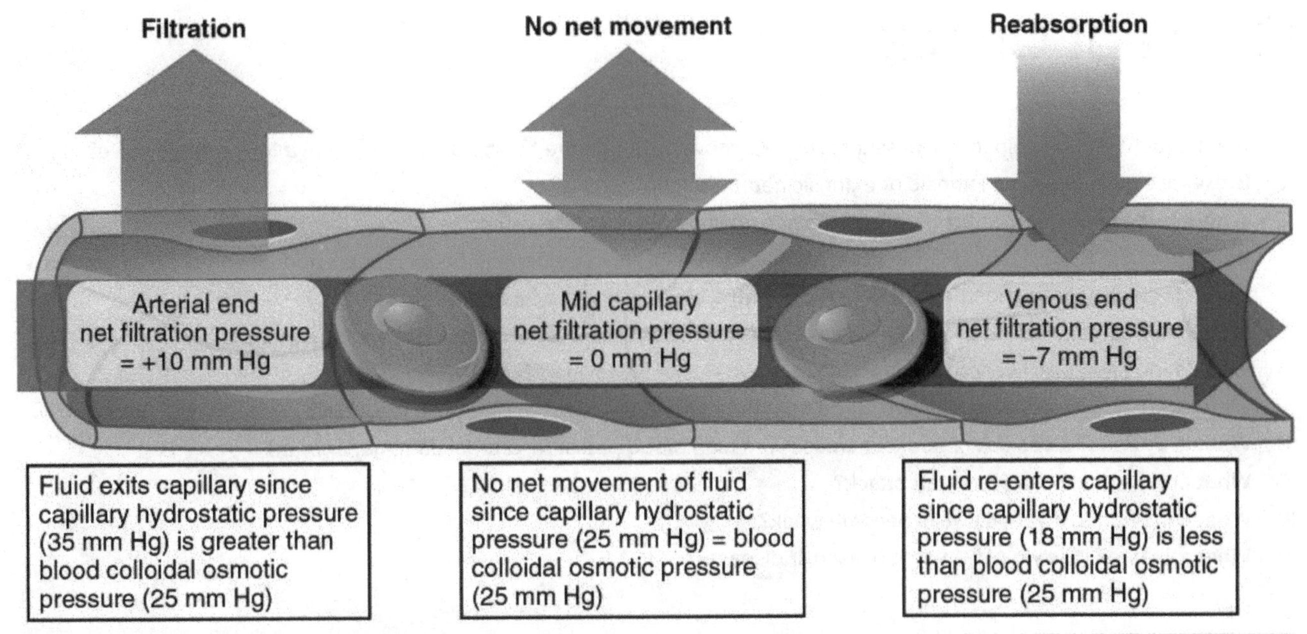

Figure 1. Net filtration occurs near the arterial end of the capillary since capillary hydrostatic pressure (CHP) is greater than blood colloidal osmotic pressure (BCOP). There is no net movement of fluid near the midpoint since CHP = BCOP. Net reabsorption occurs near the venous end since BCOP is greater than CHP.

Concept Check

1. **What** structures will a blood cell pass through as it moves through the heart from the vena cava to the aortic valve?
2. **Is** exercise considered an intrinsic or extrinsic heart rate control?
3. **Where** is the impulse to contract collected before being passed to the ventricles?
4. **What** does the P wave of an ECG indicate? **What** is the heart doing when the QRS complex appears on the ECG? **What** is the correct term for the heart as the ECG is drawing the S-T interval: systole or diastole?
5. **Which** vessels can withstand the greatest pressure? **Which** vessels allow for nutrient and gas exchange?
6. **What** is the function of venous valves? Capillary beds?
7. **Which** cardiovascular circuit delivers blood from the heart to the lungs and back to the heart?
8. **What** is a stroke? **Does** risk of stroke increase with high blood pressure, arterial damage, or both?
9. **What** are the risk factors for heart attack?
10. **What** can happen if venous walls become weak?
11. **What** role does lifestyle play in cardiovascular disease?

CHAPTER 12 — THE RESPIRATORY SYSTEM

Organs and Structures of the Respiratory System

LEARNING OBJECTIVES

- List the major functions of the respiratory system
- List the structures of the respiratory system
- Outline the forces that allow for air movement into and out of the lungs
- Compare and contrast the functions of upper respiratory tract with the lower respiratory tract
- Discuss factors that can influence the respiratory rate
- Discuss the process of external respiration
- Describe the process of internal respiration
- Describe the function of hemoglobin

Hold your breath. Really! See how long you can hold your breath as you continue reading. . . . How long can you do it? Chances are you are feeling uncomfortable already. A typical human cannot survive without breathing for more than 3 minutes, and even if you wanted to hold your breath longer, your autonomic nervous system would take control. This is because every cell in the body needs to run the oxidative stages of cellular respiration, the process by which energy is produced in the form of adenosine triphosphate (ATP). For oxidative phosphorylation to occur, oxygen is used as a reactant and carbon dioxide is released as a waste product.

You may be surprised to learn that although oxygen is a critical need for cells, it is actually the accumulation of carbon dioxide that primarily drives your need to breathe. Carbon dioxide is exhaled and oxygen is inhaled through the respiratory system, which includes muscles to move air into and out of the lungs, passageways through which air moves, and microscopic gas exchange surfaces covered by capillaries. The circulatory system transports gases from the lungs to tissues throughout the body and vice versa. A variety of diseases can affect the respiratory system, such as asthma, emphysema, chronic obstruction pulmonary disorder (COPD), and lung cancer. All of these conditions affect the gas exchange process and result in labored breathing and other difficulties.

The major organs of the respiratory system function primarily to provide oxygen to body tissues for cellular respiration, remove the waste product carbon dioxide, and help to maintain acid-base balance. Portions of the respiratory system are also used for non-vital functions, such as sensing odors, speech production, and for straining, such as during childbirth or coughing.

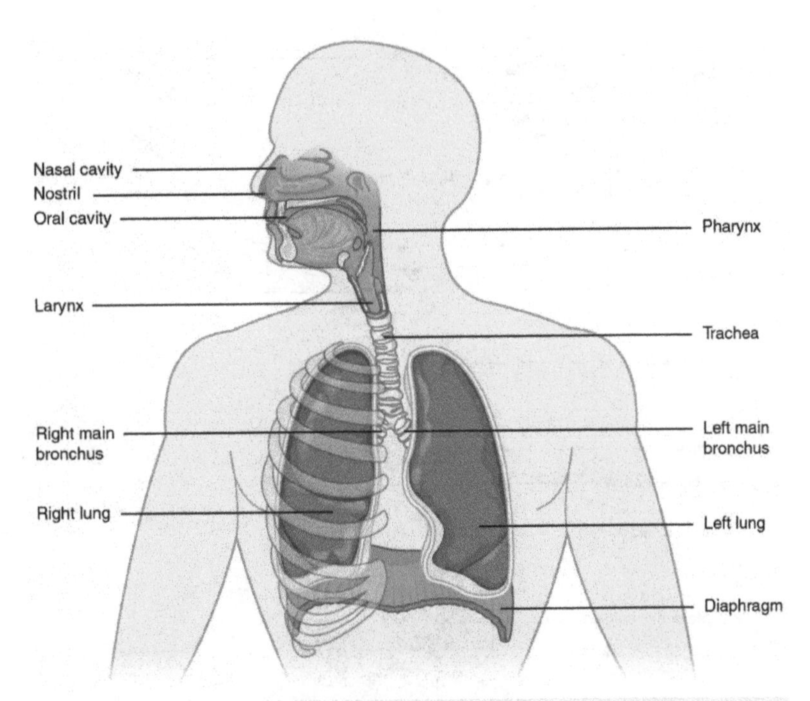

Figure 1. The major respiratory structures span the nasal cavity to the diaphragm.

Structurally, the respiratory system can be divided into the upper and lower respiratory tracts. The upper respiratory tract has direct access to the environment and conditions air as it is inhaled. The lower respiratory tract provides the surface area necessary for efficient gas exchange between the inhaled air and your blood. Functionally, the entire upper and a portion of the lower respiratory tracts make up the conducting zone while the deepest regions of the lower respiratory tract comprise the respiratory zone. The **conducting zone** of the respiratory system includes the organs and structures not directly involved in gas exchange. The gas exchange occurs in the **respiratory zone**.

Conducting Zone

The major functions of the conducting zone are to provide a route for incoming and outgoing air, remove debris and pathogens from the incoming air, and warm and humidify the incoming air. Several structures within the conducting zone perform other functions as well. The epithelium of the nasal passages, for example, is essential to sensing odors, and the bronchial epithelium that lines the lungs can metabolize some airborne carcinogens.

The Nose and its Adjacent Structures

The major entrance and exit for the respiratory system is through the nose. When discussing the nose, it is helpful to divide it into two major sections: the external nose, and the nasal cavity or internal nose.

The external nose consists of the surface and skeletal structures that result in the outward appearance of the nose and contribute to its numerous functions. The root is the region of the nose located between the eyebrows. The bridge is the part of the nose that connects the root to the rest of the nose. The dorsum nasi is the length of the nose. The apex is the tip of the nose. On either side of the apex, the nostrils are formed by the alae (singular = ala). An ala is a cartilaginous structure that forms the lateral side of each naris (plural = nares), or nostril opening. The philtrum is the concave surface that connects the apex of the nose to the upper lip.

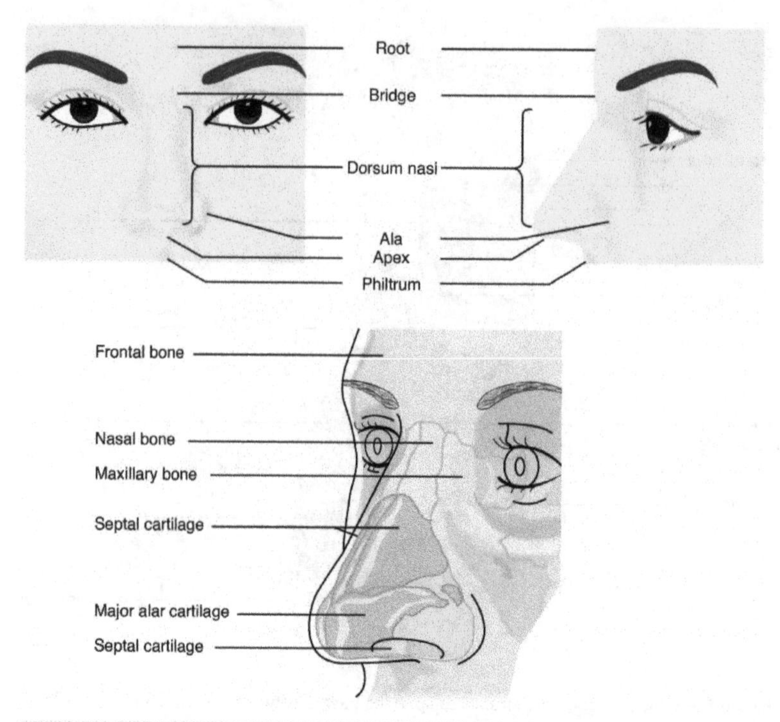

Figure 2. This illustration shows features of the external nose (top) and skeletal features of the nose (bottom).

Underneath the thin skin of the nose are its skeletal features. While the root and bridge of the nose consist of bone, the protruding portion of the nose is composed of cartilage. As a result, when looking at a skull, the nose is missing. The nasal bone is one of a pair of bones that lies under the root and bridge of the nose. The nasal bone articulates superiorly with the frontal bone and laterally with the maxillary bones. Septal cartilage is flexible hyaline cartilage connected to the nasal bone, forming the dorsum nasi. The alar cartilage consists of the apex of the nose; it surrounds the naris.

The nares open into the nasal cavity, which is separated into left and right sections by the nasal septum. The nasal septum is formed anteriorly by a portion of the septal cartilage (the flexible portion you can touch with your fingers) and posteriorly by the perpendicular plate of the ethmoid bone (a cranial bone located just posterior to the nasal bones) and the thin vomer bones (whose name refers to its plough shape).

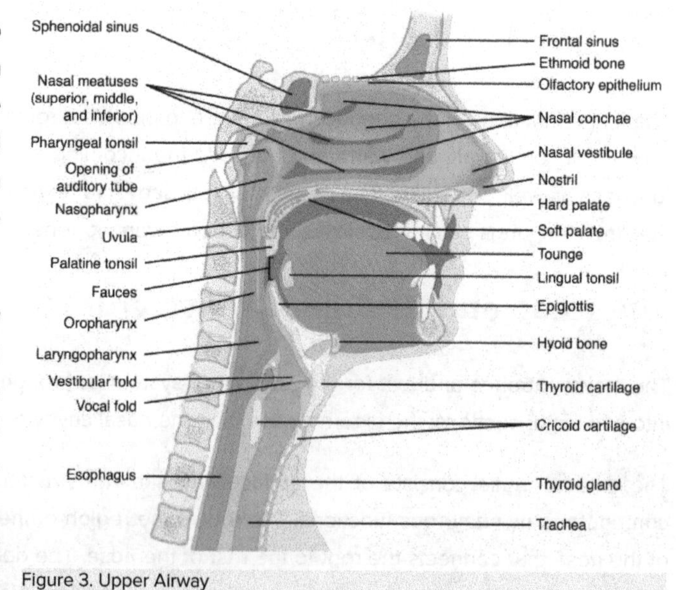

Figure 3. Upper Airway

Each lateral wall of the nasal cavity has three bony projections, called the superior, middle, and inferior nasal conchae. The inferior conchae are separate bones, whereas the superior and middle conchae are portions of the ethmoid bone. Conchae serve to increase the surface area of the nasal cavity and to disrupt the flow of air as it enters the nose, causing air to bounce along the epithelium, where it is cleaned and warmed. The conchae also conserve water and prevent dehydration of the nasal epithelium by trapping water during exhalation. The floor of the nasal cavity is composed of the palate. The hard palate at the anterior region of the nasal cavity is composed of bone. The soft palate at the posterior portion of the nasal cavity consists of muscle tissue. Air exits the nasal cavities via the internal nares and moves into the pharynx.

Several bones that help form the walls of the nasal cavity have air-containing spaces called the paranasal sinuses, which serve to warm and humidify incoming air. Sinuses are lined with a mucosa. Each paranasal sinus is named for its associated bone: frontal sinus, maxillary sinus, sphenoidal sinus, and ethmoidal sinus. The sinuses produce mucus and lighten the weight of the skull.

The nares and anterior portion of the nasal cavities are lined with mucous membranes, containing sebaceous glands and hair follicles that serve to prevent the passage of large debris, such as dirt, through the nasal cavity. An olfactory epithelium used to detect odors is found deeper in the nasal cavity.

The conchae and paranasal sinuses are lined by respiratory epithelium composed of pseudostratified ciliated columnar epithelium. The epithelium contains goblet cells, one of the specialized, columnar epithelial cells that produce mucus to trap debris. The cilia of the respiratory epithelium help remove the mucus and debris from the nasal cavity with a constant beating motion, sweeping materials towards the throat to be swallowed. Interestingly, cold air slows the movement of the cilia, resulting in accumulation of mucus that may in turn lead to a runny nose during cold weather. This moist epithelium functions to warm and humidify incoming air. Capillaries located just beneath the nasal epithelium warm the air by convection. Serous and mucus-producing cells also secrete the lysozyme enzyme and proteins called defensins, which have antibacterial properties. Immune cells that patrol the connective tissue deep to the respiratory epithelium provide additional protection.

The pharynx is a tube formed by skeletal muscle and lined with a mucous membrane that is continuous with that of the nasal cavities. The pharynx is divided into three major regions: the nasopharynx, the oropharynx, and the laryngopharynx.

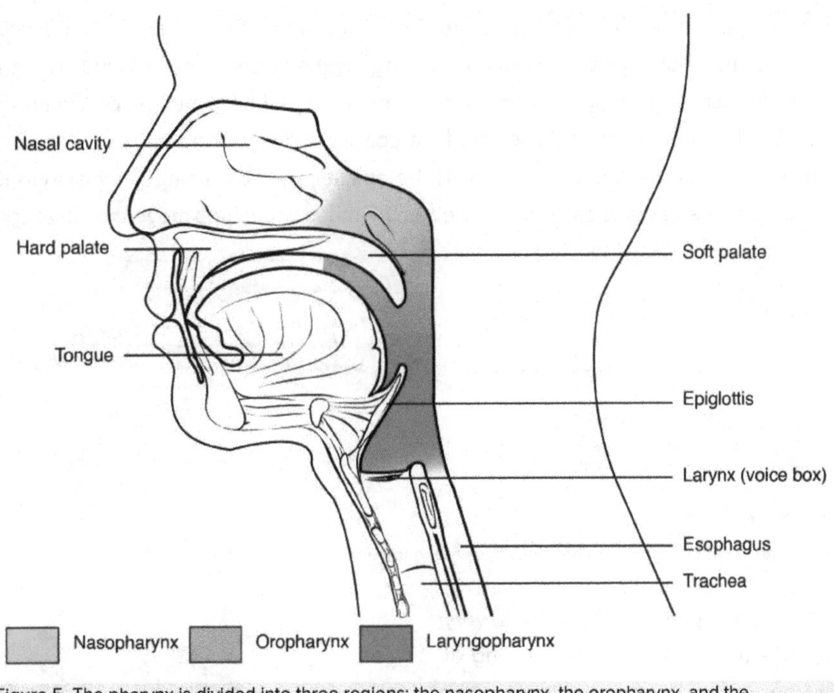

Figure 5. The pharynx is divided into three regions: the nasopharynx, the oropharynx, and the laryngopharynx.

The nasopharynx is flanked by the conchae of the nasal cavity, and it serves only as an airway. At the top of the nasopharynx are the pharyngeal tonsils. A pharyngeal tonsil, also called an adenoid, is an aggregate of lymphoid reticular tissue similar to a lymph node that lies at the superior portion of the nasopharynx. The function of the pharyngeal tonsil is not well understood, but it contains a rich supply of lymphocytes and is covered with ciliated epithelium that traps and destroys invading pathogens that enter during inhalation. The pharyngeal tonsils are large in children, but interestingly, tend to regress with age and may even disappear. The uvula is a small bulbous, teardrop-shaped structure located at the apex of the soft palate. Both the uvula and soft palate move like a pendulum during swallowing, swinging upward to close off the nasopharynx to prevent ingested materials from entering the nasal cavity. In addition, auditory (Eustachian) tubes that connect to each middle ear cavity open into the nasopharynx. This connection is why colds often lead to ear infections.

The oropharynx is a passageway for both air and food. The oropharynx is bordered superiorly by the nasopharynx and anteriorly by the oral cavity. As the nasopharynx becomes the oropharynx, the epithelium changes from pseudostratified ciliated columnar epithelium to stratified squamous epithelium. The oropharynx contains two distinct sets of tonsils, the palatine and lingual tonsils. A palatine tonsil is one of a pair of structures located laterally in the oropharynx. The lingual tonsil is located at the base of the tongue. Similar to the pharyngeal tonsil, the palatine and lingual tonsils are composed of lymphoid tissue, and trap and destroy pathogens entering the body through the oral or nasal cavities.

The laryngopharynx is inferior to the oropharynx and posterior to the larynx. It continues the route for ingested material and air until its inferior end, where the digestive and respiratory systems diverge. The stratified squamous epithelium of the oropharynx is continuous with the laryngopharynx. Anteriorly, the laryngopharynx opens into the larynx, whereas posteriorly, it enters the esophagus.

Larynx

The larynx divides the upper and lower respiratory tracts and is a cartilaginous structure connecting the pharynx to the trachea helping to regulate the volume of air that enters and leaves the lungs. The structure of the larynx is formed by several pieces of cartilage. Three large cartilage pieces—the thyroid cartilage (anterior), epiglottis (superior), and cricoid cartilage (inferior)—form the major structure of the larynx. The thyroid cartilage is the largest piece of cartilage that makes up the larynx. The thyroid cartilage has a distinct shield shape and consists of the laryngeal prominence, or "Adam's apple," which tends to be more prominent in males. The thick cricoid cartilage forms a complete ring around the superior border of the trachea and has a wide posterior region and a thinner anterior region. Three smaller, paired cartilages—the arytenoids, corniculates, and cuneiforms—attach to the epiglottis and the vocal cords as well as the muscle that help move the vocal cords to produce speech.

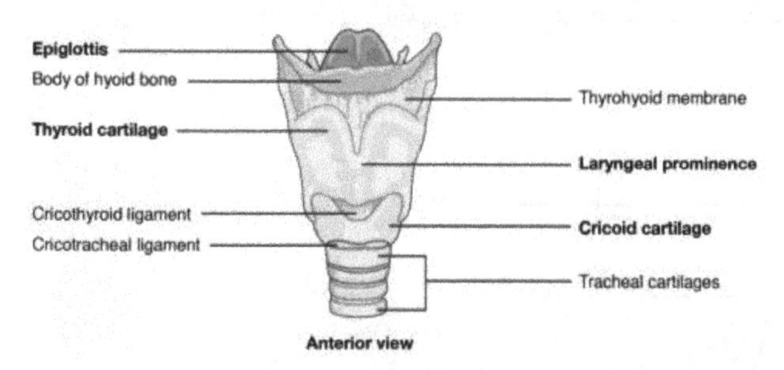

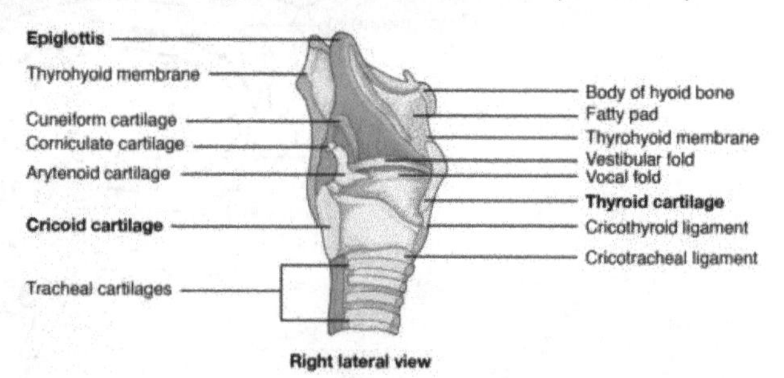

Figure 6. The larynx extends from the laryngopharynx and the hyoid bone to the trachea.

The epiglottis, attached to the thyroid cartilage, is a very flexible piece of elastic cartilage that covers the opening of the trachea. When in the "closed" position during the swallowing of food, the unattached end of the epiglottis presses down over the vestibular folds and true vocal cords covering the open space between them. A vestibular fold, or false vocal cord, is one of a pair of folded sections of mucous membrane. A true vocal cord is one of the white, membranous folds attached by muscle to the thyroid and arytenoid cartilages of the larynx on their outer edges. The inner edges of the true vocal cords are free, allowing vibration to produce sound. The size of the membranous folds of the true vocal cords differs between individuals, producing voices

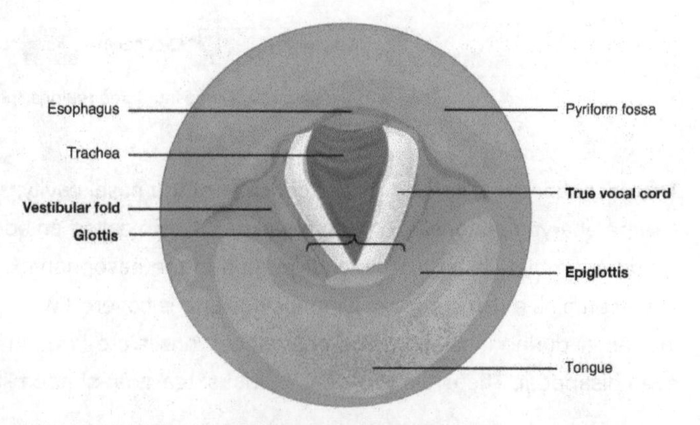

Figure 7. The true vocal cords and vestibular folds of the larynx are viewed inferiorly from the laryngopharynx.

with different pitch ranges. Folds in males tend to be larger than those in females, which create a deeper voice. The act of swallowing causes the pharynx and larynx to lift upward, allowing the pharynx to expand and the epiglottis of the larynx to swing downward, closing the opening to the trachea. These movements produce a larger area for food to pass through, while preventing food and beverages from entering the trachea.

Continuous with the laryngopharynx, the superior portion of the larynx is lined with stratified squamous epithelium, transitioning into pseudostratified ciliated columnar epithelium that contains goblet cells. Similar to the nasal cavity and nasopharynx, this specialized epithelium produces mucus to trap debris and pathogens as they enter the trachea. The cilia beat the mucus upward towards the laryngopharynx, where it can be swallowed down the esophagus.

Trachea

The trachea (windpipe) extends from the larynx toward the lungs. The trachea is formed by 16 to 20 stacked, C-shaped pieces of hyaline cartilage that are connected by dense connective tissue. The trachealis muscle (smooth muscle) and elastic connective tissue form a flexible membrane that closes the posterior surface of the trachea, connecting the C-shaped cartilages. This membrane allows the trachea to stretch and expand slightly during inhalation and exhalation, whereas the rings of cartilage provide structural support and prevent the trachea from collapsing. In addition, the trachealis muscle can be contracted to force air through the trachea during exhalation. The trachea is lined with pseudostratified ciliated columnar epithelium, which is continuous with the larynx. The esophagus borders the trachea posteriorly.

Bronchial Tree

The trachea branches into the right and left primary bronchi at the carina. These bronchi are also lined by pseudostratified ciliated columnar epithelium containing mucus-producing goblet cells. The carina is a raised structure that contains specialized nervous tissue that induces violent coughing if a foreign body, such as food, is present. Rings of cartilage, similar to those of the trachea, support the structure of the bronchi and prevent their collapse. The primary bronchi (singular: bronchus) enter the lungs at the **hilum**, a concave region where blood vessels, lymphatic vessels, and nerves also enter the lungs. Each bronchus divides into secondary bronchi, then into tertiary bronchi, which in turn divide, creating smaller and smaller diameter bronchioles as they split and spread through the lung. A bronchial tree (or respiratory tree) is the collective term used for these multiple-branched bronchi. The main function of the bronchi, like other conducting zone structures, is to provide a passageway for air to move into and out of each lung. In addition, the mucous membrane traps debris and pathogens.

Like the trachea, the bronchi are made of cartilage and smooth muscle. As the passageways decrease in diameter, the relative amount of smooth muscle increases and cartilage is replaced with elastic fibers. Bronchi are innervated by nerves of both the parasympathetic and sympathetic nervous systems that control muscle contraction (parasympathetic) or relaxation (sympathetic) in the bronchi and bronchioles, depending on the nervous system's cues.

Respiratory Zone

In contrast to the conducting zone, the respiratory zone includes structures that are directly involved in gas exchange. The respiratory zone begins where the terminal bronchioles join a respiratory bronchiole, the smallest type of bronchiole, which then leads to an alveolar duct, opening into a cluster of alveoli.

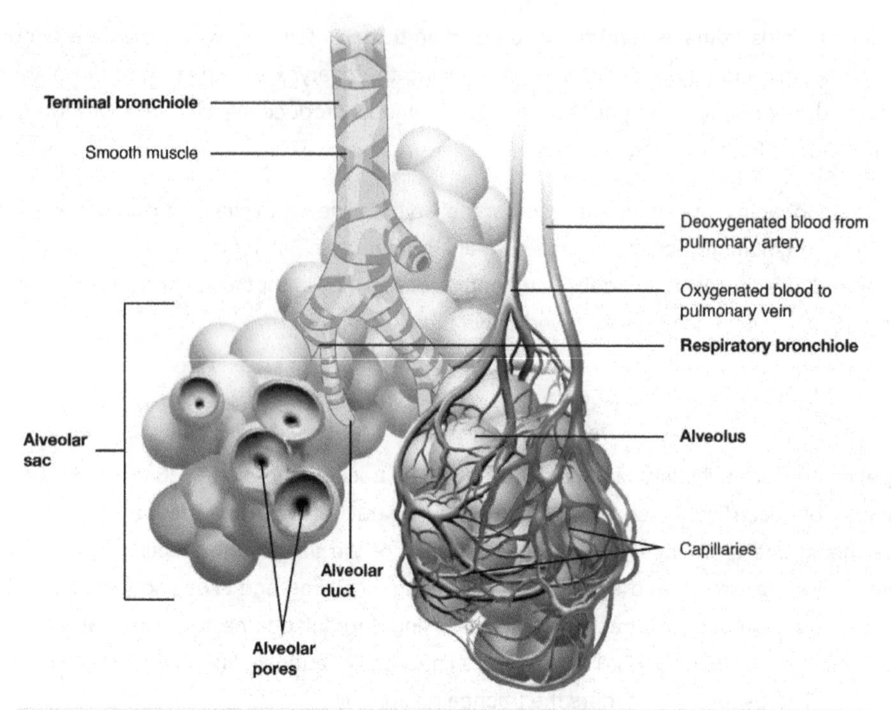

Figure 9. Bronchioles lead to alveolar sacs in the respiratory zone, where gas exchange occurs.

Alveoli

An alveolar duct is a tube composed of smooth muscle and connective tissue, which opens into a cluster of alveoli. An alveolus (plural: alveoli) is one of the many small, grape-like sacs that are attached to the alveolar ducts.

An alveolar sac is a cluster of many individual alveoli that are responsible for gas exchange. An alveolus is approximately 200 μm in diameter with elastic walls that allow the alveolus to stretch during air intake, which greatly increases the surface area available for gas exchange. Alveoli are connected to their neighbors by alveolar pores, which help maintain equal air pressure throughout the alveoli and lung.

The alveolar wall consists of three major cell types: type I alveolar cells, type II alveolar cells, and alveolar macrophages. A type I alveolar cell is a squamous epithelial cell of the alveoli, which constitute up to 97 percent of the alveolar surface area. These cells are about 25 nm thick and are highly permeable to gases. A type II alveolar cell is interspersed among the type I cells and secretes pulmonary surfactant, a substance composed of phospholipids and proteins that reduces the surface tension of the alveoli. Roaming around the alveolar wall is the alveolar macrophage, a phagocytic cell of the immune system that removes debris and pathogens that have reached the alveoli.

The simple squamous epithelium formed by type I alveolar cells is attached to a thin, elastic basement membrane. This epithelium is extremely thin and borders the endothelial membrane of capillaries. Taken together, the alveoli and capillary membranes form a **respiratory membrane** that is approximately 0.5 mm thick. The respiratory membrane allows gases to cross by simple diffusion, allowing oxygen to be picked up by the blood for transport and CO_2 to be released into the air of the alveoli.

DISEASES OF THE RESPIRATORY SYSTEM: ASTHMA

Asthma is common condition that affects the lungs in both adults and children. Approximately 8.2 percent of adults (18.7 million) and 9.4 percent of children (7 million) in the United States suffer from asthma. In addition, asthma is the most frequent cause of hospitalization in children.

Asthma is a chronic disease characterized by inflammation and edema of the airway, and bronchospasms (that is, constriction of the bronchioles), which can inhibit air from entering the lungs. In addition, excessive mucus secretion can occur, which further contributes to airway occlusion. Cells of the immune system, such as eosinophils and mononuclear cells, may also be involved in infiltrating the walls of the bronchi and bronchioles.

Bronchospasms occur periodically and lead to an "asthma attack." An attack may be triggered by environmental factors such as dust, pollen, pet hair, or dander, changes in the weather, mold, tobacco smoke, and respiratory infections, or by exercise and stress.

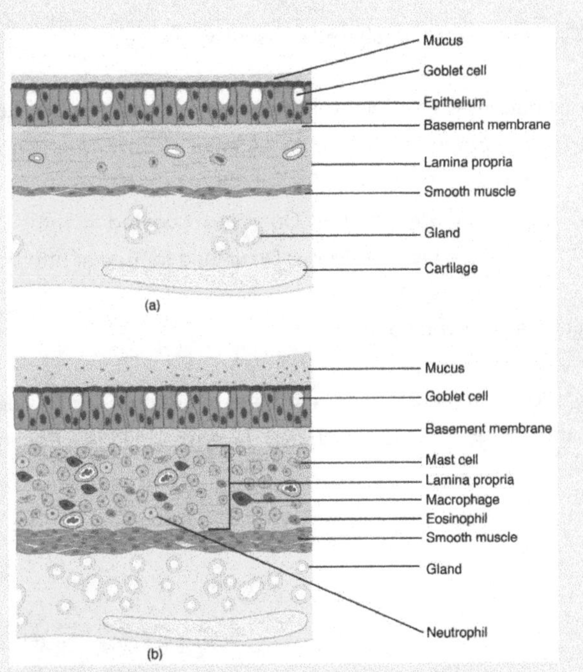

Figure 11. (a) Normal lung tissue does not have the characteristics of lung tissue during (b) an asthma attack, which include thickened mucosa, increased mucus-producing goblet cells, and eosinophil infiltrates.

Symptoms of an asthma attack involve coughing, shortness of breath, wheezing, and tightness of the chest. Symptoms of a severe asthma attack that requires immediate medical attention would include difficulty breathing that results in blue (cyanotic) lips or face, confusion, drowsiness, a rapid pulse, sweating, and severe anxiety. The severity of the condition, frequency of attacks, and identified triggers influence the type of medication that an individual may require. Longer-term treatments are used for those with more severe asthma. Short-term, fast-acting drugs that are used to treat an asthma attack are typically administered via an inhaler. For young children or individuals who have difficulty using an inhaler, asthma medications can be administered via a nebulizer.

In many cases, the underlying cause of the condition is unknown. However, recent research has demonstrated that certain viruses, such as human rhinovirus C (HRVC), and the bacteria *Mycoplasma pneumoniae* and *Chlamydia pneumoniae* that are contracted in infancy or early childhood, may contribute to the development of many cases of asthma.

References

Bizzintino J, Lee WM, Laing IA, Vang F, Pappas T, Zhang G, Martin AC, Khoo SK, Cox DW, Geelhoed GC, et al. Association between human rhinovirus C and severity of acute asthma in children. Eur Respir J [Internet]. 2010 [cited 2013 Mar 22]; 37(5):1037–1042. Available from: http://erj.ersjournals.com/gca?submit=Go&gca=erj%3B37%2F5%2F1037&allch=

Kumar V, Ramzi S, Robbins SL. Robbins Basic Pathology. 7th ed. Philadelphia (PA): Elsevier Ltd; 2005.

Martin RJ, Kraft M, Chu HW, Berns, EA, Cassell GH. A link between chronic asthma and chronic infection. J Allergy Clin Immunol [Internet]. 2001 [cited 2013 Mar 22]; 107(4):595-601. Available from: http://erj.ersjournals.com/gca?submit=Go&gca=erj%3B37%2F5%2F1037&allch=

The Lungs

A major organ of the respiratory system, each **lung** houses structures of both the conducting and respiratory zones. The main function of the lungs is to perform the exchange of oxygen and carbon dioxide between blood and the atmosphere. To this end, the lungs exchange respiratory gases across a very large epithelial surface area—about 70 square meters—that is highly permeable to gases.

Gross Anatomy of the Lungs

The lungs are pyramid-shaped, paired organs that are connected to the trachea by the right and left bronchi; on the inferior surface, the lungs are bordered by the diaphragm. The diaphragm is the flat, dome-shaped muscle located at the base of the lungs and thoracic cavity. The lungs are enclosed by the pleurae, which are attached to the mediastinum. The right lung is shorter and wider than the left lung, and the left lung occupies a smaller volume than the right. The **cardiac notch** is an indentation on the surface of the left lung, and it allows space for the heart (Figure 1). The apex of the lung is the superior region, whereas the base is the opposite region near the diaphragm.

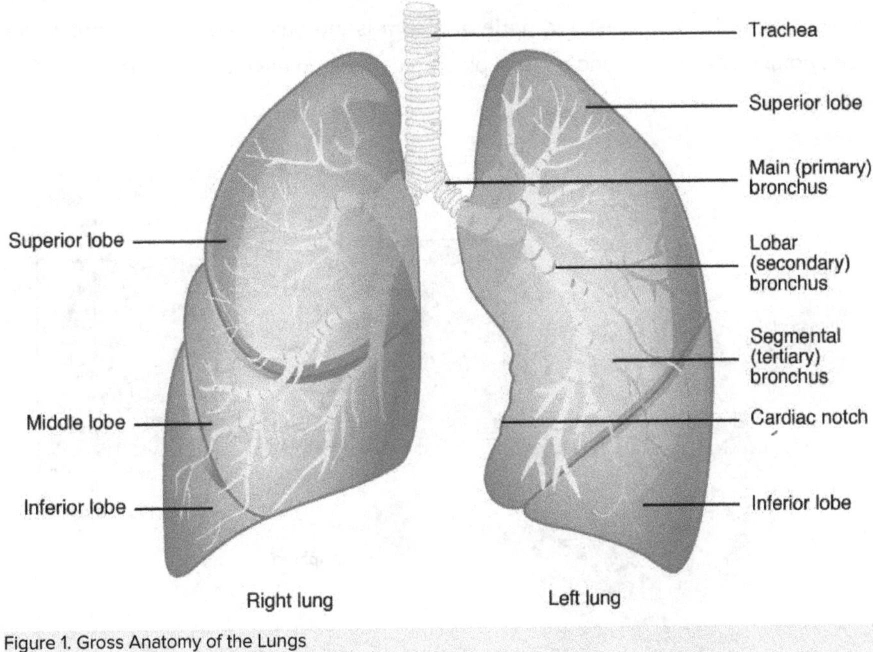

Figure 1. Gross Anatomy of the Lungs

Each lung is composed of smaller units called lobes. Fissures separate these lobes from each other. The right lung consists of three lobes: the superior, middle, and inferior lobes. The left lung consists of two lobes: the superior and inferior lobes. Each lobe can be further divided into bronchopulmonary segments. Each bronchopulmonary segment receives air from its own tertiary bronchus and is supplied with blood by its own artery. Some diseases of the lungs typically affect one or more bronchopulmonary segments, and in some cases, the diseased segments can be surgically removed with little influence on neighboring segments.

Blood Supply and Nervous Innervation of the Lungs

The blood supply of the lungs plays an important role in gas exchange and serves as a transport system for gases throughout the body. In addition, innervation by the both the parasympathetic and sympathetic nervous systems provides an important level of control through dilation and constriction of the airway.

Blood Supply

The major function of the lungs is to perform gas exchange, which requires blood from the pulmonary circulation. This blood supply contains deoxygenated blood and travels to the lungs where erythrocytes, also known as red blood cells, pick up oxygen to be transported to tissues throughout the body. The **pulmonary artery** is an artery that arises from the pulmonary trunk and carries *deoxygenated*, arterial blood to the alveoli. The pulmonary artery branches multiple times as it follows the bronchi, and each branch becomes progressively smaller in diameter. As they near the alveoli, the pulmonary arteries become the pulmonary capillary network. The pulmonary capillary network consists of tiny vessels with very thin walls that lack smooth muscle fibers. The capillaries branch and follow the bronchioles and structure of the alveoli. It is at this point that the capillary wall meets the alveolar wall, creating the **respiratory membrane**. Once the blood is oxygenated, it drains from the alveoli by way of multiple pulmonary veins, which exit the lungs through the **hilum**.

Pleura of the Lungs

Each lung is enclosed within a cavity that is surrounded by the pleura. The pleura (plural = pleurae) is a serous membrane that surrounds the lung. The right and left pleurae, which enclose the right and left lungs, respectively, are separated by the mediastinum. The pleurae consist of two layers. The **visceral pleura** is the layer that is superficial to the lungs, and extends into and lines the lung fissures (Figure 2). In contrast, the **parietal pleura** is the outer layer that connects to the thoracic wall, the mediastinum, and the diaphragm. The visceral and parietal pleurae connect to each other at the hilum. The **pleural cavity** is the space between the visceral and parietal layers.

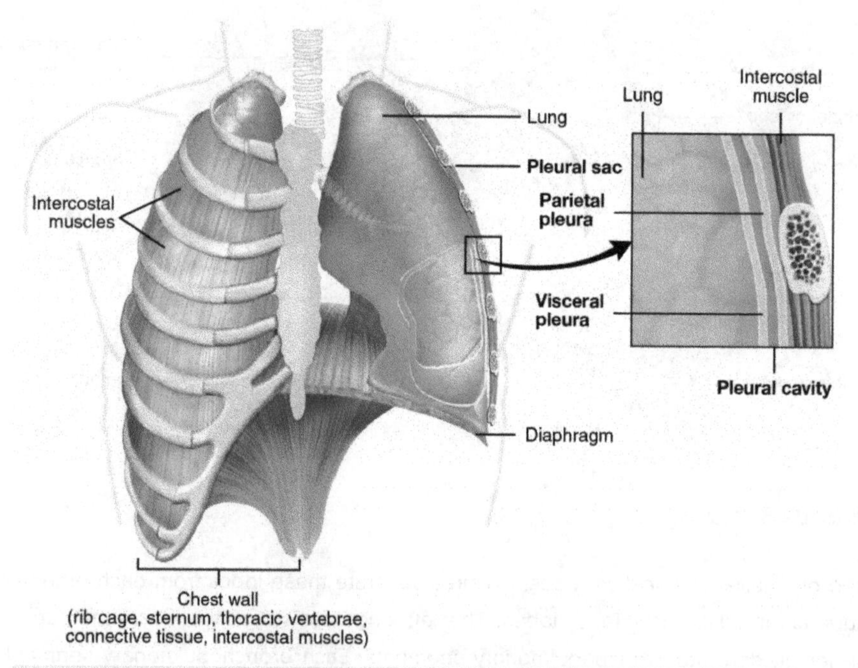

Figure 2. Parietal and Visceral Pleurae of the Lungs

The pleurae perform two major functions: They produce pleural fluid and create cavities that separate the major organs. **Pleural fluid** is secreted by mesothelial cells from both visceral and parietal layers and acts to lubricate their surfaces. This lubrication reduces friction between the two layers to prevent trauma during breathing, and creates surface tension that helps maintain the position of the lungs against the thoracic wall. This adhesive characteristic of the pleural fluid causes the lungs to enlarge when the thoracic wall expands during ventilation, allowing the lungs to fill with air. The pleurae also create a division between major organs that prevents interference due to the movement of the organs, while preventing the spread of infection.

EVERYDAY CONNECTIONS: THE EFFECTS OF SECOND-HAND TOBACCO SMOKE

The burning of a tobacco cigarette creates multiple chemical compounds that are released through mainstream smoke, which is inhaled by the smoker, and through sidestream smoke, which is the smoke that is given off by the burning cigarette. Second-hand smoke, which is a combination of sidestream smoke and the mainstream smoke that is exhaled by the smoker, has been demonstrated by numerous scientific studies to cause disease. At least 40 chemicals in sidestream smoke have been identified that negatively impact human health, leading to the development of cancer or other conditions, such as immune system dysfunction, liver toxicity, cardiac arrhythmias, pulmonary edema, and neurological dysfunction. Furthermore, second-hand smoke has been found to harbor at least 250 compounds that are known to be toxic, carcinogenic, or both. Some major classes of carcinogens in second-hand smoke are polyaromatic hydrocarbons (PAHs), N-nitrosamines, aromatic amines, formaldehyde, and acetaldehyde.

Tobacco and second-hand smoke are considered to be carcinogenic. Exposure to second-hand smoke can cause lung cancer in individuals who are not tobacco users themselves. It is estimated that the risk of developing lung cancer is increased by up to 30 percent in nonsmokers who live with an individual who smokes in the house, as compared to nonsmokers who are not regularly exposed to second-hand smoke. Children are especially affected by second-hand smoke. Children who live with an individual who smokes inside the home have a larger number of lower respiratory infections, which are associated with hospitalizations, and higher risk of sudden infant death syndrome (SIDS). Second-hand smoke in the home has also been linked to a greater number of ear infections in children, as well as worsening symptoms of asthma.

The Process of Breathing

Pulmonary ventilation is the act of breathing, which can be described as the movement of air into and out of the lungs. The major mechanisms that drive pulmonary ventilation are atmospheric pressure (P_{atm}); the air pressure within the alveoli, called alveolar pressure (P_{alv}); and the pressure within the pleural cavity, called intrapleural pressure (P_{ip}).

Mechanisms of Breathing

The alveolar and intrapleural pressures are dependent on certain physical features of the lung. However, the ability to breathe—to have air enter the lungs during inspiration and air leave the lungs during expiration—is dependent on the air pressure of the atmosphere and the air pressure within the lungs.

Pressure Relationships

Inspiration (or inhalation) and expiration (or exhalation) are dependent on the differences in pressure between the atmosphere and the lungs. In a gas, pressure is a force created by the movement of gas molecules that are confined. For example, a certain number of gas molecules in a two-liter container has more room than the same number of gas molecules in a one-liter container. In this case, the force exerted by the movement of the gas molecules against the walls of the two-liter container is lower than the force exerted by the gas molecules in the one-liter container. Therefore, the pressure is lower in the two-liter container and higher in the one-liter container.

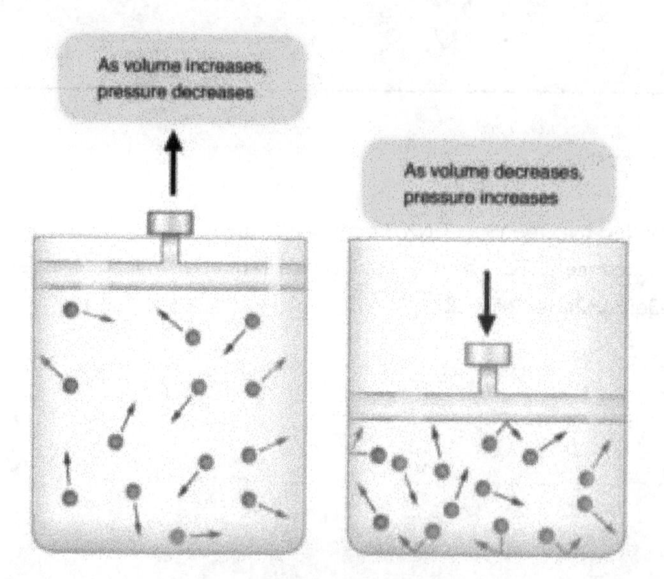

Atmospheric pressure is the amount of force that is exerted by gases in the air surrounding any given surface, such as the body. Atmospheric pressure can be expressed in terms of the unit atmosphere, abbreviated atm, or in millimeters of mercury (mm Hg). One atm is equal to 760 mm Hg, which is the atmospheric pressure at sea level. Typically, for respiration, other pressure values are discussed in relation to atmospheric pressure. Therefore, negative pressure is pressure lower than the atmospheric pressure, whereas positive pressure is pressure that it is greater than the atmospheric pressure. A pressure that is equal to the atmospheric pressure is expressed as zero.

Intra-alveolar pressure is the pressure of the air within the alveoli, which changes during the different phases of breathing (Figure 2). Because the alveoli are connected to the atmosphere via the bronchial tree the intrapulmonary pressure of the alveoli always equalizes with the atmospheric pressure.

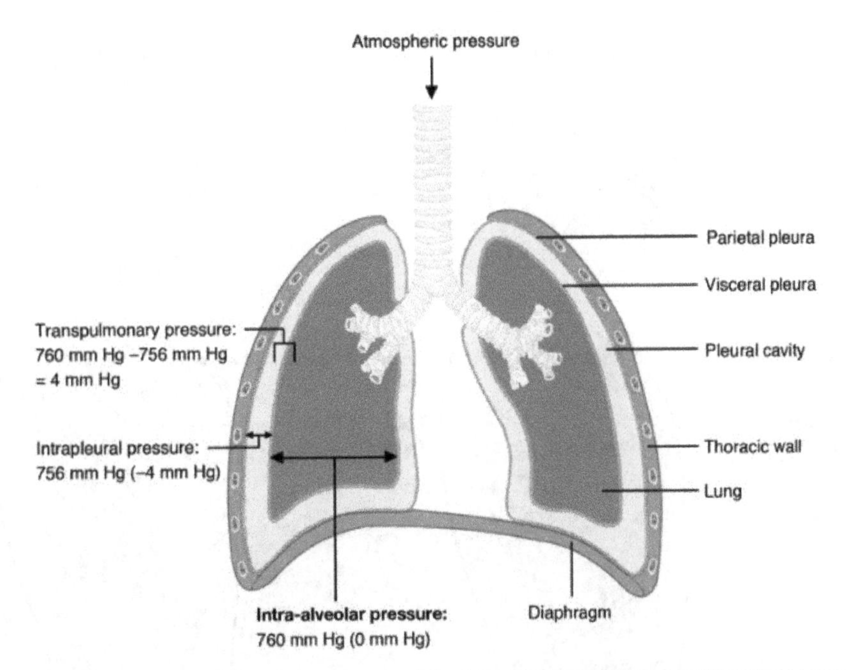

Figure 2. Alveolar pressure changes during the different phases of the cycle. It equalizes at 760 mm Hg but does not remain at 760 mm Hg.

Physical Factors Affecting Ventilation

In addition to the differences in pressures, breathing is also dependent upon the contraction and relaxation of muscle fibers of both the diaphragm and thorax. The lungs themselves are passive during breathing, meaning they are not involved in creating the movement that helps inspiration and expiration. This is because of the adhesive nature of the pleural fluid, which allows the lungs to be pulled outward when the thoracic wall moves during inspiration. The recoil of the thoracic wall during expiration causes compression of the lungs. Contraction and relaxation of the diaphragm and intercostal muscles (found between the ribs) cause most of the pressure changes that result in inspiration and expiration. These muscle movements and subsequent pressure changes cause air to either rush in or be forced out of the lungs.

Pulmonary Ventilation

The difference in pressures drives pulmonary ventilation because air flows down a pressure gradient, that is, air flows from an area of *higher pressure* to an area of *lower pressure*. Air flows into the lungs largely due to a difference in pressure; atmospheric pressure is greater than intra-alveolar pressure. Air flows out of the lungs during expiration based on the same principle; pressure within the lungs becomes greater than the atmospheric pressure.

Pulmonary ventilation comprises two major steps: inspiration and expiration. **Inspiration** is the process that causes air to enter the lungs, and **expiration** is the process that causes air to leave the lungs (Figure 3). A **respiratory cycle** is one sequence of inspiration and expiration. In general, two muscle groups are used during normal inspiration: the diaphragm and the external intercostal muscles. Additional muscles can be used if a bigger breath is required. When the diaphragm contracts, it moves inferiorly toward the abdominal cavity, creating a larger thoracic cavity and more space for the lungs. Contraction of the external intercostal muscles moves the ribs upward and outward, causing the rib cage to expand, which increases the volume of the thoracic cavity. Due to the adhesive force of the pleural fluid, the expansion of the thoracic cavity forces the lungs to stretch and expand as well. This increase in volume leads to a decrease in intra-alveolar pressure, creating a pressure lower than atmospheric pressure. As a result, a pressure gradient is created that drives air into the lungs.

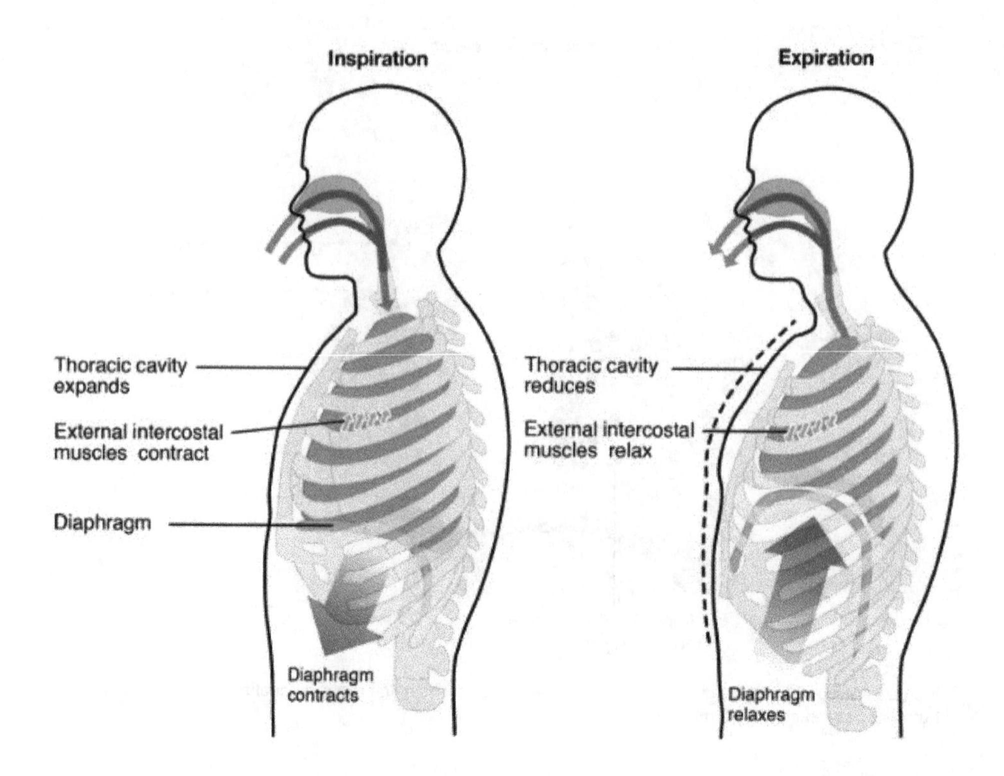

Figure 3. Inspiration and expiration occur due to the expansion and contraction of the thoracic cavity, respectively.

The process of normal expiration is passive, meaning that energy is not required to push air out of the lungs. Instead, the elasticity of the lung tissue causes the lung to recoil, as the diaphragm and intercostal muscles relax following inspiration. In turn, the thoracic cavity and lungs decrease in volume, causing an increase in interpulmonary pressure. The interpulmonary pressure rises above atmospheric pressure, creating a pressure gradient that causes air to leave the lungs.

There are different types, or modes, of breathing that require a slightly different process to allow inspiration and expiration. **Quiet breathing**, also known as eupnea, is a mode of breathing that occurs at rest and does not require the cognitive thought of the individual. During quiet breathing, the diaphragm and external intercostals must contract.

A deep breath, called diaphragmatic breathing, requires the diaphragm to contract. As the diaphragm relaxes, air passively leaves the lungs. A shallow breath, called costal breathing, requires contraction of the intercostal muscles. As the intercostal muscles relax, air passively leaves the lungs.

In contrast, **forced breathing**, also known as hyperpnea, is a mode of breathing that can occur during exercise or actions that require the active manipulation of breathing, such as singing. During forced breathing, inspiration and expiration both occur due to muscle contractions. In addition to the contraction of the diaphragm and intercostal muscles, other accessory muscles must also contract. During forced inspiration, muscles of the neck, including the scalenes, contract and lift the thoracic wall, increasing lung volume. During forced expiration, accessory muscles of the abdomen, including the obliques, contract, forcing abdominal organs upward against the diaphragm. This helps to push the diaphragm further into the thorax, pushing more air out. In addition, accessory muscles (primarily the internal intercostals) help to compress the rib cage, which also reduces the volume of the thoracic cavity.

Respiratory Volumes and Capacities

Respiratory volume is the term used for various volumes of air moved by or associated with the lungs at a given point in the respiratory cycle. There are four major types of respiratory volumes: tidal, residual, inspiratory reserve, and expiratory reserve (Figure 4).

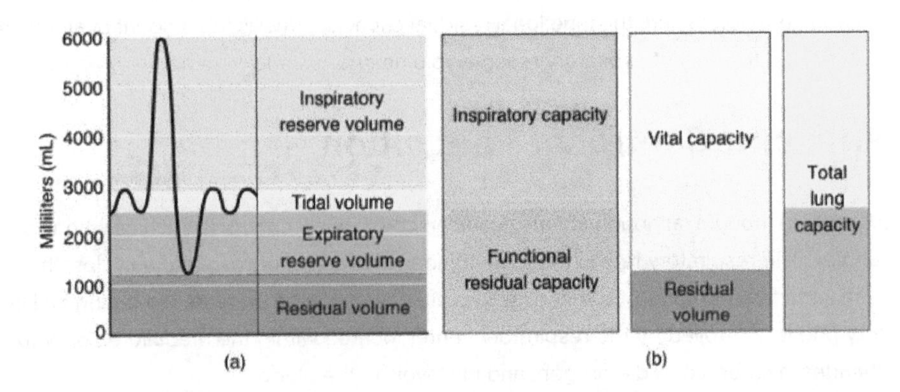

Figure 4. These two graphs show (a) respiratory volumes and (b) the combination of volumes that results in respiratory capacity.

Tidal volume (TV) is the amount of air that normally enters the lungs during quiet breathing, which is about 500 milliliters. **Expiratory reserve volume (ERV)** is the amount of air you can forcefully exhale past a normal tidal expiration, up to 1200 milliliters for men. **Inspiratory reserve volume (IRV)** is produced by a deep inhalation, past a tidal inspiration. This is the extra volume that can be brought into the lungs during a forced inspiration. **Residual volume (RV)** is the air left in the lungs if you exhale as much air as possible. The residual volume makes breathing easier by preventing the alveoli from collapsing. Respiratory volume is dependent on a variety of factors, and measuring the different types of respiratory volumes can provide important clues about a person's respiratory health (see Table 1).

Table 1. Pulmonary Function Testing

Pulmonary Function Test	Instrument	Measures	Function
Spirometry	Spirometer	Forced vital capacity (FVC)	Volume of air that is exhaled after maximum inhalation
		Forced expiratory volume (FEV)	Volume of air exhaled in one breath
		Forced expiratory flow, 25–75	Air flow in the middle of exhalation
		Peak expiratory flow (PEF)	Rate of exhalation
		Maximum voluntary ventilation (MVV)	Volume of air that an be inspired and expired in 1 minute
		Slow vital capacity (SVC)	Volume of air that can be slowly exhaled after inhaling past the tidal volume
		Total lung capacity (TLC)	Volume of air in the lungs after maximum inhalation
		Functional residual capacity (FRC)	Volume of air left in the lungs after normal expiration
		Residual volume (RV)	Volume of air in the lungs after maximum exhalation
		Total lung capacity (TLC)	Maximum volume of air that the lungs can hold
		Expiratory reserve volume (ERV)	The volume of air that can be exhaled beyond normal exhalation
Gas diffusion	Blood gas analyzer	Arterial blood gases	Concentration of oxygen and carbon dioxide in the blood

Respiratory capacity is the combination of two or more selected volumes, which further describes the amount of air in the lungs during a given time. For example, total lung capacity (TLC) is the sum of all of the lung volumes (TV, ERV, IRV, and RV), which represents the total amount of air a person can hold in the lungs after a forceful inhalation. TLC is about 6000 mL air for men, and about 4200 mL for women. Vital capacity (VC) is the amount of air a person can move into or out of his or her lungs, and is the sum of all of the volumes except residual volume (TV, ERV, and IRV), which is between 4000 and 5000 milliliters. Inspiratory capacity (IC) is the maximum amount of air that can be inhaled past a normal tidal expiration, is the sum of the tidal volume and inspiratory reserve volume. On the other hand, the functional residual capacity (FRC) is the amount of air that remains in the lung after a normal tidal expiration; it is the sum of expiratory reserve volume and residual volume.

Respiratory Rate and Control of Ventilation

Breathing usually occurs without thought, although at times you can consciously control it, such as when you swim under water, sing a song, or blow bubbles. The respiratory rate is the total number of breaths, or respiratory cycles, that occur each minute. Respiratory rate can be an important indicator of disease, as the rate may increase or decrease during an illness or in a disease condition. The respiratory rate is controlled by the respiratory center located within the *medulla oblongata* in the brain, which responds primarily to changes in carbon dioxide, oxygen, and pH levels in the blood.

The normal respiratory rate of a child decreases from birth to adolescence. A child under 1 year of age has a normal respiratory rate between 30 and 60 breaths per minute, but by the time a child is about 10 years old, the normal rate is closer to 18 to 30. By adolescence, the normal respiratory rate is similar to that of adults, 12 to 18 breaths per minute.

Factors That Affect the Rate and Depth of Respiration

The respiratory rate and the depth of inspiration are regulated by the respiration centers of the brain: the medulla oblongata and pons. These centers respond to systemic stimuli utilizing a positive-feedback relationship in which the greater the stimulus, the greater the response. Thus, increasing stimuli results in forced breathing. Multiple systemic factors are involved in stimulating the brain to produce pulmonary ventilation.

The major factor that stimulates the medulla oblongata and pons to produce respiration is surprisingly not oxygen concentration, but rather the concentration of carbon dioxide in the blood. As you recall, carbon dioxide is a waste product of cellular respiration and can be toxic. Concentrations of chemicals are sensed by chemoreceptors. Central chemoreceptors are located in the brain and brainstem, whereas peripheral chemoreceptors are located in the carotid arteries and aortic arch. Concentration changes in certain substances, such as carbon dioxide or hydrogen ions, stimulate these receptors, which in turn signal the respiration centers of the brain. In the case of carbon dioxide, as the concentration of CO_2 in the blood increases, it readily diffuses across the blood-brain barrier, where it collects in the extracellular fluid. As will be explained in more detail later, increased carbon dioxide levels lead to increased levels of hydrogen ions, decreasing pH. The increase in hydrogen ions in the brain triggers the central chemoreceptors to stimulate the respiratory centers to initiate contraction of the diaphragm and intercostal muscles. As a result, the rate and depth of respiration increase, allowing more carbon dioxide to be expelled, which brings more air into and out of the lungs promoting a reduction in the blood levels of carbon dioxide, and therefore hydrogen ions, in the blood. In contrast, low levels of carbon dioxide in the blood cause low levels of hydrogen ions in the brain, leading to a decrease in the rate and depth of pulmonary ventilation, producing shallow, slow breathing.

Another factor involved in influencing the respiratory activity of the brain is systemic arterial concentrations of hydrogen ions. Increasing carbon dioxide levels can lead to increased H^+ levels, as mentioned above, as well as other metabolic activities, such as lactic acid accumulation after strenuous exercise. Peripheral chemoreceptors of the aortic arch and carotid arteries sense arterial levels of hydrogen ions. When peripheral chemoreceptors sense decreasing, or more acidic, pH levels, they stimulate an increase in ventilation to remove carbon dioxide from the blood at a quicker rate. Removal of carbon dioxide from the blood helps to reduce hydrogen ions, thus increasing systemic pH.

Blood levels of oxygen are also important in influencing respiratory rate. The peripheral chemoreceptors are responsible for sensing large changes in blood oxygen levels. If blood oxygen levels become quite low—about 60 mm Hg or less—then peripheral chemoreceptors stimulate an increase in respiratory activity. The chemoreceptors are only able to sense dissolved oxygen molecules, not the oxygen that is bound to hemoglobin. As you recall, the majority of oxygen is bound by hemoglobin; when dissolved levels of oxygen drop, hemoglobin releases oxygen. Therefore, a large drop in oxygen levels is required to stimulate the chemoreceptors of the aortic arch and carotid arteries.

The hypothalamus and other brain regions associated with the limbic system also play roles in influencing the regulation of breathing by interacting with the respiratory centers. The hypothalamus and other regions associated with the limbic system are involved in regulating respiration in response to emotions, pain, and temperature. For example, an increase in body temperature causes an increase in respiratory rate. Feeling excited or the fight-or-flight response will also result in an increase in respiratory rate.

DISORDERS OF THE RESPIRATORY SYSTEM: SLEEP APNEA

Sleep apnea is a chronic disorder that can occur in children or adults, and is characterized by the cessation of breathing during sleep. These episodes may last for several seconds or several minutes, and may differ in the frequency with which they are experienced. Sleep apnea leads to poor sleep, which is reflected in the symptoms of fatigue, evening napping, irritability, memory problems, and morning headaches. In addition, many individuals with sleep apnea experience a dry throat in the morning after waking from sleep, which may be due to excessive snoring.

There are two types of sleep apnea: obstructive sleep apnea and central sleep apnea. Obstructive sleep apnea is caused by an obstruction of the airway during sleep, which can occur at different points in the airway, depending on the underlying cause of the obstruction. For example, the tongue and throat muscles of some individuals with obstructive sleep apnea may relax excessively, causing the muscles to push into the airway. Another example is obesity, which is a known risk factor for sleep apnea, as excess adipose tissue in the neck region can push the soft tissues towards the lumen of the airway, causing the trachea to narrow.

In central sleep apnea, the respiratory centers of the brain do not respond properly to rising carbon dioxide levels and therefore do not stimulate the contraction of the diaphragm and intercostal muscles regularly. As a result, inspiration does not occur and breathing stops for a short period. In some cases, the cause of central sleep apnea is unknown. However, some medical conditions, such as stroke and congestive heart failure, may cause damage to the pons or medulla oblongata. In addition, some pharmacologic agents, such as morphine, can affect the respiratory centers, causing a decrease in the respiratory rate. The symptoms of central sleep apnea are similar to those of obstructive sleep apnea.

A diagnosis of sleep apnea is usually done during a sleep study, where the patient is monitored in a sleep laboratory for several nights. The patient's blood oxygen levels, heart rate, respiratory rate, and blood pressure are monitored, as are brain activity and the volume of air that is inhaled and exhaled. Treatment of sleep apnea commonly includes the use of a device called a continuous positive airway pressure (CPAP) machine during sleep. The CPAP machine has a mask that covers the nose, or the nose and mouth, and forces air into the airway at regular intervals. This pressurized air can help to gently force the airway to remain open, allowing more normal ventilation to occur. Other treatments include lifestyle changes to decrease weight, eliminate alcohol and other sleep apnea–promoting drugs, and changes in sleep position. In addition to these treatments, patients with central sleep apnea may need supplemental oxygen during sleep.

Gas Exchange

Gas Exchange

Gas exchange occurs at two sites in the body: in the lungs, where oxygen is picked up and carbon dioxide is released at the respiratory membrane, and at the tissues, where oxygen is released and carbon dioxide is picked up. **External respiration** is the exchange of gases with the external environment, and occurs in the alveoli of the lungs. **Internal respiration** is the exchange of gases with the internal environment, and occurs in the tissues. The actual exchange of gases occurs due to simple diffusion. Energy is not required to move oxygen or carbon dioxide across membranes. Instead, these gases follow pressure gradients that allow them to diffuse. The anatomy of the lung maximizes the diffusion of gases: The respiratory membrane is highly permeable to gases; the respiratory and blood capillary membranes are very thin; and there is a large surface area throughout the lungs.

External Respiration

The pulmonary artery carries deoxygenated blood into the lungs from the heart where it branches and eventually becomes the capillary network composed of pulmonary capillaries. These pulmonary capillaries create the respiratory membrane with the alveoli. As the blood is pumped through this capillary network, gas exchange occurs. Although a small amount of the oxygen is able to dissolve directly into plasma from the alveoli, most of the oxygen is picked up by erythrocytes (red blood cells) and binds to a protein called hemoglobin, a process described later in this chapter. Oxygenated hemoglobin is red, causing the overall appearance of bright red oxygenated blood, which returns to the heart through the pulmonary veins. Carbon dioxide is released in the opposite direction of oxygen, from the blood to the alveoli. Some of the carbon dioxide is returned on hemoglobin, but can also be dissolved in plasma or is present as a converted form, also explained in greater detail later in this chapter.

External respiration occurs as a function of partial pressure differences in oxygen and carbon dioxide between the alveoli and the blood in the pulmonary capillaries.

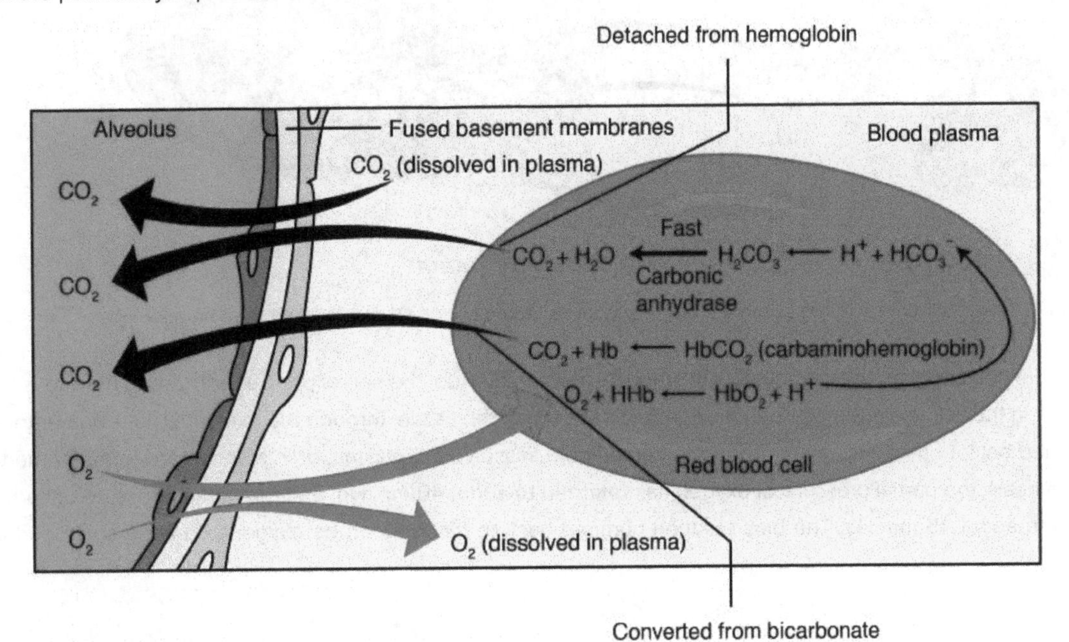

Figure 2. In external respiration, oxygen diffuses across the respiratory membrane from the alveolus to the capillary, whereas carbon dioxide diffuses out of the capillary into the alveolus.

Although the solubility of oxygen in blood is not high, there is a drastic difference in the partial pressure of oxygen in the alveoli versus in the blood of the pulmonary capillaries. This difference is about 64 mm Hg: The partial pressure of oxygen in the alveoli is about 104 mm Hg, whereas its partial pressure in the blood of the capillary is about 40 mm Hg. This large difference in partial pressure creates a very strong pressure gradient that causes oxygen to rapidly cross the respiratory membrane from the alveoli into the blood.

The partial pressure of carbon dioxide is also different between the alveolar air and the blood of the capillary. However, the partial pressure difference is less than that of oxygen, about 5 mm Hg. The partial pressure of carbon dioxide in the blood of the capillary is about 45 mm Hg, whereas its partial pressure in the alveoli is about 40 mm Hg. However, the solubility of carbon dioxide is much greater than that of oxygen—by a factor of about 20—in both blood and alveolar fluids. As a result, the relative concentrations of oxygen and carbon dioxide that diffuse across the respiratory membrane are similar.

Internal Respiration

Internal respiration is gas exchange that occurs at the level of body tissues (Figure 3). Similar to external respiration, internal respiration also occurs as simple diffusion due to a partial pressure gradient. However, the partial pressure gradients are opposite of those present at the respiratory membrane. The partial pressure of oxygen in tissues is low, about 40 mm Hg, because oxygen is continuously used for cellular respiration. In contrast, the partial pressure of oxygen in the blood is about 100 mm Hg. This creates a pressure gradient that causes oxygen to dissociate from hemoglobin, diffuse out of the blood, cross the interstitial space, and enter the tissue. Hemoglobin that has little oxygen bound to it loses much of its brightness, so that blood returning to the heart is more burgundy in color.

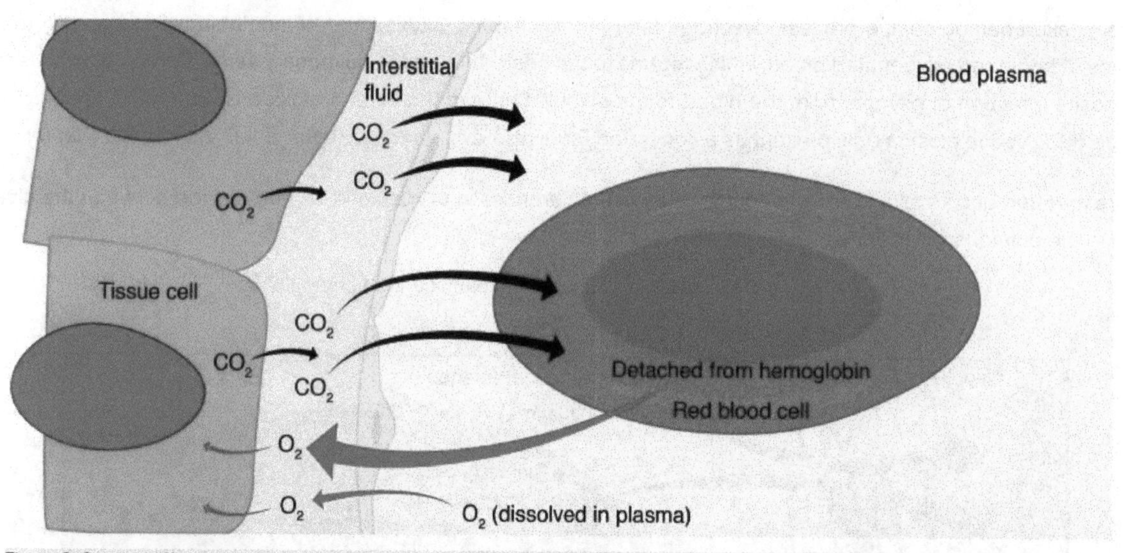

Figure 3. Oxygen diffuses out of the capillary and into cells, whereas carbon dioxide diffuses out of cells and into the capillary.

Considering that cellular respiration continuously produces carbon dioxide, the partial pressure of carbon dioxide is lower in the blood than it is in the tissue, causing carbon dioxide to diffuse out of the tissue, through the interstitial fluid, and enter the blood. It is then carried back to the lungs either bound to hemoglobin, dissolved in plasma, or in a converted form. By the time blood returns to the heart, the partial pressure of oxygen has returned to about 40 mm Hg, and the partial pressure of carbon dioxide has returned to about 45 mm Hg. The blood is then pumped back to the lungs to be oxygenated once again during external respiration.

EVERYDAY CONNECTIONS: HYPERBARIC CHAMBER TREATMENT

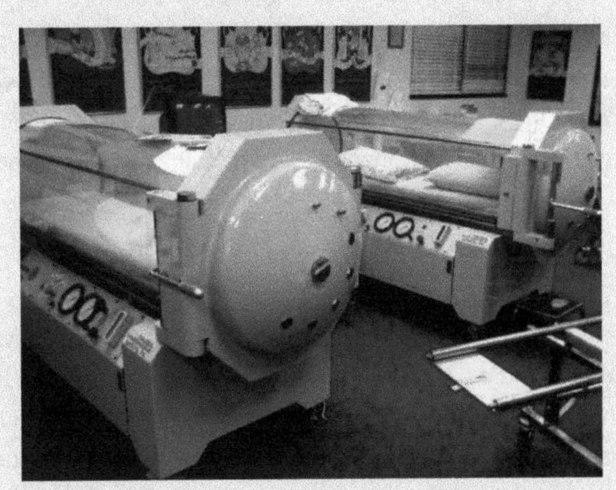

Figure 4. Hyperbaric Chamber (credit: "komunews"/flickr.com)

A type of device used in some areas of medicine that exploits the behavior of gases is hyperbaric chamber treatment. A hyperbaric chamber is a unit that can be sealed and expose a patient to either 100 percent oxygen with increased pressure or a mixture of gases that includes a higher concentration of oxygen than normal atmospheric air, also at a higher partial pressure than the atmosphere. There are two major types of chambers: monoplace and multiplace. Monoplace chambers are typically for one patient, and the staff tending to the patient observes the patient from outside of the chamber. Some facilities have special monoplace hyperbaric chambers that allow multiple patients to be treated at once, usually in a sitting or reclining position, to help ease feelings of isolation or claustrophobia. Multiplace chambers are large enough for multiple patients to be treated at one time, and the staff attending these patients is present inside the chamber. In a multiplace chamber, patients are often treated with air via a mask or hood, and the chamber is pressurized.

Hyperbaric chamber treatment is based on the behavior of gases. As you recall, gases move from a region of higher partial pressure to a region of lower partial pressure. In a hyperbaric chamber, the atmospheric pressure is increased, causing a greater amount of oxygen than normal to diffuse into the bloodstream of the patient. Hyperbaric chamber therapy is used to treat a variety of medical problems, such as wound and graft healing, anaerobic bacterial infections, and carbon monoxide poisoning. Exposure to and poisoning by carbon monoxide is difficult to reverse, because hemoglobin's affinity for carbon monoxide is much stronger than its affinity for oxygen, causing carbon monoxide to replace oxygen in the blood. Hyperbaric chamber therapy can treat carbon monoxide poisoning, because the increased atmospheric pressure causes more oxygen to diffuse into the bloodstream. At this increased pressure and increased concentration of oxygen, carbon monoxide is displaced from hemoglobin. Another example is the treatment of anaerobic bacterial infections, which are created by bacteria that cannot or prefer not to live in the presence of oxygen. An increase in blood and tissue levels of oxygen helps to kill the anaerobic bacteria that are responsible for the infection, as oxygen is toxic to anaerobic bacteria. For wounds and grafts, the chamber stimulates the healing process by increasing energy production needed for repair. Increasing oxygen transport allows cells to ramp up cellular respiration and thus ATP production, the energy needed to build new structures.

Transport of Gases

The other major activity in the lungs is the process of respiration, the process of gas exchange. The function of respiration is to provide oxygen for use by body cells during cellular respiration and to eliminate carbon dioxide, a waste product of cellular respiration, from the body. In order for the exchange of oxygen and carbon dioxide to occur, both gases must be transported between the external and internal respiration sites. Although carbon dioxide is more soluble than oxygen in blood, both gases require a specialized transport system for the majority of the gas molecules to be moved between the lungs and other tissues.

Oxygen Transport in the Blood

Even though oxygen is transported via the blood, you may recall that oxygen is not very soluble in liquids. A small amount of oxygen does dissolve in the blood and is transported in the bloodstream, but it is only about 1.5% of the total amount. The majority of oxygen molecules are carried from the lungs to the body's tissues by a specialized transport system, which relies on the erythrocyte—the red blood cell. Erythrocytes contain hemoglobin, which serves to bind oxygen molecules to the erythrocyte (Figure 1). Heme is the portion of hemoglobin that contains iron, hemoglobin's metal ion cofactor, and it is heme that binds oxygen. One erythrocyte contains four iron ions, and because of this, each erythrocyte is capable of carrying up to four molecules of oxygen. As oxygen diffuses across the respiratory membrane from the alveolus to the capillary, it also diffuses into the red blood cell and is bound by hemoglobin. The following reversible chemical reaction describes the production of the final product, **oxyhemoglobin** ($Hb–O_2$), which is formed when oxygen binds to hemoglobin. Oxyhemoglobin is a bright red-colored molecule that contributes to the bright red color of oxygenated blood.

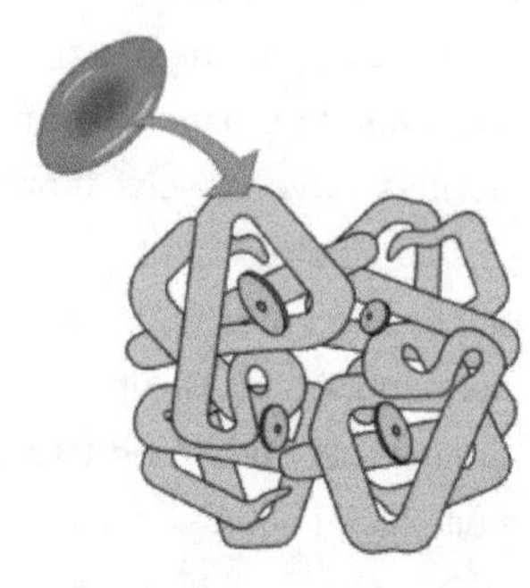

Figure 1. Hemoglobin consists of four subunits, each of which contains one molecule of iron.

$$Hb + O_2 \leftrightarrow Hb - O_2$$

In this formula, Hb represents reduced hemoglobin, that is, hemoglobin that does not have oxygen bound to it. There are multiple factors involved in how readily heme binds to and dissociates from oxygen, which will be discussed in the subsequent sections.

Function of Hemoglobin

Hemoglobin is composed of four subunits, a protein structure referred to as a quaternary structure. Each of the four subunits that make up hemoglobin is arranged in a ring-like fashion with an iron atom covalently bound to the heme in the center of each subunit. Binding of the first oxygen molecule causes a conformational change in hemoglobin that allows the second molecule of oxygen to bind more readily. As each molecule of oxygen is bound, it further facilitates the binding of the next molecule, until all four heme sites are occupied by oxygen. The opposite occurs as well: After the first oxygen molecule dissociates and is "dropped off" at the tissues, the next oxygen molecule dissociates more readily. When all four heme sites are occupied, the hemoglobin is said to be *saturated*. When one to three heme sites are occupied, the hemoglobin is said to be *partially saturated*. Therefore, when considering the blood as a whole, the percent of the available heme units that are bound to oxygen at a given time is called hemoglobin saturation. Hemoglobin saturation of 100 percent means that every heme unit in all of the erythrocytes of the body is bound to oxygen. In a healthy individual with normal hemoglobin levels, hemoglobin saturation generally ranges from 95 percent to 99 percent.

Carbon Dioxide Transport in the Blood

Carbon dioxide is transported by three major mechanisms. The first mechanism of carbon dioxide transport is by blood plasma, as some carbon dioxide molecules dissolve in the blood. The second mechanism is transport in the form of bicarbonate (HCO_3^-), which also dissolves in plasma. The third mechanism of carbon dioxide transport is similar to the transport of oxygen by erythrocytes.

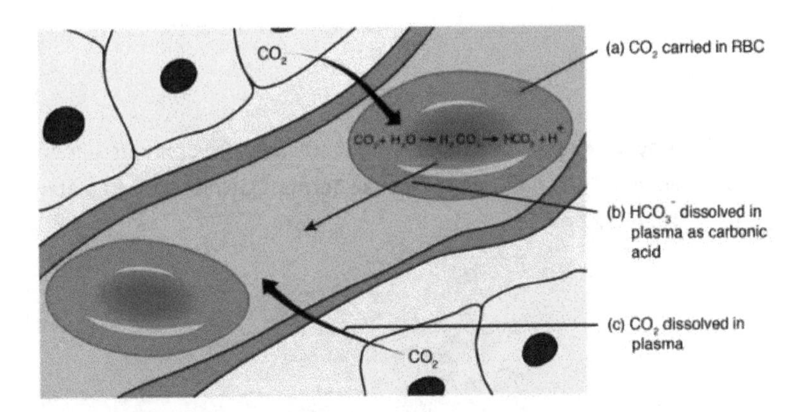

Figure 4. Carbon dioxide is transported by three different methods: (a) in erythrocytes; (b) after forming carbonic acid (H_2CO_3), which is dissolved in plasma; (c) and in plasma.

Dissolved Carbon Dioxide

Although carbon dioxide is not considered to be highly soluble in blood, a small fraction—about 7 to 10 percent—of the carbon dioxide that diffuses into the blood from the tissues dissolves in plasma. The dissolved carbon dioxide then travels in the bloodstream and when the blood reaches the pulmonary capillaries, the dissolved carbon dioxide diffuses across the respiratory membrane into the alveoli, where it is then exhaled during pulmonary ventilation.

Bicarbonate Buffer

A large fraction—about 70 percent—of the carbon dioxide molecules that diffuse into the blood is transported to the lungs as bicarbonate. Most bicarbonate is produced in erythrocytes after carbon dioxide diffuses into the capillaries, and subsequently into red blood cells. **Carbonic anhydrase (CA)** causes carbon dioxide and water to form carbonic acid (H_2CO_3), which dissociates into two ions: bicarbonate (HCO_3^-) and hydrogen (H^+). The following formula depicts this reaction:

$$CO_2 + H_2O \leftrightarrow H_2CO_3 \leftrightarrow H^+ + HCO_3^-$$

At the pulmonary capillaries, the chemical reaction that produced bicarbonate (shown above) is reversed, and carbon dioxide and water are the products. Hydrogen ions and bicarbonate ions join to form carbonic acid, which is converted into carbon dioxide and water by carbonic anhydrase. Carbon dioxide diffuses out of the erythrocytes and into the plasma, where it can further diffuse across the respiratory membrane into the alveoli to be exhaled during pulmonary ventilation.

Carbaminohemoglobin

About 20 percent of carbon dioxide is bound by hemoglobin and is transported to the lungs. Carbon dioxide does not bind to iron as oxygen does; instead, carbon dioxide binds to amino acids of the globin portions of hemoglobin to form **carbaminohemoglobin**, which forms when hemoglobin and carbon dioxide bind. When hemoglobin is not transporting oxygen, it tends to have a darker maroon tone to it, creating the color typical of deoxygenated blood.

Similar to the transport of oxygen by heme, the binding and dissociation of carbon dioxide to and from hemoglobin is dependent on the partial pressure of carbon dioxide. Because carbon dioxide is released from the lungs, blood that leaves the lungs and reaches body tissues has a lower partial pressure of carbon dioxide than is found in the tissues. As a result, carbon dioxide leaves the tissues because of its higher partial pressure, enters the blood, and then moves into red blood cells, binding to hemoglobin. In contrast, in the pulmonary capillaries, the partial pressure of carbon dioxide is high compared to within the alveoli. As a result, carbon dioxide dissociates readily from hemoglobin and diffuses across the respiratory membrane into the air.

Concept Check

1. **What** are the main functions of the respiratory system? **What** are the minor functions of this system?
2. **What** structure marks the break point between the upper and lower respiratory tracts? **What** are the functions of the upper respiratory tract?
3. **Which** laryngeal cartilage is responsible for preventing solids from entering the lower respiratory tract?
4. **How** are the conducting zone and the respiratory zone different?
5. **How** does the structure of the alveolar sac relate directly to its function?
6. **How** does muscle activity promote inhalation and exhalation?
7. **Which** dissolved gas is responsible for triggering breathing?
8. **What** is the relationship between ERV, IRV, and VC?
9. **Which** type of respiration is responsible for delivering oxygen to the tissues of the body: external or internal respiration?
10. **What** is the force that moves oxygen from air to blood, and from blood to tissue cells?
11. **What** characteristics of hemoglobin make it ideal for oxygen transport? In other words, **when** does hemoglobin pick up oxygen, and under **what** conditions does it release it?
12. **What** is one positive role of carbon dioxide in the blood?
13. **How** does asthma interfere with respiration?

CHAPTER 13 — THE NERVOUS SYSTEM

Neurons and Glial Cells

LEARNING OBJECTIVES

- List and describe the functions of the structural components of a neuron
- Compare the functions of different types of glial cells
- Explain the stages of an action potential and how action potentials are propagated
- Explain the similarities and differences between chemical and electrical synapses
- Identify the spinal cord, cerebral lobes, and other brain areas on a diagram of the brain
- Describe the basic functions of the spinal cord, cerebral lobes, and other brain areas
- Describe the organization and functions of the sympathetic and parasympathetic nervous systems
- Compare the structure of somatic and autonomic reflex arcs
- Describe the symptoms, potential causes, and treatment of several examples of nervous system disorders

The nervous system is made up of neurons, specialized cells that can receive and transmit chemical or electrical signals, and glia, cells that provide support functions for the neurons by playing an information processing role that is complementary to neurons. A neuron can be compared to an electrical wire—it transmits a signal from one place to another. Glia can be compared to the workers at the electric company who make sure wires go to the right places, maintain the wires, and take down wires that are broken. Although glia have been compared to workers, recent evidence suggests that they also usurp some of the signaling functions of neurons.

There is great diversity in the types of neurons and glia that are present in different parts of the nervous system. There are four major types of neurons, and they share several important cellular components.

Neurons

The nervous system of the common laboratory fly, *Drosophila melanogaster*, contains around 100,000 neurons, the same number as a lobster. This number compares to 75 million in the mouse and 300 million in the octopus. A human brain contains around 86 billion neurons. Despite these very different numbers, the nervous systems of these animals control many of the same behaviors—from basic reflexes to more complicated behaviors like finding food and courting mates. The ability of neurons to communicate with each other as well as with other types of cells underlies all of these behaviors.

Most neurons share the same cellular components. But neurons are also highly specialized—different types of neurons have different sizes and shapes that relate to their functional roles.

Parts of a Neuron

Like other cells, each neuron has a cell body (or soma) that contains a nucleus, smooth and rough endoplasmic reticulum, Golgi apparatus, mitochondria, and other cellular components. Neurons also contain unique structures, illustrated in Figure 2 for receiving and sending the electrical signals that make neuronal communication possible. Dendrites are tree-like structures that extend away from the cell body to receive messages from other neurons at specialized junctions called synapses. Although some neurons do not have any dendrites, some types of neurons have multiple dendrites. Dendrites can have small protrusions called dendritic spines, which further increase surface area for possible synaptic connections.

Once a signal is received by the dendrite, it then travels to the cell body. The cell body contains a specialized structure, the axon hillock that integrates signals from multiple synapses and serves as a junction between the cell body and an axon. An axon is a tube-like structure that propagates the integrated signal to specialized endings called axon terminals. These terminals in turn synapse on other neurons, muscle, or target organs. Chemicals released at axon terminals allow signals to be communicated to these other cells. Neurons usually have one or two axons, but some neurons, like amacrine cells in the retina, do not contain any axons. Some axons are covered with myelin, which acts as an insulator to minimize dissipation of the electrical signal as it travels down the axon, greatly increasing the speed of conduction. This insulation is important as the axon from a human motor neuron can be as long as a meter—from the base of the spine to the toes. The myelin sheath is not actually part of the neuron. Myelin is produced by glial cells. Along the axon there are periodic gaps in the myelin sheath. These gaps are called nodes of Ranvier and are sites where the signal is "recharged" as it travels along the axon.

It is important to note that a single neuron does not act alone—neuronal communication depends on the connections that neurons make with one another (as well as with other cells, like muscle cells). Dendrites from a single neuron may receive synaptic contact from many other neurons. For example, dendrites from a Purkinje cell in the cerebellum are thought to receive contact from as many as 200,000 other neurons.

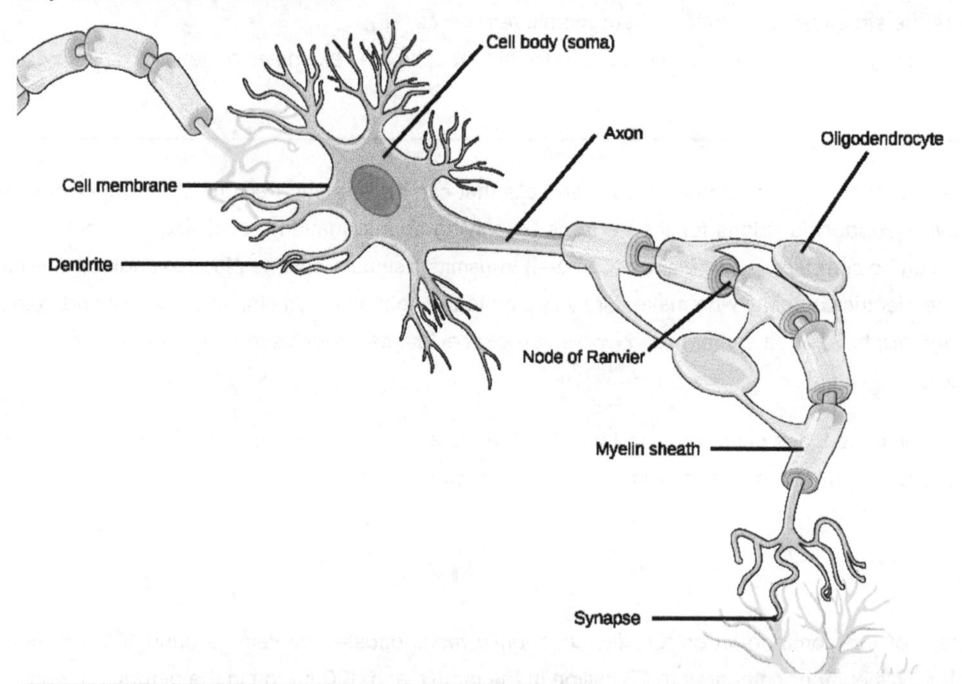

Figure 2. Neurons contain organelles common to many other cells, such as a nucleus and mitochondria. They also have more specialized structures, including dendrites and axons.

Types of Neurons

There are different types of neurons, and the functional role of a given neuron is intimately dependent on its structure. There is an amazing diversity of neuron shapes and sizes found in different parts of the nervous system (and across species).

While there are many defined neuron cell subtypes, neurons are broadly divided into four basic types: unipolar, bipolar, multipolar, and pseudounipolar. Figure 3 illustrates these four basic neuron types. Unipolar neurons have only one structure that extends away from the soma. These neurons are not found in vertebrates but are found in insects where they stimulate muscles or glands. A bipolar neuron has one axon and one dendrite extending from the soma. An example of a bipolar neuron is a retinal bipolar cell, which receives signals from photoreceptor cells that are sensitive to light and transmits these signals to ganglion cells that carry the signal to the brain. Multipolar neurons are the most common type of neuron. Each multipolar neuron contains one axon and multiple dendrites. Multipolar neurons can be found in the central nervous system (brain and spinal cord). An example of a multipolar neuron is a Purkinje cell in the cerebellum, which has many branching dendrites but only one axon. Pseudounipolar cells share characteristics with both unipolar and bipolar cells. A pseudounipolar cell has a single process that extends from the soma, like a unipolar cell, but this process later branches into two distinct structures, like a bipolar cell. Most sensory neurons are pseudounipolar and have an axon that branches into two extensions: one connected to dendrites that receive sensory information and another that transmits this information to the spinal cord.

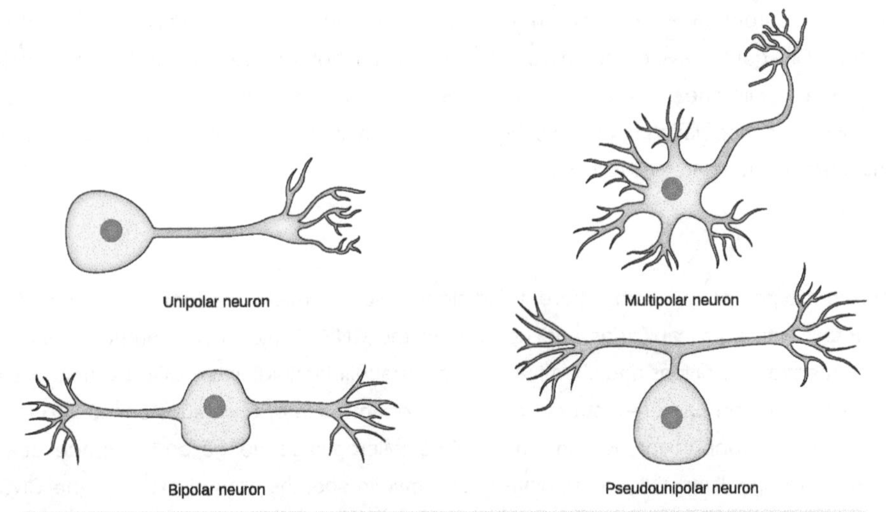

Figure 3. Neurons are broadly divided into four main types based on the number and placement of axons: (1) unipolar, (2) bipolar, (3) multipolar, and (4) pseudounipolar.

EVERYDAY CONNECTION

Neurogenesis

At one time, scientists believed that people were born with all the neurons they would ever have. Research performed during the last few decades indicates that neurogenesis, the birth of new neurons, continues into adulthood. Neurogenesis was first discovered in songbirds that produce new neurons while learning songs. For mammals, new neurons also play an important role in learning: about 1000 new neurons develop in the hippocampus (a brain structure involved in learning and memory) each day. While most of the new neurons will die, researchers found that an increase in the number of surviving new neurons in the hippocampus correlated with how well rats learned a new task. Interestingly, both exercise and some antidepressant medications also promote neurogenesis in the hippocampus. Stress has the opposite effect. While neurogenesis is quite limited compared to regeneration in other tissues, research in this area may lead to new treatments for disorders such as Alzheimer's, stroke, and epilepsy.

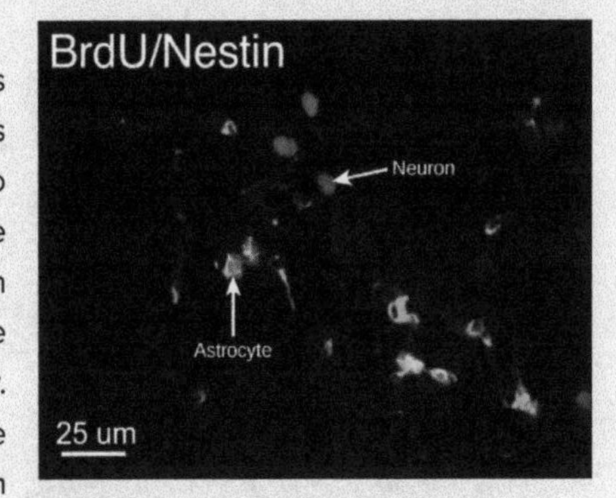

Figure 4. This micrograph shows fluorescently labeled new neurons in a rat hippocampus. Cells that are actively dividing have bromodoxyuridine (BrdU) incorporated into their DNA and are labeled in red. Cells that express glial fibrillary acidic protein (GFAP) are labeled in green. Astrocytes, but not neurons, express GFAP. Thus, cells that are labeled both red and green are actively dividing astrocytes, whereas cells labeled red only are actively dividing neurons. (credit: modification of work by Dr. Maryam Faiz, et. al., University of Barcelona; scale-bar data from Matt Russell)

How do scientists identify new neurons? A researcher can inject a compound called bromodeoxyuridine (BrdU) into the brain of an animal. While all cells will be exposed to BrdU, BrdU will only be incorporated into the DNA of newly generated cells that are in S phase. A technique called immunohistochemistry can be used to attach a fluorescent label to the incorporated BrdU, and a researcher can use fluorescent microscopy to visualize the presence of BrdU, and thus new neurons, in brain tissue. (Figure 4) is a micrograph which shows fluorescently labeled neurons in the hippocampus of a rat.

Glia

While glia are often thought of as the supporting cast of the nervous system, the number of glial cells in the brain actually outnumbers the number of neurons by a factor of ten. Neurons would be unable to function without the vital roles that are fulfilled by these glial cells. Glia guide developing neurons to their destinations, buffer ions and chemicals that would otherwise harm neurons, and provide myelin sheaths around axons. Scientists have recently discovered that they also play a role in responding to nerve activity and modulating communication between nerve cells. When glia do not function properly, the result can be disastrous—most brain tumors are caused by mutations in glia.

Types of Glia

There are several different types of glia with different functions, two of which are shown in Figure 6. Astrocytes, shown in Figure 5 **a** make contact with both capillaries and neurons in the CNS. They provide nutrients and other substances to neurons, regulate the concentrations of ions and chemicals in the extracellular fluid, and provide structural support for synapses. Astrocytes also form the blood-brain barrier—a structure that blocks entrance of toxic substances into the brain. Satellite glia provide nutrients and structural support for neurons in the PNS. Microglia scavenge and degrade dead cells and protect the brain from invading microorganisms. Oligodendrocytes form myelin sheaths around axons in the CNS. One axon can be myelinated by several oligodendrocytes, and one oligodendrocyte can provide myelin for multiple neurons. This is distinctive from the PNS where a single Schwann cell provides myelin for only one axon as the entire Schwann cell surrounds the axon. Ependymal cells line fluid-filled ventricles of the brain and the central canal of the spinal cord. They are involved in the production of cerebrospinal fluid, which serves as a cushion for the brain, moves the fluid between the spinal cord and the brain, and is a component for the choroid plexus.

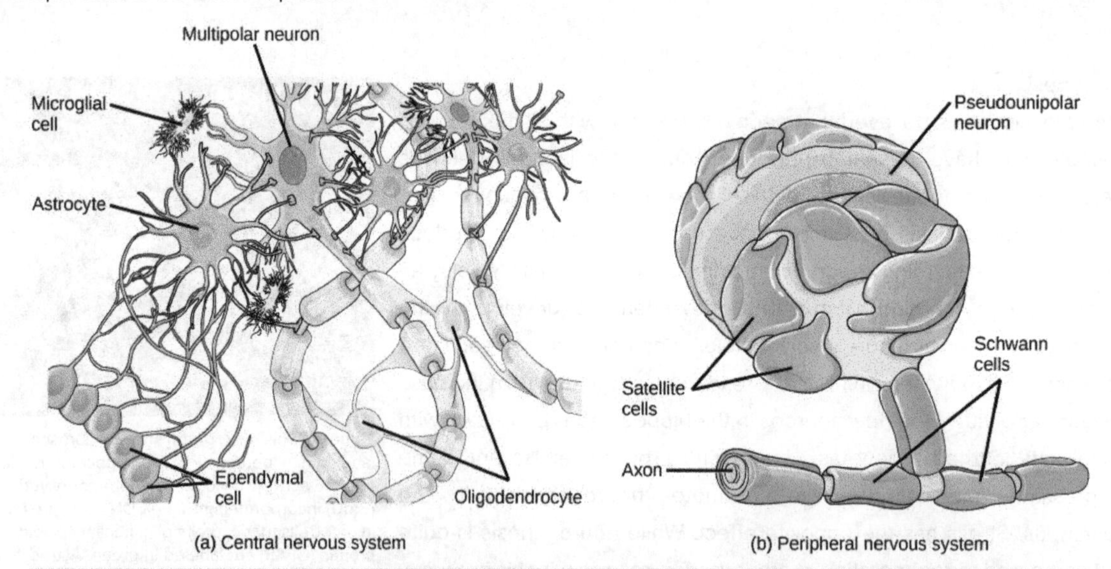

Figure 5. Glial cells support neurons and maintain their environment. Glial cells of the (a) central nervous system include oligodendrocytes, astrocytes, ependymal cells, and microglial cells. Oligodendrocytes form the myelin sheath around axons. Astrocytes provide nutrients to neurons, maintain their extracellular environment, and provide structural support. Microglia scavenge pathogens and dead cells. Ependymal cells produce cerebrospinal fluid that cushions the neurons. Glial cells of the (b) peripheral nervous system include Schwann cells, which form the myelin sheath, and satellite cells, which provide nutrients and structural support to neurons.

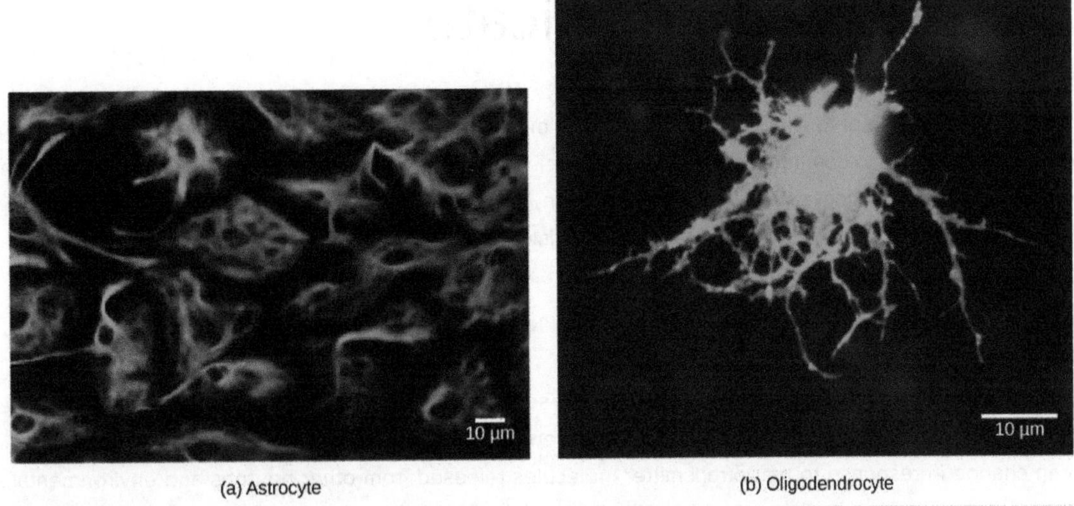

(a) Astrocyte (b) Oligodendrocyte

Figure 6. (a) Astrocytes and (b) oligodendrocytes are glial cells of the central nervous system. (credit a: modification of work by Uniformed Services University; credit b: modification of work by Jurjen Broeke; scale-bar data from Matt Russell)

How Neurons Communicate

All functions performed by the nervous system—from a simple motor reflex to more advanced functions like making a memory or a decision—require neurons to communicate with one another. While humans use words and body language to communicate, neurons use electrical and chemical signals. Just like a person in a committee, one neuron usually receives and synthesizes messages from multiple other neurons before "making the decision" to send the message on to other neurons.

Nerve Impulse Transmission within a Neuron

For the nervous system to function, neurons must be able to send and receive signals. These signals are possible because each neuron has a charged cellular membrane (a voltage difference between the inside and the outside), and the charge of this membrane can change in response to neurotransmitter molecules released from other neurons and environmental stimuli. To understand how neurons communicate, one must first understand the basis of the baseline or 'resting' membrane charge.

Neuronal Charged Membranes

The phospholipid bilayer membrane that surrounds a neuron is impermeable to charged molecules or ions. To enter or exit the neuron, ions must pass through special proteins called ion channels that span the membrane. Ion channels have different configurations: open, closed, and inactive, as illustrated in Figure 1. Some ion channels need to be activated in order to open and allow ions to pass into or out of the cell. These ion channels are sensitive to the environment and can change their shape accordingly. Ion channels that change their structure in response to voltage changes are called voltage-gated ion channels. Voltage-gated ion channels regulate the relative concentrations of different ions inside and outside the cell. The difference in total charge between the inside and outside of the cell is called the membrane potential.

Voltage-gated Na^+ Channels

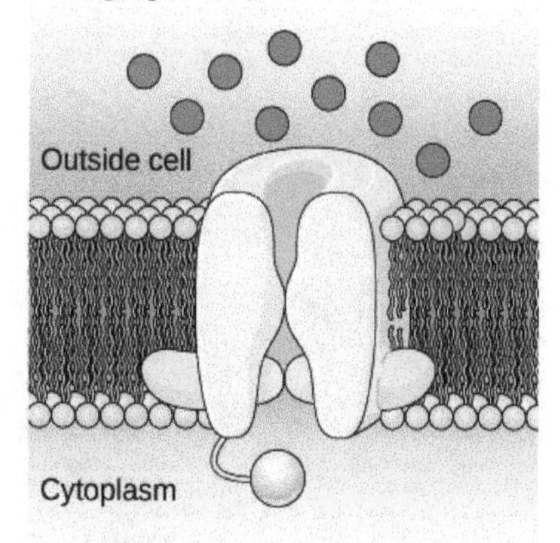

Closed At the resting potential, the channel is closed.

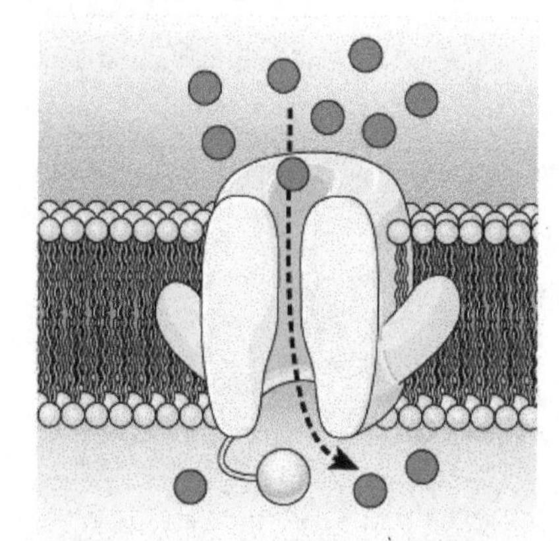

Open In response to a nerve impulse, the gate opens and Na^+ enters the cell.

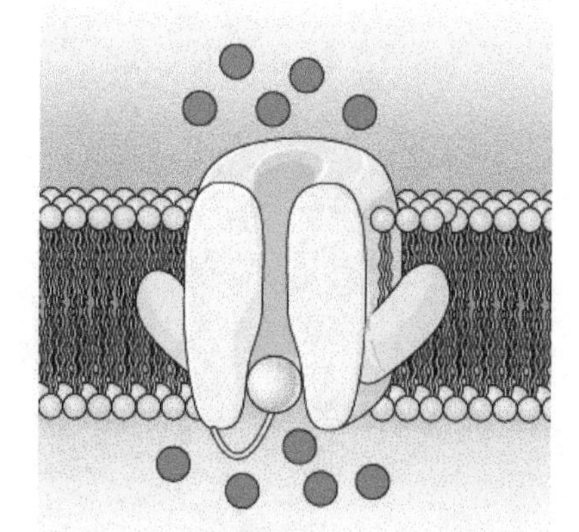

Inactivated For a brief period following activation, the channel does not open in response to a new signal.

Figure 1. Voltage-gated ion channels open in response to changes in membrane voltage. After activation, they become inactivated for a brief period and will no longer open in response to a signal.

Resting Membrane Potential

A neuron at rest is negatively charged: the inside of a cell is approximately 70 millivolts more negative than the outside (−70 mV, note that this number varies by neuron type and by species). This voltage is called the resting membrane potential; it is caused by differences in the concentrations of ions inside and outside the cell. If the membrane were equally permeable to all ions, each type of ion would diffuse across the membrane and the system would reach equilibrium. Because ions cannot simply cross the membrane at will, there are different concentrations of several ions inside and outside the cell, as shown in Figure 2. The difference in the number of positively charged potassium ions (K^+) inside and outside the cell dominates the resting membrane potential. When the membrane is at rest, K^+ ions accumulate inside the cell due to a net movement with the concentration gradient. The negative resting membrane potential is created and maintained by increasing the concentration of cations outside the cell (in the extracellular fluid) relative to inside the cell (in the cytoplasm). The negative charge within the cell is created by the cell membrane being more permeable to potassium ion movement than sodium ion movement. In neurons, potassium ions are maintained at high concentrations within the cell while sodium ions are maintained at high concentrations outside of the cell. The cell possesses potassium and sodium leakage channels that allow the two cations to diffuse down their concentration gradient. However, the neurons have far more potassium leakage channels than sodium leakage channels. Therefore, potassium diffuses out of the cell at a much faster rate than sodium leaks in. Because more cations are leaving the cell than are entering, this causes the interior of the cell to be negatively charged relative to the outside of the cell. The actions of the sodium potassium pump help to maintain the resting potential, once established. Sodium potassium pumps brings two K^+ ions into the cell while removing three Na^+ ions per ATP consumed. As more cations are expelled from the cell than taken in, the inside of the cell remains negatively charged relative to the extracellular fluid. It should be noted that chloride ions (Cl^-) tend to accumulate outside of the cell because they are repelled by negatively-charged proteins within the cytoplasm.

The resting membrane potential is a result of different concentrations inside and outside the cell.

Ion Concentration Inside and Outside Neurons			
Ion	**Extracellular concentration (mM)**	**Intracellular concentration (mM)**	**Ratio outside/inside**
Na^+	145	12	12
K+	4	155	0.026
Cl^-	120	4	30
Organic anions (A−)	—	100	

(a) Resting potential

At the resting potential, all voltage-gated Na^+ channels and most voltage-gated K^+ channels are closed. The Na^+/K^+ transporter pumps K^+ ions into the cell and Na^+ ions out.

(b) Depolarization

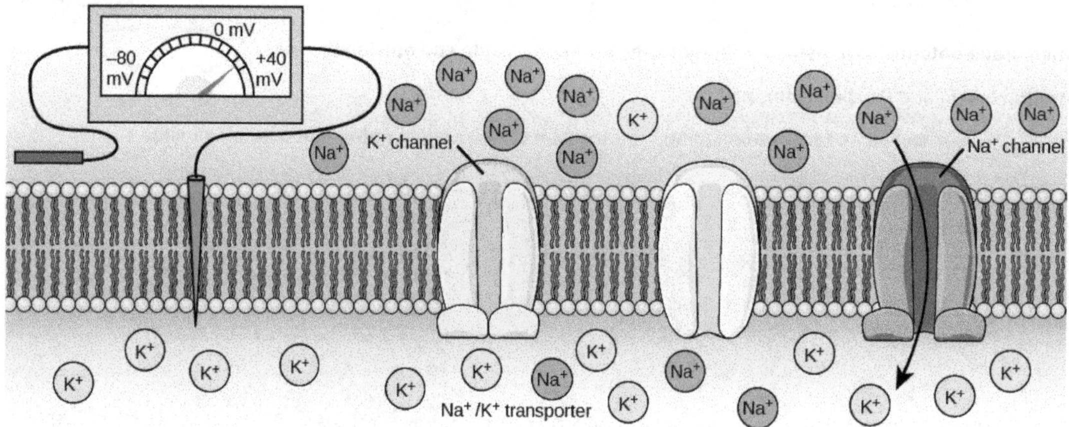

In response to a depolarization, some Na^+ channels open, allowing Na^+ ions to enter the cell. The membrane starts to depolarize (the charge across the membrane lessens). If the threshold of excitation is reached, all the Na^+ channels open.

(c) Hyperpolarization

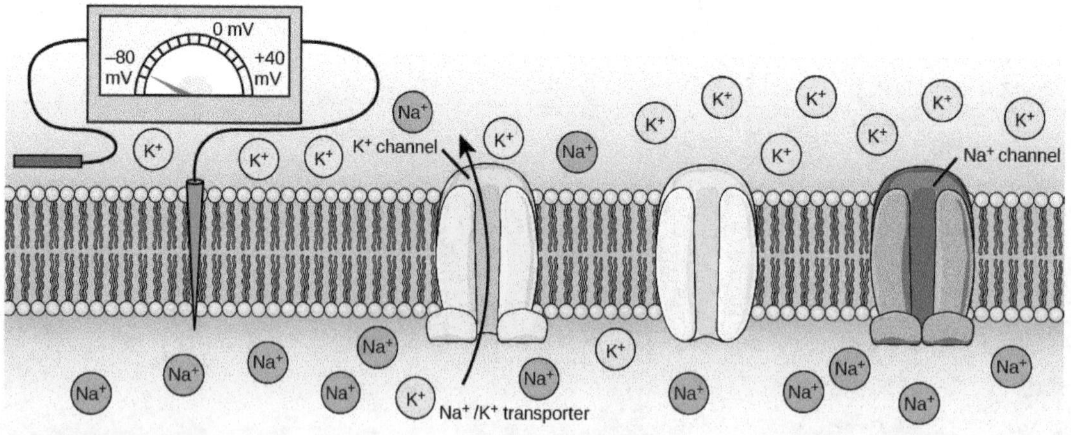

At the peak action potential, Na^+ channels close while K^+ channels open. K^+ leaves the cell, and the membrane eventually becomes hyperpolarized.

Figure 2. The (a) resting membrane potential is a result of different concentrations of Na+ and K+ ions inside and outside the cell. A nerve impulse causes Na+ to enter the cell, resulting in (b) depolarization. At the peak action potential, K+ channels open and the cell becomes (c) hyperpolarized.

Action Potential

A neuron can receive input from other neurons and, if this input is strong enough, send the signal to downstream neurons. Transmission of a signal between neurons is generally carried by a chemical called a neurotransmitter. Transmission of a signal within a neuron (from dendrite to axon terminal) is carried by a brief reversal of the resting membrane potential called an action potential. When neurotransmitter molecules bind to receptors located on a neuron's dendrites, ion channels open. At excitatory synapses, this opening allows positive ions to enter the neuron and results in depolarization of the membrane—a decrease in the difference in voltage between the inside and outside of the neuron. A stimulus from a sensory cell or another neuron depolarizes the target neuron to its threshold potential (-55 mV). Na^+ channels in the axon hillock open, allowing positive ions to enter the cell (Figure 2). Once the sodium channels open, the neuron completely depolarizes to a membrane potential of about +40 mV. Action potentials are considered an "all-or nothing" event, in that, once the threshold potential is reached, the neuron always completely depolarizes. Once depolarization is complete, the cell must now "reset" its membrane voltage back to the resting potential. To accomplish this, the Na^+ channels close and cannot be opened. This begins the neuron's refractory period, in which it cannot produce another action potential because its sodium channels will not open. At the same time, voltage-gated K^+ channels open, allowing K^+ to leave the cell. As K^+ ions leave the cell, the membrane potential once again becomes negative. The diffusion of K^+ out of the cell actually hyperpolarizes the cell, in that the membrane potential becomes more negative than the cell's normal resting potential. At this point, the sodium channels will return to their resting state, meaning they are ready to open again if the membrane potential again exceeds the threshold potential. Eventually the extra K^+ ions diffuse out of the cell through the potassium leakage channels, bringing the cell from its hyperpolarized state, back to its resting membrane potential.

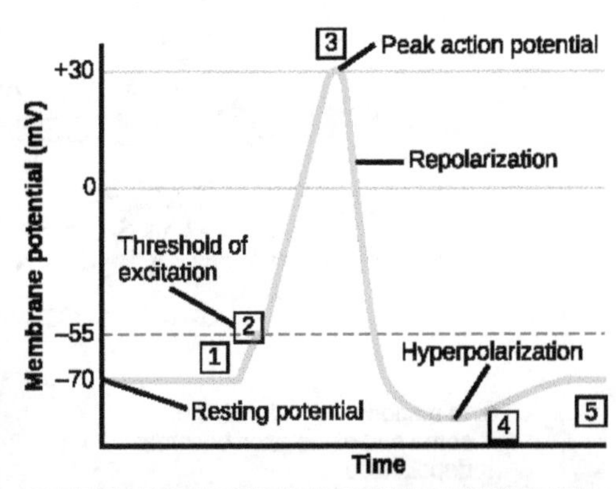

Figure 3. The formation of an action potential can be divided into five steps: (1) A stimulus from a sensory cell or another neuron causes the target cell to depolarize toward the threshold potential. (2) If the threshold of excitation is reached, all Na+ channels open and the membrane depolarizes. (3) At the peak action potential, K+ channels open and K+ begins to leave the cell. At the same time, Na+ channels close. (4) The membrane becomes hyperpolarized as K+ ions continue to leave the cell. The hyperpolarized membrane is in a refractory period and cannot fire. (5) The K+ channels close and the Na+/K+ transporter restores the resting potential.

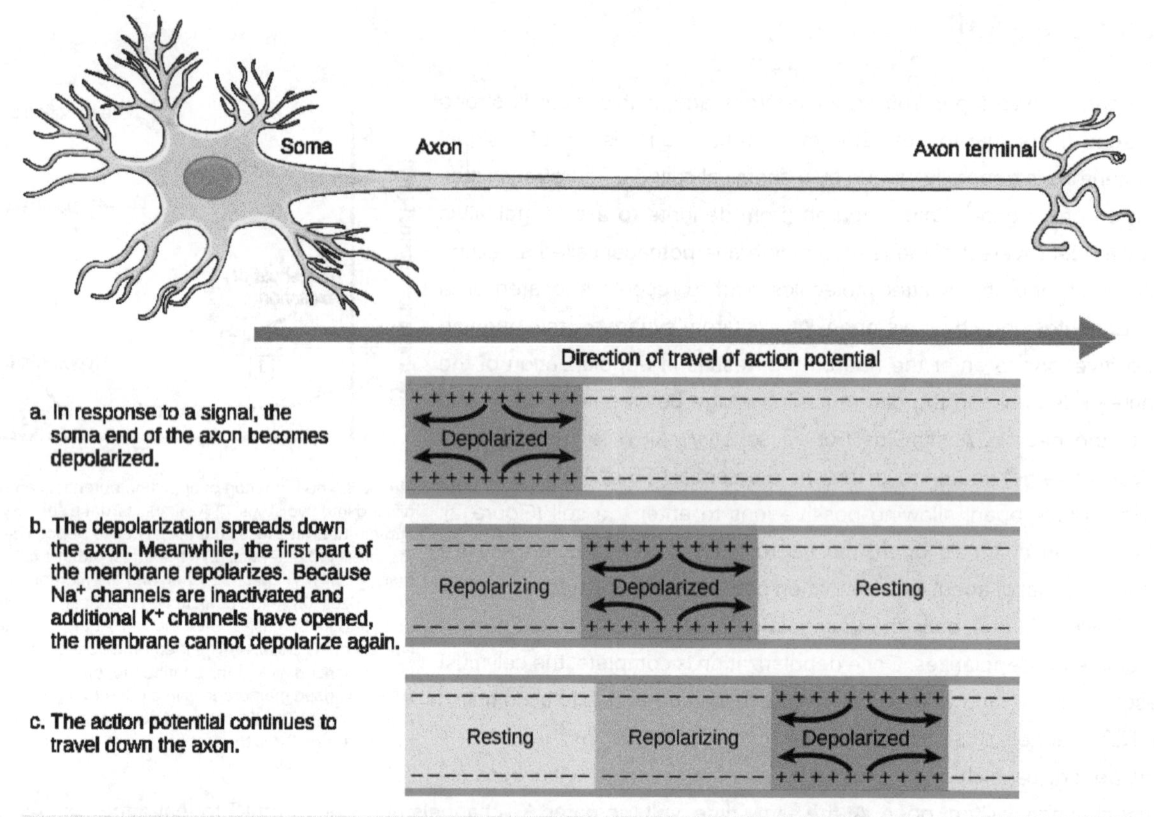

Figure 4. The action potential is conducted down the axon as the axon membrane depolarizes, then repolarizes.

Myelin and the Propagation of the Action Potential

For an action potential to communicate information to another neuron, it must travel along the axon and reach the axon terminals where it can initiate neurotransmitter release. The speed of conduction of an action potential along an axon is influenced by both the diameter of the axon and the axon's resistance to current leak. Myelin acts as an insulator that prevents current from leaving the axon; this increases the speed of action potential conduction. In demyelinating diseases like multiple sclerosis, action potential conduction slows because current leaks from previously insulated axon areas. The nodes of Ranvier, illustrated in Figure 5 are gaps in the myelin sheath along the axon. These unmyelinated spaces are about one micrometer long and contain voltage-gated Na^+ and K^+ channels. Flow of ions through these channels, particularly the Na^+ channels, regenerates the action potential over and over again along the axon. This 'jumping' of the action potential from one node to the next is called saltatory conduction. If nodes of Ranvier were not present along an axon, the action potential would propagate very slowly since Na^+ and K^+ channels would have to continuously regenerate action potentials at every point along the axon instead of at specific points. Nodes of Ranvier also save energy for the neuron since the channels only need to be present at the nodes and not along the entire axon.

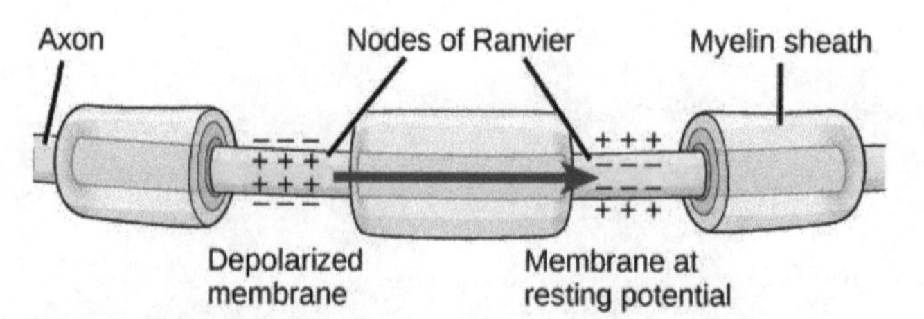

Figure 5. Nodes of Ranvier are gaps in myelin coverage along axons. Nodes contain voltage-gated K+ and Na+ channels. Action potentials travel down the axon by jumping from one node to the next.

Synaptic Transmission

The synapse or "gap" is the place where information is transmitted from one neuron to another. Synapses usually form between axon terminals and dendritic spines, but this is not universally true. There are also axon-to-axon, dendrite-to-dendrite, and axon-to-cell body synapses. The neuron transmitting the signal is called the presynaptic neuron, and the neuron receiving the signal is called the postsynaptic neuron. Note that these designations are relative to a particular synapse—most neurons are both presynaptic and postsynaptic. There are two types of synapses: chemical and electrical.

Chemical Synapse

When an action potential reaches the axon terminal it depolarizes the membrane and opens voltage-gated Na^+ channels. Na^+ ions enter the cell, further depolarizing the presynaptic membrane. This depolarization causes voltage-gated Ca^{2+} channels to open. Calcium ions entering the cell initiate a signaling cascade that causes small membrane-bound vesicles, called synaptic vesicles, containing neurotransmitter molecules to fuse with the presynaptic membrane. Synaptic vesicles are shown in Figure 6, which is an image from a scanning electron microscope.

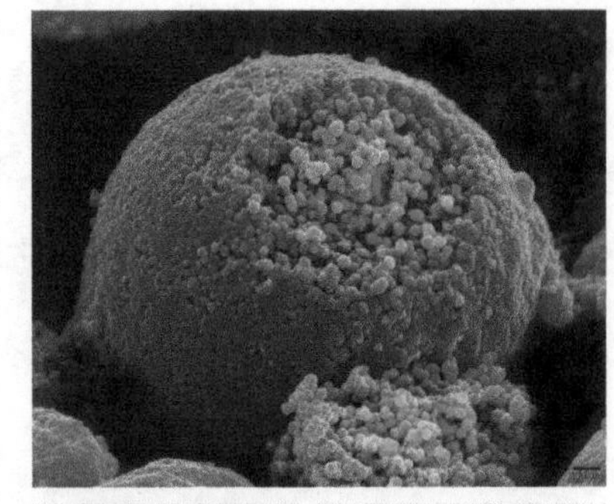

Figure 6. This pseudocolored image taken with a scanning electron microscope shows an axon terminal that was broken open to reveal synaptic vesicles (blue and orange) inside the neuron. (credit: modification of work by Tina Carvalho, NIH-NIGMS; scale-bar data from Matt Russell)

Fusion of a vesicle with the presynaptic membrane causes neurotransmitter to be released into the synaptic cleft, the extracellular space between the presynaptic and postsynaptic membranes, as illustrated in Figure 7. The neurotransmitter diffuses across the synaptic cleft and binds to receptor proteins on the postsynaptic membrane.

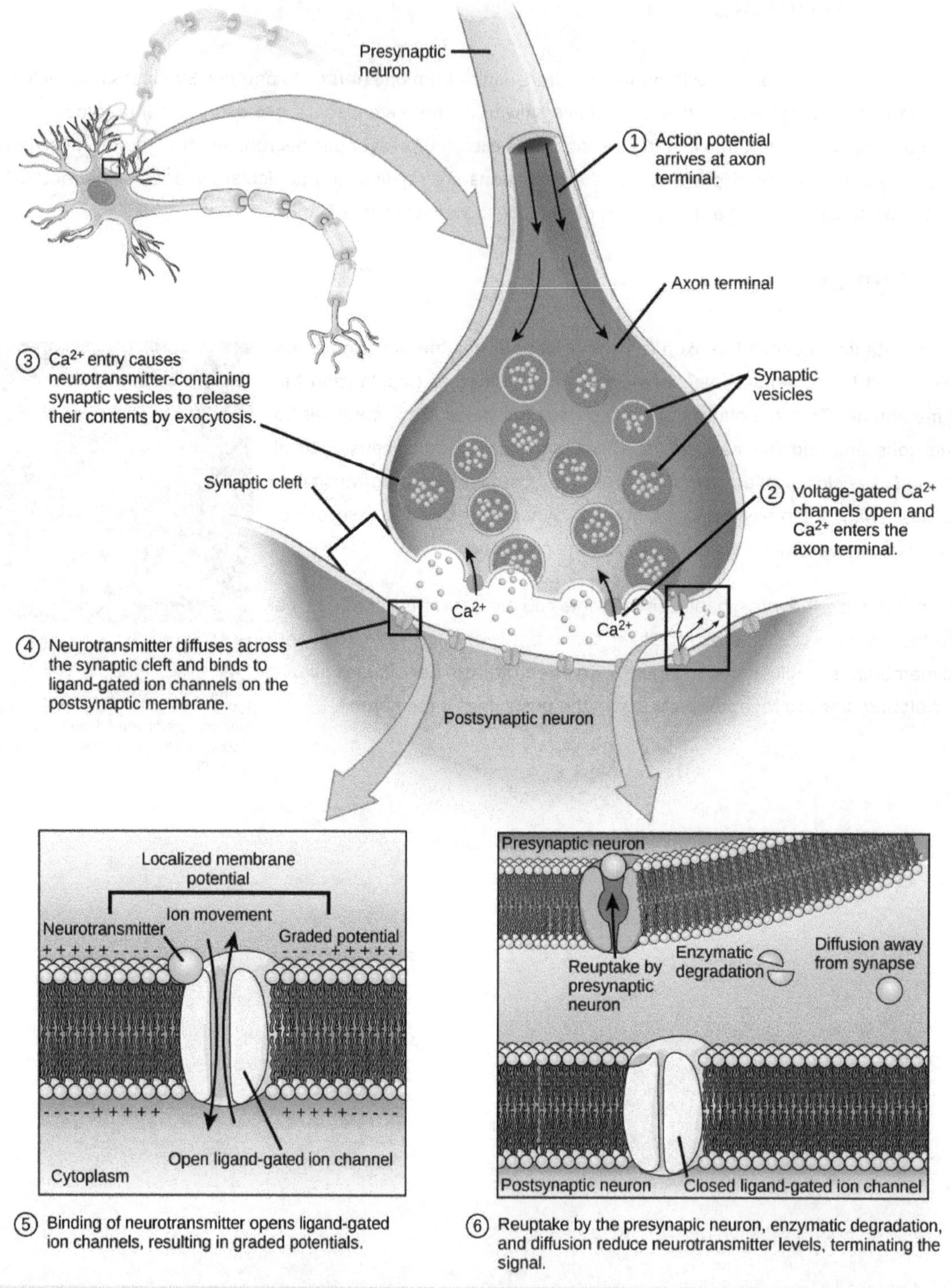

Figure 7. Communication at chemical synapses requires release of neurotransmitters. When the presynaptic membrane is depolarized, voltage-gated Ca2+ channels open and allow Ca2+ to enter the cell. The calcium entry causes synaptic vesicles to fuse with the membrane and release neurotransmitter molecules into the synaptic cleft. The neurotransmitter diffuses across the synaptic cleft and binds to ligand-gated ion channels in the postsynaptic membrane, resulting in a localized depolarization or hyperpolarization of the postsynaptic neuron.

The binding of a specific neurotransmitter causes particular ion channels, in this case ligand-gated channels, on the postsynaptic membrane to open. Neurotransmitters can either have excitatory or inhibitory effects on the postsynaptic membrane, as detailed in. For example, when acetylcholine is released at the synapse between a nerve and muscle (called the neuromuscular junction) by a presynaptic neuron, it causes postsynaptic Na^+ channels to open. Na^+ enters the postsynaptic cell and causes the postsynaptic membrane to depolarize. This depolarization is called an excitatory postsynaptic potential (EPSP) and makes the postsynaptic neuron more likely to fire an action potential. Release of neurotransmitter at inhibitory synapses causes inhibitory postsynaptic potentials (IPSPs), a hyperpolarization of the presynaptic membrane. For example, when the neurotransmitter GABA (gamma-aminobutyric acid) is released from a presynaptic neuron, it binds to and opens Cl^- channels. Cl^- ions enter the cell and hyperpolarizes the membrane, making the neuron less likely to fire an action potential.

Once neurotransmission has occurred, the neurotransmitter must be removed from the synaptic cleft so the postsynaptic membrane can "reset" and be ready to receive another signal. This can be accomplished in three ways: the neurotransmitter can diffuse away from the synaptic cleft, it can be degraded by enzymes in the synaptic cleft, or it can be recycled (sometimes called reuptake) by the presynaptic neuron. Several drugs act at this step of neurotransmission. For example, some drugs that are given to Alzheimer's patients work by inhibiting acetylcholinesterase, the enzyme that degrades acetylcholine. This inhibition of the enzyme essentially increases neurotransmission at synapses that release acetylcholine. Once released, the acetylcholine stays in the cleft and can continually bind and unbind to postsynaptic receptors.

Neurotransmitter Function and Location

Neurotransmitter	Example	Location
Acetylcholine	—	CNS and/or PNS
Biogenic amine	Dopamine, serotonin, norepinephrine	CNS and/or PNS
Amino acid	Glycine, glutamate, aspartate, gamma aminobutyric acid	CNS
Neuropeptide	Substance P, endorphins	CNS and/or PNS

Electrical Synapse

While electrical synapses are fewer in number than chemical synapses, they are found in all nervous systems and play important and unique roles. The mode of neurotransmission in electrical synapses is quite different from that in chemical synapses. In an electrical synapse, the presynaptic and postsynaptic membranes are very close together and are actually physically connected by channel proteins forming gap junctions. Gap junctions allow current to pass directly from one cell to the next. In addition to the ions that carry this current, other molecules, such as ATP, can diffuse through the large gap junction pores.

There are key differences between chemical and electrical synapses. Because chemical synapses depend on the release of neurotransmitter molecules from synaptic vesicles to pass on their signal, there is an approximately one millisecond delay between when the axon potential reaches the presynaptic terminal and when the neurotransmitter leads to opening of postsynaptic ion channels. Additionally, this signaling is unidirectional. Signaling in electrical synapses, in contrast, is virtually instantaneous (which is important for synapses involved in key reflexes), and some electrical synapses are bidirectional. Electrical synapses are also more reliable as they are less likely to be blocked, and they are important for synchronizing the electrical activity of a group of neurons. For example, electrical synapses in the thalamus are thought to regulate slow-wave sleep, and disruption of these synapses can cause seizures.

Signal Summation

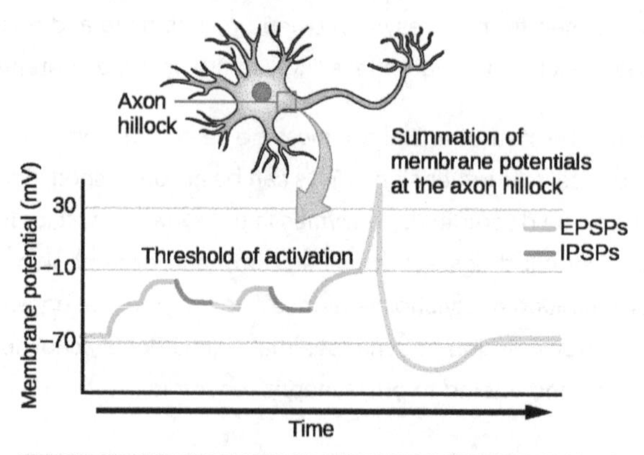

Figure 8. A single neuron can receive both excitatory and inhibitory inputs from multiple neurons, resulting in local membrane depolarization (EPSP input) and hyperpolarization (IPSP input). All these inputs are added together at the axon hillock. If the EPSPs are strong enough to overcome the IPSPs and reach the threshold of excitation, the neuron will fire.

Sometimes a single EPSP is strong enough to induce an action potential in the postsynaptic neuron, but often multiple presynaptic inputs must create EPSPs around the same time for the postsynaptic neuron to be sufficiently depolarized to fire an action potential. This process is called summation and occurs at the axon hillock, as illustrated in Figure 8. Additionally, one neuron often has inputs from many presynaptic neurons—some excitatory and some inhibitory—so IPSPs can cancel out EPSPs and vice versa. It is the net change in postsynaptic membrane voltage that determines whether the postsynaptic cell has reached its threshold of excitation needed to fire an action potential. Together, synaptic summation and the threshold for excitation act as a filter so that random "noise" in the system is not transmitted as important information.

EVERYDAY CONNECTION

Brain-computer interface

Amyotrophic lateral sclerosis (ALS, also called Lou Gehrig's Disease) is a neurological disease characterized by the degeneration of the motor neurons that control voluntary movements. The disease begins with muscle weakening and lack of coordination and eventually destroys the neurons that control speech, breathing, and swallowing; in the end, the disease can lead to paralysis. At that point, patients require assistance from machines to be able to breathe and to communicate. Several special technologies have been developed to allow "locked-in" patients to communicate with the rest of the world. One technology, for example, allows patients to type out sentences by twitching their cheek. These sentences can then be read aloud by a computer.

A relatively new line of research for helping paralyzed patients, including those with ALS, to communicate and retain a degree of self-sufficiency is called brain-computer interface (BCI) technology and is illustrated in Figure 9. This technology sounds like something out of science fiction: it allows paralyzed patients to control a computer using only their thoughts. There are several forms of BCI. Some forms use EEG recordings from electrodes taped onto the skull. These recordings contain information from large populations of neurons that can be decoded by a computer. Other forms of BCI require the implantation of an array of electrodes smaller than a postage stamp in the arm and hand area of the motor cortex. This form of BCI, while more invasive, is very powerful as each electrode can record actual action potentials from one or more neurons. These signals are then sent to a computer, which has been trained to decode the signal and feed it to a tool—such as a cursor on a computer screen. This means that a patient with ALS can use e-mail, read the Internet, and communicate with others by thinking of moving his or her hand or arm (even though the paralyzed patient cannot make that bodily movement). Recent advances have allowed a paralyzed locked-in patient who suffered a stroke 15 years ago to control a robotic arm and even to feed herself coffee using BCI technology.

Despite the amazing advancements in BCI technology, it also has limitations. The technology can require many hours of training and long periods of intense concentration for the patient; it can also require brain surgery to implant the devices.

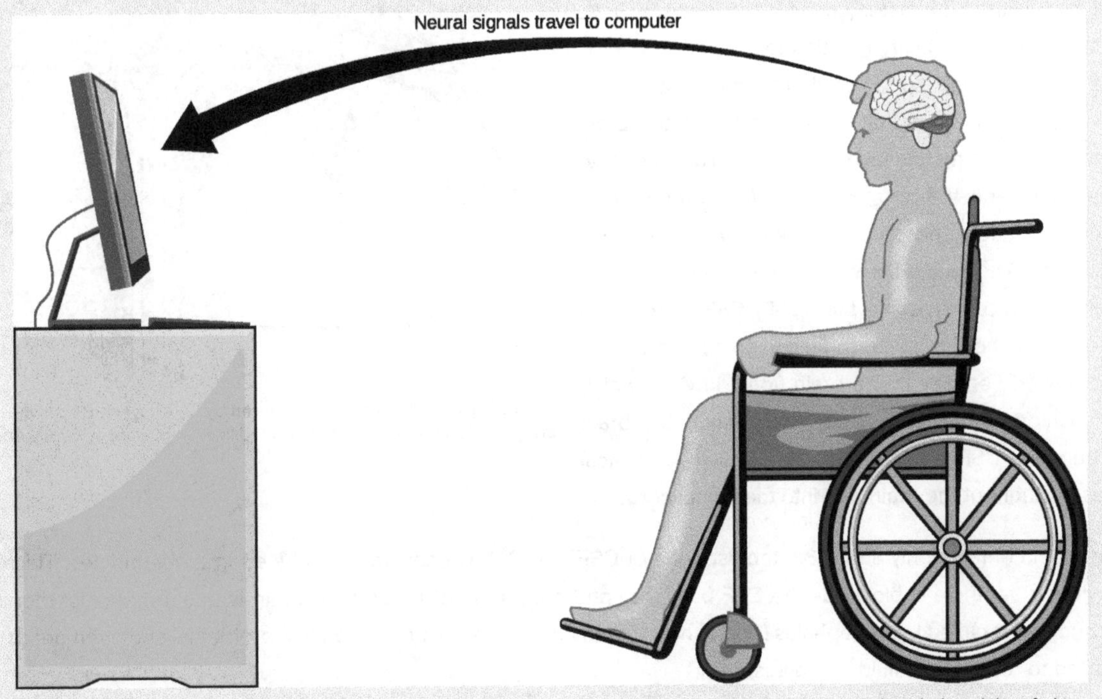

Figure 9. With brain-computer interface technology, neural signals from a paralyzed patient are collected, decoded, and then fed to a tool, such as a computer, a wheelchair, or a robotic arm.

The Central Nervous System

The central nervous system (CNS) is made up of the brain, a part of which is shown in Figure 1 and spinal cord and is covered with three layers of protective coverings called meninges (from the Greek word for membrane). The outermost layer is the dura mater (Latin for "hard mother"). As the Latin suggests, the primary function for this thick layer is to protect the brain and spinal cord. The dura mater also contains vein-like structures that carry blood from the brain back to the heart. The middle layer is the web-like arachnoid mater. The last layer is the pia mater (Latin for "soft mother"), which directly contacts and covers the brain and spinal cord like plastic wrap. The space between the arachnoid and pia maters is filled with cerebrospinal fluid (CSF). CSF is produced by a tissue called choroid plexus in fluid-filled compartments in the CNS called ventricles. The brain floats in CSF, which acts as a cushion and shock absorber and makes the brain neutrally buoyant. CSF also functions to circulate chemical substances throughout the brain and into the spinal cord.

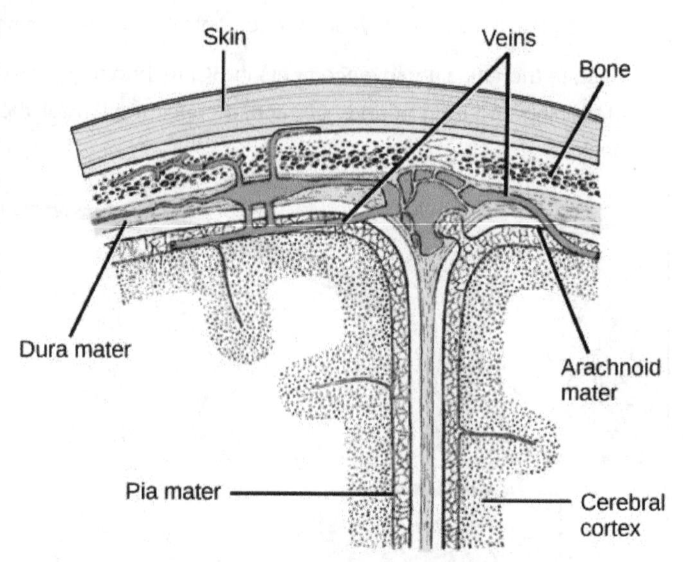

Figure 1. The cerebral cortex is covered by three layers of meninges: the dura, arachnoid, and pia maters. (credit: modification of work by Gray's Anatomy)

The entire brain contains only about 8.5 tablespoons of CSF, but CSF is constantly produced in the ventricles. This creates a problem when a ventricle is blocked—the CSF builds up and creates swelling and the brain is pushed against the skull. This swelling condition is called hydrocephalus ("water head") and can cause seizures, cognitive problems, and even death if a shunt is not inserted to remove the fluid and pressure.

Brain

The brain is the part of the central nervous system that is contained in the cranial cavity of the skull. It includes the cerebral cortex, limbic system, basal ganglia, thalamus, hypothalamus, and cerebellum. There are three different ways that a brain can be sectioned in order to view internal structures: a sagittal section cuts the brain left to right, as shown in Figure **2b**, a coronal section cuts the brain front to back, as shown in Figure **2a**, and a horizontal section cuts the brain top to bottom.

Cerebral Cortex

The outermost part of the brain is a thick piece of nervous system tissue called the cerebral cortex, which is folded into hills called gyri (singular: gyrus) and valleys called sulci (singular: sulcus). The cortex is made up of two hemispheres—right and left—which are separated by a large sulcus. A thick fiber bundle called the corpus callosum (Latin: "tough body") connects the two hemispheres and allows information to be passed from one side to the other. Although there are some brain functions that are localized more to one hemisphere than the other, the functions of the two hemispheres are largely redundant. In fact, sometimes (very rarely) an entire hemisphere is removed to treat severe epilepsy. While patients do suffer some deficits following the surgery, they can have surprisingly few problems, especially when the surgery is performed on children who have very immature nervous systems.

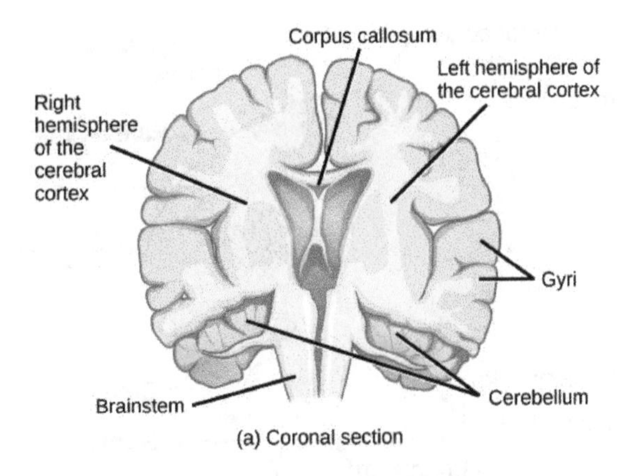

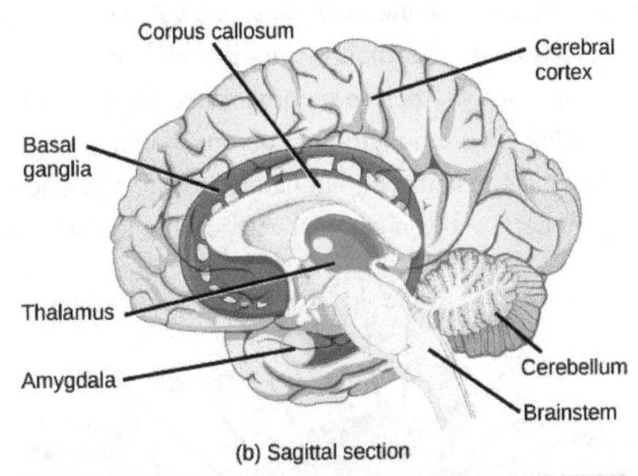

Figure 2. These illustrations show the (a) coronal and (b) sagittal sections of the human brain.

In other surgeries to treat severe epilepsy, the corpus callosum is cut instead of removing an entire hemisphere. This causes a condition called split-brain, which gives insights into unique functions of the two hemispheres. For example, when an object is presented to patients' left visual field, they may be unable to verbally name the object (and may claim to not have seen an object at all). This is because the visual input from the left visual field crosses and enters the right hemisphere and cannot then signal to the speech center, which generally is found in the left side of the brain. Remarkably, if a split-brain patient is asked to pick up a specific object out of a group of objects with the left hand, the patient will be able to do so but will still be unable to vocally identify it.

Each cortical hemisphere contains regions called lobes that are involved in different functions. Scientists use various techniques to determine what brain areas are involved in different functions: they examine patients who have had injuries or diseases that affect specific areas and see how those areas are related to functional deficits. They also conduct animal studies where they stimulate brain areas and see if there are any behavioral changes. They use a technique called transmagnetic stimulation (TMS) to temporarily deactivate specific parts of the cortex using strong magnets placed outside the head; and they use functional magnetic resonance imaging (fMRI) to look at changes in oxygenated blood flow in particular brain regions that correlate with specific behavioral tasks. These techniques, and others, have given great insight into the functions of different brain regions but have also showed that any given brain area can be involved in more than one behavior or process, and any given behavior or process generally involves neurons in multiple brain areas. That being said, each hemisphere of the mammalian cerebral cortex can be broken down into four functionally and spatially defined lobes: frontal, parietal, temporal, and occipital. Figure 3 illustrates these four lobes of the human cerebral cortex.

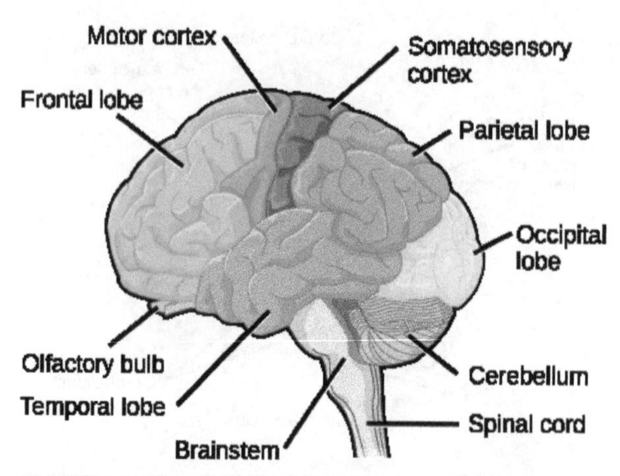

Figure 3. The human cerebral cortex includes the frontal, parietal, temporal, and occipital lobes.

The frontal lobe is located at the front of the brain, over the eyes. This lobe contains the olfactory bulb, which processes smells. The frontal lobe also contains the motor cortex, which is important for planning and implementing movement. Areas within the motor cortex map to different muscle groups, and there is some organization to this map, as shown in Figure 4. For example, the neurons that control movement of the fingers are next to the neurons that control movement of the hand. Neurons in the frontal lobe also control cognitive functions like maintaining attention, speech, and decision-making. Studies of humans who have damaged their frontal lobes show that parts of this area are involved in personality, socialization, and assessing risk.

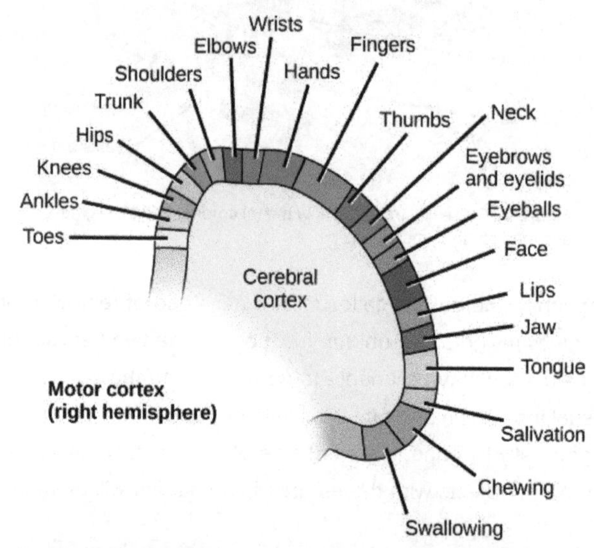

Figure 4. Different parts of the motor cortex control different muscle groups. Muscle groups that are neighbors in the body are generally controlled by neighboring regions of the motor cortex as well. For example, the neurons that control finger movement are near the neurons that control hand movement.

The parietal lobe is located at the top of the brain. Two of the parietal lobe's main functions are processing somatosensation—touch sensations like pressure, pain, heat, cold—and processing proprioception—the sense of how parts of the body are oriented in space. The parietal lobe contains a somatosensory map of the body similar to the motor cortex.

The occipital lobe is located at the back of the brain. It is primarily involved in vision—seeing, recognizing, and identifying the visual world.

The temporal lobe is located at the base of the brain by your ears and is primarily involved in processing and interpreting sounds. It also contains the hippocampus (Greek for "seahorse")—a structure that processes memory formation. The role of the hippocampus in memory was partially determined by studying one famous epileptic patient, HM, who had both sides of his hippocampus removed in an attempt to cure his epilepsy. His seizures went away, but he could no longer form new memories (although he could remember some facts from before his surgery and could learn new motor tasks).

Basal Ganglia

Interconnected brain areas called the basal ganglia (or basal nuclei) play important roles in movement control and posture. Damage to the basal ganglia, as in Parkinson's disease, leads to motor impairments like a shuffling gait when walking. The basal ganglia also regulate motivation. For example, when a wasp sting led to bilateral basal ganglia damage in a 25-year-old businessman, he began to spend all his days in bed and showed no interest in anything or anybody. But when he was externally stimulated—as when someone asked to play a card game with him—he was able to function normally. Interestingly, he and other similar patients do not report feeling bored or frustrated by their state.

Thalamus

The thalamus (Greek for "inner chamber"), illustrated in Figure 5, acts as a gateway to and from the cortex. It receives sensory and motor inputs from the body and also receives feedback from the cortex. This feedback mechanism can modulate conscious awareness of sensory and motor inputs depending on the attention and arousal state of the animal. The thalamus helps regulate consciousness, arousal, and sleep states. A rare genetic disorder called fatal familial insomnia causes the degeneration of thalamic neurons and glia. This disorder prevents affected patients from being able to sleep, among other symptoms, and is eventually fatal.

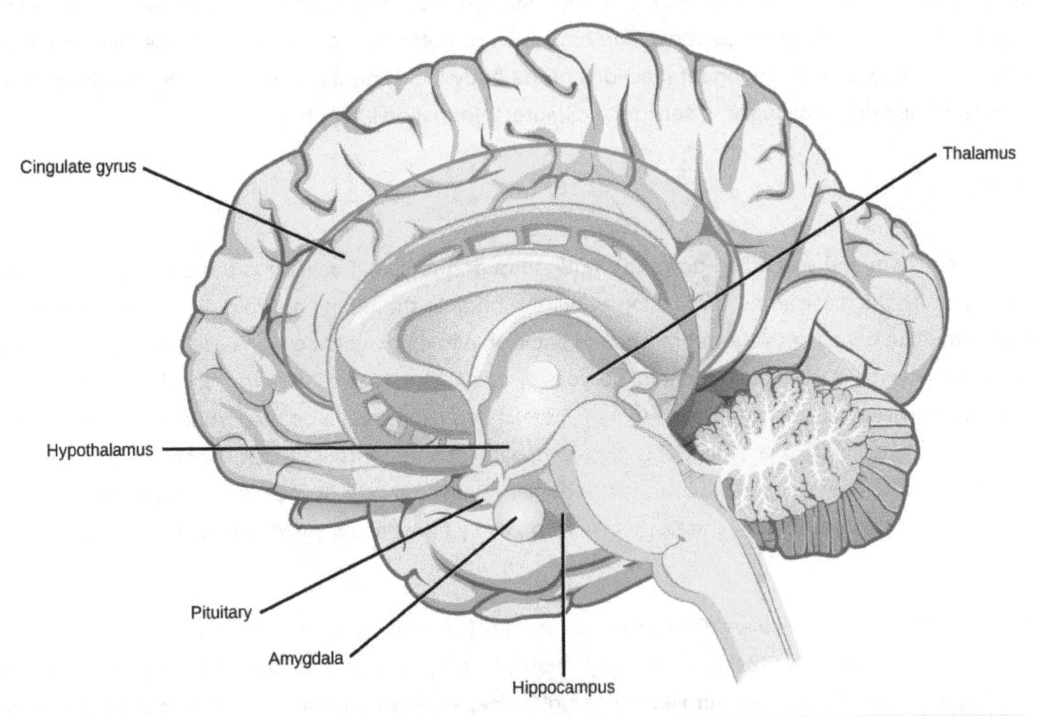

Figure 5. The limbic system regulates emotion and other behaviors. It includes parts of the cerebral cortex located near the center of the brain, including the cingulate gyrus and the hippocampus as well as the thalamus, hypothalamus, and amygdala.

Hypothalamus

Below the thalamus is the hypothalamus (Figure 5). The hypothalamus controls the endocrine system by sending signals to the pituitary gland, a pea-sized endocrine gland that releases several different hormones that affect other glands as well as other cells. This relationship means that the hypothalamus regulates important behaviors that are controlled by these hormones. The hypothalamus is the body's thermostat—it makes sure key functions like food and water intake, energy expenditure, and body temperature are kept at appropriate levels. Neurons within the hypothalamus also regulate circadian rhythms, sometimes called sleep cycles.

Limbic System

The limbic system is a connected set of structures that regulates emotion, as well as behaviors related to fear and motivation. It plays a role in memory formation and includes parts of the thalamus and hypothalamus as well as the hippocampus. One important structure within the limbic system is a temporal lobe structure called the amygdala (Greek for "almond"). The two amygdala are important both for the sensation of fear and for recognizing fearful faces. The cingulate gyrus helps regulate emotions and pain.

Cerebellum

The cerebellum (Latin for "little brain"), shown in Figure 3, sits at the base of the brain on top of the brainstem. The cerebellum controls balance and aids in coordinating movement and learning new motor tasks.

Brainstem

The brainstem, illustrated in Figure 3, connects the rest of the brain with the spinal cord. It consists of the midbrain, medulla oblongata, and the pons. Motor and sensory neurons extend through the brainstem allowing for the relay of signals between the brain and spinal cord. Ascending neural pathways cross in this section of the brain allowing the left hemisphere of the cerebrum to control the right side of the body and vice versa. The brainstem coordinates motor control signals sent from the brain to the body. The brainstem controls several important functions of the body including alertness, arousal, breathing, blood pressure, digestion, heart rate, swallowing, walking, and sensory and motor information integration.

Spinal Cord

Connecting to the brainstem and extending down the body through the spinal column is the spinal cord. The spinal cord is a thick bundle of nerve tissue that carries information about the body to the brain and from the brain to the body. The spinal cord is contained within the bones of the vertebral column but is able to communicate signals to and from the body through its connections with spinal nerves (part of the peripheral nervous system). A cross-section of the spinal cord looks like a white oval containing a gray butterfly-shape, as illustrated in Figure 6. Myelinated axons make up the "white matter" and neuron and glial cell bodies make up the "gray matter." Gray matter is also composed of interneurons, which connect two neurons each located in different parts of the body. Axons and cell bodies in the dorsal spinal cord convey mostly sensory information from the body to the brain. Axons and cell bodies in the ventral spinal cord primarily transmit signals controlling movement from the brain to the body.

The spinal cord also controls motor reflexes. These reflexes are quick, unconscious movements—like automatically removing a hand from a hot object. Reflexes are so fast because they involve local synaptic connections. For example, the knee reflex that a doctor tests during a routine physical is controlled by a single synapse between a sensory neuron and a motor neuron. While a reflex may only require the involvement of one or two synapses, synapses with interneurons in the spinal column transmit information to the brain to convey what happened (the knee jerked, or the hand was hot).

In the United States, there around 10,000 spinal cord injuries each year. Because the spinal cord is the information superhighway connecting the brain with the body, damage to the spinal cord can lead to paralysis. The extent of the paralysis depends on the location of the injury along the spinal cord and whether the spinal cord was completely severed. For example, if the spinal cord is damaged at the level of the neck, it can cause paralysis from the neck down, whereas damage to the spinal column further down may limit paralysis to the legs. Spinal cord injuries are notoriously difficult to treat because spinal nerves do not regenerate, although ongoing research suggests that stem cell transplants may be able to act as a bridge to reconnect severed nerves. Researchers are also looking at ways to prevent the inflammation that worsens nerve damage after injury. One such treatment is to pump the body with cold saline to induce hypothermia. This cooling can prevent swelling and other processes that are thought to worsen spinal cord injuries.

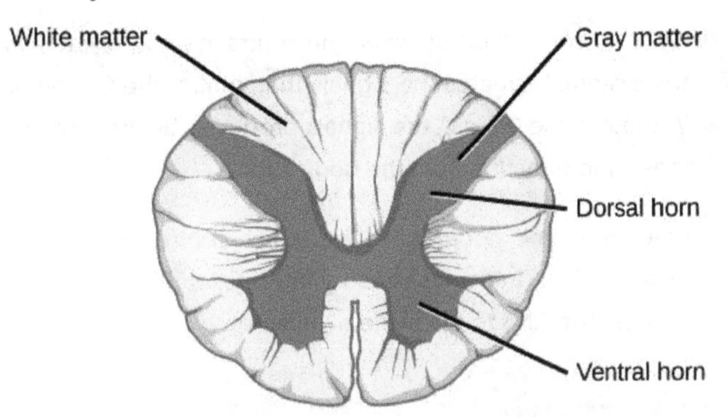

Figure 6. A cross-section of the spinal cord shows gray matter (containing cell bodies and interneurons) and white matter (containing axons).

The Peripheral Nervous System

The peripheral nervous system (PNS) is the connection between the central nervous system and the rest of the body. The CNS is like the power plant of the nervous system. It creates the signals that control the functions of the body. The PNS is like the wires that go to individual houses. Without those "wires," the signals produced by the CNS could not control the body (and the CNS would not be able to receive sensory information from the body either).

The PNS can be broken down into the autonomic nervous system, which controls bodily functions without conscious control, and the sensory-somatic nervous system, which transmits sensory information from the skin, muscles, and sensory organs to the CNS and sends motor commands from the CNS to the skeletal muscles.

Autonomic Nervous System

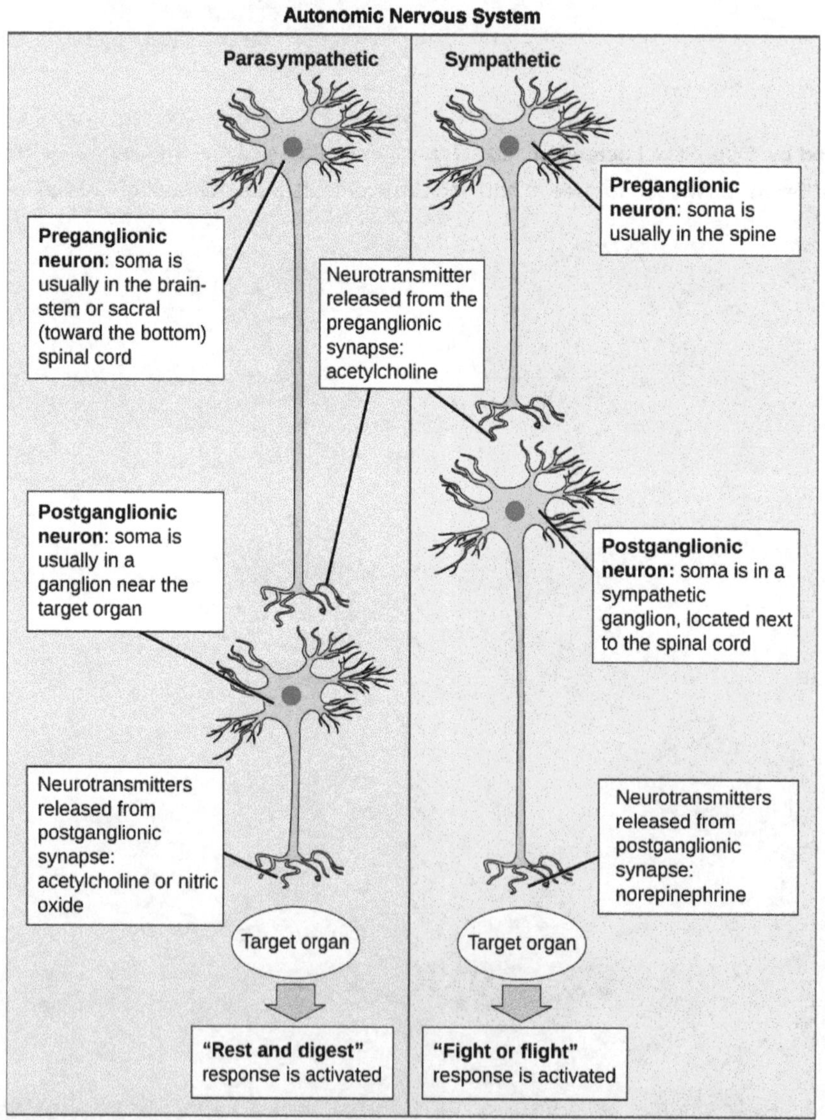

Figure 1. In the autonomic nervous system, a preganglionic neuron of the CNS synapses with a postganglionic neuron of the PNS. The postganglionic neuron, in turn, acts on a target organ. Autonomic responses are mediated by the sympathetic and the parasympathetic systems, which are antagonistic to one another. The sympathetic system activates the "fight or flight" response, while the parasympathetic system activates the "rest and digest" response.

The autonomic nervous system serves as the relay between the CNS and the internal organs. It controls the lungs, the heart, smooth muscle, and exocrine and endocrine glands. The autonomic nervous system controls these organs largely without conscious control; it can continuously monitor the conditions of these different systems and implement changes as needed. Signaling to the target tissue usually involves two synapses: a preganglionic neuron (originating in the CNS) synapses to a neuron in a ganglion that, in turn, synapses on the target organ, as illustrated in Figure 1. There are two divisions of the autonomic nervous system that often have opposing effects: the sympathetic nervous system and the parasympathetic nervous system.

Sympathetic Nervous System

The sympathetic nervous system is responsible for the "fight or flight" response that occurs when an animal encounters a dangerous situation. One way to remember this is to think of the surprise a person feels when encountering a snake ("snake" and "sympathetic" both begin with "s"). Examples of functions controlled by the sympathetic nervous system include an accelerated heart rate and inhibited digestion. These functions help prepare an organism's body for the physical strain required to escape a potentially dangerous situation or to fend off a predator.

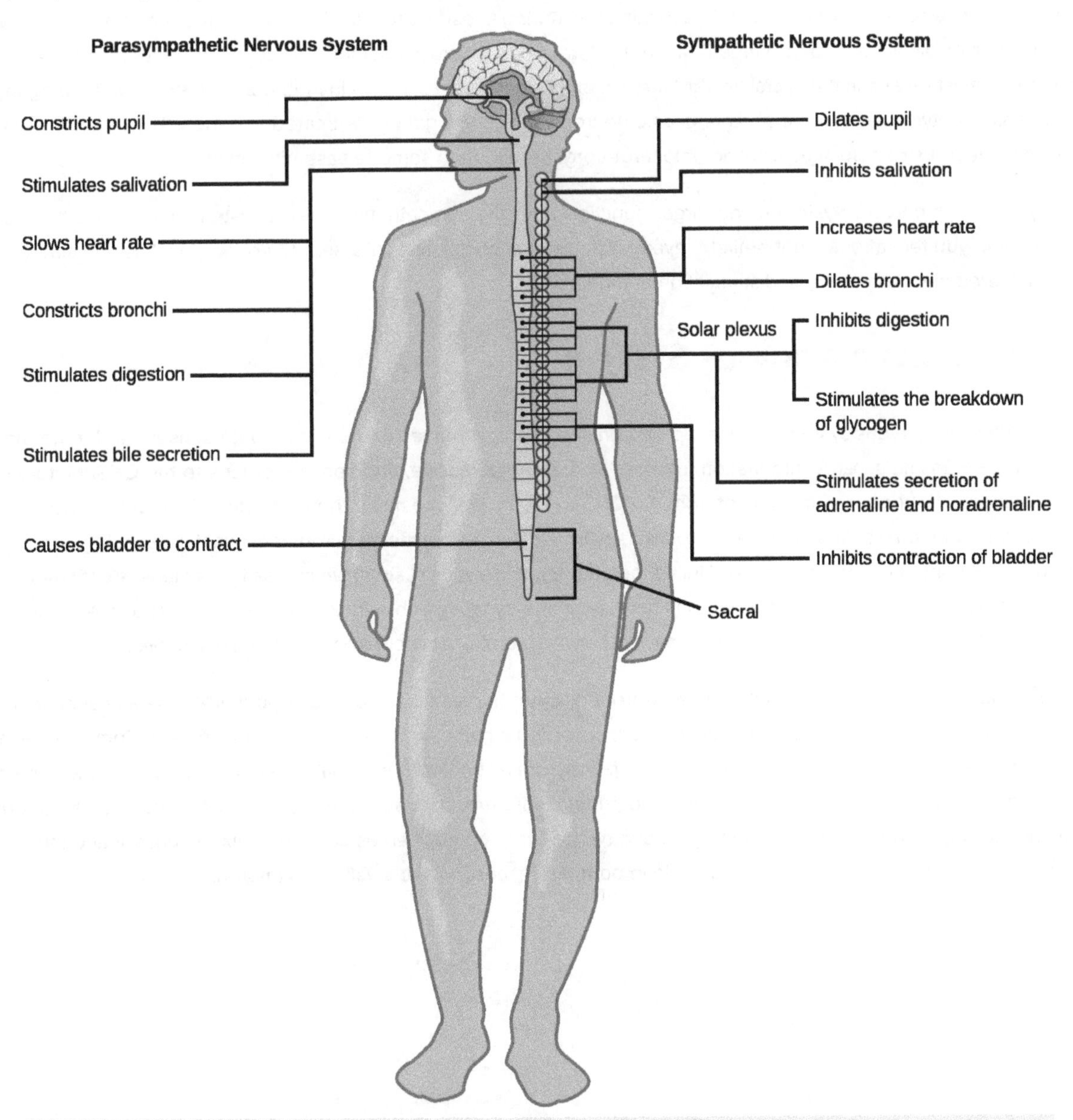

Figure 2. The sympathetic and parasympathetic nervous systems often have opposing effects on target organs.

Most preganglionic neurons in the sympathetic nervous system originate in the spinal cord, as illustrated in Figure 2. The axons of these neurons release acetylcholine on postganglionic neurons within sympathetic ganglia (the sympathetic ganglia form a chain that extends alongside the spinal cord). The acetylcholine activates the postganglionic neurons. Postganglionic neurons then release norepinephrine onto target organs. As anyone who has ever felt a rush before a big test, speech, or athletic event can attest, the effects of the sympathetic nervous system are quite pervasive. This is both because one preganglionic neuron synapses on multiple postganglionic neurons, amplifying the effect of the original synapse, and because the adrenal gland also releases norepinephrine (and the closely related hormone epinephrine) into the bloodstream. The physiological effects of this norepinephrine release include dilating the trachea and bronchi (making it easier to breathe), increasing heart rate, and moving blood from the skin to the heart, muscles, and brain making it easier to process information and run). The strength and speed of the sympathetic response helps an organism avoid danger, and scientists have found evidence that it may also increase LTP—allowing the animal to remember the dangerous situation and avoid it in the future.

Parasympathetic Nervous System

While the sympathetic nervous system is activated in stressful situations, the parasympathetic nervous system allows an animal to "rest and digest." One way to remember this is to think that during a restful situation like a picnic, the parasympathetic nervous system is in control ("picnic" and "parasympathetic" both start with "p"). Parasympathetic preganglionic neurons have cell bodies located in the brainstem and in the sacral (toward the bottom) spinal cord, as shown in Figure 2. The axons of the preganglionic neurons release acetylcholine on the postganglionic neurons, which are generally located very near the target organs. Most postganglionic neurons release acetylcholine onto target organs, although some release nitric oxide.

The parasympathetic nervous system resets organ function after the sympathetic nervous system is activated (the common adrenaline dump you feel after a 'fight-or-flight' event). Effects of acetylcholine release on target organs include slowing of the heart rate, lowered blood pressure, and stimulation of digestion.

Sensory-Somatic Nervous System

The sensory-somatic nervous system is made up of cranial and spinal nerves and contains both sensory and motor neurons. Sensory neurons transmit sensory information from the skin, skeletal muscle, and sensory organs to the CNS. Motor neurons transmit messages about desired movement from the CNS to the muscles to make them contract. Without its sensory-somatic nervous system, a person would be unable to process any information about their environment (what is seen, felt, heard, and so on) and could not control motor movements. Unlike the autonomic nervous system, which has two synapses between the CNS and the target organ, sensory and motor neurons have only one synapse—one ending of the neuron is at the organ and the other directly contacts a CNS neuron. Acetylcholine is the main neurotransmitter released at these synapses.

Humans have 12 cranial nerves, nerves that emerge from or enter the skull (cranium), as opposed to the spinal nerves, which emerge from the vertebral column. Each cranial nerve is accorded a name, which are detailed in Figure 3. Some cranial nerves transmit only sensory information. For example, the olfactory nerve transmits information about smells from the nose to the brainstem. Other cranial nerves transmit almost solely motor information. For example, the oculomotor nerve controls the opening and closing of the eyelid and some eye movements. Other cranial nerves contain a mix of sensory and motor fibers. For example, the glossopharyngeal nerve has a role in both taste (sensory) and swallowing (motor).

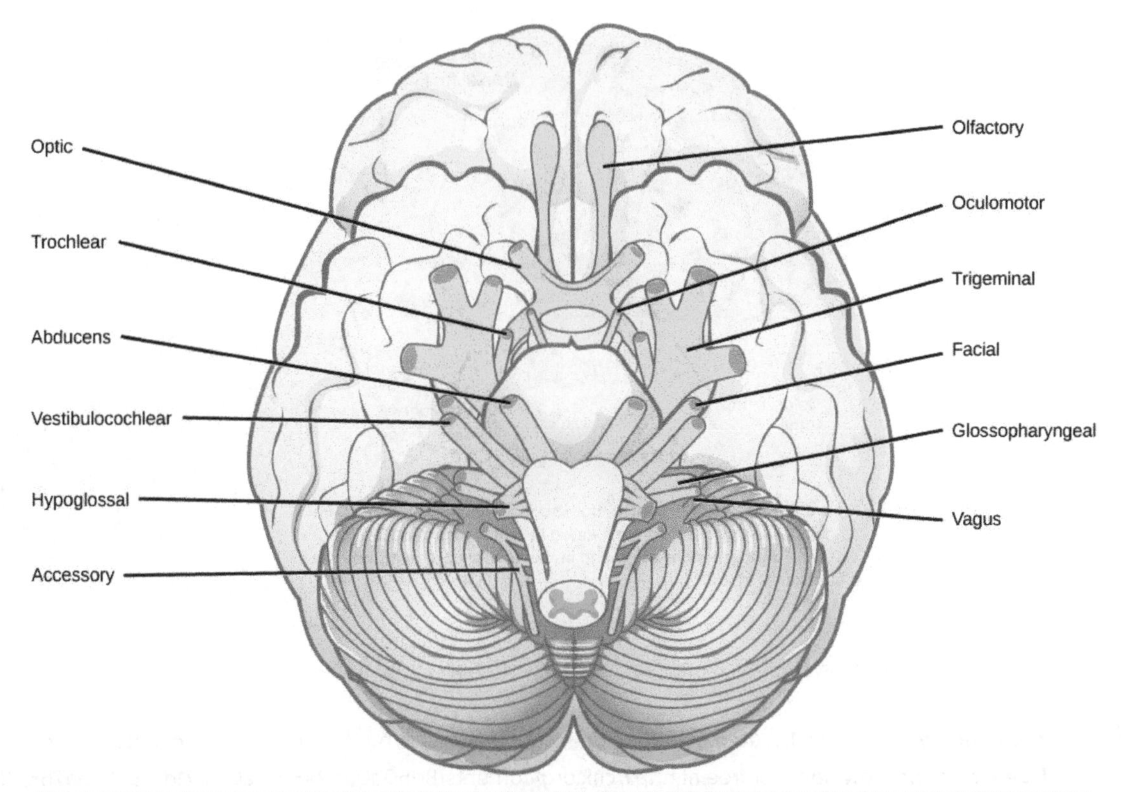

Figure 3. The human brain contains 12 cranial nerves that receive sensory input and control motor output for the head and neck.

Cranial nerves

1. Olfactory: sensory for smell
2. Optic: sensory, process visual information
3. Oculomotor: motor, movement of eyes and smooth muscles controlling pupil and lens
4. Trochlear: motor, eye movements
5. Trigeminal: sensory of upper, and mid face and upper jaw; motor for muscles of chewing
6. Abducens: motor, eye movements
7. Facial: motor for facial expression, tears and salivary glands; sensory for taste
8. Vestibulocochlear: sensory, hearing and equilibrium
9. Glossopharyngeal: motor for mouth (swallowing) and for regulation of blood pressure; sensory for tongue and pharynx and outer ear
10. Vagus: motor for swallowing, speech, cardivascular and digestive regulation; hunger and fullness; sensory from visceral organs and taste. Main parasympathetic nerve
11. Accessory: swallowing, and head, neck, shoulder movement
12. Hypoglossal: tongue movements

Spinal nerves transmit sensory and motor information between the spinal cord and the rest of the body. Each of the 31 spinal nerves contains both sensory and motor axons. The sensory neuron cell bodies are grouped in structures called dorsal root ganglia and are shown in Figure 4. Each sensory neuron has one projection—with a sensory receptor ending in skin, muscle, or sensory organs—and another that synapses with a neuron in the dorsal spinal cord. Motor neurons have cell bodies in the ventral gray matter of the spinal cord that project to muscle through the ventral root. These neurons are usually stimulated by interneurons within the spinal cord but are sometimes directly stimulated by sensory neurons.

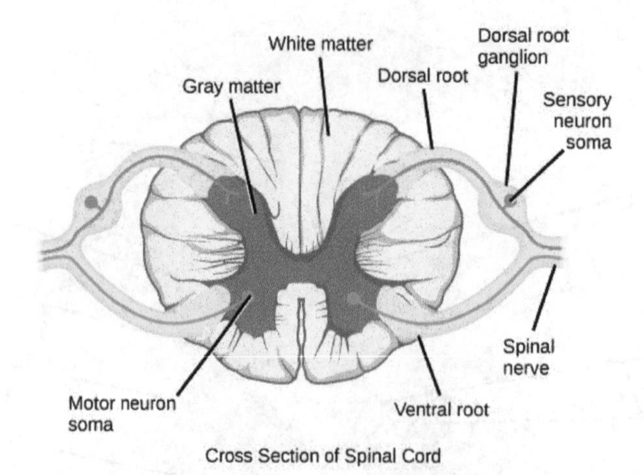

Figure 4. Spinal nerves contain both sensory and motor axons. The somas of sensory neurons are located in dorsal root ganglia. The somas of motor neurons are found in the ventral portion of the gray matter of the spinal cord.

Reflexes

The central nervous system is able to make conscious decisions about how to react to information, such as deciding how to change our behavior if we are cold or in pain. Another option is to use the sensory information to inform a visceral reflex arc, like those involved in the regulation of blood pressure or body temperature. Here we will consider using sensory information to inform somatic reflexes, where automatic motor responses occur as a result of the sensory stimuli.

No matter which type of CNS mediated reflex or response we are referring to, the general model is the same. Sensory information activates a receptor that sends information to the CNS via an afferent neuron, some level of synaptic or higher level processing occurs, and, if a response is necessary or appropriate, the response is initiated through efferent neurons. You might recognize this as the same model used to maintain homeostasis. Reflexes are a unique category of responses because they do not require the higher centers used for conscious or voluntary responses. Instead reflexes are involuntary, stereotyped (they are repeatable under the same stimulus conditions) responses that occur quickly.

Categories of Reflexes

As mentioned, reflexes can either be visceral or somatic. Visceral reflexes involve a glandular or non-skeletal muscular response carried out in internal organs such as the heart, blood vessels, or structures of the GI tract. They utilize neurons of the autonomic nervous system to elicit their actions. Visceral reflexes have been more fully discussed in the section on the autonomic nervous system. In contrast, somatic reflexes involve unconscious skeletal muscle motor responses. In doing so, these reflexes utilize some of the same lower motor neurons (alpha motor neurons) used to control skeletal muscle during conscious movement. Because reflexes are quick, it makes sense that somatic reflexes are often meant to protect us from injury. As examples, reflexes contribute to the maintenance of balance and rapid withdrawal of the hand or foot from damaging stimuli.

Somatic reflexes can either be intrinsic (present at birth) or learned. We will be focusing on intrinsic reflexes, which occur as the result of normal human development. Learned reflexes are much more complicated in their anatomical structure and result from repetitive actions, such as athletic training. Reflexes can also be categorized by the number of synapses they involve (monosynaptic reflex versus polysynaptic reflex) or the relative position of the sensory receptors to the responding muscles (ipsilateral = same side of the body, contralateral = opposite sides of the body).

Examples of Somatic Reflexes

Stretch Reflex

One of the simplest reflexes is a stretch reflex. In this reflex, when a skeletal muscle is stretched, a muscle spindle in the belly of the muscle is activated. The axon from this receptor travels to the spinal cord where it synapses with the motor neuron controlling the muscle, stimulating it to contract. This is a rapid, monosynaptic, ipsilateral reflex that helps to maintain the length of muscles and contributes to joint stabilization. A common example of this reflex is the knee jerk reflex that is elicited by a rubber hammer striking against the patellar tendon, such as during a physical exam. When the hammer strikes, it stretches the tendon, which pulls on the quadriceps femoris muscle. Because bones and tendons do not typically pull muscles, the muscle "thinks" it is stretching very rapidly, and the reflex acts to counteract this stretch. In doing so, the "knee jerk" occurs.

Along with the monosynaptic activation of the alpha motor neuron, this reflex also includes the activation of an interneuron that inhibits the alpha motor neuron of the antagonistic muscle. This aspect of the reflex ensures that contraction of the agonist muscle occurs unopposed.

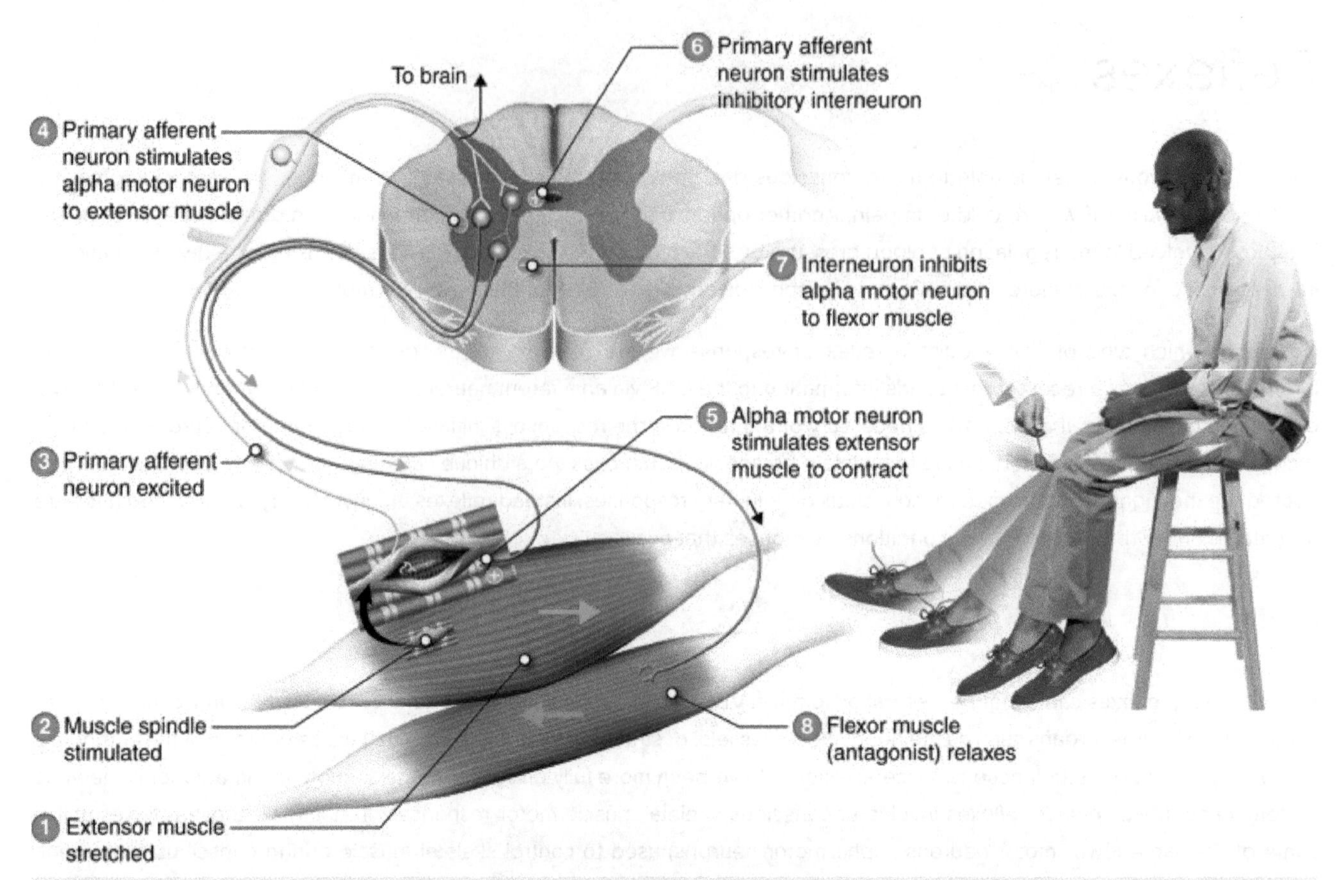

Stretch Reflex. When a muscle is stretched (1), muscle spindles (2) send information to the spinal cord (3) where it synapses on motor neuron of the same muscle (4) causing it to contract (5). At the same time, stimulation of an inhibitory interneuron (6) prevents contraction of the antagonistic muscle (7 and 8). This work by Cenveo is licensed under a Creative Commons Attribution 3.0 United States (http://creativecommons.org/licenses/by/3.0/us/).

Flexor (Withdrawal) Reflex

Recall from the beginning of this unit that when you touch a hot stove, you reflexively pull your hand away. Sensory receptors in the skin sense extreme temperature and the early signs of tissue damage. To avoid further damage, information travels along the sensory fibers from the skin and into the posterior (dorsal) horn of the spinal cord. Once in the spinal cord, the sensory fibers synapse with a variety of interneurons that mediate the responses of the reflex. These responses included a strong initial withdrawal of the flexor muscle (caused by activation of the alpha motor neurons), inhibition of the extensor muscle (mediated through inhibitory interneurons), and sustained contraction of the flexor (mediated by a spinal cord neuronal circuit). And as already discussed, the sensory information will also travel to the brain to develop a conscious awareness of the situation such that conscious decision-making can take over immediately after the reflex occurs.

Crossed-Extensor Reflex

Imagine what would happen if, when you stepped on a sharp object, it elicited a strong withdrawal reflex of your leg. You would likely topple over. In order to prevent this from happening, as the flexor (withdrawal) reflex involving the injured leg happens, an extension reflex of the opposite (contralateral) leg occurs at the same time, creating a crossed-extensor reflex. In this case, the ipsilateral limb reacts with a withdrawal reflex (stimulating flexor muscles and inhibiting extensor muscles on same side), but the contralateral extensor muscles contract so that the person can appropriately shift balance to the opposite foot during the reflex.

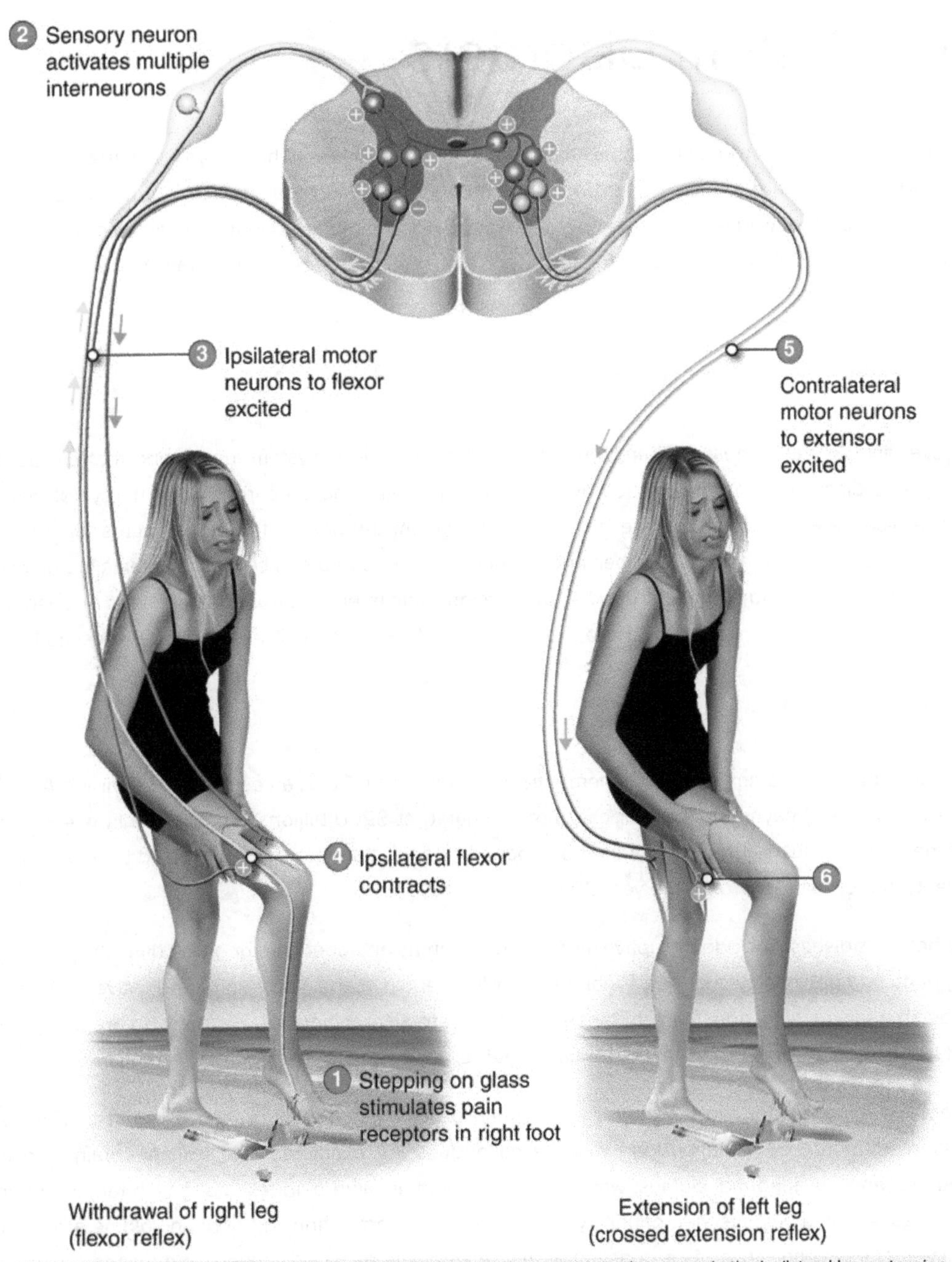

Crossed-Extensor Reflex. In this reflex, as withdrawal from the damaging stimulus occurs in the ipsilateral leg, extension occurs in the contralateral leg as a way of maintaining balance. This work by Cenveo is licensed under a Creative Commons Attribution 3.0 United States (http://creativecommons.org/licenses/by/3.0/us/).

Nervous System Disorders

A nervous system that functions correctly is a fantastically complex, well-oiled machine—synapses fire appropriately, muscles move when needed, memories are formed and stored, and emotions are well regulated. Unfortunately, each year millions of people in the United States deal with some sort of nervous system disorder. While scientists have discovered potential causes of many of these diseases, and viable treatments for some, ongoing research seeks to find ways to better prevent and treat all of these disorders.

Neurodegenerative Disorders

Neurodegenerative disorders are illnesses characterized by a loss of nervous system functioning that are usually caused by neuronal death. These diseases generally worsen over time as more and more neurons die. The symptoms of a particular neurodegenerative disease are related to where in the nervous system the death of neurons occurs. Spinocerebellar ataxia, for example, leads to neuronal death in the cerebellum. The death of these neurons causes problems in balance and walking. Neurodegenerative disorders include Huntington's disease, amyotrophic lateral sclerosis, Alzheimer's disease and other types of dementia disorders, and Parkinson's disease. Here, Alzheimer's and Parkinson's disease will be discussed in more depth.

Alzheimer's Disease

Alzheimer's disease is the most common cause of dementia in the elderly. In 2012, an estimated 5.4 million Americans suffered from Alzheimer's disease, and payments for their care are estimated at $200 billion. Roughly one in every eight people age 65 or older has the disease. Due to the aging of the baby-boomer generation, there are projected to be as many as 13 million Alzheimer's patients in the United States in the year 2050.

Symptoms of Alzheimer's disease include disruptive memory loss, confusion about time or place, difficulty planning or executing tasks, poor judgment, and personality changes. Problems smelling certain scents can also be indicative of Alzheimer's disease and may serve as an early warning sign. Many of these symptoms are also common in people who are aging normally, so it is the severity and longevity of the symptoms that determine whether a person is suffering from Alzheimer's.

Alzheimer's disease was named for Alois Alzheimer, a German psychiatrist who published a report in 1911 about a woman who showed severe dementia symptoms. Along with his colleagues, he examined the woman's brain following her death and reported the presence of abnormal clumps, which are now called amyloid plaques, along with tangled brain fibers called neurofibrillary tangles. Amyloid plaques, neurofibrillary tangles, and an overall shrinking of brain volume are commonly seen in the brains of Alzheimer's patients. Loss of neurons in the hippocampus is especially severe in advanced Alzheimer's patients. (Figure 1) compares a normal brain to the brain of an Alzheimer's patient. Many research groups are examining the causes of these hallmarks of the disease.

One form of the disease is usually caused by mutations in one of three known genes. This rare form of early onset Alzheimer's disease affects fewer than five percent of patients with the disease and causes dementia beginning between the ages of 30 and 60. The more prevalent, late-onset form of the disease likely also has a genetic component. One particular gene, apolipoprotein E (APOE) has a variant (E4) that increases a carrier's likelihood of getting the disease. Many other genes have been identified that might be involved in the pathology.

Unfortunately, there is no cure for Alzheimer's disease. Current treatments focus on managing the symptoms of the disease. Because decrease in the activity of cholinergic neurons (neurons that use the neurotransmitter acetylcholine) is common in Alzheimer's disease, several drugs used to treat the disease work by increasing acetylcholine neurotransmission, often by inhibiting the enzyme that breaks down acetylcholine in the synaptic cleft. Other clinical interventions focus on behavioral therapies like psychotherapy, sensory therapy, and cognitive exercises. Since Alzheimer's disease appears to hijack the normal aging process, research into prevention is prevalent. Smoking, obesity, and cardiovascular problems may be risk factors for the disease, so treatments for those may also help to prevent Alzheimer's disease. Some studies have shown that people who remain intellectually active by playing games, reading, playing musical instruments, and being socially active in later life have a reduced risk of developing the disease.

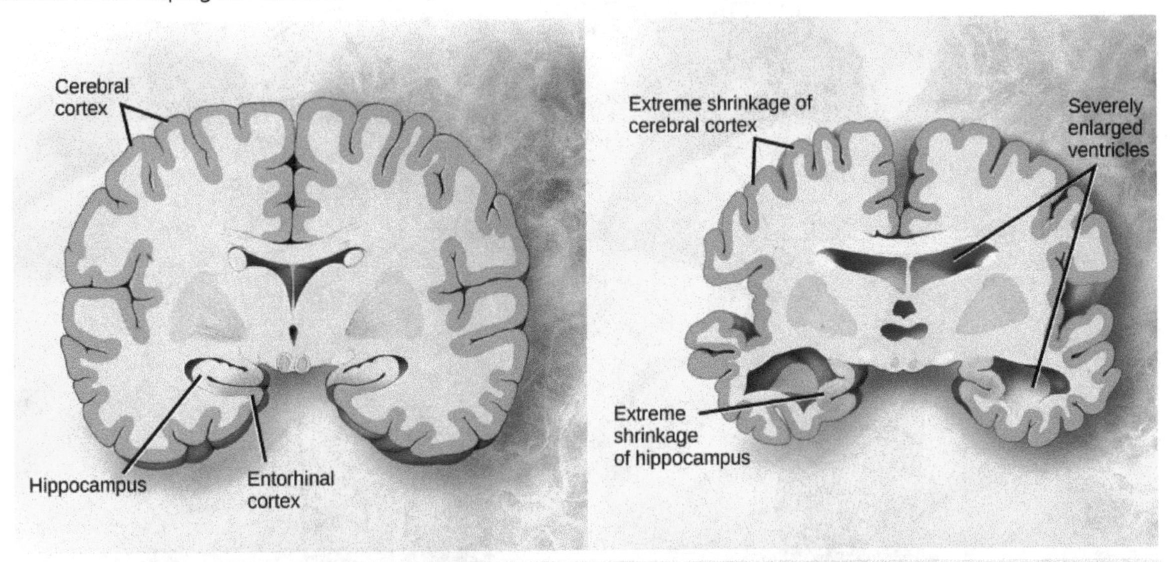

Figure 1. Compared to a normal brain (left), the brain from a patient with Alzheimer's disease (right) shows a dramatic neurodegeneration, particularly within the ventricles and hippocampus. (credit: modification of work by "Garrando"/Wikimedia Commons based on original images by ADEAR: "Alzheimer's Disease Education and Referral Center, a service of the National Institute on Aging")

Parkinson's Disease

Like Alzheimer's disease, Parkinson's disease is a neurodegenerative disease. It was first characterized by James Parkinson in 1817. Each year, 50,000-60,000 people in the United States are diagnosed with the disease. Parkinson's disease causes the loss of dopamine neurons in the substantia nigra, a midbrain structure that regulates movement. Loss of these neurons causes many symptoms including tremor (shaking of fingers or a limb), slowed movement, speech changes, balance and posture problems, and rigid muscles. The combination of these symptoms often causes a characteristic slow hunched shuffling walk, illustrated in (Figure 2). Patients with Parkinson's disease can also exhibit psychological symptoms, such as dementia or emotional problems.

Although some patients have a form of the disease known to be caused by a single mutation, for most patients the exact causes of Parkinson's disease remain unknown: the disease likely results from a combination of genetic and environmental factors (similar to Alzheimer's disease). Post-mortem analysis of brains from Parkinson's patients shows the presence of Lewy bodies—abnormal protein clumps—in dopaminergic neurons. The prevalence of these Lewy bodies often correlates with the severity of the disease.

There is no cure for Parkinson's disease, and treatment is focused on easing symptoms. One of the most commonly prescribed drugs for Parkinson's is L-DOPA, which is a chemical that is converted into dopamine by neurons in the brain. This conversion increases the overall level of dopamine neurotransmission and can help compensate for the loss of dopaminergic neurons in the substantia nigra. Other drugs work by inhibiting the enzyme that breaks down dopamine.

Figure 2. Parkinson's patients often have a characteristic hunched walk.

Neurodevelopmental Disorders

Neurodevelopmental disorders occur when the development of the nervous system is disturbed. There are several different classes of neurodevelopmental disorders. Some, like Down Syndrome, cause intellectual deficits. Others specifically affect communication, learning, or the motor system. Some disorders like autism spectrum disorder and attention deficit/hyperactivity disorder have complex symptoms.

Autism

Autism spectrum disorder (ASD) is a neurodevelopmental disorder. Its severity differs from person to person. Estimates for the prevalence of the disorder have changed rapidly in the past few decades. Current estimates suggest that one in 88 children will develop the disorder. ASD is four times more prevalent in males than females.

A characteristic symptom of ASD is impaired social skills. Children with autism may have difficulty making and maintaining eye contact and reading social cues. They also may have problems feeling empathy for others. Other symptoms of ASD include repetitive motor behaviors (such as rocking back and forth), preoccupation with specific subjects, strict adherence to certain rituals, and unusual language use. Up to 30 percent of patients with ASD develop epilepsy, and patients with some forms of the disorder (like Fragile X) also have intellectual disability. Because it is a spectrum disorder, other ASD patients are very functional and have good-to-excellent language skills. Many of these patients do not feel that they suffer from a disorder and instead think that their brains just process information differently.

Except for some well-characterized, clearly genetic forms of autism (like Fragile X and Rett's Syndrome), the causes of ASD are largely unknown. Variants of several genes correlate with the presence of ASD, but for any given patient, many different mutations in different genes may be required for the disease to develop. At a general level, ASD is thought to be a disease of "incorrect" wiring. Accordingly, brains of some ASD patients lack the same level of synaptic pruning that occurs in non-affected people. In the 1990s, a research paper linked autism to a common vaccine given to children. This paper was retracted when it was discovered that the author falsified data, and follow-up studies showed no connection between vaccines and autism.

Treatment for autism usually combines behavioral therapies and interventions, along with medications to treat other disorders common to people with autism (depression, anxiety, obsessive compulsive disorder). Although early interventions can help mitigate the effects of the disease, there is currently no cure for ASD.

Attention Deficit Hyperactivity Disorder (ADHD)

Approximately three to five percent of children and adults are affected by attention deficit/hyperactivity disorder (ADHD). Like ASD, ADHD is more prevalent in males than females. Symptoms of the disorder include inattention (lack of focus), executive functioning difficulties, impulsivity, and hyperactivity beyond what is characteristic of the normal developmental stage. Some patients do not have the hyperactive component of symptoms and are diagnosed with a subtype of ADHD: attention deficit disorder (ADD). Many people with ADHD also show comorbitity, in that they develop secondary disorders in addition to ADHD. Examples include depression or obsessive compulsive disorder (OCD). (Figure 3) provides some statistics concerning comorbidity with ADHD.

The cause of ADHD is unknown, although research points to a delay and dysfunction in the development of the prefrontal cortex and disturbances in neurotransmission. According to studies of twins, the disorder has a strong genetic component. There are several candidate genes that may contribute to the disorder, but no definitive links have been discovered. Environmental factors, including exposure to certain pesticides, may also contribute to the development of ADHD in some patients. Treatment for ADHD often involves behavioral therapies and the prescription of stimulant medications, which paradoxically cause a calming effect in these patients.

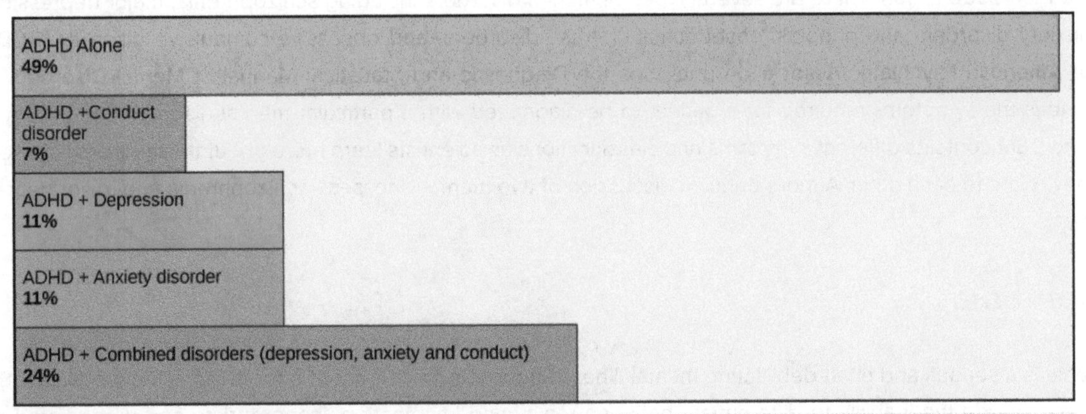

Figure 3. Many people with ADHD have one or more other neurological disorders. (credit "chart design and illustration": modification of work by Leigh Coriale; credit "data": Drs. Biederman and Faraone, Massachusetts General Hospital).

CAREER CONNECTION

Neurologist

Neurologists are physicians who specialize in disorders of the nervous system. They diagnose and treat disorders such as epilepsy, stroke, dementia, nervous system injuries, Parkinson's disease, sleep disorders, and multiple sclerosis. Neurologists are medical doctors who have attended college, medical school, and completed three to four years of neurology residency.

When examining a new patient, a neurologist takes a full medical history and performs a complete physical exam. The physical exam contains specific tasks that are used to determine what areas of the brain, spinal cord, or peripheral nervous system may be damaged. For example, to check whether the hypoglossal nerve is functioning correctly, the neurologist will ask the patient to move his or her tongue in different ways. If the patient does not have full control over tongue movements, then the hypoglossal nerve may be damaged or there may be a lesion in the brainstem where the cell bodies of these neurons reside (or there could be damage to the tongue muscle itself).

Neurologists have other tools besides a physical exam they can use to diagnose particular problems in the nervous system. If the patient has had a seizure, for example, the neurologist can use electroencephalography (EEG), which involves taping electrodes to the scalp to record brain activity, to try to determine which brain regions are involved in the seizure. In suspected stroke patients, a neurologist can use a computerized tomography (CT) scan, which is a type of X-ray, to look for bleeding in the brain or a possible brain tumor. To treat patients with neurological problems, neurologists can prescribe medications or refer the patient to a neurosurgeon for surgery.

Mental Illnesses

Mental illnesses are nervous system disorders that result in problems with thinking, mood, or relating with other people. These disorders are severe enough to affect a person's quality of life and often make it difficult for people to perform the routine tasks of daily living. Debilitating mental disorders plague approximately 12.5 million Americans (about 1 in 17 people) at an annual cost of more than $300 billion. There are several types of mental disorders including schizophrenia, major depression, bipolar disorder, anxiety disorders and phobias, post-traumatic stress disorders, and obsessive-compulsive disorder (OCD), among others. The American Psychiatric Association publishes the Diagnostic and Statistical Manual of Mental Disorders (or DSM), which describes the symptoms required for a patient to be diagnosed with a particular mental disorder. Each newly released version of the DSM contains different symptoms and classifications as scientists learn more about these disorders, their causes, and how they relate to each other. A more detailed discussion of two mental illnesses—schizophrenia and major depression—is given below.

Schizophrenia

Schizophrenia is a serious and often debilitating mental illness affecting one percent of people in the United States. Symptoms of the disease include the inability to differentiate between reality and imagination, inappropriate and unregulated emotional responses, difficulty thinking, and problems with social situations. People with schizophrenia can suffer from hallucinations and hear voices; they may also suffer from delusions. Patients also have so-called "negative" symptoms like a flattened emotional state, loss of pleasure, and loss of basic drives. Many schizophrenic patients are diagnosed in their late adolescence or early 20s. The development of schizophrenia is thought to involve malfunctioning dopaminergic neurons and may also involve problems with glutamate signaling. Treatment for the disease usually requires antipsychotic medications that work by blocking dopamine receptors and decreasing dopamine neurotransmission in the brain. This decrease in dopamine can cause Parkinson's disease-like symptoms in some patients. While some classes of antipsychotics can be quite effective at treating the disease, they are not a cure, and most patients must remain medicated for the rest of their lives.

Depression

Major depression affects approximately 6.7 percent of the adults in the United States each year and is one of the most common mental disorders. To be diagnosed with major depressive disorder, a person must have experienced a severely depressed mood lasting longer than two weeks along with other symptoms including a loss of enjoyment in activities that were previously enjoyed, changes in appetite and sleep schedules, difficulty concentrating, feelings of worthlessness, and suicidal thoughts. The exact causes of major depression are unknown and likely include both genetic and environmental risk factors. Some research supports the "classic monoamine hypothesis," which suggests that depression is caused by a decrease in norepinephrine and serotonin neurotransmission. One argument against this hypothesis is the fact that some antidepressant medications cause an increase in norepinephrine and serotonin release within a few hours of beginning treatment—but clinical results of these medications are not seen until weeks later. This has led to alternative hypotheses: for example, dopamine may also be decreased in depressed patients, or it may actually be an increase in norepinephrine and serotonin that causes the disease, and antidepressants force a feedback loop that decreases this release. Treatments for depression include psychotherapy, electroconvulsive therapy, deep-brain stimulation, and prescription medications. There are several classes of antidepressant medications that work through different mechanisms. For example, monoamine oxidase inhibitors (MAO inhibitors) block the enzyme that degrades many neurotransmitters (including dopamine, serotonin, norepinephrine), resulting in increased neurotransmitter in the synaptic cleft. Selective serotonin reuptake inhibitors (SSRIs) block the reuptake of serotonin into the presynaptic neuron. This blockage results in an increase in serotonin in the synaptic cleft. Other types of drugs such as norepinephrine-dopamine reuptake inhibitors and norepinephrine-serotonin reuptake inhibitors are also used to treat depression.

Other Neurological Disorders

There are several other neurological disorders that cannot be easily placed in the above categories. These include chronic pain conditions, cancers of the nervous system, epilepsy disorders, and stroke. Epilepsy and stroke are discussed below.

Epilepsy

Estimates suggest that up to three percent of people in the United States will be diagnosed with epilepsy in their lifetime. While there are several different types of epilepsy, all are characterized by recurrent seizures. Epilepsy itself can be a symptom of a brain injury, disease, or other illness. For example, people who have intellectual disability or ASD can experience seizures, presumably because the developmental wiring malfunctions that caused their disorders also put them at risk for epilepsy. For many patients, however, the cause of their epilepsy is never identified and is likely to be a combination of genetic and environmental factors. Often, seizures can be controlled with anticonvulsant medications. However, for very severe cases, patients may undergo brain surgery to remove the brain area where seizures originate.

Stroke

A stroke results when blood fails to reach a portion of the brain for a long enough time to cause damage. Without the oxygen supplied by blood flow, neurons in this brain region die. This neuronal death can cause many different symptoms—depending on the brain area affected— including headache, muscle weakness or paralysis, speech disturbances, sensory problems, memory loss, and confusion. Stroke is often caused by blood clots and can also be caused by the bursting of a weak blood vessel. Strokes are extremely common and are the third most common cause of death in the United States. On average one person experiences a stroke every 40 seconds in the United States. Approximately 75 percent of strokes occur in people older than 65. Risk factors for stroke include high blood pressure, diabetes, high cholesterol, and a family history of stroke. Smoking doubles the risk of stroke. Because a stroke is a medical emergency, patients with symptoms of a stroke should immediately go to the emergency room, where they can receive drugs that will dissolve any clot that may have formed. These drugs will not work if the stroke was caused by a burst blood vessel or if the stroke occurred more than three hours before arriving at the hospital. Treatment following a stroke can include blood pressure medication (to prevent future strokes) and (sometimes intense) physical therapy.

Concept Check

1. **What** are the functions of the nervous system?
2. **What** are the differences between the somatic and autonomic divisions of the efferent peripheral nervous system?
3. **What** is the difference between action potential and membrane potential?
4. **What** type of channel is the sodium channel found on neuron cell membranes?
5. **What** are the main steps in an action potential?
6. **How** do neurotransmitters reach the post-synaptic neuron?
7. **What** is the covering of the CNS that lies directly against the nervous tissue of the brain and spinal cord? **Describe** the structure of the arachnoid mater.
8. **Where** is CSF produced?
9. **Which** part of the brain controls muscle movement and coordination? **Which** part contains the vital centers?
10. **Which** cerebral hemisphere is more involved in logic and quantitative thought processes?
11. **How** does the placement of white matter in the spinal cord differ from that in the brain?
12. **What** are the steps in a typical reflex?
13. **What** is the difference between spinal and cranial nerves?

CHAPTER 14 — SENSORY SYSTEMS

Sensory Processes

LEARNING OBJECTIVES

- Identify the general and special senses in humans
- Describe four important mechanoreceptors in human skin
- Explain in what way smell and taste stimuli differ from other sensory stimuli
- Trace the path of sound through the auditory system to the site of transduction of sound
- Identify the structures of the vestibular system that respond to gravity
- Trace the path of light through the eye to the point of the optic nerve

Senses provide information about the body and its environment. Humans have five special senses: olfaction (smell), gustation (taste), equilibrium (balance and body position), vision, and hearing. Additionally, we possess general senses, also called somatosensation, which respond to stimuli like temperature, pain, pressure, and vibration. Vestibular sensation, which is an organism's sense of spatial orientation and balance, proprioception (position of bones, joints, and muscles), and the sense of limb position that is used to track kinesthesia (limb movement) are part of somatosensation. Although the sensory systems associated with these senses are very different, all share a common function: to convert a stimulus (such as light, or sound, or the position of the body) into an electrical signal in the nervous system. This process is called sensory transduction.

There are two broad types of cellular systems that perform sensory transduction. In one, a neuron works with a sensory receptor, a cell, or cell process that is specialized to engage with and detect a specific stimulus. Stimulation of the sensory receptor activates the associated afferent neuron, which carries information about the stimulus to the central nervous system. In the second type of sensory transduction, a sensory nerve ending responds to a stimulus in the internal or external environment: this neuron constitutes the sensory receptor. Free nerve endings can be stimulated by several different stimuli, thus showing little receptor specificity. For example, pain receptors in your gums and teeth may be stimulated by temperature changes, chemical stimulation, or pressure.

Reception

The first step in sensation is reception, which is the activation of sensory receptors by stimuli such as mechanical stimuli (being bent or squished, for example), chemicals, or temperature. The receptor can then respond to the stimuli. The region in space in which a given sensory receptor can respond to a stimulus, be it far away or in contact with the body, is that receptor's receptive field. Think for a moment about the differences in receptive fields for the different senses. For the sense of touch, a stimulus must come into contact with the body. For the sense of hearing, a stimulus can be a moderate distance away. For vision, a stimulus can be very far away; for example, the visual system perceives light from stars at enormous distances.

Transduction

The most fundamental function of a sensory system is the translation of a sensory signal to an electrical signal in the nervous system. This takes place at the sensory receptor, and the change in electrical potential that is produced is called the receptor potential. How is sensory input, such as pressure on the skin, changed to a receptor potential? In this example, a type of receptor called a mechanoreceptor (as shown in Figure 1) possesses specialized membranes that respond to pressure. Disturbance of these dendrites by compressing them or bending them opens gated ion channels in the plasma membrane of the sensory neuron, changing its electrical potential. Recall that in the nervous system, a positive change of a neuron's electrical potential (also called the membrane potential), depolarizes the neuron. Receptor potentials are graded potentials: the magnitude of these graded (receptor) potentials varies with the strength of the stimulus. If the magnitude of depolarization is sufficient (that is, if membrane potential reaches a threshold), the neuron will fire an action potential. In most cases, the correct stimulus impinging on a sensory receptor will drive membrane potential in a positive direction, although for some receptors, such as those in the visual system, this is not always the case.

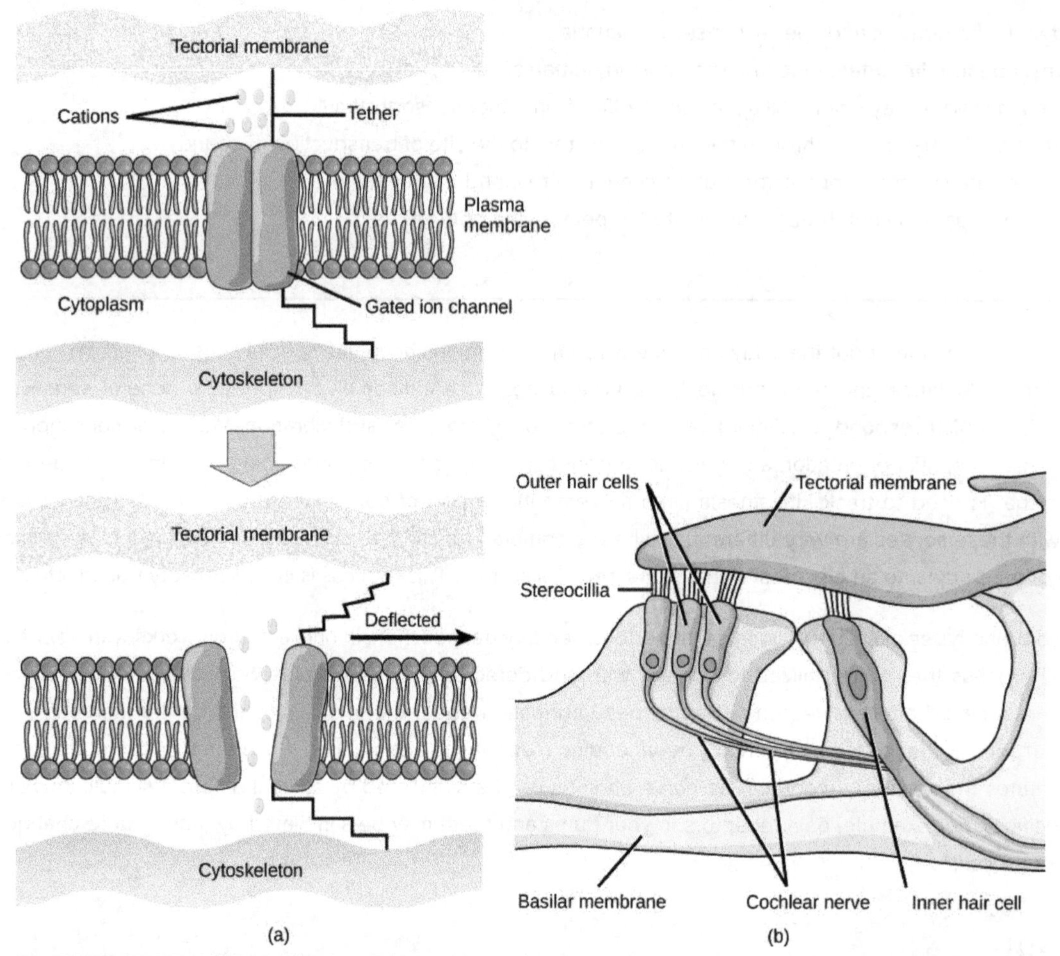

Figure 1. (a) Mechanosensitive ion channels are gated ion channels that respond to mechanical deformation of the plasma membrane. A mechanosensitive channel is connected to the plasma membrane and the cytoskeleton by hair-like tethers. When pressure causes the extracellular matrix to move, the channel opens, allowing ions to enter or exit the cell. (b) Stereocilia in the human ear are connected to mechanosensitive ion channels. When a sound causes the stereocilia to move, mechanosensitive ion channels transduce the signal to the cochlear nerve.

Sensory receptors for different senses are very different from each other, and they are specialized according to the type of stimulus they sense: they have receptor specificity. For example, touch receptors, light receptors, and sound receptors are each activated by different stimuli. Touch receptors are not sensitive to light or sound; they are sensitive only to touch or pressure. However, stimuli may be combined at higher levels in the brain, as happens with olfaction, contributing to our sense of taste.

Encoding and Transmission of Sensory Information

Four aspects of sensory information are encoded by sensory systems: the type of stimulus, the location of the stimulus in the receptive field, the duration of the stimulus, and the relative intensity of the stimulus. Thus, action potentials transmitted over a sensory receptor's afferent axons encode one type of stimulus, and this segregation of the senses is preserved in other sensory circuits. For example, auditory receptors transmit signals over their own dedicated system, and electrical activity in the axons of the auditory receptors will be interpreted by the brain as an auditory stimulus—a sound.

The intensity of a stimulus is often encoded in the rate of action potentials produced by the sensory receptor. Thus, an intense stimulus will produce a more rapid train of action potentials, and reducing the stimulus will likewise slow the rate of production of action potentials. A second way in which intensity is encoded is by the number of receptors activated. An intense stimulus might initiate action potentials in a large number of adjacent receptors, while a less intense stimulus might stimulate fewer receptors. Integration of sensory information begins as soon as the information is received in the CNS, and the brain will further process incoming signals.

Perception

Perception is an individual's interpretation of a sensation. Although perception relies on the activation of sensory receptors, perception happens not at the level of the sensory receptor, but at higher levels in the nervous system, in the brain. The brain distinguishes sensory stimuli through a sensory pathway: action potentials from sensory receptors travel along neurons that are dedicated to a particular stimulus. These neurons are dedicated to that particular stimulus and synapse with particular neurons in the brain or spinal cord.

All sensory signals, except those from the olfactory system, are transmitted though the central nervous system and are routed to the thalamus and to the appropriate region of the cortex. Recall that the thalamus is a structure in the forebrain that serves as a clearinghouse and relay station for sensory (as well as motor) signals. When the sensory signal exits the thalamus, it is conducted to the specific area of the cortex (Figure 2) dedicated to processing that particular sense.

How are neural signals interpreted? Interpretation of sensory signals between individuals is largely similar; however, there are some individual differences. A good example of this is individual tolerances to a painful stimulus, such as dental pain, which certainly differ.

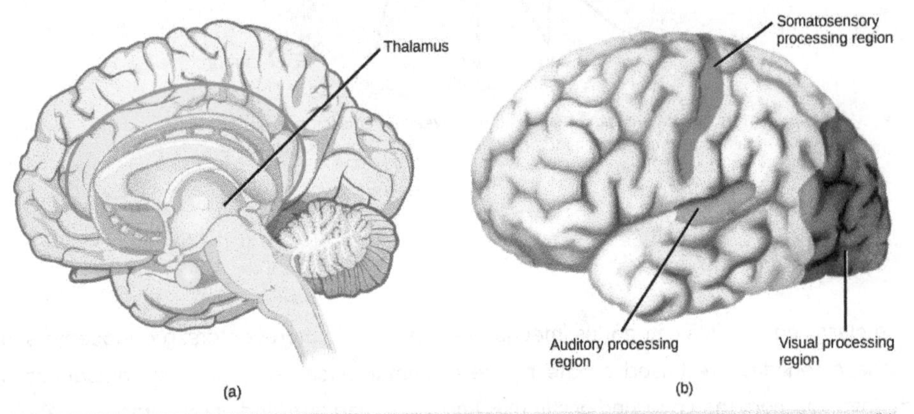

Figure 2. In humans, with the exception of olfaction, all sensory signals are routed from the (a) thalamus to (b) final processing regions in the cortex of the brain. (credit b: modification of work by Polina Tishina)

Somatosensation

Somatosensation is a mixed sensory category and includes all sensation received from the skin and mucous membranes, as well from as the limbs and joints. Somatosensation is also known as tactile sense, or more familiarly, as the sense of touch. Somatosensation occurs all over the exterior of the body and at some interior locations as well. A variety of receptor types—embedded in the skin, mucous membranes, muscles, joints, internal organs, and cardiovascular system—play a role.

Recall that the epidermis is the outermost layer of skin. It is relatively thin, composed of keratin-filled cells, and has no blood supply. The epidermis serves as a barrier to water and to invasion by pathogens. Below this, the much thicker dermis contains blood vessels, sweat glands, hair follicles, lymph vessels, and lipid-secreting sebaceous glands (Figure 1). Below the epidermis and dermis is the subcutaneous tissue, or hypodermis, the fatty layer that contains blood vessels, connective tissue, and the axons of sensory neurons. The hypodermis, which holds about 50 percent of the body's fat, attaches the dermis to the bone and muscle, and supplies nerves and blood vessels to the dermis.

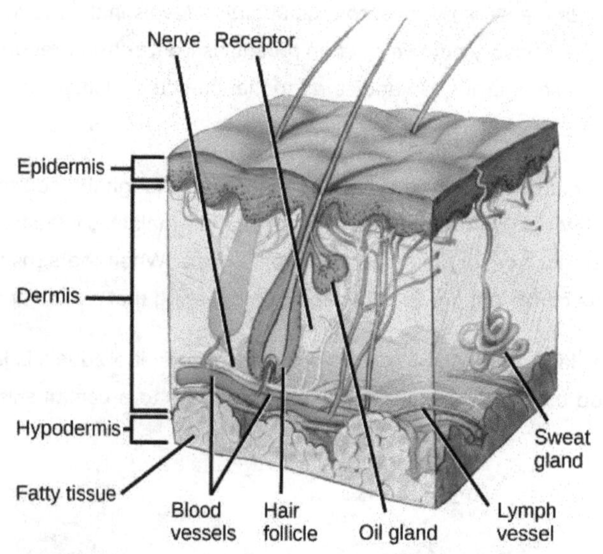

Figure 1. Mammalian skin has three layers: an epidermis, a dermis, and a hypodermis. (credit: modification of work by Don Bliss, National Cancer Institute)

Somatosensory Receptors

Sensory receptors are classified into five categories: mechanoreceptors, thermoreceptors, proprioceptors, pain receptors, and chemoreceptors. These categories are based on the nature of stimuli each receptor class transduces. What is commonly referred to as "touch" involves more than one kind of stimulus and more than one kind of receptor. Mechanoreceptors in the skin are described as encapsulated (that is, surrounded by a capsule) or unencapsulated (a group that includes free nerve endings). A free nerve ending, as its name implies, is an unencapsulated dendrite of a sensory neuron. Free nerve endings are the most common nerve endings in skin, and they extend into the middle of the epidermis. Free nerve endings are sensitive to painful stimuli, to hot and cold, and to light touch. They are slow to adjust to a stimulus and so are less sensitive to abrupt changes in stimulation.

There are three classes of mechanoreceptors: tactile, proprioceptors, and baroreceptors. Mechanoreceptors sense stimuli due to physical deformation of their plasma membranes. They contain mechanically gated ion channels whose gates open or close in response to pressure, touch, stretching, and sound." There are four primary tactile mechanoreceptors in human skin: Merkel's disks, Meissner's corpuscles, Ruffini endings, and Pacinian corpuscles; two are located toward the surface of the skin and two are located deeper. A fifth type of mechanoreceptor, Krause end bulbs, are found only in specialized regions. Merkel's disks (shown in Figure 2) are found in the upper layers of skin near the base of the epidermis, both in skin that has hair and on glabrous skin, that is, the hairless skin found on the palms and fingers, the soles of the feet, and the lips of humans and other primates. Merkel's disks are densely distributed in the fingertips and lips. They are slow-adapting, encapsulated nerve endings, and they respond to light touch. Light touch, also known as discriminative touch, is a light pressure that allows the location of a stimulus to be pinpointed. The receptive fields of Merkel's disks are small with well-defined borders. That makes them finely sensitive to edges and they come into use in tasks such as typing on a keyboard.

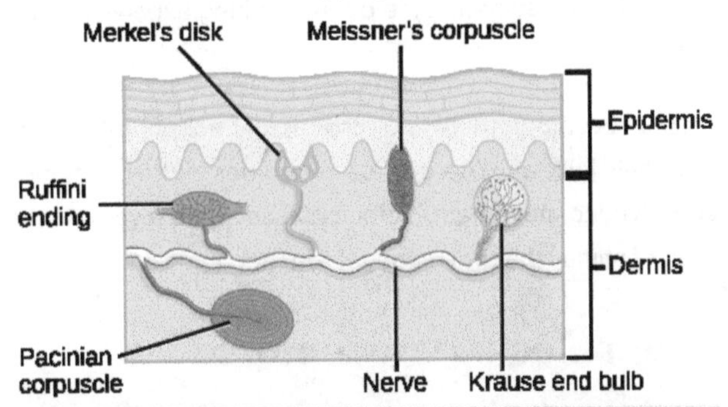

Figure 2. Four of the primary mechanoreceptors in human skin are shown. Merkel's disks, which are unencapsulated, respond to light touch. Meissner's corpuscles, Ruffini endings, Pacinian corpuscles, and Krause end bulbs are all encapsulated. Meissner's corpuscles respond to touch and low-frequency vibration. Ruffini endings detect stretch, deformation within joints, and warmth. Pacinian corpuscles detect transient pressure and high-frequency vibration. Krause end bulbs detect cold.

Meissner's corpuscles, also known as tactile corpuscles, are found in the upper dermis, but they project into the epidermis. They, too, are found primarily in the glabrous skin on the fingertips and eyelids. They respond to fine touch and pressure, but they also respond to low-frequency vibration or flutter. They are rapidly adapting, fluid-filled, encapsulated neurons with small, well-defined borders and are responsive to fine details. Like Merkel's disks, Meissner's corpuscles are not as plentiful in the palms as they are in the fingertips.

Deeper in the epidermis, near the base, are Ruffini endings, which are also known as bulbous corpuscles. They are found in both glabrous and hairy skin. These are slow-adapting, encapsulated mechanoreceptors that detect skin stretch and deformations within joints, so they provide valuable feedback for gripping objects and controlling finger position and movement. Thus, they also contribute to proprioception and kinesthesia. Ruffini endings also detect warmth. Note that these warmth detectors are situated deeper in the skin than are the cold detectors. It is not surprising, then, that humans detect cold stimuli before they detect warm stimuli.

Pacinian corpuscles (seen in Figure 4) are located deep in the dermis of both glabrous and hairy skin and are structurally similar to Meissner's corpuscles; they are found in the bone periosteum, joint capsules, pancreas and other viscera, breast, and genitals. They are rapidly adapting mechanoreceptors that sense deep transient (but not prolonged) pressure and high-frequency vibration. Pacinian receptors detect pressure and vibration by being compressed, stimulating their internal dendrites. There are fewer Pacinian corpuscles and Ruffini endings in skin than there are Merkel's disks and Meissner's corpuscles.

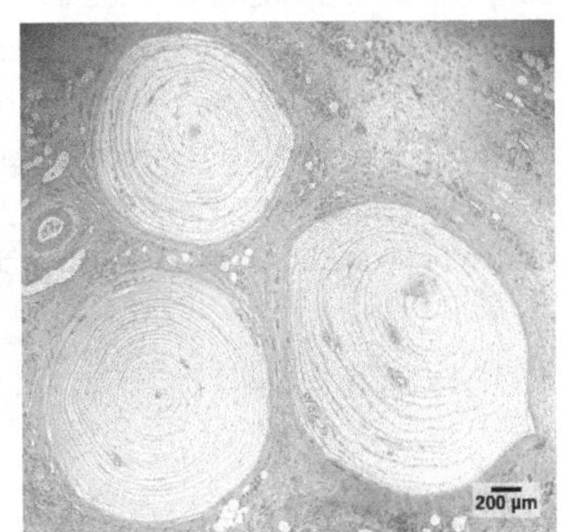

Figure 4. Pacinian corpuscles, such as these visualized using bright field light microscopy, detect pressure (touch) and high-frequency vibration. (credit: modification of work by Ed Uthman; scale-bar data from Matt Russell)

In proprioception, proprioceptive and kinesthetic signals travel through myelinated afferent neurons running from the spinal cord to the medulla. Neurons are not physically connected, but communicate via neurotransmitters secreted into synapses or "gaps" between communicating neurons. Once in the medulla, the neurons continue carrying the signals to the thalamus.

Muscle spindles are stretch receptors that detect the amount of stretch, or lengthening of muscles. Related to these are Golgi tendon organs, which are tension receptors that detect the force of muscle contraction. Unconscious proprioceptive signals run from the spinal cord to the cerebellum, the brain region that coordinates muscle contraction, rather than to the thalamus, like most other sensory information.

Baroreceptors detect pressure changes in an organ. They are found in the walls of the carotid artery and the aorta where they monitor blood pressure, and in the lungs where they detect the degree of lung expansion. Stretch receptors are found at various sites in the digestive and urinary systems.

In addition to these two types of deeper receptors, there are also rapidly adapting hair receptors, which are found on nerve endings that wrap around the base of hair follicles. There are a few types of hair receptors that detect slow and rapid hair movement, and they differ in their sensitivity to movement. Some hair receptors also detect skin deflection, and certain rapidly adapting hair receptors allow detection of stimuli that have not yet touched the skin.

Integration of Signals from Mechanoreceptors

The configuration of the different types of receptors working in concert in human skin results in a very refined sense of touch. The nociceptive receptors—those that detect pain—are located near the surface. Small, finely calibrated mechanoreceptors—Merkel's disks and Meissner's corpuscles—are located in the upper layers and can precisely localize even gentle touch. The large mechanoreceptors—Pacinian corpuscles and Ruffini endings—are located in the lower layers and respond to deeper touch. (Consider that the deep pressure that reaches those deeper receptors would not need to be finely localized.) Both the upper and lower layers of the skin hold rapidly and slowly adapting receptors. Both primary somatosensory cortex and secondary cortical areas are responsible for processing the complex picture of stimuli transmitted from the interplay of mechanoreceptors.

Density of Mechanoreceptors

The distribution of touch receptors in human skin is not consistent over the body. In humans, touch receptors are less dense in skin covered with any type of hair, such as the arms, legs, torso, and face. Touch receptors are denser in glabrous skin (the type found on fingertips and lips, for example), which is typically more sensitive and is thicker than hairy skin (4 to 5 mm versus 2 to 3 mm).

How is receptor density estimated in a human subject? The relative density of pressure receptors in different locations on the body can be demonstrated experimentally using a two-point discrimination test. In this demonstration, two sharp points, such as two thumbtacks, are brought into contact with the subject's skin (though not hard enough to cause pain or break the skin). The subject reports if he or she feels one point or two points. If the two points are felt as one point, it can be inferred that the two points are both in the receptive field of a single sensory receptor. If two points are felt as two separate points, each is in the receptive field of two separate sensory receptors. The points could then be moved closer and retested until the subject reports feeling only one point, and the size of the receptive field of a single receptor could be estimated from that distance.

Thermoreception

In addition to Krause end bulbs that detect cold and Ruffini endings that detect warmth, there are different types of cold receptors on some free nerve endings: thermoreceptors, located in the dermis, skeletal muscles, liver, and hypothalamus, that are activated by different temperatures. Their pathways into the brain run from the spinal cord through the thalamus to the primary somatosensory cortex. Warmth and cold information from the face travels through one of the cranial nerves to the brain.

You know from experience that a tolerably cold or hot stimulus can quickly progress to a much more intense stimulus that is no longer tolerable. Any stimulus that is too intense can be perceived as pain because temperature sensations are conducted along the same pathways that carry pain sensations.

Pain

Pain is the name given to nociception, which is the neural processing of injurious stimuli in response to tissue damage. Pain is caused by true sources of injury, such as contact with a heat source that causes a thermal burn or contact with a corrosive chemical. But pain also can be caused by harmless stimuli that mimic the action of damaging stimuli, such as contact with capsaicins, the compounds that cause peppers to taste hot and which are used in self-defense pepper sprays and certain topical medications. Peppers taste "hot" because the protein receptors that bind capsaicin open the same calcium channels that are activated by warm receptors.

Nociception starts at the sensory receptors, but pain, inasmuch as it is the perception of nociception, does not start until it is communicated to the brain. There are several nociceptive pathways to and through the brain. Most axons carrying nociceptive information into the brain from the spinal cord project to the thalamus (as do other sensory neurons) and the neural signal undergoes final processing in the primary somatosensory cortex. Interestingly, one nociceptive pathway projects not to the thalamus but directly to the hypothalamus in the forebrain, which modulates the cardiovascular and neuroendocrine functions of the autonomic nervous system. Recall that threatening—or painful—stimuli stimulate the sympathetic branch of the visceral sensory system, readying a fight-or-flight response.

Taste and Smell

Taste, also called gustation, and smell, also called olfaction, are the most interconnected senses in that both involve molecules of the stimulus entering the body and bonding to receptors. Smell lets an animal sense the presence of food or other animals—whether potential mates, predators, or prey—or other chemicals in the environment that can impact their survival. Similarly, the sense of taste allows animals to discriminate between types of foods. While the value of a sense of smell is obvious, what is the value of a sense of taste? Different tasting foods have different attributes, both helpful and harmful. For example, sweet-tasting substances tend to be highly caloric, which could be necessary for survival in lean times. Bitterness is associated with toxicity, and sourness is associated with spoiled food. Salty foods are valuable in maintaining homeostasis by helping the body retain water and by providing ions necessary for cells to function.

Tastes and Odors

Both taste and odor stimuli are molecules taken in from the environment. The primary tastes detected by humans are sweet, sour, bitter, salty, and umami. The first four tastes need little explanation. The identification of umami as a fundamental taste occurred fairly recently—it was identified in 1908 by Japanese scientist Kikunae Ikeda while he worked with seaweed broth, but it was not widely accepted as a taste that could be physiologically distinguished until many years later. The taste of umami, also known as savoriness, is attributable to the taste of the amino acid L-glutamate. In fact, monosodium glutamate, or MSG, is often used in cooking to enhance the savory taste of certain foods. What is the adaptive value of being able to distinguish umami? Savory substances tend to be high in protein.

All odors that we perceive are molecules in the air we breathe. If a substance does not release molecules into the air from its surface, it has no smell. And if a human or other animal does not have a receptor that recognizes a specific molecule, then that molecule has no smell. Humans have about 350 olfactory receptor subtypes that work in various combinations to allow us to sense about 10,000 different odors. Compare that to mice, for example, which have about 1,300 olfactory receptor types, and therefore probably sense more odors. Both odors and tastes involve molecules that stimulate specific chemoreceptors. Although humans commonly distinguish taste as one sense and smell as another, they work together to create the perception of flavor. A person's perception of flavor is reduced if he or she has congested nasal passages.

Reception and Transduction

Odorants (odor molecules) enter the nose and dissolve in the olfactory epithelium, the mucosa at the back of the nasal cavity (as illustrated in Figure 1). The olfactory epithelium is a collection of specialized olfactory receptors in the back of the nasal cavity that spans an area about 5 cm^2 in humans. Recall that sensory cells are neurons. An olfactory receptor, which is a dendrite of a specialized neuron, responds when it binds certain molecules inhaled from the environment by sending impulses directly to the olfactory bulb of the brain. Humans have about 12 million olfactory receptors, distributed among hundreds of different receptor types that respond to different odors. Twelve million seems like a large number of receptors, but compare that to other animals: rabbits have about 100 million, most dogs have about 1 billion, and bloodhounds—dogs selectively bred for their sense of smell—have about 4 billion. The overall size of the olfactory epithelium also differs between species, with that of bloodhounds, for example, being many times larger than that of humans.

Olfactory neurons are bipolar neurons (neurons with two processes from the cell body). Each neuron has a single dendrite buried in the olfactory epithelium, and extending from this dendrite are 5 to 20 receptor-laden, hair-like cilia that trap odorant molecules. The sensory receptors on the cilia are proteins, and it is the variations in their amino acid chains that make the receptors sensitive to different odorants. Each olfactory sensory neuron has only one type of receptor on its cilia, and the receptors are specialized to detect specific odorants, so the bipolar neurons themselves are specialized. When an odorant binds with a receptor that recognizes it, the sensory neuron associated with the receptor is stimulated. Olfactory stimulation is the only sensory information that directly reaches the cerebral cortex, whereas other sensations are relayed through the thalamus.

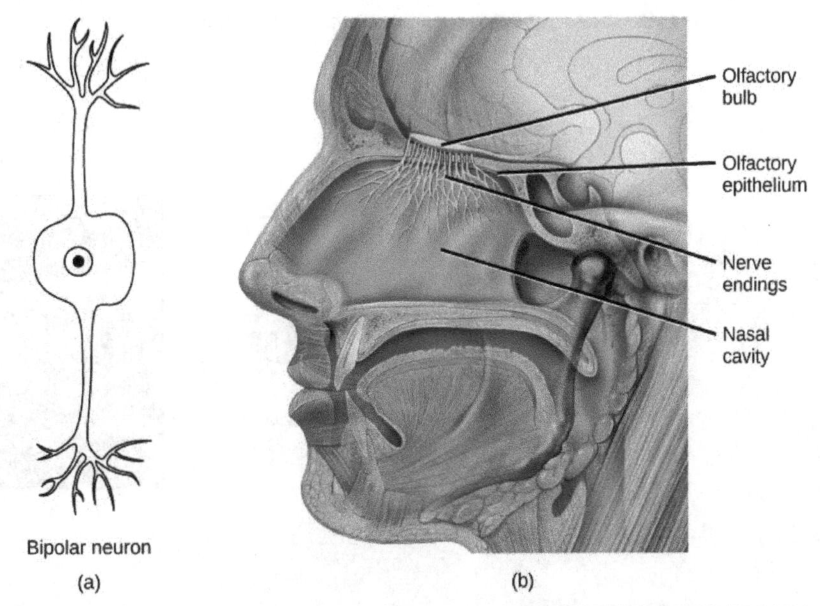

Figure 1. In the human olfactory system, (a) bipolar olfactory neurons extend from (b) the olfactory epithelium, where olfactory receptors are located, to the olfactory bulb. (credit: modification of work by Patrick J. Lynch, medical illustrator; C. Carl Jaffe, MD, cardiologist)

Taste

Detecting a taste (gustation) is fairly similar to detecting an odor (olfaction), given that both taste and smell rely on chemical receptors being stimulated by certain molecules. The primary organ of taste is the taste bud. A taste bud is a cluster of gustatory receptors (taste cells) that are located within the bumps on the tongue called papillae (singular: papilla) (illustrated in Figure 4). There are several structurally distinct papillae. Filiform papillae, which are located across the tongue, are tactile, providing friction that helps the tongue move substances, and contain no taste cells. In contrast, fungiform papillae, which are located mainly on the anterior two-thirds of the tongue, each contain one to eight taste buds and also have receptors for pressure and temperature. The large circumvallate papillae contain up to 100 taste buds and form a V near the posterior margin of the tongue.

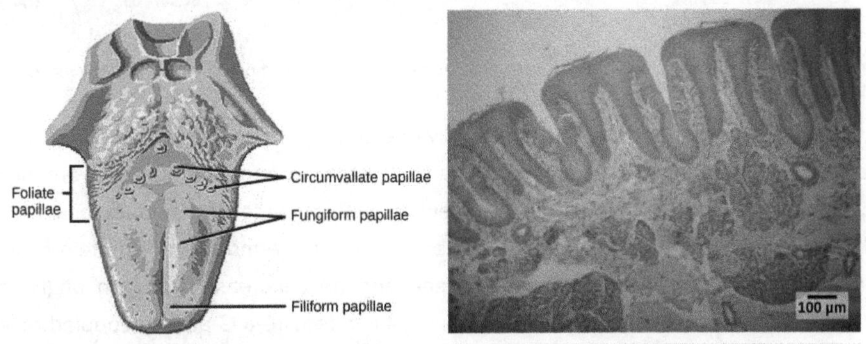

Figure 3. (a) Foliate, circumvallate, and fungiform papillae are located on different regions of the tongue. (b) Foliate papillae are prominent protrusions on this light micrograph. (credit a: modification of work by NCI; scale-bar data from Matt Russell)

In addition to those two types of chemically and mechanically sensitive papillae are foliate papillae—leaf-like papillae located in parallel folds along the edges and toward the back of the tongue, as seen in the Figure 3 micrograph. Foliate papillae contain about 1,300 taste buds within their folds. Finally, there are circumvallate papillae, which are wall-like papillae in the shape of an inverted "V" at the back of the tongue. Each of these papillae is surrounded by a groove and contains about 250 taste buds.

Each taste bud's taste cells are replaced every 10 to 14 days. These are elongated cells with hair-like processes called microvilli at the tips that extend into the taste bud pore (illustrated in Figure 4). Food molecules (tastants) are dissolved in saliva, and they bind with and stimulate the receptors on the microvilli. The receptors for tastants are located across the outer portion and front of the tongue, outside of the middle area where the filiform papillae are most prominent.

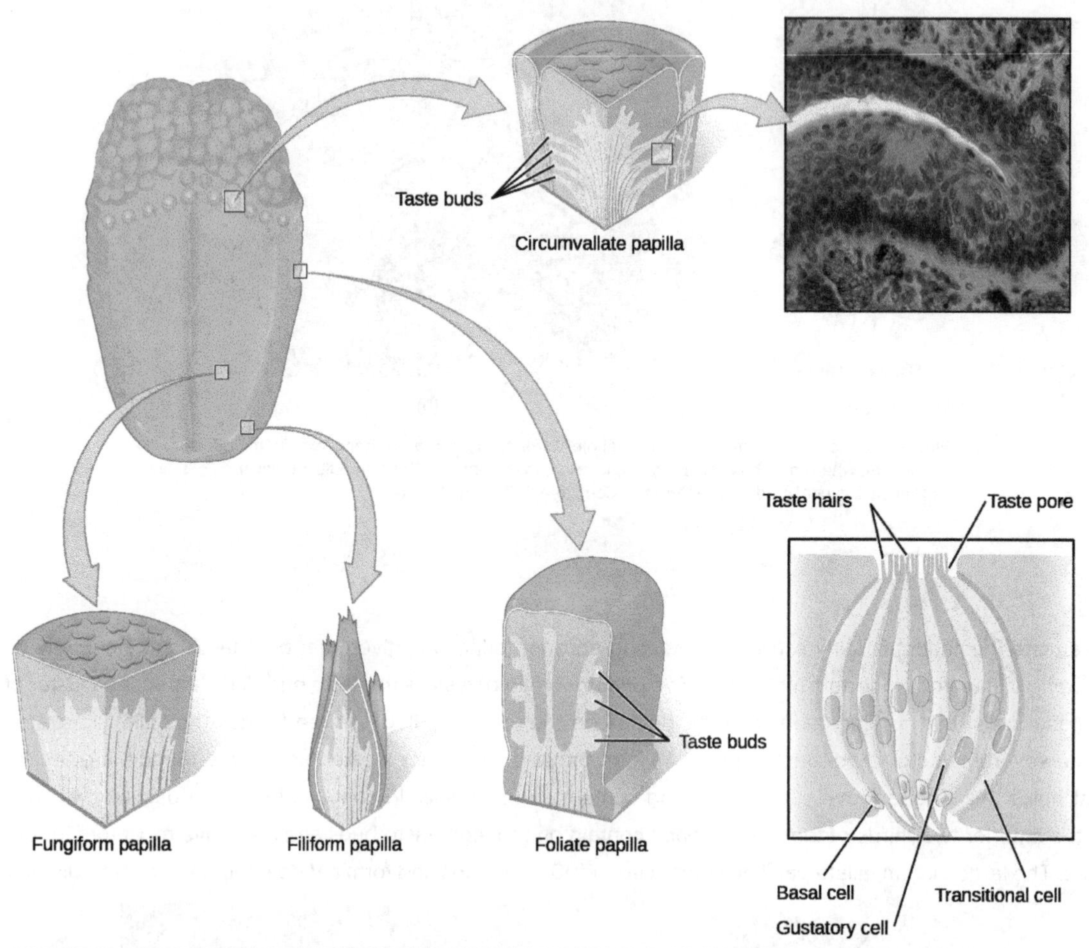

Figure 4. Pores in the tongue allow tastants to enter taste pores in the tongue. (credit: modification of work by Vincenzo Rizzo)

In humans, there are five primary tastes, and each taste has only one corresponding type of receptor. Thus, like olfaction, each receptor is specific to its stimulus (tastant). Transduction of the five tastes happens through different mechanisms that reflect the molecular composition of the tastant. A salty tastant (containing NaCl) provides the sodium ions (Na^+) that enter the taste neurons and excite them directly. Sour tastants are acids and belong to the thermoreceptor protein family. Binding of an acid or other sour-tasting molecule triggers a change in the ion channel and these increase hydrogen ion (H^+) concentrations in the taste neurons, thus depolarizing them. Sweet, bitter, and umami tastants require a G-protein coupled receptor. These tastants bind to their respective receptors, thereby exciting the specialized neurons associated with them.

Both tasting abilities and sense of smell change with age. In humans, the senses decline dramatically by age 50 and continue to decline. A child may find a food to be too spicy, whereas an elderly person may find the same food to be bland and unappetizing.

Smell and Taste in the Brain

Olfactory neurons project from the olfactory epithelium to the olfactory bulb as thin, unmyelinated axons. The olfactory bulb is composed of neural clusters called glomeruli, and each glomerulus receives signals from one type of olfactory receptor, so each glomerulus is specific to one odorant. From glomeruli, olfactory signals travel directly to the olfactory cortex and then to the frontal cortex and the thalamus. Recall that this is a different path from most other sensory information, which is sent directly to the thalamus before ending up in the cortex. Olfactory signals also travel directly to the amygdala, thereafter reaching the hypothalamus, thalamus, and frontal cortex. The last structure that olfactory signals directly travel to is a cortical center in the temporal lobe structure important in spatial, autobiographical, declarative, and episodic memories. Olfaction is finally processed by areas of the brain that deal with memory, emotions, reproduction, and thought.

Taste neurons project from taste cells in the tongue, esophagus, and palate to the medulla, in the brainstem. From the medulla, taste signals travel to the thalamus and then to the primary gustatory cortex. Information from different regions of the tongue is segregated in the medulla, thalamus, and cortex.

Hearing and Vestibular Sensation

Audition, or hearing, is important to humans and to other animals for many different interactions. It enables an organism to detect and receive information about danger, such as an approaching predator, and to participate in communal exchanges like those concerning territories or mating. On the other hand, although it is physically linked to the auditory system, the vestibular system is not involved in hearing. Instead, an animal's vestibular system detects its own movement, both linear and angular acceleration and deceleration, and balance.

Sound

Auditory stimuli are sound waves, which are mechanical, pressure waves that move through a medium, such as air or water. There are no sound waves in a vacuum since there are no air molecules to move in waves. The speed of sound waves differs, based on altitude, temperature, and medium, but at sea level and a temperature of 20° C (68° F), sound waves travel in the air at about 343 meters per second.

As is true for all waves, there are four main characteristics of a sound wave: frequency, wavelength, period, and amplitude. Frequency is the number of waves per unit of time, and in sound is heard as pitch. High-frequency (≥15.000Hz) sounds are higher-pitched (short wavelength) than low-frequency (long wavelengths; ≤100Hz) sounds. Frequency is measured in cycles per second, and for sound, the most commonly used unit is hertz (Hz), or cycles per second. Most humans can perceive sounds with frequencies between 30 and 20,000 Hz. Women are typically better at hearing high frequencies, but everyone's ability to hear high frequencies decreases with age. Dogs detect up to about 40,000 Hz; cats, 60,000 Hz; bats, 100,000 Hz; and dolphins 150,000 Hz, and American shad (*Alosa sapidissima*), a fish, can hear 180,000 Hz. Those frequencies above the human range are called ultrasound.

Amplitude, or the dimension of a wave from peak to trough, in sound is heard as volume and is illustrated in Figure 1. The sound waves of louder sounds have greater amplitude than those of softer sounds. For sound, volume is measured in decibels (dB). The softest sound that a human can hear is the zero point. Humans speak normally at 60 decibels.

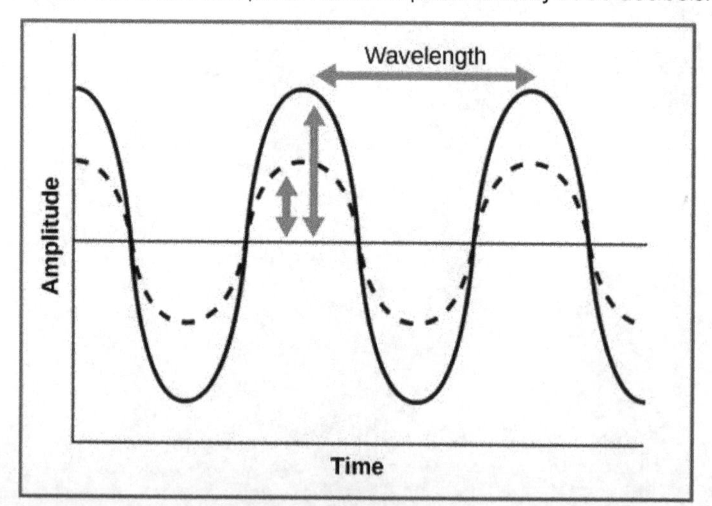

Figure 1. For sound waves, wavelength corresponds to pitch. Amplitude of the wave corresponds to volume. The sound wave shown with a dashed line is softer in volume than the sound wave shown with a solid line. (credit: NIH)

Reception of Sound

Sound waves are collected by the external, cartilaginous part of the ear called the pinna, then travel through the auditory canal and cause vibration of the thin diaphragm called the tympanum or ear drum, the innermost part of the outer ear (illustrated in Figure 2). Interior to the tympanum is the middle ear. The middle ear holds three small bones called the ossicles, which transfer energy from the moving tympanum to the inner ear. The three ossicles are the malleus (also known as the hammer), the incus (the anvil), and stapes (the stirrup). The aptly named stapes looks very much like a stirrup. The three ossicles are unique to mammals, and each plays a role in hearing. The malleus attaches at three points to the interior surface of the tympanic membrane. The incus attaches the malleus to the stapes. In humans, the stapes is not long enough to reach the tympanum. If we did not have the malleus and the incus, then the vibrations of the tympanum would never reach the inner ear. These bones also function to collect force and amplify sounds.

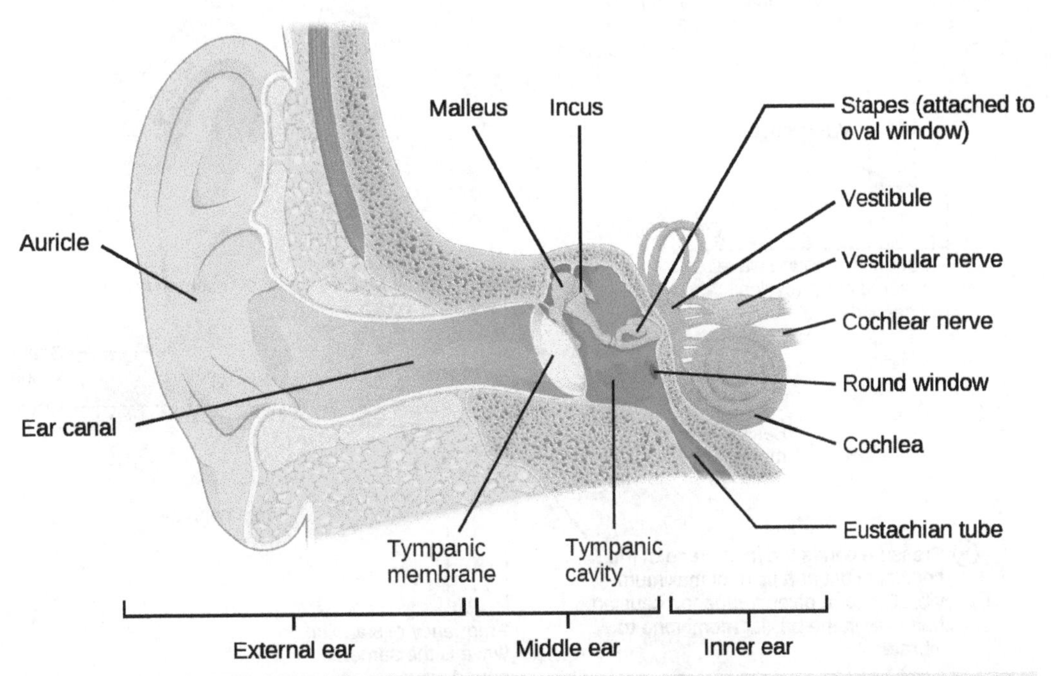

Figure 2. Sound travels through the outer ear to the middle ear, which is bounded on its exterior by the tympanic membrane. The middle ear contains three bones called ossicles that transfer the sound wave to the oval window, the exterior boundary of the inner ear. The organ of Corti, which is the organ of sound transduction, lies inside the cochlea.

Transduction of Sound

Vibrating objects, such as vocal cords, create sound waves or pressure waves in the air. When these pressure waves reach the ear, the ear transduces this mechanical stimulus (pressure wave) into a nerve impulse (electrical signal) that the brain perceives as sound. The pressure waves strike the tympanum, causing it to vibrate. The mechanical energy from the moving tympanum transmits the vibrations to the three bones of the middle ear. The stapes transmits the vibrations to a thin diaphragm called the oval window, which is the outermost structure of the inner ear. The structures of the inner ear are found in the labyrinth, a bony, hollow structure that is the most interior portion of the ear. Here, the energy from the sound wave is transferred from the stapes through the flexible oval window and to the fluid of the cochlea. The vibrations of the oval window create pressure waves in the fluid (perilymph) inside the cochlea. The cochlea is a whorled structure, like the shell of a snail, and it contains receptors for transduction of the mechanical wave into an electrical signal (as illustrated in Figure 3). Inside the cochlea, the basilar membrane is a mechanical analyzer that runs the length of the cochlea, curling toward the cochlea's center.

The mechanical properties of the basilar membrane change along its length, such that it is thicker, tauter, and narrower at the outside of the whorl (where the cochlea is largest), and thinner, floppier, and broader toward the apex, or center, of the whorl (where the cochlea is smallest). Different regions of the basilar membrane vibrate according to the frequency of the sound wave conducted through the fluid in the cochlea. For these reasons, the fluid-filled cochlea detects different wave frequencies (pitches) at different regions of the membrane. When the sound waves in the cochlear fluid contact the basilar membrane, it flexes back and forth in a wave-like fashion. Above the basilar membrane is the tectorial membrane.

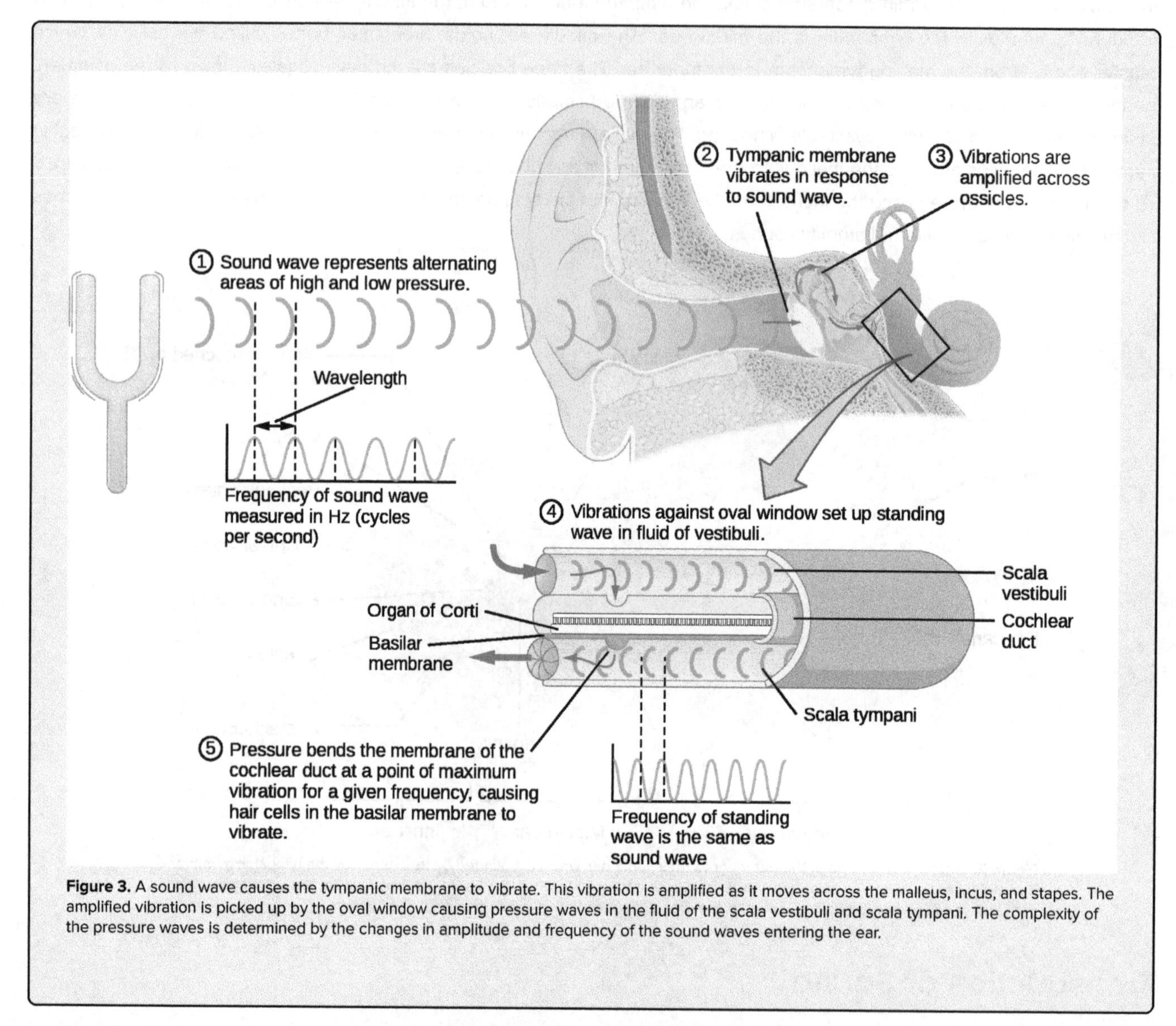

Figure 3. A sound wave causes the tympanic membrane to vibrate. This vibration is amplified as it moves across the malleus, incus, and stapes. The amplified vibration is picked up by the oval window causing pressure waves in the fluid of the scala vestibuli and scala tympani. The complexity of the pressure waves is determined by the changes in amplitude and frequency of the sound waves entering the ear.

The site of transduction is in the organ of Corti (spiral organ). It is composed of hair cells held in place above the basilar membrane like flowers projecting up from soil, with their exposed short, hair-like stereocilia contacting or embedded in the tectorial membrane above them. The inner hair cells are the primary auditory receptors and exist in a single row, numbering approximately 3,500. The stereocilia from inner hair cells extend into small dimples on the tectorial membrane's lower surface. The outer hair cells are arranged in three or four rows. They number approximately 12,000, and they function to fine tune incoming sound waves. The longer stereocilia that project from the outer hair cells actually attach to the tectorial membrane. All of the stereocilia are mechanoreceptors, and when bent by vibrations they respond by opening a gated ion channel. As a result, the hair cell membrane is depolarized, and a signal is transmitted to the chochlear nerve. Intensity (volume) of sound is determined by how many hair cells at a particular location are stimulated.

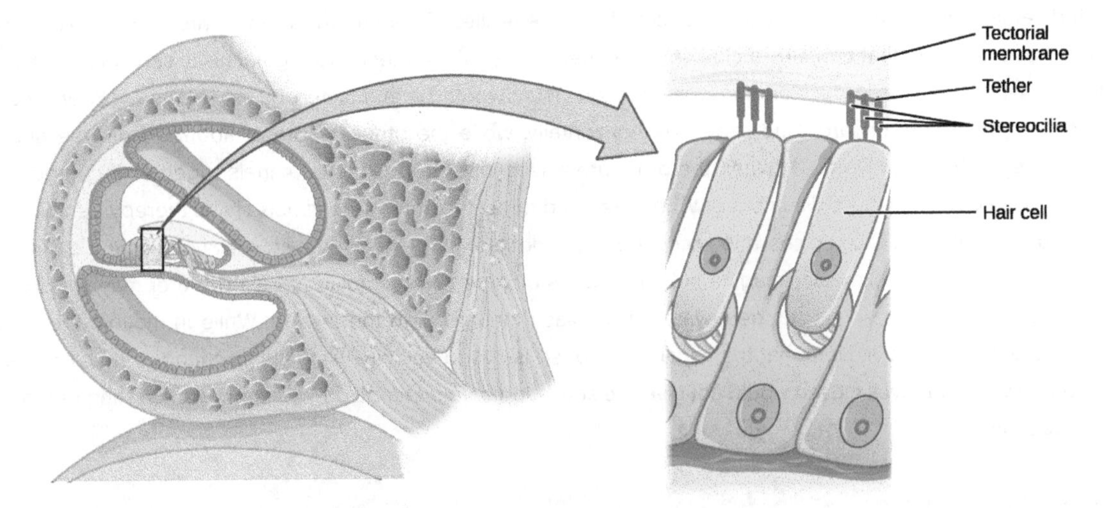

Figure 4. The hair cell is a mechanoreceptor with an array of stereocilia emerging from its apical surface. The stereocilia are tethered together by proteins that open ion channels when the array is bent toward the tallest member of their array, and closed when the array is bent toward the shortest member of their array.

When sound waves produce fluid waves inside the cochlea, the basilar membrane flexes, bending the stereocilia that attach to the tectorial membrane. Their bending results in action potentials in the hair cells, and auditory information travels along the neural endings of the bipolar neurons of the hair cells (collectively, the auditory nerve) to the brain. When the hairs bend, they release an excitatory neurotransmitter at a synapse with a sensory neuron, which then conducts action potentials to the central nervous system. The cochlear branch of the vestibulocochlear cranial nerve sends information on hearing. The auditory system is very refined, and there is some modulation or "sharpening" built in. The brain can send signals back to the cochlea, resulting in a change of length in the outer hair cells, sharpening or dampening the hair cells' response to certain frequencies.

Vestibular Information

The stimuli associated with the vestibular system are linear acceleration (gravity) and angular acceleration and deceleration. Gravity, acceleration, and deceleration are detected by evaluating the inertia on receptive cells in the vestibular system. Gravity is detected through head position. Angular acceleration and deceleration are expressed through turning or tilting of the head.

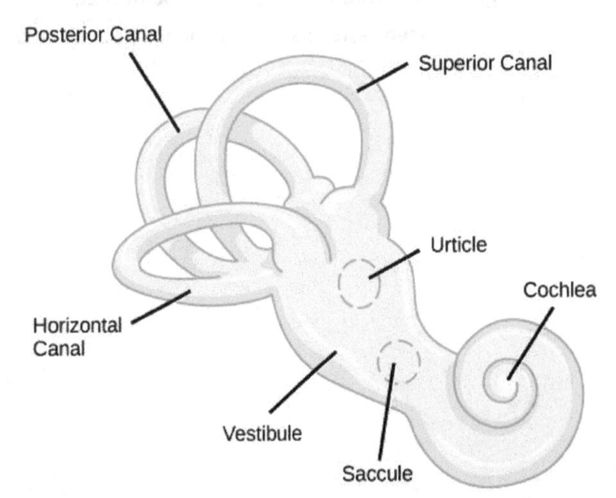

Figure 5. The structure of the vestibular labyrinth is shown. (credit: modification of work by NIH)

The vestibular system has some similarities with the auditory system. It utilizes hair cells just like the auditory system, but it excites them in different ways. There are five vestibular receptor organs in the inner ear: the utricle, the saccule, and three semicircular canals shown in Figure 4. The utricle and saccule respond to acceleration in a straight line, such as gravity. The roughly 30,000 hair cells in the utricle and 16,000 hair cells in the saccule lie below a gelatinous layer, with their stereocilia projecting into the gelatin. Embedded in this gelatin are calcium carbonate crystals—like tiny rocks. When the head is tilted, the crystals continue to be pulled straight down by gravity, but the new angle of the head causes the gelatin to shift, thereby bending the stereocilia. The bending of the stereocilia stimulates the neurons, and they signal to the brain that the head is tilted, allowing the maintenance of balance. It is the vestibular branch of the vestibulocochlear cranial nerve that deals with balance.

The fluid-filled semicircular canals are tubular loops set at oblique angles. They are arranged in three spatial planes. The base of each canal has a swelling that contains a cluster of hair cells. The hairs project into a gelatinous cap called the cupula and monitor angular acceleration and deceleration from rotation. They would be stimulated by driving your car around a corner, turning your head, or falling forward. One canal lies horizontally, while the other two lie at about 45 degree angles to the horizontal axis, as illustrated in Figure 4. When the brain processes input from all three canals together, it can detect angular acceleration or deceleration in three dimensions. When the head turns, the fluid in the canals shifts, thereby bending stereocilia and sending signals to the brain. Upon cessation accelerating or decelerating—or just moving—the movement of the fluid within the canals slows or stops. For example, imagine holding a glass of water. When moving forward, water may splash backwards onto the hand, and when motion has stopped, water may splash forward onto the fingers. While in motion, the water settles in the glass and does not splash. Note that the canals are not sensitive to velocity itself, but to changes in velocity, so moving forward at 60mph with your eyes closed would not give the sensation of movement, but suddenly accelerating or braking would stimulate the receptors.

Higher Processing

Hair cells from the utricle, saccule, and semicircular canals also communicate through bipolar neurons to the cochlear nucleus in the medulla. Cochlear neurons send descending projections to the spinal cord and ascending projections to the pons, thalamus, and cerebellum. Connections to the cerebellum are important for coordinated movements. There are also projections to the temporal cortex, which account for feelings of dizziness; projections to autonomic nervous system areas in the brainstem, which account for motion sickness; and projections to the primary somatosensory cortex, which monitors subjective measurements of the external world and self-movement. People with lesions in the vestibular area of the somatosensory cortex see vertical objects in the world as being tilted. Finally, the vestibular signals project to certain optic muscles to coordinate eye and head movements.

Vision

Vision is the ability to detect light patterns from the outside environment and interpret them into images. Animals are bombarded with sensory information, and the sheer volume of visual information can be problematic. Fortunately, the visual systems of species have evolved to attend to the most-important stimuli. The importance of vision to humans is further substantiated by the fact that about one-third of the human cerebral cortex is dedicated to analyzing and perceiving visual information.

Light

As with auditory stimuli, light travels in waves. The compression waves that compose sound must travel in a medium—a gas, a liquid, or a solid. In contrast, light is composed of electromagnetic waves and needs no medium; light can travel in a vacuum (Figure 1). The behavior of light can be discussed in terms of the behavior of waves and also in terms of the behavior of the fundamental unit of light—a packet of electromagnetic radiation called a photon. A glance at the electromagnetic spectrum shows that visible light for humans is just a small slice of the entire spectrum, which includes radiation that we cannot see as light because it is below the frequency of visible red light and above the frequency of visible violet light.

Certain variables are important when discussing perception of light. Wavelength (which varies inversely with frequency) manifests itself as hue. Light at the red end of the visible spectrum has longer wavelengths (and is lower frequency), while light at the violet end has shorter wavelengths (and is higher frequency). The wavelength of light is expressed in nanometers (nm); one nanometer is one billionth of a meter. Humans perceive light that ranges between approximately 380 nm and 740 nm. Some other animals, though, can detect wavelengths outside of the human range. For example, bees see near-ultraviolet light in order to locate nectar guides on flowers, and some non-avian reptiles sense infrared light (heat that prey gives off).

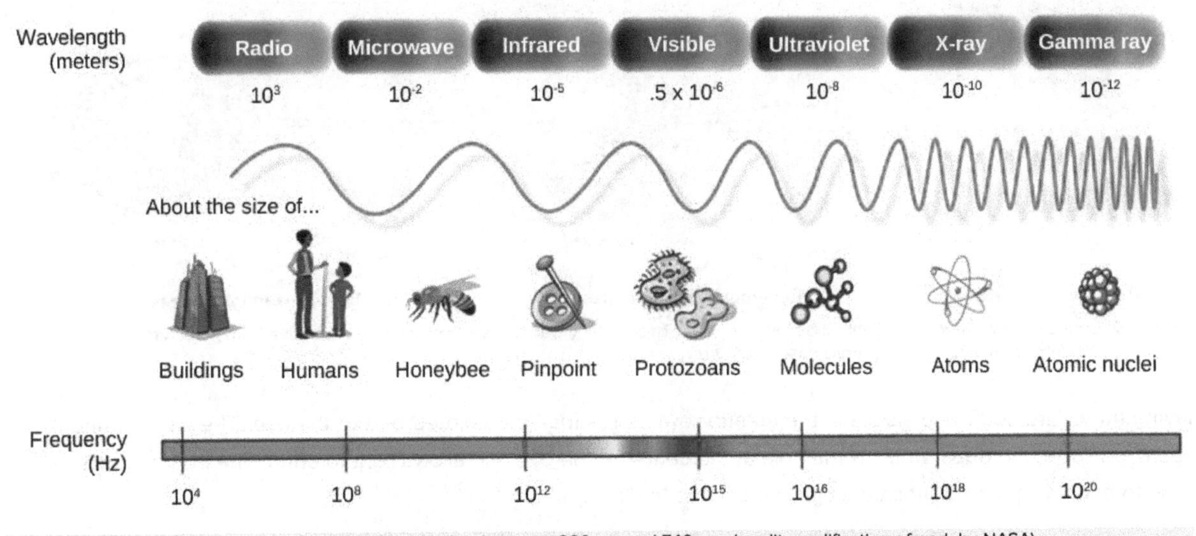

Figure 1. In the electromagnetic spectrum, visible light lies between 380 nm and 740 nm. (credit: modification of work by NASA)

Wave amplitude is perceived as luminous intensity, or brightness. The standard unit of intensity of light is the candela, which is approximately the luminous intensity of one common candle.

Light waves travel 299,792 km per second in a vacuum, (and somewhat slower in various media such as air and water), and those waves arrive at the eye as long (red), medium (green), and short (blue) waves. What is termed "white light" is light that is perceived as white by the human eye. This effect is produced by light that stimulates equally the color receptors in the human eye. The apparent color of an object is the color (or colors) that the object reflects. Thus a red object reflects the red wavelengths in mixed (white) light and absorbs all other wavelengths of light.

Anatomy of the Eye

The eye itself is a hollow sphere composed of three layers of tissue. The outermost layer is the **fibrous tunic**, which includes the white **sclera** and clear **cornea**. The sclera accounts for five sixths of the surface of the eye, most of which is not visible, though humans are unique compared with many other species in having so much of the "white of the eye" visible (Figure 2). The transparent cornea covers the anterior tip of the eye and allows light to enter the eye.

The middle layer of the eye is the **vascular tunic**, which is mostly composed of the choroid, ciliary body, and iris. The **choroid** is a layer of highly vascularized connective tissue that provides a blood supply to the eyeball. The choroid is posterior to the **ciliary body**, a muscular structure that is attached to the **lens** by **zonule fibers**. These two structures bend the lens, allowing it to focus light on the back of the eye. The lens is dynamic, focusing and re-focusing light as the eye rests on near and far objects in the visual field. The ciliary body stretches the lens flat or allows it to thicken, changing the focal length of light coming through it to focus it sharply on the retina. With age comes the loss of the flexibility of the lens, and a form of farsightedness called presbyopia results. Presbyopia occurs because the image focuses behind the retina. Presbyopia is a deficit similar to a different type of farsightedness called hyperopia caused by an eyeball that is too short. For both defects, images in the distance are clear but images nearby are blurry. Myopia (nearsightedness) occurs when an eyeball is elongated and the image focus falls in front of the retina. In this case, images in the distance are blurry but images nearby are clear.

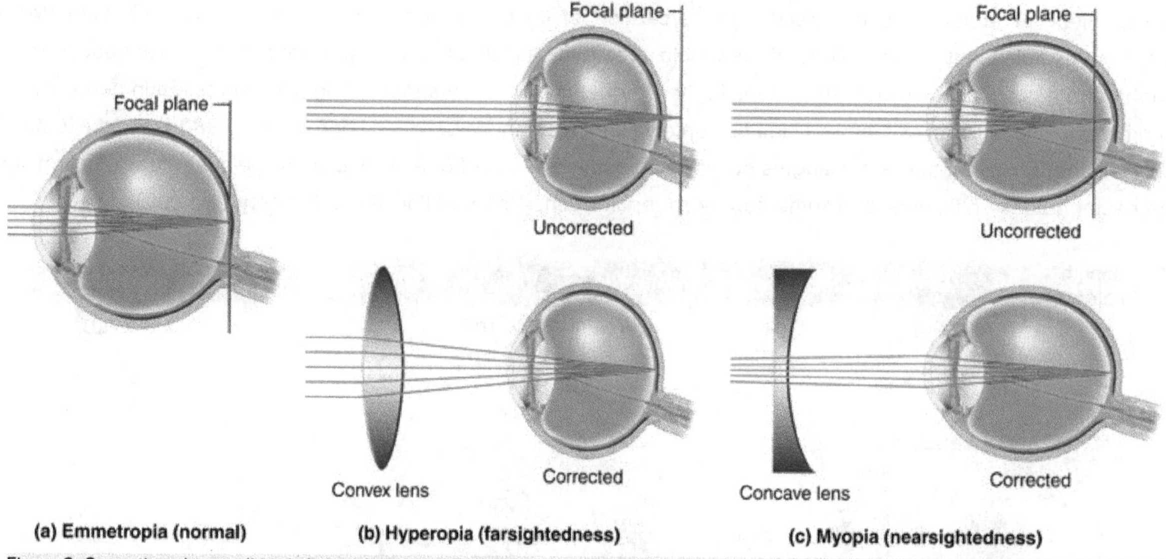

Figure 2: Correcting abnormalities in light refraction in the eye. This work by Cenveo is licensed under a Creative Commons Attribution 3.0 United States (http://creativecommons.org/licenses/by/3.0/us/).

Overlaying the ciliary body, and visible in the anterior eye, is the **iris**—the colored part of the eye. The iris is a smooth muscle that opens or closes the **pupil**, which is the hole at the center of the eye that allows light to enter. The iris constricts the pupil in response to bright light and dilates the pupil in response to dim light.

The innermost layer of the eye is the **neural tunic**, or **retina**, which contains the nervous tissue responsible for photoreception. The eye is also divided into two cavities: the anterior cavity and the posterior cavity. The anterior cavity is the space between the cornea and lens, including the iris and ciliary body. It is filled with a watery fluid called the **aqueous humor**. The posterior cavity is the space behind the lens that extends to the posterior side of the interior eyeball, where the retina is located. The posterior cavity is filled with a more viscous fluid called the **vitreous humor**. The retina is composed of several layers and contains specialized cells for the initial processing of visual stimuli. The photoreceptors (rods and cones) change their membrane potential when stimulated by light energy. The change in membrane potential alters the amount of neurotransmitter that the photoreceptor cells release onto **bipolar cells** in the **outer synaptic layer**. It is the bipolar cell in the retina that connects a photoreceptor to a **retinal ganglion cell (RGC)** in the **inner synaptic layer**. There, **amacrine cells** additionally contribute to retinal processing before an action potential is produced by the RGC. The axons of RGCs, which lie at the innermost layer of the retina, collect at the **optic disc** and leave the eye as the **optic nerve** (see Figure 3). Because these axons pass through the retina, there are no photoreceptors at the very back of the eye, where the optic nerve begins. This creates a "blind spot" in the retina, and a corresponding blind spot in our visual field.

Note that the photoreceptors in the retina (rods and cones) are located behind the axons, RGCs, bipolar cells, and retinal blood vessels. A significant amount of light is absorbed by these structures before the light reaches the photoreceptor cells. However, at the exact center of the retina is a small area known as the **fovea**. At the fovea, the retina lacks the supporting cells and blood vessels, and only contains photoreceptors. Therefore, **visual acuity**, or the sharpness of vision, is greatest at the fovea. This is because the fovea is where the least amount of incoming light is absorbed by other retinal structures (see Figure 3).

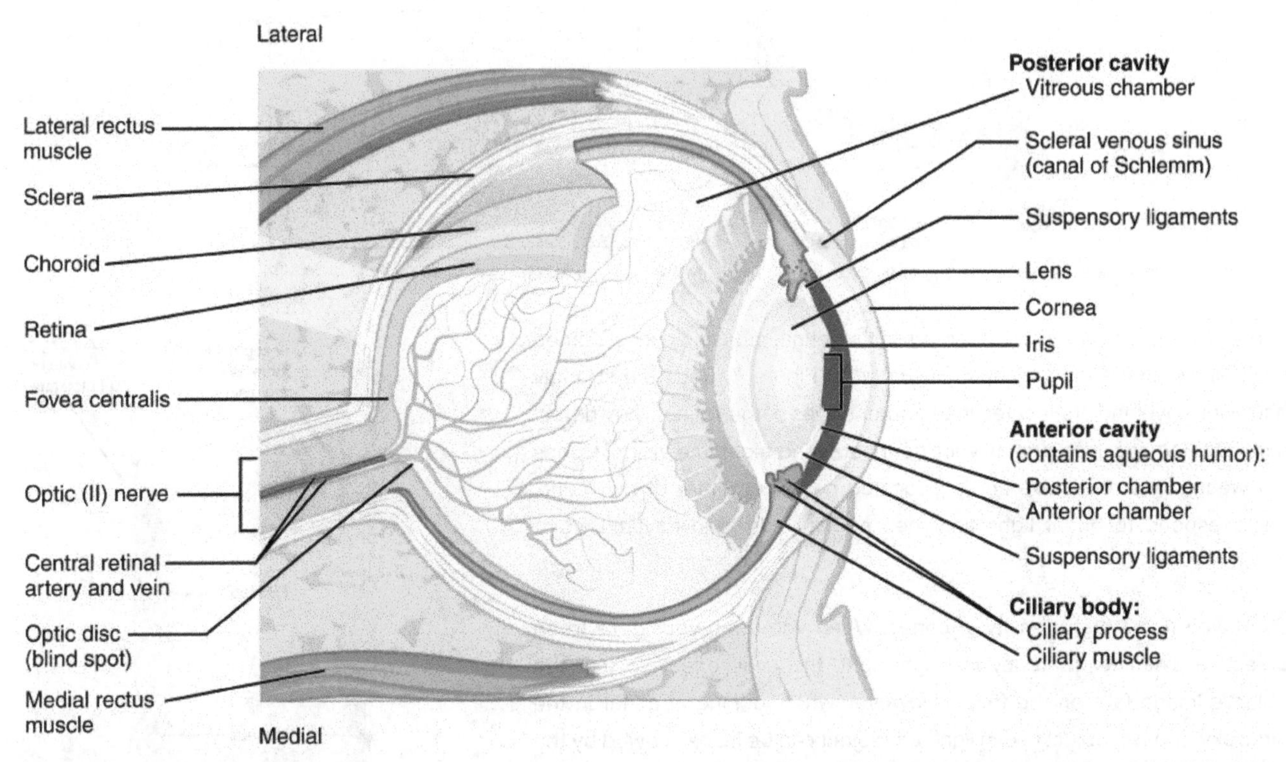

Figure 3. Structure of the Eye The sphere of the eye can be divided into anterior and posterior chambers. The wall of the eye is composed of three layers: the fibrous tunic, vascular tunic, and neural tunic. Within the neural tunic is the retina, with three layers of cells and two synaptic layers in between. The center of the retina has a small indentation known as the fovea.

As one moves in either direction from this central point of the retina, visual acuity drops significantly. In addition, each photoreceptor cell of the fovea is connected to a single RGC. Therefore, this RGC does not have to integrate inputs from multiple photoreceptors, which reduces the accuracy of visual transduction. Toward the edges of the retina, several photoreceptors converge on RGCs (through the bipolar cells) up to a ratio of 50 to 1.

The difference in visual acuity between the fovea and peripheral retina is easily evidenced by looking directly at a word in the middle of this paragraph. The visual stimulus in the middle of the field of view falls on the fovea and is in the sharpest focus. Without moving your eyes off that word, notice that words at the beginning or end of the paragraph are not in focus. The images in your peripheral vision are focused by the peripheral retina, and have vague, blurry edges and words that are not as clearly identified. As a result, a large part of the neural function of the eyes is concerned with moving the eyes and head so that important visual stimuli are centered on the fovea. Light falling on the retina causes chemical changes to pigment molecules in the photoreceptors, ultimately leading to a change in the activity of the RGCs.

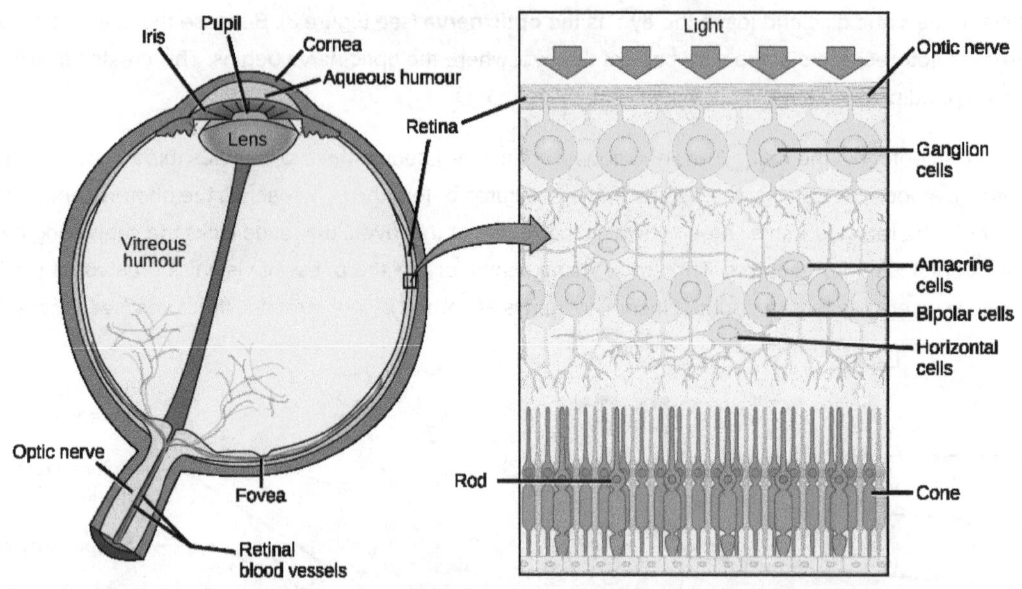

Figure 4. (a) The human eye is shown in cross section. (b) A blowup shows the layers of the retina.

There are two types of photoreceptors in the retina: rods and cones, named for their general appearance as illustrated in Figure 5. Rods are strongly photosensitive and are located in the outer edges of the retina. They detect dim light and are used primarily for peripheral and nighttime vision. Cones are weakly photosensitive and are located near the center of the retina. They respond to bright light, and their primary role is in daytime, color vision.

The fovea has a high density of cones. When you bring your gaze to an object to examine it intently in bright light, the eyes orient so that the object's image falls on the fovea. However, when looking at a star in the night sky or other object in dim light, the object can be better viewed by the peripheral vision because it is the rods at the edges of the retina, rather than the cones at the center, that operate better in low light. In humans, cones far outnumber rods in the fovea.

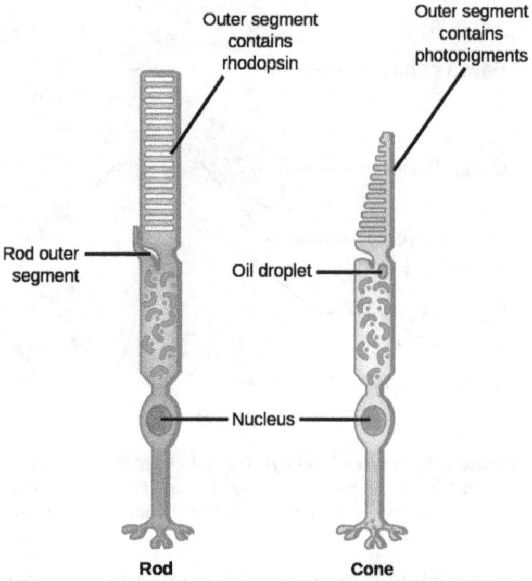

Figure 5. Rods and cones are photoreceptors in the retina. Rods respond in low light and can detect only shades of gray. Cones respond in intense light and are responsible for color vision. (credit: modification of work by Piotr Sliwa)

Transduction of Light

Photoreceptor cells have two parts, the **inner segment** and the **outer segment**. The inner segment contains the nucleus and other common organelles of a cell, whereas the outer segment is a specialized region in which photoreception takes place. There are two types of photoreceptors—rods and cones—which differ in the shape of their outer segment. The rod-shaped outer segments of the **rod photoreceptor** contain a stack of membrane-bound discs that contain the photosensitive pigment **rhodopsin**. The cone-shaped outer segments of the **cone photoreceptor** contain their photosensitive pigments in infoldings of the cell membrane. There are three cone photopigments, called **opsins**, which are each sensitive to a particular wavelength of light. The wavelength of visible light determines its color. The pigments in human eyes are specialized in perceiving three different primary colors: red, green, and blue.

At the molecular level, visual stimuli cause changes in the photopigment molecule that lead to changes in membrane potential of the photoreceptor cell. A single unit of light is called a **photon**, which is described in physics as a packet of energy with properties of both a particle and a wave. The energy of a photon is represented by its wavelength, with each wavelength of visible light corresponding to a particular color. Visible light is electromagnetic radiation with a wavelength between 380 and

720 nm. Longer wavelengths of less than 380 nm fall into the infrared range, whereas shorter wavelengths of more than 720 nm fall into the ultraviolet range. Light with a wavelength of 380 nm is blue whereas light with a wavelength of 720 nm is dark red. All other colors fall between red and blue at various points along the wavelength scale.

Opsin pigments are actually transmembrane proteins that contain a cofactor known as **retinal**. Retinal is a hydrocarbon molecule related to vitamin A. When a photon hits retinal, the long hydrocarbon chain of the molecule is biochemically altered. Specifically, photons cause some of the double-bonded carbons within the chain to switch from a *cis* to a *trans* conformation. This process is called **photoisomerization**. Before interacting with a photon, retinal's flexible double-bonded carbons are in the *cis* conformation. This molecule is referred to as 11-*cis*-retinal. A photon interacting with the molecule causes the flexible double-bonded carbons to change to the *trans*– conformation, forming all-*trans*-retinal, which has a straight hydrocarbon chain (Figure 7).

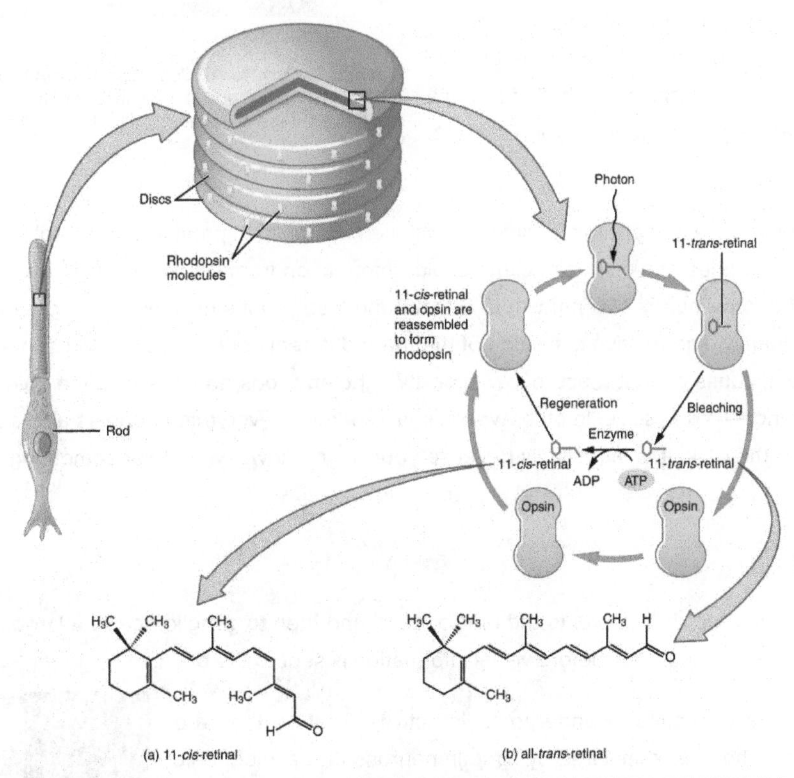

Figure 7. Retinal Isomers The retinal molecule has two isomers, (a) one before a photon interacts with it and (b) one that is altered through photoisomerization.

The shape change of retinal in the photoreceptors initiates visual transduction in the retina. Activation of retinal and the opsin proteins result in activation of a G protein. The G protein changes the membrane potential of the photoreceptor cell, which then releases less neurotransmitter into the outer synaptic layer of the retina. Until the retinal molecule is changed back to the 11-*cis*-retinal shape, the opsin cannot respond to light energy, which is called bleaching. When a large group of photopigments is bleached, the retina will send information as if opposing visual information is being perceived. After a bright flash of light, afterimages are usually seen in negative. The photoisomerization is reversed by a series of enzymatic changes so that the retinal responds to more light energy.

The opsins are sensitive to limited wavelengths of light. Rhodopsin, the photopigment in rods, is most sensitive to light at a wavelength of 498 nm. The three color opsins have peak sensitivities of 564 nm, 534 nm, and 420 nm corresponding roughly to the primary colors of red, green, and blue (Figure 8). The absorbance of rhodopsin in the rods is much more sensitive than in the cone opsins; specifically, rods are sensitive to vision in low light conditions, and cones are sensitive to brighter conditions.

In normal sunlight, rhodopsin will be constantly bleached while the cones are active. In a darkened room, there is not enough light to activate cone opsins, and vision is entirely dependent on rods. Rods are so sensitive to light that a single photon can result in an action potential from a rod's corresponding RGC.

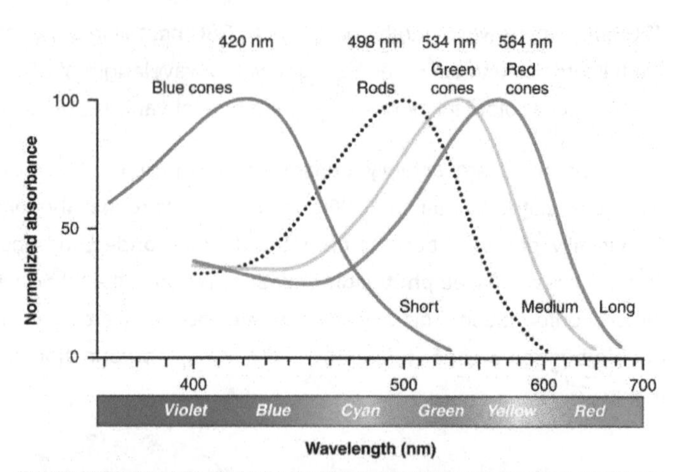

Figure 8. Comparison of Color Sensitivity of Photopigments Comparing the peak sensitivity and absorbance spectra of the four photopigments suggests that they are most sensitive to particular wavelengths.

The three types of cone opsins, being sensitive to different wavelengths of light, provide us with color vision. By comparing the activity of the three different cones, the brain can extract color information from visual stimuli. For example, a bright blue light that has a wavelength of approximately 450 nm would activate the "red" cones minimally, the "green" cones marginally, and the "blue" cones predominantly. The relative activation of the three different cones is calculated by the brain, which perceives the color as blue. However, cones cannot react to low-intensity light, and rods do not sense the color of light. Therefore, our low-light vision is—in essence—in grayscale. In other words, in a dark room, everything appears as a shade of gray. If you think that you can see colors in the dark, it is most likely because your brain knows what color something is and is relying on that memory.

Retinal Processing

Visual signals leave the cones and rods, travel to the bipolar cells, and then to ganglion cells. A large degree of processing of visual information occurs in the retina itself, before visual information is sent to the brain.

Photoreceptors in the retina continuously undergo tonic activity. That is, they are always slightly active even when not stimulated by light. In neurons that exhibit tonic activity, the absence of stimuli maintains a firing rate at a baseline; while some stimuli increase firing rate from the baseline, and other stimuli decrease firing rate. In the absence of light, the bipolar neurons that connect rods and cones to ganglion cells are continuously and actively inhibited by the rods and cones. Exposure of the retina to light hyperpolarizes the rods and cones and removes their inhibition of bipolar cells. The now active bipolar cells in turn stimulate the ganglion cells, which send action potentials along their axons (which leave the eye as the optic nerve). Thus, the visual system relies on change in retinal activity, rather than the absence or presence of activity, to encode visual signals for the brain. Sometimes horizontal cells carry signals from one rod or cone to other photoreceptors and to several bipolar cells. When a rod or cone stimulates a horizontal cell, the horizontal cell inhibits more distant photoreceptors and bipolar cells, creating lateral inhibition. This inhibition sharpens edges and enhances contrast in the images by making regions receiving light appear lighter and dark surroundings appear darker. Amacrine cells can distribute information from one bipolar cell to many ganglion cells.

Figure 9. View this flag to understand how retinal processing works. Stare at the center of the flag (indicated by the white dot) for 45 seconds, and then quickly look at a white background, noticing how colors appear.

You can demonstrate this using an easy demonstration to "trick" your retina and brain about the colors you are observing in your visual field. Look fixedly at Figure 9 for about 45 seconds. Then quickly shift your gaze to a sheet of blank white paper or a white wall. You should see an afterimage of the Norwegian flag in its correct colors. At this point, close your eyes for a moment, then reopen them, looking again at the white paper or wall; the afterimage of the flag should continue to appear as red, white, and blue. What causes this? According to an explanation called opponent process theory, as you gazed fixedly at the green, black, and yellow flag, your retinal ganglion cells that respond positively to green, black, and yellow increased their firing dramatically. When you shifted your gaze to the neutral white ground, these ganglion cells abruptly decreased their activity and the brain interpreted this abrupt downshift as if the ganglion cells were responding now to their "opponent" colors: red, white, and blue, respectively, in the visual field. Once the ganglion cells return to their baseline activity state, the false perception of color will disappear.

Higher Processing

The myelinated axons of ganglion cells make up the optic nerves. Within the nerves, different axons carry different qualities of the visual signal. Some axons constitute the magnocellular (big cell) pathway, which carries information about form, movement, depth, and differences in brightness. Other axons constitute the parvocellular (small cell) pathway, which carries information on color and fine detail. Some visual information projects directly back into the brain, while other information crosses to the opposite side of the brain. This crossing of optical pathways produces the distinctive optic chiasma (Greek, for "crossing") found at the base of the brain and allows us to coordinate information from both eyes.

Once in the brain, visual information is processed in several places, and its routes reflect the complexity and importance of visual information to humans and other animals. One route takes the signals to the thalamus, which serves as the routing station for all incoming sensory impulses except olfaction. In the thalamus, the magnocellular and parvocellular distinctions remain intact, and there are different layers of the thalamus dedicated to each. When visual signals leave the thalamus, they travel to the primary visual cortex at the rear of the brain. From the visual cortex, the visual signals travel in two directions. One stream that projects to the parietal lobe, in the side of the brain, carries magnocellular ("where") information. A second stream projects to the temporal lobe and carries both magnocellular ("where") and parvocellular ("what") information.

Another important visual route is a pathway from the retina to the superior colliculus in the midbrain, where eye movements are coordinated and integrated with auditory information. Finally, there is the pathway from the retina to the suprachiasmatic nucleus (SCN) of the hypothalamus. The SCN is a cluster of cells that is considered to be the body's internal clock, which controls our circadian (day-long) cycle. The SCN sends information to the pineal gland, which is important in sleep/wake patterns and annual cycles.

Concept Check

1. **What** are special senses?
2. **How** do neurons involved in the sense of taste obtain information?
3. **What** three organs are responsible for converting sound waves to mechanical waves?
4. **What** are the common traits in the physiology of balance and hearing?
5. **What** are the three layers of the eye, and **what** does each consist of?
6. **What** structures does light pass through in the eye as it reaches the retina?
7. **What** causes nearsightedness and farsightedness, and **how** are they treated?

CHAPTER 15 — THE LYMPHATIC AND IMMUNE SYSTEM

Anatomy of the Lymphatic and Immune Systems

LEARNING OBJECTIVES

- Describe the structure and function of the lymphatic tissue (lymph fluid, vessels, ducts, and organs)
- Discuss the cells of the immune system, how they function, and their relationship with the lymphatic system
- Explain the advantages of the adaptive immune response over the innate immune response
- Describe the major T cell types and their functions
- Discuss how B cells are activated and differentiate into plasma cells
- Describe the structure of the antibody classes and their functions
- Discuss inherited and acquired immunodeficiencies
- Explain why blood typing is important and what happens when mismatched blood is used in a transfusion

The **immune system** is the complex collection of cells and organs that destroys or neutralizes pathogens that would otherwise cause disease or death. The lymphatic system, for most people, is associated with the immune system to such a degree that the two systems are virtually indistinguishable. The **lymphatic system** is the system of vessels, cells, and organs that carries excess fluids to the bloodstream and filters pathogens from the blood. The swelling of lymph nodes during an infection and the transport of lymphocytes via the lymphatic vessels are but two examples of the many connections between these critical organ systems.

Functions of the Lymphatic System

A major function of the lymphatic system is to drain body fluids and return them to the bloodstream. Blood pressure causes leakage of fluid from the capillaries, resulting in the accumulation of fluid in the interstitial space—that is, spaces between individual cells in the tissues. In humans, 20 liters of plasma is released into the interstitial space of the tissues each day due to capillary filtration. Once this filtrate is out of the bloodstream and in the tissue spaces, it is referred to as interstitial fluid. Of this, 17 liters is reabsorbed directly by the blood vessels. But what happens to the remaining three liters? This is where the lymphatic system comes into play. It drains the excess fluid and empties it back into the bloodstream via a series of vessels, trunks, and ducts. **Lymph** is the term used to describe interstitial fluid once it has entered the lymphatic system. When the lymphatic system is damaged in some way, such as by being blocked by cancer cells or destroyed by injury, protein-rich interstitial fluid accumulates (sometimes "backs up" from the lymph vessels) in the tissue spaces. This inappropriate accumulation of fluid referred to as lymphedema may lead to serious medical consequences.

As the vertebrate immune system evolved, the network of lymphatic vessels became convenient avenues for transporting the cells of the immune system. Additionally, the transport of dietary lipids and fat-soluble vitamins absorbed in the gut uses this system.

Cells of the immune system not only use lymphatic vessels to make their way from interstitial spaces back into the circulation, but they also use lymph nodes as major staging areas for the development of critical immune responses. A **lymph node** is one of the small, bean-shaped organs located throughout the lymphatic system.

Structure of the Lymphatic System

The lymphatic vessels begin as open-ended capillaries, which feed into larger and larger lymphatic vessels, and eventually empty into the bloodstream by a series of ducts. Along the way, the lymph travels through the lymph nodes, which are commonly found near the groin, armpits, neck, chest, and abdomen. Humans have about 500–600 lymph nodes throughout the body.

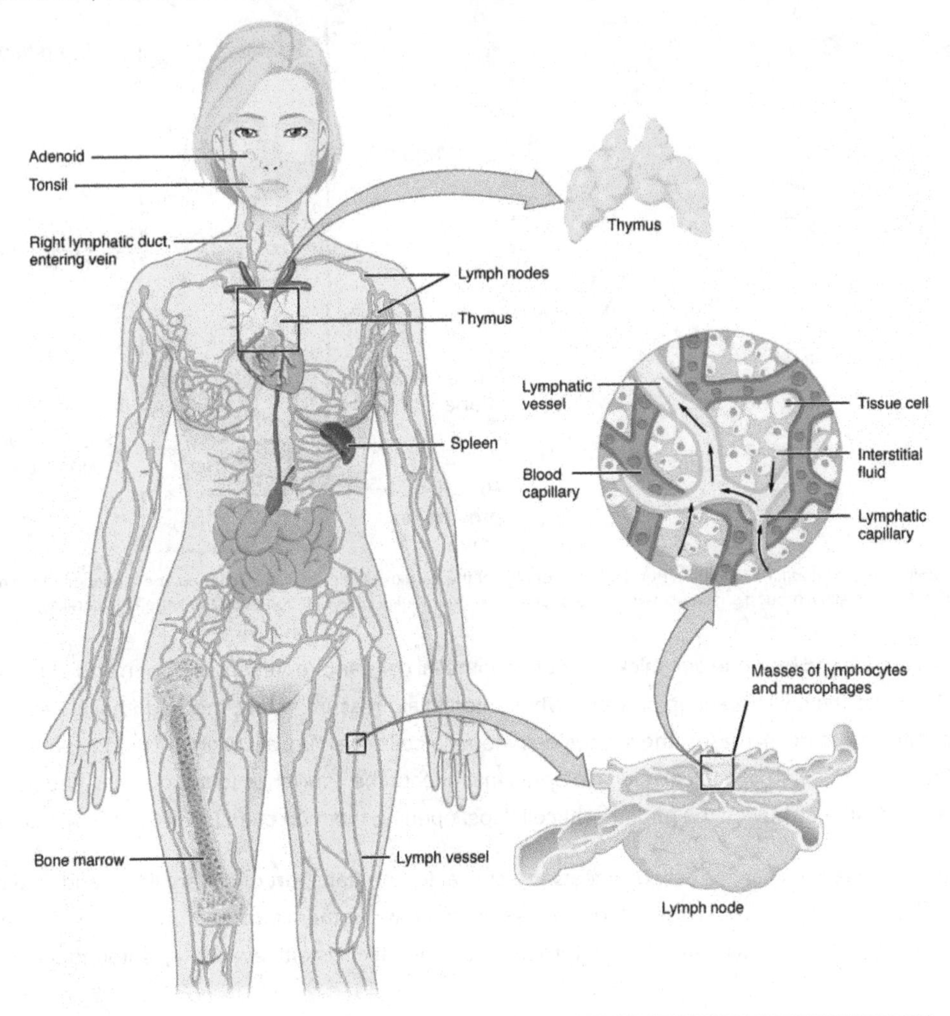

Figure 1. Lymphatic vessels in the arms and legs convey lymph to the larger lymphatic vessels in the torso.

A major distinction between the lymphatic and cardiovascular systems in humans is that lymph is not actively pumped by the heart, but is forced through the vessels by the movements of the body, the contraction of skeletal muscles during body movements, and breathing. One-way valves (semi-lunar valves) in lymphatic vessels keep the lymph moving toward the heart. Lymph flows from the lymphatic capillaries, through lymphatic vessels, and then is dumped into the circulatory system via the lymphatic ducts located at the junction of the jugular and subclavian veins in the neck.

Lymphatic Capillaries

Lymphatic capillaries, also called the terminal lymphatics, are vessels where interstitial fluid enters the lymphatic system to become lymph fluid. Located in almost every tissue in the body, these vessels are interlaced among the arterioles and venules of the circulatory system in the soft connective tissues of the body. Exceptions are the central nervous system, bone marrow, bones, teeth, and the cornea of the eye, which do not contain lymph vessels.

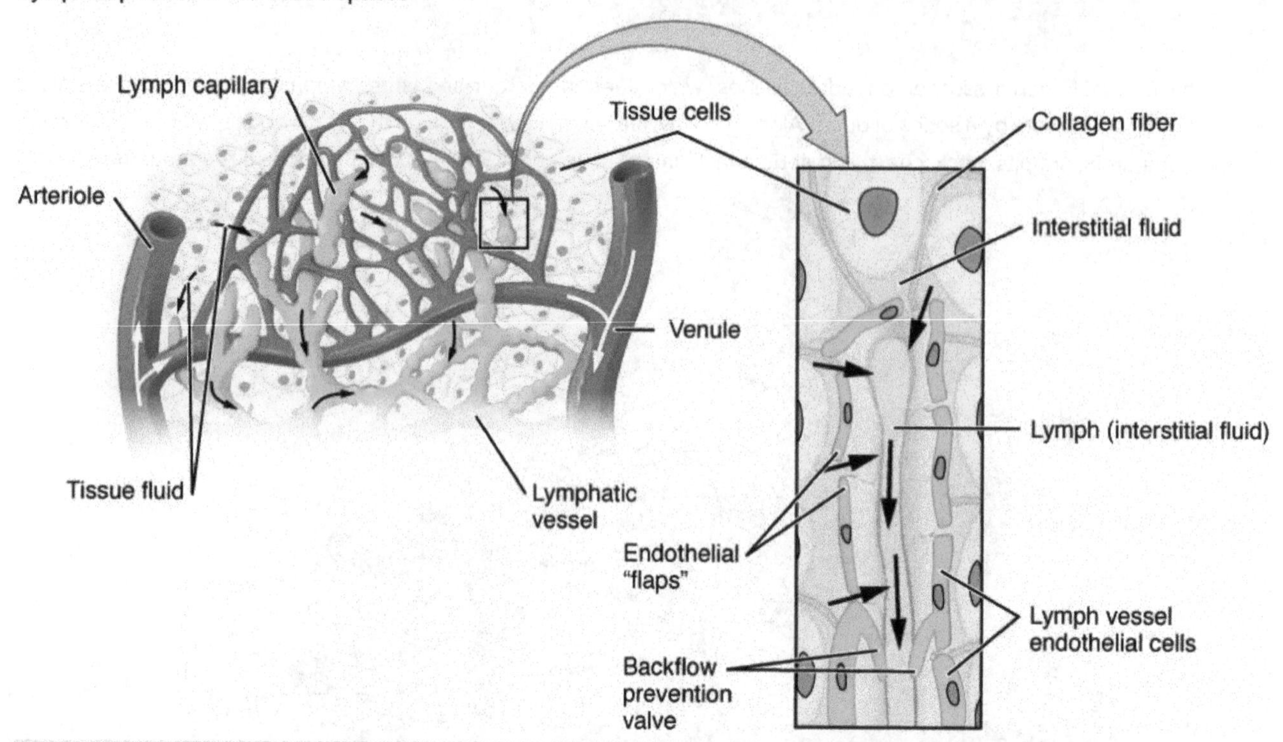

Figure 2. Lymphatic capillaries are interlaced with the arterioles and venules of the cardiovascular system. Collagen fibers anchor a lymphatic capillary in the tissue (inset). Interstitial fluid slips through spaces between the overlapping endothelial cells that compose the lymphatic capillary.

Lymphatic capillaries are formed by a one cell-thick layer of endothelial cells and represent the open end of the system, allowing interstitial fluid to flow into them via overlapping cells. When interstitial pressure is low, the endothelial flaps close to prevent "backflow." As interstitial pressure increases, the spaces between the cells open up, allowing the fluid to enter. Entry of fluid into lymphatic capillaries is also enabled by the collagen filaments that anchor the capillaries to surrounding structures. As interstitial pressure increases, the filaments pull on the endothelial cell flaps, opening them up even further to allow easy entry of fluid.

In the small intestine, lymphatic capillaries called lacteals are critical for the transport of dietary lipids and lipid-soluble vitamins to the bloodstream. In the small intestine, dietary triglycerides combine with other lipids and proteins, and enter the lacteals to form a milky fluid called **chyle**. The chyle then travels through the lymphatic system, eventually entering the liver and then the bloodstream.

Larger Lymphatic Vessels, Trunks, and Ducts

The lymphatic capillaries empty into larger lymphatic vessels, which are similar to veins in terms of their three-tunic structure and the presence of valves. These one-way valves are located fairly close to one another, and each one causes a bulge in the lymphatic vessel, giving the vessels a beaded appearance.

The superficial and deep lymphatics eventually merge to form larger lymphatic vessels known as **lymphatic trunks**. On the right side of the body, the right sides of the head, thorax, and right upper limb drain lymph fluid into the right subclavian vein via the right lymphatic duct. On the left side of the body, the remaining portions of the body drain into the larger thoracic duct, which drains into the left subclavian vein. The thoracic duct itself begins just beneath the diaphragm in the **cisterna chyli**, a sac-like chamber that receives lymph from the lower abdomen, pelvis, and lower limbs by way of the left and right lumbar trunks and the intestinal trunk.

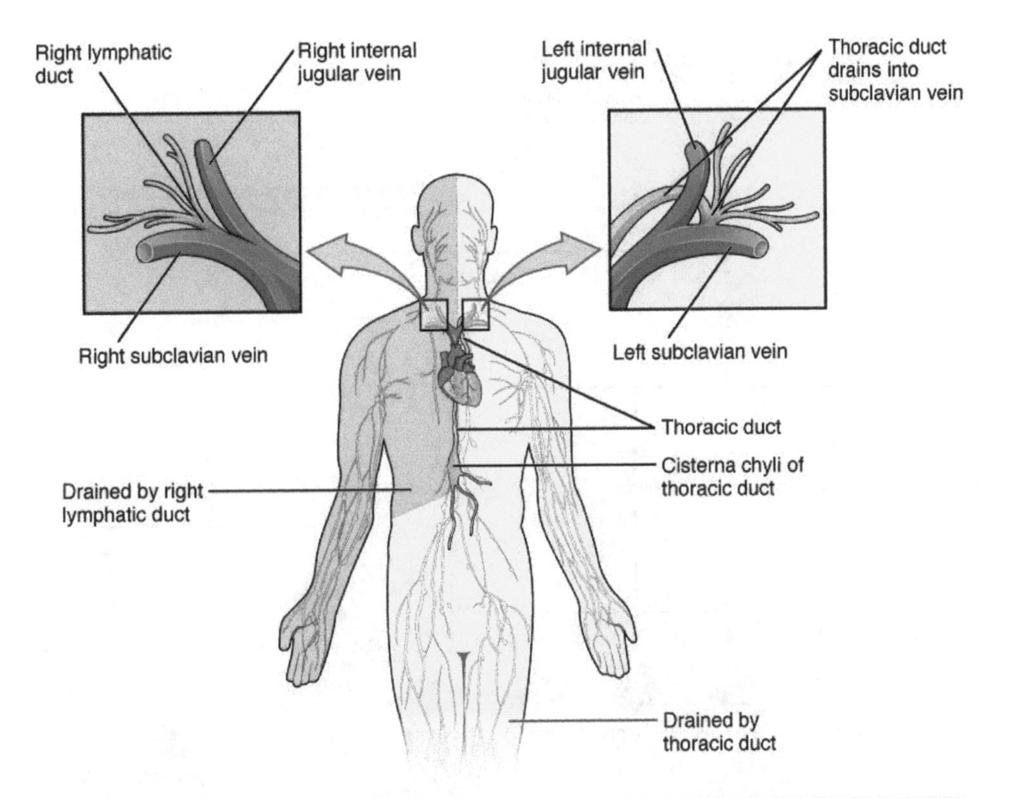

Figure 3. The thoracic duct drains a much larger portion of the body than does the right lymphatic duct.

The overall drainage system of the body is asymmetrical. The **right lymphatic duct** receives lymph from only the upper right side of the body. The lymph from the rest of the body enters the bloodstream through the **thoracic duct** via all the remaining lymphatic trunks.

The Organization of Immune Function

The immune system is a collection of barriers, cells, and soluble proteins that interact and communicate with each other in extraordinarily complex ways. The modern model of immune function is organized into three phases based on the timing of their effects. The three temporal phases consist of the following:

- **Barrier defenses** such as the skin and mucous membranes, which act instantaneously to prevent pathogenic invasion into the body tissues
- The rapid but nonspecific **innate immune response**, which consists of a variety of specialized cells and soluble factors
- The slower but more specific and effective **adaptive immune response**, which involves many cell types and soluble factors, but is primarily controlled by white blood cells (leukocytes) known as **lymphocytes**, which help control immune responses

The cells of the blood, including all those involved in the immune response, arise in the bone marrow via various differentiation pathways from hematopoietic stem cells. In contrast with embryonic stem cells, hematopoietic stem cells are present throughout adulthood and allow for the continuous differentiation of blood cells to replace those lost to age or function. These cells can be divided into three classes based on function:

- Phagocytic cells, which ingest pathogens to destroy them
- Lymphocytes, which specifically coordinate the activities of adaptive immunity
- Cells containing cytoplasmic granules, which help mediate immune responses against parasites and intracellular pathogens such as viruses

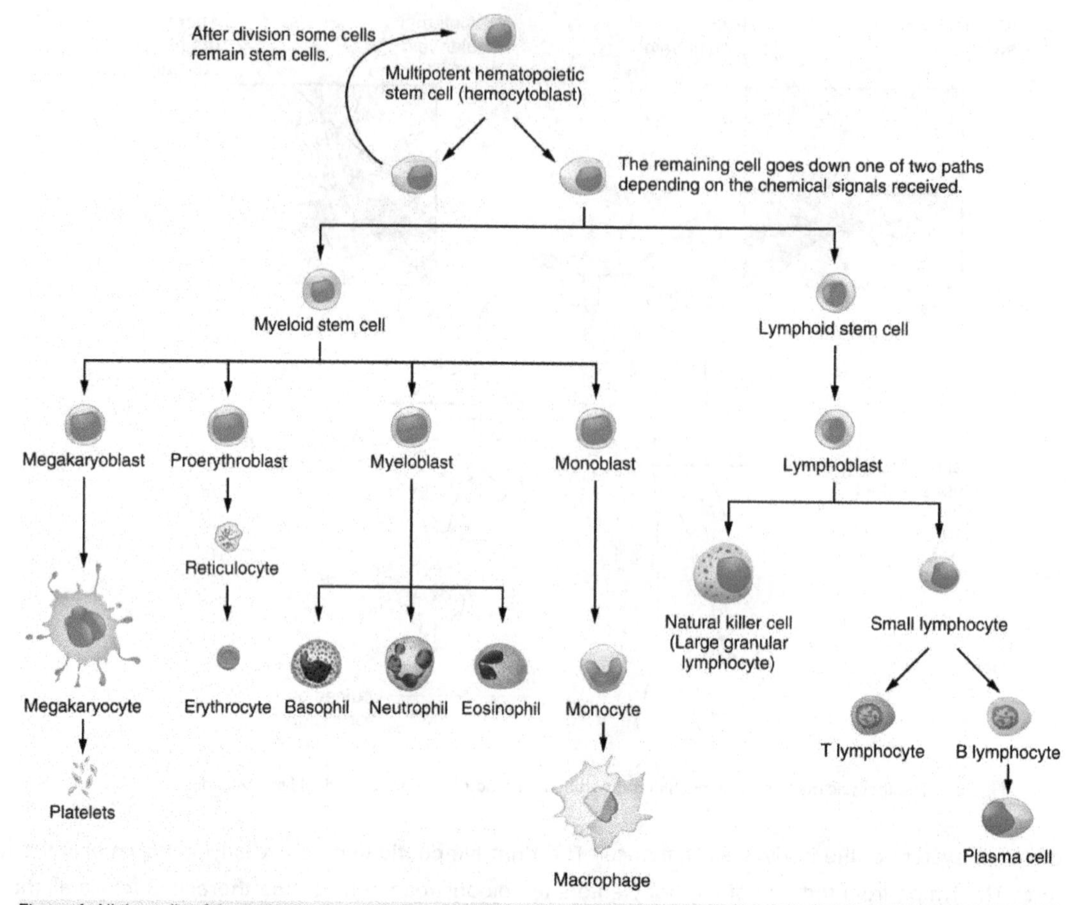

Figure 4. All the cells of the immune response as well as of the blood arise by differentiation from hematopoietic stem cells. Platelets are cell fragments involved in the clotting of blood.

Lymphocytes

As stated above, lymphocytes are the primary cells of adaptive immune responses (see Table 1 for more details). The two basic types of lymphocytes, B cells and T cells, are identical morphologically with a large central nucleus surrounded by a thin layer of cytoplasm. They are distinguished from each other by their surface protein markers as well as by the molecules they secrete. While B cells mature in red bone marrow and T cells mature in the thymus, they both initially develop from bone marrow. T cells migrate from bone marrow to the thymus gland where they further mature. B cells and T cells are found in many parts of the body, circulating in the bloodstream and lymph, and residing in secondary lymphoid organs, including the spleen and lymph nodes, which will be described later in this section. The human body contains approximately 10^{12} lymphocytes.

B Cells

B cells are immune cells that function primarily by producing antibodies. An **antibody** is any of the group of proteins that binds specifically to pathogen-associated molecules known as antigens. An **antigen** is a chemical structure on the surface of a pathogen that binds to T or B lymphocyte antigen receptors. Once activated by binding to antigen, B cells differentiate into cells that secrete a soluble form of their surface antibodies. These activated B cells are known as plasma cells.

T Cells

T cells, on the other hand, do not secrete antibodies but perform a variety of functions in the adaptive immune response. Different T cell types have the ability to either secrete soluble factors that communicate with other cells of the adaptive immune response or destroy cells infected with intracellular pathogens. The roles of T and B lymphocytes in the adaptive immune response will be discussed further in this chapter.

Plasma Cells

Another type of lymphocyte of importance is the plasma cell. A **plasma cell** is a B cell that has differentiated in response to antigen binding, and has thereby gained the ability to secrete soluble antibodies. These cells differ in morphology from standard B and T cells in that they contain a large amount of cytoplasm packed with the protein-synthesizing machinery known as rough endoplasmic reticulum.

Natural Killer Cells

A fourth important lymphocyte is the natural killer cell, a participant in the innate immune response. A **natural killer cell (NK)** is a circulating blood cell that contains cytotoxic (cell-killing) granules in its extensive cytoplasm. It shares this mechanism with the cytotoxic T cells of the adaptive immune response. NK cells are among the body's first lines of defense against viruses and certain types of cancer.

Table 1. Lymphocytes

Type of lymphocyte	Primary function
B lymphocyte	Generates diverse antibodies
T lymphocyte	Secretes chemical messengers
Plasma cell	Secretes antibodies
NK cell	Destroys virally infected cells

Primary Lymphoid Organs and Lymphocyte Development

Understanding the differentiation and development of B and T cells is critical to the understanding of the adaptive immune response. It is through this process that the body (ideally) learns to destroy only pathogens and leaves the body's own cells relatively intact. The **primary lymphoid organs** are the bone marrow, spleen, and thymus gland. The lymphoid organs are where lymphocytes mature, proliferate, and are selected, which enables them to attack pathogens without harming the cells of the body.

Bone Marrow

In the embryo, blood cells are made in the yolk sac. As development proceeds, this function is taken over by the spleen, lymph nodes, and liver. Later, the bone marrow takes over most hematopoietic functions, although the final stages of the differentiation of some cells may take place in other organs. The red **bone marrow** is a loose collection of cells where hematopoiesis occurs, and the yellow bone marrow is a site of energy storage, which consists largely of fat cells. The B cell undergoes nearly all of its development in the red bone marrow, whereas the immature T cell, called a **thymocyte**, leaves the bone marrow and matures largely in the thymus gland.

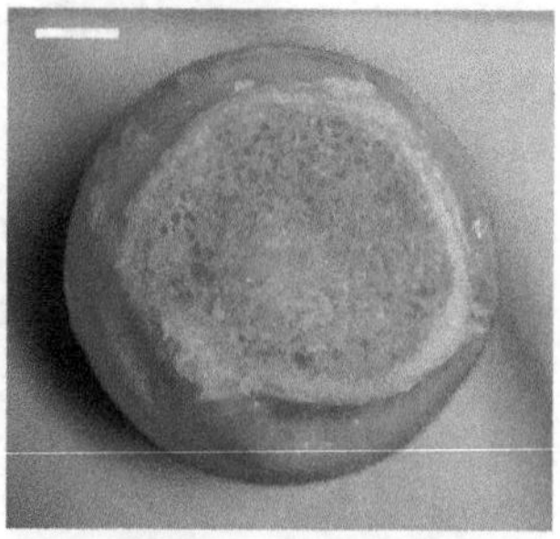

Figure 5. Red bone marrow fills the head of the femur, and a spot of yellow bone marrow is visible in the center. The white reference bar is 1 cm.

Thymus

The **thymus** gland is a bilobed organ found in the space between the sternum and the aorta of the heart. Connective tissue holds the lobes closely together but also separates them and forms a capsule.

The connective tissue capsule further divides the thymus into lobules via extensions called trabeculae. The outer region of the organ is known as the cortex and contains large numbers of thymocytes with some epithelial cells, macrophages, and dendritic cells (two types of phagocytic cells that are derived from monocytes). The cortex is densely packed so it stains more intensely than the rest of the thymus. The medulla, where thymocytes migrate before leaving the thymus, contains a less dense collection of thymocytes, epithelial cells, and dendritic cells.

Secondary Lymphoid Organs and their Roles in Active Immune Responses

Lymphocytes develop and mature in the primary lymphoid organs, but they mount immune responses from the **secondary lymphoid organs**. A **naïve lymphocyte** is one that has left the primary organ and entered a secondary lymphoid organ. Naïve lymphocytes are fully functional immunologically, but have yet to encounter an antigen to respond to. In addition to circulating in the blood and lymph, lymphocytes concentrate in secondary lymphoid organs, which include the lymph nodes, spleen, and lymphoid nodules. All of these tissues have many features in common, including the following:

- The presence of lymphoid follicles, the sites of the formation of lymphocytes, with specific B cell-rich and T cell-rich areas
- An internal structure of reticular fibers with associated fixed macrophages
- **Germinal centers**, which are the sites of rapidly dividing B lymphocytes and plasma cells, with the exception of the spleen
- Specialized post-capillary vessels known as **high endothelial venules**; the cells lining these venules are thicker and more columnar than normal endothelial cells, which allow cells from the blood to directly enter these tissues

Lymph Nodes

Lymph nodes function to remove debris and pathogens from the lymph, and are thus sometimes referred to as the "filters of the lymph". Any bacteria that infect the interstitial fluid are taken up by the lymphatic capillaries and transported to a regional lymph node. Dendritic cells and macrophages within this organ internalize and kill many of the pathogens that pass through, thereby removing them from the body. The lymph node is also the site of adaptive immune responses mediated by T cells, B cells, and accessory cells of the adaptive immune system. Like the thymus, the bean-shaped lymph nodes are surrounded by a tough capsule of connective tissue and are separated into compartments by trabeculae, the extensions of the capsule. In addition to the structure provided by the capsule and trabeculae, the structural support of the lymph node is provided by a series of reticular fibers laid down by fibroblasts.

The major routes into the lymph node are via **afferent lymphatic vessels**. Cells and lymph fluid that leave the lymph node may do so by another set of vessels known as the **efferent lymphatic vessels**. Within the cortex of the lymph node are lymphoid follicles, which consist of germinal centers of rapidly dividing B cells surrounded by a layer of T cells and other accessory cells. As the lymph continues to flow through the node, it enters the medulla, which consists of medullary cords of B cells and plasma cells, and the medullary sinuses where the lymph collects before leaving the node via the efferent lymphatic vessels.

Spleen

In addition to the lymph nodes, the **spleen** is a major secondary lymphoid organ. It is about 12 cm (5 in) long and is attached to the lateral border of the stomach. The spleen is a fragile organ without a strong capsule, and is dark red due to its extensive vascularization. The spleen is sometimes called the "filter of the blood" because of its extensive vascularization and the presence of macrophages and dendritic cells that remove microbes and other materials from the blood, including dying red blood cells. The spleen also functions as the location of immune responses to blood-borne pathogens.

The spleen is also divided by trabeculae of connective tissue, and within each splenic nodule is an area of red pulp, consisting of mostly red blood cells, and white pulp, which resembles the lymphoid follicles of the lymph nodes. Upon entering the spleen, the splenic artery splits into several arterioles (surrounded by white pulp) and eventually into sinusoids. Blood from the capillaries subsequently collects in the venous sinuses and leaves via the splenic vein. The red pulp consists of reticular fibers with fixed macrophages attached, free macrophages, and all of the other cells typical of the blood, including some lymphocytes. The white pulp surrounds a central arteriole and consists of germinal centers of dividing B cells surrounded by T cells and accessory cells, including macrophages and dendritic cells. Thus, the red pulp primarily functions as a filtration system of the blood, using cells of the relatively nonspecific immune response, and white pulp is where adaptive T and B cell responses are mounted.

Lymphoid Nodules

The other lymphoid tissues, the **lymphoid nodules**, have a simpler architecture than the spleen and lymph nodes in that they consist of a dense cluster of lymphocytes without a surrounding fibrous capsule. These nodules are located in the respiratory and digestive tracts, areas routinely exposed to environmental pathogens.

Tonsils are lymphoid nodules located along the inner surface of the pharynx and are important in developing immunity to oral pathogens. The tonsil located at the back of the throat, the pharyngeal tonsil, is sometimes referred to as the adenoid when swollen. Such swelling is an indication of an active immune response to infection. Tonsils do not contain a complete capsule, and the epithelial layer folds deeply into the interior of the tonsil to form tonsillar crypts. These structures, which accumulate all sorts of materials taken into the body through eating and breathing, actually "encourage" pathogens to penetrate deep into the tonsillar tissues where they are acted upon by numerous lymphoid follicles and eliminated. This seems to be the major function of tonsils—to help children's bodies recognize, destroy, and develop immunity to common environmental pathogens so that they will be protected in their later lives. Tonsils are often removed in those children who have recurring throat infections, especially those involving the palatine tonsils on either side of the throat, whose swelling may interfere with their breathing and/or swallowing.

Mucosa-associated lymphoid tissue (MALT) consists of an aggregate of lymphoid follicles directly associated with the mucous membrane epithelia. MALT makes up dome-shaped structures found underlying the mucosa of the gastrointestinal tract, breast tissue, lungs, and eyes. Peyer's patches, a type of MALT in the small intestine, are especially important for immune responses against ingested substances. Peyer's patches contain specialized endothelial cells called M (or microfold) cells that sample material from the intestinal lumen and transport it to nearby follicles so that adaptive immune responses to potential pathogens can be mounted.

Bronchus-associated lymphoid tissue (BALT) consists of lymphoid follicular structures with an overlying epithelial layer found along the bifurcations of the bronchi, and between bronchi and arteries. They also have the typically less-organized structure of other lymphoid nodules. These tissues, in addition to the tonsils, are effective against inhaled pathogens.

Barrier Defenses and the Innate Immune Response

The immune system can be divided into two overlapping mechanisms to destroy pathogens: the innate immune response, which is relatively rapid but nonspecific and thus not always effective, and the adaptive immune response, which is slower in its development during an initial infection with a pathogen, but is highly specific and effective at attacking a wide variety of pathogens (see Figure 1).

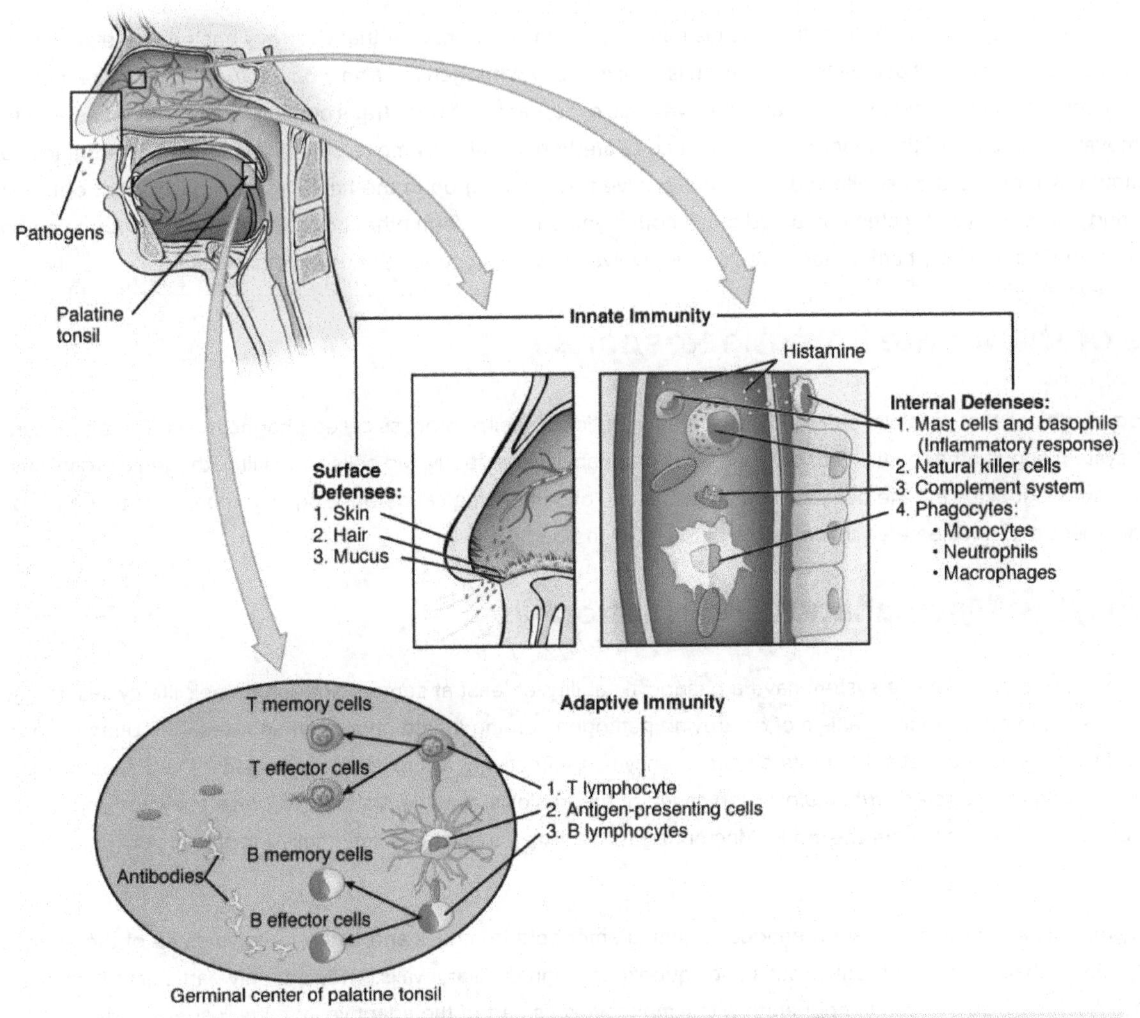

Figure 1. The innate immune system enhances adaptive immune responses so they can be more effective.

Any discussion of the innate immune response usually begins with the physical barriers that prevent pathogens from entering the body, destroy them after they enter, or flush them out before they can establish themselves in the hospitable environment of the body's soft tissues. Barrier defenses are part of the body's most basic defense mechanisms. The barrier defenses are not a response to infections, but they are continuously working to protect against a broad range of pathogens.

The different modes of barrier defenses are associated with the external surfaces of the body, where pathogens may try to enter (see Table 1). The primary barrier to the entrance of microorganisms into the body is the skin. Not only is the skin covered with a layer of dead, keratinized epithelium that is too dry for bacteria to grow in, but as these cells are continuously sloughed off from the skin, they carry bacteria and other pathogens with them. Additionally, sweat and other skin secretions may lower pH, contain toxic lipids, and physically wash microbes away.

Table 1. Barrier Defenses

Site	Specific defense	Protective aspect
Skin	Epidermal surface	Keratinized cells of surface, Langerhans cells
Skin (sweat/secretions)	Sweat glands, sebaceous glands	Low pH, washing action
Oral cavity	Salivary glands	Lysozyme
Stomach	Gastrointestinal tract	Low pH
Mucosal surfaces	Mucosal epithelium	Nonkeratinized epithelial cells
Normal flora (nonpathogenic bacteria)	Mucosal tissues	Prevent pathogens from growing on mucosal surfaces

Another barrier is the saliva in the mouth, which is rich in lysozyme—an enzyme that destroys bacteria by digesting their cell walls. The acidic environment of the stomach, which is fatal to many pathogens, is also a barrier. Additionally, the mucus layer of the gastrointestinal tract, respiratory tract, reproductive tract, eyes, ears, and nose traps both microbes and debris, and facilitates their removal. In the case of the upper respiratory tract, ciliated epithelial cells move potentially contaminated mucus upwards to the mouth, where it is then swallowed into the digestive tract, ending up in the harsh acidic environment of the stomach. Considering how often you breathe compared to how often you eat or perform other activities that expose you to pathogens, it is not surprising that multiple barrier mechanisms have evolved to work in concert to protect this vital area.

Cells of the Innate Immune Response

A phagocyte is a cell that is able to surround and engulf a particle or cell, a process called **phagocytosis**. The phagocytes of the immune system engulf other particles or cells, either to clean an area of debris, old cells, or to kill pathogenic organisms such as bacteria. The phagocytes are the body's fast acting, first line of immunological defense against organisms that have breached barrier defenses and have entered the vulnerable tissues of the body.

Phagocytes: Macrophages and Neutrophils

Many of the cells of the immune system have a phagocytic ability, at least at some point during their life cycles. Phagocytosis is an important and effective mechanism of destroying pathogens during innate immune responses. The phagocyte takes the organism inside itself and fuses it with its digestive enzymes, effectively killing many pathogens. On the other hand, some bacteria including *Mycobacteria tuberculosis*, the cause of tuberculosis, may be resistant to these enzymes and are therefore much more difficult to clear from the body. Macrophages, neutrophils, and dendritic cells are the major phagocytes of the immune system.

A **macrophage** is an irregularly shaped phagocyte that is amoeboid in nature and is the most versatile of the phagocytes in the body. Macrophages move through tissues and squeeze through capillary walls. They not only participate in innate immune responses but have also evolved to cooperate with lymphocytes as part of the adaptive immune response. Macrophages exist in many tissues of the body, either freely roaming through connective tissues or fixed to reticular fibers within specific tissues such as lymph nodes. When pathogens breach the body's barrier defenses, macrophages are the first line of defense.

A **neutrophil** is a phagocytic cell that is attracted from the bloodstream to infected tissues. These spherical cells contain cytoplasmic granules, which in turn contain a variety of molecules capable of affecting the diameter of blood vessels such as histamine. In contrast, macrophages have few or no cytoplasmic granules. Whereas macrophages act like sentries, always on guard against infection, neutrophils can be thought of as military reinforcements that are called into a battle to hasten the destruction of the enemy. Although, usually thought of as the primary pathogen-killing cell of the inflammatory process of the innate immune response, new research has suggested that neutrophils play a role in the adaptive immune response as well, just as macrophages do.

A **monocyte** is a circulating precursor cell that differentiates into either a macrophage or dendritic cell, which can be rapidly attracted to areas of infection by signal molecules of inflammation.

Table 2. Phagocytic Cells of the Innate Immune System

Cell	Cell type	Primary location	Function in the innate immune response
Macrophage	Agranulocyte	Body cavities/organs	Phagocytosis
Neutrophil	Granulocyte	Blood	Phagocytosis
Monocyte	Agranulocyte	Blood	Precursor of macrophage/dendritic cell

Natural Killer Cells

NK cells are a type of lymphocyte that have the ability to induce apoptosis, that is, programmed cell death, in cells infected with intracellular pathogens such as bacteria and viruses. NK cells recognize these cells by mechanisms that are still not well understood, but that presumably involve their surface receptors. NK cells can induce apoptosis, in which a cascade of events inside the cell causes its own death by either of two mechanisms:

1. NK cells are able to respond to chemical signals and express the fas ligand. The **fas ligand** is a surface molecule that binds to the fas molecule on the surface of the infected cell, sending it apoptotic signals, thus killing the cell and the pathogen within it
2. The granules of the NK cells release perforins and granzymes. A **perforin** is a protein that forms pores in the membranes of infected cells. A **granzyme** is a protein-digesting enzyme that enters the cell via the perforin pores and triggers apoptosis intracellularly.

Both mechanisms are especially effective against virally infected cells. If apoptosis is induced before the virus has the ability to synthesize and assemble all its components, no infectious virus will be released from the cell, thus preventing further infection.

Recognition of Pathogens

Cells of the innate immune response, the phagocytic cells, and the cytotoxic NK cells recognize patterns of pathogen-specific molecules, such as bacterial cell wall components or bacterial flagellar proteins, using pattern recognition receptors. A **pattern recognition receptor (PRR)** is a membrane-bound receptor that recognizes characteristic features of a pathogen and molecules released by stressed or damaged cells.

These receptors are present on the cell surface whether they are needed or not. Should the cells of the innate immune system come into contact with a species of pathogen they recognize, the cell will bind to the pathogen and initiate phagocytosis (or cellular apoptosis in the case of an intracellular pathogen) in an effort to destroy the offending microbe. Receptors vary somewhat according to cell type, but they usually include receptors for bacterial components and for complement, discussed below.

Soluble Mediators of the Innate Immune Response

The previous discussions have alluded to chemical signals that can induce cells to change various physiological characteristics, such as the expression of a particular receptor. These soluble factors are secreted during innate or early induced responses, and later during adaptive immune responses.

Cytokines and Chemokines

A **cytokine** is signaling molecule that allows cells to communicate with each other over short distances. Cytokines are secreted into the intercellular space, and the action of the cytokine induces the receiving cell to change its physiology. A **chemokine** is a soluble chemical mediator similar to cytokines except that its function is to attract cells (chemotaxis) from longer distances.

Early Induced Proteins

Early induced proteins are those that are not constitutively present in the body, but are made as they are needed early during the innate immune response. **Interferons** are an example of early induced proteins. Cells infected with viruses secrete interferons that travel to adjacent cells and induce them to make antiviral proteins. Thus, even though the initial cell is sacrificed, the surrounding cells are protected.

Complement System

The **complement** system is a series of proteins constitutively found in the blood plasma. As such, these proteins are not considered part of the **early induced immune response.** Made in the liver, they have a variety of functions in the innate immune response, using what is known as the "alternate pathway" of complement activation. Additionally, complement functions in the adaptive immune response as well, in what is called the classical pathway. The complement system consists of several proteins that enzymatically alter and fragment later proteins in a series, which is why it is termed cascade. Once activated, the series of reactions is irreversible, and releases fragments that have the following actions:

- Bind to the cell membrane of the pathogen that activates it, labeling it for phagocytosis (opsonization)
- Diffuse away from the pathogen and act as chemotactic agents to attract phagocytic cells to the site of inflammation
- Form damaging pores in the plasma membrane of the pathogen

Figure 2 shows the classical pathway, which requires antibodies of the adaptive immune response. The alternate pathway does not require an antibody to become activated.

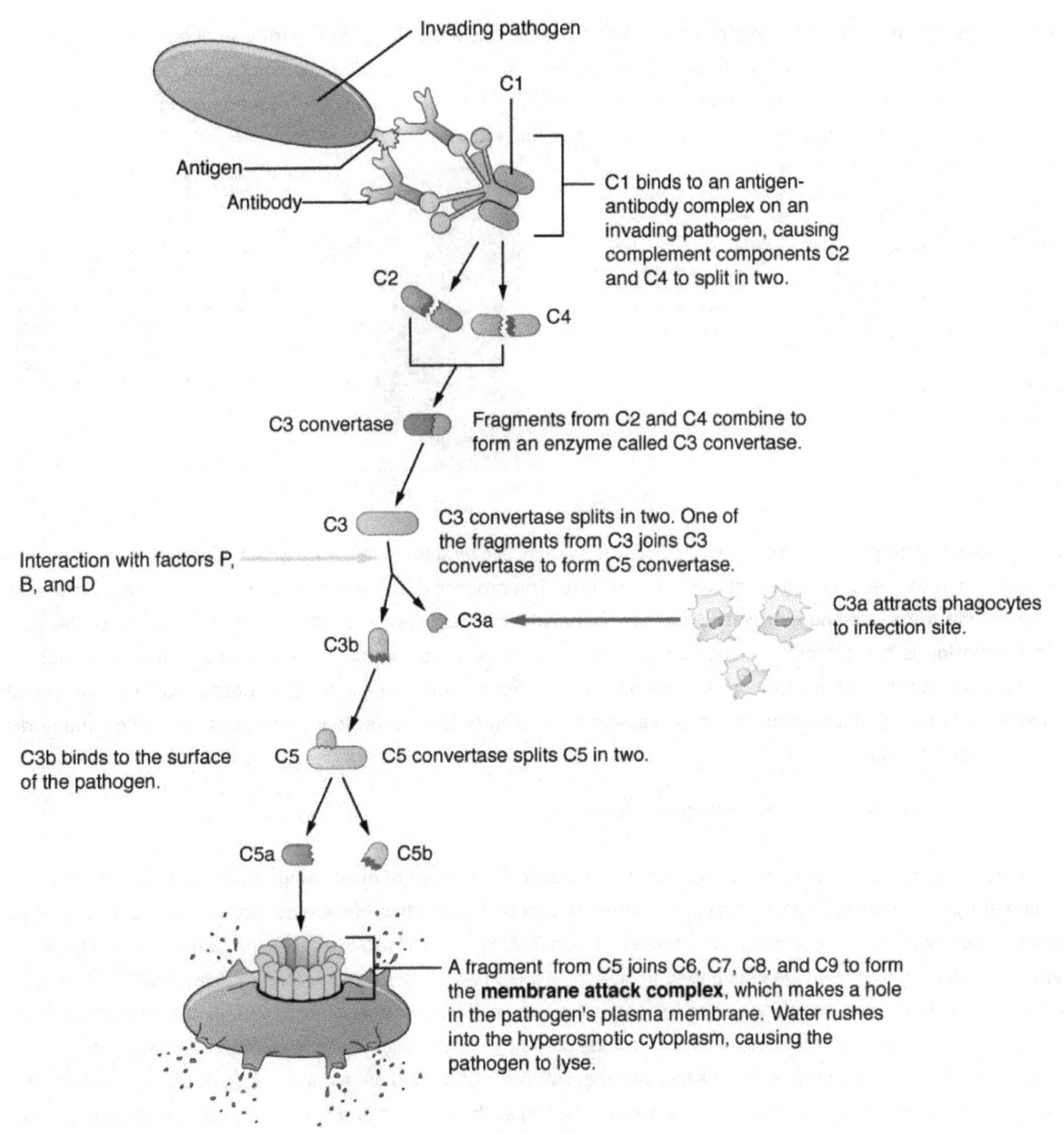

Figure 2. The classical pathway, used during adaptive immune responses, occurs when C1 reacts with antibodies that have bound an antigen.

The splitting of the C3 protein is the common step to both pathways. In the alternate pathway, C3 is activated spontaneously and, after reacting with the molecules factor P, factor B, and factor D, splits apart. The classical pathway is similar, except the early stages of activation require the presence of antibody bound to antigen, and thus is dependent on the adaptive immune response. The earlier fragments of the cascade also have important functions. Phagocytic cells such as macrophages and neutrophils are attracted to an infection site by chemotactic attraction to smaller complement fragments. Additionally, once they arrive, their receptors for surface-bound C3b opsonize the pathogen for phagocytosis and destruction.

Inflammatory Response

The hallmark of the innate immune response is **inflammation**. Inflammation is something everyone has experienced. Stub a toe, cut a finger, or do any activity that causes tissue damage and inflammation will result, with its four characteristics: heat, redness, pain, and swelling ("loss of function" is sometimes mentioned as a fifth characteristic). It is important to note that inflammation does not have to be initiated by an infection, but can also be caused by tissue injuries. The release of damaged cellular contents into the site of injury is enough to stimulate the response, even in the absence of breaks in physical barriers that would allow pathogens to enter (by hitting your thumb with a hammer, for example). The inflammatory reaction brings in phagocytic cells to the damaged area to clear cellular debris and to set the stage for wound repair (Figure 3).

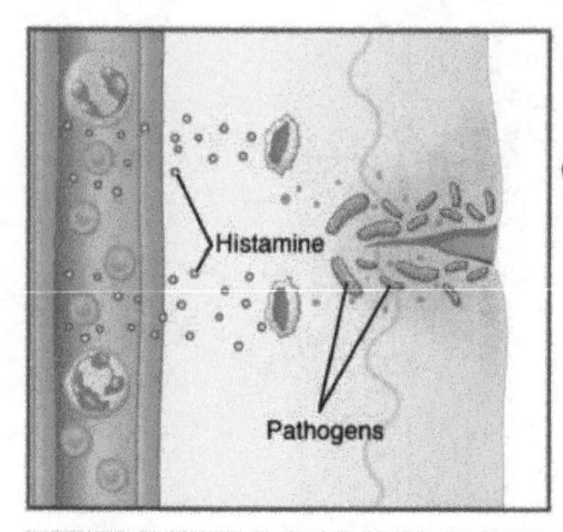

① Mast cells detect injury to nearby cells and release histamine, initiating inflammatory response.

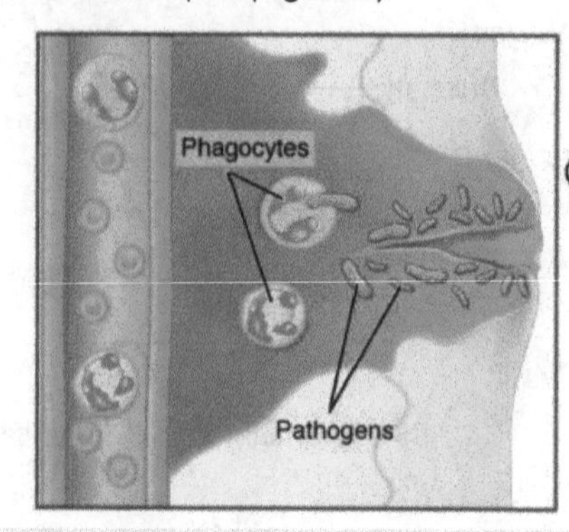

② Histamine increases blood flow to the wound sites, bringing in phagocytes and other immune cells that neutralize pathogens. The blood influx causes the wound to swell, redden, and become warm and painful.

Figure 3. The Inflammatory Process

This reaction also brings in the cells of the innate immune system, allowing them to get rid of the sources of a possible infection. Inflammation is part of a very basic form of immune response. The process not only brings fluid and cells into the site to destroy the pathogen and remove it and debris from the site, but also helps to isolate the site, limiting the spread of the pathogen. **Acute inflammation** is a short-term inflammatory response to an insult to the body. If the cause of the inflammation is not resolved, however, it can lead to chronic inflammation, which is associated with major tissue destruction and fibrosis. **Chronic inflammation** is ongoing inflammation. It can be caused by foreign bodies, persistent pathogens, and autoimmune diseases such as rheumatoid arthritis.

There are four important parts to the inflammatory response:

- *Tissue Injury.* The released contents of injured cells stimulate the release of **mast cell** granules and their potent inflammatory mediators such as histamine, leukotrienes, and prostaglandins. **Histamine** increases the diameter of local blood vessels (vasodilation), causing an increase in blood flow. Histamine also increases the permeability of local capillaries, causing plasma to leak out and form interstitial fluid. This causes the swelling associated with inflammation. Additionally, injured cells, phagocytes, and basophils are sources of inflammatory mediators, including prostaglandins and leukotrienes. Leukotrienes attract neutrophils from the blood by chemotaxis and increase vascular permeability. Prostaglandins cause vasodilation by relaxing vascular smooth muscle and are a major cause of the pain associated with inflammation. Nonsteroidal anti-inflammatory drugs (NSAIDS) such as aspirin and ibuprofen relieve pain by inhibiting prostaglandin production.
- *Vasodilation.* Many inflammatory mediators such as histamine are vasodilators that increase the diameters of local capillaries. This causes increased blood flow and is responsible for the heat and redness of inflamed tissue. It allows greater access of the blood to the site of inflammation.
- *Increased Vascular Permeability.* At the same time, inflammatory mediators increase the permeability of the local vasculature, causing leakage of fluid into the interstitial space, resulting in the swelling, or edema, associated with inflammation.
- *Recruitment of Phagocytes.* Leukotrienes are particularly good at attracting neutrophils from the blood to the site of infection by chemotaxis. Following an early neutrophil infiltrate stimulated by macrophage cytokines, more macrophages are recruited to clean up the debris left over at the site. When local infections are severe, neutrophils are attracted to the sites of infections in large numbers, and as they phagocytose the pathogens and subsequently die, their accumulated cellular remains are visible as pus at the infection site.

Overall, inflammation is valuable for many reasons. Not only are the pathogens killed and debris removed, but the increase in vascular permeability encourages the entry of clotting factors, the first step towards wound repair. Inflammation also facilitates the transport of antigen to lymph nodes by dendritic cells for the development of the adaptive immune response.

The Adaptive Immune Response: T lymphocytes and Their Functional Types

Innate immune responses (and early induced responses) are in many cases ineffective at completely controlling pathogen growth. However, they slow pathogen growth and allow time for the adaptive immune response to strengthen and either control or eliminate the pathogen. The innate immune system also sends signals to the cells of the adaptive immune system, guiding them in how to attack the pathogen. Thus, these are the two important arms of the immune response.

The Benefits of the Adaptive Immune Response

The specificity of the adaptive immune response—its ability to specifically recognize and make a response against a wide variety of pathogens—is its great strength. Antigens, the small chemical groups often associated with pathogens, are recognized by receptors on the surface of B and T lymphocytes. The adaptive immune response to these antigens is so versatile that it can respond to nearly any pathogen. This increase in specificity comes because the adaptive immune response has a unique way to develop as many as 100 trillion different receptors to recognize nearly every conceivable pathogen. How could so many different types of antibodies be encoded? And what about the many specificities of T cells? There is not nearly enough DNA in a cell to have a separate gene for each specificity. The mechanism was finally worked out in the 1970s and 1980s using the new tools of molecular genetics

Primary Disease and Immunological Memory

The immune system's first exposure to a pathogen is called a **primary adaptive response**. Symptoms of a first infection, called primary disease, are always relatively severe because it takes time for an initial adaptive immune response to a pathogen to become effective.

Upon re-exposure to the same pathogen, a secondary adaptive immune response is generated, which is stronger and faster than the primary response. The **secondary adaptive response** often eliminates a pathogen before it can cause significant tissue damage or any symptoms. Without symptoms, there is no disease, and the individual is not even aware of the infection. This secondary response is the basis of **immunological memory**, which protects us from getting diseases repeatedly from the same pathogen. By this mechanism, an individual's exposure to pathogens early in life spares the person from these diseases later in life.

Self Recognition

A third important feature of the adaptive immune response is its ability to distinguish between self-antigens, those that are normally present in the body, and foreign antigens, those that might be on a potential pathogen. As T and B cells mature, there are mechanisms in place that prevent them from recognizing self-antigen, preventing a damaging immune response against the body. These mechanisms are not 100 percent effective, however, and their breakdown leads to autoimmune diseases, which will be discussed later in this chapter.

T Cell-Mediated Immune Responses

The primary cells that control the adaptive immune response are the lymphocytes, the T and B cells. T cells are particularly important, as they not only control a multitude of immune responses directly, but also control B cell immune responses in many cases as well. Thus, many of the decisions about how to attack a pathogen are made at the T cell level, and knowledge of their functional types is crucial to understanding the functioning and regulation of adaptive immune responses as a whole.

T lymphocytes recognize antigens based on a two-chain protein receptor. The most common and important of these are the alpha-beta T cell receptors (Figure 1).

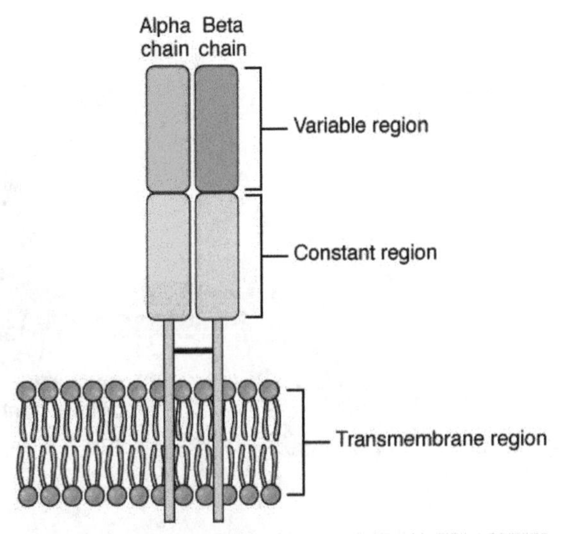

Figure 1. Notice the constant and variable regions of each chain, anchored by the transmembrane region.

There are two chains in the T cell receptor, and each chain consists of two domains. The **variable region domain** is furthest away from the T cell membrane and is so named because its amino acid sequence varies between receptors. In contrast, the **constant region domain** has less variation. The differences in the amino acid sequences of the variable domains are the molecular basis of the diversity of antigens the receptor can recognize. Thus, the antigen-binding site of the receptor consists of the terminal ends of both receptor chains, and the amino acid sequences of those two areas combine to determine its antigenic specificity. Each T cell produces only one type of receptor and thus is specific for a single particular antigen.

Antigens

Antigens on pathogens are usually large and complex, and consist of many antigenic determinants. An **antigenic determinant** (epitope) is one of the small regions within an antigen to which a receptor can bind. They usually consist of six or fewer amino acid residues in a protein, or one or two sugar moieties in a carbohydrate antigen. Carbohydrate antigens are found on bacterial cell walls and on red blood cells (the ABO blood group antigens). Protein antigens are complex because of the variety of three-dimensional shapes that proteins can assume, and are especially important for the immune responses to viruses and worm parasites. It is the interaction of the shape of the antigen and the complementary shape of the amino acids of the antigen-binding site that accounts for the chemical basis of specificity.

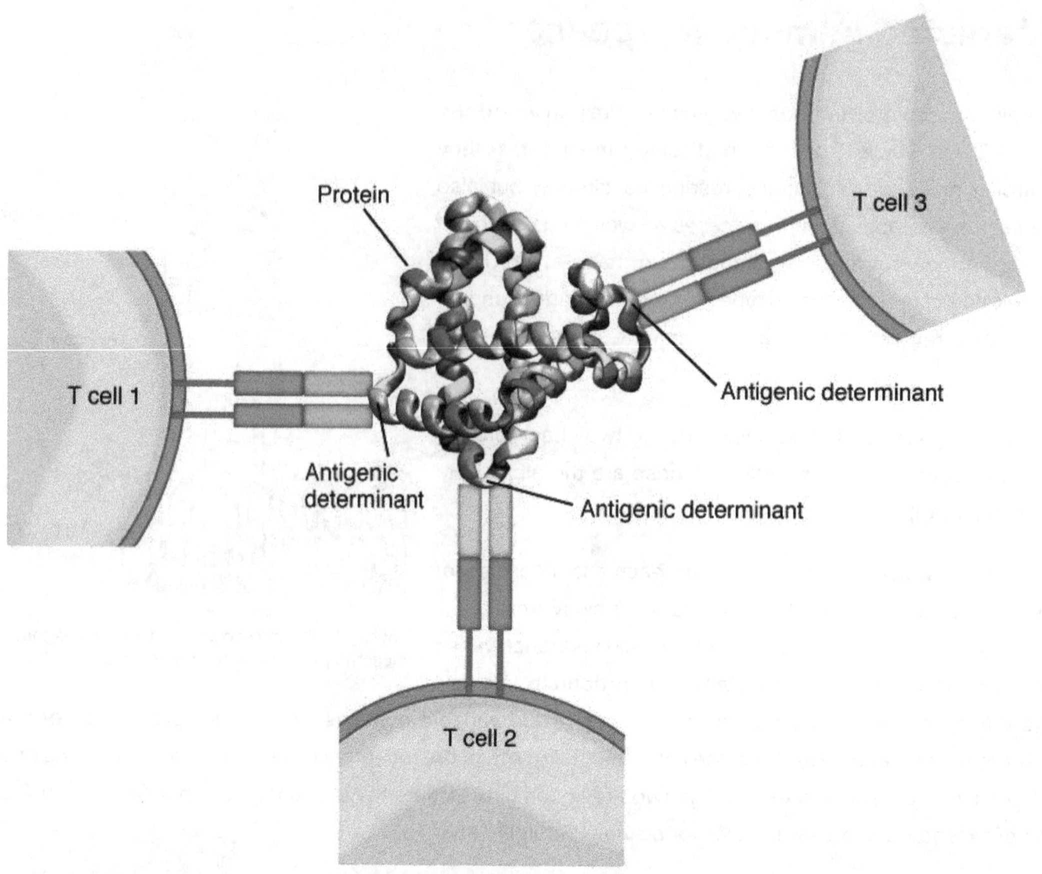

Figure 2. A typical protein antigen has multiple antigenic determinants, shown by the ability of T cells with three different specificities to bind to different parts of the same antigen.

Antigen Processing and Presentation

Although Figure 2 shows T cell receptors interacting with antigenic determinants directly, the mechanism that T cells use to recognize antigens is, in reality, much more complex. T cells do not recognize free-floating or cell-bound antigens as they appear on the surface of the pathogen – they only recognize antigen on the surface of specialized cells called antigen-presenting cells (APCs). Antigens are internalized by these antigen-presenting cells.

Antigen processing is a mechanism that enzymatically cleaves the antigen into smaller pieces once it has been internalized. The antigen fragments are then brought to the APC's surface and associated with a specialized type of antigen-presenting protein known as a **major histocompatibility complex (MHC)** molecule. The association of the antigen fragments with an MHC molecule on the surface of a cell is known as **antigen presentation** and results in the recognition of antigen by a T cell. It is the combination of the MHC molecule and the fragment of the original peptide or carbohydrate that is actually physically recognized by the T cell receptor.

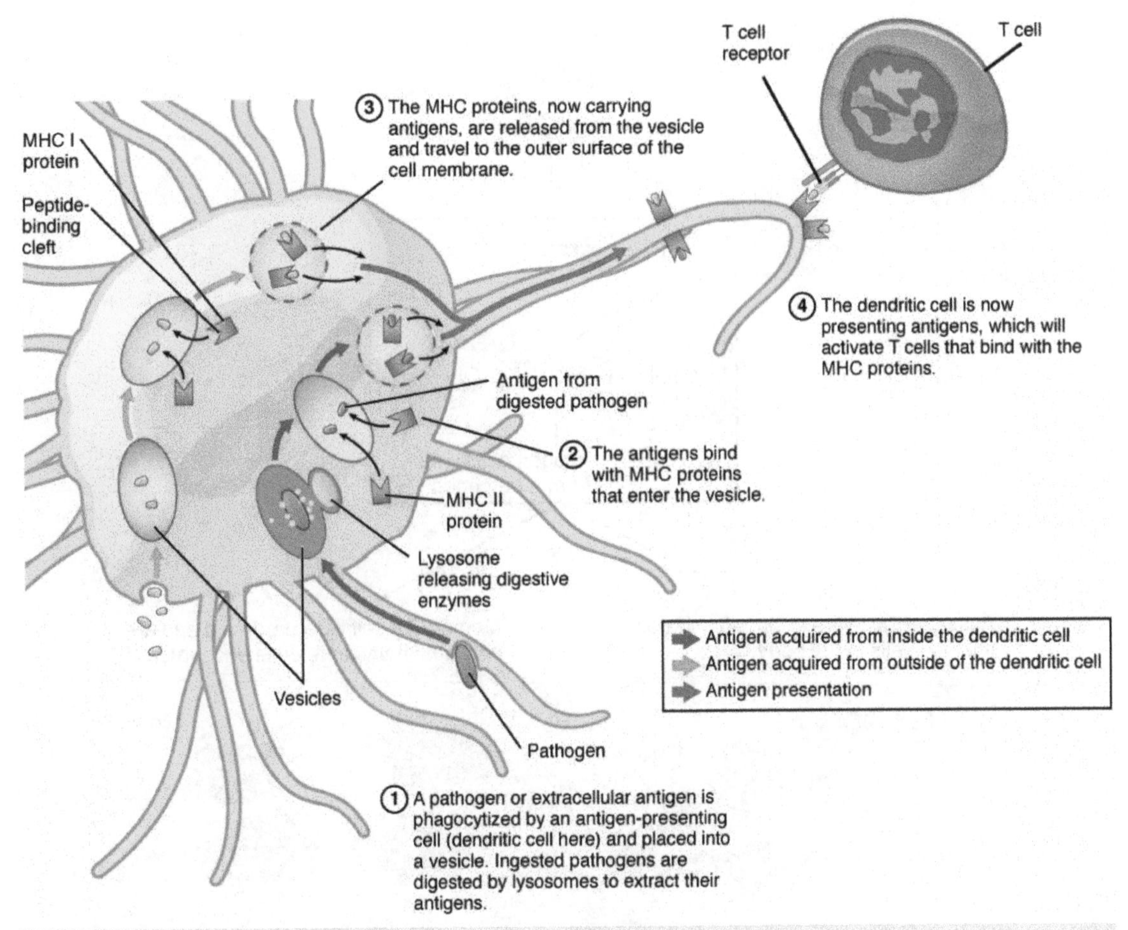

Figure 3. Antigen Processing and Presentation

Mechanisms of T Cell-mediated Immune Responses

Mature T cells become activated by recognizing processed foreign antigen in association with a self-MHC molecule and begin dividing rapidly by mitosis. This proliferation of T cells is called **clonal expansion** and is necessary to make the immune response strong enough to effectively control a pathogen. How does the body select only those T cells that are needed against a specific pathogen? Again, the specificity of a T cell is based on the amino acid sequence and the three-dimensional shape of the antigen-binding site formed by the variable regions of the two chains of the T cell receptor. **Clonal selection** is the process of antigen binding only to those T cells that have receptors specific to that antigen. Each T cell that is activated has a specific receptor "hard-wired" into its DNA, and all of its progeny will have identical DNA and T cell receptors, forming clones of the original T cell.

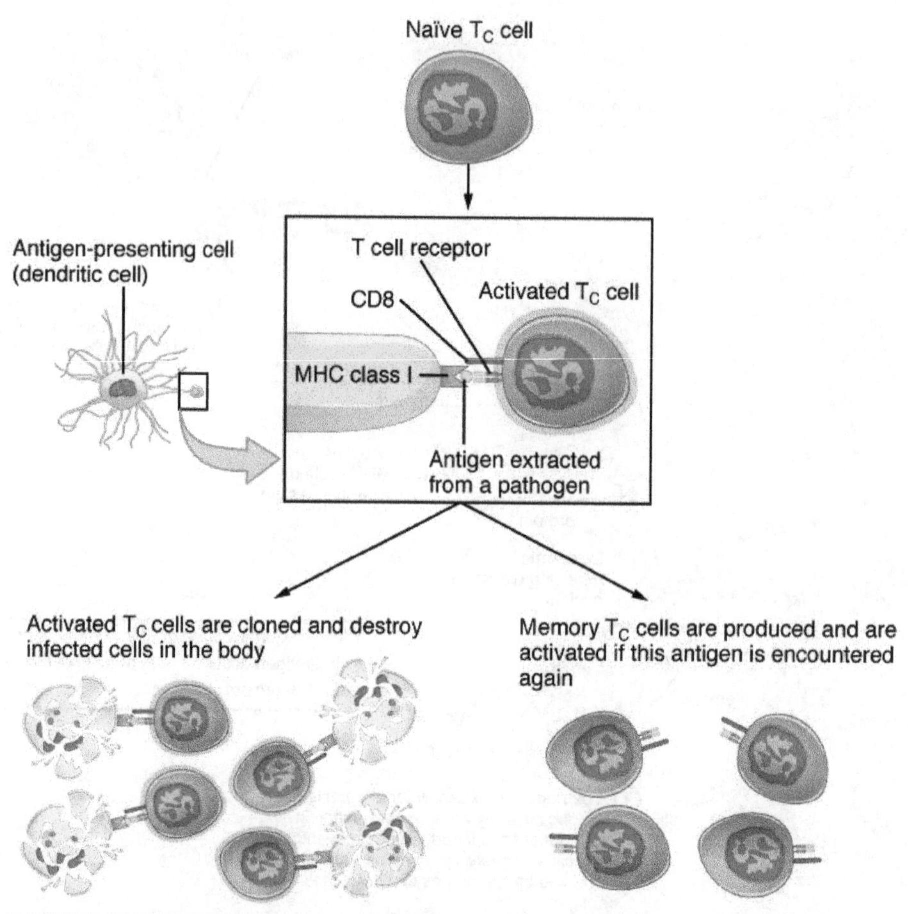

Figure 4. Stem cells differentiate into T cells with specific receptors, called clones. The clones with receptors specific for antigens on the pathogen are selected for and expanded.

Only those clones of lymphocytes whose receptors are activated by the antigen are stimulated to proliferate. Keep in mind that most antigens have multiple antigenic determinants, so a T cell response to a typical antigen involves a polyclonal response. A **polyclonal response** is the stimulation of multiple T cell clones. Once activated, the selected clones increase in number and make many copies of each cell type, each clone with its unique receptor. By the time this process is complete, the body will have large numbers of specific lymphocytes available to fight the infection.

The Cellular Basis of Immunological Memory

As already discussed, one of the major features of an adaptive immune response is the development of immunological memory.

During a primary adaptive immune response, both **memory T cells** and **effector T cells** are generated. Memory T cells are long-lived and can even persist for a lifetime. Memory cells are primed to act rapidly. Thus, any subsequent exposure to the pathogen will elicit a very rapid T cell response. This rapid, secondary adaptive response generates large numbers of effector T cells so fast that the pathogen is often overwhelmed before it can cause any symptoms of disease. This is what is meant by immunity to a disease. The same pattern of primary and secondary immune responses occurs in B cells and the antibody response, as will be discussed later in the chapter.

T Cell Types and their Functions

Mature T cells express cell adhesion molecules (CD4 or CD8) that keep the T cell in close contact with the antigen-presenting cell by directly binding to the MHC molecule (to a different part of the molecule than does the antigen). Thus, T cells and antigen-presenting cells are held together in two ways: by CD4 or CD8 attaching to MHC and by the T cell receptor binding to antigen.

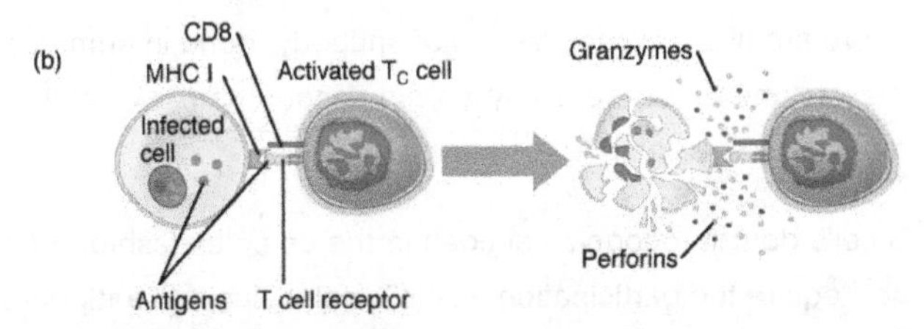

Figure 5. (a) CD4 is associated with helper and regulatory T cells. An extracellular pathogen is processed and presented in the binding cleft of a class II MHC molecule, and this interaction is strengthened by the CD4 molecule. (b) CD8 is associated with cytotoxic T cells. An intracellular pathogen is presented by a class I MHC molecule, and CD8 interacts with it.

Helper T Cells and their Cytokines

Helper T cells (Th) function by secreting cytokines that act to enhance other immune responses. There are two classes of Th cells, and they act on different components of the immune response. These cells are not distinguished by their surface molecules but by the characteristic set of cytokines they secrete.

- **Th1 cells** are a type of helper T cell that secretes cytokines that regulate the immunological activity and development of a variety of cells, including macrophages and other types of T cells.
- **Th2 cells** are cytokine-secreting cells that act on B cells to drive their differentiation into plasma cells that make antibody. T cell help is required for antibody responses to most protein antigens.

Cytotoxic T cells

Cytotoxic T cells (Tc) are T cells that kill target cells by inducing apoptosis using the same mechanism as NK cells. They either express Fas ligand, which binds to the fas molecule on the target cell, or act by using perforins and granzymes contained in their cytoplasmic granules. As was discussed earlier with NK cells, killing a virally infected cell before the virus can complete its replication cycle results in the production of no infectious particles. As more Tc cells are developed during an immune response, they overwhelm the ability of the virus to cause disease. In addition, each Tc cell can kill more than one target cell, making them especially effective.

The Adaptive Immune Response: B-lymphocytes and Antibodies

Antibodies were the first component of the adaptive immune response to be characterized by scientists working on the immune system. It was already known that individuals who survived a bacterial infection were immune to re-infection with the same pathogen. Early microbiologists took serum from an immune patient and mixed it with a fresh culture of the same type of bacteria, then observed the bacteria under a microscope. The bacteria became clumped in a process called agglutination. When a different bacterial species was used, the agglutination did not happen. Thus, there was something in the serum of immune individuals that could specifically bind to and agglutinate bacteria.

Scientists now know the cause of the agglutination is an antibody molecule, also called an **immunoglobulin**. What is an antibody? An antibody protein is essentially a secreted form of a B cell receptor. (In fact, surface immunoglobulin is another name for the B cell receptor.) Not surprisingly, the same genes encode both the secreted antibodies and the surface immunoglobulins.

There are five different classes of antibody found in humans: IgM, IgD, IgG, IgA, and IgE. Each of these has specific functions in the immune response, so by learning about them, researchers can learn about the great variety of antibody functions critical to many adaptive immune responses.

B cells do not recognize antigen in the complex fashion of T cells. B cells can recognize native, unprocessed antigen and do not require the participation of MHC molecules and antigen-presenting cells.

B Cell Differentiation and Activation

B cells differentiate in the bone marrow. During the process of maturation, up to 100 trillion different clones of B cells are generated, which is similar to the diversity of antigen receptors seen in T cells.

After B cells are activated by their binding to antigen, they differentiate into plasma cells. Plasma cells often leave the secondary lymphoid organs, where the response is generated, and migrate back to the bone marrow, where the whole differentiation process started. After secreting antibodies for a specific period, they die, as most of their energy is devoted to making antibodies and not to maintaining themselves. Thus, plasma cells are said to be terminally differentiated.

Memory B cells function in a way similar to memory T cells. They lead to a stronger and faster secondary response when compared to the primary response, as illustrated below.

Antibody Structure

Antibodies are glycoproteins consisting of two types of polypeptide chains with attached carbohydrates. The **heavy chain** and the **light chain** are the two polypeptides that form the antibody. The main differences between the classes of antibodies are in the differences between their heavy chains, but as you shall see, the light chains have an important role, forming part of the antigen-binding site on the antibody molecules.

Four-chain Models of Antibody Structures

All antibody molecules have two identical heavy chains and two identical light chains. (Some antibodies contain multiple units of this four-chain structure.) The **Fc region** of the antibody is formed by the two heavy chains coming together, usually linked by disulfide bonds. The Fc portion of the antibody is important in that many effector cells of the immune system have Fc receptors. Cells having these receptors can then bind to antibody-coated pathogens, greatly increasing the specificity of the effector cells. At the other end of the molecule are two identical antigen-binding sites.

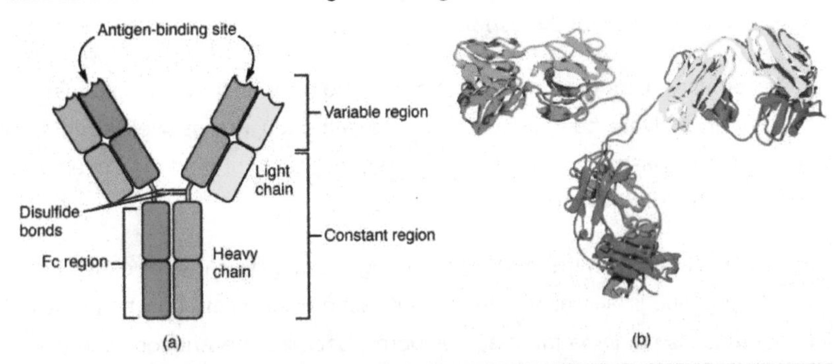

Figure 1. The typical four chain structure of a generic antibody (a) and the corresponding three-dimensional structure of the antibody IgG2 (b). (credit b: modification of work by Tim Vickers)

Five Classes of Antibodies and their Functions

In general, antibodies have two basic functions. They can act as the B cell antigen receptor or they can be secreted, circulate, and bind to a pathogen, often labeling it for identification by other forms of the immune response. Of the five antibody classes, notice that only two can function as the antigen receptor for naïve B cells: IgM and **IgD** (see Table 1). Mature B cells that leave the bone marrow express both IgM and IgD, but both antibodies have the same antigen specificity. Only IgM is secreted, however, and no other nonreceptor function for IgD has been discovered.

Table 1. The Five Immunoglobulin (Ig) Classes

	IgM	IgG	IgA	IgE	IgD
Diagram			Secretory component		
Crosses placenta	no	yes	no	no	no
Fixes complement	yes	yes	no	no	no
Fc binds to		phagocytes		mast cells and basophils	
Function	Main antibody of primary responses, best at fixing complement; the monomer form of IgM serves as the B cell receptor	Main blood antibody of secondary responses, neutralizes toxins, opsonization	Secreted into mucus, tears, saliva, colostrum	Antibody of allergy and antiparasitic activity	B cell receptor

IgM is usually the first antibody made during a primary response. Its 10 antigen-binding sites and large shape allow it to bind well to many bacterial surfaces promoting agglutination. It is excellent at binding complement proteins and activating the complement cascade, consistent with its role in promoting chemotaxis, opsonization, and cell lysis. It is a very effective antibody against bacteria at early stages of a primary antibody response.

IgG is a major antibody of late primary responses and the main antibody of secondary responses in the blood. IgG clears pathogens from the blood and can activate complement proteins (although not as well as IgM), taking advantage of its antibacterial activities. Furthermore, this class of antibody crosses the placenta to protect the developing fetus from disease and exits the blood to the interstitial fluid to fight extracellular pathogens.

IgA exists in two forms, a four-chain structure in the blood and an eight-chain structure, or dimer, in exocrine gland secretions of the mucous membranes, including mucus, saliva, and tears. Thus, dimeric IgA is the only antibody to leave the interior of the body to protect body surfaces. IgA is also of importance to newborns, because this antibody is present in mother's breast milk (colostrum), which serves to protect the infant from disease.

IgE is usually associated with allergies and anaphylaxis. IgE makes mast cell degranulation very specific, such that if a person is allergic to peanuts, there will be peanut-specific IgE bound to his or her mast cells. In this person, eating peanuts will cause the mast cells to degranulate, sometimes causing severe allergic reactions, including anaphylaxis, a severe, systemic allergic response that can cause death.

Clonal Selection of B Cells

Clonal selection and expansion work much the same way in B cells as in T cells. Only B cells with appropriate antigen specificity are selected for and expanded. Eventually, the plasma cells secrete antibodies with antigenic specificity identical to those that were on the surfaces of the selected B cells. Notice in the figure that both plasma cells and memory B cells are generated simultaneously.

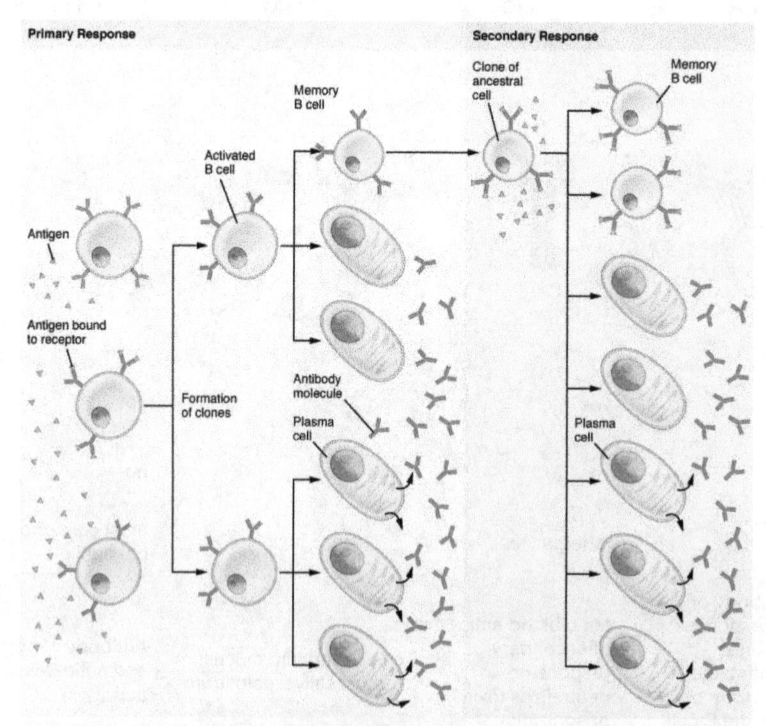

Figure 2. During a primary B cell immune response, both antibody-secreting plasma cells and memory B cells are produced. These memory cells lead to the differentiation of more plasma cells and memory B cells during secondary responses.

Primary versus Secondary B Cell Responses

Primary and secondary responses as they relate to T cells were discussed earlier. This section will look at these responses with B cells and antibody production. Because antibodies are easily obtained from blood samples, they are easy to follow and graph (Figure 3). As you will see from the figure, the primary response to an antigen (representing a pathogen) is delayed by several days. This is the time it takes for the B cell clones to expand and differentiate into plasma cells. The level of antibody produced is low, but it is sufficient for immune protection. The second time a person encounters the same antigen, there is no time delay, and the amount of antibody made is much higher. Thus, the secondary antibody response overwhelms the pathogens quickly and, in most situations, no symptoms are felt. When a different antigen is used, another primary response is made with its low antibody levels and time delay.

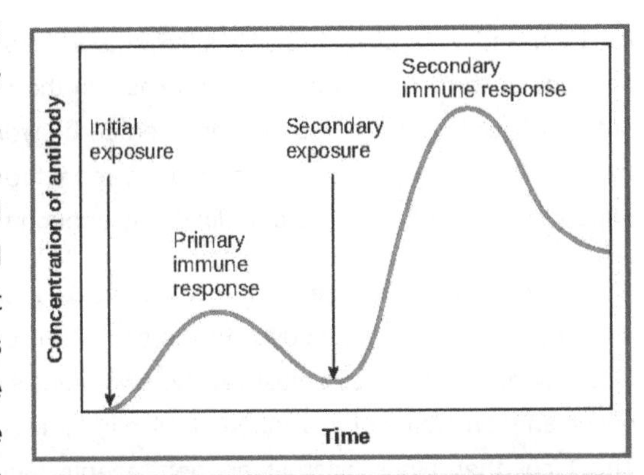

Figure 3. Antigen A is given once to generate a primary response and later to generate a secondary response. When a different antigen is given for the first time, a new primary response is made.

Active versus Passive Immunity

Immunity to pathogens, and the ability to control pathogen growth so that damage to the tissues of the body is limited, can be acquired by (1) the active development of an immune response in the infected individual or (2) the passive transfer of immune components from an immune individual to a nonimmune one. Both active and passive immunity have examples in the natural world and as part of medicine.

Active immunity is the resistance to pathogens acquired during an adaptive immune response within an individual. Naturally acquired active immunity, the response to a pathogen, is the focus of this chapter. Artificially acquired active immunity involves the use of vaccines. A vaccine is a killed or weakened pathogen or its components that, when administered to a healthy individual, leads to the development of immunological memory (a weakened primary immune response) without causing much in the way of symptoms. Thus, with the use of vaccines, one can avoid the damage from disease that results from the first exposure to the pathogen, yet reap the benefits of protection from immunological memory. The advent of vaccines was one of the major medical advances of the twentieth century and led to the eradication of smallpox and the control of many infectious diseases, including polio, measles, and whooping cough.

Table 2. Active versus Passive Immunity

	Natural	Artificial
Active	Adaptive immune response	Vaccine response
Passive	Trans-placental antibodies/breastfeeding	Immune globulin injections

Passive immunity arises from the transfer of antibodies to an individual without requiring them to mount their own active immune response. Naturally acquired passive immunity is seen during fetal development. IgG is transferred from the maternal circulation to the fetus via the placenta, protecting the fetus from infection and protecting the newborn for the first few months of its life. As already stated, a newborn benefits from the IgA antibodies it obtains from milk during breastfeeding. The fetus and newborn thus benefit from the immunological memory of the mother to the pathogens to which she has been exposed. In medicine, artificially acquired passive immunity usually involves injections of immunoglobulins, taken from animals previously exposed to a specific pathogen. This treatment is a fast-acting method of temporarily protecting an individual who was possibly exposed to a pathogen. The downside to both types of passive immunity is the lack of the development of immunological memory. Once the antibodies are transferred, they are effective for only a limited time before they degrade.

T cell-dependent versus T cell-independent Antigens

As discussed previously, Th2 cells secrete cytokines that drive the production of antibodies in a B cell, responding to complex antigens such as those made by proteins. On the other hand, some antigens are T cell independent. A **T cell-independent antigen** usually is in the form of repeated carbohydrate moieties found on the cell walls of bacteria. Each antibody on the B cell surface has two binding sites, and the repeated nature of T cell-independent antigen leads to crosslinking of the surface antibodies on the B cell. The crosslinking is enough to activate it in the absence of T cell cytokines.

A **T cell-dependent antigen**, on the other hand, usually is not repeated to the same degree on the pathogen and thus does not crosslink surface antibody with the same efficiency. To elicit a response to such antigens, the B and T cells must come close together. The B cell must receive two signals to become activated. First, its surface immunoglobulin must recognize a native antigen. Some of this antigen is also internalized, processed, and presented to the Th2 cells on a class II MHC molecule. The T cell then binds using its antigen receptor and is activated to secrete cytokines that diffuse to the B cell creating the second signal, finally activating it completely. Thus, the B cell receives signals from both its surface antibody and the T cell via its cytokines, and acts as an antigen-presenting cell in the process.

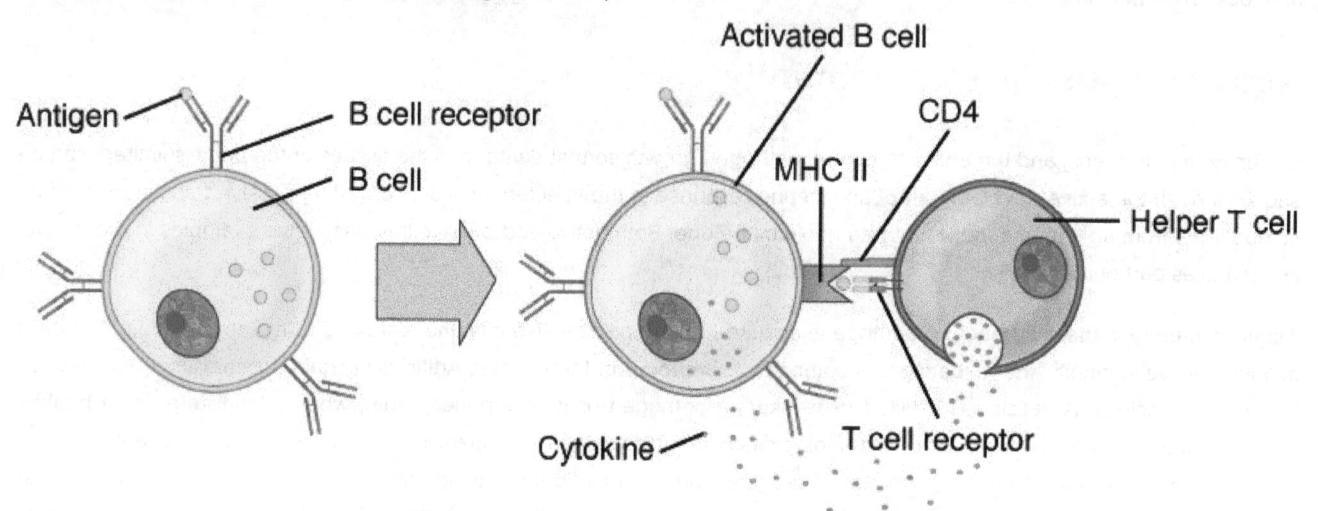

Figure 4. To elicit a response to a T cell-dependent antigen, the B and T cells must come close together. To become fully activated, the B cell must receive two signals from the native antigen and the T cell's cytokines.

The Immune Response against Pathogens

Now that you understand the development of mature, naïve B cells and T cells, and some of their major functions, how do all of these various cells, proteins, and cytokines come together to actually resolve an infection? Ideally, the immune response will rid the body of a pathogen entirely. The adaptive immune response, with its rapid clonal expansion, is well suited to this purpose. Think of a primary infection as a race between the pathogen and the immune system. The pathogen bypasses barrier defenses and starts multiplying in the host's body. During the first 4 to 5 days, the innate immune response will partially control, but not stop, pathogen growth. As the adaptive immune response gears up, however, it will begin to clear the pathogen from the body, while at the same time becoming stronger and stronger.

EVERYDAY CONNECTION: DISINFECTANTS—FIGHTING THE GOOD FIGHT?

"Wash your hands!" Parents have been telling their children this for generations. Dirty hands can spread disease. But is it possible to get rid of enough pathogens that children will never get sick? Are children who avoid exposure to pathogens better off? The answers to both these questions appears to be no.

Antibacterial wipes, soaps, gels, and even toys with antibacterial substances embedded in their plastic are ubiquitous in our society. Still, these products do not rid the skin and gastrointestinal tract of bacteria, and it would be harmful to our health if they did. We need these nonpathogenic bacteria on and within our bodies to keep the pathogenic ones from growing. The urge to keep children perfectly clean is thus probably misguided. Children will get sick anyway, and the later benefits of immunological memory far outweigh the minor discomforts of most childhood diseases. In fact, getting diseases such as chickenpox or measles later in life is much harder on the adult and are associated with symptoms significantly worse than those seen in the childhood illnesses. Of course, vaccinations help children avoid some illnesses, but there are so many pathogens, we will never be immune to them all.

Could over-cleanliness be the reason that allergies are increasing in more developed countries? Some scientists think so. Allergies are based on an IgE antibody response. Many scientists think the system evolved to help the body rid itself of worm parasites. The hygiene theory is the idea that the immune system is geared to respond to antigens, and if pathogens are not present, it will respond instead to inappropriate antigens such as allergens and self-antigens. This is one explanation for the rising incidence of allergies in developed countries, where the response to nonpathogens like pollen, shrimp, and cat dander cause allergic responses while not serving any protective function.

The Mucosal Immune Response

Mucosal tissues are major barriers to the entry of pathogens into the body. The IgA (and sometimes IgM) antibodies in mucus and other secretions can bind to the pathogen, and in the cases of many viruses and bacteria, neutralize them. **Neutralization** is the process of coating a pathogen with antibodies, making it physically impossible for the pathogen to bind to receptors. Neutralization, which occurs in the blood, lymph, and other body fluids and secretions, protects the body constantly. Neutralizing antibodies are the basis for the disease protection offered by vaccines. Vaccinations for diseases that commonly enter the body via mucous membranes, such as influenza, are usually formulated to enhance IgA production.

Immune responses in some mucosal tissues such as the Peyer's patches in the small intestine take up particulate antigens by specialized cells known as microfold or M cells (Figure 2). These cells allow the body to sample potential pathogens from the intestinal lumen. Dendritic cells then take the antigen to the regional lymph nodes, where an immune response is mounted.

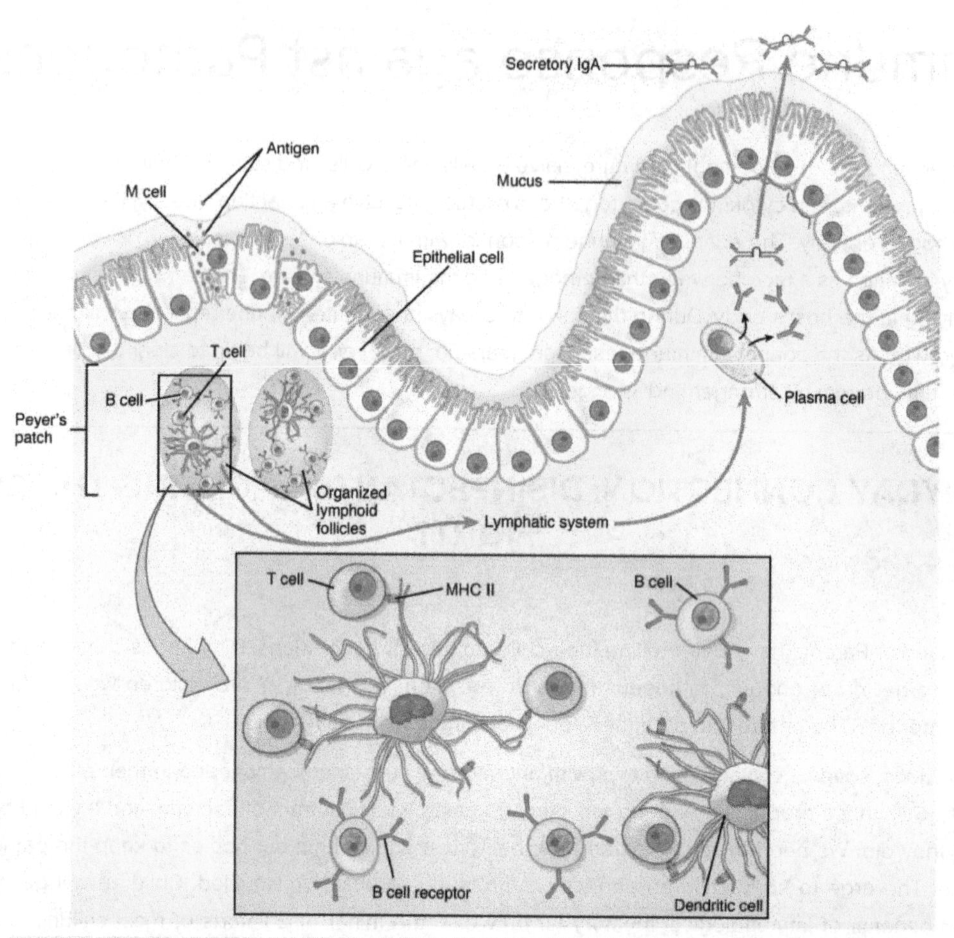

Figure 2. The nasal-associated lymphoid tissue and Peyer's patches of the small intestine generate IgA immunity. Both use M cells to transport antigen inside the body so that immune responses can be mounted.

Defenses against Bacteria and Fungi

The body fights bacterial pathogens with a wide variety of immunological mechanisms, essentially trying to find one that is effective. Bacteria such as *Mycobacterium leprae*, the cause of leprosy, are resistant to lysosomal enzymes and can persist in macrophage organelles or escape into the cytosol. In such situations, infected macrophages receiving cytokine signals from Th1 cells turn on special metabolic pathways. **Macrophage oxidative metabolism** is hostile to intracellular bacteria, often relying on the production of nitric oxide to kill the bacteria inside the macrophage.

Fungal infections, such as those from *Aspergillus*, *Candida*, and *Pneumocystis*, are largely opportunistic infections that take advantage of suppressed immune responses. Most of the same immune mechanisms effective against bacteria have similar effects on fungi, both of which have characteristic cell wall structures that protect their cells.

Defenses against Parasites

Worm parasites such as helminths are seen as the primary reason why the mucosal immune response, IgE-mediated allergy and asthma, and eosinophils evolved. These parasites were at one time very common in human society. When infecting a human, often via contaminated food, some worms take up residence in the gastrointestinal tract. Eosinophils are attracted to the site by T cell cytokines, which release their granule contents upon their arrival. Mast cell degranulation also occurs, and the fluid leakage caused by the increase in local vascular permeability is thought to have a flushing action on the parasite, expelling its larvae from the body. Furthermore, if IgE labels the parasite, the eosinophils can bind to it by its Fc receptor.

Defenses against Viruses

The primary mechanisms against viruses are NK cells, interferons, and cytotoxic T cells. Antibodies are effective against viruses mostly during protection, where an immune individual can neutralize them based on a previous exposure. Antibodies have no effect on viruses or other intracellular pathogens once they enter the cell, since antibodies are not able to penetrate the plasma membrane of the cell. Many cells respond to viral infections by downregulating their expression of MHC class I molecules. This is to the advantage of the virus, because without MHC expression, cytotoxic T cells have no activity. NK cells, however, can recognize virally infected MHC-negative cells and destroy them. Thus, NK and cytotoxic T cells have complementary activities against virally infected cells.

Interferons have activity in slowing viral replication and are used in the treatment of certain viral diseases, such as hepatitis B and C, but their ability to eliminate the virus completely is limited. The cytotoxic T cell response, though, is key, as it eventually overwhelms the virus and kills infected cells before the virus can complete its replicative cycle. Clonal expansion and the ability of cytotoxic T cells to kill more than one target cell make these cells especially effective against viruses. In fact, without cytotoxic T cells, it is likely that humans would all die at some point from a viral infection (if no vaccine were available).

Evasion of the Immune System by Pathogens

It is important to keep in mind that although the immune system has evolved to be able to control many pathogens, pathogens themselves have evolved ways to evade the immune response. An example already mentioned is in *Mycobactrium tuberculosis*, which has evolved a complex cell wall that is resistant to the digestive enzymes of the macrophages that ingest them, and thus persists in the host, causing the chronic disease tuberculosis. This section briefly summarizes other ways in which pathogens can "outwit" immune responses. But keep in mind, although it seems as if pathogens have a will of their own, they do not. All of these evasive "strategies" arose strictly by evolution, driven by selection.

Bacteria sometimes evade immune responses because they exist in multiple strains, such as different groups of *Staphylococcus aureus*. *S. aureus* is commonly found in minor skin infections, such as boils, and some healthy people harbor it in their nose. One small group of strains of this bacterium, however, called methicillin-resistant *Staphylococcus aureus*, has become resistant to multiple antibiotics and is essentially untreatable. Different bacterial strains differ in the antigens on their surfaces. The immune response against one strain (antigen) does not affect the other; thus, the species survives.

Another method of immune evasion is mutation. Because viruses' surface molecules mutate continuously, viruses like influenza change enough each year that the flu vaccine for one year may not protect against the flu common to the next. New vaccine formulations must be derived for each flu season.

Genetic recombination—the combining of gene segments from two different pathogens—is an efficient form of immune evasion. For example, the influenza virus contains gene segments that can recombine when two different viruses infect the same cell. Recombination between human and pig influenza viruses led to the 2010 H1N1 swine flu outbreak.

Pathogens can produce immunosuppressive molecules that impair immune function, and there are several different types. Viruses are especially good at evading the immune response in this way, and many types of viruses have been shown to suppress the host immune response in ways much more subtle than the wholesale destruction caused by HIV.

Diseases Associated with Depressed or Overactive Immune Responses

This section is about how the immune system goes wrong. When it goes haywire, and becomes too weak or too strong, it leads to a state of disease. The factors that maintain immunological homeostasis are complex and incompletely understood.

Immunodeficiencies

As you have seen, the immune system is quite complex. It has many pathways using many cell types and signals. Because it is so complex, there are many ways for it to go wrong. Inherited immunodeficiencies arise from gene mutations that affect specific components of the immune response. There are also acquired immunodeficiencies with potentially devastating effects on the immune system, such as HIV.

Inherited Immunodeficiencies

A list of all inherited immunodeficiencies is well beyond the scope of this book. The list is almost as long as the list of cells, proteins, and signaling molecules of the immune system itself. Certainly, the most serious of the inherited immunodeficiencies is **severe combined immunodeficiency disease (SCID)**. This disease is complex because it is caused by many different genetic defects. What groups them together is the fact that both the B cell and T cell arms of the adaptive immune response are affected.

Children with this disease usually die of opportunistic infections within their first year of life unless they receive a bone marrow transplant. Such a procedure had not yet been perfected for David Vetter, the "boy in the bubble," who was treated for SCID by having to live almost his entire life in a sterile plastic cocoon for the 12 years before his death from infection in 1984. One of the features that make bone marrow transplants work as well as they do is the proliferative capability of hematopoietic stem cells of the bone marrow. Only a small amount of bone marrow from a healthy donor is given intravenously to the recipient. It finds its own way to the bone where it populates it, eventually reconstituting the patient's immune system, which is usually destroyed beforehand by treatment with radiation or chemotherapeutic drugs.

New treatments for SCID using gene therapy, inserting nondefective genes into cells taken from the patient and giving them back, have the advantage of not needing the tissue match required for standard transplants. Although not a standard treatment, this approach holds promise, especially for those in whom standard bone marrow transplantation has failed.

Human Immunodeficiency Virus/AIDS

Although many viruses cause suppression of the immune system, only Human Immunodeficiency Virus (HIV) wipes it out completely. It is worth discussing the biology of this virus, which can lead to the well-known AIDS, so that its full effects on the immune system can be understood. The virus is transmitted through semen, vaginal fluids, and blood, and can be caught by risky sexual behaviors and the sharing of needles by intravenous drug users. There are sometimes, but not always, flu-like symptoms in the first 1 to 2 weeks after infection.

Treatment for the disease consists of drugs that target virally encoded proteins that are necessary for viral replication but are absent from normal human cells. By targeting the virus itself and sparing the cells, this approach has been successful in significantly prolonging the lives of HIV-positive individuals. On the other hand, an HIV vaccine has been 30 years in development and is still years away. Because the virus mutates rapidly to evade the immune system, scientists have been looking for parts of the virus that do not change and thus would be good targets for a vaccine candidate.

Hypersensitivities

The word "hypersensitivity" simply means sensitive beyond normal levels of activation. Allergies and inflammatory responses to nonpathogenic environmental substances have been observed since the dawn of history. Hypersensitivity is a medical term describing symptoms that are now known to be caused by unrelated mechanisms of immunity. Still, it is useful for this discussion to use the four types of hypersensitivities as a guide to understand these mechanisms. Figure 1 and Table 1 provide an overview of four types of hypersensitivity:

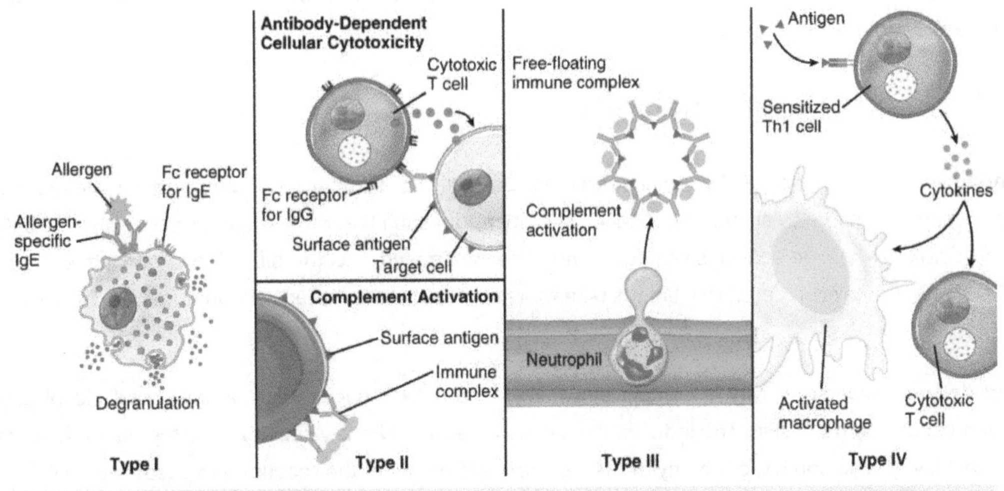

Figure 1. Four types of hypersensitivity.

Table 1. Components of the immune system cause four types of hypersensitivity.

Type I: IgE-Mediated Hypersensitivity	Type II: IgG-Mediated Cytotoxic Hypersensitivity	Type III: Immune Complex-Mediated Hypersensitivity	Type IV: Cell-Mediated Hypersensitivity
IgE is bound to mast cells via its Fc portion. When an allergen binds to these antibodies, crosslinking of IgE induces degranulation.	Cells are destroyed by bound antibody, either by activation of complement or by a cytotoxic T cell with an Fc receptor for the antibody (ADCC).	Antigen–antibody complexes are deposited in tissues, causing activation of complement, which attracts neutrophils to the site.	Th1 cells secrete cytokines, which activate macrophages and cytotoxic T cells and can cause macrophage accumulation at the site.
Causes localized and systemic anaphylaxis, seasonal allergies including hay fever, food allergies such as those to shellfish and peanuts, hives, and eczema.	Red blood cells destroyed by complement and antibody during a transfusion of mismatched blood types or during erythroblastosis fetalis	Most common forms of immune complex disease are seen in glomerulonephritis, rheumatoid arthritis, and systemic lupus erythematosus.	Most common forms are contact dermatitis, tuberculin reaction, and autoimmune diseases such as diabetes mellitus type I, multiple sclerosis, and rheumatoid arthritis.

Notice that types I–III are B cell mediated, whereas type IV hypersensitivity is exclusively a T cell phenomenon.

Immediate (Type I) Hypersensitivity

Antigens that cause allergic responses are often referred to as allergens. The specificity of the **immediate hypersensitivity** response is based on the binding of allergen-specific IgE to the mast cell surface. The process of producing allergen-specific IgE is called sensitization, and is a necessary prerequisite for the symptoms of immediate hypersensitivity to occur. Allergies and allergic asthma are mediated by mast cell degranulation that is caused by the crosslinking of the antigen-specific IgE molecules on the mast cell surface. The mediators released have various vasoactive effects already discussed, but the major symptoms of inhaled allergens are the nasal edema and runny nose caused by the increased vascular permeability and increased blood flow of nasal blood vessels. As these mediators are released with mast cell degranulation, **type I hypersensitivity** reactions are usually rapid and occur within just a few minutes, hence the term immediate hypersensitivity.

Type II and Type III Hypersensitivities

Type II hypersensitivity, which involves IgG-mediated lysis of cells by complement proteins, occurs during mismatched blood transfusions and blood compatibility diseases.

Type III hypersensitivity occurs with diseases such as systemic lupus erythematosus, where soluble antigens, mostly DNA and other material from the nucleus, and antibodies accumulate in the blood to the point that the antigen and antibody precipitate along blood vessel linings. These immune complexes often lodge in the kidneys, joints, and other organs where they can activate complement proteins and cause inflammation.

Delayed (Type IV) Hypersensitivity

Delayed hypersensitivity, or type IV hypersensitivity, is basically a standard cellular immune response. In delayed hypersensitivity, the first exposure to an antigen is called **sensitization**, such that on re-exposure, a secondary cellular response results, secreting cytokines that recruit macrophages and other phagocytes to the site. These sensitized T cells, of the Th1 class, will also activate cytotoxic T cells. The time it takes for this reaction to occur accounts for the 24- to 72-hour delay in development.

Another type of delayed hypersensitivity is contact sensitivity, where substances such as the metal nickel cause a red and swollen area upon contact with the skin. The individual must have been previously sensitized to the metal. A much more severe case of contact sensitivity is poison ivy, but many of the harshest symptoms of the reaction are associated with the toxicity of its oils and are not T cell mediated.

Autoimmune Responses

The worst cases of the immune system over-reacting are autoimmune diseases. Somehow, tolerance breaks down and the immune systems in individuals with these diseases begin to attack their own bodies, causing significant damage. The trigger for these diseases is, more often than not, unknown, and the treatments are usually based on resolving the symptoms using immunosuppressive and anti-inflammatory drugs such as steroids. These diseases can be localized and crippling, as in rheumatoid arthritis, or diffuse in the body with multiple symptoms that differ in different individuals, as is the case with systemic lupus erythematosus.

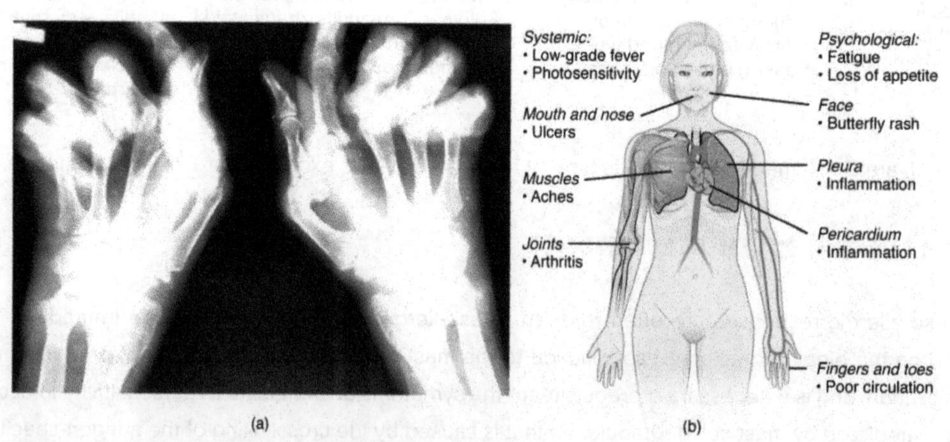

Figure 2. (a) Extensive damage to the right hand of a rheumatoid arthritis sufferer is shown in the x-ray. (b) The diagram shows a variety of possible symptoms of systemic lupus erythematosus.

Environmental triggers seem to play large roles in autoimmune responses. One explanation for the breakdown of tolerance is that, after certain bacterial infections, an immune response to a component of the bacterium cross-reacts with a self-antigen. This mechanism is seen in rheumatic fever, a result of infection with *Streptococcus* bacteria, which causes strep throat. The antibodies to this pathogen's M protein cross-react with an antigenic component of heart myosin, a major contractile protein of the heart that is critical to its normal function. The antibody binds to these molecules and activates complement proteins, causing damage to the heart, especially to the heart valves. On the other hand, some hypotheses propose that having multiple common infectious diseases actually prevents autoimmune responses. The fact that autoimmune diseases are rare in countries that have a high incidence of infectious diseases supports this idea, another example of the hygiene hypothesis discussed earlier in this chapter.

There are genetic factors in autoimmune diseases as well. Some diseases are associated with the MHC genes that an individual expresses. The reason for this association is likely because if one's MHC molecules are not able to present a certain self-antigen, then that particular autoimmune disease cannot occur. Overall, there are more than 80 different autoimmune diseases, which are a significant health problem in the elderly. The table below lists several of the most common autoimmune diseases, the antigens that are targeted, and the segment of the adaptive immune response that causes the damage.

Table 2. Autoimmune Diseases

Disease	Autoantigen	Symptoms
Celiac disease	Tissue transglutaminase	Damage to small intestine
Diabetes mellitus type I	Beta cells of pancreas	Low insulin production; inability to regulate serum glucose
Graves' disease	Thyroid-stimulating hormone receptor (antibody blocks receptor)	Hyperthyroidism
Hashimoto's thyroiditis	Thyroid-stimulating hormone receptor (antibody mimics hormone and stimulates receptor)	Hypothyroidism
Lupus erythematosus	Nuclear DNA and proteins	Damage of many body systems
Myasthenia gravis	Acetylcholine receptor in neuromuscular junctions	Debilitating muscle weakness
Rheumatoid arthritis	Joint capsule antigens	Chronic inflammation of joints

Transplantation and Cancer Immunology

The immune responses to transplanted organs and to cancer cells are both important medical issues. With the use of tissue typing and anti-rejection drugs, transplantation of organs and the control of the anti-transplant immune response have made huge strides in the past 50 years. Today, these procedures are commonplace. **Tissue typing** is the determination of MHC molecules in the tissue to be transplanted to better match the donor to the recipient. The immune response to cancer, on the other hand, has been more difficult to understand and control. Although it is clear that the immune system can recognize some cancers and control them, others seem to be resistant to immune mechanisms.

The Rh Factor

Red blood cells can be typed based on their surface antigens. ABO blood type, in which individuals are type A, B, AB, or O according to their genetics, is one example. A separate antigen system seen on red blood cells is the Rh antigen. When someone is "A positive" for example, the positive refers to the presence of the Rh antigen, whereas someone who is "A negative" would lack this molecule.

An interesting consequence of Rh factor expression is seen in **erythroblastosis fetalis**, a hemolytic disease of the newborn. This disease occurs when mothers negative for Rh antigen have multiple Rh-positive children. During the birth of a first Rh-positive child, the mother makes a primary anti-Rh antibody response to the fetal blood cells that enter the maternal bloodstream. If the mother has a second Rh-positive child, IgG antibodies against Rh-positive blood mounted during this secondary response cross the placenta and attack the fetal blood, causing anemia. This is a consequence of the fact that the fetus is not genetically identical to the mother, and thus the mother is capable of mounting an immune response against it. This disease is treated with antibodies specific for Rh factor. These are given to the mother during the subsequent births, destroying any fetal blood that might enter her system and preventing the immune response.

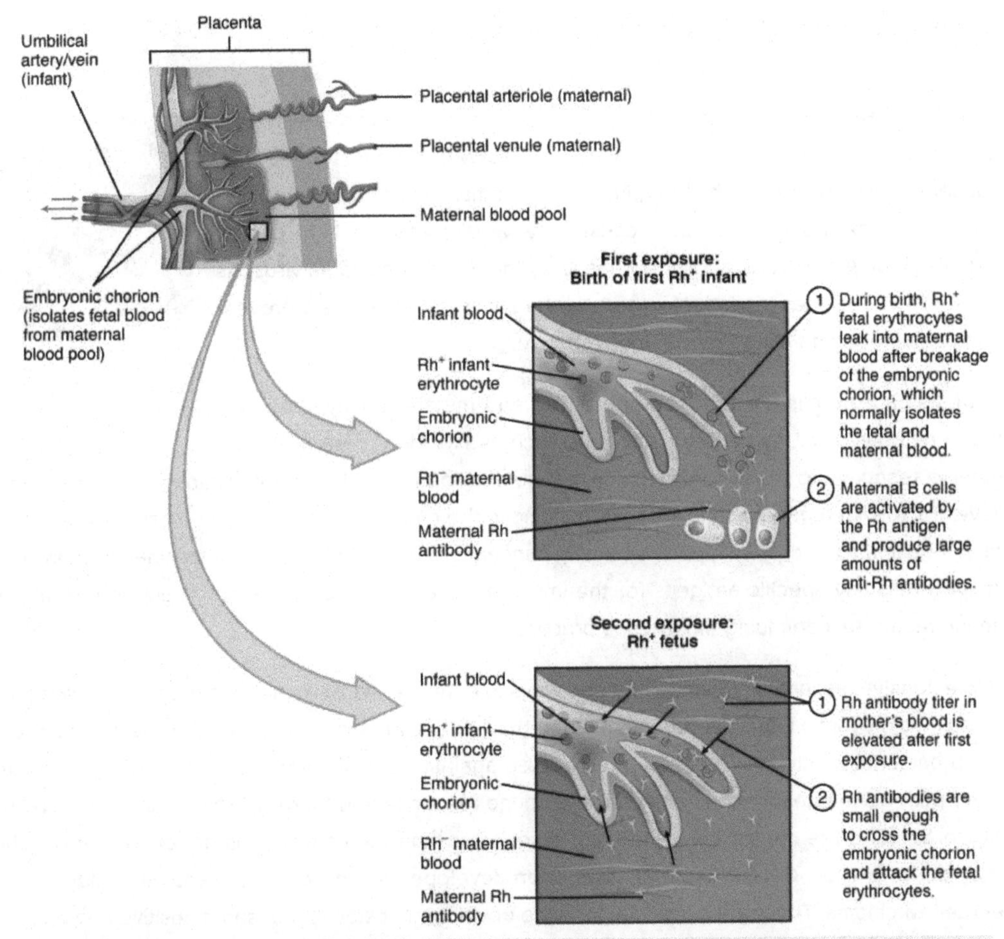

Figure 1. Erythroblastosis fetalis (hemolytic disease of the newborn) is the result of an immune response in an Rh-negative mother who has multiple children with an Rh-positive father. During the first birth, fetal blood enters the mother's circulatory system, and anti-Rh antibodies are made. During the gestation of the second child, these antibodies cross the placenta and attack the blood of the fetus. The treatment for this disease is to give the mother anti-Rh antibodies (RhoGAM) during the first pregnancy to destroy Rh-positive fetal red blood cells from entering her system and causing the anti-Rh antibody response in the first place.

Tissue Transplantation

Tissue transplantation is more complicated than blood transfusions because of two characteristics of MHC molecules. These molecules are the major cause of transplant rejection (hence the name "histocompatibility"). When a donor organ expresses MHC molecules that are different from the recipient, the latter will often mount a cytotoxic T cell response to the organ and reject it. If a biopsy of a transplanted organ exhibits massive infiltration of T lymphocytes within the first weeks after transplant, it is a sign that the transplant is likely to fail. The response is a classical, and very specific, primary T cell immune response. As far as medicine is concerned, the immune response in this scenario does the patient no good at all and causes significant harm.

Immunosuppressive drugs such as cyclosporine A have made transplants more successful, but matching the MHC molecules is still key. Family members, since they share a similar genetic background, are much more likely to share MHC molecules than unrelated individuals do. In fact, due to the extensive polymorphisms in these MHC molecules, unrelated donors are found only through a worldwide database. The system is not foolproof however, as there are not enough individuals in the system to provide the organs necessary to treat all patients needing them.

Immune Responses Against Cancer

It is clear that with some cancers, for example Kaposi's sarcoma, a healthy immune system does a good job at controlling them. This disease, which is caused by the human herpesvirus, is almost never observed in individuals with strong immune systems, such as the young and immunocompetent. Other examples of cancers caused by viruses include liver cancer caused by the hepatitis B virus and cervical cancer caused by the human papilloma virus. As these last two viruses have vaccines available for them, getting vaccinated can help prevent these two types of cancer by stimulating the immune response.

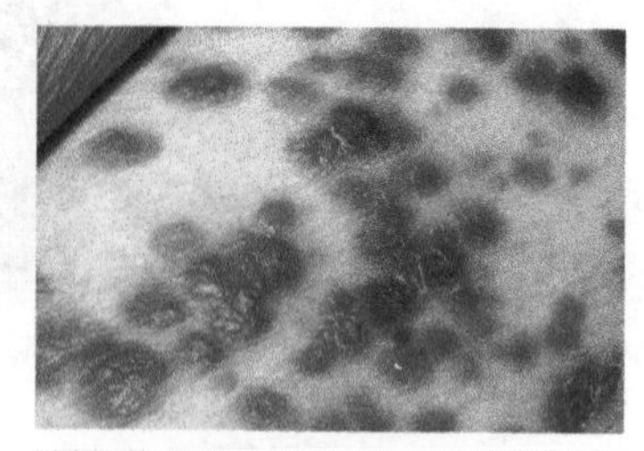

Figure 2. Karposi's Sarcoma Lesions (credit: National Cancer Institute)

On the other hand, as cancer cells are often able to divide and mutate rapidly, they may escape the immune response, just as certain pathogens such as HIV do. There are three stages in the immune response to many cancers: elimination, equilibrium, and escape. Elimination occurs when the immune response first develops toward tumor-specific antigens specific to the cancer and actively kills most cancer cells, followed by a period of controlled equilibrium during which the remaining cancer cells are held in check. Unfortunately, many cancers mutate, so they no longer express any specific antigens for the immune system to respond to, and a subpopulation of cancer cells escapes the immune response, continuing the disease process.

This fact has led to extensive research in trying to develop ways to enhance the early immune response to completely eliminate the early cancer and thus prevent a later escape. One method that has shown some success is the use of cancer vaccines, which differ from viral and bacterial vaccines in that they are directed against the cells of one's own body. Treated cancer cells are injected into cancer patients to enhance their anti-cancer immune response and thereby prolong survival. The immune system has the capability to detect these cancer cells and proliferate faster than the cancer cells do, overwhelming the cancer in a similar way as they do for viruses. Cancer vaccines have been developed for malignant melanoma, a highly fatal skin cancer, and renal (kidney) cell carcinoma. These vaccines are still in the development stages, but some positive and encouraging results have been obtained clinically.

EVERYDAY CONNECTION: HOW STRESS AFFECTS THE IMMUNE RESPONSE

The Connections between the Immune, Nervous, and Endocrine Systems of the Body

The immune system cannot exist in isolation. After all, it has to protect the entire body from infection. Therefore, the immune system is required to interact with other organ systems, sometimes in complex ways. Thirty years of research focusing on the connections between the immune system, the central nervous system, and the endocrine system have led to a new science with the unwieldy name of called **psychoneuroimmunology**. The physical connections between these systems have been known for centuries: All primary and secondary organs are connected to sympathetic nerves. What is more complex, though, is the interaction of neurotransmitters, hormones, cytokines, and other soluble signaling molecules, and the mechanism of "crosstalk" between the systems. For example, white blood cells, including lymphocytes and phagocytes, have receptors for various neurotransmitters released by associated neurons. Additionally, hormones such as cortisol (naturally produced by the adrenal cortex) and prednisone (synthetic) are well known for their abilities to suppress T cell immune mechanisms, hence, their prominent use in medicine as long-term, anti-inflammatory drugs.

One well-established interaction of the immune, nervous, and endocrine systems is the effect of stress on immune health. In the human vertebrate evolutionary past, stress was associated with the fight-or-flight response, largely mediated by the central nervous system and the adrenal medulla. This stress was necessary for survival. The physical action of fighting or running, whichever the animal decides, usually resolves the problem in one way or another. On the other hand, there are no physical actions to resolve most modern day stresses, including short-term stressors like taking examinations and long-term stressors such as being unemployed or losing a spouse. The effect of stress can be felt by nearly every organ system, and the immune system is no exception.

Table 3. Effects of Stress on Body Systems

System	Stress-related illness
Integumentary system	Acne, skin rashes, irritation
Nervous system	Headaches, depression, anxiety, irritability, loss of appetite, lack of motivation, reduced mental performance
Muscular and skeletal systems	Muscle and joint pain, neck and shoulder pain
Circulatory system	Increased heart rate, hypertension, increased probability of heart attacks
Digestive system	Indigestion, heartburn, stomach pain, nausea, diarrhea, constipation, weight gain or loss
Immune system	Depressed ability to fight infections
Male reproductive system	Lowered sperm production, impotence, reduced sexual desire
Female reproductive system	Irregular menstrual cycle, reduced sexual desire

At one time, it was assumed that all types of stress reduced all aspects of the immune response, but the last few decades of research have painted a different picture. First, most short-term stress does not impair the immune system in healthy individuals enough to lead to a greater incidence of diseases. However, older individuals and those with suppressed immune responses due to disease or immunosuppressive drugs may respond even to short-term stressors by getting sicker more often. It has been found that short-term stress diverts the body's resources towards enhancing innate immune responses, which have the ability to act fast and would seem to help the body prepare better for possible infections associated with the trauma that may result from a fight-or-flight exchange. The diverting of resources away from the adaptive immune response, however, causes its own share of problems in fighting disease.

Chronic stress, unlike short-term stress, may inhibit immune responses even in otherwise healthy adults. The suppression of both innate and adaptive immune responses is clearly associated with increases in some diseases, as seen when individuals lose a spouse or have other long-term stresses, such as taking care of a spouse with a fatal disease or dementia. The new science of psychoneuroimmunology, while still in its relative infancy, has great potential to make exciting advances in our understanding of how the nervous, endocrine, and immune systems have evolved together and communicate with each other.

References

Robinson J, Mistry K, McWilliam H, Lopez R, Parham P, Marsh SG. Nucleic acid research. IMGT/HLA Database [Internet]. 2011 [cited 2013 Apr 1]; 39:D1171–1176. Available from: http://europepmc.org/abstract/MED/21071412

Robinson J, Malik A, Parham P, Bodmer JG, Marsh SG. Tissue antigens. IMGT/HLA Database [Internet]. 2000 [cited 2013 Apr 1]; 55(3):280–287. Available from: http://europepmc.org/abstract/MED/10777106/reload=0;jsessionid=otkdw3M0TIVSa2zhvimg.6

Concept Check

1. **Which** organ system is most responsible for our first line of defense?
2. **Which** type of immunity can be trained to recognize new pathogens?
3. **Which** defensive chemical is used to ward off viral infections?
4. **Why** is it important to eat properly when combating pathogens?
5. **How** do phagocytes assist in disease prevention?
6. **Why** does lymph flow slowly through a lymph node?
7. **How** do lymphatic and blood vessels differ?
8. **Which** lymphatic organ(s) produce mature lymphocytes?
9. **Which** lymphocyte is responsible for producing antibodies?
10. **How** are inactive helper and cytotoxic T cells stimulated to begin cloning?
11. **What** is the function of the natural killer (NK) cell?
12. **Why** is the secondary immune response so much more effective than the primary immune response?
13. **How** do active and passive immunity differ?
14. **What** is the common feature of all autoimmune diseases?

CPSIA information can be obtained
at www.ICGtesting.com
Printed in the USA
BVHW060903120920
588363BV00005B/233

9 781641 760652